Surveys on Geometry and Integrable Systems

ADVANCED STUDIES
IN PURE MATHEMATICS 51

Chief Editor: Toshitake Kohno (The University of Tokyo)

Surveys on Geometry and Integrable Systems

Edited by

Martin Guest (Tokyo Metropolitan University)
Reiko Miyaoka (Tohoku University)
Yoshihiro Ohnita (Osaka City University)

Mathematical Society of Japan

This book was typeset by AMS-TEX and AMS-LATEX, the TEX macro systems of the American Mathematical Society, together with the style files aspm.sty *and* aspmfm.sty *for AMS-TEX written by Dr. Chiaki Tsukamoto and* aspmproc.sty *for AMS-LATEX written by Dr. Akihiro Munemasa and* aspm.cls *for AMS-LATEX provided by Tokyo Shoseki Printing Co., Ltd.*

TEX is a trademark of the American Mathematical Society.

Edited by the Mathematical Society of Japan.

Published by the Mathematical Society of Japan.

Distributed by the Mathematical Society of Japan, the American Mathematical Society and the World Scientific Publishing Co., Ltd.
Distributed exclusively in North America by the American Mathematical Society.

ISBN 978-4-931469-46-4

2000 Mathematics Subject Classification.
Primary 70H06.
Secondary 35Q53, 37K05, 37K10, 53A05, 53A07, 53A10, 58D25, 58D27, 58E20.

PRINTED IN JAPAN
by Tokyo Shoseki Printing Co., Ltd.

Preface

As the 9th International Research Institute of the Mathematical Society of Japan, an international conference on "Integrable Systems in Differential Geometry" was held at the University of Tokyo, during the period 17-21 July 2000. This volume consists primarily of survey articles based on the theme of the conference.

Two volumes, consisting primarily of research articles, have already been published. "Differential Geometry and Integrable Systems" (Contemporary Mathematics, vol. 308, American Mathematical Society, 2002) contains articles whose main focus is differential geometry. "Integrable Systems, Topology, and Physics" (Contemporary Mathematics, vol. 309, American Mathematical Society, 2002) contains articles related to geometry in a wider sense: topology, algebraic geometry, symplectic geometry, and mathematical physics.

The articles in this volume provide a panoramic view of the role of geometry in integrable systems, firmly rooted in surface theory but currently branching out in all directions. The longer articles are substantial surveys of several different aspects of the subject, while the shorter ones indicate more briefly how far the classical ideas have developed.

The members of the Organising Committee (with their affiliations at the time of the conference) were Takushiro Ochiai, Chairman (University of Tokyo), Martin Guest (Tokyo Metropolitan University), Mitsuhiro Itoh (University of Tsukuba), Reiko Miyaoka (Sophia University), Tohru Morimoto (Nara Women's University), Yoshihiro Ohnita (Tokyo Metropolitan University), Simon Salamon (Oxford University), Takeshi Sasaki (Kobe University), Chuu-Lian Terng (Northeastern University), and Gudlaugur Thorbergsson (Cologne University). On behalf of the entire Organising Committee we would like to thank the institutions and many individuals whose cooperation was essential for the success of the conference. Financial support was provided by the Mathematical Society of Japan, the Japan Association for Mathematical Sciences, and from individual research grants of the Japan Society for Promotion of Science. We thank Carolyn Dong and Sanae Kurosu for assistance in the preparation of the TeX files.

Martin Guest, Reiko Miyaoka, and Yoshihiro Ohnita

Note: The first-named editor offers his sincere apologies for the extremely long delay at the final stage of publication of this volume. All articles were received by 2002, and the refereeing process was completed by 2003. However, a combination of personal circumstances and technical problems delayed the final English and TeX editing, for which he accepts sole responsibility

All papers in this volume have been refereed and are in final form. No version of any of them will be submitted for publication elsewhere.

CONTENTS

Alexander I. Bobenko — Exploring surfaces through methods from the theory of integrable systems: The Bonnet problem 1

Josef Dorfmeister — Generalized Weierstraß representations of surfaces 55

Atsushi Fujioka and Jun-ichi Inoguchi — Timelike surfaces with harmonic inverse mean curvature 113

Chaohao Gu — Darboux transformations and generalized self-dual Yang-Mills flows 143

Frédéric Hélein and Pascal Romon — From CMC surfaces to Hamiltonian stationary Lagrangian surfaces 163

Dominic Joyce — Special Lagrangian 3-folds and integrable systems 189

Xiaobo Liu — Quantum product, topological recursion relations, and the Virasoro conjecture 235

Shigeki Matsutani — A generalized Weierstrass representation for a submanifold S in $\mathbb{E}^n$ arising from the submanifold Dirac operator 259

Ian McIntosh — Harmonic tori and their spectral data 285

Reiko Miyaoka — Isoparametric geometry and related fields 315

Henrik Pedersen — Kähler Ricci solitons 339

Wayne Rossman, Masaaki Umehara and Kotaro Yamada — Period problems for mean curvature one surfaces in H^3 (with applications to surfaces of low total curvature) 347

Chuu-Lian Terng — Geometries and symmetries of soliton equations and integrable elliptic equations 401

Peter Topalov — A-integrability of geodesic flows and geodesic equivalence 489

Advanced Studies in Pure Mathematics 51, 2008
Surveys on Geometry and Integrable Systems
pp. 1–53

Exploring surfaces through methods from the theory of integrable systems: The Bonnet problem

Alexander I. Bobenko

A generic surface in Euclidean 3-space is determined uniquely by its metric and curvature. Classification of all special surfaces where this is not the case, i.e. of surfaces possessing isometries which preserve the mean curvature, is known as the Bonnet problem. Regarding the Bonnet problem, we show how analytic methods of the theory of integrable systems — such as finite-gap integration, isomonodromic deformation, and loop group description — can be applied for studying global properties of special surfaces.

§1. Quaternionic description of surfaces. Bonnet problem

1.1. Differential equations of surfaces

Let $\mathcal{F}$ be a smooth orientable surface in 3-dimensional Euclidean space. The Euclidean metric induces a metric Ω on this surface, which in turn generates the complex structure of a Riemann surface $\mathcal{R}$. Under such a parametrization, which is called *conformal*, the surface $\mathcal{F}$ is given by an immersion

$$F = (F_1, F_2, F_3) : \mathcal{R} \to \mathbb{R}^3,$$

and the metric is conformal: $\Omega = e^u \, dzd\bar{z}$, where z is a local coordinate on $\mathcal{R}$.

One should keep in mind that a complex coordinate is defined up to holomorphic $z \to w(z)$ transformation. This freedom will be used to simplify the corresponding equations.

Received March 15, 2001.
Revised March 25, 2001.
2000 *Mathematics Subject Classification.* Primary 53A05.

The conformal parametrization gives the following normalization of $F(z,\bar{z})$:

$$< F_z, F_z > = < F_{\bar{z}}, F_{\bar{z}} > = 0, \ < F_z, F_{\bar{z}} > = \frac{1}{2}e^u, \tag{1}$$

where the brackets denote the scalar product

$$< a, b > = a_1 b_1 + a_2 b_2 + a_3 b_3,$$

and F_z and $F_{\bar{z}}$ are the partial derivatives $\frac{\partial F}{\partial z}$ and $\frac{\partial F}{\partial \bar{z}}$, where

$$\frac{\partial}{\partial z} = \frac{1}{2}\left(\frac{\partial}{\partial x} - i\frac{\partial}{\partial y}\right), \qquad \frac{\partial}{\partial \bar{z}} = \frac{1}{2}\left(\frac{\partial}{\partial x} + i\frac{\partial}{\partial y}\right).$$

The vectors $F_z, F_{\bar{z}}$, as well as the normal N, with

$$< F_z, N > = < F_{\bar{z}}, N > = 0, \quad < N, N > = 1, \tag{2}$$

define a moving frame on the surface, which due to (1, 2) satisfies the following Gauss-Weingarten equations:

$$\sigma_z = \mathcal{U}\sigma, \ \sigma_{\bar{z}} = \mathcal{V}\sigma, \ \ \sigma = (F_z, F_{\bar{z}}, N)^T, \tag{3}$$

$$\mathcal{U} = \begin{pmatrix} u_z & 0 & Q \\ 0 & 0 & \frac{1}{2}He^u \\ -H & -2e^{-u}Q & 0 \end{pmatrix}, \tag{4}$$
$$\mathcal{V} = \begin{pmatrix} 0 & 0 & \frac{1}{2}He^u \\ 0 & u_{\bar{z}} & \bar{Q} \\ -2e^{-u}\bar{Q} & -H & 0 \end{pmatrix},$$

where

$$Q = < F_{zz}, N >, \ \ < F_{z\bar{z}}, N > = \frac{1}{2}He^u. \tag{5}$$

The quadratic differential Qdz^2 is called the *Hopf differential.* The first and the second fundamental forms

$$\begin{aligned} < dF, dF > &= < I\begin{pmatrix} dx \\ dy \end{pmatrix}, \begin{pmatrix} dx \\ dy \end{pmatrix} >, \ z = x + iy, \\ - < dF, dN > &= < II\begin{pmatrix} dx \\ dy \end{pmatrix}, \begin{pmatrix} dx \\ dy \end{pmatrix} > \end{aligned}$$

are given by the matrices

(6)
$$I = e^u \begin{pmatrix} 1 & 0 \\ 0 & 1 \end{pmatrix}, \; II = \begin{pmatrix} Q + \bar{Q} + He^u & i(Q - \bar{Q}) \\ i(Q - \bar{Q}) & -(Q + \bar{Q}) + He^u \end{pmatrix}.$$

The *principal curvatures* k_1 and k_2 are the eigenvalues of the matrix $II \cdot I^{-1}$. This gives the following expressions for the *mean* and the *Gaussian curvatures*:

$$H = \tfrac{1}{2}\,(k_1 + k_2) = \tfrac{1}{2}\,\mathrm{tr}\,(II \cdot I^{-1}),$$
$$K = k_1 k_2 = \det\,(II \cdot I^{-1}) = H^2 - 4Q\bar{Q}e^{-2u}.$$

A point P of the surface $\mathcal{F}$ is called *umbilic* if the principal curvatures at this point coincide $k_1(P) = k_2(P)$. The Hopf differential vanishes $Q(P) = 0$ exactly at umbilic points of the surface.

Coordinates in which both fundamental forms are diagonal are called *curvature line coordinates* and the corresponding parametrization (not necessarily conformal) is called a *curvature line parametrization.* A curvature line parametrization always exists in a neighborhood of a non-umbilic point. Near umbilic points, curvature lines form more complicated patterns.

The Gauss–Codazzi equations, which are the compatibility conditions of equations (3, 4),

$$\mathcal{U}_{\bar{z}} - \mathcal{V}_z + [\mathcal{U}, \mathcal{V}] = 0,$$

have the following form:

(7)
$$\begin{array}{ll} \text{Gauss equation} & u_{z\bar{z}} + \frac{1}{2}\,H^2\,e^u - 2|Q|^2\,e^{-u} = 0, \\ \text{Codazzi equation} & Q_{\bar{z}} = \frac{1}{2}\,H_z\,e^u. \end{array}$$

These equations are necessary and sufficient for existence of the corresponding surface.

Theorem 1. (Bonnet theorem). *Given a metric $e^u\,dzd\bar{z}$, a quadratic differential $Q\,dz^2$, and a function H on $\mathcal{R}$ satisfying the Gauss–Codazzi equations, there exists an immersion*

$$F : \tilde{\mathcal{R}} \to \mathbb{R}^3$$

with the fundamental forms (6). Here $\tilde{\mathcal{R}}$ is the universal covering of $\mathcal{R}$. The immersion F is unique up to Euclidean motions in $\mathbb{R}^3$.

We finish this section with some basic facts about a special class of surfaces. A conformal curvature line parametrization is called *isothermic*. In this case the preimages of the curvature lines are the lines $x =$ constant and $y =$ constant on the parameter domain, where $z = x + iy$ is a conformal coordinate. A surface is called *isothermic* if it allows isothermic parametrization. Isothermic surfaces are divided by their curvature lines into infinitesimal squares. Written in terms of an isothermic coordinate z the Hopf differential $Q(z, \bar{z})dz^2$ of an isothermic surface is real, i.e. $Q(z, \bar{z}) \in \mathbb{R}$.

In terms of arbitrary conformal coordinates, isothermic surfaces can be characterized as follows.

Lemma 1. *Let $F : \mathcal{R} \to \mathbb{R}^3$ be a conformal immersion of an umbilic free surface in $\mathbb{R}^3$. The surface is isothermic if and only if there exists a holomorphic non-vanishing differential $f(z)dz^2$ on $\mathcal{R}$ and a function $q : \mathcal{R} \to \mathbb{R}_*$ such that the Hopf differential is of the form*

$$Q(z, \bar{z}) = f(z)\, q(z, \bar{z}). \tag{8}$$

It is easy to see that $w = \int \sqrt{f(z)}\, dz$ is an isothermic coordinate.

1.2. Quaternionic description of surfaces

We construct and investigate surfaces in $\mathbb{R}^3$ by analytic methods. For this purpose it is convenient to use the Lie algebra isomorphism $so(3) = su(2)$ and to rewrite the equations (3, 4) for the moving frame in terms of 2 by 2 matrices. This quaternionic description turns out to be useful for analytic studies of general curves and surfaces in 3- and 4-spaces as well as for investigation of special classes of surfaces [Bob1, KS2, DPW, Bob2, KPP, PP].

Let us denote the algebra of quaternions by $\mathbb{H}$, the multiplicative quaternion group by $\mathbb{H}_* = \mathbb{H}\setminus\{0\}$, and their standard basis by $\{\mathbf{1}, \mathbf{i}, \mathbf{j}, \mathbf{k}\}$, where

$$\mathbf{ij} = \mathbf{k}, \ \mathbf{jk} = \mathbf{i}, \ \mathbf{ki} = \mathbf{j}. \tag{9}$$

This basis can be represented by the Pauli matrices σ_α as follows:

$$\sigma_1 = \begin{pmatrix} 0 & 1 \\ 1 & 0 \end{pmatrix} = i\,\mathbf{i}, \ \ \sigma_2 = \begin{pmatrix} 0 & -i \\ i & 0 \end{pmatrix} = i\,\mathbf{j},$$

$$\sigma_3 = \begin{pmatrix} 1 & 0 \\ 0 & -1 \end{pmatrix} = i\,\mathbf{k}, \ \ \mathbf{1} = \begin{pmatrix} 1 & 0 \\ 0 & 1 \end{pmatrix}. \tag{10}$$

We identify $\mathbb{H}$ with 4-dimensional Euclidean space

$$q = q_0\mathbf{1} + q_1\mathbf{i} + q_2\mathbf{j} + q_3\mathbf{k} \longleftrightarrow \ q = (q_0, q_1, q_2, q_3) \in \mathbb{R}^4.$$

The sphere $S^3 \subset \mathbb{R}^4$ is then naturally identified with the group of unitary quaternions $\mathbb{H}_1 = SU(2)$. 3-dimensional Euclidean space is identified with the space of imaginary quaternions Im $\mathbb{H}$

$$(11)\ X = -i\sum_{\alpha=1}^{3} X_\alpha\sigma_\alpha \in \operatorname{Im}\mathbb{H} \ \longleftrightarrow \ X = (X_1, X_2, X_3) \in \mathbb{R}^3.$$

The scalar product of vectors in terms of quaternions and matrices is then

$$(12)\qquad < X, Y >= -\frac{1}{2}(XY + YX) = -\frac{1}{2}\operatorname{tr} XY.$$

We will also denote by F and N the matrices obtained in this way from the vectors F and N.

Let us take $\Phi \in \mathbb{H}_*$ which transforms the basis $\mathbf{i}, \mathbf{j}, \mathbf{k}$ into the frame F_x, F_y, N:

$$(13)\qquad F_x = e^{u/2}\Phi^{-1}\mathbf{i}\Phi,\ F_y = e^{u/2}\Phi^{-1}\mathbf{j}\Phi,\ N = \Phi^{-1}\mathbf{k}\Phi.$$

Then

$$(14)\ F_z = -ie^{u/2}\Phi^{-1}\begin{pmatrix} 0 & 0 \\ 1 & 0 \end{pmatrix}\Phi,\quad F_{\bar z} = -ie^{u/2}\Phi^{-1}\begin{pmatrix} 0 & 1 \\ 0 & 0 \end{pmatrix}\Phi,$$

and all the conditions (1) are automatically satisfied.

The quaternion Φ satisfies linear differential equations. To derive them we introduce matrices

$$(15)\qquad U = \Phi_z\Phi^{-1}, \qquad V = \Phi_{\bar z}\Phi^{-1}.$$

The compatibility condition $F_{z\bar z} = F_{\bar z z}$ for (14) implies

$$V_{22} - V_{11} = \frac{u_{\bar z}}{2},\ \ U_{11} - U_{22} = \frac{u_z}{2},\ \ U_{21} = -V_{12},$$

where U_{kl} and V_{kl} are the matrix elements of U and V. In the same way one obtains from (14)

$$\begin{aligned} F_{z\bar z} &= \frac{1}{2}He^uN &\rightarrow&\quad U_{21} = -V_{12} = \frac{1}{2}He^{u/2} \\ F_{zz} &= u_zF_z + QN &\rightarrow&\quad U_{12} = -Qe^{-u/2} \\ F_{\bar z\bar z} &= u_{\bar z}F_{\bar z} + \bar QN &\rightarrow&\quad V_{21} = \bar Qe^{-u/2}. \end{aligned}$$

Recall that Φ is defined up to multiplication by a scalar factor. We normalize this factor by the condition

$$\det \Phi = e^{\frac{u}{2}}, \tag{16}$$

for reasons which will be clarified in the next section. For the traces of U and V this implies

$$\operatorname{tr} U = \frac{u_z}{2}, \qquad \operatorname{tr} V = \frac{u_{\bar{z}}}{2}.$$

Finally one arrives at the following

Theorem 2. *By the isomorphism (11), the moving frame $F_z, F_{\bar{z}}, N$ of a conformally parametrized surface (z is a conformal coordinate) is described by formulae (13),(14), where $\Phi \in \mathbb{H}_*$ satisfies the equations (15) with U, V of the form*

$$U = \begin{pmatrix} \frac{u_z}{2} & -Qe^{-u/2} \\ \frac{1}{2}He^{u/2} & 0 \end{pmatrix}, \tag{17}$$

$$V = \begin{pmatrix} 0 & -\frac{1}{2}He^{u/2} \\ \bar{Q}e^{-u/2} & \frac{u_{\bar{z}}}{2} \end{pmatrix}.$$

Corollary 1. *The conformal frame Φ satisfies the Dirac equation*

$$e^{-u/2} \begin{pmatrix} 0 & \partial_z \\ -\partial_{\bar{z}} & 0 \end{pmatrix} \Phi = \frac{1}{2} H \Phi. \tag{18}$$

It turnes out that at this point the whole construction can be reversed. Namely, starting with a solution to the Dirac equation one can derive a Weierstrass type representation (see (21) below) for conformally parametrized surfaces. This idea was recently developed by Konopelchenko [Kon] and further in [Tai, PP, KS2], although in other forms the Weierstrass representation of surfaces was known already to Eisenhart [Eis] and Kenmotsu [Ken].

Theorem 3. *Let $D \subset \mathbb{C}$ be a simply connected domain and $(s_1, \bar{s}_2)^T : D \to \mathbb{C}^2$ be a solution to the Dirac equation with the potential $p \in C^\infty(D)$*

$$\begin{pmatrix} 0 & \partial_z \\ -\partial_{\bar{z}} & 0 \end{pmatrix} \begin{pmatrix} s_1 \\ \bar{s}_2 \end{pmatrix} = p \begin{pmatrix} s_1 \\ \bar{s}_2 \end{pmatrix}. \tag{19}$$

Then

$$\Phi = \begin{pmatrix} s_1 & -s_2 \\ \bar{s}_2 & \bar{s}_1 \end{pmatrix} : D \to \mathbb{H}_* \tag{20}$$

is a conformal frame (14) of the conformally immersed surface

$$\begin{aligned} F_1 + iF_2 &= \int s_1^2\, dz - \bar{s}_2^2\, d\bar{z} \\ F_3 &= \int s_1 s_2\, dz + \bar{s}_1 \bar{s}_2\, d\bar{z}. \end{aligned} \tag{21}$$

The metric and the mean curvature of the surface are given by

$$e^u\, dz d\bar{z} = (|\, s_1\,|^2 + |\, s_2\,|^2)^2\, dz d\bar{z}, \quad H = 2pe^{-u/2}. \tag{22}$$

Proof. Note that $(-s_2, \bar{s}_1)^T$ is also a solution to (19) due to the symmetry of the Dirac equation. At this point Φ given by (20) can be identified with the conformal frame Φ of Corollary 1. The formula for the metric (22) follows from (16). Substituting it into our previous formulae (14) for conformal frame one defines

$$F_z := -i \begin{pmatrix} s_1 s_2 & -s_2^2 \\ s_1^2 & -s_1 s_2 \end{pmatrix}, \quad F_{\bar{z}} := -i \begin{pmatrix} \bar{s}_2 \bar{s}_1 & \bar{s}_1^2 \\ -\bar{s}_2^2 & -\bar{s}_2 \bar{s}_1 \end{pmatrix}.$$

These formulae are automatically compatible. Integrating them one arrives at (21).

1.3. Spinor description of surfaces

As shown in [Bob2], the quaternionic description of the previous section is actually a global one. Let $\cup_i D_i = \mathcal{R}$ be an open covering of $\mathcal{R}$ with local coordinates $z_i : D_i \to \mathbb{C}$. Conditions (14, 16) determine a quaternionic valued smooth $\Phi(z_i, \bar{z}_i)$ uniquely up to sign on each D_i. To establish the global nature of Φ recall that a holomorphic line bundle S is called a *spin bundle* if it satisfies $S \otimes S = K$, where K is the canonical bundle.

Denote the first column of Φ by $\begin{pmatrix} S_1 \\ \bar{S}_2 \end{pmatrix}$.

Lemma 2. *S_1 and S_2 are smooth sections of the same holomorphic spin bundle S.*

Proof. Consider two intersecting $D_i \cap D_j \neq \emptyset$ with corresponding $\Phi_i(z_i, \bar{z}_i)$ and $\Phi_j(z_j, \bar{z}_j)$. Identifying the representations for the Gauss map in terms of Φ_i and Φ_j one obtains on $D_i \cap D_j$

$$\Phi_i = \begin{pmatrix} c_{ij} & 0 \\ 0 & \bar{c}_{ij} \end{pmatrix} \Phi_j.$$

with some $c_{ij} : D_i \cap D_j \to \mathbb{C}_*$. Further, identifying the tangent frames $F_{z_i} = F_{z_j} \frac{dz_j}{dz_i}$ and using $\Phi \sigma_2 \Phi^T \sigma_2 = \det \Phi$ one obtains

$$\Phi_i^T \begin{pmatrix} 1 & 0 \\ 0 & 0 \end{pmatrix} \Phi_i = \frac{dz_j}{dz_i} \Phi_j^T \begin{pmatrix} 1 & 0 \\ 0 & 0 \end{pmatrix} \Phi_j,$$

which finally implies $c_{ij}^2 = \frac{dz_j}{dz_i}$. The transition functions $c_{ij} : D_i \cap D_j \to \mathbb{C}_*$ defined through Φ_i obviously satisfy the cocycle condition $c_{ij}c_{jk} = c_{ik}$ and thus define a line bundle S with $S \otimes S = K$.

In local coordinates the S_n may be written as $S_n = s_n(z_i, \bar{z}_i)\sqrt{dz_i}$.

Using the equivalence of spinor representation of conformal frames of surfaces and solutions of the Dirac equation, proven in Corollary 1 and Theorem 3, we arrive at the following global reformulation [Tai, PP] of Theorem 3.

Theorem 4. *A half-density p (i.e. a smooth section of $K^{\frac{1}{2}} \otimes \bar{K}^{\frac{1}{2}}$) and two not simultaneously vanishing spinors S_1, S_2 (i.e. smooth sections of $S \cong K^{\frac{1}{2}}$ with $(S_1, S_2) \neq (0,0)\ \forall P \in \mathcal{R}$) satisfying the Dirac equation (19) determine through*

$$\begin{aligned} F_1 + iF_2 &= \int S_1^2 - \bar{S}_2^2 \\ F_3 &= \int S_1 S_2 + \bar{S}_1 \bar{S}_2 \end{aligned} \tag{23}$$

a conformal immersion $F : \tilde{\mathcal{R}} \to \mathbb{R}^3$, where $\tilde{\mathcal{R}}$ is the universal covering of $\mathcal{R}$. The metric and the mean curvature of the immersion are given by

$$e^u\, dz d\bar{z} = (\mid S_1 \mid^2 + \mid S_2 \mid^2)^2, \quad H = 2pe^{-u/2}.$$

Remark. In the case of minimal surfaces $H = 0$ the spinors S_1 and S_2 are holomorphic and the representation (23) is known as the spinor Weierstrass representation [Sul, Bob2, KS1].

On a Riemann surface of genus g there exist 2^{2g} non-isomorphic spin bundles which are distinguished by different *spin structures*. For a geometric interpretation of the spin structure of the spin bundle S in terms of the immersion (23) we refer to [Bob2]. Spin structures classify regular homotopies of immersions [Pin].

1.4. Alternative descriptions of surfaces and the Bonnet problem

Bonnet Theorem 1 characterizes surfaces via the coefficients e^u, H, Q of their fundamental forms. These coefficients are not independent and are subject to the Gauss–Codazzi equations (7). A natural question is whether some of these data are superfluous. The following natural candidates for more "economic" characterization of surfaces were studied.

(i) The most geometric setting of the problem is the oldest one and is due to Bonnet. He posed the question whether one can eliminate the Hopf differential from the description of surfaces, i.e. whether the *metric* e^u and the *mean curvature function* H alone suffice to describe a surface completely. Generic surfaces are determined uniquely by the metric and the mean curvature function. Bonnet himself [Bon] made the initial progress in the investigation of the special surfaces where it is not the case, i.e. which possess non-congruent isometric "relatives" with the same mean curvature function. The rest of these lectures is devoted to this problem, which is fairly named the *Bonnet problem.*

(ii) The *conformal Hopf differential* $q := Qe^{-\frac{u}{2}}$. Note that whereas the Hopf differential is a quadratic differential, i.e. a section of the line bundle K^2, the conformal Hopf differential is more exotic - it is a section of $K^{\frac{3}{2}} \otimes \bar{K}^{-\frac{1}{2}}$. The reason for its introduction by U. Pinkall is that q is invariant with respect to the Möbius transformations of the ambient $\mathbb{R}^3$. A non-isothermic surface is uniquely determined by q up to Möbius transformations. Counting dimensions, one immediately observes that generic sections of $K^{\frac{3}{2}} \otimes \bar{K}^{-\frac{1}{2}}$ do not correspond to surfaces in $\mathbb{R}^3$. A proper equation for q of surfaces in $\mathbb{R}^3$ is still unknown.

(iii) The *Dirac potential* or *mean-curvature half-density* $p = \frac{1}{2}He^{-\frac{u}{2}}$. As one can see from its definition, this potential is a half-density, i.e. a section of the line bundle $K^{\frac{1}{2}} \otimes \bar{K}^{\frac{1}{2}}$. Recently, description of surfaces through Dirac spinors attracted much attention [KS2, Tai, PP]. Unfortunately, one has neither existence nor uniqueness in this description. A generic Dirac operator (with generic potential) has trivial kernel, thus generic half-densities do not yield surfaces. On the other hand, there may exist many immersions with the same potential p, for example all special surfaces appearing in the Bonnet problem.

Returning to the Bonnet problem, note that already Bonnet indicated all special surfaces which possess non-congruent isometric "relatives" with the same mean curvature function. There are three cases when this happens.

1. Constant mean curvature surfaces. Let $\mathcal{F}$ be a surface with constant mean curvature H. The Gauss–Codazzi equations (7) are obviously invariant with respect to the transformation $Q \to Q_t = e^{it}Q, t \in \mathbb{R}$. Applying the Bonnet theorem one obtains the one parameter family $\mathcal{F}_t$, $\mathcal{F} = \mathcal{F}_0$ of isometric surfaces with the same constant mean curvature H. In the last ten years there was much interest in studying global properties of surfaces with constant mean curvature and now they are rather well investigated by various methods (see for example [Wen, PS, Kap, Bob1, GKS]) including methods of the theory of integrable systems.

2. Bonnet pairs are exactly two non-congruent isometric surfaces $\mathcal{F}'$ and $\mathcal{F}''$ with the same mean curvature function. The theory of Bonnet pairs is very closely related [Bia2, KPP] to the theory of isothermic surfaces and as such belongs also to geometry described by integrable systems. Up to now, global theory of Bonnet pairs is not well developed, in particular it is unknown whether there exist compact Bonnet pairs, a question first posed in [LT].

3. Bonnet families. In [Bon], Bonnet himself was able to show that besides the surfaces with constant mean curvature there exists a class of surfaces, depending on finitely many parameters which possess one-parameter family of isometries preserving the mean curvature. These surfaces were studied by many authors [Haz, Gra, Car, Che, BE1, Rou] and recently global classification [BE2] of them was obtained using methods from the theory of integrable systems.

The remaining three sections are devoted to consideration of these three cases.

§2. Constant mean curvature surfaces

2.1. Associated family

If the mean curvature of $\mathcal{F}$ is constant, then the Gauss-Codazzi equations

$$u_{z\bar{z}} + \frac{1}{2}H^2 e^u - 2Q\bar{Q}e^{-u} = 0, \qquad Q_{\bar{z}} = 0,$$

are invariant with respect to the transformation

$$Q \to Q^t = \Lambda Q, \quad |\Lambda| = 1, \tag{24}$$

Integrating the equations for the moving frame with the coefficient Q replaced by $Q^t = \Lambda Q$ we obtain a one-parameter family $\mathcal{F}^t$ of surfaces. All the surfaces $\mathcal{F}^t$ are isometric and have the same constant mean curvature. Treating t as a deformation parameter we obtain the first family of special surfaces indicated by Bonnet (see Section 1.4).

Theorem 5. *Every constant mean curvature surface has a one-parameter family of isometric deformations preserving the mean curvature. The deformation is described by the transformations (24).*

Without loss of generality we normalize $H = 1$. The quaternion $\Phi(z, \bar{z}, \Lambda)$ solving the system (15, 17) with $Q^t = \Lambda Q$ describes the moving frame $F_z, F_{\bar{z}}, N$ (13, 14) of the corresponding surface. Knowing the family $\Phi(z, \bar{z}, \Lambda)$ in a neighbourhood of $\Lambda = e^{2it}$ allows us to derive an immersion formula without integration the frame with respect to $z, \bar{z}$, but just by differentiation by t. Before presenting this important formula we pass to a gauge equivalent frame function

$$\Phi_0 = e^{-u/4} \begin{pmatrix} \frac{1}{\sqrt{i\lambda}} & 0 \\ 0 & \sqrt{i\lambda} \end{pmatrix} \Phi, \qquad \Lambda = \lambda^2, \tag{25}$$

normalized by

$$\Phi_0(z, \bar{z}, \lambda = e^{it}) \in SU(2), \qquad t \in \mathbb{R}. \tag{26}$$

Although it is known that compact CMC surfaces exist for any genus g [Kap], their analytic description remains an open problem. Its solution requires a development of new analytic methods. Until now the theory of integrable systems was successfully applied for description of planes, cylinders, tori [PS, Bob1] ($g = 1$) and punctured spheres [KMS]. In this section we are dealing essentially with the theory of CMC tori whhich can be completely classified through analytic methods from the theory of integrable systems. Since the canonical bundle in this case is trivial, introducing a global complex coordinate, one can describe spinors in terms of doubly-periodic functions (see Section 2.3).

Theorem 6. [Bob1] *Let $\Phi_0(z, \bar{z}, \lambda = e^{it})$ be a solution of the system*

$$\Phi_{0\,z} = U_0(\lambda)\Phi_0, \qquad \Phi_{0\,\bar{z}} = V_0(\lambda)\Phi_0, \tag{27}$$

(28)

$$U_0(\lambda) = \begin{pmatrix} \frac{u_z}{4} & i\lambda Q e^{-u/2} \\ \lambda \frac{i}{2} e^{u/2} & -\frac{u_z}{4} \end{pmatrix}, \quad V_0(\lambda) = \begin{pmatrix} -\frac{u_{\bar z}}{4} & \frac{i}{2\lambda} e^{u/2} \\ \frac{i}{\lambda} \bar Q e^{-u/2} & \frac{u_{\bar z}}{4} \end{pmatrix}$$

normalized by (26). Then F and N, defined by the formulae

$$(29) \quad F = -\Phi_0^{-1} \frac{\partial}{\partial t} \Phi_0 + \frac{i}{2} \Phi_0^{-1} \sigma_3 \Phi_0, \qquad N = -i\Phi_0^{-1} \sigma_3 \Phi_0,$$

describe a CMC surface and its Gauss map, with metric e^u, mean curvature $H = 1$, and Hopf differential $Q^t = e^{2it} Q$.

Conversely, let F be a conformal parametrization of a CMC surface with metric e^u, mean curvature $H = 1$, and Hopf differential Q^t. Then F is given by formula (29) where Φ_0 is a solution of (27, 29) as above.

Proof. First we note that both F and N are imaginary quaternions and therefore can be identified with vectors in $\mathbb{R}^3$. By identification (25) the system (29) coincides with the quaternionic representation (17) for the equations for the moving frame with the Hopf differential λQ. Differentiating (29) we get

$$\begin{aligned} F_z &= -\Phi_0^{-1} \frac{\partial U_0(\lambda)}{\partial t} \Phi_0 + \frac{i}{2} \Phi_0^{-1} [\sigma_3, U_0(\lambda)] \Phi_0 = -ie^{u/2} \Phi^{-1} \begin{pmatrix} 0 & 0 \\ 1 & 0 \end{pmatrix} \Phi, \\ F_{\bar z} &= -ie^{u/2} \Phi^{-1} \begin{pmatrix} 0 & 1 \\ 0 & 0 \end{pmatrix} \Phi, \end{aligned}$$

which coincides with (14).

Remark. In a neighborhood of a non-umbilic point $Q \neq 0$ by a conformal change of coordinate $z \to w(z)$ one can always normalize $Q = \frac{1}{2}$. Thus, umbilic free CMC surfaces are isothermic. In this parametrization the Gauss equation becomes the elliptic sinh-Gordon equation

$$(30) \qquad u_{z\bar z} + \sinh u = 0.$$

2.2. Loop group formulation

The matrices $A = U_0 + V_0, B = i(U_0 - V_0)$ corresponding to real vector fields $\partial_x = \partial_z + \partial_{\bar z}$ and $\partial_y = i(\partial_z - \partial_{\bar z})$ belong to the loop algebra

$$g_H[\lambda] = \{\xi : S^1 \to su(2) : \xi(-\lambda) = \sigma_3 \xi(\lambda) \sigma_3\},$$

and Φ_0 in (27) lies in the corresponding loop group

$$(31) \qquad G_H[\lambda] = \{\phi : S^1 \to SU(2) : \phi(-\lambda) = \sigma_3 \phi(\lambda) \sigma_3\}.$$

Here S^1 is the set $|\lambda| = 1$. When defined for general complex λ, elements of $g_H[\lambda]$ and $G_H[\lambda]$ satisfy the real reduction

$$\xi(\lambda) = \sigma_2 \overline{\xi(\frac{1}{\bar\lambda})} \sigma_2, \qquad \phi(\lambda) = \sigma_2 \overline{\phi(\frac{1}{\bar\lambda})} \sigma_2.$$

For applying analytic methods of the theory of integrable systems it is crucial that CMC surfaces can be characterized in terms of this loop group completely without referring to the special geometric nature of the coefficients of A and B. It is not difficult to prove the following

Theorem 7. *Let $\phi : D \to G_H[\lambda]$ be a smooth map on $D \subset \mathbb{C}$ satisfying $\phi_z \phi^{-1} = A\lambda + B$ with $A : D \to GL(2, \mathbb{C})$. Then the gauge equivalent*

$$\Phi_0 = \exp(\frac{i}{2} \arg A_{21}\, \sigma_3) \phi$$

satisfies (27) with U_0, V_0 of the form (29) and describes the conformal frame of the immersion

$$F = -\phi^{-1} \frac{\partial}{\partial t} \phi + \frac{i}{2} \phi^{-1} \sigma_3 \phi, \qquad \lambda = e^{it}$$

of D in $\mathbb{R}^3$ with the mean curvature $H = 1$.

2.3. CMC tori. Analytic formulation

Methods of Section 2.2 can be used not only in local but also in global studies of CMC surfaces. It is a classical result of Hopf [Hop] that the only CMC surface of genus zero is a round sphere. Indeed the holomorphic quadratic differential $Q\,dz^2$ on a sphere must vanish identically. Then (4) implies in particular $N + F = C =$ constant, which yields $< F - C, F - C >= 1$.

Classification of CMC tori is not as simple as of spheres but analytic tools enable us to achieve success in this case also. The reason for a simplification in the case $g = 1$ is the fact that, unlike the case of Riemann surfaces of genus $g \geq 2$, on a torus it is possible to introduce a global complex coordinate.

Any Riemann surface of genus 1 is conformally equivalent to the factor of the complex plane by a lattice $\mathbb{C}/\mathcal{L}$. The corresponding conformal parametrization of a torus is given by a doubly-periodic mapping

$F : \mathbb{C}/\mathcal{L} \to \mathbb{R}^3$. The metric and the Hopf differential in this parametrization are described by doubly-periodic functions $u(z,\bar{z})$, $Q(z,\bar{z})$. Note that $H =$ constant implies $Q_{\bar{z}} = 0$ and $Q(z)$ is a bounded elliptic function, thus a constant. This constant is not zero, otherwise, as follows from the consideration above the surface is a sphere. Thus CMC tori have no umbilic points. As before we normalize the Gauss equation to (30) by $Q = \frac{1}{2}$.

Denoting the generators of $\mathcal{L}$ by $Z_1 = X_1 + iY_1, Z_2 = X_2 + iY_2$ one obtains the following

Proposition 1. *Any torus with mean curvature $H = 1$ can be conformally parametrized by a doubly-periodic immersion $F : \mathbb{C} \to \mathbb{R}^3$*

$$F(z + Z_i, \bar{z} + \bar{Z}_i) = F(z, \bar{z}), \qquad i = 1, 2$$

with the Hopf differential $Q = \frac{1}{2}$. In this parametrization the metric $u(z,\bar{z})$ is a doubly-periodic solution to the elliptic sinh-Gordon equation (30).

Note that due to the ellipticity of equation (30) all CMC tori are real analytic.

To describe all CMC tori one should solve the following problems.

1. Describe all doubly-periodic solutions $u(z,\bar{z})$ of the elliptic sinh-Gordon equation (30).

2. Integrate linear system (27) with $U_0(\lambda), V_0(\lambda)$ respectively given by

$$(32) \qquad \frac{1}{2}\begin{pmatrix} \frac{u_z}{2} & i\lambda e^{-u/2} \\ i\lambda e^{u/2} & -\frac{u_z}{2} \end{pmatrix}, \quad \frac{1}{2}\begin{pmatrix} -\frac{u_{\bar{z}}}{2} & \frac{i}{\lambda} e^{u/2} \\ \frac{i}{\lambda} e^{-u/2} & \frac{u_{\bar{z}}}{2} \end{pmatrix}$$

to find $\Phi_0(z, \bar{z}, \lambda = e^{it})$.

3. Formula (29) for F describes the corresponding CMC immersion. In general, this immersion is not doubly-periodic. One should specify parameters of the solution $u(z,\bar{z})$, which yield doubly-periodic $F(z,\bar{z})$.

These three problems can be solved simultaneously using methods of the finite-gap integration theory. In the rest of the lecture we give an idea of how this solution is found.

2.4. Higher flows and the fundamental theorem

Let $u(z,\bar{z})$ be a solution of the sinh-Gordon equation. The perturbation $u_\epsilon(z,\bar{z}) = u(z,\bar{z}) + \epsilon v(z,\bar{z})$ of $u(z,\bar{z})$ satisfies (30) up to the terms of order $O(\epsilon^2)$ if and only if $v(z,\bar{z})$ is a solution of the linearized elliptic sinh-Gordon equation

$$(\partial_{z\bar{z}} + \cosh u(z,\bar{z}))v(z,\bar{z}) = 0. \tag{33}$$

The elliptic sinh-Gordon equation is one of the possible real versions of the sine-Gordon equation, which is one of the basic models of the theory of integrable systems. Integrable systems possess infinitely many conservation laws, which induce infinitely many commuting flows of the corresponding dynamical system. In particular, applying standard algebraic tools of the theory to the sine-Gordon equation one can prove that there exists $v(u_z,\ldots,u_z^{(k)})$, which solves (33) and is a polynomial in all its arguments. Such a polynomial can be treated as a tangential vector field to the space of solutions of the elliptic sinh-Gordon equation. These vector fields induce flows on the phase space of the dynamical system (30), which in the theory of solitons are called *higher flows*.

There exists a regular algebraic description of these commuting flows through *formal Killing field* (see [FPPS]), which is in our case a symmetric $K_0(-\lambda) = \sigma_3 K_0(\lambda)\sigma_3$ formal power series solution

$$K_0(\lambda) = \sum_{m=1}^{\infty} K_m \lambda^{-m} \tag{34}$$

of

$$K_0(\lambda)_z = \big[U_0(\lambda), K_0(\lambda)\big], \qquad K_0(\lambda)_{\bar{z}} = \big[V_0(\lambda), K_0(\lambda)\big].$$

Coefficients K_m can be computed recursively.

Lemma 3. *The diagonal terms $K_{2n} = v_n\sigma_3$, $n = 1,\ldots$ of the formal Killing field (34) define tangential vector fields v_n*

$$(\partial_{z\bar{z}} + \cosh u)v_n = 0. \tag{35}$$

v_n are polynomials in $u_z,\ldots,u_z^{(2n-1)}$.

Any complex vector field v_n generates two real tangential vector fields

$$w_{2n-1} = v_n + \bar{v}_n, \qquad w_{2n} = i(v_n - \bar{v}_n), \quad n = 1,\ldots$$

Lemma 4. *Let $u(z,\bar{z})$ be a doubly periodic solution*

$$u(z + Z_i, \bar{z} + \bar{Z}_i) = u(z,\bar{z}), \quad i = 1,2 \quad \operatorname{Im} Z_1/Z_2 \neq 0$$

of the elliptic sinh-Gordon equation and $w_n, n = 1,\ldots$ be the corresponding tangential real vector fields. Only finitely many tangential vectors w_n are linearly independent.

Proof. All w_n are also doubly-periodic. Equation (35) determines an elliptic operator L on the torus T :

$$Lw_n = (\partial_z \partial_{\bar{z}} + \cosh u) w_n = 0.$$

It is well known that the spectra of this operator is discrete, which implies in particular that all the eigenspaces are finite dimensional. All tangential vectors w_n belong to the kernel of L. This observation proves the lemma: dim span$\{w_n\}_{n=1,\ldots} < \infty$.

This lemma is the reason for the existence of a *polynomial Killing field.*

Theorem 8. *Let $u(z, \bar{z})$ be a doubly-periodic solution of the elliptic sinh-Gordon equation (30), and U_0, V_0 are given by (32). Then in the loop algebra $g_H[\lambda]$ there exists a polynomial Killing field*

$$W_0(\lambda) = \sum_{n=-(2N-1)}^{2N-1} W_n \lambda^n \tag{36}$$

which satisfies

$$\begin{aligned} W_0(\lambda)_z &= \left[U_0(\lambda), W_0(\lambda)\right], \\ W_0(\lambda)_{\bar{z}} &= \left[V_0(\lambda), W_0(\lambda)\right]. \end{aligned} \tag{37}$$

This fundamental theorem in different forms appeared first in [PS, Hit]. For a recent elegant proof see [FLPPS].

The coefficients of $W_n, n > 0$ are polynomials in $u_z, \ldots, u_z^{(2N-1-n)}$ and $e^{\pm u/2}$, those of $W_n, n < 0$ are polynomials in $u_{\bar{z}}, \ldots, u_{\bar{z}}^{(2N-1+n)}$ and $e^{\pm u/2}$, W_0 is a polynominal of $u_z, u_{\bar{z}}, \ldots, u_z^{(2N-1)}, u_{\bar{z}}^{(2N-1)}$. The leading coefficient is of the form

$$W_{2N-1} = \alpha \begin{pmatrix} 0 & e^{-u/2} \\ e^{u/2} & 0 \end{pmatrix}, \qquad 0 \neq \alpha \in \mathbb{C}.$$

Solutions possessing polynomial Killing fields are called *solutions of finite type* or *finite-gap solutions.* The theory of finite-gap solutions is a well established branch [DKN, BBEIM] of the theory of integrable equations. Due to Theorem 8, all doubly-periodic solutions of the elliptic sinh-Gordon equation are finite-gap.

2.5. The spectral curve and Baker-Akhiezer function

Let $u(z,\bar{z})$ be a solution of the elliptic sinh-Gordon equation with the polynomial Killing field $W_0(\lambda)$. The curve

$$\det(W_0(\lambda) - \mu I) = 0 \tag{38}$$

is called the *spectral curve* of the solution $u(z,\bar{z})$. The spectral curve is independent of $z, \bar{z}$.

Compactified at $\mu = \infty$ the hyperelliptic curve (38) determines a compact Riemann surface $\hat{C}$ of genus $\hat{g}$. Due to symmetries of the loop algebra $g_H[\lambda]$, besides the hyperelliptic involution $(\mu,\lambda) \to (-\mu,\lambda)$ it possesses two more involutions: a holomorphic

$$\pi : (\mu,\lambda) \to (\mu, -\lambda) \tag{39}$$

and an anti-holomorphic $\hat{\tau} : (\mu,\lambda) \to (\bar{\mu}, \frac{1}{\bar{\lambda}})$.

The factor Riemann surface $C = \hat{C}/\pi$ plays central role for the explicit construction of section 2.6. The covering $\hat{C} \to C$, $(\mu,\lambda) \mapsto (\mu,\Lambda)$, $\Lambda := \lambda^2$ is unramified and C is a Riemann surface of genus g, where $\hat{g} = 2g-1$. The anti-holomorphic involution

$$\tau : (\mu,\Lambda) \to (\bar{\mu}, \frac{1}{\bar{\Lambda}}). \tag{40}$$

acts on C.

Due to (37) the system

$$\phi_z = U_0\phi, \qquad \phi_{\bar{z}} = V_0\phi, \qquad W_0\phi = \mu\phi$$

has a common vector valued solution $\phi(P,z,\bar{z})$, which is called the *Baker-Akhiezer function*. Here $P = (\mu,\lambda)$ is a point on $\hat{C}$. In the finite-gap integration theory of the sine-Gordon equation, usually a gauge equivalent function

$$\psi = \begin{pmatrix} e^{u/4} & 0 \\ 0 & e^{-u/4} \end{pmatrix} \phi$$

is used. Immersion formula (29) is obviously invariant with respect to this transformation.

Suitably normalized as $\psi(P,0,0) = \begin{pmatrix} 1 \\ * \end{pmatrix}$, this function satisfies

$$\psi(\pi P, z, \bar{z}) = \sigma_3 \psi(P, z, \bar{z}). \tag{41}$$

The Baker-Akhiezer function ψ has essential singularities at the points $\infty^\pm, 0^\pm \in \hat{C}$ defined by $\lambda(\infty^\pm) = \infty, \lambda(0^\pm) = 0$. The involution π interchanges these points $\pi(\infty^+) = \infty^-, \pi(0^+) = 0^-$. Denote their projection on C by ∞ and 0 respectively. Due to the symmetry (41) the pole divisor of ψ on $\hat{C} \setminus \{\infty^\pm, 0^\pm\}$ is the lift of a divisor $\mathcal{D}$ on $C \setminus \{\infty, 0\}$.

Finally after some computations one can prove the following analytic properties of ψ.

Theorem 9. *The Baker-Akhiezer function ψ possesses the following analytic properties:*

1. ψ is transformed by (41) under the action of the involution π,

2. ψ is meromorphic on $\hat{C} \setminus \{\infty^\pm, 0^\pm\}$. The pole divisor $\mathcal{D}$ of ψ on $C \setminus \{\infty, 0\}$ is independent of $z, \bar{z}$, and is a non-special divisor of degree g. The Abel map $\mathcal{A}(\mathcal{D})$ of $\mathcal{D}$ on C satisfies

$$\mathcal{A}(\mathcal{D} - \tau\mathcal{D}) = \mathcal{A}(0 - \infty), \tag{42}$$

3. ψ has essential singularities at the points $\infty^\pm, 0^\pm$ of the form

$$\begin{aligned} \psi(P, z, \bar{z}) &= \left(\begin{pmatrix} 1 \\ \pm 1 \end{pmatrix} + o(1) \right) \exp(\pm \frac{i\lambda z}{2}), \qquad P \to \infty^\pm. \\ \psi(P, z, \bar{z}) &= O(1) \exp(\mp \frac{i\bar{z}}{2\lambda}), \qquad P \to 0^\pm. \end{aligned}$$

2.6. Baker-Akhiezer function. Formulae

Due to the symmetry (41) the Baker-Akhiezer function ψ can be described in terms of the data $\{C, \mathcal{D}\}$. Here C is a hyperelliptic Riemann surface of genus g with the anti-holomorphic involution (40) and branch points at $\lambda = 0, \infty$ and $\mathcal{D}$ is a non-special divisor of degree g on C satisfying (42). We call these data admissible. It is crucial that the construction of Section 2.5 can be reversed and a result similar to Theorem 7 holds.

Theorem 10. *Let $\{C, \mathcal{D}\}$ be admissible data. There exists a Baker-Akhiezer function ψ with these data and ψ is uniquely characterized by the analytic properties listed in Theorem 9.*

Admissible $\{C, \mathcal{D}\}$ generate a finite-gap solution of the elliptic sinh-Gordon equation and thus a surface with constant mean curvature $H = 1$, which we call a CMC surface of *finite type.* It follows from Sections 2.4, 2.5 that all CMC tori are CMC surfaces of finite type.

The Baker-Akhiezer functions and hence CMC surfaces of finite type can be described explicitly. Let

$$M^2 = \Lambda \prod_{i=1}^{g} (\Lambda - \Lambda_i)(\Lambda - \frac{1}{\bar{\Lambda}_i}), \qquad | \Lambda_i |< 1 \quad \forall i$$

be a non-singular hyperelliptic curve C with an anti-holomorphic involution $\tau : \Lambda \to \frac{1}{\bar{\Lambda}}$. Choose a canonical homology basis $a_1, b_1, \ldots, a_g, b_g$ with a_i-cycles surrounding the cuts $[\Lambda_i, \frac{1}{\bar{\Lambda}_i}]$, i.e. $\tau a_i = -a_i$. Let $\omega_1, \ldots, \omega_g$ be the dual basis $\int_{a_n} \omega_m = 2\pi i \delta_{nm}$ of holomorphic differentials. The period matrix $B_{nm} = \int_{b_n} \omega_m$ determines the Riemann theta function

$$\theta(u) = \sum_{k \in \mathbb{Z}^g} \exp(\frac{1}{2}(Bk, k) + (u, k)), \qquad u \in \mathbb{C}^g.$$

We need also the Abelian differentials of the second kind Ω_∞, Ω_0 normalized by the condition

$$\int_{a_n} \Omega_\infty = \int_{a_n} \Omega_0 = 0, \qquad n = 1, \ldots, g$$

and the following asymptotics at the singularities

$$\begin{array}{ll} \Omega_\infty \to d\sqrt{\Lambda}, & \Lambda \to \infty \\ \Omega_0 \to d\frac{1}{\sqrt{\Lambda}}, & \Lambda \to 0. \end{array}$$

Denote the vector of b-periods of Ω_∞ by

$$U = (U_1, \ldots, U_g), \qquad U_n = \int_{b_n} \Omega_\infty.$$

Finally note that in explicit description one can replace the divisor $\mathcal{D}$ of admissible data by its Abel map $D \in Jac(C)$. One can show that in the chosen normalizations the reality condition (42) is equivalent to the condition $D \in i\mathbb{R}^g$.

Theorem 11. *The Baker-Akhiezer function with the data $\{C, D\}$ is given by the formulae*

$$\begin{aligned} \psi_1(P, z, \bar{z}) &= \frac{\theta(\int_\infty^P \omega + W)\theta(D)}{\theta(\int_\infty^P \omega + D)\theta(W)} \exp(\frac{i}{2}\int_\infty^P (z\Omega_\infty + \bar{z}\Omega_0)) \\ \psi_2(P, z, \bar{z}) &= \frac{\theta(\int_\infty^P \omega + W + \Delta)\theta(D)}{\theta(\int_\infty^P \omega + D)\theta(W + \Delta)} \exp(\frac{i}{2}\int_\infty^P (z\Omega_\infty + \bar{z}\Omega_0)). \end{aligned}$$

Here $\Delta = \pi i(1, \ldots, 1)$, *the vector* $D \in i\mathbb{R}^g$ *is arbitrary,*

$$W = i\mathrm{Re}\,(Uz) + D$$

and the integration paths in all the integrals are identical. The corresponding solution to the sinh-Gordon equation is given by

$$u(z, \bar{z}) = 2\log\frac{\theta(W + \Delta)}{\theta(W)}. \tag{43}$$

Applying now Theorem 6 to $\psi(P, z, \bar{z})$ with $P = P_0 = (M_0, \Lambda_0)$, $|\Lambda_0| = 1$ one arrives at the following final formulae for CMC immersions of finite type [Bob1].

Theorem 12. *The quaternion valued solution* $\Phi(z, \bar{z}, \lambda_0)$ *of the linear system (27, 32) with the finite-gap coefficient (43) is given by* $\Phi =$

$$\frac{i}{\sqrt{\theta(W)\theta(W + \Delta)}}\begin{pmatrix} \theta(W + l) & \theta(W - l) \\ \theta(W + \Delta + l) & -\theta(W + \Delta - l) \end{pmatrix}\exp(i\sigma_3\mathrm{Re}\,(zL)),$$

where $l = \int_\infty^{P_0}\omega$ *is the Abel map of* $P_0 = (M_0, \Lambda_0)$ *chosen on the unit circle* $\Lambda_0 = \lambda^2 = e^{2it}$ *and* $L = \int_\infty^{P_0}\Omega_\infty$. *The matrix* Φ *is normalized by*

$$\det\Phi = 2\frac{\theta(l)\theta(l + \Delta)}{\theta(0)\theta(\Delta)}.$$

The corresponding CMC immersion is given by (29). This immersion is doubly-periodic granted a lattice $\mathcal{L}$ *with the basic vectors* Z_1, Z_2 *exists such that*

$$\mathrm{Re}\,(Z_k U) \in 2\pi\mathbb{Z}^g, \quad \mathrm{Re}\,(2Z_k\int_\infty^{P_0}\Omega_\infty) \in 2\pi\mathbb{Z}, \qquad k = 1, 2, \tag{44}$$

and the differential Ω_∞ *vanishes at the point* P_0

$$\Omega_\infty(\lambda_0) = 0. \tag{45}$$

CMC tori are singled out from general quasiperiodic immersions of finite type by the periodicity conditions (44, 45), which are in fact conditions on the corresponding hyperelliptic curve C of genus g only. One can show [Jag, Bob1] that there are no CMC tori with $g = 1$ and that for $g > 1$ there exists a discrete set of spectral curves C generating CMC tori. The parameter $D \in i\mathbb{R}^g$ remains arbitrary. So the CMC tori with $g > 2$ (changes of D in the plane span$\{\mathrm{Re}\,U, \mathrm{Im}\,U\}$ are equivalent to reparametrization of the torus) possess commuting deformation flows. These deformations are area preserving [Bob1].

2.7. Examples of CMC tori

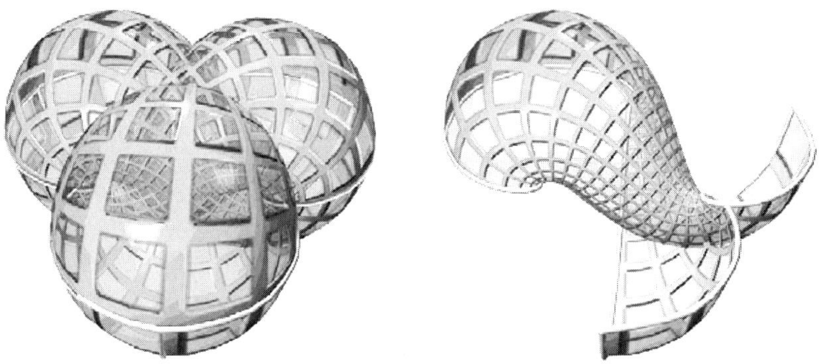

(a) (b)

Fig. 1. Wente torus having a threefold symmetry, and one fundamental piece

All finite-gap solutions of the sinh-Gordon equation of genus $g = 1$ and $g = 2$ are doubly-periodic. There are no CMC tori with $g = 1$. The simplest CMC tori were found by Wente [Wen] and analytically studied by Abresch [Abr] and Walter [Wal]. These tori presented in Figures 1a, 1b possess a family of plane curvature lines. This implies the additional symmetry $\Lambda \to 1/\Lambda$ of the corresponding spectral curve of genus $g = 2$. The Wente torus in Figure 1a comprises three congruent fundamental domains shown in Figure 1b.

Spectral curves of genus $g = 2$ without additional symmetries also generate CMC tori. An example is presented in Figure 2.

Taking spectral curves with $g = 3$ and the symmetry $\Lambda \to 1/\Lambda$ one obtains all CMC tori with spherical curvature lines. The fundamental domain of such an example is shown in Figure 3.

Figure 4 visualizes a CMC torus corresponding to a curve of genus $g = 5$. This torus possesses a 3-parameter family of area preserving deformations. Finally Figure 5 presents classical surfaces of Delaunay which correspond to spectral curves of genus $g = 1$ and are CMC surfaces of revolution.

The figures of this section are produced by Matthias Heil using formulae presented in this lecture and the software for calculations on

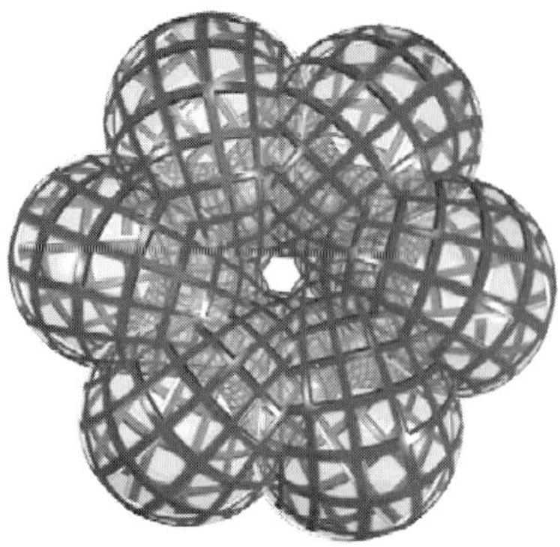

Fig. 2. Twisted Wente torus

Fig. 3. One fundamental piece of a torus with spherical curvature lines

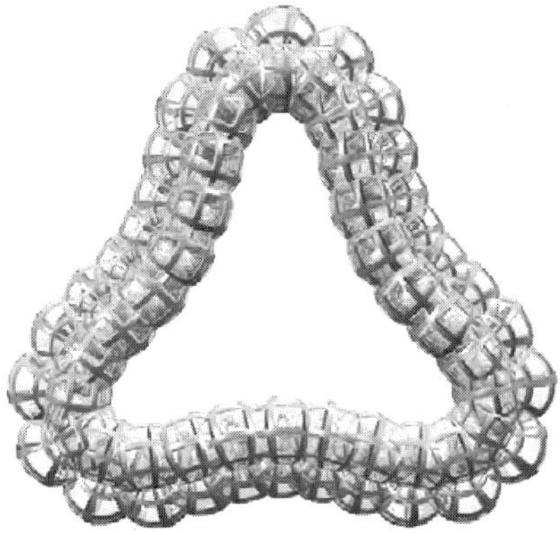

Fig. 4. A torus with spectral curve of genus $g = 5$

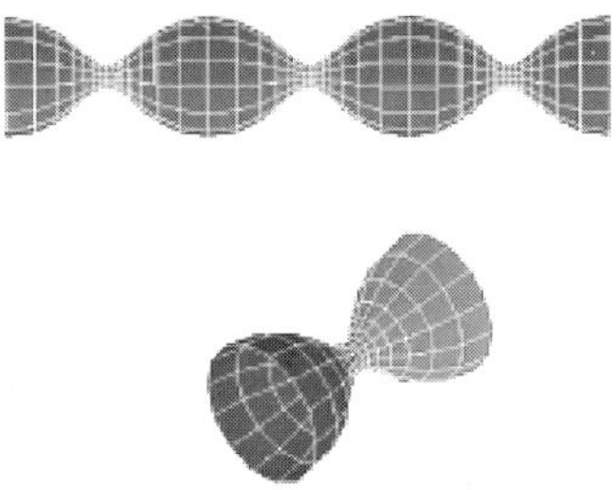

Fig. 5. Delaunay surface

hyperelliptic Riemann surfaces developed by him for Sfb288 in Berlin. Further examples can be found in [Hei].

§3. Bonnet pairs

In this section we present some preliminary results on local and global geometry of Bonnet pairs.

3.1. Basic facts about Bonnet pairs

Let $\mathcal{F}_1, \mathcal{F}_2 \subset \mathbb{R}^3$ be a smooth Bonnet pair (Bonnet mates), i.e. two isometric non-congruent surfaces with coinciding mean curvatures at the corresponding points. As conformal immersions of the same Riemann surface

$$F_1 : \mathcal{R} \to \mathbb{R}^3, \qquad F_2 : \mathcal{R} \to \mathbb{R}^3$$

they are described by the corresponding Hopf differentials Q_1, Q_2, the common metric $e^u\, dzd\bar{z}$ and the mean curvature function H. Since the surfaces are non-congruent the Hopf differentials differ $Q_1 \not\equiv Q_2$.

The Gauss-Codazzi equations immediately imply

Proposition 2. *Let Q_1 and Q_2 be the Hopf differentials of a Bonnet pair F_1 and $F_2 \to \mathbb{R}^3$. Then*

$$h = Q_2 - Q_1 \tag{46}$$

is a holomorphic quadratic differential $h\, dz^2$ on $\mathcal{R}$ and

$$\mid Q_1 \mid = \mid Q_2 \mid . \tag{47}$$

Due to the second statement of Proposition 2 the umbilic points of $\mathcal{F}_1$ and $\mathcal{F}_2$ correspond. Denote by

$$\mathcal{U} = \{P \in \mathcal{R} : \ Q_k(P) = 0\}$$

the corresponding set of umbilic points on $\mathcal{R}$.

Proposition 3. *Let Q_1 and Q_2 be the Hopf differentials of a Bonnet pair F_1 and $F_2 \to \mathbb{R}^3$. Then there exist a holomorhic quadratic differential h on $\mathcal{R}$ and a smooth real valued function $\alpha : \mathcal{R} \to \mathbb{R}$ such that*

$$Q_1 = \frac{1}{2}h(i\alpha - 1), \qquad Q_2 = \frac{1}{2}h(i\alpha + 1). \tag{48}$$

Proof. Define a smooth quadratic differential $g\, dz^2$ by $g = Q_1 + Q_2$. Identity (47) implies $h\bar{g} + g\bar{h} = 0$. Thus the quotient $\alpha = -i\frac{g}{h}$ is a real valued smooth function $\alpha : \mathcal{R}\backslash\mathcal{U}_h \to \mathbb{R}$, where $\mathcal{U}_h = \{P \in \mathcal{R} : \ h(P) = 0\}$

is the zero set of h. Let us show that α can be smoothly extended to the whole of $\mathcal{R}$. At any $z_0 \in \mathcal{U}_h$ the holomorphic differential h is of the form

$$h(z) = (z - z_0)^J h_0(z), \quad h_0(z_0) \neq 0, \quad J \in \mathbb{N}.$$

Real-valuedness of α near z_0 implies $g(z) = (z - z_0)^J g_0(z)$ with g_0 smooth, which in its turn implies smoothness of α at z_0.

Corollary 2. *Umbilic points of a Bonnet pair are isolated. The umbilic set coincides with the zero set of h, $\mathcal{U} = \mathcal{U}_h$.*

The number $-J$ where J is defined above is called the *index* of the umbilic point. We call the zero divisor $D_u = (h)$ of h the *umbilic divisor* of a Bonnet pair.

For compact Riemann surfaces $\mathcal{R}$, Propositions 2, 3 imply the following

Theorem 13.

(i) There exist no Bonnet pairs of genus $g = 0$.

(ii) Bonnet pairs of genus $g = 1$ have no umbilic points.

(iii) The umbilic divisor D_u of a Bonnet pair of genus $g \geq 1$ is of degree $4g - 4$ and its class is $D_u \equiv 2K$, where K is the canonical divisor.

Proof. A holomorphic quadratic differential on a sphere vanishes identically $h \equiv 0$, which means $Q_1 = Q_2$, and the surfaces are congruent. A holomorphic quadratic differential on a torus does not have zeros, thus $\mathcal{U} = \emptyset$ for tori.

The point (i) of Theorem 13 was proven in [LT].

Taking into account the similarity of the analytic description of Bonnet surfaces and CMC surfaces and the progress in the investigation of CMC surfaces achieved by methods from the theory of integrable systems (see Section 2), the most promising open problem to attack by these methods seems to be the problem of existence and description of Bonnet tori mates.

For tori one has $\mathcal{R} = \mathbb{C}/\mathcal{L}$. Scaling the lattice $\mathcal{L}$ appropriately one can always normalize $h = -i$, i.e.

$$Q_1 = \frac{1}{2}(\alpha + i), \qquad Q_2 = \frac{1}{2}(\alpha - i). \tag{49}$$

The corresponding Gauss-Codazzi equations of Bonnet mates become

$$
\begin{aligned}
2u_{z\bar{z}} + H^2 e^u - (1+\alpha^2)e^{-u} &= 0, \\
\alpha_{\bar{z}} - e^u H_z &= 0, \\
\alpha_z - e^u H_{\bar{z}} &= 0.
\end{aligned} \tag{50}
$$

Note that the Gauss-Codazzi equations of isothermic surfaces $Q = \alpha/2 \in \mathbb{R}$ differ only slightly

$$
\begin{aligned}
2u_{z\bar{z}} + H^2 e^u - \alpha^2 e^{-u} &= 0, \\
\alpha_{\bar{z}} - e^u H_z &= 0, \\
\alpha_z - e^u H_{\bar{z}} &= 0.
\end{aligned} \tag{51}
$$

3.2. Lax representation and connection to isothermic surfaces

Looking for a Lax representation for Bonnet pairs, it is natural to try to merge the frame equations of two Bonnet mates. For tori, cylinders, or simply connected domains it is enough to consider the case of $\mathcal{R}$ being a domain D in $\mathbb{C}$. Cylinders and tori are distinguished by the corresponding periodicity lattices $\mathcal{L}$. Since our main interest lies in the investigation of tori let us restrict ourselves to the case of umbilic free Bonnet pairs. As in Section 2, introducing a global complex variable z on D we normalize the corresponding frame matrices traceless and the Hopf differentials as in (49). The following theorem can be checked directly.

Theorem 14. *Normalized by (49), conformal frames $\Phi_1, \Phi_2 : D \to SU(2)$ of a Bonnet pair*

$$
\Phi_{k\,z} = U_k \Phi_k, \qquad \Phi_{k\,\bar{z}} = V_k \Phi_k
$$

$$
U_k = \begin{pmatrix} \frac{u_z}{4} & -Q_k e^{-u/2} \\ \frac{H}{2} e^{u/2} & -\frac{u_z}{4} \end{pmatrix}, \qquad V_k = \begin{pmatrix} -\frac{u_{\bar{z}}}{4} & -\frac{H}{2} e^{u/2} \\ \bar{Q}_k e^{-u/2} & \frac{u_{\bar{z}}}{4} \end{pmatrix}
$$

can be extended

$$
\Phi(z, \bar{z}, \lambda = 0) = \begin{pmatrix} \Phi_1 & 0 \\ 0 & \Phi_2 \end{pmatrix} (z, \bar{z}) \tag{52}
$$

to $\Phi(z, \bar{z}, \lambda)$ satisfying

$$
\Phi_z = U\Phi, \qquad \Phi_{\bar{z}} = V\Phi \tag{53}
$$

with

$$(54)\qquad U = \begin{pmatrix} \begin{matrix} & U_1 & \\ \end{matrix} & \begin{matrix} 0 & 0 \\ -i\lambda e^{u/2} & 0 \end{matrix} \\ \begin{matrix} 0 & -i\lambda e^{-u/2} \\ 0 & 0 \end{matrix} & \begin{matrix} & U_2 & \end{matrix} \end{pmatrix},$$

$$V = \begin{pmatrix} \begin{matrix} & V_1 & \end{matrix} & \begin{matrix} 0 & -i\lambda e^{u/2} \\ 0 & 0 \end{matrix} \\ \begin{matrix} 0 & 0 \\ -i\lambda e^{-u/2} & 0 \end{matrix} & \begin{matrix} & V_2 & \end{matrix} \end{pmatrix}.$$

Conversely, the linear system (54) with $Q_1 = Q \neq Q_2 = \bar{Q}$ is compatible if and only if the metric e^u, the mean curvature function H, and the Hopf differentials Q_1, Q_2 on D satisfy the Gauss-Codazzi equations of Bonnet pairs. Conformal frames of the Bonnet mates are determined through (52) by a suitably normalized common solution $\Phi(z, \bar{z}, \lambda)$ of (53) evaluated at $\lambda = 0$.

Remark. In the general case of an arbitrary Riemann surface $\mathcal{R}$ and holomorphic quadratic differential h, a spinor form of the Lax representation for Bonnet pairs similar to (14) can be easily made by merging the spinor frames (see Section 1) of the corresponding surfaces.

Remark. In the case $Q_1 = Q_2 = Q = \bar{Q}$ system (53, 54) becomes a Lax representation for isothermic surfaces in $\mathbb{R}^3$ in isothermic coordinates.

The matrices $U + V$ and $i(U - V)$ corresponding to real vector fields ∂_x and ∂_y possess the symmetries

$$(55)\qquad A(-\lambda) = \begin{pmatrix} \mathbf{1} & 0 \\ 0 & -\mathbf{1} \end{pmatrix} A(\lambda) \begin{pmatrix} \mathbf{1} & 0 \\ 0 & -\mathbf{1} \end{pmatrix},$$

$$(56)\qquad \overline{A(\bar{\lambda})} = \begin{pmatrix} \sigma_2 & 0 \\ 0 & \sigma_2 \end{pmatrix} A(\lambda) \begin{pmatrix} \sigma_2 & 0 \\ 0 & \sigma_2 \end{pmatrix}.$$

Denote by

$$g_B[\lambda] = \{A : \mathbb{R} \to gl(2, \mathbb{H}) \mid A(-\lambda) = \begin{pmatrix} \mathbf{1} & 0 \\ 0 & -\mathbf{1} \end{pmatrix} A(\lambda) \begin{pmatrix} \mathbf{1} & 0 \\ 0 & -\mathbf{1} \end{pmatrix}\}$$

the corresponding loop algebra, and by

$$G_B[\lambda] = \{\phi : \mathbb{R} \to GL(2, \mathbb{H}) \mid \phi(-\lambda) = \begin{pmatrix} \mathbf{1} & 0 \\ 0 & -\mathbf{1} \end{pmatrix} \phi(\lambda) \begin{pmatrix} \mathbf{1} & 0 \\ 0 & -\mathbf{1} \end{pmatrix}\}$$

the corresponding loop group.

By the normalization (52) the solution $\Phi(z,\bar z,\lambda)$ is determined uniquely up to right multiplication by a matrix depending only on λ

$$\Phi(z,\bar z,\lambda) \to \Phi(z,\bar z,\lambda)G(\lambda), \qquad G(0)=1. \tag{57}$$

$\Phi(z,\bar z,\lambda)$ can be chosen to lie in $G_B[\lambda]$. Then the matrix

$$\begin{pmatrix} 0 & S \\ T & 0 \end{pmatrix} := \Phi^{-1}\Phi_\lambda \mid_{\lambda=0} \tag{58}$$

is off-diagonal and its coefficients are quaternion valued functions of z and $\bar z$.

For a description of the geometry of S and T, the notion of isothermic surfaces in $\mathbb{R}^4$ and of the dual isothermic surface is required. An immersion $f : D \to \mathbb{R}^4$ is called *isothermic* if it is conformal and the vector f_{xy} lies in the tangent plane $f_{xy} \in \operatorname{span}\{f_x, f_y\}$. It is convenient to describe isothermic surfaces in $\mathbb{R}^4 = \mathbb{H}$ in quaternionic form, i.e. as mappings $f : D \to \mathbb{H}$ with the coordinates

$$f = f_0\mathbf{1} + f_1\mathbf{i} + f_2\mathbf{j} + f_3\mathbf{k}.$$

Its differential is $df = f_x\,dx + f_y\,dy$. An important property of an isothermic immersion $f : D \to \mathbb{H}$ is the closedness of the form

$$df^* := -f_x^{-1}dx + f_y^{-1}dy \tag{59}$$

The corresponding immersion determined up to translation by this form and denoted by $f^* : D \to \mathbb{H}$ is also isothermic and is called the *dual isothermic surface.* Note that the dual isothermic surface is defined through one-forms and therefore the periodicity properties of f are not respected. The relation (59) is an involution. Moreover one can check that

Lemma 5. *The transformation (59) is characteristic for isothermic surfaces.*

Proposition 4.

(i) Let Q_1 and Q_2, normalized by (49), be Hopf differentials of a Bonnet pair F_1 and $F_2 : D \to \operatorname{Im}\mathbb{H}$. Then $T : D \to \mathbb{H}$ defined by (58) is an isothermic surface in the three-dimensional sphere $S^3 \subset \mathbb{R}^4 = \mathbb{H}$ and $S : D \to \mathbb{H}$ in (58) is its dual $S = T^*$. The isothermic surfaces S and T are related to the Bonnet pair by

$$\begin{aligned} dF_1 &= dS\,T = dT^*\,T \\ dF_2 &= T\,dS = T\,dT^*. \end{aligned} \tag{60}$$

(ii) Let $Q_1 = Q_2 = Q$ in (54) be real. Then $S : D \to \operatorname{Im}\mathbb{H}$ given by (58) is the isothermic surface determined by the fundamental forms with the coefficients e^u, H, Q, and $T : D \to \operatorname{Im}\mathbb{H}$ is its dual $T = S^*$.

Proof. Let us prove the first statement. Formula (58) implies

$$d\begin{pmatrix} 0 & S \\ T & 0 \end{pmatrix} = \Phi^{-1}\left(U_\lambda dz + V_\lambda d\bar{z}\right)\Phi\mid_{\lambda=0}$$

or equivalently

$$(61)\quad dS = e^{u/2}\Phi_1^{-1}(\mathbf{i}dx + \mathbf{j}dy)\Phi_2, \qquad dT = e^{-u/2}\Phi_2^{-1}(\mathbf{i}dx - \mathbf{j}dy)\Phi_1.$$

These frames are obviously conformal and are related by (59). Lemma 5 implies that T is isothermic. Moreover dT can be integrated explicitly

$$(62)\qquad T = \Phi_2^{-1}\Phi_1.$$

Indeed, differentiating the last expression one obtains

$$dT = \Phi_2^{-1}(d\Phi_1\Phi_1^{-1} - d\Phi_2\Phi_2^{-1})\Phi_1 =$$
$$e^{-u/2}\Phi_2^{-1}\left((Q_2 - Q_1)\begin{pmatrix} 0 & 1 \\ 0 & 0 \end{pmatrix}dz + (\bar{Q}_1 - \bar{Q}_2)\begin{pmatrix} 0 & 0 \\ 1 & 0 \end{pmatrix}d\bar{z}\right)\Phi_1$$

which coincides with the previous expression for dT. Integrating, one obtains $T = \Phi_2^{-1}\Phi_1 + \text{constant}$. The constant can be normalized to zero by transformation (57) with an appropriate $G(\lambda) \in G_B[\lambda]$. Obviously the surface given by (62) lies in the three sphere $S^3 = \mathbb{H}_1$. Using (62) we obtain

$$\begin{aligned} dS\,T &= -ie^{-u/2}\Phi_1^{-1}\begin{pmatrix} 0 & d\bar{z} \\ dz & 0 \end{pmatrix}\Phi_1 \\ T\,dS &= -ie^{-u/2}\Phi_2^{-1}\begin{pmatrix} 0 & d\bar{z} \\ dz & 0 \end{pmatrix}\Phi_2 \end{aligned}$$

which coincides with (60).

The proof of the second claim is even simpler (see [BP]).

The next theorem is essentially due to Bianchi [Bia2]. A modern version of it in terms of quaternions is derived in [KPP].

Theorem 15. *F_1 and $F_2 : D \to \operatorname{Im}\mathbb{H} = \mathbb{R}^3$ build a Bonnet pair if and only if there exists an isothermic surface $T : D \to \mathbb{H}_1 = S^3 \subset \mathbb{R}^4$*

(or equivalently an isothermic surface $R : D \to \operatorname{Im}\mathbb{H} = \mathbb{R}^3$) such that

$$(63) \qquad \begin{aligned} dF_1 &= dT^* T = \frac{1}{2}(1-R)dR^*(1+R) \\ dF_2 &= TdT^* = \frac{1}{2}(1+R)dR^*(1-R). \end{aligned}$$

Proof. Let us show first the equivalence of the representations in terms of T and R. The class of isothermic surfaces is invariant under Möbius transformations. In particular, isothermic surfaces in S^3 and in $\mathbb{R}^3$ are related by stereographic projection, which in quaternionic form can be represented by $T = \frac{1+R}{1-R}$ with $R \in \operatorname{Im}\mathbb{H} = \mathbb{R}^3, T \in \mathbb{H}_1 = S^3$. For the frames this implies

$$\begin{aligned} dT &= 2(1-R)^{-1}dR(1-R)^{-1} \\ dT^* &= \frac{1}{2}(1-R)dR^*(1-R), \end{aligned}$$

which proves the equivalence of the two representations of the theorem. The passing from Bonnet pairs to isothermic surfaces is proven in Proposition 4. Conversely, given R, the representation (63) shows that dF_1 and dF_2 are conformal and lie in $\operatorname{Im}\mathbb{H}$. Due to $dF_2 = TdF_1T^{-1}$ the immersions are isometric.

Similarly one can show (see [KPP]) that the forms dF_1, dF_2 defined by (63) are closed and that the mean curvature functions of the corresponding surfaces coincide.

To study a global version of Theorem 15 in the case of Bonnet tori mates let F_1 and $F_2 : \mathbb{C} \to \mathbb{R}^3$ be a Bonnet pair with doubly periodic frames dF_1 and dF_2 with the same period lattice $\mathcal{L}$. The frames Φ_1 and Φ_2 are periodic up to a sign

$$\Phi_k(z + Z_i) = (-1)^{p_{ik}}\Phi_k(z),$$

where Z_1, Z_2 are generators of $\mathcal{L}$ and $p_{ik} \in \mathbb{Z}_2$ characterize the spin structures of the immersions. The isothermic surface given by (62) is a torus in S^3 if the spin structures of F_1 and F_2 coincide ($p_{i1} = p_{i2}, i = 1, 2$) and is a torus in three dimensional real projective space $\mathbb{R}P^3 = S^3/\{-1\}$ if the spin structures differ. Conversely, an isothermic immersion $T : \mathbb{C}/\mathcal{L} \to S^3/\{-1\}$ generates by (63) a Bonnet pair with frames dF_1 and dF_2 defined on $\mathbb{C}/\mathcal{L}$. The spin structures of these two surfaces are the same iff T is an immersion to S^3 with the lattice $\mathcal{L}$.

Corollary 3. *Bonnet mates with doubly periodic frames dF_1 and dF_2 are in one to one correspondence with isothermic tori in $\mathbb{R}P^3$. Bonnet mates with doubly periodic frames $dF_{1,2}$ and coinciding spin structure are in one to one correspondence with isothermic tori in S^3. The corresponding relations are given by formulae (63).*

Formula (63) allows us to control the periodicity of the frame of a Bonnet pair. To be able to control the periodicity of the immersion one needs an analog of formula (29) describing the corresponding immersion without integration. We call a solution

$$\begin{array}{rcl} \Phi : D \times \mathbb{R} & \to & G_B[\lambda] \\ (z, \lambda) & \mapsto & \Phi(z, \bar{z}, \lambda) \end{array}$$

of (53, 54) *normalized* (see proof of Proposition 4) if the coefficient T of its decomposition (58) at $\lambda = 0$ is a unitary quaternion $T : D \to \mathbb{H}_1$, i.e. (62) holds. Obviously this solution can be extended to all $\lambda \in \mathbb{C}$.

Proposition 5. *Let $\Phi(z, \bar{z}, \lambda)$ a normalized solution of (54) with Q of the form (49). Then the corresponding Bonnet pair $F_1(z, \bar{z})$ and $F_2(z, \bar{z})$ is restored by the following coefficients of quaternionic 2 by 2 matrices*

$$\begin{pmatrix} F_1 & 0 \\ 0 & * \end{pmatrix} = \frac{1}{2}\Phi^{-1}\Phi_{\lambda\lambda} \,|_{\lambda=0}, \tag{64}$$

$$\begin{pmatrix} * & 0 \\ 0 & F_2 \end{pmatrix} = \frac{1}{2}\Phi^{-1}\Phi_{\lambda\lambda} - \left(\Phi^{-1}\Phi_{\lambda}\right)_{\lambda} |_{\lambda=0}. \tag{65}$$

Proof. To prove the formula for dF_1 let us differentiate it by z and $\bar{z}$ to obtain

$$d(\frac{1}{2}\Phi^{-1}\Phi_{\lambda\lambda}) = \Phi^{-1}(U_{\lambda}dz + V_{\lambda}d\bar{z})\Phi_{\lambda},$$

where we used $U_{\lambda\lambda} = V_{\lambda\lambda} = 0$. Evaluating this expression at $\lambda = 0$ using (52), (62) and (61), one finally obtains

$$d(\frac{1}{2}\Phi^{-1}\Phi_{\lambda\lambda})_{|\lambda=0} = \begin{pmatrix} dS\,T & 0 \\ 0 & dT\,S \end{pmatrix}.$$

Now the formula for dF_1 follows from (60). The formula for dF_2 is proven by an analogous computation

$$d(\frac{1}{2}\Phi^{-1}\Phi_{\lambda\lambda} - \left(\Phi^{-1}\Phi_{\lambda}\right)_{\lambda})_{|\lambda=0} = \begin{pmatrix} S\,dT & 0 \\ 0 & T\,dS \end{pmatrix}.$$

3.3. Loop group description

As we have seen already in Proposition 4 the theory developed in Section 3.2 includes two different cases: the case of Bonnet pairs when the Hopf differentials $Q_1 = Q$ and $Q_2 = \bar{Q}$ are different and thus generate two non-congruent surfaces, and the case $Q = \bar{Q}$ of isothermic surfaces in $\mathbb{R}^3$. The loop group $G_B[\lambda]$ and the loop algebra $g_B[\lambda]$ of Bonnet pairs are described in Section 3.2. In the case of isothermic surfaces $Q = \bar{Q}$, the corresponding loop group and algebra are specialized further as follows:

$$\begin{aligned} G_I[\lambda] &= \{\phi \in G_B[\lambda] \mid \phi^T(\lambda)\begin{pmatrix} 0 & \sigma_2 \\ \sigma_2 & 0 \end{pmatrix}\phi(\lambda) = \begin{pmatrix} 0 & \sigma_2 \\ \sigma_2 & 0 \end{pmatrix}\} \\ g_I[\lambda] &= \{A \in g_B[\lambda] \mid A^T(\lambda) = -\begin{pmatrix} 0 & \sigma_2 \\ \sigma_2 & 0 \end{pmatrix} A(\lambda) \begin{pmatrix} 0 & \sigma_2 \\ \sigma_2 & 0 \end{pmatrix}\}. \end{aligned}$$

As in the previous sections the main strategy for applying analytic methods of the theory of integrable systems consists of two steps: first, to characterize the frame equations through analytic properties of Φ as a function of λ without referring to the special geometric nature of the coefficients of the frame equations and, second, to construct those $\Phi(\lambda)$ explicitly. For this purpose it is more convenient to pass to a gauge equivalent

$$\Psi = \begin{pmatrix} e^{-u/4}\mathbf{1} & 0 \\ 0 & e^{u/4}\mathbf{1} \end{pmatrix} \Phi \tag{66}$$

to (53), linear problem

$$\Psi_z = U_0 \Psi, \qquad \Psi_{\bar{z}} = V_0 \Psi \tag{67}$$

$$U_0 = \begin{pmatrix} 0 & -Qe^{-u/2} & 0 & 0 \\ \frac{H}{2}e^{u/2} & -\frac{u_z}{2} & -i\lambda & 0 \\ 0 & -i\lambda & \frac{u_z}{2} & -\bar{Q}e^{-u/2} \\ 0 & 0 & \frac{H}{2}e^{u/2} & 0 \end{pmatrix}, \tag{68}$$

$$V_0 = \begin{pmatrix} -\frac{u_{\bar{z}}}{2} & -\frac{H}{2}e^{u/2} & 0 & -i\lambda \\ \bar{Q}e^{-u/2} & 0 & 0 & 0 \\ 0 & 0 & 0 & -\frac{H}{2}e^{u/2} \\ -i\lambda & 0 & Qe^{-u/2} & \frac{u_{\bar{z}}}{2} \end{pmatrix},$$

normalizing the solution at $\lambda = \infty$. Note that all immersion formulae (64, 58) are preserved under this gauge transformation.

Proposition 6. *Let*

$$\begin{array}{rcl}\Psi : D \times \mathbb{R} & \to & G_B[\lambda] \\ z, \lambda & \mapsto & \Psi(z, \bar{z}, \lambda)\end{array}$$

be a smooth mapping satisfying

$$\Psi_z \Psi^{-1} = A(z, \bar{z})\lambda + B(z, \bar{z}) \tag{69}$$

with some $A(z, \bar{z})$ and $B(z, \bar{z})$, having the asymptotics

$$\Psi(z, \bar{z}, \lambda) = (L + M(z, \bar{z})\lambda^{-1} + o(\lambda^{-1})) \exp(-i\lambda(J_1 z + J_2 \bar{z}))C(\lambda) \tag{70}$$

at $\lambda \to \infty$ with

$$L = \begin{pmatrix} 1 & 0 & 0 & 1 \\ 0 & 1 & 1 & 0 \\ 0 & 1 & -1 & 0 \\ 1 & 0 & 0 & -1 \end{pmatrix}$$

and

$$J_1 = \begin{pmatrix} 0 & 0 & 0 & 0 \\ 0 & 1 & 0 & 0 \\ 0 & 0 & -1 & 0 \\ 0 & 0 & 0 & 0 \end{pmatrix}, \quad J_2 = \begin{pmatrix} 1 & 0 & 0 & 0 \\ 0 & 0 & 0 & 0 \\ 0 & 0 & 0 & 0 \\ 0 & 0 & 0 & -1 \end{pmatrix}$$

for some $M(z, \bar{z})$ and $C(\lambda)$. If, for the coefficients of M, one has

$$M_{31}(z, \bar{z}) = -M_{42}(z, \bar{z}), \tag{71}$$

then

$$U_0 := \Psi_z \Psi^{-1}, \qquad V_0 := \Psi_{\bar{z}} \Psi^{-1} \tag{72}$$

can be parametrized as in (68) with some real valued $u(z, \bar{z}), H(z, \bar{z})$. If in addition

$$M_{21}(z, \bar{z}) = -M_{12}(z, \bar{z}), \tag{73}$$

then $Q(z, \bar{z})$ in (68) is also real valued.

Proof. Due to the assumption of the theorem, U_0 is of the form $U_0 = A\lambda + B$. Substituting the asymptotics (70) one obtains for the coefficients

$$A = -iLJ_1L^{-1}, \qquad B = [ML^{-1}, A].$$

Similarly the symmetry (55) and the asymptotics (70) imply $V_0 = C\lambda + D$ with

$$C = -iLJ_2L^{-1}, \qquad D = [ML^{-1}, C].$$

The matrices A and C are of the required form. For the matrices B and D a simple computation gives

$$(74)\quad \begin{aligned} B &= -i\begin{pmatrix} 0 & M_{12} & 0 & 0 \\ -M_{31} & M_{22}-M_{32} & 0 & 0 \\ 0 & 0 & M_{32}-M_{22} & -M_{21} \\ 0 & 0 & M_{42} & 0 \end{pmatrix} \\ D &= -i\begin{pmatrix} M_{11}-M_{41} & -M_{42} & 0 & 0 \\ M_{21} & 0 & 0 & 0 \\ 0 & 0 & 0 & M_{31} \\ 0 & 0 & -M_{12} & M_{41}-M_{11} \end{pmatrix}, \end{aligned}$$

where we have used the symmetry of M

$$\begin{pmatrix} 1 & 0 & 0 & 0 \\ 0 & 1 & 0 & 0 \\ 0 & 0 & -1 & 0 \\ 0 & 0 & 0 & -1 \end{pmatrix} M \begin{pmatrix} 0 & 0 & 0 & 1 \\ 0 & 0 & 1 & 0 \\ 0 & 1 & 0 & 0 \\ 1 & 0 & 0 & 0 \end{pmatrix} = -M.$$

The anti-holomorphic involution $\lambda \to \bar{\lambda}$ of the loop group implies

$$M_{12} = \bar{M}_{21},\ M_{42} = \bar{M}_{31},\ M_{11} = -\bar{M}_{22},\ M_{32} = -\bar{M}_{41}.$$

Finally, comparing the coefficients and using the compatibility conditions one shows that U_0, V_0 are of the form (68).

The result of Proposition 6 is not optimal. Whereas the conditions (69) and (70) are quite standard for the loop group description of integrable systems, the constraint (71) is difficult to take into account by analytic construction of the Ψ-function. For isothermic surfaces the situation is more fortunate. Ψ-functions satisfying both constraints (71), (73) can be characterized in terms of the corresponding loop group $G_I[\lambda]$.

Theorem 16. *Let*

$$\begin{aligned} \Psi : D \times \mathbb{R} &\to G_I[\lambda] \\ z, \lambda &\mapsto \Psi(z, \bar{z}, \lambda) \end{aligned}$$

be a smooth mapping satisfying (69) and (70). Then its logarithmic derivatives (72) can be parametrized as in (68) with some real valued $u(z,\bar{z}), H(z,\bar{z}), Q(z,\bar{z})$. *The corresponding isothermic surface* $R : D \to \mathbb{R}^3$ *and its dual* $R^* : D \to \mathbb{R}^3$ *are given by*

$$(75)\qquad \begin{pmatrix} 0 & R \\ R^* & 0 \end{pmatrix} = \Psi^{-1}\Psi_\lambda \mid_{\lambda=0}.$$

Proof. Both constraints (71, 73) follow from the condition $U_0 = A\lambda + B \in g_I[\lambda]$ and formula (74). The immersion formula for isothermic surfaces mentioned as part (ii) of Proposition 4 is preserved by the gauge transformation (66).

The Kamberov-Pedit-Pinkall formula (63) for the frame of a Bonnet pair also can be explicitly integrated in terms of the Ψ-function of the corresponding isothermic surface in $\mathbb{R}^3$. The corresponding formula was obtained jointly with P. Grinevich.

Theorem 17. *Let $\Psi : D \times \mathbb{R} \to G_I[\lambda]$ be a Ψ-function of an isothermic surface $R : D \to \mathbb{R}^3$, i.e. it satisfies (67) with $Q = \bar{Q}$. Then the Bonnet pair $F_{1,2} : D \to \mathbb{R}^3$ corresponding to it by (63) is given by*

$$
\begin{aligned}
F_1 &= \frac{1}{2}(R^* - RB + C + A + B - RR^*) \\
F_2 &= \frac{1}{2}(R^* - RB + C - A - B + RR^*),
\end{aligned}
\tag{76}
$$

where the quaternionic coefficients are defined by (75) and

$$
\begin{pmatrix} A & 0 \\ 0 & B \end{pmatrix} = \frac{1}{2}\Psi^{-1}\Psi_{\lambda\lambda}\,|_{\lambda=0} \qquad \begin{pmatrix} 0 & C \\ D & 0 \end{pmatrix} = \frac{1}{6}\Psi^{-1}\Psi_{\lambda\lambda\lambda}\,|_{\lambda=0}\,. \tag{77}
$$

Proof. Since all the immersion formulae are invariant with respect to the gauge transformation (66) one can use any of the extended frames Φ or Ψ of the isothermic surface performing computations. The diagonal or off-diagonal structure of the matrices (75) and (77) follows from the symmetry (55). Introducing $W := Udz + Vd\bar{z}$ and using $W_{\lambda\lambda} = W_{\lambda\lambda\lambda} = 0$ one obtains

$$
\begin{aligned}
d\Psi &= W\Psi \\
d(\Psi^{-1}\Psi_\lambda) &= \Psi^{-1}W_\lambda\Psi \\
d(\Psi^{-1}\Psi_{\lambda\lambda}) &= 2\Psi^{-1}W_\lambda\Psi_\lambda = 2d(\Psi^{-1}\Psi_\lambda)\Psi^{-1}\Psi_\lambda \\
d(\Psi^{-1}\Psi_{\lambda\lambda\lambda}) &= 3\Psi^{-1}W_\lambda\Psi_{\lambda\lambda} = 3d(\Psi^{-1}\Psi_\lambda)\Psi^{-1}\Psi_{\lambda\lambda}.
\end{aligned}
$$

For the coefficients at $\lambda = 0$ this implies

$$
\begin{aligned}
dA &= (dR)R^* \\
dB &= (dR^*)R \\
dC &= (dR)B = d(RB) - R(dR^*)R \\
dD &= (dR^*)A = d(R^*A) - R^*(dR)R^*.
\end{aligned}
$$

Using these expressions all the terms in the formula (63) can be transformed to differentials

$$\begin{aligned} (dR^*)R - RdR^* &= dA + dB - d(RR^*) \\ R(dR^*)R &= d(RB - C). \end{aligned}$$

Integrating one arrives at (76).

3.4. Surfaces of finite type

We have shown that isothermic surfaces and Bonnet pairs can be studied in frames of the theory of integrable systems. Applying standard methods one can define for these surfaces higher flows, the Bäcklund-Darboux transformations[1], finite-gap solutions etc.

In contrast to the case of CMC surfaces, the linearizations of the Gauss-Codazzi equations of isothermic surfaces (51) and of Bonnet pairs (50) are not elliptic. This fact prevents us from applying the arguments of Section 2 and claiming that all corresponding immersions with doubly periodic fundamental forms are generated by finite-gap solutions. We call surfaces corresponding to the finite-gap solutions of the Gauss-Codazzi equations or equivalently surfaces with a polynomial Killing field (see Section 2) surfaces of *finite type*. This class of surfaces is worth studying, in particular, since it may contain Bonnet tori mates.

A polynomial Killing field $W(\lambda)$ of an isothermic surface or a Bonnet pair of finite type is an element of the loop algebra $g_I[\lambda]$. The coresponding spectral curve $\hat{C}$

$$\det(W(\lambda) - \mu I) = 0$$

is a four-sheeted covering $\hat{C} \to \bar{\mathbb{C}} \ni \mu$. Due to symmetries of the loop algebra it possesses the holomorphic involutions $\pi : (\mu, \lambda) \to (\mu, -\lambda)$ and $\hat{\sigma} : (\mu, \lambda) \to (-\mu, \lambda)$, and an anti-holomorphic involution $\hat{\tau} : (\mu, \lambda) \to (\bar{\mu}, \bar{\lambda})$. The Riemann surface $C = \hat{C}/\pi$ is an algebraic curve of μ and $\Lambda = \lambda^2$. It also possesses the symmetries $\sigma : (\mu, \Lambda) \to (-\mu, \Lambda)$ and $\tau : (\mu, \Lambda) \to (\bar{\mu}, \bar{\Lambda})$. The factor curve $C_\sigma := C/\sigma$ is quadratic in μ^2 and thus a hyperelliptic one.

Proposition 7. *The spectral curve C of an isothermic surface and of a Bonnet pair is a double covering of a hyperelliptic curve.*

---

[1]The Bäcklund-Darboux transformations of isothermic surfaces are known already in local differential geometry [Dar, Bia3] and in theory of solitons [Cie]

Let us indicate the steps of construction of the finite-gap solutions in this case. Although the considerations are similar to those of Section 2, technically they are more involved. By more or less standard technique one describes explicitly finite-gap solutions of the complexified system (51), i.e. of the system corresponding to the loop algebra without the real reduction (56). The spectral curve in this case remains to be a double covering of a hyperelliptic curve, but does not necessarily possess the real involution τ. The differentials analogous to the differentials Ω_1 and Ω_2 turn out to be Abelian differentials of the second kind with singularities at the branch points of the covering $C \to C_\sigma$. They are odd with respect to the involution σ. Their vectors of b-periods describing the velocity of the linear dynamics on the Jacobian lie in the odd part of the Jacobian with respect to the involution σ. The dynamic of the corresponding nonlinear system lies on the Prym variety $Prym_\sigma(C)$ of the covering $C \to C_\sigma$.

The real reduction (56) leads to constraints on parameters of the finite-gap solution. It is a technical but rather involved problem to classify all possible cases leading to real finite-gap solutions of (51) and thus to isothermic surfaces and Bonnet pairs of finite type. This is not yet done. Note that due to explicit formulae (75, 76) for the corresponding immersions, the isolation of tori from general surfaces of finite type is then straightforward.

§4. Bonnet families

4.1. Definition of Bonnet surfaces and simplest properties

A natural question is whether there may exist more than two isometric surfaces with the same mean curvature function, for example three such surfaces (Bonnet triple). It was known already to Bonnet that a Bonnet triple implies the existence of a one-parameter family of isometric surfaces with the same mean curvature function. We will prove this result in Section 4.4 after we learn more about these families.

Let $\mathcal{F}$ be a smooth surface in $\mathbb{R}^3$ with non-constant mean curvature function. $\mathcal{F}$ is called a *Bonnet surface* if it possesses a one-parameter family

$$\mathcal{F}_\tau, \qquad \tau \in (-\epsilon, \epsilon),\ \epsilon > 0,\ \mathcal{F}_0 = \mathcal{F}$$

of non-trivial[2] isometric deformations preserving the mean curvature function. The family $(\mathcal{F}_\tau)_{\tau\in(-\epsilon,\epsilon)}$ is called a *Bonnet family*. A Bonnet family can be described as a conformal mapping

$$F: \begin{array}{ccc} (-\epsilon,\epsilon)\times\mathcal{R} & \longrightarrow & \mathbb{R}^3 \\ (\tau,z) & \mapsto & F(\tau,z,\bar{z}) \end{array} \qquad \epsilon>0, \tag{78}$$

where z is a local coordinate z on the Riemann surface $\mathcal{R}$.

The set $\mathcal{U}\subset\mathcal{R}$ of preimages of the umbilic points on $\mathcal{F}_\tau$ is independent of τ (see Section 3). Obviously, the set $\mathcal{V}=\{P\in\mathcal{R}\,:\,dH(P)=0\}$ of preimages of critical points of the mean curvature function on $\mathcal{F}_\tau$ is also τ-independent.

Similarly to Proposition 3 one can prove

Proposition 8. *The holomorphic quadratic differential*

$$\varphi(z,\tau)\,dz^2 := \frac{\partial}{\partial\tau}\,Q(z,\bar{z},\tau)\,dz^2. \tag{79}$$

vanishes exactly at the umbilic points.

In a neighbourhood of a non-umbilic point $Q(P)\neq 0$ there exist smooth real-valued functions $\psi(z,\bar{z},\tau)$, $q(z,\bar{z})$ such that $Q(z,\bar{z},\tau)=e^{\mathrm{i}\,\psi(z,\bar{z},\tau)}\,q(z,\bar{z})$. Differentiating we obtain

$$Q(z,\bar{z},\tau) = -\mathrm{i}\,\frac{\varphi(z,\tau)}{\psi_\tau(z,\bar{z},\tau)} \tag{80}$$

This representation is a special case of (8), which implies

Corollary 4. *Bonnet surfaces are isothermic away from umbilic points.*

Representation (80) implies that ψ_τ is a harmonic function on $\mathcal{R}\setminus\mathcal{U}$. Moreover, from a more detailed analysis of the local behaviour at an umbilic point one can deduce [BE2] that the function ψ_τ can be extended to a nowhere vanishing harmonic function $\psi_\tau:\mathcal{R}\to\mathbb{R}_*$.

With ψ_τ defined this way, the identity

$$\varphi = \mathrm{i}\,\psi_\tau\,Q \tag{81}$$

holds on all $\mathcal{R}$.

---

[2]We call an isometry of a surface non-trivial if it is not induced by an isometry of the ambient space.

Theorem 18. *Let $\mathcal{F}$ be a Bonnet surface. Then*

(i) Umbilic points are critical points $\mathcal{U} \subset \mathcal{V}$ of the mean curvature function.

(ii) The set $\mathcal{V}$ of critical points of the mean curvature function is discrete in $\mathcal{R}$.

Proof. From the Codazzi equations it follows that $dH = 0$ if and only if $Q_{\bar z} = 0$. Differentiating (81) with respect to $\bar z$ one obtains

$$\mathrm{i}\,\psi_{T\bar z}\,Q + \mathrm{i}\,Q_{\bar z} \equiv 0.$$

Thus $Q = 0$ implies $Q_{\bar z} = 0$. To show that $\mathcal{V}$ is a discrete subset of $\mathcal{R}$, we use

$$Q_{\bar z} = 0 \Longleftrightarrow \mathrm{i}\,\psi_{T\bar z}\,Q = 0 \Longleftrightarrow \mathrm{i}\,\psi_{Tz}\,Q = 0 \Longleftrightarrow \frac{\psi_{Tz}\,Q_T}{\psi_T} = 0$$

where we use that ψ_T is a non-vanishing harmonic function on $\mathcal{R}$. Since $\psi_{Tz}\,Q_T\,dz^3$ is a cubic holomorphic differential, its zeros (which comprise the set $\mathcal{V}$) are discrete.

4.2. Local theory away from critical points

In this section we develop local theory of Bonnet surfaces in $\mathbb{R}^3$ away from possible critical points of the mean curvature function $F : (-\epsilon, \epsilon) \times \mathcal{R} \setminus \mathcal{V} \to \mathbb{R}^3$. The preimages of holomorphic local charts $z : U \subset \mathcal{R} \setminus \mathcal{V} \to \mathbb{C}$ are always assumed to be simply connected.

The following "stationary" characterisation of Bonnet surfaces is classical [Gra].

Theorem 19. *An umbilic free surface $\mathcal{F}$ is a Bonnet surface if and only if*

(i) $\mathcal{F}$ is isothermic.

(ii) $1/Q$ is harmonic, i.e.

$$\left(\frac{1}{Q(z,\bar z)}\right)_{z\bar z} = 0, \tag{82}$$

where z is an isothermic coordinate and $Q\,dz^2$ is the Hopf differential.

Proof. "$\Rightarrow$": follows from Corollary 4, (80), and harmonicity of ψ_T. "$\Leftarrow$": Let z be an isothermic coordinate and $e^{u(z,\bar z)}, H(z,\bar z), Q(z,\bar z)$ be a solution to the Gauss-Codazzi equations (7) with Q satisfying (82). Locally there exists a holomorphic function $h(z)$ such that $Q(z,\bar{z}) = 1/(h(z)+\bar{h}(\bar{z}))$. Define $Q(z,\bar{z},\tau)$ via

$$Q(z,\bar{z},\tau) = \left(\frac{1 - \mathrm{i}T\,\overline{h(z)}}{1+\mathrm{i}T\,h(z)}\right)\frac{1}{h(z)+\overline{h(z)}}, \tag{83}$$

where T is a deformation parameter equivalent to τ. One can easily check that $|Q(z,\bar{z},\tau)| = |Q(z,\bar{z})|$ and $Q_{\bar{z}}(z,\bar{z},\tau) = Q_{\bar{z}}(z,\bar{z})$ holds. Thus $e^{u(z,\bar{z})}$, $H(z,\bar{z})$, $Q(z,\bar{z},\tau)$ is also a solution of (7) for all T. The surfaces corresponding to different T are isometric and have the same mean curvature. They form a Bonnet family.

It is easy to see that the Codazzi equations imply

$$h'(z)\,H_z(z,\bar{z}) = \overline{h'(z)}\,H_{\bar{z}}(z,\bar{z}).$$

Introducing locally the new conformal coordinate

$$w = \int \frac{1}{h'(z)}\,dz \tag{84}$$

one finds that the mean curvature function depends on $t = w + \bar{w}$ only. This finally leads to the fact that the Gauss-Codazzi equations can be reduced to an ordinary differential equation, which is derived below.

One can directly check that Q satisfies

$$Q_{\bar{w}}(w,\bar{w},T) = \bar{Q}_w(w,\bar{w},T) = -|Q(w,\bar{w},T)|^2.$$

Inserting

$$e^u = \frac{2\,Q_{\bar{w}}}{H_w} = -\frac{2\,|Q|^2}{H'} \tag{85}$$

into the Gauss equation one obtains

$$\left(\frac{H''(t)}{H'(t)}\right)' - H'(t) = |Q|^2\left(2 - \frac{H^2(t)}{H'(t)}\right), \quad ' = \frac{d}{dt}. \tag{86}$$

For a general holomorphic function $h(w)$, equation (86) does not possess a solution depending on t only.

The identity $|Q|^2_w = |Q|^2_{\bar{w}}$ implies

$$\left(\frac{h''(w)}{h'(w)} - \frac{\overline{h''(w)}}{\overline{h'(w)}}\right)\left(h(w)+\overline{h(w)}\right) = 2\,(h'(w) - \overline{h'(w)}).$$

Type	$Q(w,\bar{w})$	*H solves*	e^u
A_1	$-2\left(\dfrac{\sin(2\bar{w})}{\sin(2w)}\right)\dfrac{1}{\sin(2t)}$	$\left(\left(\dfrac{H''}{H'}\right)' - H'\right)\dfrac{\sin^2(2t)}{4} = 2 - \dfrac{H^2}{H'}$	$-\dfrac{8}{\sin^2(2t)H'}$
A_2	$2\left(\dfrac{\cos(2\bar{w})}{\cos(2w)}\right)\dfrac{1}{\sin(2t)}$		
B	$-2\left(\dfrac{\sinh(2\bar{w})}{\sinh(2w)}\right)\dfrac{1}{\sinh(2t)}$	$\left(\left(\dfrac{H''}{H'}\right)' - H'\right)\dfrac{\sinh^2(2t)}{4} = 2 - \dfrac{H^2}{H'}$	$-\dfrac{8}{\sinh^2(2t)H'}$
C	$-\dfrac{\bar{w}}{w}\dfrac{1}{t}$	$\left(\left(\dfrac{H''}{H'}\right)' - H'\right)t^2 = 2 - \dfrac{H^2}{H'}$	$-\dfrac{2}{t^2H'}$

Table 1. Table of fundamental functions

This equation can be solved explicitly. Up to appropriate normalizations $h(w)$ is one of the following five forms

$$\begin{array}{rclrclrcl} h_1(w) &=& w, & h_2(w) &=& e^{4\,i\,w}, & h_3(w) &=& -\dfrac{1}{w}, \\ h_4(w) &=& \tanh(2\,w), & h_5(w) &=& \tan(2\,w). & & & \end{array}$$

Finally one arrives at the following classical result of E. Cartan [Car] (see also [BE1, BE2] for detail).

Theorem 20. *Away from umbilic points there are 3 types of Bonnet families, which are characterized by the modulus of the Hopf differential:*

$$\textbf{Type A}: |Q^A(w,\bar{w},T)|^2 = \frac{4}{\sin^2(2t)}.$$
$$\textbf{Type B}: |Q^B(w,\bar{w},T)|^2 = \frac{4}{\sinh^2(2t)}.$$
$$\textbf{Type C}: |Q^C(w,\bar{w},T)|^2 = \frac{1}{t^2}\,.$$

The families are given by the surfaces with the fundamental forms presented in Table 1.

Here $H(t)$ is any smooth solution with $H' < 0$ of the corresponding ordinary differential equation in Table 1. The corresponding one-parameter families of isometries are intrinsic isometries of the surface *described by imaginary translations of the coordinate* w

$$w \to w + \mathrm{i}\,\rho(T). \tag{87}$$

The surfaces of type A_1 *and* A_2 *are isometric with the same mean curvature function.*

Corollary 5. *Bonnet surfaces are real analytic.*

The equations for the mean curvature function in Table 1 were first derived by N. Hazzidakis in [Haz]. We call them Hazzidakis equations.

A Bonnet surface in $\mathbb{R}^3$ is said to be *of type* **A**, **B**, or **C**, respectively, if away from critical points of the mean curvature function it is of the corresponding type. One can show that a Bonnet surface is exactly of one of the types **A**, **B**, or **C**.

4.3. Local theory at critical points

In this section we derive a differential equation describing Bonnet surfaces near (isolated) critical points.

The identity (81) at an umbilic point implies the following local representation of the Hopf differential.

Proposition 9. *Let* $P \in \mathcal{U} \subset \mathcal{R}$ *be an umbilic point of a Bonnet surface. Then there exists a neighbourhood* U *of* P*, a local conformal chart* $z : U \to \mathbb{C}$ *with* $z(P) = 0$*, a holomorphic non-vanishing function* $\varphi : U \to \mathbb{C}_*$*, and an integer* $J > 0$ *such that the Hopf differential on* U *is*

$$Q(z, \bar{z}, \tau) = -\mathrm{i}\, z^J \left(\frac{\varphi(z,\tau)}{\psi_\tau(z,\bar{z},\tau)} \right). \tag{88}$$

The function $\psi_{\tau\, z}$ is holomorphic on U and therefore can be represented as $\psi_{\tau z} = z^M\, \theta(z)$, $\theta(0) \neq 0$ with some $M \geq 0$. Analysing the Codazzi equations

$$\bar{Q}_z = \bar{z}^J\, z^M\, \theta\, \overline{\left(\frac{\mathrm{i}\varphi}{\psi_\tau^2}\right)} = \frac{1}{2}\, H_{\bar{z}}\, e^u, \quad Q_{\bar{z}} = z^J\, \bar{z}^M\, \bar{\theta}\, \left(\frac{\mathrm{i}\varphi}{\psi_\tau^2}\right) = \frac{1}{2}\, H_z\, e^u. \tag{89}$$

one obtains

$$\psi_{\tau z} = z^{J+1}\, \theta(z), \qquad \theta(0) \neq 0. \tag{90}$$

Non-umbilic critical points correspond to $J = 0$. We call $-J$ with J given by (88, 90) the *index of the critical point.*

The derivation of an ordinary differential equation at a critical point is similar to the one of section 4.2. The Codazzi equations (89) imply

$$(91)\qquad \left(\frac{\theta}{i\varphi}\right) z\,H_z = \overline{\left(\frac{\theta}{i\varphi}\right)}\,\bar z\,H_{\bar z} \in \mathbb{R}.$$

Introducing a new conformal parameter $w(z)$ by

$$(92)\qquad \frac{z}{f}\,\frac{\partial}{\partial z} = \frac{w}{f(0)}\,\frac{\partial}{\partial w},$$

with $w(0) = 0$ and $f(z) = -i\,\varphi(z)/\theta(z)$, one arrives at $w\,H_w = \bar w\,H_{\bar w}$. The last identity shows that the mean curvature function is a function of $s = |w|^2$ only. Proceeding further as in Section 4.2, after apropriate normalizations (see [BE2] for detail) one obtains the following

Theorem 21. *Let $\mathcal{F}$ be a Bonnet surface in $\mathbb{R}^3$ with an (isolated) critical point of index $-J$. Then there exists a local conformal chart w at the critical point $w = 0$ such that the coefficients of its fundamental forms are given by*

(93)

$$\begin{aligned} Q(w,\bar w)\,dw^2 &= (J+2)\left(\frac{1-\bar w^{J+2}}{1-w^{J+2}}\right)\frac{w^J}{1-s^{J+2}}\,dw^2, \\ e^{u(w,\bar w)}\,|dw|^2 &= -2\,(J+2)^2\,\frac{s^J}{(1-s^{J+2})^2\,H'(s)}\,|dw|^2, \end{aligned} \qquad s = |w|^2\,,$$

and $H(w,\bar w) = H(|w|^2)$ is a solution of

$$(94)\qquad \left(\frac{sH''(s)}{H'(s)}\right)' - H'(s) = \frac{(J+2)^2\,s^{J+1}}{(1-s^{J+2})^2}\left(2 - \frac{H^2(s)}{s\,H'(s)}\right), \qquad ' = \frac{d}{d\,s},$$

with the asymptotics

$$(95)\qquad H(s) = H(0) + s^{J+1}\,B(s)$$

at $s = 0$, where $B(s)$ is a non-vanishing, smooth function. Conversely, any solution of (94) with the asymptotics (95) at $s = 0$ via (93) determines a Bonnet surface with critical point of index $-J$. The Bonnet family is given by the intrinisic isometry

$$(96)\qquad w \to e^{i\alpha}\,w, \quad \alpha \in \mathbb{R}.$$

The existence of Bonnet surfaces in $\mathbb{R}^3$ with critical points of the mean curvature function will be proven in Section 4.6.

Finally note that Bonnet surfaces with critical points are necessarily of type **B**. The relation between the corresponding coordinates w of this section and $w_{\mathbf{B}} = w$ of Section 4.2 is given by

$$w = e^{-\frac{4}{J+2} w_{\mathbf{B}}}. \tag{97}$$

We call a Bonnet surface with an isolated critical point a Bonnet surface of type $\mathbf{B}_{\mathcal{V}}$.

4.4. Bonnet triple implies Bonnet family

Theorem 22. *Let F_1, F_2, F_3 be a Bonnet triple, i.e. isometric non-congruent immersions $F_i : U \to \mathbb{R}^3$, $i = 1, 2, 3$ with the same mean curvature. Then there exists a one-parameter family of such surfaces F_t, $t \in (-\epsilon, \epsilon)$ with $F_{t=0} = F_1$.*

The proof is divided into several Lemmas. The statement is local, and since the umbilic points of Bonnet pairs are isolated (Corollary 2) it is enough to consider the case of umbilic free surfaces $F_i : U \to \mathbb{R}^3$. Denote by Q_i the corresponding Hopf differentials. Proposition 3 implies that there exist smooth functions $a, b, c : U \to \mathbb{R}^3$ and holomorphic non-vanishing h_1, h_2, h_3 such that

$$\begin{aligned} Q_1 &= h_2(1 + ic) = h_3(-1 + ib), \\ Q_2 &= h_3(1 + ib) = h_1(-1 + ia), \\ Q_3 &= h_2(-1 + ic) = h_1(1 + ia). \end{aligned}$$

By change of the conformal variable the differential Q_1 can be brought to the form

$$Q_1(z) = 1 + ic = h(-1 + ib), \tag{98}$$

with holomorphic non-vanishing h.

Lemma 6. *Function $c(z, \bar{z})$ in (98) satisfies*

$$c_{z\bar{z}}(1 + c^2) = 2c\, c_z c_{\bar{z}}. \tag{99}$$

Proof. Eliminating b from (98) yields $h^{-1}(1+ic)+\bar{h}^{-1}(1-ic) = -2$. Differentiating by z and substituting $\bar{h}$ one obtains that

$$\frac{2ic_z}{1 + c^2} = \frac{h_z}{(h + 1)h}$$

is holomorphic, which implies (99).

Lemma 7. *Let $c : U \to \mathbb{R}$ be a smooth solution of (99). Then there exists harmonic $r : U \to \mathbb{R}_+$ such that*

$$f := (1 + ic)r \tag{100}$$

is holomorphic.

Proof. The Cauchy-Riemann equations for (100) with real valued r read

$$r_x(1 + c^2) + r(cc_x - c_y) = 0$$
$$r_y(1 + c^2) + r(cc_y + c_x) = 0,$$

where $z = x + iy$. The compatibility condition for this system is exactly (99). By direct computation one can check $r_{xx} + r_{yy} = 0$.

Lemma 8. *A surface with the Hopf differential $Q_1(z)dz^2$ given by (98) is isothermic. The isothermic parametrization w is determined by*

$$\left(\frac{dw}{dz}\right)^2 = f$$

with f given by (100). In the isothermic coordinate w the Hopf differential of F_1 equals $\frac{1}{r}dw^2$.

Proof. Written in the coordinate z the Hopf differential equals

$$Q_1(z)dz^2 = \frac{f}{r}dz^2.$$

The conformal coordinate w of Lemma 8 is isothermic since the differential in this coordinate is real.

Since r is harmonic, the statement of Theorem 22 now follows from the characterization of the Bonnet families in Theorem 19.

4.5. Bonnet surfaces via Painlevé transcendents

In this section we give an explicit description of Bonnet surfaces in terms of solutions of Painlevé equations, which are certain ordinary differential equations of the second order

$$y'' = R(y', y, x). \tag{101}$$

Solutions of the Painlevé equations – Painlevé transcendents – are treated nowadays as non-linear special functions. The theory of these special functions is rather well developed (see for example [IN, IKSY, Its]). In Section 4.6 we use elements of this theory for global classification of Bonnet surfaces. Of special importance for us is the *Painlevé property* of these equations. Recall that a differential equation (101) is said to possess the Painlevé property if it is free of movable branch and essential singular points, i.e. the only singularities of the solutions which change their position if one varies the initial data are poles.

Regarding examples of the Bonnet surfaces of type **B**, let us expain how the quaternionic representation of Bonnet surfaces naturally leads to the above mentioned remarkable connection. Note that the Hazzidakis equations of types **A** and **B** are analytically equivalent $H_{\mathbf{A}}(it) = -iH_{\mathbf{B}}(t)$.

Since the mean curvature and the metric of a Bonnet surfaces of type **B** depend on $t = w + \bar{w}$ only, it is natural to change the parametrization once more using t (or a function of t) as one of the variables. After some computations one arrives to the idea of considering

$$x = e^{-4\,(w+\bar{w})}, \qquad \lambda = e^{-4\,w} \tag{102}$$

as new variables. In these coordinates the Hazzidakis equation is

$$4\left(x\,\frac{\mathcal{H}''(x)}{\mathcal{H}'(x)}\right)' + \mathcal{H}'(x) = \frac{4}{(x-1)^2}\left(2 + \frac{\mathcal{H}^2(x)}{4\,x\,\mathcal{H}'(x)}\right), \tag{103}$$

where $\mathcal{H}(x) \equiv H(t)$. The frame equations (15, 17) in parametrization (102) are now

$$\frac{\partial\Phi}{\partial\lambda}\,\Phi^{-1} = \frac{B_0(x)}{\lambda} + \frac{B_1(x)}{\lambda-1} + \frac{B_x(x)}{\lambda-x}, \tag{104}$$

$$\frac{\partial\Phi}{\partial x}\,\Phi^{-1} = -\frac{B_x(x)}{\lambda-x} + C(x), \tag{105}$$

where all the coefficients of the matrices are given by some explicit formulae through $\mathcal{H}(x)$, $\mathcal{H}'(x)$ and $\mathcal{H}''(x)$. Ignore for the moment the complicated formulae for the coefficients in the matrices (104). What is more important is the special dependence of (104) on λ. Equation (104) is a 2 by 2 matrix dimensional Fuchsian system with four regular singularities (at $\lambda = 0, 1, x, \infty$). Equation (105) describes deformations preserving the monodromy group of the system. It is well known (see for example [JM]) that such isomonodromic deformations are the issue of the Painlevé VI equation. In particular, all the coefficients of the matrices can be expressed in terms of solutions of this equation – Painlevé transcendents. A rather involved calculation (see [BE1]) of the corresponding gauge transformation identifying the corresponding descriptions yields finally the following

Theorem 23. *Equation (103) possesses the first integral*

$$x^2 \left(\frac{\mathcal{H}''(x)}{\mathcal{H}'(x)} + \frac{2}{x-1} \right)^2 + \frac{x\,\mathcal{H}'(x)}{2} + \frac{\mathcal{H}^2(x)}{2\,(x-1)^2\,\mathcal{H}'(x)} + \frac{\mathcal{H}(x)\,(x+1)}{2\,(x-1)} = \theta^2. \tag{106}$$

Let $\mathcal{H}(x)$ be a solution of (103) with $\mathcal{H}'(x) \neq 0$ and θ be a fixed root of (106). Then the function $y(x)$ defined by

$$y(x) \equiv -\frac{2}{\mathcal{H}'(x)} \left(\frac{x\,(x-1)\,\mathcal{H}''(x) + (\theta - x\,(\theta-2))\,\mathcal{H}'(x)}{\mathcal{H}(x) + (x-1)\,\mathcal{H}'(x)} \right)^2 \tag{107}$$

solves the Painlevé VI equation of the following form

$$\begin{aligned} \frac{d^2 y}{d x^2} &= \frac{1}{2}\left(\frac{1}{y} + \frac{1}{y-1} + \frac{1}{y-x}\right) y'^2 - \left(\frac{1}{x} + \frac{1}{x-1} + \frac{1}{y-x}\right) y' + \\ &\quad \frac{y\,(y-1)\,(y-x)}{2\,x^2\,(x-1)^2} \left(\theta^2 \frac{(x-1)}{(y-1)^2} - \theta\,(\theta+2)\,\frac{x\,(x-1)}{(y-x)^2}\right). \end{aligned} \tag{108}$$

Conversely, for any solution of the Painlevé VI (108) which is not of the form $y(x) = \text{constant} \cdot x^{-\theta}$, the function

$$\mathcal{H}(x) \equiv -2\,\frac{(x-1)\,(\theta^2\,y(x)^2 - x^2 y'^2(x))}{y(x)\,(y(x)-1)(y(x)-x)} \tag{109}$$

is a solution of (103) with first integral $\pm\theta$ (106).

The mappings (107) and (109) with the same θ are inverse one to another.

In a similar way [BE1] Bonnet surfaces of type **C** are described through solutions of the Painlevé V equation.

Corollary 6. *The Hazzidakis equations for Bonnet surfaces of all types possess the Painlevé property.*

4.6. Global classification of Bonnet surfaces

In this section *maximal* or *global* Bonnet immersions $F : \mathcal{R} \to \mathbb{R}^3$ are classified. These surfaces are characterized by the following natural "maximality" property: given an immersed Bonnet surface $\mathcal{F} \subset \mathbb{R}^3$ there exists a global Bonnet immersion containing $\mathcal{F}$, i.e.

$$\mathcal{F} = F(U), \quad U \subset \mathcal{R}.$$

Let us first prove the existence of critical points.

Theorem 24. *For arbitrary $H(0) \in \mathbb{R}$, $H_0 < 0$ there exists a real analytic Bonnet surface of type $\mathbf{B}_\mathcal{V}$ with a critical point of index $-J$ and with the mean curvature function and the metric at the critical point given by*

$$H(0) \text{ and } -2\,\frac{(J+2)^2}{(J+1)\,H_0}\,dw\,d\bar{w}.$$

Proof. Substituting the ansatz

$$H(s) = H(0) + s^{J+1}\sum_{i=0}^{\infty} H_i\, s^i, \quad H_0 \neq 0$$

into equation (94) and using the corresponding recurrence formulae for H_i, one can prove that the series converges absolutely in a neighbourhood of $s = 0$ (see [BE2]). Thus, for arbitrary $H(0) \in \mathbb{R}$, $H_0 \in \mathbb{R}_*$ there exists a real analytic solution of equation (94) at $s = 0$ with the asymptotic

$$H(s) = H(0) + s^{J+1} H_0 + O(s^{J+2}). \tag{110}$$

The claim follows now from Theorem 21.

Let the local coordinate w be defined by (84) for the Bonnet surfaces of type **A**, **B**, and **C** and by (92) for the Bonnet surfaces of type $\mathbf{B}_\mathcal{V}$. We denote by $\mathbf{U}$ the largest connected domain in the w-plane, for which Q (see Table **1** and (93) respectively) is bounded. Furthermore, let $\mathcal{D} = \{t = w + \bar{w}\,|w \in \mathbf{U}\}$ for the types **A**, **B**, and **C**, and $\mathcal{D} = \{s = |w|^2\,|w \in \mathbf{U}\}$ for the type $\mathbf{B}_\mathcal{V}$. In the following Table **2**, all the cases $\mathbf{U}$, $\mathcal{D}$, and the harmonic function ψ_τ, parametrized by w, are listed:

Proposition 10. *Let H be a solution of one of the Hazzidakis equations of types **A**, **B**, **C**, or $\mathbf{B}_\mathcal{V}$ (see Table **1** and (94)) with $H'(t) < 0$ at some point $t \in \mathcal{D}$ (the corresponding domains $\mathcal{D}$ are listed in Table **2**). Then H is real analytic on $\mathcal{D}$.*

The proof of this proposition is based on using the Painlevé property (see Corollary 6) together with the following

Lemma 9. *Let $\mathcal{D} \ni t$ be an open interval with smooth positive-valued functions f, g, $|Q|^2 : \mathcal{D} \to \mathbb{R}_+$ on it. Let $H = H(t)$ be a real-valued solution of the generalized Hazzidakis equation*

$$\left(f(t)\,\frac{H''(t)}{H'(t)}\right)' - H'(t) = |Q|^2\left(2 - \frac{H^2(t)}{g(t)\,H'(t)}\right), \tag{111}$$

Type	$\mathbf{U}$	$\mathcal{D}$	$\psi_\mathcal{T}$
$\mathbf{A}_1$, $\mathbf{A}_2$	$\{w \in \mathbb{C} \mid 0 < \mathrm{Re}(w) < \frac{\pi}{4}\}$	$(0, \frac{\pi}{2})$	$2\,\mathrm{Re}(\tan(2w))$
$\mathbf{B}$	$\{w \in \mathbb{C} \mid \mathrm{Re}(w) > 0\}$	$(0, \infty)$	$2\,\mathrm{Re}(\tanh(2w))$
$\mathbf{B}_\mathcal{V}$	$\{w \in \mathbb{C} \mid \ \lvert w\rvert < 1\}$	$[0, 1)$	$2\,\mathrm{Re}\left(\frac{1 + w^{J+2}}{1 - w^{J+2}}\right)$
$\mathbf{C}$	$\{w \in \mathbb{C} \mid \mathrm{Re}(w) > 0\}$	$(0, \infty)$	$2\,\mathrm{Re}(w)$

Table 2. Global description of Bonnet surfaces

smooth on $\mathcal{D}\backslash\mathcal{P}$, where $\mathcal{P}$ is a discrete set of poles of $H(t)$. If $H'(t_0) < 0$ at some $t_0 \in \mathcal{D}$, then H is smooth everywhere on $\mathcal{D}$, i.e. $\mathcal{P} = \emptyset$, with $H'(t) < 0$ for all $t \in \mathcal{D}$.

To prove smoothness of $H(t)$ one shows that all poles of $H(t)$ are necessarily simple with negative residues and that in addition $H'(t)$ never changes its sign. These two properties contradict one another.

It is not difficult to show that immersions $F : \mathbf{U} \to \mathbb{R}^3$ of Bonnet surfaces of type **A**, **B**, **C**, or $\mathbf{B}_\mathcal{V}$ given in Tables 1 and 2 are maximal. The function $\psi_\mathcal{T}$ defined in Table 2 is a non-vanishing function on $\mathbf{U}$. The continuity of $\psi_\mathcal{T}$ yields that this function can not be extended beyond $\mathbf{U}$.

Finally using the arguments of Section 4.2 it is easy to show that the classified global Bonnet surfaces are all different, i.e. for two global Bonnet immersions $F_i : \mathcal{R}_i \to \mathbb{R}^3$, $i = 1, 2$ there exist no open sets $U_i \subset \mathcal{R}_i$, $i = 1, 2$ on which the surfaces coincide.

Theorem 25. *Any Bonnet surface in $\mathbb{R}^3$ can be given a conformal parametrization $F : \mathcal{R} \to \mathbb{R}^3$ by a corresponding global Bonnet immersion $F : \mathbf{U} \to \mathbb{R}^3$, $\mathcal{R} \subset \mathbf{U}$. The latter are of one of the types* **A**, **B**, **C**, $\mathbf{B}_\mathcal{V}$. *The corresponding domains* $\mathbf{U}$ *are listed in Table 2. Given $t_0 \in \mathcal{D}$ (see Table 2) and arbitrary $H(t_0)$, $H'(t_0) < 0$, $H''(t_0)$ there exists a unique solution $H(t)$ of the Hazzidakis equation of type* **A**, **B**, **C** *(see Table 1), real analytic on $\mathcal{D}$. This function determines the fundamental forms (Table 1) of the corresponding global Bonnet immersions of the type* **A**, **B**, **C**. *Given $H(0)$, $H_0 < 0$ there exists a unique solution $H(s)$ of the Hazzidakis equation (94), real analytic on $\mathcal{D}$, with the asymptotics*

(110). It determines by (93) the fundamental forms of the corresponding global Bonnet surface of type $\mathbf{B}_\mathcal{V}$.

4.7. Examples of Bonnet surfaces

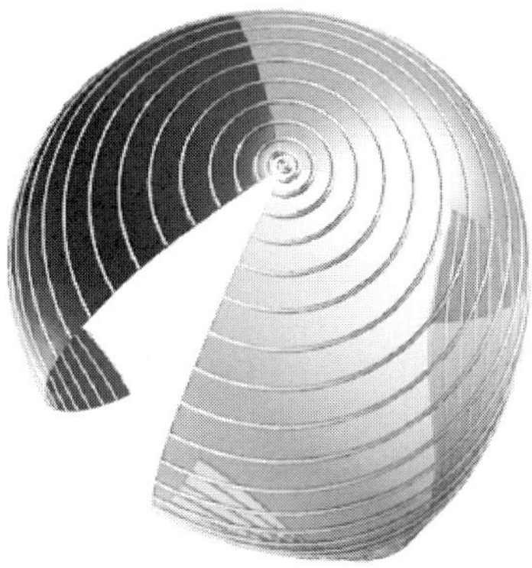

Fig. 6. A branched Bonnet surface of type **B**

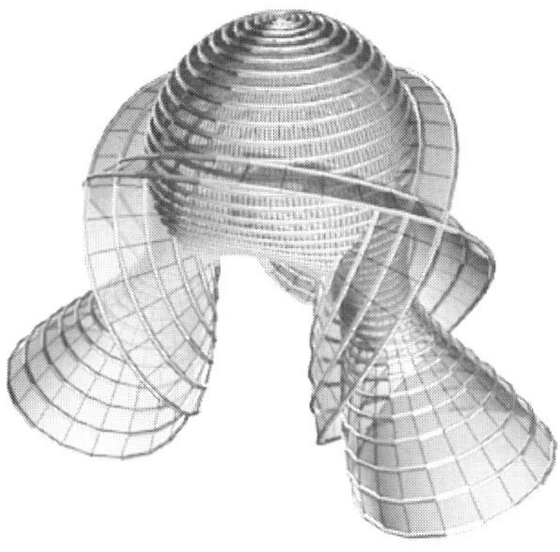

Fig. 7. Bonnet surface of type $\mathbf{B}_\mathcal{V}$

Let us present some figures of Bonnet surfaces of types **B** and $\mathbf{B}_\mathcal{V}$. Tubes correspond to parameter lines t = constant, i.e. to the trajectories of the isometric flow preserving the mean curvature function. Both the mean curvature function and the metric are preserved along these lines. The last fact can be clearly observed - the strips bounded by two consequent parameter lines $t = t_1$ and $t = t_2$ are of constant width. The isometry is intrinsic, i.e. is not induced by a Euclidean motion of the ambient $\mathbb{R}^3$. The immersion domain **U** of Bonnet surfaces of type **B** is naturally split into fundamental domains

$$\mathbf{U}_n = \{w \in \mathbb{C} \,|\, (n-1)\frac{\pi}{2} < \mathrm{Im}(w) < n\,\frac{\pi}{2}\}.$$

Indeed the fundamental forms (see Table **1**) are invariant with respect to the shift

$$w \to w + \mathrm{i}\,\frac{\pi}{2},$$

and thus immersed $\mathbf{U}_n$ with different n's are congruent in $\mathbb{R}^3$. A Bonnet surface comprising three fundamental domains is shown in Figure 6. For an appropriate choice of parameters, several copies of the fundamental domain can close up and thus comprise a closed surface with a critical point. Figure 7 shows such a case with three fundamental domains. It is worth mentioning that it was this figure which led us to suggest the existence of Bonnet surfaces with critical points.

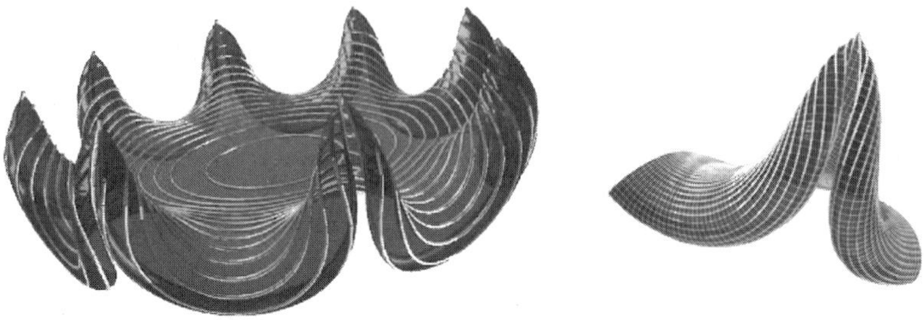

(a) (b)

Fig. 8. Bonnet disk with a critical point with $J = 6$, and its cusp

Figures 8a and 8b present another example of type $\mathbf{B}_{\mathcal{V}}$. The surface in Figure 8a is an immersed disk. The index of the critical point is $J = 6$. A more detailed view of one of the cusps of this surface is shown in Figure 8b. As before, the tubed curves are integral curves of the isometry field. The surface is probably embedded.

Further examples can be found in [BE1, BE2].

References

[Abr] U. Abresch, Constant mean curvature tori in terms of elliptic functions, J. Reine Angew. Math., **394** (1987), 169–192.

[BBEIM] E. D. Belokolos, A. I. Bobenko, V. Z. Enolskii, A. R. Its and V. B. Matveev, Algebro-Geometric Approach to Nonlinear Integrable Equations, Springer-Verlag, 1994.

[Bia1] L. Bianchi, Vorlesung über Differentialgeometrie, Leipzig, Berlin, 1910, p. 454.

[Bia2] L. Bianchi, Sulle superficie a linee di curvature isotherme, Rend. Acc. Naz. dei Lincei, (5), **12** (1903), 511–520.

[Bia3] L. Bianchi, Ricerche sulle superficie isotherme and sulla deformatione delle quadriche, Annali di Math. Ser. III, **11** (1905), 93–157.

[Bob1] A. I. Bobenko, Constant mean curvature surfaces and integrable equations, Russ. Math. Surveys, **46** (1991), 1–45.

[Bob2] A. I. Bobenko, Surfaces in terms of 2 by 2 matrices. Old and new integrable cases, In: Harmonic Maps and Integrable Systems, (eds. A. P. Fordy and J. Wood), Aspects Math., **E23**, Braunschweig Wiesbaden, Vieweg, 1994, pp. 83–127.

[BE1] A. I. Bobenko and U. Eitner, Bonnet surfaces and Painlevé equations, J. Reine Angew. Math., **499** (1998), 47–79.

[BE2] A. I. Bobenko and U. Eitner, Painlevé Equations in the Differential Geometry of Surfaces, Lecture Notes in Math., **1753**, Springer-Verlag, 2000.

[BP] A. Bobenko and U. Pinkall, Discrete isothermic surfaces, J. Reine Angew. Math., **475** (1996), 187–208.

[Bon] O. Bonnet, Mémoire sur la théorie des surfaces applicables, J. Ec. Polyt., **42** (1867), 72–92.

[Car] E. Cartan, Sur les couples de surfaces applicables avec conversation des courbieres principles, Bull. Sc. Math., **66** (1942), 1–30.

[Che] S. S. Chern, Deformation of Surfaces Preserving Principal Curvatures, In: Differential Geometry and Complex Analysis, (eds. I. Chavel and H. Farkas), Springer-Verlag, 1985, pp. 155–163.

[Cie] J. Cieslinski, The Darboux-Bianchi transformation for isothermic surfaces. Classical results versus the soliton approach, Diff. Geom. Appl., **7** (1997), 1–28.

[Dar] G. Darboux, Sur les surfaces isothermiques, Comptes Rendus, **122** (1899), 1299–1305, 1483–1487, p. 1538.

[DKN] B. A. Dubrovin, I. M. Krichever and S. P. Novikov, Integrable systems I, In: Contemprorary problems of mathematics, Fundamental Directions, Itogi nauki i Tekhniki, VINITI AN SSSR, Moscow, **4** (1985), 210–315.

[DPW] J. Dorfmeister, F. Pedit and H. Wu, Weierstrass type representation of harmonic maps into symmetric spaces, Commun. Anal. Geom., **6** (1998), 633–668.

[Eis] L. P. Eisenhart, A Treatise on the Differential Geometry of Curves and Surfaces, Allyn and Bacon, Boston, 1909.

[FLPPS] D. Ferus, K. Leschke, F. Pedit and U. Pinkall, Quaternionic Holomorphic Geometry: Plücker Formula, Dirac Eigenvalue Estimates and Energy Estimates of Harmonic 2-Tori, Invent. Math., **146** (2001), 507–593.

[FPPS] D. Ferus, F. Pedit and U. Pinkall, Sterling, I.: Minimal tori in S^4, J. Reine Angew. Math., **429** (1992), 1–47.

[Gra] W. C. Graustein, Applicability with preservation of both curvatures, Bull. Amer. Math. Soc., **30** (1924), 19–23.

[GKS] K. Grosse-Braukmann, R. Kusner and J. Sullivan, Classification of embedded constant mean curvature surfaces with genus zero and three ends, preprint, 1997.

[Haz] J. N. Hazzidakis, Biegung mit Erhaltung der Hauptkrümmungsradien, J. Reine Angew. Math., **117** (1887), 42–56.

[Hei] M. Heil, Numerical tools for the study of finite gap solutions of integrable systems, Ph.D. thesis, TU Berlin, 1995.

[Hit] N. J. Hitchin, Harmonic maps from a 2-torus to the 3-sphere, J. Diff. Geom., **31** (1991), 627–710.

[Hop] H. Hopf, Differential geometry in the large, Lecture Notes in Math., **1000**, Springer-Verlag, 1983.

[Its] A. R. Its, The Painlevé transcendents as nonlinear special functions, In: The Painlevé Transcendents, Their Asymptotics and Physical Applications, (eds. D. Levi and P. Winternitz), New York, Plenum, 1992, pp. 40–60.

[IN] A. R. Its and V. Y. Novokshenov, The Isomonodromic Deformation Method in the Theory of Painlevé Equations, Lectures Notes in Math., **1191**, Springer-Verlag, 1986.

[IKSY] K. Iwasaki, H. Kimura, S. Shimomura and M. Yoshida, From Gauss to Painlevé. A Modern Theory of Special Functions, Braunschweig, Vieweg, 1991.

[Jag] C. Jaggy, On the classification of constant mean curvature tori in $\mathbb{R}^3$, Comment. Math. Helvetici, **69** (1994), 640–658.

[JM] M. T. Jimbo and T. Miwa, Monodromy preserving deformation of linear ordinary differential equations with rational coefficients, II, Physica, **2 D** (1981), 407–448.

[KPP] G. Kamberov, F. Pedit and U. Pinkall, Bonnet pairs and isothermic surfaces, Duke Math. J., **92** (1998), 637–644.

[Kap] N. Kapouleas, Compact constant mean curvature surfaces in Euclidean three-space, J. Differential Geom., **33** (1991), 683–715.

[Ken] K. Kenmotsu, Weierstrass formula for surfaces of prescribed mean curvature, Math. Ann., **245** (1979), 89–99.

[KMS] M. Kilian, I. McIntosh and N. Schmitt, New constant mean curvature surfaces, Experiment. Math., **9** (2000), 595–611.

[Kon] B. G. Konopelchenko, Induced surfaces and their integrable dynamics, Stud. Appl. Math., **96** (1996), 9–51.

[KS1] R. Kusner and N. Schmidt, The spinor representation of minimal surfaces in space, GANG, preprint III.27, 1993.

[KS2] R. Kusner and N. Schmidt, The spinor representation of surfaces in space, GANG, preprint IV.18, 1996.

[LT] H. B. Lawson, Tribuzy, R. de, On the mean curvature function for compact surfaces, J. Differential Geom., **16** (1981), 179–183.

[PP] F. Pedit and U. Pinkall, Quaternionic analysis on Riemann surfaces and differential geometry, Doc. Math., Extra Volume ICM 1998, 389–400.

[Pin] U. Pinkall, Regular homotopy classes of immersed surfaces, Topology, **24** (1985), 421–434.

[PS] U. Pinkall and I. Sterling, On the classification of constant mean curvature tori, Ann. of Math. (2), **130** (1989), 407–451.

[Rou] I. M. Roussos, Global Results on Bonnet surfaces, J. Geom., **65** (1999), 151–168.

[Sul] D. Sullivan, The spinor representation of minimal surfaces in space, notes, 1989.

[Tai] I. Taimanov, The Weierstrass representation of closed surfaces in $\mathbb{R}^3$, Funct. Anal. Appl., **32** (1998), 49–62.

[Wal] R. Walter, Explicit examples to the H-problem of Heinz Hopf, Geom. Dedicata, **23** (1987), 187–213.

[Wen] H. C. Wente, Counterexample to a conjecture of H. Hopf, Pacific J. Math., **121** (1986), 193–243.

Mathematisches Institut
Technische Universität Berlin
Strasse des 17. Juni 136
10623 Berlin
GERMANY

E-mail address:
`bobenko@math.TU-Berlin.DE`

Advanced Studies in Pure Mathematics 51, 2008
Surveys on Geometry and Integrable Systems
pp. 55–111

Generalized Weierstraß representations of surfaces

Josef Dorfmeister

§1. Introduction

The classical Weierstraß representation

$$\phi(z,\bar{z}) = \mathrm{Re}\left\{\int_{z_0}^{z} (\frac{1}{2}f(1-g^2), \frac{i}{2}f(1+g^2), fg)dz\right\}$$

has been a very useful tool for the construction and the investigation of minimal surfaces in $\mathbb{R}^3$. While the differential equation describing these surfaces is highly nonlinear, the Weierstraß data, a pair consisting of a holomorphic function f and a meromorphic function g, is completely unconstrained. Moreover, the relation between the Weierstraß data (f, g) and ϕ is sufficiently direct that it is possible to relate geometric properties of the surface to the properties of the Weierstraß data.

In recent years a generalized Weierstraß representation was found, which applies to the construction of all surfaces of constant mean curvature in $\mathbb{R}^3$. If the mean curvature H vanishes, i.e. if the surface actually is a minimal surface, then the new procedure leads to the classical Weierstraß representation in a straightforward fashion. If $H \neq 0$, then the generalized Weierstraß representation is a new tool for the construction of conformal immersions of these surfaces. The Weierstraß data consists again of a pair of functions (Q, f), where Q is holomorphic and f is meromorphic. More precisely, Qdz^2 is the Hopf differential of the surface to be constructed and f is closely related with the conformal factor of the induced metric. In spite of this, the relation between the Weierstraß data and the geometry of the conformally immersed surface is much less direct than in the classical case. However, it turns out that at least some

Received July 11, 2001.
Revised October 5, 2001; December 13, 2007 (updated).
2000 *Mathematics Subject Classification.* Primary 58E20.

features of surfaces, such as the number of umbilical points and also the number of embedded ends, are well reflected in the Weierstraß data.

In this paper we will attempt to give a survey of the use of the generalized Weierstraß representation for the construction of constant mean curvature surfaces in $\mathbb{R}^3$. Since the original paper [26] introducing this method actually discussed harmonic maps into arbitrary compact symmetric spaces and since the notion of harmonic maps also exists for pseudo-Riemannian spaces, it is not surprising that there are generalized Weierstraß representations also for other classes of surfaces, such as the surfaces in $\mathbb{R}^3$ of constant Gauß curvature $K = -1$, timelike surfaces of constant mean curvature in three dimensional Minkowski space, and affine spheres in $\mathbb{R}^3$.

While the number of papers using methods similar to the one used in [26] is still reasonably small, we have still been unable to present all results in sufficient detail. We would like to apologize sincerely to everyone whose work we failed to represent adequately.

Acknowledgements: Over the years I have profited greatly from communication with many people. In particular I would like to mention Franz Pedit, Fran Burstall, Ivan Sterling, Martin Guest, and Junichi Inoguchi. I would also like to thank Martin Guest, Junichi Inoguchi, Magdalena Toda, and the referee for thoughtful and pertinent questions and comments on a preliminary version of this article. Finally, I would like to thank Professors M. Guest, R. Miyaoka and Y. Ohnita for their generous invitation to participate in the 9th MSJ-IRI in Tokyo and for the opportunity to publish this survey in these conference proceedings.

§2. Constructing potentials from surfaces

2.1 Let M be a connected, smooth, orientable surface and $\phi : M \longrightarrow \mathbb{R}^3$ an immersion of constant mean curvature. We can assume that $\phi \circ \pi$ is conformal, where $\pi : \mathbb{D} \longrightarrow M$ realizes the universal cover of M. In this article we will always assume that $\mathbb{D}$ is noncompact. For the case $\mathbb{D} \cong S^2$ we refer to [56],[9],[36].

By a result of Ruh and Vilms [48] we know that ϕ has constant mean curvature if and only if the Gauß map $N : M \longrightarrow S^2$ is harmonic. Kenmotsu [41] has shown how one can construct ϕ from N. In the method discussed below we will construct both N and ϕ from an "extended moving frame". To this end we write $S^2 = SU(2)/U(1)$, where $U(1)$ is represented as the subgroup $U(1) \cong \{ \operatorname{diag}(\alpha, \alpha^{-1}) \mid |\alpha| = 1\}$ of diagonal matrices in $SU(2)$. Then there exists some frame $F : \mathbb{D} \longrightarrow SU(2)$, such that the diagram

$$\begin{array}{ccc} & & SU(2) \\ & \nearrow & \\ \mathbb{D} & & \downarrow \\ & \searrow & \\ & & S^2 \end{array} \tag{2.1.1}$$

commutes, i.e. $N \equiv F \bmod U(1)$.

One can characterize the harmonicity of N by differential equations for the Maurer-Cartan form of F. More precisely, let $\mathfrak{g} = \mathfrak{p} + \mathfrak{k}$ denote the Cartan decomposition of $\mathfrak{g} = su(2, \mathbb{C}) = \operatorname{Lie}(SU(2))$ associated with the symmetric space $S^2 \cong SU(2)/U(1)$. In particular, $\mathfrak{k} = \operatorname{Lie}(U(1))$. Then the Maurer-Cartan form

$$\alpha = F^{-1}dF \tag{2.1.2}$$

of F decomposes as $\alpha = \alpha_{\mathfrak{p}} + \alpha_{\mathfrak{k}}$. Decomposing $\alpha_{\mathfrak{p}}$ further into the (1,0)-part $\alpha'_{\mathfrak{p}}$ and the (0,1)-part $\alpha''_{\mathfrak{p}}$, where $\alpha_{\mathfrak{p}} = \alpha'_{\mathfrak{p}} + \alpha''_{\mathfrak{p}}$, one introduces a "loop" parameter $\lambda \in S^1$ as in [56] by

$$\alpha_\lambda = \lambda^{-1}\alpha'_{\mathfrak{p}} + \alpha_{\mathfrak{k}} + \lambda\alpha''_{\mathfrak{p}}. \tag{2.1.3}$$

For later reference we state the explicit form of α_λ as it is used in this article (see e.g. the Appendix of [17]): Let z be a conformal coordinate and denote by $ds^2 = e^u dz d\bar{z}$ the metric, by H the mean curvature and by $Q(z)dz^2$ the Hopf differential, so that α_λ is given by $\alpha_\lambda =$

$$\begin{pmatrix} \frac{1}{4}u_z & -\frac{1}{2}\lambda^{-1}He^{\frac{u}{2}} \\ \lambda^{-1}Qe^{-\frac{u}{2}} & -\frac{1}{4}u_z \end{pmatrix} dz + \begin{pmatrix} -\frac{1}{4}u_{\bar{z}} & -\lambda\bar{Q}e^{-\frac{u}{2}} \\ \frac{1}{2}\lambda He^{\frac{u}{2}} & \frac{1}{4}u_{\bar{z}} \end{pmatrix} d\bar{z}. \tag{2.1.4}$$

Then one has the easy but crucial

Theorem 2.1.1. *([26]) The map $N : \mathbb{D} \longrightarrow S^2$ is harmonic if and only if α is integrable for all $\lambda \in S^1$, i.e.*

$$d\alpha_\lambda + \frac{1}{2}[\alpha_\lambda \wedge \alpha_\lambda] = 0 \quad \textit{for all } \lambda \in S^1. \tag{2.1.5}$$

It is well known that the integrability of α is equivalent to the existence of some map $\hat{F} : \mathbb{D} \times S^1 \longrightarrow SU(2)$ such that $\hat{F}^{-1}d\hat{F} = \alpha_\lambda$

and $\hat{F}(z,1) = F(z)$. *Such a map $\hat{F}$ is called an extended framing for N. From now on we will simply write F for the extended framing.*

As an immediate consequence of the theorem above we obtain

Corollary 2.1.2. *Let F be the extended framing associated with some harmonic map $N : \mathbb{D} \longrightarrow S^2$. Then the map, denoted by N again, $N : \mathbb{D} \times S^1 \longrightarrow S^2$, defined by $N \equiv F \bmod U(1)$, forms an S^1-family of harmonic maps.*

2.2 At this point one needs to observe that by the introduction of the loop parameter λ the extended framing $F(\cdot, \lambda)$ is, for fixed λ, actually a framing for the harmonic map $N(\cdot, \lambda)$. From a more technical point of view, F can be considered as a map from $\mathbb{D}$ into the group of loops in $Sl(2,\mathbb{C})$, i.e. the "loop group" $\Lambda Sl(2,\mathbb{C})$. Actually, from the definition of α_λ we see that odd powers of λ have coefficient matrices in $\mathfrak{p}$ and even powers of λ have coefficient matrices in $\mathfrak{k}$ (for the Lie algebra generated by $\alpha_\lambda(z,\bar{z}), z \in \mathbb{D}$). Therefore, α_λ is an element of the *twisted loop algebra*

$$\Lambda sl(2,\mathbb{C})_\sigma = \{\beta(\lambda) \in sl(2,\mathbb{C}) \mid \beta(-\lambda) = \sigma_3 \beta(\lambda) \sigma_3\}, \tag{2.2.1}$$

where $\sigma_3 = \mathrm{diag}(1,-1)$ denotes the third Pauli matrix. And F is in the twisted loop group

$$\Lambda Sl(2,\mathbb{C})_\sigma = \{g(\lambda) \in Sl(2,\mathbb{C}) \mid g(-\lambda) = \sigma_3 g(\lambda) \sigma_3\}. \tag{2.2.2}$$

In this case the twisting amounts to having even functions of λ on the diagonal and odd functions of λ off the diagonal. Since topological questions are not completely irrelevant for the splitting theorems mentioned below, we will tacitly assume that all the functions of λ are elements of the *Wiener algebra*

$$\mathcal{A} = \{f(\lambda) = \sum_{n\in\mathbb{Z}} f_n \lambda^n \mid \sum_{n\in\mathbb{Z}} | f_n | < \infty\}. \tag{2.2.3}$$

To get the loop group technicalities out of the way we note that for our purposes there are three subgroups of special importance:

$$\Lambda^+ Sl(2,\mathbb{C})_\sigma = \{f(\lambda) = \sum_{0\le n<\infty} f_n \lambda^n \in \Lambda Sl(2,\mathbb{C})_\sigma\} \tag{2.2.4}$$

$$\Lambda^- Sl(2,\mathbb{C})_\sigma = \{f(\lambda) = \sum_{-\infty<n\le 0} f_n \lambda^n \in \Lambda Sl(2,\mathbb{C})_\sigma\} \tag{2.2.5}$$

$$\Lambda SU(2)_\sigma = \{f \in \Lambda Sl(2,\mathbb{C})_\sigma \mid f(\lambda) \in SU(2)\ \forall \lambda \in S^1\}. \tag{2.2.6}$$

Since we have assumed that all coefficient functions are in the Wiener algebra, all the groups mentioned so far are Banach Lie groups (and Banach subgroups, where applicable). The corresponding Banach Lie algebras will be denoted by lower case letters, as in $\Lambda sl(2,\mathbb{C})_\sigma$. Crucial ingredients for the generalized Weierstraß representation are the following two group splitting results.

To ensure uniqueness in the first of these results we need to use the group $\Lambda_*^- Sl(2,\mathbb{C})_\sigma$, which is defined by the requirement that the coefficient of λ^0 is the identity matrix I.

Theorem 2.2.1. *(Birkhoff Splitting)*

$$\Lambda Sl(2,\mathbb{C})_\sigma = \bigcup_{n=0}^{\infty} \Lambda^- Sl(2,\mathbb{C})_\sigma \cdot \operatorname{diag}(\lambda^n, \lambda^{-n}) \cdot \Lambda^+ Sl(2,\mathbb{C})_\sigma \tag{2.2.7}$$

Moreover, the group multiplication map

$$\Lambda_*^- Sl(2,\mathbb{C})_\sigma \times \Lambda^+ Sl(2,\mathbb{C})_\sigma \longrightarrow \Lambda_*^- Sl(2,\mathbb{C})_\sigma \cdot \Lambda^+ Sl(2,\mathbb{C})_\sigma \tag{2.2.8}$$

is a complex analytic diffeomorphism onto an open and dense subset of $\Lambda Sl(2,\mathbb{C})_\sigma$. For + and − interchanged the analogous statements hold.

For a proof of this theorem see [47],[26],[16]. Note that if one ignores the subscript * above, then, in the case $n = 0$, the two representations $a_- \cdot b_+ = \hat{a}_- \cdot \hat{b}_+$ are equivalent to $a_- = \hat{a}_- \cdot c, b_+ = c^{-1} \cdot \hat{b}_+$, where c does not depend on λ, i.e. c is a λ-independent diagonal matrix.

For the second basic splitting result we need the group $\Lambda_P^+ Sl(2,\mathbb{C})_\sigma$, which is defined by the requirement that the (diagonal) coefficient matrix of λ^0 has only real, non-negative entries.

Theorem 2.2.2. *(Iwasawa Splitting)*

$$\Lambda Sl(2,\mathbb{C})_\sigma = \Lambda SU(2)_\sigma \cdot \Lambda^+ Sl(2,\mathbb{C})_\sigma \tag{2.2.9}$$

Moreover, the group multiplication map

$$\Lambda SU(2)_\sigma \times \Lambda_P^+ Sl(2,\mathbb{C})_\sigma \longrightarrow \Lambda Sl(2,\mathbb{C})_\sigma \tag{2.2.10}$$

is a real analytic diffeomorphism. For + replaced by − the analogous statement holds.

For a proof of this theorem see [47],[26],[16]. If one ignores the restriction "P" above, then two representations $a \cdot b_+ = \hat{a} \cdot \hat{b}_+$ are equivalent to $a = \hat{a} \cdot k$ and $\hat{b}_+ = k^{-1} \cdot b_+$, where k does not depend on λ and is unitary, i.e. $k = \operatorname{diag}(e^{it}, e^{-it})$, for $t \in \mathbb{R}$.

Remark 2.2.3. 1. Usually we call any decomposition of the form $g = u \cdot v_+$ with $u \in \Lambda SU(2)_\sigma$ and $v_+ \in \Lambda^+ Sl(2, \mathbb{C})_\sigma$ an Iwasawa splitting of g. The freedom just explained will then be referred to as the *freedom in the Iwasawa splitting.*

2. We would like to point out that the prototype of most splitting theorems used at present in geometric contexts seems to be Theorem 4.1 of [3]. Notwithstanding, of course, Birkhoff's original contribution.

3. The two theorems above work so well in the generalized Weierstraß representation because they are in some sense "complementary", namely, a relation of the form $V_- = F \cdot V_+$ can be interpreted as an (almost) 1-1 correspondence between V_- and F :

a) Given V_-, one obtains F (almost uniquely) via the Iwasawa splitting.

b) Given F, one obtains V_- (almost uniquely) via the Birkhoff splitting.

4. The splitting theorems have been stated for the full (twisted) loop group over the Wiener algebra. It will be important for applications, like the ones involving dressing (see section 5), to have these theorems in this generality. However, the actually geometrically relevant quantities are all holomorphic in $\lambda \in \mathbb{C}^*$ and are thus contained in a much smaller loop group. The extended frames, for example, are defined and holomorphic for $\lambda \in \mathbb{C}^*$, since α_λ is holomorphic in $\lambda \in \mathbb{C}^*$

2.3 Returning to the geometry under consideration we want to replace the extended frame F, whose Maurer-Cartan form α_λ satisfies some nonlinear differential equation, by some holomorphic extended frame $C = C(z, \lambda)$, which contains the same information as F. However, while the integrability condition for α_λ is nonlinear and nontrivial, the integrability condition for the Maurer-Cartan form of C will be trivial !

To achieve this, we will introduce a gauge $V_+ : \mathbb{D} \longrightarrow \Lambda^+ Sl(2, \mathbb{C})_\sigma$ such that

$$C = F \cdot V_+ \tag{2.3.1}$$

is holomorphic in z and in λ. For our setting it is also useful to require

$$C(z_0, \lambda) = I, \tag{2.3.2}$$

where z_0 is a base point chosen once and for all.

It is straightforward to see that C is holomorphic in z if and only if the $(0,1)$-part of the Maurer-Cartan form of $F \cdot V_+$ vanishes. This is equivalent to

$$\partial_{\bar{z}} V_+ = -(\alpha_{\mathfrak{k}}^{''} + \lambda \cdot \alpha_{\mathfrak{p}}^{''}) \cdot V_+, \tag{2.3.3}$$

where $\alpha_{\mathfrak{k}}^{''}$ and $\alpha_{\mathfrak{p}}^{''}$ denote the $(0,1)-$ parts of $\alpha_{\mathfrak{k}}$ and $\alpha_{\mathfrak{p}}$ respectively. As shown in the Appendix of [26], this equation admits a global solution $V_+ : \mathbb{D} \longrightarrow \Lambda^+ Sl(2,\mathbb{C})_\sigma$. Moreover, we can assume that V_+ is holomorphic in $\lambda \in \mathbb{C}^*$ and also that $C(z_0, \lambda) = I$. This result uses the local solvability of the $\bar{\partial}-$problem and the fact that every holomorphic cocycle on $\mathbb{D}$ is a boundary cycle [7]. Note that the holomorphic extended frame C in 2.3.1 is not uniquely determined. Any holomorphic gauge $W_+ : \mathbb{D} \longrightarrow \Lambda^+ Sl(2,\mathbb{C})_\sigma$ satisfying $W_+(z_0, \lambda) = I$ produces another holomorphic extended frame

$$\tilde{C} = C \cdot W_+ = F \cdot (V_+ \cdot W_+). \tag{2.3.4}$$

We will see later (e.g. section 2.4) that for certain purposes one or the other holomorphic extended frame is preferable.

2.4 It certainly seems to be useful to look for some holomorphic extended frame C which is uniquely determined by F and which, in turn, determines F essentially uniquely. Such a curve would be F_-, if F splits as $F = F_- \cdot F_+$ for all $z \in \mathbb{D}$. In general, this is not possible. However, one can prove:

Theorem 2.4.1. *([26]) If F is an extended framing of a harmonic map $N : \mathbb{D} \longrightarrow S^2$, then there exists some discrete set $S \subset \mathbb{D}$ such that the extended framing F can be split in the form*

$$F = F_- \cdot F_+ \quad for \ \ all \ \ z \in \mathbb{D} \setminus S \tag{2.4.1}$$

with $F_- \in \Lambda_*^- Sl(2,\mathbb{C})_\sigma$. Then F_- is uniquely determined by F and, considered as a function on all of $\mathbb{D}$, meromorphic. Moreover, we have $F_-(z_0, \lambda) = I$.

Remark 2.4.2. 1. Since $F(z_0, \lambda) = I$, we know that F_- is not singular at $z = z_0$.

2. To verify that F_- is meromorphic, one observes first that the Maurer-Cartan form ξ of F_- is integrable and that by 2.4.1 it is obtained from the Maurer-Cartan form of F by some gauge transformation which does not contain any negative powers of λ. It follows from this that ξ

is a $(1,0)-$form. More precisely, it is obtained from $\alpha'_{\mathfrak{p}}$ by conjugation with some matrix independent of λ. Therefore the $\bar{z}-$derivative of F_- vanishes and F_- is holomorphic where defined.

3. It is easy to see that the argument just given cannot be applied to F_+.

The Maurer-Cartan form

$$\xi = F_-^{-1} dF_- \tag{2.4.2}$$

is called the *normalized potential* for N (or for F). This ξ is uniquely determined by the harmonic map N. In contrast, the Maurer-Cartan forms associated with the holomorphic extended frames C associated with N are called *holomorphic potentials* for N. Holomorphic potentials are not uniquely determined by N.

Theorem 2.4.3. *([26]) a) The normalized potential ξ for some harmonic map $N : \mathbb{D} \longrightarrow S^2$ is of the form*

$$\xi = \lambda^{-1} \xi_{-1} dz, \tag{2.4.3}$$

where $\xi_{-1} : \mathbb{D} \longrightarrow sl(2,\mathbb{C})$ is meromorphic.

b) The holomorphic potentials

$$\eta = C^{-1} dC \tag{2.4.4}$$

are of the form

$$\eta = \lambda^{-1}\eta_{-1} + \lambda^0 \eta_0 + \lambda^1 \eta_1 + \dots \tag{2.4.5}$$

where all $\eta_j : \mathbb{D} \longrightarrow sl(2,\mathbb{C})$, $j = -1, 0, 1, \dots$ are holomorphic.

c) If η is a holomorphic potential for the harmonic map $N : \mathbb{D} \longrightarrow G/K$ and if $W_+ : \mathbb{D} \longrightarrow \Lambda^+ Sl(2,\mathbb{C})_\sigma$ is a holomorphic map satisfying $W_+(z_0, \lambda) = I$, then the gauged potential

$$\tilde{\eta} = W_+{}^{-1} \eta W_+ + W_+{}^{-1} dW_+ \tag{2.4.6}$$

is also a holomorphic potential for N.

Recall that the coefficient matrices at even powers of λ are diagonal and the coefficient matrices at odd powers of λ are off-diagonal.

Remark 2.4.4. 1. The coefficient matrix of normalized potentials and the coefficient matrix at λ^{-1} of holomorphic potentials is in all cases of the form $\begin{pmatrix} 0 & a \\ b & 0 \end{pmatrix}$. A comparison with 2.1.4 shows that the corresponding immersion of constant mean curvature is minimal if and only if $a = 0$, and it is totally umbilical if and only if $b = 0$. For the first case see section 3.5. The second case we have excluded from our discussion.

2. By the theorem above, every normalized potential has a meromorphic coefficient function. More generally, potentials for which the coefficient functions at all powers of λ are meromorphic will be called *meromorphic potentials* in this article. In particular, normalized potentials are meromorphic potentials. In many articles (including previous articles of the author) the notion of a meromorphic potential is used synonymously with what we call normalized potential. In this article the notion of a meromorphic potential will not be so restricted (see section 6.6).

§3. Constructing surfaces from potentials

3.1 In this section we will reverse the steps carried out in the previous section and construct constant mean curvature surfaces from potentials.

We start from some holomorphic potential η, i.e. some holomorphic $(1,0)$–form on $\mathbb{D}$ of the following type

$$\eta = \lambda^{-1}\eta_{-1} + \lambda^{0}\eta_0 + \lambda^{1}\eta_1 + \dots \tag{3.1.1}$$

where the coefficient matrices at even powers of λ are diagonal and the coefficient matrices at odd powers of λ are off-diagonal.

In view of 2.4.4, to pass from the level of potentials to the level of "holomorphic extended frames" we solve the *complex ordinary differential equation*

$$dC = C \cdot \eta, \quad C(z_0, \lambda) = I. \tag{3.1.2}$$

In view of 2.3.1, the transition from the holomorphic extended frame to the extended frame is established by the Iwasawa splitting

$$C = F \cdot V_+ \tag{3.1.3}$$

where we can assume $F(z_0, \lambda) = I$. Using Lemma 4.5 of [26] and Theorem 3.1 of [22] we have

Theorem 3.1.1. *Let η be a holomorphic potential and let F be constructed as above. If b_{-1} denotes the $(1,2)-$entry of η_{-1}, and if $b_{-1} \neq 0$, then $N \equiv F \bmod U(1)$ is a harmonic map from $\mathbb{D} \setminus \hat{S}$ to S^2 with extended framing F, where $\hat{S} = \{z \in \mathbb{D} \mid b_{-1}(z) = 0\}$.*

The case $b_{-1} = 0$ will be treated in section 3.5.

If the mean curvature H of the associated immersion is different from 0, then the final step, constructing the surface of constant mean curvature associated with N, can be carried out via Kenmotsu's work [41]. But using the associated family, one can construct the immersion associated with N directly using the extended framing ([4], Theorem 1.2):

Theorem 3.1.2. *(Sym-Bobenko Formula) If F is the extended framing of some harmonic map $N : \mathbb{D} \longrightarrow S^2$, associated with an immersion ϕ of constant mean curvature $H \neq 0$, then*

$$\phi = -\frac{1}{2H}\{\partial_t F \cdot F^{-1} + F \cdot \frac{i}{2} \operatorname{diag}(1, -1) \cdot F^{-1}\}, \tag{3.1.4}$$

where $\lambda = \exp(it)$.

Here we have realized $\mathbb{R}^3$ as $su(2)$. A natural isomorphism of $\mathbb{R}^3$ with $su(2)$ is given by the spin representation (see e.g. the Appendix of [17]).

Remark 3.1.3. It is important to note that this formula yields the same immersion ϕ for the extended frames F and $F \cdot k$, where $k \in SU(2)$ is diagonal and independent of λ. In particular, the freedom in the Iwasawa splitting (see Remark 2.2.3) has *no* effect on the final immersion, if $H \neq 0$.

3.2 Starting, conversely, from some normalized potential ξ, one can proceed as above. However, due to possible monodromy in the solutions of the ODE involved, one may have to introduce "cuts" into $\mathbb{D}$. On the other hand, if one wants to construct immersions without singularities on $\mathbb{D}$, then some conditions need to be satisfied at the poles of ξ. First of all we note that smooth immersions on $\mathbb{D}$ have holomorphic Hopf differentials on $\mathbb{D}$. In addition, for the smoothness on all of $\mathbb{D}$ of the immersion produced from ξ, two more conditions need to be satisfied:

(3.2.1) The solution F_- to $dF_- = F_- \cdot \xi$, $F_-(z_0, \lambda) = I$ is meromorphic on $\mathbb{D}$

(3.2.2) The extended framing F obtained from F_- via $F_- = F \cdot V_+$ is smooth on $\mathbb{D}$.

Condition 3.2.1 is necessary by Theorem 2.4.1 and condition 3.2.2 is obviously necessary. For the case $H = 0$ see section 3.5. Assume now $H \neq 0$. Converting the matrix differential equation $dF_- = F_- \cdot \xi$, with

$$\xi = \lambda^{-1} \begin{pmatrix} 0 & f \\ Q/f & 0 \end{pmatrix} dz,$$

into a scalar second order ordinary differential equation, one obtains

$$y'' - \frac{f'}{f} y' - \lambda^{-2} Q y = 0. \tag{3.2.3}$$

One observes that only regular singular points occur. Thus for every point of $\mathbb{D}$ one of the two solutions, say y_1, can already be assumed to be meromorphic. For the second solution, y_2, to be meromorphic a simple residue condition needs to be satisfied

Theorem 3.2.1. *([17], Theorem 2.8.2) The equation*

$$dF_- = F_- \cdot \xi, \quad F_-(z_0, \lambda) = I$$

has a global meromorphic solution in $\Lambda^- Sl(2, \mathbb{C})_\sigma$ *if and only if*

$$\oint \frac{f}{{y_1}^2} dz = \mathrm{res}_p \frac{f}{{y_1}^2} = 0 \tag{3.2.4}$$

for any zero or pole p of f.

Condition 3.2.4 can be expressed in terms of the orders of the roots and poles of the coefficient functions of ξ.

Theorem 3.2.2. *([17]) Let*

$$\xi = \lambda^{-1} \begin{pmatrix} 0 & f \\ Q/f & 0 \end{pmatrix} dz, \tag{3.2.5}$$

be a normalized potential for which the solution to the differential equation $dC = C \cdot \xi$, $C(0, \lambda) = I$, is meromorphic. Then for ξ to yield a *constant mean curvature surface without branch points under the generalized Weierstraß representation it is necessary and sufficient that for every $z_0 \in \mathbb{D}$, where f has a pole or zero, the following conditions are satisfied: Let m denote the order of z_0 as a zero of the Hopf differential Q. If f has a pole of order n at z_0, then $n = 2$ or for some integer $r \geq 1$*

$$n = r(2m+4) \qquad \text{or} \qquad n = r(2m+4)+2. \tag{3.2.6}$$

If f has a zero of order n at z_0, then for some integer $r \geq 1$

$$n = r(2m+4) \qquad \text{or} \qquad n = r(2m+4)-2. \tag{3.2.7}$$

Remark 3.2.3. 1. The conditions for the smoothness of the immersion are expressed for holomorphic potentials and for normalized potentials in quite different ways. If a normalized potential ξ as above is actually holomorphic, then for the vanishing orders n and m of f and Q at some point $z_0 \in \mathbb{D}$ we have $n \neq m$. But then 3.2.7 does not make sense for positive n. Therefore only $n = 0$ is possible, if the associated immersion is supposed to have no branch points at z_0. In other words, if ξ is holomorphic, then the immersion does not have any branch points if and only if f never vanishes.

2. When considering non-umbilical points, one deals with the case $m = 0$. In this case the theorem simply states that n is even.

3. Constructing a constant mean curvature immersion ϕ from a holomorphic potential η or a normalized potential ξ is called a *generalized Weierstraß representation* of ϕ. The potentials η or ξ are called the Weierstraß data for ϕ.

4. The first, still fairly slow, computer implementation of the generalized Weierstraß representation was carried out by Lerner and Sterling [44]. Inspired by their work, Pinkall and Gunn improved the algorithm considerably, leading to a fast visualization method for surfaces of constant mean curvature. Recently, Schmitt has produced further improvements [43].

3.3 Examples 1. Consider the holomorphic potential

$$\eta = \lambda^{-1} \begin{pmatrix} 0 & 1 \\ 1 & 0 \end{pmatrix} dz. \tag{3.3.1}$$

The associated family of surfaces corresponding to η contains (for $\lambda = 1$) the round cylinder. In this case the Weierstraß representation can be computed explicitly by hand.

2. Consider the holomorphic potential

$$\eta = \lambda^{-1}\begin{pmatrix} 0 & X \\ \bar{X} & 0 \end{pmatrix} dz, \tag{3.3.2}$$

where $X = a\lambda^{-1} + \lambda b,\ \ a, b \in \mathbb{R}$. If $a = b = 0$, then the map ϕ defined in 3.1.4 is constant, and thus degenerate. If $a = 0, b \neq 0$, then the map ϕ describes a minimal surface [24]. If $a \neq 0, b = 0$, then the map ϕ describes (part of) a sphere. This follows from the fact that in this case the Hopf differential vanishes identically and thus every point of the surface is an umbilical point. Assume now $ab \neq 0$. Since the associated family does not change, if we replace λ by $-\lambda$, or by λ^{-1}, we can assume $a > 0$ and $a \geq b$. If $ab > 0$ and $a+b = \frac{1}{2}$, then ϕ describes the associated family of an unduloid (the unduloid is the surface obtained for $\lambda = 1$). In particular, this is a *Delaunay surface*, i.e. a surface of revolution of constant mean curvature. For more information about Delaunay surfaces we refer to [42],[29],[30].

3. Consider the potential

$$\eta = \lambda^{-1}\begin{pmatrix} 0 & 1 \\ z^n & 0 \end{pmatrix} dz \tag{3.3.3}$$

where n is a non-negative integer. The associated family of surfaces contains (for $\lambda = 1$) a *Smyth surface* [51]. Its metric is invariant under a one-parameter group of self-isometries. All such surfaces of constant mean curvature have been determined by Smyth [51]. The associated family of such a surface contains either a Delaunay surface or a Smyth surface.

4. Consider the potential

$$\xi = \begin{pmatrix} 0 & \lambda^{-1}\alpha + \lambda\beta \\ \lambda^{-1}\beta + \lambda\alpha & 0 \end{pmatrix} \tag{3.3.4}$$

where α and β are holomorphic 1-forms on $\mathbb{C}^*$ of the form

$$\alpha = -h\frac{dz}{z}, \tag{3.3.5}$$

and

$$\beta = (h + \sum_{n=1}^{\infty} b_n(z^{-n} + z^n))\frac{dz}{z}, \tag{3.3.6}$$

where $h, b_n \in \mathbb{R}$ and $h \neq 0$.

Theorem 3.3.1. *([42], Theorem 7.1.1) Let α and β be as above and assume that the coefficients of β satisfy*

$$\int_0^{2\pi} \cos\left(4\sum_{n=1}^{\infty} \frac{b_n}{n} \sin(nt)\right) dt = 0. \tag{3.3.7}$$

Then the potential ξ generates, for $\lambda = 1$, an immersion of a cylinder which has constant mean curvature and two planar symmetries. By choosing α and β appropriately one can obtain any number of umbilical points.

By different methods more examples, such as immersions of cylinders and tori, will be given in the following sections.

3.4 Wu's Formula In general, interest in the loop group method is concentrated on the construction of CMC surfaces with specific preassigned properties. Thus the goal is to choose potentials in order to obtain surfaces with such properties. For this purpose it is useful to compute the potentials for some known examples, and this can be achieved by a remarkably simple formula. The crucial observation behind this formula—as explained in more detail in section 7.7.6—is that the theory presented so far is in fact a "real form" of a complex theory, in which z and $\bar{z}$ become independent variables. A thorough investigation of this complex theory has not yet been carried out.

Theorem 3.4.1. *([58]) Let $\phi : \mathbb{D} \longrightarrow \mathbb{R}^3$ be an associated family of immersions of constant mean curvature $H \neq 0$. If $ds^2 = \zeta(z, \bar{z}) dz d\bar{z}$ denotes the metric and $\lambda^{-2}Q(z)dz^2$ the Hopf differential, then the conformal factor ζ is the restriction of a meromorphic map $\hat{\zeta} : \mathbb{D} \times \bar{\mathbb{D}} \longrightarrow \mathbb{C}$ to the subset $\{(z, \bar{z}) \mid z \in \mathbb{D}\}$ of $\mathbb{D} \times \bar{\mathbb{D}}$, and the normalized potential, uniquely associated with ϕ, is of the form*

$$\xi = \lambda^{-1} \begin{pmatrix} 0 & f \\ Q/f & 0 \end{pmatrix} dz, \tag{3.4.1}$$

where

$$f(z) = \frac{\zeta(z, 0)}{\sqrt{\zeta(0, 0)}}. \tag{3.4.2}$$

3.5 Minimal Surfaces The concrete expression 2.1.4 for the Maurer-Cartan form of the extended frames shows:

a) A holomorphic potential $\eta = \lambda^{-1}\eta_{-1} + \lambda^0\eta_0 + \lambda^1\eta_1 + \dots$ corresponds to a minimal surface iff $\eta_{-1} = \begin{pmatrix} 0 & 0 \\ p & 0 \end{pmatrix} dz$,

b) A normalized potential $\xi = \lambda^{-1}\xi_{-1}$ corresponds to a minimal surface iff $\xi_{-1} = \begin{pmatrix} 0 & 0 \\ p & 0 \end{pmatrix} dz$.

Let ξ be the normalized potential of some minimal surface. Then

$$F_- = \begin{pmatrix} 1 & 0 \\ \lambda^{-1}\int_{z_0}^{z} p(w)dw & 1 \end{pmatrix}. \tag{3.5.1}$$

We put

$$q(z) = \int_{z_0}^{z} p(w)dw. \tag{3.5.2}$$

Because of the very simple form of F_- one can find an Iwasawa splitting explicitly. We have $F_- = F \cdot V_+$, where

$$F = (1 + |q|^2)^{-1/2} \begin{pmatrix} 1 & -\lambda\bar{q} \\ \lambda^{-1}q & 1 \end{pmatrix}, \tag{3.5.3}$$

and

$$V_+ = (1 + |q|^2)^{-1/2} \begin{pmatrix} 1 + |q|^2 & \lambda\bar{q} \\ 0 & 1 \end{pmatrix}. \tag{3.5.4}$$

Interpreting F as the coordinate frame of some minimal surface we obtain (e.g. by comparing the Maurer-Cartan forms)

$$ds^2 = 4(1 + |q|^2)^2 dz d\bar{z}, \tag{3.5.5}$$

$$Q = q_z = p. \tag{3.5.6}$$

In contrast with the case $H \neq 0$, the freedom in the Iwasawa splitting (see Remark 2.2.3) does affect the associated minimal immersions. Let F and $\tilde{F} = F \cdot k$ be coordinate frames obtained via Iwasawa splitting:

$$F_- = \tilde{F} \cdot \tilde{V}_+ = F \cdot V_+. \tag{3.5.7}$$

Then, setting $k = \operatorname{diag}(e^{i\alpha}, e^{-i\alpha})$, we obtain for the associated conformal factor and the Hopf differential

$$d\tilde{s}^2 = (1 + |q|^2)^2 \cdot e^{Re(h)} \cdot dz d\bar{z} \tag{3.5.8}$$

$$\tilde{Q} = Qe^{\bar{h}/2}, \tag{3.5.9}$$

where h is an antiholomorphic map satisfying

$$\mathrm{Im}(h) = -4\alpha. \tag{3.5.10}$$

Since F and $\tilde{F}$ yield the same Gauß map, and since the relation between Gauß maps and meromorphic potentials is one-to-one, the freedom in the Iwasawa splitting describes the family of minimal surfaces possessing the same Gauß map. The splitting 3.5.3 and 3.5.4 given above represents one natural choice of minimal surface associated with a fixed Gauß map.

Before making a precise connection with the classical Weierstraß representation we note that the normalized potential for minimal surfaces contains exactly one meromorphic function, which defines — via the freedom in the Iwasawa splitting — many minimal surfaces and one Gauß map, which is the same for all these minimal surfaces. The second function that appears in the classical Weierstraß representation enters as the function h above, representing the freedom in the Iwasawa splitting.

Let F be the extended coordinate framing of some minimal surface and denote the entries of F by a, b and $-\bar{b}, \bar{a}$ as usual. Then from 2.1.4 we derive, since $H = 0$,

$$\bar{a}_z = -\frac{1}{4}u_z\bar{a}, \tag{3.5.11}$$

$$b_z = -\frac{1}{4}u_z b, \tag{3.5.12}$$

whence, since u is real,

$$a = e^{-\frac{1}{4}u}s, \tag{3.5.13}$$

$$b = \lambda e^{-\frac{1}{4}u}\bar{r}, \tag{3.5.14}$$

where s and r are holomorphic functions. In view of 3.5.3 and the relation $F = \tilde{F} \cdot k$ for the general frame, we see that r and s are independent of λ. If $\phi : \mathbb{D} \longrightarrow \mathbb{R}^3$ is the minimal immersion which corresponds to F, then (A.5.6) and (A.6.3) of [17] show that

$$G^{-1} \cdot J(\phi_z) \cdot G = -\frac{i}{2}\lambda^{-1}e^{u/2}F\sigma_+F^{-1}, \tag{3.5.15}$$

where

$$G = i\begin{pmatrix} 0 & \lambda^{-1/2} \\ \lambda^{1/2} & 0 \end{pmatrix} \text{ and } \sigma_+ = \begin{pmatrix} 0 & 1 \\ 0 & 0 \end{pmatrix}, \tag{3.5.16}$$

while

$$J(x,y,z) = -\frac{i}{2}\begin{pmatrix} z & x-iy \\ x+iy & -z \end{pmatrix}. \tag{3.5.17}$$

As a consequence

$$\phi_z = (\frac{1}{2}(s^2 - \lambda^{-4}r^2), -\frac{i}{2}(s^2 + \lambda^{-4}r^2), -sr\lambda^{-2}). \tag{3.5.18}$$

If we give up the requirement that F is I at z_0, we can consider the frame

$$\hat{F} = DF, \text{ where } D = \operatorname{diag}(\sqrt{i}\lambda^{-1}, \sqrt{i}^{-1}\lambda). \tag{3.5.19}$$

Then for the corresponding minimal immersion $\hat{\phi}$ we obtain

$$\hat{\phi}_z = \lambda^{-2}(\frac{1}{2}(r^2 - s^2), \frac{i}{2}(s^2 + r^2), sr), \tag{3.5.20}$$

which is a version of the classical Weierstraß representation. One obtains the usual formula for $\mu = r^2$ and $\nu = s/r$. Note that the special splitting 3.5.3 corresponds to $s = 1$ and $r = -q$, equivalently $\mu = q^2$ and $\nu = -q^{-1}$. Note that ν corresponds to the stereographically projected Gauß map. For $q = -z^{-1}$ one obtains $\nu = z$ and $\mu = z^{-2}$, which describes the catenoid. More examples are contained in [25].

From now on we will assume, unless stated otherwise, that "constant mean curvature" means that H is constant and nonzero.

§4. Surfaces of finite type

4.1 The generalized Weierstraß representation constructs all constant mean curvature surfaces. However, as it is not easy to trace properties from the input for the Iwasawa splitting to its output, the procedure is difficult to use in most concrete cases. We will see in section 5 that geometric features can be controlled in some cases via dressing. Still, the construction of the associated surfaces is not very explicit, at least from a non-numerical point of view. It is therefore very fortunate that for a large and important class of constant mean curvature surfaces an independent geometric construction procedure is known, which, in addition, is relatively explicit. We follow in essence the exposition of [10],[12] and [8].

For $d \in 2\mathbb{N}+1$ we set

$$\xi_* = \Lambda_d = \{\sum_{n=-d}^{d} \xi_{-n}\lambda^n \in \Lambda su(2)_\sigma, \xi_d \neq 0\}. \tag{4.1.1}$$

Then the basic results are Theorem 3.2 and Theorem 2.5 of [10], which we shall summarize. Note that if $\xi \in \Lambda_d$, then ξ_{d-1} is diagonal, since d is odd. Therefore $\xi_{d-1} \in \mathfrak{k}^{\mathbb{C}}$. Specializing the general procedure (see (2.5) in [10]) we set for any $\tau \in \mathfrak{k}^{\mathbb{C}}$

$$r(\tau) = \frac{1}{2}\tau \tag{4.1.2}$$

Using this notation we can state

Theorem 4.1.1. *([10]) For each $d \in 2\mathbb{N}+1$ and $\xi_* \in \Lambda_d$, there is a unique solution $\xi : \mathbb{R}^2 \longrightarrow \Lambda_d$, to the differential equation*

$$\frac{\partial \xi}{\partial z} = [\xi, \lambda^{-1}\xi_d + r(\xi_{d-1})], \quad \xi(z_0) = \xi_*. \tag{4.1.3}$$

Moreover, in this case, the $\Lambda su(2)_\sigma$-valued 1-form α given by

$$\alpha = (\lambda^{-1}\xi_d + r(\xi_{d-1}))dz + (\lambda\xi_{-d} + \overline{r(\xi_{d-1})})d\bar{z} \tag{4.1.4}$$

satisfies the Maurer-Cartan equations. In addition, the extended framing F defined by

$$dF = F \cdot \alpha, \qquad F(z_0, \lambda) = I \tag{4.1.5}$$

induces a harmonic map $N : \mathbb{D} \longrightarrow S^2$ and a surface of constant mean curvature $\phi : \mathbb{D} \longrightarrow \mathbb{R}^3$ via the Sym-Bobenko formula.

Constant mean curvature surfaces defined this way are said to be of *finite type*. A crucial observation is that from the theorem above one can construct naturally a very special holomorphic potential for surfaces of finite type (see Theorem 4.2.1 below).

4.2 Consider the potential

$$\eta = \lambda^{d-1}\xi_* dz, \tag{4.2.1}$$

where $\xi_* \in \Lambda_d$ is, as in the previous section, independent of z and $\bar{z}$. Then the holomorphic extended frame C defined by the differential equation 3.1.2 is of the form

$$C(z,\lambda) = e^{z\lambda^{d-1}\xi_*}. \tag{4.2.2}$$

Consider the Iwasawa splitting

$$C = \hat{F} \cdot \hat{V}_+. \tag{4.2.3}$$

The trivial, but important, consequence of this is that

$$\hat{F}^{-1} \cdot \xi_* \cdot \hat{F} \in \Lambda_d, \tag{4.2.4}$$

which is easy to verify, since C commutes with ξ_* and $\hat{\xi} = \hat{F}^{-1} \cdot \xi_* \cdot \hat{F}$ is skew-hermitian.

It is straightforward to verify that

$$\frac{\partial \hat{\xi}}{\partial z} = [\hat{\xi}, \hat{F}^{-1}\frac{\partial \hat{F}}{\partial z}], \quad \textit{where} \quad \hat{\xi}(0,\lambda) = \xi_*. \tag{4.2.5}$$

Since $\hat{\xi} \in \Lambda_d$, one can derive from 4.2.3 and 4.2.4 that $\hat{F}^{-1}\frac{\partial \hat{F}}{\partial z} = (\lambda^{-1}\hat{\xi}_d + r(\hat{\xi}_{d-1}))$ and thus has exactly the form of the $(1,0)$-part of α in 4.1.4. Since $\hat{F}^{-1}\frac{\partial \hat{F}}{\partial z}$ is skew-hermitian, $\hat{F}^{-1}d\hat{F}$ has the form 4.1.4. Therefore, the surface derived from a potential of the form 4.2.1 is of finite type. Conversely, if we start from a surface of finite type, then with the notation of Theorem 4.1.1, ξ satisfies the differential equation 4.1.3 with initial condition ξ_*. If the potential η in 4.2.1 is formed with the same ξ_*, then the argument above shows that $\hat{\xi}$ satisfies the same differential equation with the same initial condition as ξ. Therefore $\xi = \hat{\xi}$. From this it is easy to verify that the Maurer-Cartan forms of F and $\hat{F}$ also coincide, whence $F = \hat{F}$. Thus we obtain

Theorem 4.2.1. *([10]) If ξ satisfies 4.1.3 and ϕ is the constant mean curvature immersion of finite type defined from F, which is given by 4.1.4 and 4.1.5, then ϕ coincides with the constant mean curvature immersion induced from the constant potential $\eta = \lambda^{d-1}\xi_* dz$ via the generalized Weierstraß representation. Conversely, if $\xi_* \in \Lambda_d$ and $\eta = \lambda^{d-1}\xi_* dz$, then the associated constant mean curvature immersion is of finite type.*

Remark 4.2.2. 1. We would like to emphasize that equation 4.2.4 implies that ξ only involves finitely many powers of λ.

2. Surfaces of finite type have been investigated by many authors. We have followed closely the work of Burstall and Pedit [10],[12]. A starting point for many investigations has been the work of Pinkall and Sterling [46].

3. A beautiful and also very explicit description of all constant mean curvature surfaces of finite type has been given by algebro-geometric methods. This includes a discussion of periodicity conditions. The connection with ξ_* is given by the algebraic curve $\det(\xi_*(\lambda) - \mu I) = 0$. For details see [4],[5].

Example 4.2.3. (1) $d = 1$. *In this case the potentials have the form* $\eta = \xi_* = (\lambda^{-1}A_{-1} + A_0 + \lambda A_1)dz$. *From 3.3.2 one sees that the potentials of the Delaunay surfaces are of this type.*

(2) $d = 3$. *In this case a special choice of* ξ_* *leads to the Wente torus.*

4.3 In the last section we have obtained a very explicit and simple definition of surfaces of finite type in terms of certain holomorphic potentials. Since the Hopf differential can be obtained as the determinant of the coefficient matrix at λ^{-1} in any potential of the surface, it is clear that surfaces of finite type do not have any umbilics. In particular, compact constant mean curvature surfaces of genus > 1 are not of finite type. On the other hand, not every surface without umbilics is of finite type. As an example we mention the potentials

$$\eta = \lambda^{-1}\begin{pmatrix} 0 & 1 \\ c & 0 \end{pmatrix}$$

which do not yield surfaces of finite type, if $c \in \mathbb{R}$, $c \neq 0, \pm 1$ [44]. Despite the somewhat obscure definition of "finite type", we know

Theorem 4.3.1. *([46]) Every constant mean curvature surface with doubly periodic metric is of finite type. In particular, every constant mean curvature torus is of finite type.*

For immersions of cylinders the situation is quite different: some constant mean curvature cylinders are of finite type, while others have umbilics [22],[42] and are therefore not of finite type.

Remark 4.3.2. The notion of "finite type" as spelled out in Theorem 4.1.1 is quite roundabout. One of its consequences is that $F \cdot \xi_* \cdot F^{-1}$ is polynomial in λ (see the remark above). It would seem to be more natural to define "finite type" by this condition, namely that $F \cdot \xi_* \cdot F^{-1}$

is polynomial in λ, without assuming the reality of ξ_* [36]. It would be interesting to understand when this condition implies "finite type" in the sense of this article.

§5. Dressing

5.1 Dressing is a deformation technique which, applied to constant mean curvature surfaces, produces new surfaces from old ones. Since its invention by Zakharov and Shabat [59], dressing methods have been applied in many situations. For example, constructing solutions to the KdV equation via flows on infinite dimensional Grassmannians is in essence dressing, producing nontrivial solutions from the trivial one. For the situation discussed in this paper the basic result is

Theorem 5.1.1. *Let F be the extended frame of some constant mean curvature surface. Let $h_+ \in \Lambda^+ Sl(2,\mathbb{C})_\sigma$ and consider the Iwasawa splitting*

$$h_+ \cdot F = \hat{F} \cdot \hat{V}_+. \tag{5.1.1}$$

Then $\hat{F}$ is the extended frame of some constant mean curvature surface.

In 5.1.1 we can assume that the coefficient matrix at λ^0 of h_+ has only non-negative entries. Therefore we can assume that h_+ is in the identity component of $\Lambda^+ Sl(2,\mathbb{C})_\sigma$. Hence we can connect h_+ with I by a continuous curve h_+^s, $0 \le s \le 1$, $h_+^0 = I$, $h_+^1 = h_+$. Then the family of immersions ϕ^s associated with $\hat{F}^s$, ${h_+}^s \cdot F = \hat{F}^s \cdot \hat{V}^s$, represents a continuous deformation within the class of constant mean curvature immersions, starting from $\phi = \phi^0$ and ending at $\hat{\phi} = \phi^1$.

Remark 5.1.2. This already shows that dressing as defined above is an important procedure. It turns out, however, that for the most interesting applications the theorem above needs to be generalized.

For this one observes that the Maurer-Cartan form of F depends on λ in a holomorphic fashion, with λ varying in $\mathbb{C}^*$. Therefore, the solution F to the equation $dF = F \cdot \alpha$, $F(0,\lambda) = I$, also depends holomorphically on $\lambda \in \mathbb{C}^*$. This leads to the introduction of the group $\Lambda_r Sl(2,\mathbb{C})_\sigma$ of matrices defined on the circle $|\lambda| = r$, for $0 < r \le 1$. We also say $h_+ \in \Lambda_r^+ Sl(2,\mathbb{C})_\sigma$ if $h_+ \in \Lambda_r Sl(2,\mathbb{C})_\sigma$ is holomorphic in the interior of the disk of radius r. Note that for $h_+ \in \Lambda_r^+ Sl(2,\mathbb{C})_\sigma$ the expression $h_+ \cdot F$ does make sense on the circle $|\lambda| = r$. In this situation we have the following

Theorem 5.1.3. *(r-Iwasawa Splitting) Let $h \in \Lambda_r Sl(2,\mathbb{C})_\sigma$. Then there exist $U, V_+ \in \Lambda_r Sl(2,\mathbb{C})_\sigma$ such that $h = U \cdot V_+$ and*

(5.1.2) $V_+ \in \Lambda_r^+ Sl(2,\mathbb{C})_\sigma$

(5.1.3) *U extends holomorphically to $A_r = \{\lambda \in \mathbb{C} \mid r < |\lambda| < 1/r\}$*

(5.1.4) *U is unitary when restricted to the unit circle.*

For a proof of this theorem, see [45]. The result follows easily from [3]. The group of matrices in $\Lambda_r Sl(2,\mathbb{C})_\sigma$ satisfying the last two conditions will be denoted by $\Lambda_r SU(2)_\sigma$.

Corollary 5.1.4. *Theorem 5.1.1 remains true if h is chosen in $\Lambda_r^+ Sl(2,\mathbb{C})_\sigma$.*

If $\hat{F}$ is defined as in Theorem 5.1.1 by some $h \in \Lambda_r^+ Sl(2,\mathbb{C})_\sigma$ we will say that *the frame $\hat{F}$ and the corresponding immersion $\hat{\phi}$ is obtained by r-dressing from F and ϕ respectively.* If $r = 1$, then we will generally use "dressing" instead of "1-dressing".

Since a group splitting is involved in the r-dressing procedure, computations by hand are possible only in very special cases. For example one can prove

Theorem 5.1.5. *([17], section 3.3) a) If $\xi = \lambda^{-1} \begin{pmatrix} 0 & f \\ Q/f & 0 \end{pmatrix} dz$ is the normalized potential of some constant mean curvature immersion ϕ, then dressing with $h_+ \in \Lambda^+ Sl(2,\mathbb{C})_\sigma$ yields a constant mean curvature immersion $\hat{\phi}$ with normalized potential $\hat{\xi} = \lambda^{-1} \begin{pmatrix} 0 & \hat{f} \\ Q/\hat{f} & 0 \end{pmatrix} dz$.*

b) If $h_+ = \begin{pmatrix} s & 0 \\ 0 & s^{-1} \end{pmatrix} dz$, then $\hat{f} = s^2 \cdot f$.

c) If $h_+ = \begin{pmatrix} 1 & 0 \\ t\lambda & 1 \end{pmatrix} dz$, then $\hat{f} = (1 + tb_1)^{-2} \cdot f$.

d) If $h_+ = \begin{pmatrix} 1 & t\lambda \\ 0 & 1 \end{pmatrix} dz$, then $\hat{f} = (1 + tc_1)^2 \cdot f$.

Here $b_1(z) = \int_{z_0}^z f(w)dw$ and $c_1(z) = \int_{z_0}^z \frac{Q(w)}{f(w)} dw$.

Clearly, Q is invariant under the dressing operation. The transformation $f \longmapsto \hat{f}$ is quite explicit for the three simple dressing matrices listed above. Moreover, these three matrices generate $\Lambda_r^+ Sl(2,\mathbb{C})_\sigma$ as a

Banach Lie group. Nevertheless, since $f \longmapsto \hat{f}$ is non-linear and involves integration, it is difficult to find an explicit formula for more general h_+. It would be helpful to understand the dressing transformation $f \longrightarrow \hat{f}$ analytically in a way which reflects the group structure of dressing.

From a geometric point of view, work of Inoguchi-Toda and also Mahler seems to show that the Bäcklund transformations defined by Bianchi (see e.g. [52]) can be obtained by special dressing matrices (different from the ones listed above). One hopes that eventually this will lead to a better understanding of more general dressing matrices.

Another geometric result related to dressing is

Theorem 5.1.6. *([22], Theorem 5.3) If $\phi : \mathbb{D} \longrightarrow \mathbb{R}^3$ is a constant mean curvature surface, which induces a complete metric on $\mathbb{D}$, then all surfaces in the dressing orbit of ϕ are complete.*

5.2 In general, the dressing orbits of the group $\Lambda_r^+ Sl(2,\mathbb{C})_\sigma$ are quite large:

Theorem 5.2.1. *a) ([21]) If the surface ϕ_0 has umbilics, then the isotropy group under the dressing action is trivial.*

b) If the surface ϕ_0 is generated by the constant holomorphic potential η, then the isotropy group $\mathrm{Iso}(\phi_0)$ under the dressing action is

$$\mathrm{Iso}(\phi_0) = \{g \in \Lambda_r^+ Sl(2,\mathbb{C})_\sigma \mid [g, \eta_{-1}] = 0\}. \tag{5.2.1}$$

Proof. Only b) needs some argument. Assume $h_+ \cdot F_0 = F_0 \cdot V_+$, then for the holomorphic extended frame C_0 associated with η we have $h_+ \cdot C_0 = C_0 \cdot W_+$. Since $C_0 = \exp(z\eta)$, this is equivalent to the formula $\exp(z \cdot ad(\eta)) \cdot h_+ = W_+$, which, in turn, is equivalent to the condition

$$(ad(\eta))^m h_+ \text{ only contains } \lambda^k \text{ for } k \geq 0.$$

Writing $\eta = \lambda^{-1}\eta_{-1} + \eta_+$, it is easy to see that this statement is equivalent to the condition

$$(ad(\lambda^{-1} \cdot \eta_{-1}))^m h_+ \text{ only contains } \lambda^k \text{ for } k \geq 0.$$

Recall that both coefficients of the off-diagonal matrix η_{-1} do not vanish, as $H \neq 0$, since we do not consider totally umbilical surfaces. Now the claim follows by a straightforward induction. □

Parts a) and b) of the theorem above do not cover the case of a general, never vanishing Hopf differential. It would be interesting to investigate this case.

5.3 Dressing and r-dressing can be defined not only on the level of frames, but also on the level of holomorphic extended frames:

$$h_+ \cdot C = \hat{C} \cdot \hat{W}_+. \tag{5.3.1}$$

Here we require that C and $\hat{C}$ are defined for $|\lambda| = 1$.

There is considerable freedom in the choice of h_+. Given $\hat{C}$ as above, one can also choose $\tilde{C} = \hat{C} \cdot U_+$, where U_+ is holomorphic in z and holomorphic in λ in the disk of radius 1. On the other hand, one can consider $\hat{C} = h_+ \cdot C \cdot {h_+}^{-1}$. If η is the Maurer-Cartan form of C, then the Maurer-Cartan form of $\hat{C}$ is $h_+ \cdot \eta \cdot {h_+}^{-1}$. If this is defined on the unit circle, then we would call this latter one-form a potential dressed from η.

Let C, $\hat{C}$ and η, $\hat{\eta}$ be as above. Then one obtains for the Maurer-Cartan forms η and $\hat{\eta}$ and W_+ the equation

$$d\hat{W}_+ = \hat{W}_+ \cdot \eta - \hat{\eta} \cdot \hat{W}_+ \tag{5.3.2}$$

Thus, in order to relate η and $\hat{\eta}$, one needs only to solve a linear ODE. However, we are interested in solutions involving only non-negative powers of λ, while η and $\hat{\eta}$ also involve some negative powers of λ. Thus the existence of such a solution $\hat{W}_+$ is a delicate and difficult problem.

The existence of a solution to 5.3.2 implies that C and $\hat{C} \cdot \hat{W}_+$ solve the same linear ODE, whence $A \cdot C = \hat{C} \cdot \hat{W}_+$ with some $A(\lambda)$ independent of z. An evaluation at $z = z_0$ yields $A(\lambda) = \hat{W}_+(z_0, \lambda)$, and hence a dressing equation 5.3.1. Thus dressing with some h_+ is equivalent to solving the *gauge equation* 5.3.2 with some $\hat{W}_+$ (not necessarily normalized as to satisfy $W_+(z_0, \lambda) = I$).

One could perhaps hope that all potentials with the same Hopf differential $Q(z)dz^2$ can be dressed onto each other. This is not true. It turns out that it is not even possible in general to solve the gauge equation 5.3.2 with a formal power series $\hat{W}_+ = \sum_{n=0}^{\infty} \hat{W}_n \lambda^n$. The formal solvability has been characterized by Wu [57]. (There are two misprints in [57]: in 5.12 there should be a + sign between p'/p and q'/q and b in 5.14 should be a q.) It would be desirable to find an analytic connection between "convergent solutions" and "formal solutions".

5.4 There have been several applications of the dressing transformation. One of the first results in the context of constant mean curvature surfaces related round cylinders to constant mean curvature tori:

Theorem 5.4.1. *([27], 5.3) Let ϕ be an associated family of constant mean curvature surfaces, which contains a torus. Then there exists some $0 < r < 1$ such that ϕ can be obtained by r-dressing from the associated family ϕ_0 of the round cylinder.*

This result was generalized by Burstall and Pedit [12]. Note that for the description of tori $r = 1$ does not work. This stems from the fact that the points defining the hyperelliptic curve, which is used to describe constant mean curvature tori in the algebro-geometric approach, lie between the r-circle and the $1/r$-circle. To make this more precise we quote two results from [19], which refer to 1-dressing.

Theorem 5.4.2. *Other than round cylinders there are no surfaces with translational symmetry in the dressing orbit of the round cylinder.*

Corollary 5.4.3. *The 1-dressing action does not act transitively on the set of all constant mean curvature surfaces without umbilics.*

The following result is in the same spirit and context as the results above, but admits r-dressing.

Theorem 5.4.4. *([20], Theorem 4.9) Every constant mean curvature surface, for which the metric is invariant under a discrete group of translations, and which is in the r-dressing orbit of the round cylinder, is of finite type.*

5.5 Recently, dressing was used to construct surfaces which are not of finite type and have nontrivial topology:

Theorem 5.5.1. *([22], Theorem 5.4 and Theorem 5.5) Consider the holomorphic potential*

$$\eta = \frac{1}{2}(\lambda + \lambda^{-1}) \begin{pmatrix} 0 & A \\ B\cos(2\pi z) & 0 \end{pmatrix} dz, \tag{5.5.1}$$

with associated constant mean curvature immersion ϕ_0. Then for all sufficiently small $A, B \neq 0$ there exists some $h_+ \in \Lambda^+ Sl(2, \mathbb{C})_\sigma$ such that the constant mean curvature immersions ϕ obtained from ϕ_0 by dressing with h_+ are complete and for $\lambda = \pm i$ one obtains constant mean curvature immersions of a cylinder with umbilics at $z = 1/4$ and $z = 3/4$.

Starting from [22], Kilian [42] has found many more cylinders with umbilics. See Example 4 in section 3.3.

5.6 We have seen in section 3.5 that also for minimal surfaces there is a 1-1-relation between normalized potentials and Gauß maps. However, there are many minimal surfaces with the same Gauß map. And these surfaces are all generated from the same potential via the freedom in the Iwasawa splitting (which becomes irrelevant in the case of mean curvature $H \neq 0$). Thus the primary interest is to understand how dressing affects the Gauß map. This is equivalent to asking how dressing affects the normalized potential of a minimal surface. Let $\xi = \lambda^{-1} \begin{pmatrix} 0 & 0 \\ p & 0 \end{pmatrix} dz$ be the normalized potential of a minimal surface, and

$$C_- = \begin{pmatrix} 1 & 0 \\ \lambda^{-1} q & 1 \end{pmatrix}$$

its associated holomorphic extended frame, where $q = \int_{z_0}^{z} p(w)dw$. Then the dressing equation $h_+ \cdot C_- = \hat{C}_- \cdot \hat{V}_+$ yields after a straightforward computation that $\hat{C}_-$ is of the form

$$\hat{C}_- = \begin{pmatrix} 1 & 0 \\ \lambda^{-1} \hat{q} & 1 \end{pmatrix},$$

where $\hat{q}$ is given by the explicit linear fractional formula $\hat{q} = q/(a + bq)$, for certain complex numbers a and b. In particular, dressing preserves the class of minimal surfaces and the dressing orbit has complex dimension 2. Recall from section 3.5 that q describes the Gauß map of the associated surfaces. Thus $\hat{q}$ belongs to a surface with a Gauß map which is different from the one of the undressed surfaces. It was, of course, classically known that this fractional linear formula transforms Gauß maps of minimal surfaces to Gauß maps of minimal surfaces [25],[56],[3].

§6. Symmetries

6.1 In the previous sections we have discussed the main features of the generalized Weierstraß representation of surfaces. For the construction of specific examples fundamental groups need to be incorporated as well. First we present the general framework [18],[19].

Let $\phi : \mathbb{D} \longrightarrow \mathbb{R}^3$ be an immersion of constant mean curvature, which induces a complete metric on $\mathbb{D}$. As always in this article $\mathbb{D}$

denotes an open and simply connected subset of $\mathbb{C}$. Assume now that R is an orientation-preserving rigid motion of $\mathbb{R}^3$ such that

$$R\phi(\mathbb{D}) = \phi(\mathbb{D}). \tag{6.1.1}$$

Then it is well known that there exists a holomorphic map $\gamma : \mathbb{D} \longrightarrow \mathbb{D}$ such that for all $z \in \mathbb{D}$

$$R\phi(z) = \phi(\gamma z) \tag{6.1.2}$$

Therefore we define *symmetries* for ϕ to be pairs (γ, R), where γ is biholomorphic on $\mathbb{D}$ and R is a rigid motion as above, such that 6.1.2 holds. Each symmetry induces a transformation formula for the frame F associated with ϕ,

$$F(\gamma z) = R_0 \cdot F(z) \cdot k(z), \tag{6.1.3}$$

where R_0 is the (orientation preserving, orthogonal) linear part of the affine transformation R and k is a unitary diagonal matrix. As a consequence one obtains

$$\gamma^* \alpha = k^{-1} \cdot \alpha \cdot k + k^{-1} \cdot dk. \tag{6.1.4}$$

From this it is easy to see that, after introducing the parameter $\lambda \in S^1$,

$$\gamma^* \alpha_\lambda = k^{-1} \cdot \alpha_\lambda \cdot k + k^{-1} \cdot dk. \tag{6.1.5}$$

This implies that for the extended frame, which by abuse of notation we will also denote by F, there exists some $\chi \in \Lambda SU(2)$ such that $\chi|_{\lambda=1} = R_0$ and

$$F(\gamma z, \lambda) = \chi(\lambda) \cdot F(z, \lambda) \cdot k(z), \tag{6.1.6}$$

where k is the unitary diagonal matrix of 6.1.3, hence independent of λ.

6.2 Equation 6.1.6 can be evaluated in two "directions". On the one hand, since $H \neq 0$, we can apply the Sym-Bobenko formula, and obtain

$$\phi(\gamma z, \lambda) = \chi(\lambda) \cdot \phi(z, \lambda) \cdot \chi^{-1}(\lambda) + i\lambda\partial_\lambda \chi(\lambda) \cdot \chi^{-1}(\lambda). \tag{6.2.1}$$

On the other hand, from 6.1.6, one obtains for all holomorphic extended frames associated with F the transformation formula

$$C(\gamma z, \lambda) = \chi(\lambda) \cdot C(z, \lambda) \cdot W_+(z, \lambda). \tag{6.2.2}$$

The holomorphic potential $\eta = C^{-1} \cdot dC$ then satisfies

$$\gamma^* \eta = {W_+}^{-1} \cdot \eta \cdot W_+ + {W_+}^{-1} \cdot dW_+. \tag{6.2.3}$$

6.3 Since we are interested in constructing constant mean curvature immersions from potentials, it is natural to start with potentials η satisfying 6.2.3. Then the corresponding holomorphic extended frame C satisfies

$$C(\gamma z, \lambda) = \rho(\lambda) \cdot C(z, \lambda) \cdot W_+(z, \lambda) \tag{6.3.1}$$

with some $\rho \in \Lambda Sl(2, \mathbb{C})_\sigma$. If $\rho \in \Lambda SU(2)_\sigma$, then the extended frame F obtained from C via Iwasawa splitting satisfies 6.1.6 for $\chi = \rho$ and the Sym-Bobenko formula produces a constant mean curvature immersion satisfying 6.2.1 with $\chi = \rho$. Unfortunately, in general, ρ will *not* be unitary for all $\lambda \in S^1$. As a consequence, the transition from C to F does *not* result in a formula like 6.1.1. And therefore we do *not* know anything about the transformation behaviour of the immersion constructed from C via F and the Sym-Bobenko formula.

There seem to be at least three avenues to overcome this complication:

1. One needs to find some criterion for η which ensures that ρ will be unitary for all $\lambda \in S^1$. While this seems to be the most desirable solution, no general criterion seems to be known. However, in some special cases such criteria have been found [42].

2. Assume now that η satisfies 6.2.3, and therefore

$$C(\gamma z, \lambda) = \rho(\lambda) \cdot C(z, \lambda) \cdot W_+(z, \lambda) \tag{6.3.2}$$

for some $\rho \in \Lambda Sl(2, \mathbb{C})_\sigma$. We carry out the Iwasawa splitting $\rho = \rho_u \cdot \rho_+$ for ρ and dress the holomorphic extended frame C with ρ_+: $\rho_+ \cdot C = \hat{C} \cdot \hat{V}_+$. Then we obtain

$$C(\gamma z, \lambda) = \rho_u(\lambda) \cdot \hat{C} \cdot S_+. \tag{6.3.3}$$

This leads to

$$F(\gamma z, \lambda) = \rho_u \cdot \hat{F} \cdot \hat{k}, \tag{6.3.4}$$

with $\hat{k}$ independent of λ. Applying the Sym-Bobenko formula to this equation one finally obtains

$$\phi(\gamma z, \lambda) = \rho_u \cdot \hat{\phi}(z, \lambda) \cdot {\rho_u}^{-1} + i\lambda \partial_\lambda \rho_u(\lambda) \cdot \rho_u(\lambda)^{-1}. \tag{6.3.5}$$

Since ϕ and $\hat{\phi}$ are different surfaces, this approach does not seem to be useful for the construction of surfaces admitting specific fundamental groups.

3. Assume again that η satisfies 6.2.3, whence C satisfies 6.3.2. This third approach assumes that ρ is of the form $\rho = \rho_+ \cdot \rho_u \cdot \rho_+^{-1}$ with $\rho_+ \in \Lambda_r^+ Sl(2,\mathbb{C})_\sigma$ and $\rho_u \in \Lambda SU_r(2)_\sigma$. Then dressing C with ${\rho_+}^{-1}$ yields ${\rho_+}^{-1} \cdot C = \hat{C} \cdot \hat{V_+}$ and 6.3.2 translates into

$$\hat{C}(\gamma z, \lambda) = \rho_u \cdot \hat{C} \cdot \hat{S_+}. \tag{6.3.6}$$

Note that this equation is, in contrast to 6.3.3, an equation involving only one surface. Similarly, for the new frame $\hat{F}$ one obtains the transformation formula

$$\hat{F}(\gamma z, \lambda) = \rho_u \cdot \hat{F}(z, \lambda) \cdot \hat{k}(z) \tag{6.3.7}$$

and for the associated immersion

$$\hat{\phi}(\gamma z, \lambda) = \rho_u(\lambda) \cdot \hat{\phi}(z, \lambda) \cdot \rho_u(\lambda)^{-1} + i\lambda \partial_\lambda \rho_u(\lambda) \cdot \rho_u(\lambda)^{-1}. \tag{6.3.8}$$

Clearly, for the new immersion $\hat{\phi}$ the pair (γ, ρ_u) is a symmetry.

6.4 From 6.3.8 one can show that the invariance of $\hat{\phi}$ under the operation of γ, $\hat{\phi}(\gamma z, \lambda_0) = \hat{\phi}(z, \lambda_0)$, is equivalent to the so called *closing conditions*

$$\rho_u(\lambda_0) = \pm I, \tag{6.4.1}$$

$$\partial_\lambda \mid_{\lambda = \lambda_0} \rho_u = 0. \tag{6.4.2}$$

As a consequence, if for fixed $\lambda_0 \in S^1$ the immersion $\hat{\phi}$ factors through a Riemann surface M, then the closing conditions need to be satisfied for all $\gamma \in \pi_1(M)$.

More precisely, if $\operatorname{Ker}\hat{\phi}$ denotes the biholomorphic maps γ of $\mathbb{D}$ which fix $\hat{\phi}$ for $\lambda = \lambda_0$, then $\gamma \in \operatorname{Ker}\hat{\phi}$ if and only if the associated ρ_u satisfies the closing conditions. In particular

$$\pi_1(M) \subset \operatorname{Ker}\hat{\phi} \quad \text{if} \quad \hat{\phi} \text{ factors through M at } \lambda = \lambda_0. \tag{6.4.3}$$

In this case, one can factor $\hat{\phi}$ through the Riemann surface $M' = \mathbb{D}/\operatorname{Ker}\hat{\phi}$ [19]. In most cases, factoring through a given Riemann surface is only possible for a few values of λ. In general, $\operatorname{Ker}\hat{\phi} = \{\mathrm{id}\}$ for all but finitely many λ's. It would be interesting to determine the "degenerate cases", i.e. those where $\rho_u = I$ for all $\lambda \in S^1$. In particular, it is an open question whether there exist complete immersions with this property.

6.5 If one starts with some biholomorphic map γ of $\mathbb{D}$ and a rigid motion R satisfying $\phi(\gamma z) = R\phi(z)$, then γ is uniquely determined modulo $\operatorname{Ker}\phi$. Conversely, given some $\gamma \in \operatorname{Aut}\mathbb{D}$ and a symmetry (γ, R), then R is uniquely determined by γ as a rigid motion of $\mathbb{R}^3 \cong su(2)$ [18]. Thus the map $\gamma \longrightarrow \chi(\gamma, \lambda)$ is a homomorphism up to sign:

$$\chi(\gamma\mu, \lambda) = \pm\chi(\gamma, \lambda) \cdot \chi(\mu, \lambda). \tag{6.5.1}$$

The map $\gamma \longrightarrow W_+(\gamma, z, \lambda)$ of 6.3.1 is a cocycle up to sign:

$$W_+(\gamma\mu, z, \lambda) = \pm W_+(\mu, z, \lambda) W_+(\gamma, \mu z, \lambda). \tag{6.5.2}$$

6.6 From 6.4.1 it is clear that for the construction of constant mean curvature surfaces one needs to find good potentials to start with. In addition one needs to find the cocyles W_+. However, these can be chosen to be trivial if M is non-compact.

Theorem 6.6.1. *([22]) If M is a non-compact Riemann surface and $\phi : M \longrightarrow \mathbb{R}^3$ a constant mean curvature immersion with universal cover $\pi : \mathbb{D} \longrightarrow M$, then the associated family of ϕ can be generated from a holomorphic potential $\eta = \sum_{j=-1}^{\infty} \eta_j \lambda^j$ on $\mathbb{D}$, which is invariant under $\pi_1(M) \subset \operatorname{Aut}\mathbb{D}$:*

(6.6.1) *η is a holomorphic $(1,0)$-form on $\mathbb{D}$ and $\gamma^*\eta = \eta \ \forall \gamma \in \pi_1(M)$.*

For compact M one can show

Theorem 6.6.2. *If M is a compact Riemann surface and $\phi : M \longrightarrow \mathbb{R}^3$ a constant mean curvature immersion with universal cover $\pi : \mathbb{D} \longrightarrow M$, then the associated family of ϕ can be generated from a meromorphic potential $\eta = \sum_{j=-1}^{\infty} \eta_j \lambda^j$ on $\mathbb{D}$, which is invariant under $\pi_1(M) \subset \operatorname{Aut}\mathbb{D}$:*

(6.6.2) *η is a meromorphic (1,0)-form on $\mathbb{D}$ and $\gamma^*\eta = \eta \ \forall \gamma \in \pi_1(M)$.*

We emphasize (see also section 2.4) that the notion of a "normalized potential" involves only one power of λ, namely λ^{-1}, and a meromorphic coefficient function, while the meromorphic potentials occurring in the theorem above involve more than one power of λ by the results of section 2.3 of [21]. The proofs of the two theorems above are almost identical.

The first one uses the fact that holomorphic cocycles on a non-compact Riemann surface with values in a Banach Lie group are holomorphic boundaries. The second theorem uses the fact that meromorphic cocycles on compact Riemann surfaces are meromorphic boundaries (see [7] for both cases).

Remark 6.6.3. As mentioned in the previous section, an important goal is to construct surfaces with a given fundamental group Γ. For the non-compact case one can start, due to Theorem 6.6.1, from a holomorphic potential η satisfying 6.6.1. This implies for the holomorphic extended frame C the equation $C(\gamma.z,\lambda) = \rho(\gamma,\lambda)\cdot C(z,\lambda)$ for $\gamma \in \Gamma$, $z \in \mathbb{D}$, $\lambda \in S^1$. If ρ is unitary for all $\gamma \in \Gamma$ and all $\lambda \in S^1$, then one obtains for the extended frame F associated to C, $C = F \cdot V_+$, the equation $F(\gamma.z,\lambda) = \rho(\gamma,\lambda)\cdot F(z,\lambda)\cdot k(z)$ and the Sym-Bobenko formula implies formula 6.3.8 for $\phi = \hat{\phi}$ derived from $F = \hat{F}$. So it only remains to satisfy the closing conditions 6.4.1 and 6.4.2, which is not too hard. The crucial problem is that it is quite difficult to choose η so that ρ is unitary for all $\gamma \in \Gamma$. An instance where this task has been carried out successfully is the work of Kilian [42]. A more general situation occurs when ρ is not unitary, but can be represented in the form $\rho = h_+ \cdot \rho_u \cdot h_+{+}^{-1}$ with $h_+ \in \Lambda_r^+ Sl(2,\mathbb{C})_\sigma$ and $\rho_u \in \Lambda_r SU(2)_\sigma$. In this case the third approach outlined in section 6.3 can be applied, producing an immersion $\hat{\phi}$ of constant mean curvature. If in addition the closing conditions 6.4.1 and 6.4.2 are satisfied for $\hat{\phi}$, then $\hat{\phi}$ factors through $\Gamma\backslash\mathbb{D} = M$. This way one can produce surfaces with fundamental group Γ from $\Gamma-$invariant potentials after some dressing. This has been carried out successfully (see Theorem 5.5.1) in [22] for the construction of cylinders of constant mean curvature, initiating the study [42]. In [28] the same approach is used for the construction of trinoids and more generally N-noids. Based on the dressing idea [22] outlined above and a special representation of the holomorphic extended frame [28], Schmitt has developed a numerical algorithm which produces pictures of cylinders, trinoids and N-noids [43],[49], available at *www.gang.umass.edu.* The fact that the pictures of trinoids and N-noids actually do represent surfaces of the indicated topological type is not proved in [43], or [49]. In some cases, however, this follows from [42] or [22]. We expect that the other cases will follow from [28].

It should be noted also that by completely different methods (almost) embedded trinoids and planar N-noids have been investigated (see e.g. [34],[35], and the references listed there). Some pictures have been produced following this approach.

6.7 There are various groups associated with a constant mean curvature immersion $\phi : M \longrightarrow \mathbb{R}^3$, its lift $\psi : \mathbb{D} \longrightarrow \mathbb{R}$ and the canonical projection $\pi : \mathbb{D} \longrightarrow M$. Let

$$\text{(6.7.1)} \qquad \operatorname{Aut}\phi(M) = \{\text{orientation preserving rigid motions } R \text{ of } \mathbb{R}^3 \text{ such that } R\phi(M) = \phi(M)\}.$$

Recall that in this article $\mathbb{D}$ always denotes a simply connected open subset of $\mathbb{C}$.

Theorem 6.7.1. *([19]) Let $\phi : M \longrightarrow \mathbb{R}^3$ be a complete immersion of constant mean curvature with simply connected cover $\pi : \mathbb{D} \longrightarrow M$. Then $\operatorname{Aut}\phi(M)$ contains a one-parameter group if and only if the map $\psi : \mathbb{D} \longrightarrow \mathbb{R}^3$ is in the associated family of a Delaunay surface.*

As mentioned earlier, for complete surfaces every $R \in \operatorname{Aut}\phi(M)$ induces an automorphism $\gamma = \gamma_R$ of $\mathbb{D}$ satisfying $\phi(\gamma z) = R\phi(z)$. Thus the theorem above has thc

Corollary 6.7.2. *A complete constant mean curvature surface different from the sphere, possesses a one-parameter group of symmetries if and only if $\mathbb{D} = \mathbb{C}$ and the surface $\psi : \mathbb{D} \longrightarrow \mathbb{R}^3$ is in the associated family of a Delaunay surface.*

A very similar result holds if one only considers self-isometries:

Theorem 6.7.3. *([51]) Let $\phi : M \longrightarrow \mathbb{R}^3$ be a complete immersion of constant mean curvature with simply connected cover $\pi : \mathbb{D} \longrightarrow M$. Then $\mathbb{D}$ admits a one-parameter group of self-isometries for the induced metric if and only if $\mathbb{D} = \mathbb{C}$ and, with the induced metric, $\mathbb{D}$ is isometrically isomorphic to the simply connected cover of a Delaunay surface or of a Smyth surface. More precisely,*

a) if $\mathbb{D}$ admits a one-parameter group of translations, then $\mathbb{D}$ is isometric to the simply connected cover of a Delaunay surface,

b) if $\mathbb{D}$ admits a one-parameter group of rotations around a fixed point, then $\mathbb{D}$ is isometric to the simply connected cover of a Smyth surface.

Note that the existence of a one-parameter group of self-isometries implies that the corresponding metric only depends on one variable or is constant. Very little is known about surfaces with large but discrete groups of symmetries or self-isometries. Constant mean curvature cylinders have been investigated by Kilian [42] and constant mean curvature surfaces with doubly periodic frames, and in particular tori, have been investigated in [46],[20],[4],[5]. Recently, work on trinoids and more generally N-noids of genus zero has been started [28],[43],[49] (see also Remark 6.6.3).

§7. Weierstraß representations for other classes of surfaces

7.1 For the discussion of constant mean curvature surfaces we have used the loop group $\Lambda Sl(2,\mathbb{C})_\sigma$ and some of its subgroups. For other classes one needs to choose different groups $\mathcal{H}$ and subgroups $\mathcal{H}_+, \mathcal{H}_-$ and $\mathcal{U}$. But certain features will remain the same. In particular, for the surfaces in a given class under consideration:

1. For every surface there is a "frame" F in $\mathcal{U}$
2. There is a "Birkhoff splitting" $F = F_- \cdot F_+$ with $F_\pm \in \mathcal{H}_\pm$.

Using these features one can — in the language of the previous sections — construct immersions and normalized potentials $F_-^{-1} \cdot dF_-$. It is, of course, crucial that one can also construct frames from potentials.

3. There is an "Iwasawa splitting" $F_- = F \cdot F_+$ such that

a) in this equation the Birkhoff splitting and the Iwasawa splitting are essentially inverse to each other: starting from F the Birkhoff splitting produces F_-, starting from F_- the Iwasawa splitting produces F.

b) the frames obtained via Iwasawa splitting from normalized potentials are frames of surfaces in the class considered.

In most cases there is a "Sym-Bobenko formula" which produces the immersion from the extended frame F. However, there are some cases (e.g. see section 7.4) where such a formula does not seem to be known. We would like to point out that in the examples listed in the next few sections, the relevant Iwasawa splitting is no longer globally defined. Also, the "loops" are no longer maps from S^1, but from other curves, like the real line. This requires generalized splitting theorems. The basic results in this direction can be found in [1] and [40]. The splittings actually used in the examples below can easily be derived from these results.

7.2 Pseudospherical Surfaces In this section we discuss the class of surfaces of constant Gauß curvature $K = -1$ following [55]. We will always use asymptotic line coordinates. Frequently we will even assume that these coordinates form a Chebyshev net. The starting point for this approach is the fact that a pseudospherical surface M, endowed with the second fundamental form, is a Lorentz manifold, and the Gauß map $N : M \longrightarrow S^2$ of a Lorentz manifold M is (Lorentz) harmonic if and only if M is pseudospherical. One needs to note that "harmonic" is defined here by the vanishing of the tension field, where the "metric coefficients" involved are for M the ones of the pseudo-Riemannian structure and for S^2 the ones of the usual Riemannian structure. Viewing S^2 as a quotient of $SO(3)$, $S^2 = SO(3)/SO(2)$, we can lift the Gauß map N to a map into $SO(3)$. Among these lifts there is a "normalized" one, for which the Maurer-Cartan form looks particularly simple. Expressing the coordinate frame in terms of the angle ϕ between the asymptotic lines, one obtains the normalized frame U by rotating the coordinate frame by the angle $\Theta = \phi/2$ and then taking its transpose. (The latter step is only to obtain the form of the moving frame equations as in the constant mean curvature case.)

With these conventions the moving frame equations then read (for the coefficient matrices in asymptotic line coordinates):

$$\partial_x U = U \cdot \begin{pmatrix} 0 & -\phi_x & 0 \\ \phi_x & 0 & 1 \\ 0 & -1 & 0 \end{pmatrix}, \tag{7.2.1}$$

$$\partial_y U = U \cdot \begin{pmatrix} 0 & 0 & -\sin\phi \\ 0 & 0 & -\cos\phi \\ \sin\phi & \cos\phi & 0 \end{pmatrix}. \tag{7.2.2}$$

Introducing the *real, positive* parameter λ one obtains the extended frame equations

$$\partial_x U = U \cdot \begin{pmatrix} 0 & -\phi_x & 0 \\ \phi_x & 0 & \lambda \\ 0 & -\lambda & 0 \end{pmatrix}, \tag{7.2.3}$$

$$\partial_y U = U \cdot \lambda^{-1} \begin{pmatrix} 0 & 0 & -\sin\phi \\ 0 & 0 & -\cos\phi \\ \sin\phi & \cos\phi & 0 \end{pmatrix}. \tag{7.2.4}$$

Proposition 7.2.1. *The extended frame equations can be solved simultaneously for all λ if and only if the corresponding surface is pseudospherical.*

The family of surfaces, parametrized by λ, is called the associated family. Since λ is (for geometric applications) a real parameter, we need to be careful when defining the loop groups for pseudospherical surfaces. However, the crucial observation is that all geometrically relevant quantities actually are analytic in $\lambda \in \mathbb{C}^*$. Therefore we can carry out all loop group operations in a standard loop group, remembering that for geometric evaluations λ needs to be chosen real and positive. With this in mind we consider the group $\Lambda SO(3)_P$ whose elements $g(\lambda)$ are defined by the following four conditions:

$$g \in \Lambda Sl(3,\mathbb{C}) \tag{7.2.5}$$

$$g(\lambda)^T = g(\lambda)^{-1} \tag{7.2.6}$$

$$\overline{g(\overline{\lambda})} = g(\lambda) \tag{7.2.7}$$

$$g(-\lambda) = P \cdot g(\lambda) \cdot P^{-1} \tag{7.2.8}$$

where $P = \mathrm{diag}(1, 1, -1)$ and the subscript P will denote in this section "twisting by P", which differs from the use of the subscript P in section 2.2. Thus the elements of $\Lambda SO(3)_P$ are loops in $\Lambda Sl(3,\mathbb{C})$, which have real coefficients at all powers of λ, are in $SO(3)$ for every real λ, and are twisted by P. We set $\Lambda^+ SO(3)_P = \Lambda^+ Sl(3,\mathbb{C}) \bigcap \Lambda SO(3)_P$, and define $\Lambda^- SO(3)_P$ analogously.

Now we are ready to define the "loop group" used for pseudospherical surfaces:

$$\mathcal{H} = \Lambda SO(3)_P \times \Lambda SO(3)_P, \tag{7.2.9}$$

and the relevant subgroups

$$\mathcal{U} = \{(g,g) \mid g \in \Lambda SO(3)_P\} \tag{7.2.10}$$

$$\mathcal{H}^+ = \Lambda^+ SO(3)_P \times \Lambda^- SO(3)_P \tag{7.2.11}$$

$$\mathcal{H}^- = \Lambda^- SO(3)_P \times \Lambda^+ SO(3)_P. \tag{7.2.12}$$

As with previous conventions we write for example $(g,h) \in \mathcal{H}^+_*$, if g and h have I as the coefficient of λ^0. We will write $(g,h) \in \mathcal{H}^+_\sharp$, if we require only I for the coefficient of g of λ^0. Then the "Birkhoff splitting" means to write an element of $\mathcal{H}$ as a product of elements in $\mathcal{H}^+$ and $\mathcal{H}^-_*$. It is not difficult to see that $\mathcal{H}^-_* \cdot \mathcal{H}^+$ is open and dense in $\mathcal{H}$, and the group splitting is an analytic operation on this set [55]. Similarly, $\mathcal{U} \cdot \mathcal{H}^+_\sharp$ is open and dense and the group splitting is an analytic operation on this set. (To verify this one considers $(g,h) \in \mathcal{H}$ and forms the element $h^{-1}g \in \Lambda SO(3)_P$. By the previous result, in any neighbourhood of g

there is some $g' \in \Lambda SO(3)_P$ such that $h^{-1}g' = v_- \cdot v_+^{-1}$ with $v_+ \in \Lambda_*^+ SO(3)_P$ and $v_- \in \Lambda^- SO(3)_P$. As a consequence, setting $u = hv_- = g'v_+$ we obtain $(g', h) = (u, u)(v_+, v_-)$ and the first part of the claim follows. The second part is a consequence of the analyticity of the map $\Lambda^- SO(3)_P \cdot \Lambda_*^+ SO(3)_P \longrightarrow \Lambda^- SO(3)_P \times \Lambda_*^+ SO(3)_P$.) The "Birkhoff splitting" here is actually a pair of two standard Birkhoff splittings, while the "Iwasawa splitting" here for a pair (g, h) means writing $g = u \cdot v_+$ and $h = u \cdot v_-$, which is equivalent to $h^{-1} \cdot g = {v_-}^{-1} \cdot v_+$.

Following the outline at the beginning of this section we want to produce "normalized potentials" by splitting the "frame". For a pseudospherical surface we consider its normalized frame U and form the element $F = (U, U) \in \mathcal{U}$. This is the "frame", which will work for our purposes. We perform the Birkhoff splitting in $\mathcal{H}$:

$$F = F_- \cdot F_+, \quad \textit{where } F_\pm \in \mathcal{H}^\pm. \tag{7.2.13}$$

In this case it turns out that

$$F_- = (U_-, U_+) \tag{7.2.14}$$

is such that U_- only depends on the variable y, while U_+ only depends on x. The normalized potential is

$$\xi = (\xi_-, \xi_+) = F_-^{-1} dF_-, \tag{7.2.15}$$

where

$$\xi_- = U_-^{-1} dU_- = \lambda^{-1} \xi_{-1}(y) dy, \tag{7.2.16}$$

$$\xi_+ = U_+^{-1} dU_+ = \lambda \xi_1(x) dx, \tag{7.2.17}$$

and

$$\xi_{-1}(y) = \begin{pmatrix} 0 & 0 & \sin\phi(0,y) \\ 0 & 0 & \cos\phi(0,y) \\ -\sin\phi(0,y) & -\cos\phi(0,y) & 0 \end{pmatrix} dy, \tag{7.2.18}$$

$$\xi_1(x) = \begin{pmatrix} 0 & 0 & -\sin\hat{\phi}(x) \\ 0 & 0 & -\cos\hat{\phi}(x) \\ \sin\hat{\phi}(x) & \cos\hat{\phi}(x) & 0 \end{pmatrix} dx \tag{7.2.19}$$

where $\hat{\phi}(x) = \phi(0,0) - \phi(x, 0)$.

Remark 7.2.2. We have constructed a "normalized potential" for each pseudospherical surface. Since the group splittings used in this case are not globally defined, the normalized potentials may acquire singularities. At this point it has not yet been investigated what type of singularities can occur. At any rate, locally around a given base point, the potential can be assumed to be smooth. It would be interesting to investigate whether there are analogues of the "holomorphic potentials" of the constant mean curvature case. These would need to be defined on simply connected Lorentz surfaces and would perhaps be globally smooth.

For the converse construction, producing surfaces from potentials we start from two matrices of the form 7.2.18 and 7.2.19, where we replace the functions $\phi(x,0)$ and $\phi(0,y)$ by some arbitrary smooth functions $a(x)$ and $b(y)$. Next we solve the two ODE's 7.2.15 for U_- and U_+ with trivial initial condition. Next we perform an Iwasawa splitting of (U_-, U_+). As pointed out above, this is equivalent to performing the classical Birkhoff splitting $U_-^{-1} \cdot U_+ = W_+ \cdot W_-^{-1}$. Then one sets $U = U_- \cdot W_+ = U_+ \cdot W_-$ and shows that U is a frame associated with a pseudospherical surface, which is obtained by the *Sym formula*

$$\psi(x,y,\lambda) = \frac{\partial U}{\partial t} \cdot U^{-1}, \quad \text{where } \lambda = e^t. \tag{7.2.20}$$

For details, examples, and proofs we refer to [55].

7.3 Timelike Surfaces in Minkowski Space $\mathbb{E}^3_1$ of Constant Mean Curvature This case has similarities with both the theory of surfaces of constant mean curvature and the theory of surfaces of constant negative Gauß curvature. For details see [23]. Consider the Minkowski space $\mathbb{E}^3_1$ defined by the metric $\langle .,. \rangle = -du_1^2 + du_2^2 + du_3^2$. The starting point for this approach is the fact that the Gauß map $N : M \longrightarrow S^2_1 = Sl(2,\mathbb{R})/K$, where $K = \{\mathrm{diag}(a, a^{-1}) \mid a \neq 0\}$, of a timelike surface in $\mathbb{E}^3_1$, is harmonic if and only if the (Lorentzian) mean curvature is constant. As in the previous cases, "harmonic" is defined by the vanishing of the tension field. In the case under consideration both the timelike surface M and the Lorentz sphere S^2_1 carry a Lorentz metric. Lifting the Lorentzian frame, which defines an element of the Lorentz group, to the double cover $Sl(2,\mathbb{R})$, one obtains the moving frame equations (we will always use null coordinates):

$$\partial_y U = U \cdot \begin{pmatrix} \frac{1}{4}\omega_y & -\frac{1}{2}H\exp(\frac{1}{2}\omega) \\ R\exp(-\frac{1}{2}\omega) & -\frac{1}{4}\omega_y \end{pmatrix}, \tag{7.3.1}$$

(7.3.2)
$$\partial_x U = U \cdot \begin{pmatrix} -\frac{1}{4}\omega_x & -Q\exp(-\frac{1}{2}\omega) \\ \frac{1}{2}H\exp(\frac{1}{2}\omega) & \frac{1}{4}\omega_x \end{pmatrix},$$

where $Q = \langle \phi_{xx}, N\rangle$, $R = \langle \phi_{yy}, N\rangle$, $H = 2\exp(-\omega)\langle \phi_{xy}, N\rangle$ and the metric is $ds^2 = e^{\omega}dxdy$. Note that we have two Hopf differentials this time.

Introducing the *real, positive* parameter λ one obtains the extended frame equations

(7.3.3)
$$\partial_y U = U \cdot \begin{pmatrix} \frac{1}{4}\omega_y & -\lambda^{-1}\frac{1}{2}H\exp(\frac{1}{2}\omega) \\ \lambda^{-1}R\exp(-\frac{1}{2}\omega) & -\frac{1}{4}\omega_y \end{pmatrix},$$

(7.3.4)
$$\partial_x U = U \cdot \begin{pmatrix} -\frac{1}{4}\omega_x & -\lambda Q\exp(-\frac{1}{2}\omega) \\ \lambda\frac{1}{2}H\exp(\frac{1}{2}\omega) & \frac{1}{4}\omega_x \end{pmatrix}.$$

For our loop group approach the following result is basic:

Proposition 7.3.1. *The extended frame equations can be solved simultaneously for all λ if and only if the corresponding timelike surface in $\mathbb{E}^3_1$ has constant mean curvature.*

The family of surfaces parametrized by λ, is called the *associated family.*

As in the case of pseudospherical surfaces the parameter λ is real, but all geometrically relevant quantities are actually analytic for $\lambda \in \mathbb{C}^*$. Therefore we can carry out all loop group operations in a standard loop group, remembering that for geometric evaluations λ needs to be chosen real and positive.

With this in mind we consider the group $\Lambda Sl(2,\mathbb{R})_\sigma$, whose elements $g(\lambda)$ are defined by the following three conditions:

(7.3.5)
$$g \in \Lambda Sl(2,\mathbb{C})$$

(7.3.6)
$$\overline{g(\overline{\lambda})} = g(\lambda)$$

(7.3.7)
$$g(-\lambda) = \sigma_3 \cdot g(\lambda) \cdot \sigma_3^{-1},$$

where $\sigma_3 = \mathrm{diag}(1,-1)$. Thus the elements of $\Lambda Sl(2,\mathbb{R})_\sigma$ are loops in $Sl(2,\mathbb{C})$, which have real coefficients at all powers of λ, are in $Sl(2,\mathbb{R})$ for real λ, and are twisted by σ_3. We set $\Lambda^+ Sl(2,\mathbb{R})_\sigma = \Lambda^+ Sl(2,\mathbb{C})_\sigma \bigcap \Lambda Sl(2,\mathbb{R})$ and define $\Lambda^- Sl(2,\mathbb{R})_\sigma$ similarly.

Now we define the "loop group" used for timelike surfaces of constant mean curvature in $\mathbb{E}^3_1$:

(7.3.8)
$$\mathcal{H} = \Lambda Sl(2,\mathbb{R})_\sigma \times \Lambda Sl(2,\mathbb{R})_\sigma$$

and the relevant subgroups

$$\mathcal{U} = \{(g,g) \mid g \in \Lambda Sl(2,\mathbb{R})_\sigma\} \tag{7.3.9}$$

$$\mathcal{H}^- = \Lambda^+ Sl(2,\mathbb{R})_\sigma \times \Lambda^- Sl(2,\mathbb{R})_\sigma \tag{7.3.10}$$

$$\mathcal{H}^+ = \Lambda^- Sl(2,\mathbb{R})_\sigma \times \Lambda^+ Sl(2,\mathbb{R})_\sigma. \tag{7.3.11}$$

As usual we add a subscript * to indicate that the coefficient of λ^0 is the identity matrix. The comments concerning the Birkhoff and Iwasawa splittings in the previous section also apply here.

To produce a normalized potential from a surface we consider its extended frame $U \in \Lambda Sl(2,\mathbb{C})_\sigma$ and form the element $F = (U, U)$. Again, this is the "frame" which will work for our purposes. We perform the Birkhoff splitting in $\mathcal{H}$:

$$F = F_- \cdot F_+, \quad \text{where} \quad F_\pm \in \mathcal{H}^\pm. \tag{7.3.12}$$

In this case it turns out that

$$F_- = (U_+, U_-) \tag{7.3.13}$$

is such that U_- only depends on the variable y, while U_+ only depends on the variable x. Thus the normalized potential is

$$\xi = (\xi', \xi'') = F_-^{-1} dF_-, \tag{7.3.14}$$

where

$$\xi' = U_+^{-1} dU_+ = \lambda \begin{pmatrix} 0 & \frac{Q(x)}{f(x)} \\ -\frac{1}{2} H f(x) & 0 \end{pmatrix} dx, \tag{7.3.15}$$

$$\xi'' = U_-^{-1} dU_- = \lambda^{-1} \begin{pmatrix} 0 & \frac{1}{2} H g(y) \\ \frac{R(x)}{g(y)} & 0 \end{pmatrix} dy, \tag{7.3.16}$$

where $f(x) = \exp(\omega(x,0) - \frac{1}{2}\omega(0,0))$ and $g(y) = \exp(\omega(0,y) - \frac{1}{2}\omega(0,0))$. These are the "normalized potentials" for the case under consideration. Remark 7.2.2 applies here as well.

For the converse construction, producing surfaces from potentials, we start from two matrices of the form 7.3.15 and 7.3.16, where we replace the functions $\omega(x,0)$ and $\omega(0,y)$ by some arbitrary smooth functions $a(x)$ and $b(y)$. Next we solve the two ODE's $\partial_x U_+ = U_+ \cdot \xi'$ and $\partial_y U_- = U_- \cdot \xi''$ for U_- and U_+ with trivial initial condition. Then we perform an Iwasawa splitting of (U_-, U_+). As pointed out before, this is equivalent to performing the classical Birkhoff splitting $U_-^{-1} \cdot U_+ = W_+ \cdot W_-^{-1}$. Finally we set $U = U_- \cdot W_+ = U_+ \cdot W_-$ and show that U is a frame associated with a timelike surface in $\mathbb{E}^3_1$ of constant mean curvature, which is obtained by the *Generalized Sym formula*

$$\phi(x,y,\lambda) = -\frac{1}{H}\left\{\frac{\partial U}{\partial t} \cdot U^{-1} - \frac{1}{2} U \cdot \sigma_3 \cdot U^{-1}\right\}, \quad \lambda = e^t. \tag{7.3.17}$$

For details, examples, and proofs we refer to [23].

7.4 Affine Spheres By definition, affine spheres are Blaschke surfaces and therefore the Gauß curvature K never vanishes. According to whether $K > 0$ or $K < 0$ different coordinates are natural and different loop groups are needed. The case $K > 0$ is very similar to the case of constant mean curvature surfaces in $\mathbb{R}^3$. Therefore we will restrict here to the case $K < 0$ and use asymptotic line coordinates u and v. Since we consider proper affine spheres, the affine mean curvature H is constant and different from 0. We will assume without loss of generality that $H = -1$. Then the moving frame equations for the frame $\hat{U} = (f_u, f_v, f)$ of an affine sphere are

$$\partial_u \hat{U} = \hat{U} \begin{pmatrix} \omega_u & 0 & 1 \\ Ae^{-\omega} & 0 & 0 \\ 0 & e^{\omega} & 0 \end{pmatrix}, \tag{7.4.1}$$

$$\partial_v \hat{U} = \hat{U} \begin{pmatrix} 0 & Be^{-\omega} & 0 \\ 0 & \omega_v & 0 \\ e^{\omega} & 0 & 1 \end{pmatrix}. \tag{7.4.2}$$

Gauging the moving frame $\hat{U}$ by $D = \text{diag}(\lambda^{-1}e^{-\frac{1}{2}\omega}, \lambda e^{-\frac{1}{2}\omega}, 1)$, with λ *real and positive*, we obtain for the *modified* (extended) frame $U = \hat{U}D$ the equations

$$U^{-1} \cdot U_u = \frac{\omega_u}{2}\begin{pmatrix} 1 & 0 & 0 \\ 0 & -1 & 0 \\ 0 & 0 & 0 \end{pmatrix} + \lambda \begin{pmatrix} 0 & 0 & e^{\omega/2} \\ Ae^{-\omega} & 0 & 0 \\ 0 & e^{\omega/2} & 0 \end{pmatrix}, \tag{7.4.3}$$

$$U^{-1} \cdot U_v = \frac{\omega_v}{2}\begin{pmatrix} -1 & 0 & 0 \\ 0 & 1 & 0 \\ 0 & 0 & 0 \end{pmatrix} + \lambda^{-1} \begin{pmatrix} 0 & Be^{-\omega} & 0 \\ 0 & 0 & e^{\omega/2} \\ e^{\omega/2} & 0 & 0 \end{pmatrix}. \tag{7.4.4}$$

For the loop group approach the following result is basic:

Proposition 7.4.1. *The modified frame equations 7.4.3 and 7.4.4 can be solved simultaneously for all* λ *if and only if the corresponding affine surface is a proper affine sphere.*

The family parametrized by λ is called the *associated family.*

As in the previous two sections the parameter λ is real, but all geometrically relevant quantities are actually analytic for $\lambda \in \mathbb{C}^*$. Therefore, in this case also, we can carry out all loop group operations in a standard loop group, remembering that for geometric evaluations λ needs to be chosen real and positive. With this in mind we consider the group $G[\lambda]$, whose elements are defined by the following four conditions

$$g \in \Lambda Sl(3, \mathbb{C}) \tag{7.4.5}$$

$$\overline{g(\overline{\lambda})} = g(\lambda) \tag{7.4.6}$$

$$Q \cdot g(\epsilon\lambda) \cdot Q^{-1} = g(\lambda) \tag{7.4.7}$$

$$T \cdot [g(-\lambda)^{-1}]^t \cdot T = g(\lambda) \tag{7.4.8}$$

(7.4.9)

where $\epsilon = \exp(2\pi i/3)$, $Q = \text{diag}(\epsilon, \epsilon^2, 1)$, and

$$T = \begin{pmatrix} 0 & 1 & 0 \\ 1 & 0 & 0 \\ 0 & 0 & 1 \end{pmatrix}.$$

Thus the elements of $G[\lambda]$ are loops in $Sl(3, \mathbb{C})$, which have real coefficients at all powers of λ and satisfy two twisting conditions. It turns out that the Lie algebra $\mathfrak{g}[\lambda]$ of $G[\lambda]$ is the loop part of a Kac-Moody algebra of type $A_2^{(2)}$. Keeping with the notation of this article we set $G^{\pm}[\lambda] = G[\lambda] \bigcap \Lambda^{\pm} Sl(2, \mathbb{C})$ and we will add a subscript $*$ if we want to indicate that the coefficient at λ^0 is I. (Note that [14] uses the opposite convention.)

Next we introduce the "loop group" used for affine spheres

$$\mathcal{H} = G[\lambda] \times G[\lambda], \tag{7.4.10}$$

and the relevant subgroups

$$\mathcal{U} = \{(g, g) \mid g \in G[\lambda]\} \tag{7.4.11}$$

$$\mathcal{H}^- = G^-[\lambda] \times G^+[\lambda] \tag{7.4.12}$$

$$\mathcal{H}^+ = G^+[\lambda] \times G^-[\lambda]. \tag{7.4.13}$$

The comments made in section 7.2 about the Birkhoff and Iwasawa splittings also apply here.

To produce a normalized potential from a surface we consider its modified frame $U \in G[\lambda]$ and form the element $F = (U, U) \in \mathcal{U}$. Again, this is the "frame", which will work for our purposes. We perform the Birkhoff splitting in $\mathcal{H}$:

$$F = F_- \cdot F_+, \quad \text{where} \quad F_\pm \in \mathcal{H}^\pm. \tag{7.4.14}$$

It turns out that

$$F_- = (V_-, V_+) \tag{7.4.15}$$

is such that V_- only depends on the variable u, while V_+ only depends on the variable v. Thus the normalized potential is

$$\xi = (\xi_-, \xi_+) = F_-^{-1} dF_-, \tag{7.4.16}$$

where
(7.4.17)

$$\xi_- = V_-^{-1} dV_- = \lambda^{-1} D_-^{-1} \begin{pmatrix} 0 & Be^{-\omega} & 0 \\ 0 & 0 & e^{\omega/2} \\ e^{\omega/2} & 0 & 0 \end{pmatrix} D_- dv = \lambda^{-1} T_-,$$

(7.4.18)

$$\xi_+ = V_+^{-1} dV_+ = \lambda D_+^{-1} \begin{pmatrix} 0 & 0 & e^{\omega/2} \\ Ae^{-\omega} & 0 & 0 \\ 0 & e^{\omega/2} & 0 \end{pmatrix} D_+ du = \lambda T_+,$$

and where D_- and D_+ are independent of λ and of the form $\mathrm{diag}(a, a^{-1}, 1)$.

This way we have constructed a "normalized potential" for each proper affine sphere. The comments made in section 7.2 concerning singularities and "holomorphic potentials" apply here as well.

For the converse construction we start from an arbitrary normalized potential $\xi = (\lambda^{-1} T_-, \lambda T_+)$, where

$$T_-(v) = \begin{pmatrix} 0 & \beta_-(v) & 0 \\ 0 & 0 & \alpha_-(v) \\ \alpha_-(v) & 0 & 0 \end{pmatrix} dv \tag{7.4.19}$$

$$T_+(u) = \begin{pmatrix} 0 & 0 & \alpha_+(u) \\ \beta_+(u) & 0 & 0 \\ 0 & \alpha_+(u) & 0 \end{pmatrix} du \tag{7.4.20}$$

and α_+ and α_- never vanish. (One can show that after changing u to $-u$ and/or v to $-v$, if necessary, one can assume $\alpha_+(0) = \alpha_-(0) > 0$. Such potentials are called "normalized" in [14].)

Next we solve the ODE's 7.4.17 and 7.4.18 with initial condition I, producing some V_+ and V_-. Then we perform an Iwasawa splitting

$$(V_-, V_+) = (\tilde{U}, \tilde{U}) \cdot (L_+, L_-) \tag{7.4.21}$$

In the previous examples a Sym-type formula was applied at this point, producing the required immersion. Such a formula does not seem to be available for affine spheres. However, one can show that (after changing u into $-u$ and/or v into $-v$, if necessary) there is a $\lambda-$independent gauge $C_0 \in G[\lambda]$ such that $U = C_0(0,0)^{-1} \cdot \tilde{U} \cdot C$ is the frame of a proper affine sphere. For details, examples, and proofs we refer to [14].

7.5 Willmore Surfaces in $\mathbb{R}^3$. Part I. In the next two sections we will discuss Willmore surfaces. For details we refer to [37]. There are actually two approaches to describe Willmore surfaces via loop groups. The first one is parallel to the previous sections.

An immersion

$$\phi : M \longrightarrow \mathbb{R}^3 \tag{7.5.1}$$

is called a *Willmore surface* if it is a critical point of the Willmore functional

$$\mathcal{W}(M) = \int_M H^2 dA. \tag{7.5.2}$$

Equivalently, an immersion is a Willmore surface if the mean curvature H satisfies the non-linear elliptic equation

$$\Delta^M H + 2H(H^2 - K) = 0 \tag{7.5.3}$$

In this part we assume that M has no umbilical points and that ϕ is conformal. Under these assumptions Bryant [6] has defined a *conformal Gauß map*

$$\check{N}_\phi : M \longrightarrow S^{3,1} \subset \mathbb{E}^{4,1}, \tag{7.5.4}$$

where $\mathbb{E}^{4,1}$ is the five-dimensional Minkowski space with metric

$$\langle x, y \rangle = -x_0 y_4 - x_4 y_0 + \sum_{j=1}^{3} x_j y_j \tag{7.5.5}$$

and

$$S^{3,1} = \{y \in \mathbb{E}^{4,1} \mid \langle y, y \rangle = 1\}. \tag{7.5.6}$$

We note that $S^{3,1}$ carries naturally a pseudo-Riemannian metric induced from $\mathbb{E}^{4,1}$.

As in the previous sections one has

Theorem 7.5.1. *([6]) The immersion $\phi : M \longrightarrow \mathbb{R}^3$ is a Willmore immersion if and only if the conformal Gauß map $\check{N}_\phi : M \longrightarrow S^{3,1}$ is harmonic. Moreover, ϕ can be retrieved from $\check{N}_\phi$ by a projection and every harmonic map $\check{N} : M \longrightarrow S^{3,1}$ induces naturally a Willmore immersion.*

In this theorem harmonicity is defined as usual by the vanishing of the tension field. The metrics entering into this stress tensor are the Riemannian metric on M induced by the immersion ϕ and the pseudo-Riemannian metric on $S^{3,1}$ induced from $\mathbb{E}^{4,1}$. As a consequence, the discussion of Willmore immersions is equivalent to the discussion of harmonic maps from M to $S^{3,1}$. At this point the theory becomes parallel to [26]. We observe that

$$S^{3,1} = SO_o(4,1)/SO_o(3,1), \tag{7.5.7}$$

where

$$SO(4,1) = \{A \in \mathrm{Mat}(5,\mathbb{R}) \mid \langle Au, Av \rangle = \langle u, v \rangle \ \forall\ u, v \in \mathbb{E}^{4,1}\}. \tag{7.5.8}$$

It is easy to see that the isotropy subgroup of $e_3 = (0,0,1,0)^t$ is naturally isomorphic with $SO(3,1)$. The subscript o denotes the identity component. Moreover, $SO_o(3,1)$ is the fixed point set of the involutive automorphism $g \mapsto \tau g \tau, \tau = \mathrm{diag}(1,1,1,-1,1)$ in $SO_o(3,1)$. In particular, $S^{3,1}$ is the *semisimple symmetric space* defined by τ. Thus we can follow the general procedure of [26], which for such symmetric spaces can be found in [10].

Let $M = \mathbb{D}$ be a simply connected subset of $\mathbb{C}$. We consider the (generalized) Cartan decomposition

$$so(4,1) = so(3,1) \oplus \mathfrak{m} = \mathfrak{h} \oplus \mathfrak{m}, \tag{7.5.9}$$

where $so(4,1)$ and $so(3,1) = \mathfrak{h}$ are the Lie algebras of the corresponding Lie groups defined above, and $\mathfrak{m}$ is the (-1)-eigenspace of (the differential of) τ. We consider the *extended lift*

$$F : \mathbb{D} \longrightarrow SO_0(4,1) \tag{7.5.10}$$

and its Maurer-Cartan form

$$\alpha = F^{-1}dF. \tag{7.5.11}$$

We decompose α in conformal coordinates z and $\bar{z}$ as

$$\alpha = \alpha'_{\mathfrak{m}} + \alpha_{\mathfrak{h}} + \alpha''_{\mathfrak{m}}, \tag{7.5.12}$$

where $\alpha = \alpha_{\mathfrak{m}} + \alpha_{\mathfrak{h}}$ is the decomposition of α relative to 7.5.12 and $\alpha'_{\mathfrak{m}}$ is the (1,0)-part of α, while $\alpha''_{\mathfrak{m}}$ is the (0,1)-part of α.

Introducing the *loop parameter* $\lambda \in S^1$ we set

$$\alpha_\lambda = \lambda^{-1}\alpha'_{\mathfrak{m}} + \alpha_{\mathfrak{h}} + \lambda\alpha''_{\mathfrak{m}}. \tag{7.5.13}$$

We then have

Proposition 7.5.2. *([10]) The extended Maurer-Cartan form α_λ is integrable for all $\lambda \in S^1$, i.e.*

$$d\alpha_\lambda + \frac{1}{2}[\alpha_\lambda \wedge \alpha_\lambda] = 0 \quad \text{for all} \quad \lambda \in S^1 \tag{7.5.14}$$

if and only if $\check{N}_\phi$ is harmonic, i.e. if and only if ϕ is a Willmore immersion.

Next we introduce the loop groups

$$\Lambda SO_0(4,1)^{\mathbb{C}} = \{S^1 \longrightarrow SO_0(4,1)^{\mathbb{C}}\} \tag{7.5.15}$$

$$\Lambda^+ SO_0(4,1)^{\mathbb{C}} = \Lambda SO_0(4,1)^{\mathbb{C}} \bigcap \Lambda^+ Sl(5,\mathbb{C}) \tag{7.5.16}$$

$$\Lambda^- SO_0(4,1)^{\mathbb{C}} = \Lambda SO_0(4,1)^{\mathbb{C}} \bigcap \Lambda^- Sl(5,\mathbb{C}) \tag{7.5.17}$$

$$\Lambda SO_0(4,1) = \{S^1 \longrightarrow SO_0(4,1)\} \tag{7.5.18}$$

and the corresponding twisted loop groups twisted by τ. From 7.5.14 we know that

$$F^{-1}dF = \alpha_\lambda, \quad F(0,0,\lambda) \tag{7.5.19}$$

is solvable and

$$F \in \Lambda SO_0(4,1). \tag{7.5.20}$$

Applying the classical Birkhoff splitting we obtain

$$F = F_- \cdot F_+, \tag{7.5.21}$$

$$\xi = F_-^{-1} \cdot dF_-. \tag{7.5.22}$$

As in the case of surfaces of constant mean curvature in $\mathbb{R}^3$ one verifies that F_- only depends on z and that ξ is of the form

$$\xi = \lambda^{-1}\xi_{-1}dz, \quad \xi_{-1} \in \mathfrak{m}. \tag{7.5.23}$$

This way we have obtained normalized potentials for Willmore surfaces without umbilics.

Conversely, starting from some ξ of the form 7.5.23 we solve the ODE 7.5.22 for some F_- with initial condition $F_-(0,\lambda) = I$. Then we apply a generalized Iwasawa splitting

$$F_- = F \cdot F_+^{-1}, \tag{7.5.24}$$

where $F \in \Lambda SO_0(4,1)^{\mathbb{C}}_\tau$ and $F_+ \in \Lambda^+ SO_0(4,1)^{\mathbb{C}}$.

We note that this is possible near $z = 0$, since there $F_- = I$ and the splitting of F_- near $z = 0$ is a consequence of the Lie algebra decomposition

$$\Lambda so(4,1)^{\mathbb{C}}_\tau = \Lambda so(4,1)_\tau + \Lambda^+ so(4,1)_\tau. \tag{7.5.25}$$

For a more general statement we refer to [40],[1].

It is easy to verify that $F^{-1}dF$ is of the form

$$F^{-1}dF = \lambda^{-1}\gamma'_{\mathfrak{m}} + \gamma_{\mathfrak{h}} + \lambda\gamma''_{\mathfrak{m}}. \tag{7.5.26}$$

But this implies (see e.g. [10]) that F is the frame of a harmonic map $\check{N} : \mathbb{D} \longrightarrow S^{3,1}$. Thus we obtain a Willmore immersion from $\mathbb{D}$ to $\mathbb{R}^3$ by the theorem above.

7.6 Willmore Surfaces in $\mathbb{R}^3$. Part II. We have seen in the last section that Willmore surfaces in $\mathbb{R}^3$ without umbilics have a Weierstrass type representation in the spirit of [26]. We have pointed out that in the normalized potentials, and even in the frames and the immersions, singularities may occur due to the fact that the group splittings are not global. In the presentation of the last section, in addition umbilics show up as singularities of the harmonic maps involved. Interestingly, it turns out that in the case of Willmore surfaces there is a (somewhat) different loop group procedure, which manages to handle the umbilical points as non-singular points. Of course, the singularities which are caused by the non-global group splittings remain.

Let $\phi : \mathbb{D} \longrightarrow \mathbb{R}^3$ be a Willmore immersion and $F : \mathbb{D} \longrightarrow \Lambda SO_0(4,1)^{\mathbb{C}}$ an extended lift. Then decomposing the Maurer-Cartan form $\alpha =$

$F^{-1}dF$ according to the involution induced by τ yields the description of section 7.5. Part I which is associated with the symmetric space $S^{3,1} = SO_o(4,1)/SO_o(3,1)$. However, there is another natural involution on $\Lambda SO_0(4,1)^{\mathbb{C}}$, induced by $\sigma = \text{diag}(-1,1,1,1-1)$. This involution has the fixed point group $SO(3) \times SO(1,1)$ and yields the symmetric space

$$Gr^5_{3,0} = SO_0(4,1)/(SO(3) \times SO(1,1)), \tag{7.6.1}$$

the Grassmannian of three-dimensional spacelike subspaces of $\mathbb{E}^{4,1}$.

Decomposing α relative to the involution σ yields

$$\alpha = \delta_{\mathfrak{k}} + \delta_{\mathfrak{p}}, \tag{7.6.2}$$

where $\mathfrak{k}$ and $\mathfrak{p}$ are the eigenspaces of the involution σ for the eigenvalues 1 and -1 respectively. Decomposing further (in conformal coordinates) we obtain

$$\alpha = \delta'_{\mathfrak{p}} + \delta_{\mathfrak{k}} + \delta''_{\mathfrak{p}}, \tag{7.6.3}$$

where $\delta'_{\mathfrak{p}}$ is a (1,0)-form and $\delta''_{\mathfrak{p}}$ is a (0,1)-form. Now one can introduce a parameter $\mu \in S^1$ producing

$$\alpha^{(\mu)} = \mu^{-1}\delta'_{\mathfrak{p}} + \delta_{\mathfrak{k}} + \delta''_{\mathfrak{p}}. \tag{7.6.4}$$

It is crucial for this part of the discussion that the Willmore property can be rephrased not only relative to the extended Maurer-Cartan frame α_λ defined in 7.5.13, but also relative to $\alpha^{(\mu)}$.

Theorem 7.6.1. *([37], Theorem 2) The map $\phi : \mathbb{D} \longrightarrow \mathbb{R}^3$ is a Willmore immersion if and only if $\alpha^{(\mu)}$ is integrable for all $\mu \in S^1$:*

$$d\alpha^{(\mu)} + \frac{1}{2}[\alpha^{(\mu)} \wedge \alpha^{(\mu)}] = 0 \quad \textit{for all} \quad \mu \in S^1. \tag{7.6.5}$$

So far in this article the parameter λ was always used in relation with harmonic Gauß maps. Therefore *the roles of λ and μ are interchanged compared to* [37].

Just as in the previous section, the integrability condition 7.6.5 allows us to solve the differential equation $F^{-1}dF = \alpha^{(\mu)}, F(0,\mu) = I$. The classical Birkhoff splitting then yields

$$F = F_- \cdot F_+, \tag{7.6.6}$$

where F, F_- and F_+ are in the twisted loop groups which are defined relative to σ. To indicate our use of the parameter μ in constrast to λ we will use M in place of Λ, as in $MSO_o(4,1)^{\mathbb{C}}_\sigma$. Then the Maurer-Cartan form of F_- is of the form ([37], section 4.1):

$$F_-^{-1}dF_- = \mu^{-1}\begin{pmatrix} 0 & a^t & 0 \\ b & 0 & a \\ 0 & b^t & 0 \end{pmatrix}. \tag{7.6.7}$$

Here a and b are 1-forms with values in $\mathbb{R}^3$ satisfying

$$b^t \cdot b = b^t \cdot a = 0 \text{ and } b \neq 0. \tag{7.6.8}$$

Moreover, a and b are closed 1-forms of the type

$$b = \beta dz \quad \text{and} \quad a = \gamma dz + \zeta d\bar{z}. \tag{7.6.9}$$

These are the normalized potentials $\xi^{(\mu)}$ used in this part.

So far things are "as usual". The converse construction, however, will involve some "unusual" steps. Starting from some normalized potential $\xi^{(\mu)}$, which is of the form 7.6.7 and satisfies 7.6.8 and 7.6.9, the right side of 7.6.7 defines an integrable 1-form. Solving the ODE $F^{-1}dF = \xi^{(\mu)}$ with trivial initial condition we obtain some matrix function $F_-^{(\mu)} : \mathbb{D} \longrightarrow M^-SO_o(4,1)_\sigma$. The generalized Iwasawa splitting [40] yields

$$F_-^{(\mu)} = F^{(\mu)} \cdot F_+^{(\mu)}, \tag{7.6.10}$$

with $F^{(\mu)} \in MSO_0(4,1)_\sigma$ and $F_+^{(\mu)} \in MSO_0(4,1)^{\mathbb{C}}_\sigma$.

The crucial point is whether one is able to construct a Willmore immersion from the map $F^{(\mu)}$. It is shown in [37] that the freedom in the splitting 7.6.10 can used so that the resulting extended lift produces a harmonic map via projection onto $S^{3,1}$.

Remark 7.6.2. The case of Willmore immersions has some interesting aspects:

1. The property of being a Willmore immersion is rephrased equivalently by the harmonicity of a map which is *not* the usual Gauß map. A similar situation was encountered by Inoguchi [39] in his investigation of surfaces of constant mean curvature in hyperbolic space $\mathbb{H}^3$, and most recently in [11]. It would be interesting to look for surface classes which are defined by the harmonicity of some "natural map" and to investigate

to what extent such classes can be treated by some loop group approach akin to the ones presented in this article.

2. The transition from the frame to the harmonic map or the immersion is not given by a "Sym-type formula". However, as in the case of affine spheres the freedom in the generalized Iwasawa splitting permits one to choose the splitting so that the transition can be made. It would be interesting to understand for which surface classes a Sym-type formula does exist.

7.7 Remarks 7.7.1. *Other surface classes*: There are several other classes of surfaces which are natural candidates for a "loop group approach". Among these are surfaces of constant mean curvature in the sphere S^3 and the hyperbolic space $\mathbb{H}^3$, as well as surfaces of positive constant Gauß curvature in $\mathbb{R}^3$ and surfaces of constant Gauß curvature in $\mathbb{E}^3_1$. Spacelike surfaces of constant mean curvature in $\mathbb{E}^3_1$ fall into this class as well [54].

There are several types of "loop group involvement". On the one hand, surfaces like the ones of constant positive Gauß curvature in $\mathbb{R}^3$ are parallel to surfaces of constant mean curvature, and thus admit a "loop group approach". Next, surfaces like the surfaces of constant mean curvature in hyperbolic space $\mathbb{H}^3$ have been investigated quite successfully by introducing "loop parameters" and "extended frames" [33],[32],[54]. From the point of view of this article, however, it would be most interesting to see whether the surfaces of the classes listed above (and possibly others) can be characterized directly by the harmonicity of some "generalized Gauß map" into some symmetric (?) space and whether it is possible to construct all these Gauß maps (and thus all the surfaces in the class considered) from unconstrained "potentials". Work in this direction is presently being conducted by several mathematicians (Inoguchi, Inoguchi-Toda [39], Rossman and others).

7.7.2. *More powers of* λ: Hamiltonian stationary Lagrangian surfaces in $\mathbb{C}^2$ have been investigated via a loop group approach in [38]. An interesting feature of this paper is the use of potentials in which not only λ^{-1} occurs, but also λ^{-2}. On the other hand, the integration of the relevant differential equations can be carried out by hand. In this sense the loop group approach is only a guide and not essential for the theory. In contrast, in the investigation of harmonic maps into k-symmetric spaces [10], potentials involving negative powers $\lambda^{-k}, \ldots, \lambda^{-1}$ occur. In this case, however, only special harmonic maps ("primitive harmonic maps") can be treated.

7.7.3. *More features* of the theory of constant mean curvature in $\mathbb{R}^3$ can be carried over to the surface classes discussed in the last few sections. This certainly applies to dressing. *Wu's Formula* seems to work in all cases. *Bäcklund transformations* also appear. These transformations usually represent very specific geometric constructions of new surfaces from known ones (such as using line congruences). In all cases known to the author they can be interpreted as dressing transformations, and are usually associated with very simple dressing matrices. This is well known for the pseudospherical case [53], but seems to apply also to the Tzitzeica transformation of affine spheres and to "Date's direct method" [13]. It would be interesting to see this made explicit.

7.7.4. *A general theory* providing a framework for all known loop group approaches to the description of surfaces (and other geometric quantities) is still missing. At this point the method seems to work best for maps $\phi : M \longrightarrow M'$ for which there exists a "Gauß map" $N : M \longrightarrow S$ from M into some (not necessarily Riemannian) symmetric space S, such that N is " harmonic" if and only if the map ϕ is in the class considered. The paper [11] may be of particular influence in this context.

7.7.5. *Higher Dimensions*: The presentation so far has addressed exclusively maps from surfaces. In view of the fact that the theories presented so far all relate to "harmonic" maps, one should expect that one can extend the theory to include "pluriharmonic" maps. This seems to be the case [15]. There does exist, however, one paper which uses a loop group approach for maps defined on higher dimensional manifolds [31] in a substantially different way. In this case isometric immersions of space forms are discussed. The paper concentrates on immersions of "finite type". It seems that also "potentials" could have been considered. It would be interesting to clarify this and to relate these potentials to the parameter space of the Cartan-Kähler theory. In general it would be very interesting to find loop group approaches to higher dimensional "completely integrable soliton" equations in a direct geometric context.

7.7.6 *Complex Theory*: While a general theory of classes of surfaces which can be treated by some unified loop group approach seems to be out of reach at this point, it seems to be at least feasible to expect that some "complexified" theory should be able to provide a "higher point of view", at least for some cases. Most of the surface classes discussed in the previous sections are treated in a double-loop-group setting and their potentials consist of pairs of Lie algebra valued differential forms. The only exception is the class of surfaces of constant mean curvature in $\mathbb{R}^3$. It is easy to see that also this case has naturally a double-loop-group setting. For $g \in \Lambda Sl(2,\mathbb{C})_\sigma$ we put $\hat{g} = (g, (\bar{g}^t)^{-1})$. Thus $\Lambda Sl(2,\mathbb{C})_\sigma$ is identified with a subgroup of the double loop group $\mathcal{H} = \Lambda Sl(2,\mathbb{C})_\sigma \times \Lambda Sl(2,\mathbb{C})_\sigma$. If F is the frame of a constant mean curvature surface, then $\hat{F} = (F, F) \in \mathcal{U}$, the diagonal subgroup in $\mathcal{H}$. As in the previous cases we set $\mathcal{H}^- = \Lambda^- Sl(2,\mathbb{C})_\sigma \times \Lambda^+ Sl(2,\mathbb{C})_\sigma$ and $\mathcal{H}^+ = \Lambda^+ Sl(2,\mathbb{C})_\sigma \times \Lambda^- Sl(2,\mathbb{C})_\sigma$. Then $\mathcal{H}^- \cdot \mathcal{H}^+$ is open and dense in $\mathcal{H}$. The Birkhoff splitting $\hat{F} = \hat{F}_- \cdot \hat{F}_+$, which is $F = F_- \cdot F_+$ and $F = (\bar{F_-}^t)^{-1} \cdot (\bar{F_+}^t)^{-1}$, yields the pair of potentials $\hat{\xi} = (\xi, -\bar{\xi}^t) \in \mathrm{Lie}\,\mathcal{H}_-$, the Lie algebra of $\mathcal{H}_-$ [24]. Here the first factor depends only on z, while the second factor depends only on $\bar{z}$. Conversely, starting from $\hat{\xi}$, one solves the pair of ODE's $d\hat{F} = \hat{F} \cdot \hat{\xi}$ with initial condition $\hat{F}(z_0) = (I, I)$, and splits $\hat{F} = (F, F) \cdot (V_+, V_-)$ in $\mathcal{U} \cdot \mathcal{H}_+$. It is easy to see that, because of the special form of $\hat{\xi}$, the matrix F is unitary. Clearly, the latter splitting is a double-loop-group version of the classical Iwasawa splitting. It is natural to generalize the procedure above. Let $\Omega = (\xi(z), \eta(w))$ be a differential form in two independent variables z and w which takes values in the Lie algebra of $\mathcal{H}_-$. Then, solving the pair of ODE's $d\hat{F} = \hat{F} \cdot \Omega$ with initial condition I, one obtains a pair of functions $\hat{F}_- = (U_-(z), U_+(w))$. The generalized Iwasawa splitting $\mathcal{H} \approx \mathcal{U} \cdot \mathcal{H}_+$ then produces some $U(z, w)$, such that $(U_-(z), U_+(w)) = (U, U) \cdot (V_+, V_-)$, where U, V_+, V_- depend on z and on w. Thus we obtain the familiar formula

$$U = U_- \cdot V_+^{-1} = U_+ \cdot V_-^{-1}. \tag{7.7.1}$$

It is tempting to predict that U plays the role of some (perhaps modified) frame of some complex surface. This surface should have many interesting features. In particular, it should contain the surfaces of constant mean curvature in $\mathbb{R}^3$, the pseudospherical surfaces in $\mathbb{R}^3$, and the timelike surfaces of constant mean curvature in $\mathbb{E}^3_1$ as "real forms".

Update added, December 2007: Since this article was written, progress has been made on several fronts. A list of additional references ([60]-[108]) is provided below concerning matters closely related to the questions discussed in this article and in particular to the remarks above. This list is by no means complete and we apologize to everyone whose work is not mentioned.

References

[1] V. Balan and J. Dorfmeister, Birkhoff decompositions and Iwasawa decompositions for Loop Groups, Tohoku Math. J., **53** (2001), 593–615.

[2] V. Balan and J. Dorfmeister, A Weierstraß-type representation for harmonic maps from Riemann surfaces to general Lie groups, Balkan J. Geom. Appl., **5** (2000), 7–37.

[3] M. Bergvelt and M. Guest, Actions of loop groups on harmonic maps, Trans. Amer. Math. Soc., **326** (1991), 861–886.

[4] A. Bobenko, Constant mean curvature surfaces and integrable equations, Russian Math. Surveys, **46** (1991), 1–45.

[5] A. Bobenko, All constant mean curvature tori in R^3, S^3, H^3 in terms of theta-functions, Math. Ann., **290** (1991), 209–245.

[6] R. Bryant, A duality theorem for Willmore surfaces, J. Differential Geom., **20** (1984), 23–53.

[7] L. Bungart, On analytic fiber bundles, Topology, **7** (1968), 55–68.

[8] F. Burstall, D. Ferus, F. Pedit and U. Pinkall, Harmonic tori in symmetric spaces and commuting Hamiltonian systems on loop algebras, Ann. of Math. (2), **138** (1993), 173–212.

[9] F. Burstall and M. Guest, Harmonic two-spheres in compact symmetric spaces, revisited, Math. Ann., **309** (1997), 541–572.

[10] F. Burstall and F. Pedit, Harmonic maps via Adler-Kostant-Symes theory, In: Harmonic maps and integrable systems, Aspects Math., **E23**, Vieweg, Braunschweig, 1994, pp. 221–272.

[11] F. Burstall and U. Hertrich-Jeromin, Harmonic maps in unfashionable geometries, Manuscripta Math., **108** (2002), 171–189.

[12] F. Burstall and F. Pedit, Dressing orbits of harmonic maps, Duke Math. J., **80** (1995), 353–382.

[13] E. Date, Multi-soliton solutions and quasi-periodic solutions of nonlinear equations of Sine-Gordon type, Osaka J. Math., **19** (1982), 125–158.

[14] J. Dorfmeister and U. Eitner, Weierstraß-type representation of affine spheres, Abh. Math. Univ. Hamburg, **71** (2001), 225–250.

[15] J. Dorfmeister and J. Eschenburg, Pluriharmonic maps, loop groups and twistor theory, Ann. Global Anal. Geom., **24** (2003), 301–321.

[16] J. Dorfmeister, H. Gradl and J. Szmigielski, Systems of PDE's obtained from factorization in loop groups, Acta Appl. Math. **53** (1998), 1–58.

[17] J. Dorfmeister and G. Haak, Meromorphic potentials and smooth CMC surfaces, Math. Z., **224** (1997), 603–640.

[18] J. Dorfmeister and G. Haak, On symmetries of constant mean curvature surfaces. I. General theory Tohoku Math. J., **50** (1998), 437–454.

[19] J. Dorfmeister and G. Haak, On symmetries of constant mean curvature surfaces. II. Symmetries in a Weierstraß-type representation, Int. J. Math. Game Theory Algebra, **10** (2000), 121–146.

[20] J. Dorfmeister and G. Haak, On constant mean curvature surfaces with periodic metric, Pacific J. Math., **182** (1998), 229–287.

[21] J. Dorfmeister and G. Haak, Investigation and application of the dressing action on surfaces of constant mean curvature, Q. J. Math., **51** (2000), 57–73.

[22] J. Dorfmeister and G. Haak, Construction of non-simply connected CMC surfaces via dressing, J. Math. Soc. Japan, **55** (2003), 335–364.

[23] J. Dorfmeister, J. Inoguchi and M. Toda, Weierstraß type representation of timelike surfaces with constant mean curvature, In: Differential Geometry and Integrable Systems, Contemp. Math., **308**, Amer. Math. Soc., Providence, RI, 2002, pp. 77–99.

[24] J. Dorfmeister, I. McIntosh, F. Pedit and H. Wu, On the meromorphic potential for a harmonic surface in a k-symmetric space, Manuscripta Math., **92** (1997), 143–152.

[25] J. Dorfmeister, F. Pedit and M. Toda, Minimal surfaces via loop groups, Balkan J. Geom. Appl., **2** (1997), 25–40.

[26] J. Dorfmeister, F. Pedit and H. Wu, Weierstraß type representations of harmonic maps into symmetric spaces, Comm. Anal. Geom., **6** (1998), 633–668.

[27] J. Dorfmeister and H. Wu, Constant mean curvature surfaces and loop groups, J. Reine Angew. Math., **440** (1993), 43–76.

[28] J. Dorfmeister and H. Wu, Construction of constant mean curvature N-noids from holomorphic potentials, in preparation.

[29] J. Eells, On the surfaces of Delaunay and their Gauß maps, In: Proceedings of the IV International Colloquium on Differential Geometry, Santiago de Compostela, 1979, Cursos Congr. Univ. Santiago de Compostela, **15**, pp. 97–116.

[30] J. Eells, The surfaces of Delaunay, Math. Intelligencer, **9** (1987), 53–57.

[31] D. Ferus and F. Pedit, Curved flats in symmetric spaces, Manuscripta Math., **91** (1996), 445–454.

[32] A. Fujioka, Harmonic maps and associated maps from simply connected Riemann surfaces into the 3-dimensional space forms, Tohoku Math. J., **47** (1995), 431–439.

[33] A. Fujioka, Actions of loop groups on simply connected H-surfaces in space forms, J. Math. Soc. Japan, **50** (1998), 819–829.

[34] K. Grosse-Brauckmann, R. Kusner and J. Sullivan, Classification of embedded constant mean curvature surfaces with genus zero and three ends, Proc. Natl. Acad. Sci. USA, **97** (2000), 14067–14068.

[35] K. Grosse-Brauckmann, R. Kusner and J. Sullivan, Triunduloids: Embedded constant mean curvature surfaces with three ends and genus zero, J. Reine Angew. Math., **564** (2003), 35–61.

[36] M. Guest, Harmonic Maps, Loop Groups, and Integrable Systems, London Math. Soc. Stud. Texts, **38**, Cambridge Univ. Press, 1997.

[37] F. Hélein, Willmore immersions and loop groups, J. Differential Geom., **50** (1998), 331–385.

[38] F. Hélein and P. Romon, Hamiltonian stationary Lagrangian surfaces in $\mathbb{C}^2$, Comm. Anal. Geom., **10** (2002), 79–126.

[39] J. Inoguchi and M. Toda, Timelike minimal surfaces via loop groups, Acta Appl. Math., **83** (2004), 313–355.

[40] P. Kellersch, Eine Verallgemeinerung der Iwasawa Zerlegung in Loop Gruppen Dissertation, Technische Universität München, 1999.

[41] K. Kenmotsu, Weierstrass formula for surfaces of prescribed mean curvature, Math. Ann., **245** (1979), 89–99.

[42] M. Kilian, Constant Mean Curvature Cylinders, Dissertation, Univ. of Massachusetts, Amherst, 2000.

[43] M. Kilian, I. McIntosh and N. Schmitt, New constant mean curvature surfaces, Experiment. Math., **9** (2000), 595–611.

[44] D. Lerner and I. Sterling, Construction of constant mean curvature surfaces using the Dorfmeister-Pedit-Wu representation of harmonic maps, Proc. Symp. Differential Geometry, Hamiltonian Systems, and Operator Theory, University of the West Indies, The University Printers, 1994.

[45] I. McIntosh, Infinite-dimensional Lie groups and the two-dimensional Toda lattice, Aspects Math., **E23** (1994), 205–220.

[46] U. Pinkall and I. Sterling, On the classification of constant mean curvature tori, Ann. of Math. (2), **130** (1989), 407–451.

[47] A. Pressley and G. Segal, Loop Groups, Clarendon Press, Oxford, 1986.

[48] E. Ruh and J. Vilms, The tension field of the Gauss map, Trans. Amer. Math. Soc., **149** (1970), 569–573.

[49] N. Schmitt, New Constant Mean Curvature Trinoids, preprint.

[50] G. Segal and G. Wilson, Loop groups and equations of KdV type, Inst. Hautes Etudes Sci. Publ. Math., **61** (1985), 5–65.

[51] B. Smyth, A generalization of a theorem of Delaunay on constant mean curvature surfaces, In: Statistical thermodynamics and differential geometry of microstructured materials, IMA Vol. Math. Appl., **51**, Springer-Verlag, 1993, pp. 123–130.

[52] I. Sterling and H. Wente, Existence and classification of constant mean curvature multibubbletons of finite and infinite type, Indiana Univ. Math. J., **42** (1993), 1239–1266.

[53] C.-L. Terng and K. Uhlenbeck, Bäcklund transformations and loop group actions, Comm. Pure Appl. Math., **53** (2000), 1–75.

[54] T. Taniguchi, The Sym-Bobenko formula and constant mean curvature surfaces in Minkowski 3-space, Tokyo J. Math., **20** (1997), 463–473.

[55] M. Toda, Pseudospherical surfaces via moving frames and loop groups, Dissertation, Univ. of Kansas, 2000.

[56] K. Uhlenbeck, Harmonic maps into Lie groups: classical solutions of the chiral model, J. Differential Geom., **30** (1989), 1–50.

[57] H. Wu, On the dressing action of loop groups on constant mean curvature surfaces, Tohoku Math. J., **49** (1997), 599–621.

[58] H. Wu, A simple way for determining the normalized potentials for harmonic maps, Ann. Global Anal. Geom., **17** (1999), 189–199.

[59] V. Zakharov and A. Shabat, Integration of non-linear equations of mathematical physics by the method of inverse scattering II, Funkts. Anal. i Prilozhen, **13** (1978), 13–22.

Added December 2007:

[60] A. Amarzaya, Notes on Baecklund transformations and loop group actions for harmonic maps, J. Mong. Math. Soc., **5** (2001), 1–13.
[61] V. Balan and J. Dorfmeister, Weierstrass-type representation for harmonic maps into general symmetric spaces via loop groups, J. Math. Soc. Japan, **57** (2005), 69–94.
[62] D. Brander, K-symmetric AKS systems and flat immersions in spheres, to appear in J. London Math. Soc.
[63] D. Brander, Curved flats, pluriharmonic maps and constant curvature immersions into pseudo-Riemannian space forms, Ann. Global Anal. Geom., **32** (2007), 253–275.
[64] D. Brander, Grassmann geometries in infinite dimensional homogeneous spaces and an application to reflective submanifolds, to appear in Int. Math. Res. Notices.
[65] D. Brander and J. Dorfmeister, Generalized DPW method and an application to space forms, preprint.
[66] D. Brander and W. Rossman, A loop group formulation for constant curvature submanifolds of pseudo-Euclidean space, to appear in Taiwanese J. Math.
[67] M. Brueck, X. Du, J. Park and C.-L. Terng, The submanifold geometries associated to Grassmannian systems, Mem. Amer. Math. Soc., **155** (2002), no. 735.
[68] F. Burstall, Isothermic surfaces: conformal geometry, Clifford algebras and integrable systems, In: Integrable systems, geometry, and topology, AMS/IP Stud. Adv. Math., 36, Amer. Math. Soc., Providence, RI, 2006, pp. 1–82.
[69] F. Burstall and M. Kilian, Equivariant harmonic cylinders, Q. J. Math., **57** (2006), 449–468.
[70] B. Dai and C.-L. Terng, Periodic and homoclinic solutions of the modified 2+1 chiral model, J. Math. Phys., **46** (2005), p. 14.
[71] B. Dai and C.-L. Terng, Baecklund transformations, Ward solitons, and unitons, J. Differential Geom., **75** (2007), 57–108.
[72] B. Dai, C.-L. Terng and K. Uhlenbeck, On the space-time monopole equation, preprint.
[73] J. Dorfmeister and M. Kilian, Dressing preserving the fundamental group, Differential Geom. Appl., **23** (2005), 176–204.
[74] J. Dorfmeister and S. Kobayashi, Coarse classification of constant mean curvature cylinders, Trans. Amer. Math. Soc., **359** (2007), 2483–2500.
[75] J. Dorfmeister, S. Kobayashi and F. Pedit, Complex constant mean curvature surfaces fibered by minimal surfaces, preprint.

[76] M. Dussan, Space-like isothermic surfaces and Grassmannian systems, Tsukuba J. Math., **30** (2006), 81–102.
[77] M. Dussan and M. Magid, Complex timelike isothermic surfaces and their geometric transformations, Balkan J. Geom. Appl., **11** (2006), 39–53.
[78] J.-H. Eschenburg, Pluriharmonic maps, twisted loops and twistors, Jahresber. Deutsch. Math. Verein., **106** (2004), 39–48.
[79] M. Guest, An update on harmonic maps of finite uniton number, via the zero curvature equation, In: Integrable Systems, Topology, and Physics, Contemp. Math., **309**, Amer. Math. Soc., 2002, pp. 85–113.
[80] F. Helein and P. Romon, Hamiltonian stationary Lagrangian surfaces in Hermitian symmetric spaces, In: Differential Geometry and Integrable Systems, Contemp. Math., **308**, Amer. Math. Soc., 2002, pp. 161–178.
[81] I. Khemar, Surfaces isotropes de $\mathbb{O}$ et systemes integrables, preprint.
[82] I. Khemar, Superharmonic maps into symmetric spaces, preprint.
[83] M. Kilian, On the associated family of Delaunay surfaces, Proc. Amer. Math. Soc., **132** (2004), 3075–3082.
[84] M. Kilian, S. Kobayashi, W. Rossman and N. Schmitt, Unitarization of monodromy representations and constant mean curvature trinoids in 3-dimensional space forms, J. London Math. Soc. (2), **75** (2007), 563–581.
[85] M. Kilian, W. Rossman and N. Schmitt, Delaunay ends of constant mean curvature surfaces, to appear in Compositio Math.
[86] S. Kobayashi, Real forms of complex constant mean curvature surfaces, preprint.
[87] S. Kobayashi, Asymptotics of ends of constant mean curvature surfaces with bubbletons, to appear in Proc. Amer. Math. Soc.
[88] S. Kobayashi and J. Inoguchi, Characterizations of Bianchi-Baecklund transformations of constant mean curvature surfaces, Internat. J. Math., **16** (2005), 101–110.
[89] S. Kobayashi, M. Kilian, W. Rossman and N. Schmitt, Constant mean curvature surfaces of any positive genus, J. London Math. Soc. (2), **72** (2005), 258–272.
[90] S. Kong, E. Wang and C.-L. Terng, Associative cones and integrable systems, Chinese Ann. Math. Ser. B, **27** (2006), 153–168.
[91] S. Kresik-Juric, On the Birkhoff factorization problem for the Heisenberg magnet and Schroedinger equations, preprint.
[92] X. Liu and C.-L. Terng, The mean curvature flow for isoparametric submanifolds, preprint.
[93] H. Ma, Hamiltonian stationary Lagrangian surfaces in $\mathbb{C}\mathbf{P}^2$, Ann. Global Anal. Geom., **27** (2005), 1–16.
[94] A. Mahler, Bianchi-Baecklund transformations for constant mean curvature surfaces with umbilics, theory and applications, Dissertation, Univ. of Toledo, 2002.
[95] I. McIntosh and P. Romon, The spectral data for Hamiltonian stationary Lagrangian tori in $\mathbb{R}^4$, preprint.

[96] D. Nadler, Matsuki correspondence for the affine Grassmannian, Duke Math. J., **124** (2004), 421–457.

[97] R. Pacheco, Harmonic two-spheres in the symplectic group Sp(n), Internat. J. Math., **17** (2006), 295–311.

[98] W. Rossman and N. Schmitt, Simultaneous unitarizability of $Sl(n, \mathbb{C})-$valued maps, and constant mean curvature k-noid monodromy, Ann. Sc. Norm. Super. Pisa Cl. Sci. (5), **5** (2006), 549–577.

[99] W. Rossman and N. Sultana, Morse index of constant mean curvature tori of revolution in the 3-sphere, to appear in Illinois J. Math.

[100] C.-L. Terng, Applications of loop group factorization to geometric soliton equations, International Congress of Mathematicians, Vol. II, Madrid, 2006, pp. 927–950.

[101] C.-L. Terng, Geometries and symmetries of soliton equations and integrable elliptic equations, this volume.

[102] C.-L. Terng and E. Wang, Curved flats, exterior differential systems, and conservation laws, In: Complex, contact and symmetric manifolds, Progr. Math., **234**, Birkhäuser Boston, 2005, pp. 235–254.

[103] C.-L. Terng and K. Uhlenbeck, Schroedinger flows on Grassmannians, to appear in Integrable Systems, Geometry and Topology.

[104] M. Toda, Weierstrass-type representation of weakly regular pseudospherical surfaces in Euclidean space, Balkan J. Geom. Appl., **7** (2002), 87–136.

[105] E. Wang, Tzitzeica transformation is a dressing action, J. Math. Phys., **47** (2006), p. 13.

[106] E. Wang and C.-L. Terng, Transformations of flat Lagrangian immersions and Egoroff nets, to appear in Asian J. Math.

[107] Q.-L. Xia and Y.-B. Shen, Pluriharmonic maps from complex manifolds into symmetric spaces, (Chinese) Acta Math. Sinica (Chin. Ser.), **45** (2002), 661–670.

[108] Q. Xia, The generalized Weierstrass representation of surfaces with positive Gauss curvature in pseudo-Riemannian space forms, preprint.

Zentrum Mathematik
Technische Universität München
Boltzmannstrasse 3
D-85747 Garching
GERMANY

E-mail address:
`dorfm@mathematik.tu-muenchen.de`

Advanced Studies in Pure Mathematics 51, 2008
Surveys on Geometry and Integrable Systems
pp. 113–141

Timelike surfaces with harmonic inverse mean curvature

Atsushi Fujioka and Jun-ichi Inoguchi

Introduction

In this paper we introduce the notion of *timelike surfaces with harmonic inverse mean curvature* in 3-dimensional Lorentzian space forms, and study their fundamental properties.

In classical differential geometry, surfaces of constant mean curvature (CMC surfaces) have been studied extensively [1]. As a generalization of CMC surfaces, Bobenko [2] introduced the notion of *surface with harmonic inverse mean curvature* (HIMC surface). He showed that HIMC surfaces admit a Lax representation with *variable* spectral parameter. In [5], Bobenko, Eitner and Kitaev showed that the Gauss equations of θ-isothermic HIMC surfaces reduce to the ordinary differential equation:

$$(*) \qquad \left(\frac{q''(t)}{q'(t)}\right)' - q'(t) = \mathcal{S}(t)\left(2 - \frac{q^2(t)+c}{q'(t)}\right), \quad q'(t) < 0,$$

with $c = \theta^2 > 0$. Here the coefficient function $\mathcal{S}(t)$ is $1/\sin^2(2t)$, $1/\sinh^2(2t)$ or $1/t^2$. This ordinary differential equation is called the *generalized Hazzidakis equation.* Bobenko, Eitner and Kitaev [5] solved $(*)$ in terms of Painlevé transcendents P_{V} and P_{VI}.

For $c < 0$, solutions to $(*)$ do not describe surfaces in Euclidean 3-space. It seems to be interesting to find "corresponding surfaces" to such solutions.

2000 *Mathematics Subject Classification.* Primary 37K25; Secondary 53C42, 53C50.

Received January 12, 2001.
Revised May 22, 2001.
Fujioka: Partially supported by Grant-in-Aid for Encouragement of Young Scientists No. 12740037, JSPS; Inoguchi: Partially supported by Grant-in-Aid for Encouragement of Young Scientists No. 12740051, JSPS.

The first author extended the notion of HIMC surface in Euclidean 3-space to that of Riemannian 3-space forms [7]. Moreover he generalized a theorem due to Lawson (the *Lawson correspondence*) to HIMC surfaces. By using the Lawson correspondence for HIMC surfaces, we have classified Bonnet surfaces with constant curvature in Riemannian 3-space forms [8]. Corresponding results for spacelike surfaces in Lorentzian 3-space forms are obtained in [10].

On the other hand, very little is known about (integrable) timelike surfaces of nonconstant mean curvature in Lorentzian 3-space forms. Timelike Bonnet surfaces were investigated by the present authors very recently [11].

In this paper we introduce the notion of timelike surfaces with harmonic inverse mean curvature (THIMC surface) in Lorentzian 3-space forms. We shall show that every solution to the generalized Hazzidakis equation with $c < 0$ describes a THIMC surface in Minkowski 3-space. This is one of the motivations to study THIMC surfaces.

Because of the indefiniteness of the metric, timelike surface geometry has many aspects different from Euclidean surface geometry. For instance, there exist timelike (HIMC) surfaces with imaginary principal curvatures. Moreover there exist non totally umbilical timelike surfaces with real repeated principal curvatures. Such surfaces have no counterparts in Euclidean surface geometry and spacelike surface geometry. Thus the geometry of THIMC surfaces has its own interest.

The second motivation of the present study is to give new examples of Lax equations with variable spectral parameter, namely, Lax equations whose spectral parameters depend on the variables. Burtsev, Zakharov and Mikhailov [6] exhibited some examples of Lax equations with variable spectral parameter which appeared in theoretical physics. In differential geometry, HIMC surfaces and Bianchi surfaces are known examples. (See [2], [15] and [16].)

We shall show that THIMC surfaces Lorentzian 3-space forms admit a Lax representation with variable spectral parameter. Moreover we shall show that in de Sitter 3-space or anti-de Sitter 3-space, THIMC surfaces admit a Lax representation with *two* independent variable spectral parameters.

This paper is organized as follows. After recalling fundamental facts on Lorentzian geometry, we introduce the notion of THIMC surface in Minkowski 3-space in Section 3. We give a Lax representation and an immersion formula (Sym formula) for THIMC surfaces. Some elementary examples will be given in Section 3. In Section 4, we introduce the notion of $\pm$ isothermic timelike surface. We shall give a duality between timelike Bonnet surfaces and $\pm$ isothermic THIMC surfaces.

In Section 5, we shall investigate the normal forms of the Gauss equations of THIMC surfaces. More precisely we show that (θ-isothermic or anti-θ-isothermic) THIMC surfaces in Minkowski 3-space are derived from solutions to the generalized Hazzidakis equation with $c = -\theta^2 < 0$.

In Section 6, we shall generalize the notion of THIMC surface to Lorentzian 3-space forms and establish a Lawson-type correspondence for THIMC surfaces.

1 Lorentzian space forms

1.1 First of all, we shall describe *Lorentzian* 3*-space forms*, i.e., complete and connected Lorentzian 3-manifolds $\mathfrak{M}^3_1(c)$ of constant curvature c explicitly. Without loss of generality, we may assume that $c = 0$ or ± 1.

We equip the Cartesian 4-space $\mathbf{R}^4$, with the following scalar product $\langle \cdot, \cdot \rangle = \langle \cdot, \cdot \rangle_c$:

$$\langle a, b \rangle_c = -a_0 b_0 + a_1 b_1 + a_2 b_2 + a_3 b_3, \quad c = 1,$$
$$\langle a, b \rangle_c = -a_0 b_0 - a_1 b_1 + a_2 b_2 + a_3 b_3, \quad c = 0, \ -1.$$

The resulting semi-Euclidean 4-space $(\mathbf{R}^4, \langle \cdot, \cdot \rangle)$ is of index 1 for $c = 1$ and of index 2 for $c = 0$ or -1 respectively. The Lorentzian 3-space forms $\mathfrak{M}^3_1(c)$ are embedded in the semi-Euclidean space $(\mathbf{R}^4, \langle \cdot, \cdot \rangle_c)$ as

$$\mathfrak{M}^3_1(0) = \{p \in (\mathbf{R}^4, \langle \cdot, \cdot \rangle_0) \mid p_0 = 0\} = \mathbf{E}^3_1, \text{Minkowski 3-space},$$
$$\mathfrak{M}^3_1(1) = \{p \in (\mathbf{R}^4, \langle \cdot, \cdot \rangle_1) \mid \langle p, p \rangle_1 = 1 \ \} = S^3_1, \text{de Sitter 3-space},$$
$$\mathfrak{M}^3_1(-1) = \{p \in (\mathbf{R}^4, \langle \cdot, \cdot \rangle_{-1}) \mid \langle p, p \rangle_1 = -1 \ \} = H^3_1, \text{anti-de Sitter 3-space}.$$

For more details on semi-Riemannian geometry, we refer to O'Neill [18].

Next we recall 2 by 2 matrix models of $\mathfrak{M}^3_1(c)$ for later use.

First the semi-Euclidean 4-space $\mathbf{E}^4_2 = (\mathbf{R}^4, \langle \cdot, \cdot \rangle_{-1})$ is identified with the linear space $\mathrm{M}_2\mathbf{R}$ of all 2 by 2 real matrices via the isomorphism:

(1.1)
$$p = (p_0, p_1, p_2, p_3) \longleftrightarrow p_0\mathbf{1} + p_1\mathbf{i} + p_2\mathbf{j}' + p_3\mathbf{k}' = \begin{pmatrix} p_0 - p_3 & -p_1 + p_2 \\ p_1 + p_2 & p_0 + p_3 \end{pmatrix}.$$

The semi-Euclidean metric of $\mathbf{E}^4_2$ corresponds to the following scalar product on $\mathrm{M}_2\mathbf{R}$.

$$(1.2)\qquad \langle X,Y\rangle=\frac{1}{2}\{\mathrm{tr}(XY)-\mathrm{tr}(X)\mathrm{tr}(Y)\},\quad X,Y\in \mathrm{M}_2\mathbf{R}.$$

Under the identification (1.1), the Minkowski 3-space $\mathbf{E}^3_1(p_1,p_2,p_3)$ is identified with the Lie algebra $\mathfrak{g}=\mathfrak{sl}_2\mathbf{R}=\{X\in \mathrm{M}_2\mathbf{R}\mid \mathrm{tr}\ X=0\}$ with metric $\langle X,Y\rangle=\frac{1}{2}\mathrm{tr}(XY),\ X,Y\in\mathfrak{g}$.

Next, since $\langle X,X\rangle=-\det X$ for all $X\in \mathrm{M}_2\mathbf{R}$, the anti-de Sitter 3-space $H^3_1\subset\mathbf{E}^4_2$ corresponds to the real special linear group:

$$G=\mathrm{SL}_2\mathbf{R}=\left\{\begin{pmatrix} a & b\\ c & d\end{pmatrix}\in \mathrm{M}_2\mathbf{R}\mid ad-bc=1\right\}.$$

Since the Lorentzian metric of G is bi-invariant, the product group $G\times G$ acts transitively and isometrically on H^3_1 as follows:

$$\mu_H:(G\times G)\times H^3_1\longrightarrow H^3_1,\quad \mu_H(g_1,g_2)X=g_1\ X\ g_2^{-1}$$

for $(g_1,g_2)\in G\times G,\ X\in H^3_1$. The isotropy subgroup Δ of $G\times G$ at $\mathbf{1}$ is the diagonal subgroup of $G\times G$, that is, $\Delta=\{(g_1,g_1)\mid g_1\in G\}$. Hence the anti-de Sitter 3-space H^3_1 is represented by $H^3_1=(G\times G)/\Delta$ as a Lorentzian symmetric space. The natural projection $p_H:G\times G\to H^3_1$ is given explicitly by $p_H(g_1,g_2)=g_1\ g_2^{-1},\ (g_1,g_2)\in G\times G$.

Moreover G acts isometrically on $\mathbf{E}^3_1$ via the Ad-action:

$$\mathrm{Ad}:G\times\mathbf{E}^3_1\to\mathbf{E}^3_1;\ \mathrm{Ad}(a)X=aXa^{-1},\ a\in G,\ X\in\mathbf{E}^3_1.$$

Finally we recall a 2 by 2 matrix model of S^3_1. The Minkowski 4-space $\mathbf{E}^4_1=(\mathbf{R}^4,\langle\cdot,\cdot\rangle_1)$ is identified with the space $\mathbb{H}$ of all Hermitian 2-matrices via the following isomorphism:

$$(1.4)\qquad p=(p_0,p_1,p_2,p_3)\longleftrightarrow\begin{pmatrix} p_0+p_1 & p_3-\sqrt{-1}p_2\\ p_3+\sqrt{-1}p_2 & p_0-p_1\end{pmatrix}\in\mathbb{H}.$$

Under the identification (1.4), the scalar product $\langle\cdot,\cdot\rangle_1$ of $\mathbf{E}^4_1$ corresponds to the following scalar product on $\mathbb{H}$:

$$(1.5)\quad \langle X,Y\rangle=-\frac{1}{2}\mathrm{tr}(\mathbf{i}'X\ \mathbf{i}'Y^t),\ X,Y\in\mathbb{H},\ \mathbf{i}'=\begin{pmatrix} 0 & -\sqrt{-1}\\ \sqrt{-1} & 0\end{pmatrix}.$$

In particular $\det X = -\langle X, X\rangle_1$ under (1.4). Thus the de Sitter 3-space S_1^3 is represented by $S_1^3 = \{X \in \mathbb{H} \mid \det X = -1\}$. The complex special linear group $\mathrm{SL}_2\mathbf{C}$ acts transitively and isometrically on S_1^3 by $\mu_S : \mathrm{SL}_2\mathbf{C} \times S_1^3 \to S_1^3$, $\mu_S(g) = gXg^*$. Here g^* denotes the transposed complex conjugate of g. The isotropy subgroup of $\mathrm{SL}_2\mathbf{C}$ at $\mathbf{i}'$ is $\mathrm{SL}_2\mathbf{R}$. Hence the de Sitter 3-space S_1^3 is represented by $S_1^3 = G^{\mathbf{C}}/G$ as a Lorentzian symmetric space. The natural projection $p_S : G^{\mathbf{C}} \to G^{\mathbf{C}}/G$ is given explicitly by $p_S(g) = \mu(g)\,\mathbf{i}' = g\,\mathbf{i}'\,g^*$, $g \in G^{\mathbf{C}}$.

2 Timelike surfaces in Lorentzian space forms

We start with some preliminaries on the geometry of timelike surfaces in Lorentzian space forms $\mathfrak{M}_1^3(c)$.

Let M be a connected 2-manifold and $F : M \to \mathfrak{M}_1^3(c)$ an immersion. The immersion F is said to be *timelike* if the induced metric I of M is Lorentzian. Hereafter we assume that M is an orientable timelike surface in $\mathfrak{M}_1^3(c)$ immersed by F. The induced Lorentzian metric I of M determines a Lorentzian conformal structure on M. We treat M as a Lorentz surface with respect to this conformal structure and F as a conformal immersion. Our general reference on Lorentz surfaces is Weinstein [21].

On a timelike surface M, there exists a local coordinate system (x, y) such that

$$I = e^{\omega}(-dx^2 + dy^2). \tag{2.1}$$

Such a local coordinate system (x, y) is called a *Lorentz isothermal coordinate system.*

Let (u, v) be the local *null coordinate system* of M derived from (x, y). Namely (u, v) is defined by $u = x + y$, $v = -x + y$. Then the induced metric I can be written as

$$I = e^{\omega}du dv. \tag{2.2}$$

Now, let N be a unit normal vector field to M. The *second fundamental form II* of (M, F) derived from N is defined by $II = -\langle dF, dN\rangle$. The *shape operator S* of (M, F) relative to N is defined by $S = -dN$. The (complex) eigenvalues of S are called *principal curvatures* of (M, F). The *mean curvature H* of (M, F) is defined by $H = \mathrm{tr}\, S/2$. The Gaussian curvature K of (M, I) is given by the formula $K = \det S$.

The Gauss-Codazzi equations of (M, F) have the following form:

$$(\mathrm{G}_c) \qquad \omega_{uv} + \frac{1}{2}(H^2 + c)e^{\omega} - 2QRe^{-\omega} = 0,$$

$$(\mathrm{C}_c) \qquad H_u = 2e^{-\omega}Q_v, \quad H_v = 2e^{-\omega}R_u.$$

Here the functions $Q = \langle F_{uu}, N\rangle$ and $R = \langle F_{vv}, N\rangle$ define global null 2-differentials $Q^{\#} = Qdu^2$ and $R^{\#} = Rdv^2$ on M. These two differentials are called the *Hopf differentials* of M. The Gauss equation implies

$$(2.3) \qquad K = -2\,\omega_{uv}\,e^{-\omega}.$$

Let us denote by $\mathcal{D}$ the discriminant of the characteristic equation $\det(t\mathrm{I} - S) = 0$ for the shape operator S. Here I is the identity transformation of the tangent bundle TM of M. Then by the Gauss equation, we have

$$(2.4) \qquad \mathcal{D} = H^2 - K + c = 4e^{-2\omega}\,QR.$$

The first and second fundamental forms are related by the formula

$$II - HI = Q^{\#} + R^{\#}.$$

This formula implies that the common zero of Q and R coincides with the umbilic point of (M, F). Even if S has real and equal eigenvalues, (M, F) is not necessarily totally umbilic. In fact, there exist timelike surfaces with $QR = 0$ but $II - HI \neq 0$. See Example 3.3.

In the study of timelike surfaces, we also use the following local coordinate system:

Lemma 2.1. *Let $F : M \to \mathfrak{M}^3_1(c)$ be a timelike surface. Then there exists a local coordinate system $(\check{u}, \check{v})$ such that*

$$(2.5) \qquad I = -e^{\check{\omega}}d\check{u}d\check{v}.$$

With respect to this coordinate system, the Gauss-Codazzi equations are

$$(\mathrm{G}^-_c) \qquad \check{\omega}_{\check{u}\check{v}} - \frac{1}{2}(H^2 + c)e^{\check{\omega}} + 2\check{Q}\check{R}e^{-\check{\omega}} = 0,$$

$$(\mathrm{C}^-_c) \qquad H_{\check{u}} = -2e^{-\check{\omega}}\check{Q}_{\check{v}}, \quad H_{\check{v}} = -2e^{-\check{\omega}}\check{R}_{\check{u}}$$

for $\check{Q} = \langle F_{\check{u}\check{u}}, N \rangle, \check{R} = \langle F_{\check{v}\check{v}}, N \rangle$.

We call the local coordinate system $(\check{u}, \check{v})$ an *anti-isothermal coordinate system.* Anti-isothermal coordinate systems will be used for introducing the notion of the Christoffel transformation of an anti-isothermic surface. See Proposition 4.14.

3 Timelike HIMC surfaces in Minkowski 3-space

In this section we shall consider a generalization of timelike CMC surfaces in Minkowski 3-space in terms of integrability theory.

We start by recalling the Lax representation for timelike surfaces in $\mathbf{E}^3_1$. Hereater we assume $H \neq 0$.

Let $F : M \to \mathbf{E}^3_1$ be a timelike surface. Let us take an $\mathrm{SL}_2\mathbf{R}$-valued framing Φ defined by $\mathrm{Ad}(\Phi)(\mathbf{i}, \mathbf{j}', \mathbf{k}') = (e^{-\frac{\omega}{2}} F_x, e^{-\frac{\omega}{2}} F_y, N)$. Thus we get the following Lax representation of the Gauss-Codazzi equations:

$$\frac{\partial}{\partial u}\Phi = \Phi U, \ \frac{\partial}{\partial v}\Phi = \Phi V, \tag{3.1}$$

$$U = \begin{pmatrix} -\frac{1}{4}\omega_u & -Qe^{-\frac{\omega}{2}} \\ \frac{H}{2}e^{\frac{\omega}{2}} & \frac{1}{4}\omega_u \end{pmatrix}, \quad V = \begin{pmatrix} \frac{1}{4}\omega_v & -\frac{H}{2}e^{\frac{\omega}{2}} \\ Re^{-\frac{\omega}{2}} & -\frac{1}{4}\omega_v \end{pmatrix}. \tag{3.2}$$

Now we shall insert a *variable spectral parameter* λ , *i.e.*, an additional *real* parameter λ which depends on the coordinates (u, v) into the Lax pair (3.2) in the following way:

$$U_\lambda = \begin{pmatrix} -\frac{1}{4}\omega_u & -Qe^{-\frac{\omega}{2}} \\ \frac{H}{2}\lambda e^{\frac{\omega}{2}} & \frac{1}{4}\omega_u \end{pmatrix}, \quad V_\lambda = \begin{pmatrix} \frac{1}{4}\omega_v & -\frac{H}{2}\lambda^{-1}e^{\frac{\omega}{2}} \\ Re^{-\frac{\omega}{2}} & -\frac{1}{4}\omega_v \end{pmatrix}. \tag{3.3}$$

Then the compatibility condition

$$\frac{\partial}{\partial u}V_\lambda - \frac{\partial}{\partial v}U_\lambda + [\, U_\lambda, V_\lambda \,] = 0 \tag{3.4}$$

for the deformed Lax pair $\{U_\lambda, V_\lambda\}$ yields

$$\omega_{uv} + \frac{1}{2}H^2 e^{\omega} - 2QRe^{-\omega} = 0, \tag{G$_0$}$$

$$Q_v = \frac{e^\omega}{2}(H\lambda^{-1})_u, \quad R_u = \frac{e^\omega}{2}(H\lambda)_v. \tag{3.5}$$

The Lax pair $\{U_\lambda, V_\lambda\}$ describes a timelike surface in $\mathbf{E}^3_1$ if and only if the equations (3.5) are consistent with the Codazzi equations (C_0). The equations (3.5) are consistent with (C_0) if and only if

$$\frac{\partial}{\partial v}\{H(1-\lambda)\} = 0, \quad \frac{\partial}{\partial u}\{H(1-\lambda^{-1})\} = 0. \tag{3.6}$$

These equations (3.6) can be solved easily as follows:

$$H = \frac{1}{f(u)+g(v)}, \quad \lambda = -\frac{g(v)}{f(u)}, \tag{3.7}$$

where $f(u)$ and $g(v)$ are smooth functions. It is easy to see that the mean curvature H is invariant under the one parametric deformation

$$f \longmapsto f + \frac{1}{2\tau}, \quad g \longmapsto g - \frac{1}{2\tau}, \quad \tau \in \mathbf{R}^*.$$

Under this deformation, the spectral parameter λ is transformed as

$$\lambda = -\frac{g}{f} \longmapsto \lambda(u,v;\tau) = \frac{1-2\tau g}{1+2\tau f}, \quad \tau \in \mathbf{R}.$$

Note that $\lambda(u,v;0) \equiv 1$. The form (3.7) of H is equivalent to the Lorentz-harmonicity of $1/H$, *i.e.*, $(1/H)_{uv} = 0$. As in Euclidean surface geometry [2] and spacelike surface geometry [10], we shall call a timelike surface M in $\mathbf{E}^3_1$, a *timelike surface with harmonic inverse mean curvature* (THIMC surface) if $1/H$ is a Lorentz-harmonic function.

Here we would like to exhibit three elementary examples of THIMC surfaces.

Example 3.1. (THIMC cylinders.) Let $a(y) = (a_2(y), a_3(y))$ be a curve in Euclidean plane $\mathbf{E}^2(\xi_2, \xi_3)$ parametrized by the arclength parameter $y \in \mathcal{I}$. Here $\mathcal{I}$ is an interval. *A timelike cylinder over the curve* a is a flat timelike surface in $\mathbf{E}^3_1$ defined by the immersion $F : \mathcal{I} \times \mathbf{R} \longrightarrow \mathbf{E}^3_1$; $F(x,y) = (x, a_2(y), a_3(y))$. It is straightforward to see that the mean curvature of the cylinder is $H = \kappa(y)/2$. Here κ is the curvature of a. Thus the cylinder F is a THIMC surface if and only if the base curve has the curvature $\frac{1}{C_1 y + C_2}$, $C_1, C_2 \in \mathbf{R}$. It is well known that curves with curvature $\frac{1}{C_1 y + C_2}$ are logarithmic spirals or circles. Hence all the THIMC cylinders over a Euclidean curve are cylinders over a logarithmic spiral or a circular cylinder.

Example 3.2. (THIMC cylinders over timelike curves.) Let $a(x) = (a_1(x), a_2(x))$ be a timelike curve in Minkowski plane $\mathbf{E}^2_1(\xi_1, \xi_2)$ parametrized by the proper time parameter x defined on an interval $\mathcal{I}$. *A timelike cylinder over the timelike curve a* is a flat timelike surface in $\mathbf{E}^3_1$ defined by the immersion $F : \mathcal{I} \times \mathbf{R} \longrightarrow \mathbf{E}^3_1$; $F(x, y) = (a_1(x), a_2(x), y)$. The mean curvature of F is $H = \kappa(x)/2$. Here κ is the curvature of a. The cylinder F is THIMC if and only if $1/\kappa = C_1 x + C_2$, $C_1, C_2 \in \mathbf{R}$.

We can see that timelike curves with curvature $\frac{1}{C_1 x + C_2}$ are logarithmic pseudo-spirals or timelike hyperbolas (*cf.* the Appendix of [10]). Hence all the THIMC cylinders are cylinders over a logarithmic pseudo-spiral or a timelike hyperbola.

Example 3.3. (B-scrolls.) A curve $\gamma(s)$ in $\mathbf{E}^3_1$ is said to be a *null Frenet curve* if it admits a frame field $\mathcal{L} = (A, B, C)$ along γ (called a *null frame field*) such that $A = \gamma'$, $\langle A, A\rangle = \langle B, B\rangle = 0$, $\langle A, B\rangle = 1$, $\langle C, C\rangle = 1$, $\langle A, C\rangle = \langle B, C\rangle = 0$ and

$$\frac{d}{ds}\mathcal{L} = \mathcal{L}\begin{pmatrix} 0 & 0 & -\tau \\ 0 & 0 & -\kappa \\ \kappa & \tau & 0 \end{pmatrix}.$$

The functions κ and τ are called the *curvature* and *torsion* of γ respectively. The ruled surface $F(s, t) = \gamma(s) + tB(s)$ is called the *B-scroll* of γ. (See Graves [12] and McNertney [17]). The mean curvature of F is the torsion $\tau(s)$. It is straightforward to check that for any null Frenet curve with $\tau \neq 0$, its B-scroll is a THIMC surface.

Remark. The Gaussian curvature of the B-scroll is τ^2. Thus every B-scroll satisfies $QR = 0$ but is not totally umbilical ($II - HI \neq 0$). The property $QR = 0$ implies that every B-scroll is a timelike Bonnet surface. Here *timelike Bonnet surfaces* are timelike surfaces which admit nontrivial isometric deformation preserving mean curvature [11]. Conversely we have proved that every timelike Bonnet surface with $QR = 0$ is a B-scroll [11].

In [14], we have obtained a one-parameter "isometric" deformation of timelike surfaces with constant mean curvature (TCMC surfaces). For THIMC surfaces in $\mathbf{E}^3_1$, we get the following one-parameter family of "conformal" deformations.

Proposition 3.4. *Let $F : M \to \mathbf{E}^3_1$ be a timelike surface with harmonic inverse mean curvature. Let us express the mean curvature H as*

$$H = \frac{1}{f(u) + g(v)}$$

in terms of null coordinate system (u, v). Here $f(u)$ and $g(v)$ are smooth functions. Then F admits the following Lax representation with variable spectral parameter $\lambda(u, v; \tau) = (1 - 2\tau g(v))/(1 + 2\tau f(u))$, $\tau \in \mathbf{R}$:

$$\frac{\partial}{\partial u}\Phi_\lambda = \Phi_\lambda U_\lambda, \ \frac{\partial}{\partial v}\Phi_\lambda = \Phi_\lambda V_\lambda, \tag{3.8}$$

$$U_\lambda = \begin{pmatrix} -\frac{1}{4}\omega_u & -Qe^{-\frac{\omega}{2}} \\ \frac{H}{2}\lambda e^{\frac{\omega}{2}} & \frac{1}{4}\omega_u \end{pmatrix}, \quad V_\lambda = \begin{pmatrix} \frac{1}{4}\omega_v & -\frac{H}{2}\lambda^{-1}e^{\frac{\omega}{2}} \\ Re^{-\frac{\omega}{2}} & -\frac{1}{4}\omega_v \end{pmatrix}.$$

Let $\Phi_\lambda(u, v)$ be a solution of (3.8). Then

$$F_\lambda = -\frac{\partial}{\partial \tau}\Phi_\lambda \cdot \Phi_\lambda^{-1} \tag{3.9}$$

describes a family of THIMC surfaces through $F = F_\lambda|_{\tau=0}$ with Gauss map $N_\lambda = \mathrm{Ad}(\Phi_\lambda)\,\mathbf{k}'$. The fundamental associated quantities of F_λ are given as follows:

$$I_\lambda = \frac{e^\omega dudv}{(1 + 2\tau f)^2(1 - 2\tau g)^2}, \tag{3.10}$$

$$\frac{1}{H_\lambda} = f_\lambda + g_\lambda, \ f_\lambda = \frac{f}{(1 + 2\tau f)}, \ g_\lambda = \frac{g}{(1 - 2\tau g)}, \tag{3.11}$$

$$Q_\lambda = \frac{Q}{(1 + 2\tau f)^2}, \quad R_\lambda = \frac{R}{(1 - 2\tau g)^2}, \tag{3.12}$$

$$K_\lambda = (1 + 2\tau f)(1 - 2\tau g)K, \tag{3.13}$$

$$H_\lambda^2/K_\lambda \equiv H^2/K. \tag{3.14}$$

The formula (3.14) implies that the members of the one parameter family F_λ have the same ratio of principal curvatures.

4 Timelike HIMC surfaces in Minkowski 3-space

In the study of HIMC surfaces in Riemannian space forms, *isothermic surfaces* play a fundamental role. In this section we shall consider such surfaces in timelike surface geometry.

Definition 4.1. Let $F : M \to \mathfrak{M}^3_1(c)$ be a timelike surface. Then (M, F) is said to be *isothermic* if there exists a local isothermal–curvature line coordinate system around any point of M.

An isothermal–curvature line coordinate system is a local Lorentz–isothermal coordinate system such that both parameter curves are curvature lines. It should be remarked that the isothermic property implies the positivity of the discriminant $\mathcal{D}$ for the characteristic equation for the shape operator S. Equivalently, every isothermic timelike surface has real distinct principal curvatures.

The isothermic property for timelike surfaces in $\mathfrak{M}^3_1(c)$ can be reformulated in terms of *associated null coordinate system* as follows.

Proposition 4.2. *A timelike surface (M, F) is isothermic if and only if there exists a local null coordinate system (u, v) around any point of M such that the Hopf differentials take the following form:*

$$(4.1)\quad Q(u,v) = \frac{1}{2}\mathfrak{q}(u,v)\varrho(u), \quad R(u,v) = \frac{1}{2}\mathfrak{q}(u,v)\sigma(v), \quad \varrho > 0, \ \sigma > 0.$$

Here $\mathfrak{q}$ is a real smooth function and ϱ and σ are positive Lorentz holomorphic and anti-holomorphic functions respectively.

Remark. On a Lorentz surface M with null coordinate system (u, v), a smooth function f on M depending only on u [resp. v] is called a *Lorentz holomorphic function* [resp. *Lorentz anti-holomorphic function*].

Hereafter we shall call a null coordinate system derived from an isothermic coordinate system simply an *isothermic coordinate system.*

Remark. Isothermic timelike surfaces in $\mathbf{E}^3_1$ correspond to solutions of the Zoomeron equation studied in soliton theory. Note that the Zoomeron equation is related to the Davey-Stewartson III-equation. See Schief [19, p. 97].

Typical examples of isothermic timelike surfaces are timelike surfaces of revolution in $\mathbf{E}^3_1$. Let us recall the notion of timelike surfaces of revolution in $\mathbf{E}^3_1$. A *revolution of* $\mathbf{E}^3_1$ is a linear isometry which lies in the identity component $\mathrm{O}^{++}_1(3)$ of the Lorentz group $\mathrm{O}_1(3)$. Every revolution fixes a line pointwise. Such a fixed line of a revolution is called the *axis of* revolution. Hence revolutions of $\mathbf{E}^3_1$ can be characterised by the causal character of the axis.

By a *timelike surface of revolution* in $\mathbf{E}^3_1$ we mean a timelike surface obtained by revolving about an axis a regular curve lying in some plane containing the axis [17].

Example 4.3. (Spacelike axis and Euclidean profile curve.) Let $F : M \longrightarrow \mathbf{E}^3_1$ be a timelike surface of revolution with spacelike axis and Euclidean profile curve. Then there exists an isothermic parametrization

$$F(x,y) = \frac{1}{a}\left(e^{\frac{\omega(y)}{2}}\sinh(ax), e^{\frac{\omega(y)}{2}}\cosh(ax), c(y)\right), \quad a \in \mathbf{R}^*$$

so that $c'(y)^2 e^{-\omega(x)} + \left(\frac{\omega'(y)}{2}\right)^2 = a^2$. With respect to this isothermic coordinate system, the mean curvature is given by

$$H(y) = \frac{1}{8c'(y)}\{4a^2 - \omega'(y)^2 - 2\omega''(y)\}, \ c''(y) = e^{\omega(y)}\omega'(y)H(y).$$

Example 4.4. (Spacelike axis and timelike profile curve.) Let $F : M \longrightarrow \mathbf{E}^3_1$ be a timelike surface of revolution with spacelike axis and timelike profile curve. Then there exists an isothermic parametrization

$$F(x,y) = \frac{1}{a}\left(e^{\frac{\omega(x)}{2}}\cosh(ay), e^{\frac{\omega(x)}{2}}\sinh(ay), c(x)\right), \quad a \in \mathbf{R}^*$$

so that $-c'(x)^2 e^{-\omega(x)} + \left(\frac{\omega'(x)}{2}\right)^2 = a^2$. With respect to this isothermic coordinate system, the mean curvature is given by

$$H(x) = \frac{1}{8c'(x)}\{4a^2 - \omega'(x)^2 - 2\omega''(x)\}, \ c''(x) = -e^{\omega(x)}\omega'(x)H(x).$$

Example 4.5. (Timelike axis.) Let $F : M \longrightarrow \mathbf{E}^3_1$ be a timelike surface of revolution with timelike axis. Then there exists an isothermic parametrization

$$F(x,y) = \frac{1}{a}\left(c(x), e^{\frac{\omega(x)}{2}}\cos(ay), e^{\frac{\omega(x)}{2}}\sin(ay),\right), \quad a \in \mathbf{R}^*$$

so that $c'(x)^2 e^{-\omega(x)} - \left(\frac{\omega'(x)}{2}\right)^2 = a^2$. With respect to this isothermic coordinate system, the mean curvature is given by

$$H(x) = -\frac{1}{8c'(x)}\{2\omega''(x) + \omega'(x)^2 + 4a^2\}, \ c''(x) = -e^{\omega(x)}\omega'(x)H(x).$$

Example 4.6. (Null axis.) Let $F : M \longrightarrow \mathbf{E}^3_1$ be a timelike surface of revolution with null axis. Then there exists a null basis $\{L_1, L_2, L_3\}$ of $\mathbf{E}^3_1$ and an isothermic parametrization

$$F(x,y) = \left(a(x), b(x) - \frac{y^2}{2}a(x), ya(x)\right), \quad a \in \mathbf{R}^*$$

relative to the null basis $\{L_1, L_2, L_3\}$ so that $2a'(x)b'(x) = -a(x)^2$. Here a linear null frame means a basis of $\mathbf{E}^3_1$ such that $\langle L_1, L_1\rangle = \langle L_2, L_2\rangle = 0$, $\langle L_1, L_2\rangle = 1$, $\langle L_3, L_3\rangle = 1$, $\langle L_1, L_3\rangle = \langle L_2, L_3\rangle = 0$. With respect to this isothermic parametrization, the mean curvature of F is given by

$$H = \frac{a''(x)a(x) + a'(x)^2}{4a(x)^2 a'(x)}.$$

Proposition 4.7. *For any THIMC surface of revolution with non-constant mean curvature, there exists an isothermic coordinate system (x, y) such that $H(x) = 1/x$ or $H(y) = 1/y$.*

Proposition 4.8. *Let $F : M \longrightarrow \mathbf{E}^3_1$ be a timelike surface of revolution with spacelike axis and Euclidean profile curve parametrized as in Example 4.3 with harmonic inverse mean curvature $1/H = y$ and $a = 2$. Then there exists a real valued function ϕ such that*

$$e^{\omega(y)} = \frac{y^2}{4}\{\phi'(y) + 2\sin\phi(y)\}^2, \quad c(y) = -\frac{y^2}{4}\{\phi'(y)^2 - 4\sin^2\phi(y)\}.$$

Furthermore ϕ is a solution to the third Painlevé equation of trigonometric form:

$$y\{\phi''(y) - 2\sin(2\phi(y))\} + \phi'(y) + 2\sin\phi(y) = 0. \tag{4.2}$$

Proposition 4.9. *Let $F : M \longrightarrow \mathbf{E}^3_1$ be a timelike surface of revolution with spacelike axis and timelike profile curve parametrized as in Example 4.4 with harmonic inverse mean curvature $1/H = x$ and $a = 2$. Then there exists a real valued function ϕ such that*

$$e^{\omega(x)} = \frac{x^2}{4}\left\{\phi'(x) - 2\sinh\phi(x)\right\}^2, \quad c(x) = \frac{x^2}{4}\{\phi'(x)^2 - 4\sinh^2\phi(x)\}.$$

Furthermore ϕ is a solution to the third Painlevé equation of hyperbolic form:

$$x\left\{\phi''(x) - 2\sinh(2\phi(x))\right\} + \phi'(x) \mp 2\sinh\phi(x) = 0. \tag{4.3}$$

Proposition 4.10. *Let $F : M \longrightarrow \mathbf{E}^3_1$ be a timelike surface of revolution with timelike axis parametrized as in Example 4.5 with harmonic inverse mean curvature $1/H = x$ and $a = 2$. Then there exists a real valued function ϕ such that*

$$e^{\omega(x)} = \frac{x^2}{4}\left\{\phi'(x) + 2\cosh\phi(x)\right\}^2, \quad c(x) = \frac{x^2}{4}\{\phi'(x)^2 - 4\cosh^2\phi(x)\}.$$

Furthermore ϕ is a solution to the ordinary differential equation

$$x\left\{\phi''(x) - 2\sinh(2\phi(x))\right\} - \phi'(x) - 2\cosh\phi(x) = 0. \tag{4.4}$$

Remark 4.11. The ordinary differential equations (4.2) and (4.3) are related to the third Painlevé equation. More precisely let $w = w(x)$ be a solution to the third Painlevé equation:

$$w'' - \frac{1}{w}(w')^2 + \frac{w'}{x} - \frac{\alpha w^2 - \alpha}{x} - \frac{\gamma}{w^3} - \frac{\gamma}{w} = 0 \tag{P_{III}}$$

with unit modulus, *i.e.*, $w(x) = e^{\sqrt{-1}\psi(x)}$ for some real valued function $\psi(x)$. Then (P_{III}) is equivalent to the following ordinary differential equation:

$$x\left\{\psi''(x) + 2\gamma\sin(2\psi(x))\right\} + \psi'(x) + 2\alpha\sin\psi(x) = 0.$$

If we choose $\alpha = \gamma = 1$ then we get (4.2). In addition, if we complexify the above third Painlevé equation in trigonometric form and put $\psi = \sqrt{-1}\phi$ then ϕ satisfies

$$x\left\{\phi''(x) + 2\gamma\sinh(2\phi(x))\right\} + \phi'(x) + 2\alpha\sinh\phi(x) = 0.$$

If we choose $\alpha = \mp 1$ and $\gamma = -1$ then we get (4.3).

Timelike HIMC surfaces of revolution with null axis can be classified as follows:

Proposition 4.12. *Let $F : M \longrightarrow \mathbf{E}^3_1$ be a timelike surface of revolution with null axis parametrized as in Example 4.6 with harmonic inverse mean curvature $1/H = 4x$. Then the function $a(x)$ is a solution to the following ordinary differential equation:*

$$x\left\{a''(x)a(x) + a'(x)^2\right\} = a^2(x)a'(x). \tag{4.5}$$

This ordinary differential equation can be explicitly solved by quadratures. In fact the solution $a(x)$ is given as follows.

$$12\int \frac{a}{2a^3 + 3a^2 + c_1}da = 2\log|x| + c_2, \;\; c_1, c_2 \in \mathbf{R}. \tag{4.6}$$

Next, to study timelike surfaces with imaginary principal curvatures we shall introduce the notion of *anti-isothermic surface.*

Definition 4.13. Let $F : M \to \mathfrak{M}^3_\nu(c)$ be a timelike surface. A null coordinate system (u, v) is said to be *anti-isothermic* if its Hopf differentials take the following form:

$$Q(u,v) = \frac{1}{2}\mathfrak{q}(u,v)\varrho(u), \quad R(u,v) = -\frac{1}{2}\mathfrak{q}(u,v)\sigma(v), \quad \varrho > 0, \sigma > 0. \tag{4.7}$$

In addition (M, F) is said to be *anti-isothermic* if there exists an anti-isothermic coordinate system around any point of M.

Note that anti-isothermic property implies that M has imaginary principal curvatures. In $\mathfrak{M}^3_1(c)$, $c \geq 0$, anti-isothermic surfaces have non negative Gaussian curvature. (See (2.4).)

The following result plays a fundamental role in the study of isothermic timelike surfaces and anti-isothermic timelike surfaces in $\mathbf{E}^3_1$. We write these alternatives together as $\pm$ isothermic.

Proposition 4.14. *Let (M, F) be a $\pm$ isothermic timelike surface in $\mathbf{E}^3_1$ and $(\mathfrak{D}; u, v)$ a simply connected $\pm$ isothermic coordinate region so that*

$$I = e^\omega du dv, \quad Q = \frac{1}{2}\mathfrak{q}(u,v)\varrho(u), \quad R = \pm\frac{1}{2}\mathfrak{q}(u,v)\sigma(v), \quad \varrho > 0, \sigma > 0.$$

Then the formulas

$$F^*_u = e^{-\omega}\varrho F_v,\ F^*_v = \pm e^{-\omega}\sigma F_u,\ N^* = N \tag{4.8}$$

define a $\pm$ isothermic timelike immersion $F^* : \mathfrak{D} \to \mathbf{E}^3_1$. The conformal structure of $\mathfrak{D}$ induced by F^* is anti-conformal to the original conformal structure determined by I. The fundamental quantities of F^* are given as follows:

(4.9)

$$I^* = \pm e^{\omega^*} dudv = \pm e^{-\omega}\varrho\sigma dudv,\quad H^* = \mathfrak{q},\quad Q^* = \varrho H/2,\quad R^* = \pm\sigma H/2.$$

The new immersion F^* is called the Christoffel transform of F or dual of F.

In particular for $\pm$ isothermic THIMC surfaces, we have the following.

Corollary 4.15. *Every $\pm$ isothermic THIMC surface in $\mathbf{E}^3_1$ is dual to a $\pm$ timelike Bonnet surface in $\mathbf{E}^3_1$ and vice versa.*

5 The Hazzidakis equation

In this section we shall investigate normal forms of the Gauss equation for THIMC surfaces.

5.1 Timelike surfaces with $\pm$ holomorphic inverse mean curvature

Let $F : M \to \mathbf{E}^3_1$ be a $\pm$ isothermic timelike surface with $\pm$ holomorphic inverse mean curvature.

Without loss of generality we may assume that $1/H = g(v)$. Take a $\pm$ isothermic coordinate system (u, v) such that $Q = \varepsilon R = \mathfrak{q}(u, v)/2$. Here ε denotes the signature $+$ or $-$. The Codazzi equations (C_0) become

$$\mathfrak{q} = \mathfrak{q}(u),\ e^{\omega} = -\frac{\varepsilon g^2 \mathfrak{q}_u}{g_v}. \tag{5.1}$$

Hence we get $\omega_{uv} = 0$ and hence M is flat by (2.3). On the other hand the Gauss equation (G_0) implies

$$e^{2\omega} = \frac{4\varepsilon Q^2}{H^2} = \frac{\varepsilon \mathfrak{q}^2}{H^2}. \tag{5.2}$$

Hence M is isothermic. Moreover (5.1) and (5.2) imply that $g^2\mathfrak{q}_u^2 = \mathfrak{q}^2 g_v^2$. Hence $g\mathfrak{q}_u = \pm\mathfrak{q}g_v$. Thus we have $g(v) = C_1 e^{\alpha v}$, $\mathfrak{q}(u) = C_2 e^{\mu\alpha u}$, $\mu = \pm 1$, $\alpha \in \mathbf{R}$, $C_1, C_2 \in \mathbf{R}^*$, $C_1C_2/\mu < 0$. These formulas show that timelike surfaces with $\pm$ holomorphic inverse mean curvature are the flat Bonnet surfaces with $\pm$holomorphic mean curvature described in [11, Theorem 3.1]. In particular the case $\alpha = 0$ corresponds to timelike CMC cylinders.

Proposition 5.1. *Let M be a timelike surface in $\mathbf{E}_1^3$ with $\pm$ holomorphic inverse mean curvature. If M is $\pm$ isothermic then M is a flat isothermic timelike Bonnet surface.*

The notion of $\pm$ isothermic surface can be generalized to the notion of "(ε, ϑ)-isothermic surface" in the following way:

Definition 5.2. A timelike surface (M, F) is *(ε, ϑ)-isothermic* if there exists a local null coordinate system (u, v) around any point of M such that the Hopf differentials Q and R have the following form:
(5.3)

$$Q(u,v) = \frac{1}{2}(\mathfrak{q}(u,v)+\vartheta)\varrho(u), \quad R(u,v) = \frac{\varepsilon}{2}(\mathfrak{q}(u,v)-\vartheta)\sigma(v), \quad \varrho > 0, \ \sigma > 0.$$

Here $\mathfrak{q}$ is a real smooth function, ϱ and σ are $\pm$ Lorentz-holomorphic functions and ϑ is a real constant. If $\varepsilon = +$ [resp. $\varepsilon = -$], then we call M a *ϑ–isothermic* surface [resp. an anti-*ϑ–isothermic* surface].

Note that the constant ϑ has no global meaning, in fact, ϑ depends on the choice of (u, v).

Proposition 5.3. *Let M be an (ε, ϑ)–isothermic timelike surface with $\vartheta \neq 0$. Then M is $\pm$ isothermic if and only if M is a timelike Bonnet surface.*

Proposition 5.1 can be generalized as follows:

Theorem 5.4. *Let M be an (ε, ϑ)–isothermic timelike surface with Lorentz anti-holomorphic inverse mean curvature $1/H = g(v)$, $\vartheta \neq 0$. Then M is flat and has real distinct principal curvatures.*

(1) *If M is ϑ–isothermic then $g(v) = Ce^{\alpha v}$, $\mathfrak{q}(u) = \vartheta\cosh(\alpha u + \beta)$,*

(2) *If M is anti-ϑ-isothermic then $g(v) = Ce^{\alpha v}$, $\mathfrak{q}(u) = \vartheta\sin(\alpha u + \beta)$, $C \in \mathbf{R}^*$, $\alpha, \beta \in \mathbf{R}$.*

For any (ε, ϑ)–isothermic THIMC surface in $\mathbf{E}^3_1$, we can consider the *dual* Bonnet surface in H^3_1 or S^3_1.

Proposition 5.5. *Let (M, F) be an (ε, ϑ)-isothermic timelike surface in $\mathbf{E}^3_1$ and $(\mathfrak{D}; u, v)$ a simply connected (ε, ϑ)-isothermic coordinate region such that the Hopf differentials take the following forms:*

$$Q = \frac{1}{2}(\mathfrak{q}(u, v) + \vartheta), \ R = \frac{\varepsilon}{2}(\mathfrak{q}(u, v) - \vartheta).$$

Then

(1) *if $\varepsilon = +$, there exists a timelike immersion*

$$F^* : \mathfrak{D} \longrightarrow \begin{cases} H^3_1(\frac{1}{|\vartheta|}), & \vartheta \neq 0, \\ \mathbf{E}^3_1, & \vartheta = 0. \end{cases}$$

(2) *if $\varepsilon = -$, there exists a timelike immersion*

$$F^* : \mathfrak{D} \longrightarrow \begin{cases} S^3_1(\frac{1}{|\vartheta|}), & \vartheta \neq 0, \\ \mathbf{E}^3_1, & \vartheta = 0. \end{cases}$$

The timelike immersion F^* is called a dual surface of F. In particular if F is a THIMC surface then F^* is a timelike Bonnet surface and vice versa.

Remark. In Section 6, we shall prove a Lawson correspondence between THIMC surfaces in Lorentzian space forms. Combining the duality in the preceding proposition and Lawson correspondence, we get a duality between THIMC surfaces and timelike Bonnet surfaces in H^3_1.

5.2 Timelike surfaces with non ± holomorphic inverse mean curvature

Let $F : M \to \mathbf{E}^3_1$ be a THIMC surface parametrized by a null coordinate system $(\bar{u}, \bar{v})$. Since the reciprocal of the mean curvature of (M, F) is harmonic, the mean curvature H can be written as

$$\frac{1}{H} = f(\bar{u}) + g(\bar{v}). \tag{5.4}$$

Inserting (5.4) in the Codazzi equation (C_0) we get

$$f_{\bar{u}} R_{\bar{u}} = g_{\bar{v}} Q_{\bar{v}}. \tag{5.5}$$

Inserting this formula into the Gauss equation (G_0) we get

$$f_{\bar{u}}\left(\frac{Q_{\bar{u}\bar{v}}}{Q_{\bar{v}}}\right)_{\bar{v}}-Q_{\bar{v}}=\frac{f_{\bar{u}}g_{\bar{v}}}{(f+g)^2}\left(2f_{\bar{u}}-\frac{QR}{R_{\bar{u}}}\right). \tag{5.6}$$

Thanks to (5.5), the equation (5.6) is equivalent to

$$g_{\bar{v}}\left(\frac{R_{\bar{u}\bar{v}}}{R_{\bar{u}}}\right)_{\bar{u}}-R_{\bar{u}}=\frac{f_{\bar{u}}g_{\bar{v}}}{(f+g)^2}\left(2g_{\bar{v}}-\frac{QR}{Q_{\bar{v}}}\right). \tag{5.7}$$

As long as $f_{\bar{u}}\neq 0$, $g_{\bar{v}}\neq 0$, we may assume $\xi:=f(\bar{u})$, $\eta:=g(\bar{v})$ is a local null coordinate system. With respect to (ξ,η), the Gauss-Codazzi equations (G_0) and (C_0) become:

$$\left(\frac{Q_{\xi\eta}}{Q_\eta}\right)_\eta-Q_\eta=\frac{1}{(\xi+\eta)^2}\left(2-\frac{QR}{R_\xi}\right),\quad Q_\eta=R_\xi. \tag{5.8}$$

We should remark that every solution $\{Q,R\}$ to

$$2-\frac{QR}{R_\xi}=0 \tag{5.9}$$

solves (5.8). Let $\{Q,R\}$ be a solution to (5.9). Then by the Codazzi equations (C_0) and the formula $1/H=\xi+\eta$, we get

$$e^{\omega(\xi,\eta)}=-2(\xi+\eta)^2R_\xi=-(\xi+\eta)^2Q(\xi,\eta)R(\xi,\eta).$$

Hence the solution $\{Q,R\}$ to (5.9) defines a THIMC surface if and only if $QR<0$. Such THIMC surfaces have no Euclidean counterparts. (Compare with the Euclidean case [5, p. 203].)

Hereafter we restrict our attention to (ε,ϑ)–isothermic THIMC surfaces. Namely we assume

$$Q(\xi,\eta)=\frac{1}{2}(\mathfrak{q}(\xi,\eta)+\vartheta)\varrho(\xi),\quad R(\xi,\eta)=\frac{\varepsilon}{2}(\mathfrak{q}(\xi,\eta)-\vartheta)\sigma(\eta),\quad \varrho>0,\sigma>0. \tag{5.10}$$

To adapt our computations to [5] and [10], and avoid unnecessary 1/2's, we shall use the following convention:

$$q(u,v):=\frac{\varepsilon}{2}\mathfrak{q}(u,v),\quad \theta:=\frac{1}{2}\vartheta.$$

We call (ξ,η) an (ε,θ)*–isothermic coordinate system.*

Inserting (5.10) into (5.5), we get

$$\varepsilon\,\sigma(\eta)q_\xi(\xi,\eta)=\varrho(\xi)q_\eta(\xi,\eta). \tag{5.11}$$

Now we introduce a new null coordinate system (u,v) by

$$u=\int\varrho(\xi)d\xi,\quad v=\int\sigma(\eta)d\eta.$$

Then the formula (5.11) implies that q depends only on $t:=\varepsilon u+v$.

We consider two cases: (1) $2-QR/R_\xi=0$, (2) $2-QR/R_\xi\neq 0$.

Case 1: $2-QR/R_\xi=0$. In this case, the Hopf differentials are given by

$$Q(\xi,\eta)=\varrho(\xi)(\varepsilon q(t)+\theta),\quad R(\xi,\eta)=\sigma(\eta)(q(t)-\varepsilon\theta).$$

$$q(t)=\begin{cases}-\theta\tanh(\theta t/2), & \theta\neq 0\\ -2/t, & \theta=0\end{cases} \tag{5.12}$$

Inserting (5.12) into (C_0), we have

$$e^{\omega(u,v)}=\begin{cases}\varepsilon\theta^2\left(\xi(u)+\eta(v)\right)^2/\cosh^2(\theta t/2), & \theta\neq 0,\\ -4\varepsilon\left(\xi(u)+\eta(v)\right)^2/t^2, & \theta=0.\end{cases}$$

These formulas imply that $\varepsilon=+$ for $\theta\neq 0$ and $\varepsilon=-$ for $\theta=0$.

Proposition 5.7. *Let (M,F) be an (ε,θ)–isothermic THIMC surface in $\mathbf{E}^3_1$ with (ε,θ)–isothermic coordinate (ξ,η) of the form (5.10) and (5.11). If $2R_\xi-QR=0$, then $\varepsilon=+$ for $\theta\neq 0$ and $\varepsilon=-$ for $\theta=0$. The fundamental quantities of (M,F) are given by*

$$Q(u,v)=\frac{\varepsilon q(t)+\theta}{\varrho(\xi(u))},\ R(u,v)=\frac{q(t)-\varepsilon\theta}{\sigma(\eta(v))},\ q(t)=\begin{cases}-\theta\tanh(\frac{\theta t}{2}), & \theta\neq 0\\ -2/t, & \theta=0,\end{cases}$$

$$H(u,v)=\frac{1}{\xi(u)+\eta(v)},\quad I=\begin{cases}\theta^2(\xi(u)+\eta(v))^2dudv/\cosh^2(\frac{\theta t}{2}), & \theta\neq 0,\\ 4(\xi(u)+\eta(v))^2\,dudv/t^2, & \theta=0.\end{cases}$$

The dual surface of (M,F) is given by the following formulas :

(1) If $\theta\neq 0$ then the dual surface F^* in $H^3_1(1/(2|\theta|))$ is defined by the data

$$e^{\omega^*(u,v)}=\frac{\cosh^2(\frac{\theta t}{2})}{\theta^2(\xi(u)+\eta(v))^2},$$

$$Q^*(u,v) = R^*(u,v) = \frac{1}{2(\xi(u)+\eta(v))}, \quad H^*(u,v) = -2\theta\tanh(\frac{\theta t}{2}).$$

The dual surface F^* is an isothermic timelike Bonnet surface in $H_1^3(1/(2|\theta|))$.

(2) If $\theta = 0$ then the dual surface F^* in $\mathbf{E}_1^3$ is defined by the data:

$$e^{\omega^*(u,v)} = \frac{t^2}{4(\xi(u)+\eta(v))^2},$$

$$Q^*(u,v) = -R^*(u,v) = \frac{1}{2(\xi(u)+\eta(v))}, \quad H^*(u,v) = \frac{-4}{t}.$$

The dual surface F^* is an anti-isothermic timelike Bonnet surface in $\mathbf{E}_1^3$.

We call a THIMC surface (M,F) *generic* if (M,F) does not correspond to a solution of $2R_\xi - QR = 0$.

Case 2: $2R_\xi - QR \neq 0$. In this case, inserting $Q(\xi,\eta) = \varrho(\xi)(\varepsilon q(t)+\theta)$, $R(\xi,\eta) = \sigma(\eta)(q(t)-\varepsilon\theta)$, into (5.8) and by the assumption $2R_\xi - QR \neq 0$, we can define the following function

$$\mathcal{S}(t) = \frac{1}{\varrho(\xi(u))\sigma(\eta(v))(\xi(u)+\eta(v))^2}. \tag{5.13}$$

The following theorem is proved as in [5] and [10].

Theorem 5.8. *There exist three classes– A, B and C– of associated families of generic (ε,θ)–isothermic THIMC surfaces in $\mathbf{E}_1^3$. The immersion function of each family is given by the Sym formula* (3.8) *and* (3.9) *in Proposition* 3.4, *where the data (ω,Q,R,H) in* (3.8) *are determined by*

$$e^{\omega(u,v)} = -2\varepsilon q'(t)(\xi(u)+\eta(v))^2,$$

$$Q(u,v) = \frac{\varepsilon q(t)+\theta}{\varrho(\xi(u))}, \quad R(u,v) = \frac{q(t)-\varepsilon\theta}{\sigma(\eta(v))}, \quad H(u,v) = \frac{1}{\xi(u)+\eta(v)}.$$

Here $q(t)$ is a solution to the generalized Hazzidakis equation:

$$\left(\frac{q''(t)}{q'(t)}\right)' - q'(t) = \mathcal{S}(t)\left(2 - \frac{q^2(t)-\theta^2}{q'(t)}\right), \quad -\varepsilon q'(t) > 0. \tag{$\bigstar^{-\varepsilon}_{-\theta^2}$}$$

Here the coefficient function $\mathcal{S}(t)$ in the generalized Hazzidakis equation is given by

Family	*Coefficient*
A-family	$\mathcal{S}(t) = 1/\sin^2(2t)$
B-family	$\mathcal{S}(t) = 1/\sinh^2(2t)$
C-family	$\mathcal{S}(t) = 1/t^2$

Any generic (ε, θ)-isothermic THIMC surface belongs to one of these families A, B or C.

Via the duality between ± 1-isothermic THIMC surfaces in $\mathbf{E}^3_1$ and isothermic timelike Bonnet surfaces in H^3_1, the generalized Hazzidakis equation $(\bigstar^-_{-1})$ coincides with that for isothermic timelike Bonnet surfaces in H^3_1 obtained in [11, Theorem 6.1].

Moreover the generalized Hazzidakis equation $(\bigstar^-_{-1})$ coincides with that for Bonnet surfaces in hyperbolic 3-space H^3. (See [4, Theorem 3.3.1] and [20].) Thus $(\bigstar^-_{-1})$ for A or B-family [respectively, C-family] is solved by the Painlevé transcendents P_{VI} [respectively, P_{V}]. See [4, Theorem 3.5.1, 3.5.2]. Hence Bonnet surfaces in H^3 of non-Willmore type, $(\varepsilon, \pm 1)$-isothermic THIMC surfaces in $\mathbf{E}^3_1$ and (generic) timelike Bonnet surfaces in H^3_1 are derived from P_{V} and P_{VI}. Note that $(\bigstar^+_{\theta^2})$ coincides with generalized Hazzidakis equation for θ-isothermic spacelike HIMC surfaces in $\mathbf{E}^3_1$ (and hence spacelike Bonnet surfaces in H^3_1) [10].

6 Timelike HIMC surfaces in $\mathfrak{M}^3_1(c)$

In this section we shall generalize the notion of THIMC surface in Minkowski 3-space to that of $\mathfrak{M}^3_1(c)$.

Proposition 6.1. *Let $I[c]$ be a 1-dimensional Riemannian manifold defined by*

$$I[c] = \begin{cases} (\mathbf{R}, g[c]) & c = 0,\ 1, \\ (\mathbf{R} \setminus \{\pm 1\}, g[c]) & c = -1, \end{cases}, \quad g[c] = \frac{dt^2}{(1+ct^2)^2}.$$

Let $\varphi : M \to I[c]$ be a smooth map from a Lorentz surface M. Then φ is a (Lorentzian) harmonic map if and only if

$$\frac{\partial^2 \varphi}{\partial u \partial v} - \frac{2c\varphi}{1+c\varphi^2}\frac{\partial \varphi}{\partial u}\frac{\partial \varphi}{\partial v} = 0 \tag{6.1}$$

with respect to any (and hence all) null coordinate systems (u,v).

The harmonic map equation (6.1) may be considered as a nonlinear generalization of the classical linear wave equation $\varphi_{uv}=0$. As is well known, the classical linear wave equation can be solved by the *d'Alembert formula.* The following may be regarded as a nonlinear d'Alembert formula for (6.1).

Proposition 6.2. *The harmonic map equation* (6.1) *can be solved as follows*:

$$\varphi(u,v)=\begin{cases} f(u)+g(v), & c=0,\\ \frac{f(u)+g(v)}{1-cf(u)g(v)} \text{ or } \frac{1-cf(u)g(v)}{f(u)+g(v)}, & c=\pm 1. \end{cases}$$

The following definition is a generalization of that in Section 3.

Definition 6.3. Let $F : M \to \mathfrak{M}^3_1(c)$ be a timelike surface. Then M is said to be a *timelike surface with harmonic inverse mean curvature* (THIMC surface) if $1/H$ is a harmonic map into $I[c]$.

Hereafter we assume that M is *simply connected.* We denote by $\mathcal{C}_H$ the *moduli space* of conformal immersions of M into $\mathfrak{M}^3_1(c)$ with prescribed mean curvature H:

$\mathcal{C}_H=\{F : M \to \mathfrak{M}^3_1(c) \mid$ a conformal timelike immersion with mean curvature $H\ \}/\mathfrak{I}_0(c)$.

Here $\mathfrak{I}_0(c)$ is the identity component of the full isometry group of $\mathfrak{M}^3_1(c)$. Then we can deduce (by the fundamental theorem of surface theory) that

$$\mathcal{C}_H \cong \{(\omega,Q,R)|\text{ a solution to } (\mathrm{G}_c) \text{ and } (\mathrm{C}_c) \text{ with mean curvature } H\ \}.$$

Theorem 6.4. (Generalized Lawson Correspondences) *Let M be a simply connected Lorentz surface, f a holomorphic function and g an anti-holomorphic function on M. We define a function H_c by $H_c := (1-cfg)/(f+g)$. Then the three moduli spaces $\mathcal{C}_{H_0}$, $\mathcal{C}_{H_1}$, $\mathcal{C}_{H_{-1}}$ are mutually isomorphic.*

Proof. Let (ω,Q,R,H_c) be a solution of (G_c) and (C_c) for $c=\pm 1$.

Then $(\tilde{\omega}, \tilde{Q}, \tilde{R}, H_0)$ defined by

$$e^{\tilde{\omega}} := (1+cf^2)(1+cg^2)\, e^{\omega}, \quad \tilde{Q} := (1+cf^2)Q, \quad \tilde{R} := (1+cg^2)R$$

is a solution to (G_c) and (C_c). Note that in the case $c=-1$, the function $(1+cf^2)(1+cg^2)$ is positive if and only if $H^2_{-1} > 1$. ■

Theorem 6.4 may be considered as a generalization of the so-called *Lawson correspondences* for timelike CMC surfaces.

Remark. In the Riemannian case, (non CMC) HIMC surfaces in H^3 have Lawson correspondents if and only if $H^2 > 1$. On the other hand, in the spacelike case, (non CMC) spacelike HIMC surfaces in S^3_1 have Lawson correspondents if and only if $H^2 > 1$. See [7], [10].

Using the Lawson correspondences described above, we can give an immersion formulas for THIMC surfaces in $\mathfrak{M}^3_1(c)$, $c = \pm 1$. Before doing so, we point out the following invariance of (6.1). Let φ be a solution to (6.1) of the form:

$$\varphi(u,v) = \frac{f(u)+g(v)}{1-cf(u)g(v)}.$$

Then the replacements $f \longmapsto 2\tau f$, $g \longmapsto 2\tau g$, $\tau \in \mathbf{R}^*$ produce a new solution to (6.1). More precisely, the function $\varphi[\tau]$ defined by

$$\varphi[\tau](u,v) = \frac{2\tau(f(u)+g(v))}{1-4c\tau^2 f(u)g(v)}$$

is still a solution to (6.1).

Let Φ_λ be a solution of the zero curvature equations (3.8) with variable spectral parameter λ. To describe immersion formulas we shall use the following notational convention.

$$\Phi[\tau] := \Phi_\lambda, \ \lambda = (1-2\tau g)/(1+2\tau f), \quad \tau \in \mathbf{R}.$$

Since the zero curvature equation (3.8) is completely integrable, (3.8) has also solutions for all $\tau \in \mathbf{C}$.

Direct computations similar to those in [1], [7] and [10] show the following.

Theorem 6.5. (Immersion formulas) *Let $\Phi[\tau] : M \times \mathbf{C} \to G^{\mathbf{C}}$ be a complexified solution to* (3.8). *Then the following hold.*

($c = 0$) *For every* $\tau \in \mathbf{R}$, $F^{(0)}(\tau) := -\frac{\partial}{\partial \tau}\Phi[\tau] \cdot \Phi[\tau]^{-1}$, $\tau \in \mathbf{R}$ *describes a THIMC surface in* $\mathbf{E}^3_1$ *given in* Proposition 3.4.

($c = -1$) *For any* $\tau \in \mathbf{R}^*$, $F^{(-1)}(\tau) := p_H(\Phi[\tau], \Phi[-\tau])$ *is a THIMC surface in* H^3_1 *with unit normal vector field* $N = -\mu_H(\Phi[\tau], \Phi[-\tau])\mathbf{k}'$.

($c = 1$) *Let* $\Phi[\sqrt{-1}\tau]$, $\tau \in \mathbf{R}$ *be a complexified solution to* (3.8). *Then for every* $\tau \in \mathbf{R}$ $F^{(1)}(\tau) := p_S(\Phi[\sqrt{-1}\tau])$ *is a THIMC surface in* S^3_1 *with unit normal vector field* $N = -\mu_S(\Phi[\sqrt{-1}\tau])\mathbf{j}'$.

The first fundamental form of $F^{(c)}$, $c = \pm 1$ *is*

$$I^{(c)}(\tau) = \frac{4\tau^2 e^{\omega}}{(1 + 4c\tau^2 f^2)(1 + 4c\tau^2 g^2)}.$$

The mean curvature of $F^{(c)}(\tau)$, $c = \pm 1$ *is*

$$H = \frac{1 - 4c\tau^2 fg}{2\tau(f + g)}.$$

Moreover the mean curvature of $F^{(-1)}(t)$ *satisfies* $H^2 > 1$. *In particular, for* $c = \pm 1$, $F^{(\pm 1)}(1/2)$ *is the Lawson correspondent of* $F = F^{(0)}$. *The conformal deformations of THIMC surfaces in* $\mathfrak{M}^3_1(c)$ *preserve* $K/(H^2 + c)$.

Remark. The conformal deformation of HIMC surfaces in Riemannian space forms [respectively, spacelike HIMC surfaces in Lorentzian space forms] preserves $K/(H^2 + c)$ [respectively, $K/(H^2 - c)$] Note that in the case $c = 0$, the constancy of $K/(H^2 - c)$ is equivalent to the constancy of the ratio of principal curvatures.

Computing the Gaussian curvature or $K/(H^2 + c)$, we have the following theorem.

Theorem 6.7. *Let* (M, F) *be an* (ε, θ)*-isothermic THIMC surface in* $\mathfrak{M}^3_1(c)$.

(1) *If* K *is constant then* $K = 0$ *or* c.

(2) *If* $K/(H^2 + c)$ *is constant then* (M, F) *is a flat timelike Bonnet surface.*

As an application of the Lawson correspondence above, one can classify $\pm$ isothermic flat timelike Bonnet surfaces in Lorentzian space forms. In fact, since the Lawson correspondence preserves the $\pm$ isothermic property or flatness, we have obtained the following ([11, Theorem 6.2]).

Theorem 6.8. *Flat simply connected $\pm$ isothermic timelike Bonnet surfaces in one Lorentzian 3-space form correspond to those in another Lorentzian 3-space form.*

In [11], timelike Bonnet surfaces in $\mathfrak{M}^3_1(c)$ with constant Gaussian curvature are classified.

Finally we consider THIMC surfaces in H^3_1 with mean curvature $H^2 < 1$. To investigate such surfaces, we use the following invariance of (6.1) with $c = \pm 1$. Let H be a solution of (6.1) of the form:

$$H(u,v) = \frac{f(u) + g(v)}{1 - cf(u)g(v)}.$$

Then for any $\tau \in \mathbf{R}^*$, the replacements $f \longmapsto \tau f, \quad g \longmapsto \tau^{-1} g$ produce a new solution of (6.1). Namely the function $H[\tau]$ defined by

$$H[\tau](u,v) = \frac{\tau f(u) + \tau^{-1} g(v)}{1 - cf(u)g(v)}$$

is also a solution of (6.1). Based on this deformation, we define two auxiliary functions (*variable spectral parameters*):

$$\lambda(u,\tau) := \frac{\tau(1 - cf(u)^2)}{\tau^2 - cf(u)^2}, \quad \nu(v,\tau) := \frac{\tau(1 - cg(v)^2)}{\tau^2 - cg(v)^2}.$$

Then we have the following.

Theorem 6.9. *Let $\Psi[\tau]$ be a solution to*

$$\frac{\partial}{\partial u}\Psi[\tau] = \Psi[\tau]U[\tau], \quad \frac{\partial}{\partial v}\Psi[\tau] = \Psi[\tau]V[\tau], \tag{6.2}$$

$$U[\tau] = \begin{pmatrix} -\frac{1}{4}\omega_u & -Qe^{-\omega/2} \\ \frac{1}{2}(H[\tau] + c)\lambda e^{\omega/2} & \frac{1}{4}\omega_u \end{pmatrix},$$

$$V[\tau] = \begin{pmatrix} \frac{1}{4}\omega_v & -\frac{1}{2}(H[\tau] - c)\nu e^{\omega/2} \\ Re^{-\omega/2} & -\frac{1}{4}\omega_v \end{pmatrix}.$$

Then for any $\tau \in \mathbf{R}^*$, $F^{(-1)}[\tau](u,v) := p_H(\Psi[\tau], \Psi[-\tau])$ *is a THIMC surface in* H^3_1 *with unit normal vector field* $N = -\mu_H(\Psi[\tau], \Psi[-\tau])\mathbf{k}'$ *and mean curvature* $H[\tau]$. *The first fundamental form of* $F[\tau]$ *is given by*

$$I^{(-1)}[\tau] = \frac{\tau^2(1-cf(u)^2)(1-cg(v)^2)e^{\omega}}{(\tau^2-cf(u)^2)(\tau^2-cg(v)^2)}dudv.$$

Since the Lax equation (6.2) with two variable spectral parameters λ and ν is completely integrable, (6.2) has solutions for all $\tau \in \mathbf{C}$. Such complexified solutions $\Psi[\tau]$ to (6.2) describe another kind of surface in S^3_1.

Theorem 6.10. *Let* $\Psi[\tau] : M \times \mathbf{C} \to G^{\mathbf{C}}$ *be a complexified solution to* (6.2). *Then for any* $\tau \in \mathbf{R}^*$, $F^{(1)}[\tau](u,v) = p_S(\Psi[\sqrt{-1}\tau])$ *is a timelike surface in* S^3_1 *with unit normal vector field* $N = \mu_S(\Psi[\sqrt{-1}\tau])\mathbf{j}'$ *and mean curvature* $(\tau f - \tau^{-1}g)/(1-cfg)$. *The first fundamental form of* $F^{(1)}[\tau]$ *is given by*

$$I^{(1)}[\tau] = \frac{\tau^2(1-cf(u)^2)(1-cg(v)^2)e^{\omega}}{(\tau^2+cf(u)^2)(\tau^2+cg(v)^2)}dudv.$$

The inverse mean curvature of $F^{(1)}[\tau]$ in Theorem 6.10 is a harmonic map into $I[-1]$.

References

[1] A. I. Bobenko, Constant mean curvature surfaces and integrable equations, Russian Math. Surveys, **46** (1991), 1–45.

[2] A. I. Bobenko, Surfaces in terms of 2 by 2 matrices. Old and new integrable cases, In: Harmonic Maps and Integrable Systems, (eds. A. P. Fordy and J. C. Wood), Aspects of Math., **E 23**, Vieweg, Braunschweig.

[3] A. I. Bobenko and U. Eitner, Bonnet surfaces and Painlevé equations, J. Reine Angew. Math., **499** (1998), 47–79.

[4] A. I. Bobenko and U. Eitner, Painlevé Equations in Differential Geometry of Surfaces, Lecture Notes in Math., **1753** (2000), Springer-Verlag.

[5] A. I. Bobenko, U. Eitner and A. Kitaev, Surfaces with harmonic inverse mean curvature and Painlevé equations, Geom. Dedicata, **68** (1998), 187–227.

[6] S. P. Burtsev, V. E. Zakharov and A. V. Mikhairov, Inverse scattering method with variable spectral parameter, Theoret. and Math. Phys., **70** (1987), 227–240.

[7] A. Fujioka, Surfaces with harmonic inverse mean curvature in space forms, Proc. Amer. Math. Soc., **127** (1999), 3021–3025.
[8] A. Fujioka and J. Inoguchi, Bonnet surfaces with constant curvature, Results Math., **33** (1998), 288–293.
[9] A. Fujioka and J. Inoguchi, On some generalisations of constant mean curvature surfaces, Lobachevskii J. Math., **3** (1999), 73–95, `http:ljm.ksu.ru/vol3/fujioka.htm`.
[10] A. Fujioka and J. Inoguchi, Spacelike surfaces with harmonic inverse mean curvature, J. Math. Sci. Univ. Tokyo, **7** (2000), 657–698..
[11] A. Fujioka and J. Inoguchi, Timelike Bonnet surfaces in Lorentzian space forms, Diff. Geom. Appl., **18** (2003), 103–111.
[12] L. K. Graves, Codimension one isometric immersions between Lorentz spaces, Trans. Amer. Math. Soc., **252** (1979), 367–392.
[13] J. N. Hazzidakis, Biegung mit Erhaltung der Hauptkrümmungsradien, J. Reine Angew. Math., **117** (1897), 42–56.
[14] J. Inoguchi, Timelike surfaces of constant mean curvature in Minkowski 3-space, Tokyo J. Math., **21** (1998), 141–152.
[15] D. A. Korotkin, On some integrable cases in surface theory, J. Math. Sci. (New York), **94** (1999), 1177–1217.
[16] D. A. Korotkin and V. A. Reznik, Bianchi surfaces in $\mathbf{R}^3$ and deformation of hyperelliptic curves, Math. Notes, **52** (1992), 930–937.
[17] L. McNertney, One-parameter families of surfaces with constant curvature in Lorentz 3-space, Ph.D. thesis., Brown Univ., 1980.
[18] B. O'Neill, Semi–Riemannian Geometry with Applications to Relativity, Pure and Applied Math., **130**, Academic Press, 1983.
[19] W. K. Schief, Isothermic surfaces in spaces of arbitrary dimension: integrablity, discretization, and Bäcklund transformations–a discrete Calapso equation, Stud. Appl. Math., **116** (2001), 85–137.
[20] K. Voss, Bonnet surfaces in spaces of constant curvature In: First MSJ International Research Institute on Geometry and Global Analysis, Lecture Notes, **2** (1993), Tôhoku Univ., Sendai.
[21] T. Weinstein, An Introduction to Lorentz Surfaces, de Gruyter Exp. Math., Walter de Gruyter, Berlin.

Atsushi Fujioka
Department of Mathematics
Faculty of Science
Kanazawa University
Kakuma-machi, Kanazawa
Ishikawa 920–1192
JAPAN

Current address:
Graduate School of Economics
Hitotsubashi University
2-1 Naka, Kunitachi
Tokyo 186–8601
JAPAN

E-mail address:
fujioka@math.hit-u.ac.jp

Jun-ichi Inoguchi
Department of Applied Mathematics
Fukuoka University
Fukuoka 814-0180
JAPAN

Current address:
Department of Mathematics Education
Utsunomiya University
Utsunomiya 321-8505
JAPAN

E-mail address:
inoguchi@cc.utsunomiya-u.ac.jp

Advanced Studies in Pure Mathematics 51, 2008
Surveys on Geometry and Integrable Systems
pp. 143–161

Darboux transformations and generalized self-dual Yang-Mills flows

Gu Chaohao

§1. Introduction

The self-dual Yang-Mills equations are important partial differential equations in mathematical physics and have significant applications in mathematics [2, 3]. Besides the outstanding results related to 4-dimensional topology [5], it was known long ago that the self-dual Yang-Mills equations are an integrable system in the sense that they are the integrability condition of an overdetermined system of partial differential equations with a spectral parameter λ [12, 16]. Consequently, the rapidly developing theory of integrable systems can be applied to the self-dual Yang-Mills equations. Moreover, it has been found that many known soliton equations are reductions of the self-dual Yang-Mills equations [1, 17].

The self-dual Yang-Mills equations are equations in 4 dimensional space. It is interesting to generalize the equations to higher dimensions [13, 14, 18]. In [14], the generalization to $\mathbf{R}^{4n}$ is a simple case. In [8], more general integrable systems called generalized self-dual Yang-Mills flows (GSYMF) in $\mathbf{R}^{2n,1}$ were introduced. These are dynamical systems on the moduli space of the generalized self-dual Yang-Mills equations. The systems are very general in the sense that almost all known soliton equations of all dimensions are the reduction of these systems. It has been found that the Darboux transformation method is applicable to the GSYMF, giving new solutions explicitly. For the multi-soliton solutions, the phenomena of separation of solitons and confinement of solitons occurs.

Received February 17, 2001.
2000 *Mathematics Subject Classification.* Primary 35Q51; Secondary 35Q58, 70S15.

In the present paper we give an overview of the generalized self-dual Yang-Mills flows and the Darboux transformation method for obtaining explicit solutions.

In §2, we give a very brief sketch of some basic facts on Yang-Mills fields and self-duality. §3 is devoted to the derivation of the GSYMF from the Lax equations. In §4, the Darboux transformations (DT) for the GSYMF, as a method to obtain explicit GSYMF from a known one, are described. The algorithm is purely algebraic, depending on the choice of several arbitrary holomorphic functions. The most important cases of $U(N)$ and $SU(N)$ are emphasized. §5 is devoted to the explicit construction of solitons. The asymptotic behavior of solitons, which is a generalization of the usual KdV solitons, is much more complicated. In particular the multi-solitons may exhibit separation or confinement of single solitons as $|t| \to \infty$. In §6 we point out that the AKNS system in $\mathbf{R}^{n,1}$ can be a reduction of the GSYMF. Hence we can say the GSYMF may be the most general integrable system known to date, since almost all soliton equations for spaces of lower dimensions (≤ 4) are the reduction of the self-dual Yang-Mills equations.

§2. Yang-Mills fields, self-duality

First we recall some well-known basic facts on self-dual Yang-Mills fields. Let G be a matrix Lie group, g its Lie algebra and $\mathbf{R}^N$ an N-dimensional flat space $\{(x_1, \cdots, x_N)\}$ with a metric of the form $\mathrm{d}s^2 = \sum_{\alpha,\beta}^N \eta_{\alpha\beta}\,\mathrm{d}x^\alpha\,\mathrm{d}x^\beta$, where $\eta_{\alpha\alpha} = \pm 1$, $\eta_{\alpha\beta} = 0$ $(\alpha \neq \beta)$. The Yang-Mills potential is a 1-form valued in the Lie algebra g

$$\mathcal{A} = A_\alpha \mathrm{d}x^\alpha \tag{1}$$

and the field strength is

$$\mathcal{F} = \mathrm{d}\mathcal{A} + \mathcal{A} \wedge \mathcal{A} = \frac{1}{2} F_{\lambda\mu} \mathrm{d}x^\lambda \wedge \mathrm{d}x^\mu \tag{2}$$

where

$$F_{\lambda\mu} = \partial_\lambda A_\mu - \partial_\mu A_\lambda - [A_\mu, A_\lambda]. \tag{3}$$

Differentiating $\mathcal{F}$, we obtain the Bianchi identity

$$F_{(\lambda\mu|\nu)} = 0. \tag{4}$$

Here $F_{\lambda\mu|\nu}$ is the gauge derivative of $F_{\lambda\mu}$ defined by

$$F_{\lambda\mu|\nu} = \partial_\nu F_{\lambda\mu} + [A_\nu, F_{\lambda\mu}] \tag{5}$$

and $(\lambda\mu|\nu)$ is the symbol for the sum of cyclic permutations, i.e.

$$F_{(\lambda\mu|\nu)} = F_{\lambda\mu|\nu} + F_{\mu\nu|\lambda} + F_{\nu\lambda|\mu}.$$

The Yang-Mills equations are

$$\eta^{\mu\nu} F_{\lambda\mu|\nu} = 0. \tag{6}$$

A transformation from a potential $\mathcal{A}$ to $\mathcal{A}'$ defined by

$$\mathcal{A}' = \phi\mathcal{A}\phi^{-1} - (\mathrm{d}\phi)\phi^{-1} \qquad (\phi : \mathbf{R}^N \to G) \tag{7}$$

is called a gauge transformation; $\mathcal{A}'$ and $\mathcal{A}$ are called gauge equivalent. Gauge equivalent potentials are regarded as the same physical entity. The Yang-Mills theory is the most important theory in modern physics and has had great influence on mathematics. In mathematical terms, $\mathcal{A}$ is the connection form and $\mathcal{F}$ is the curvature of a G-bundle over $\mathbf{R}^N$. The case $N = 4$ is the most important, but the study of the higher dimensional case is also interesting [9, 13, 14, 18].

If $N = 4$ and $\mathbf{R}^4$ is Euclidean, a duality operator $*$ (Hodge operator) can be applied to $\mathcal{F}$:

$$(*\mathcal{F})_{\alpha\beta} = \frac{1}{2}\epsilon_{\alpha\beta\gamma\delta}\eta^{\gamma\lambda}\eta^{\delta\mu}\mathcal{F}_{\lambda\mu} \tag{8}$$

where

$$\epsilon_{\alpha\beta\gamma\delta} = \begin{cases} 1 & (\alpha,\beta,\gamma,\delta) \text{ is an even permutation of } (1,2,3,4), \\ -1 & (\alpha,\beta,\gamma,\delta) \text{ is an odd permutation of } (1,2,3,4), \end{cases} \tag{9}$$

A Yang-Mills potential is called self-dual (respectively, anti-self-dual) if

$$*\mathcal{F} = \mathcal{F} \qquad (\text{respectively, } *\mathcal{F} = -\mathcal{F}) \tag{10}$$

or

$$F_{12} = \pm F_{34}, \qquad F_{23} = \pm F_{14}, \qquad F_{31} = \pm F_{24}, \tag{10$'$}$$

where "+" (respectively, "−") is for the self-dual (respectively, anti-self-dual) case.

It is well-known that a self-dual (respectively, anti-self-dual) Yang-Mills potential satisfies the Yang-Mills equation. This can be seen from the definitions if the Bianchi identity is used.

The concept of self-duality can be extended to $\mathbf{R}^{2,2}$ where the metric can be written as

$$ds^2 = 2(dx_1\,dx_4 - dx_2\,dx_3). \tag{11}$$

In this case the self-dual condition is

$$F_{12} = 0, \qquad F_{34} = 0, \qquad F_{14} - F_{23} = 0. \tag{12}$$

Here we have changed the sign of $\epsilon_{\alpha\beta\gamma\delta}$, e.g. $\epsilon_{1234} = -1$ (or we consider the anti-self-dual case).

From $F_{12} = 0$ we see that the equations

$$\partial_1\phi = \phi A_1, \qquad \partial_2\phi = \phi A_2 \tag{13}$$

are completely integrable on any plane x_3 = constant, x_4 = constant. Hence we can find $\phi(x_1, x_2, x_3, x_4) \in G$ such that after the gauge transformation defined by ϕ,

$$A_1' = \phi A_1\phi^{-1} - (\partial_1\phi)\phi^{-1} = 0, \qquad A_2' = \phi A_2\phi^{-1} - (\partial_2\phi)\phi^{-1} = 0. \tag{14}$$

This special gauge is called the R-gauge [19].

Let $p_2 = x_4$, $p_1 = x_3$. Under an R-gauge, the equations (12) become

$$\frac{\partial A_{p_1}}{\partial p_2} - \frac{\partial A_{p_2}}{\partial p_1} - [A_{p_1}, A_{p_2}] = 0, \qquad \frac{\partial A_{p_1}}{\partial x_2} - \frac{\partial A_{p_2}}{\partial x_1} = 0 \tag{15}$$

and $A_{x_1} = A_{x_2} = 0$. Hence the self-dual Yang-Mills equations in $\mathbf{R}^{2,2}$ can be simplified to (15) in which A_{x_1} and A_{x_2} disappear.

§3. Generalized self-dual Yang-Mills fields and their flows

We consider the generalization of the self-dual Yang-Mills fields in the higher dimensional flat space $\mathbf{R}^{n,n}$.

Let $(x, p) = (x_1, \cdots, x_n; p_1, \cdots, p_n)$ be the coordinates of points in $\mathbf{R}^{n,n}$ $(n \geq 2)$ $(i = 1, 2, \cdots, n)$. We start with the Lax equations

$$L\Psi = \left(\frac{\partial}{\partial p_i} - \lambda\frac{\partial}{\partial x_i}\right)\Psi = -A_i\Psi. \tag{16}$$

Here Ψ is an $N \times N$ matrix in G, $A_i = A_i(x, p)$ take values in the Lie algebra g of G, and λ is the spectral parameter.

By calculation, we see that the integrability condition for (16) is

$$\frac{\partial A_i}{\partial p_j} - \frac{\partial A_j}{\partial p_i} - [A_i, A_j] = 0 \tag{17}$$

and

$$\frac{\partial A_i}{\partial x_j} - \frac{\partial A_j}{\partial x_i} = 0. \tag{18}$$

If $n = 2$, these are just the self-dual Yang-Mills equations in the R-gauge. Thus (17) and (18) may be considered as the generalized self-dual Yang-Mills equations with $A_{p_i} = A_i$, $A_{x_i} = 0$. The generalization was initiated by Takasaki [13] for even n. However, as an integrable system, it makes sense for odd n too [7].

From (17) we see that there is a matrix-valued function J such that

$$A_i = -\frac{\partial J}{\partial p_i} J^{-1}, \tag{19}$$

and (18) becomes

$$\frac{\partial}{\partial x_i}\left(\frac{\partial J}{\partial p_j} J^{-1}\right) - \frac{\partial}{\partial x_j}\left(\frac{\partial J}{\partial p_i} J^{-1}\right) = 0. \tag{20}$$

From (18) we see that there is a matrix-valued function K such that

$$A_i = \frac{\partial K}{\partial x_i} \tag{21}$$

and (17) becomes

$$\frac{\partial^2 K}{\partial p_j \partial x_i} - \frac{\partial^2 K}{\partial p_i \partial x_j} - \left[\frac{\partial K}{\partial x_i}, \frac{\partial K}{\partial x_j}\right] = 0. \tag{22}$$

Consequently (20) or (22) is an equivalent form of the generalized self-dual Yang-Mills equations.

The metric of $\mathbf{R}^{n,n}$ is defined as

$$\mathrm{d}s^2 = \sum_{k=0}^{\frac{n}{2}-1} (\mathrm{d}x_{2k+1}\,\mathrm{d}p_{2k+2} - \mathrm{d}x_{2k+2}\,\mathrm{d}p_{2k+1}) \ (n \text{ even}) \tag{23}$$

and

$$\mathrm{d}s^2 = \sum_{k=0}^{[\frac{n}{2}]-1} (\mathrm{d}x_{2k+1}\,\mathrm{d}p_{2k+2} - \mathrm{d}x_{2k+2}\,\mathrm{d}p_{2k+1}) + \mathrm{d}x_n\,\mathrm{d}p_n \ (n \text{ odd}). \tag{24}$$

As in the 4-dimensional case, it can be shown that:

Theorem 1. *If n is even, a solution of generalized self-dual Yang-Mills equations satisfies the Yang-Mills equations.*

If Ψ is a solution to (16) and $\det\Psi \neq 0$, it is called a Lax representation of the generalized self-dual Yang-Mills field.

Remark 1. *If we replace (16) by*

$$L_i\Psi = (-A_i + \lambda B_i)\Psi, \tag{25}$$

then the integrability condition is still given by the equations

$$F_{p_ip_j} = 0, \qquad F_{p_ix_j} - F_{p_jx_i} = 0, \qquad F_{x_ix_j} = 0. \tag{26}$$

Here $A_{x_i} = B_i$, $A_{p_i} = A_i$ are the gauge potentials in a general gauge.

In the general case, let ϕ be a function valued in G and $\Psi' = \phi\Psi$, then

$$\begin{aligned} L_i\Psi' &= (L_i\phi)\Psi + \phi(-A_i + \lambda B_i)\Psi \\ &= (L_i\phi)\phi^{-1}\Psi' + \phi(-A_i + \lambda B_i)\phi^{-1}\Psi' = -A_i'\Psi' + \lambda B_i'\Psi' \end{aligned}$$

with

$$\begin{aligned} A_i' &= \phi A_i\phi^{-1} - (\partial_{p_i}\phi)\phi^{-1}, \\ B_i' &= \phi B_i\phi^{-1} - (\partial_{x_i}\phi)\phi^{-1}. \end{aligned} \tag{27}$$

This is exactly a gauge transformation.

We turn to the flows of the generalized self-dual Yang-Mills equations under the R-gauge in $\mathbf{R}^{n,n}$.

Suppose the functions Ψ and A_i depend on (x, p) and a "time" t and an evolution equation for Ψ

$$\frac{\partial\Psi}{\partial t} = V\Psi = \sum_{a=0}^{m+q} V_a\lambda^{m-a}\Psi, \quad (q \geq 0), \tag{28}$$

is satisfied. Besides, (16) and (28) constitute an integrable system in $\mathbf{R}^{2n,1}$. Here the V_a's are independent of the spectral parameter λ and can be determined in the following way.

Considering the condition of integrability for (16) and (28) we obtain a system of equations for the V_a's:

$$(29)\qquad \begin{aligned} &\frac{\partial V_0}{\partial x_i} = 0, \\ &\frac{\partial V_a}{\partial x_i} = \frac{\partial V_{a-1}}{\partial p_i} + [A_i, V_{a-1}], \quad (a = 1, 2, \cdots, m), \end{aligned}$$

and

$$(30)\qquad \begin{aligned} &\frac{\partial V_{m+q}}{\partial p_i} = [V_{m+q}, A_i], \\ &\frac{\partial V_{a-1}}{\partial p_i} = \frac{\partial V_a}{\partial x_i} + [V_{a-1}, A_i], \quad (a = m+q, \cdots, m+2) \end{aligned}$$

and a system of evolution equations for the A_i's:

$$(31)\qquad \frac{\partial A_i}{\partial t} + \frac{\partial V_m}{\partial p_i} - \frac{\partial V_{m+1}}{\partial x_i} + [A_i, V_m] = 0.$$

Theorem 2. *If the A_i's satisfy the generalized self-dual Yang-Mills equations (17), (18), then (29) and (30) are completely integrable and V_a $(a = 0, 1, \cdots, m, m+1, \cdots, m+q)$ can be expressed as integro-differential expressions of the A_i's.*

Proof. From the first equation of (30) we have

$$(32)\qquad V_0 = V_0(p, t).$$

Assume that $V_1, V_2, \cdots, V_{a-1}$ $(a < m)$ are constructed already. By differentiating (29) with respect to x_j, we have

$$\frac{\partial^2 V_a}{\partial x_i \partial x_j} = \frac{\partial^2 V_a}{\partial x_j \partial x_i}$$

if (17) and (18) hold. Hence $V_a(x, p)$ can be obtained by integrating the right hand sides of (29) along any paths connecting $(0, p)$ and (x, p) in the plane $p_i =$ constant, i.e.

$$(33)\qquad V_a = \sum_i \int_{(0,p)}^{(x,p)} \left(\frac{\partial V_{a-1}}{\partial p_i} + [A_i, V_{a-1}] \right) \mathrm{d}x_i + V_a^0(p, t).$$

The integral is independent of the choice of paths and $V_0(p, t)$, $V_a^0(p, t)$ are arbitrary functions.

Let

$$V_a = JW_aJ^{-1}, \qquad a = m+q, \cdots, m+1. \tag{34}$$

Equation (30) becomes

$$W_{m+q} = W_{m+q}(x,t), \tag{35}$$

$$\frac{\partial W_{a-1}}{\partial p_i} = \frac{\partial W_a}{\partial x_i} + \left[J^{-1}\frac{\partial J}{\partial x_i}, W_a\right], \qquad (a = m+q, \cdots, m+2). \tag{36}$$

When (17) and (18) are satisfied, we have

$$W_{a-1} = \sum_i \int_{(x,0)}^{(x,p)} \left(\frac{\partial W_a}{\partial x_i} + \left[J^{-1}\frac{\partial J}{\partial x_i}, W_a\right]\right) \mathrm{d}p_i + W^0_{a-1}(x,t). \tag{37}$$

Here, the path of integration is any curve on the plane x_i = constant connecting $(x,0)$ and (x,p) and $W_{m+q}(x,t)$, $W^0_a(x,t)$ $(m < a < m+q)$ are arbitrary functions of integration. The theorem is proved.

Equations (31) can be considered as dynamical equations on the moduli space of generalized self-dual Yang-Mills equations.

§4. Darboux transformations

The Darboux transformation of GSYMF is actually a modified gauge transformation

$$\widetilde{\Psi} = S\Psi \tag{38}$$

of the solutions of Lax equations with

$$S = \lambda I + \alpha \tag{39}$$

such that $(L_i\widetilde{\Psi})\widetilde{\Psi}^{-1}$ is independent of λ. S is called a Darboux matrix. Here I is the unit matrix and α is to be determined.

We can write

$$L_i\widetilde{\Psi} = \left(\frac{\partial}{\partial p_i} - \lambda\frac{\partial}{\partial x_i}\right)\widetilde{\Psi} = -\widetilde{A}_i\widetilde{\Psi}, \tag{40}$$

$$\frac{\partial\widetilde{\Psi}}{\partial t} = \sum V'_a\lambda^{m-a}\widetilde{\Psi}. \tag{41}$$

Substituting (38) and (39) in (40), we obtain

$$(42)\quad L_i\widetilde{\Psi} = \left(\left(\frac{\partial}{\partial p_i} - \lambda\frac{\partial}{\partial x_i}\right)\alpha - (\lambda I + \alpha)A_i\right)\Psi = -\widetilde{A}_i(\lambda I + \alpha)\Psi.$$

Hence

$$(43)\qquad \widetilde{A}_i = A_i + \frac{\partial\alpha}{\partial x_i},$$

and

$$(44)\qquad \frac{\partial\alpha}{\partial p_i} = \alpha A_i - \widetilde{A}_i\alpha = [\alpha, A_i] - \frac{\partial\alpha}{\partial x_i}\alpha.$$

Similarly, from (41) we obtain

$$(45)\qquad \begin{aligned} &\widetilde{V}_0 = V_0,\\ &\widetilde{V}_a = V_a + \alpha V_{a-1} - \widetilde{V}_{a-1}\alpha \ (a = 1, 2, \cdots, m),\\ &\widetilde{V}_{m+q} = \alpha V_{m+q}\alpha^{-1},\\ &\widetilde{V}_{m+k} = \alpha V_{m+k}\alpha^{-1} + (V_{m+k+1} - \widetilde{V}_{m+k+1})\alpha^{-1} \ (k = q-1, \cdots, 1) \end{aligned}$$

and

$$(46)\qquad \frac{\partial\alpha}{\partial t} + \alpha V_m - \widetilde{V}_m\alpha + V_{m+1} - \widetilde{V}_{m+1} = 0.$$

Since the generalized self-dual Yang-Mills equations (17), (18) and the evolution equation (31) are consequences of the Lax equations (16) and (28) (provided that $\det\Psi \neq 0$), the $\widetilde{A}_i$'s are solutions to (17), (18) and (31) as well. We have

Theorem 3. *Let A_i be a solution to the generalized self-dual Yang-Mills equations (17), (18) and the evolution equations (31), and let Ψ be the Lax representation. If α is a solution to (44) and (46), then the*

$$(47)\qquad \widetilde{A}_i = A_i + \frac{\partial\alpha}{\partial x_i}$$

satisfy (17), (18) and (31), in which the V_a are replaced by the $\widetilde{V}_a$ in (45). Moreover,

$$\widetilde{\Psi} = (\lambda I + \alpha)\Psi$$

is the Lax representation of $\widetilde{A}_i$.

The transformation $(\Psi, A_i) \to (\widetilde{\Psi}, \widetilde{A}_i)$ defined by (38), (39) and (47) is called a Darboux transformation.

System (44) and (46) is nonlinear and very complicated. However, we are able to construct explicit solutions as follows:

Let λ_α $(\alpha = 1, 2, \cdots, N)$ be N numbers (at least two of them not equal, $\lambda_\alpha \neq 0$) and $h_\alpha(\lambda_\alpha, x, p)$ be N column vectors, satisfying (16) and (28) for $\lambda = \lambda_\alpha$ $(\alpha = 1, 2, \cdots, N)$, i.e.

$$\frac{\partial h_\alpha}{\partial p_i} = \lambda_\alpha \frac{\partial h_\alpha}{\partial x_i} - A_i h_\alpha, \quad \frac{\partial h_\alpha}{\partial t} = \sum_a V_a \lambda_\alpha^{m-a} h_\alpha. \tag{48}$$

Define

$$H = [h_1, h_2, \cdots, h_N]. \tag{49}$$

Here we require that $\det H \neq 0$.

Theorem 4. *The matrix α defined by*

$$\alpha = -H\Lambda H^{-1} \text{ with } \Lambda = \begin{pmatrix} \lambda_1 & & & \\ & \lambda_2 & & \\ & & \ddots & \\ & & & \lambda_N \end{pmatrix} \tag{50}$$

satisfies equations (44) and (46).

Proof. From (48) it is easily seen that

$$\frac{\partial H}{\partial p_i} = \frac{\partial H}{\partial x_i}\Lambda - A_i H. \tag{51}$$

Hence

$$\begin{aligned} \frac{\partial \alpha}{\partial p_i} &= -\frac{\partial H}{\partial x_i}\Lambda^2 H^{-1} + H\Lambda H^{-1}\frac{\partial H}{\partial x_i}\Lambda H^{-1} - A_i\alpha + \alpha A_i, \\ \frac{\partial \alpha}{\partial x_i} &= -\frac{\partial H}{\partial x_i}\Lambda H^{-1} + H\Lambda H^{-1}\frac{\partial H}{\partial x_i}H^{-1}. \end{aligned} \tag{52}$$

By using (50), the definition of α, it is easily seen that (44) is satisfied.

Similarly, we have

$$\frac{\partial H}{\partial t} = \sum_{a=0}^{m+q} V_a H \Lambda^{m-a}. \tag{53}$$

Hence

$$\begin{aligned}\frac{\partial\alpha}{\partial t} &= -\sum_{a=0}^{m+q} V_a H\Lambda^{m-a+1}H^{-1} + \sum_{a=0}^{m+q} H\Lambda H^{-1}V_a H\Lambda^{m-a}H^{-1} \\ &= \sum_{a=0}^{m+q} V_a(-\alpha)^{m-a}\alpha - \alpha\sum_{a=0}^{m+q} V_a(-\alpha)^{m-a}. \qquad (54)\end{aligned}$$

From (45) it can be seen that the right hand side of (54) is just $\widetilde{V}_m\alpha - \alpha V_m + \widetilde{V}_{m+1} - V_{m+1}$. Thus (46) is satisfied.

Remark 2. *The column solutions h_α in (48) can be obtained from $\Psi(\lambda)$ by setting $h_\alpha = \Psi(\lambda_\alpha)l_\alpha$, where l_α satisfies*

$$L_i l_\alpha = \frac{\partial l_a}{\partial p_i} - \lambda\frac{\partial l_\alpha}{\partial x_i} = 0,$$

i.e. $l_\alpha = l_\alpha(\lambda_\alpha p + x)$. It is easily seen that the l_α's are holomorphic functions of $\lambda_\alpha p + x$.

Remark 3. *As in many other cases, we can apply Darboux transformations successively to obtain an infinite sequence of solutions by the purely algebraic algorithm indicated in Theorem 4, provided that a seed solution (Ψ, A_i) is known.*

Remark 4. *It can be seen that $L_i(\Psi(\bar{\lambda})^*\Psi(\lambda)) = 0$, $\partial_t(\Psi(\bar{\lambda})^*\Psi(\lambda)) = 0$. If we take a suitable initial condition for $\Psi(\lambda)$, the $\Psi(\lambda)$ can be normalized so that $\Psi(\bar{\lambda})^*\Psi(\lambda) = I$.*

For the case of $U(N)$, we consider (x,p) as real coordinates of $\mathbf{R}^{n,n}$. The potential A_i's should take values in the Lie algebra $u(N)$, i.e.

$$A_i^* + A_i = 0. \qquad (55)$$

Moreover, we can choose the functions of integration $V_0(p)$, $V_a^0(p)$ $(a = 1,2,\cdots,m)$ such that V_0, V_1, $\cdots$, V_m take values in $u(N)$ too. In this case we can apply the Darboux transformation as before. However, we should consider the additional requirement

$$\widetilde{A}_i^* + \widetilde{A}_i = 0 \qquad (56)$$

which is equivalent to

$$\frac{\partial\alpha^*}{\partial x_i} + \frac{\partial\alpha}{\partial x_i} = 0 \qquad (57)$$

since (55) and (43) hold. For this purpose we choose the parameters λ_α's and l_α's as follows. Let μ be a complex number ($\mu \neq \bar{\mu}$) and

$$\lambda_\alpha = \mu \text{ or } \bar{\mu} \qquad (\alpha = 1, 2, \cdots, N). \tag{58}$$

We assume that Ψ is normalized and choose the l_α such that

$$l_\alpha^* l_\beta = 0 \text{ if } \lambda_\alpha \neq \lambda_\beta. \tag{59}$$

Then we have

$$h_\alpha^* h_\beta = 0 \text{ if } \lambda_\alpha \neq \lambda_\beta. \tag{60}$$

From (50), the definition of α, we have

$$\alpha h_\beta = -\lambda_\beta h_\beta, \qquad h_\beta^* \alpha^* = -\lambda_\beta^* h_\beta^*. \tag{61}$$

Hence

$$h_\beta^* (\alpha^* + \alpha) h_\gamma = -h_\beta^* (\mu^* + \mu) h_\gamma \tag{62}$$

holds for all $\beta, \gamma = 1, 2, \cdots, N$. Since the columns $\{h_\alpha\}$ are linearly independent, we have

$$\alpha^* + \alpha = -(\mu + \bar{\mu}) I \tag{63}$$

and (57) holds. Thus we obtain:

Theorem 5. *If $A_i \in u(N)$, $V_a \in u(N)$ and parameters $\lambda_1, \cdots, \lambda_N$; $l_1, \cdots, l_N$ for constructing the Darboux matrix satisfy (58) and (59) and the condition of linear independence, then $\widetilde{A}_i \in u(N)$.*

Remark 5. *The matrices α and S can be expressed in a geometric way. Without loss of generality, let $\lambda_a = \mu$ ($a = 1, 2, \cdots, k$), $\lambda_b = \bar{\mu}$ ($b = k+1, \cdots, N$). At each point of $\mathbf{R}^{n,n}$, the l_a's span a k-dimensional subspace L_1 of $\mathbf{C}_n$ and the l_b's span the orthogonal complement L_2. They are holomorphic functions of $\mu p + x$ and $\bar{\mu} p + x$, respectively. Define Σ_1 to be the bundle of k-dimensional spaces on $\mathbf{R}^{n,n} \times \mathbf{R}^1$ spanned by the $\Psi(\mu) l_a$, and let π be the Hermitian projection from $\mathbf{C}^n \times (\mathbf{R}^{n,n} \times \mathbf{R}^1)$ on Σ_1 and $\pi^\perp$ the complement of π. Then from (61) we see that*

$$\alpha = -(\mu \pi + \bar{\mu} \pi^\perp) \tag{64}$$

and

$$S = (\lambda - \mu)\pi + (\lambda - \bar{\mu})\pi^\perp = (\lambda - \mu)\left(I + \frac{\mu - \bar{\mu}}{\lambda - \mu}\pi^\perp\right). \tag{65}$$

Hence the Darboux transformation of Ψ *can be written as*

$$\Psi_1 = \left(I + \frac{\mu - \bar{\mu}}{\lambda - \mu} \pi^{\perp} \right) \Psi. \tag{66}$$

Here both Ψ_1 *and* Ψ *are normalized. Consequently, a Darboux transformation is defined by a complex number* μ *and a holomorphic map* ρ *from* $\mathbf{C}^n$ *to the Grassmannian* $G_{N,k}$ *such that the subspace* L_1 *is the image of* $\rho(\mu p + x)$.

Remark 6. *The above statements are valid for the case of* $SU(N)$, *since we have* $\mathrm{tr}\alpha = k\mu + (N-k)\bar{\mu} =$ *constant and* $\mathrm{tr}\widetilde{A}_i = 0$, *from* $\mathrm{tr}A_i = 0$.

§5. Solitons and interactions

We start with the trivial solution $A_i = 0$. In order to obtain the solution $\Psi_0(\lambda)$ to the Lax equations, we have to determine V_0, V_1, $\cdots$, V_m. For $A_i = 0$, (29), (30) become

$$V_0 = V_0(p),$$

$$\frac{\partial V_a}{\partial x_i} = \frac{\partial V_{a-1}}{\partial p_i} \quad (a = 1, 2, \cdots, m+q), \tag{67}$$

$$\frac{\partial V_{m+q}}{\partial p_i} = 0.$$

We assume that V_0, V_a^0, W_{m+k}^0, W_{m+q} are independent of t. It is easily seen that

$$V(\lambda) = U_1(\lambda p + x) + U_2\left(p + \frac{x}{\lambda}\right), \tag{68}$$

where U_1, U_2 are matrix-valued polynomials of degrees m and q, and $U_1, U_2 \in u(N)$ (or $su(N)$) for real λ. Thus

$$\Psi_0 = \exp[V(\lambda)t]. \tag{69}$$

We can find explicit expressions for H and α. We assume $N = 2$ in order to simplify the expressions, and let

$$V(\lambda) = \begin{pmatrix} \mathrm{i}v(\lambda) & 0 \\ 0 & -\mathrm{i}v(\lambda) \end{pmatrix} \tag{70}$$

with

$$v(\lambda) = u_1(\lambda p + x) + u_2\left(p + \frac{x}{\lambda}\right). \tag{71}$$

Here u_1, u_2 are polynomials of degrees m and q with real coefficients.

The matrix H takes the form

$$H = \begin{pmatrix} e^{\mathrm{i}v(\mu)t} & -\overline{g(\mu)}e^{\mathrm{i}v(\bar{\mu})t} \\ g(\mu)e^{-\mathrm{i}v(\mu)t} & e^{-\mathrm{i}v(\bar{\mu})t} \end{pmatrix}. \tag{72}$$

Here $g(\mu)$ is a holomorphic function of $\mu p + x$ and $\mu = \sigma + \mathrm{i}\tau$ $(\tau \neq 0)$. The condition $h_2^* h_1 = 0$ holds. By calculation we have

$$\begin{aligned} \Delta &= \det H = e^{wt} + |g(\mu)|^2 e^{-wt}, \\ \alpha &= -\frac{1}{\Delta}\begin{pmatrix} \mu e^{wt} + \bar{\mu}|g|^2 e^{-wt} & (\mu - \bar{\mu})\bar{g}e^{\mathrm{i}ut} \\ (\mu - \bar{\mu})g e^{-\mathrm{i}ut} & \mu|g|^2 e^{-wt} + \bar{\mu}e^{wt} \end{pmatrix}. \end{aligned} \tag{73}$$

Here

$$w = \mathrm{i}(v(\mu) - v(\bar{\mu})), \qquad u = v(\mu) + v(\bar{\mu}) \tag{74}$$

are real functions of p and x. Consequently,

$$A_i = \frac{\partial \alpha}{\partial x_i} = \begin{pmatrix} a_{i11} & a_{i12} \\ -\bar{a}_{i12} & -a_{i11} \end{pmatrix} \tag{75}$$

with

$$\begin{aligned} a_{i11} &= 2\tau\mathrm{i}\frac{1}{(e^{wt} + |g(\mu)|^2 e^{-wt})^2}\left(\frac{\partial |g|^2}{\partial x_i} - 2|g|^2 \frac{\partial w}{\partial x_i} t\right), \\ a_{i12} &= -2\tau\mathrm{i}\frac{\left(\frac{\partial \bar{g}}{\partial x_i} + \bar{g}\frac{\partial u}{\partial x_i} t\mathrm{i}\right) e^{\mathrm{i}ut}}{e^{wt} + |g(\mu)|^2 e^{-wt}} \\ &\quad + 2\tau\mathrm{i}\frac{(e^{wt} - |g(\mu)|^2 e^{-wt})\frac{\partial w}{\partial x_i} + e^{-wt}\frac{\partial |g|^2}{\partial x_i}}{(e^{wt} + |g(\mu)|^2 e^{-wt})^2}\bar{g}e^{\mathrm{i}ut}. \end{aligned} \tag{76}$$

These are the explicit formulae for single solitons. The asymptotic behavior of the single solitons as $t \to \pm\infty$ depends on μ, g and v. For the general case it is quite complicated.

We assume

$$g(\mu) = P(\mu)\exp\{Q(\mu)\} \tag{77}$$

where P and Q are polynomials of $\mu p + x$. Imagine an observer who moves in the space $\mathbf{R}^{n,n}$ with constant velocity. His world line l in $\mathbf{R}^{n,n,1}$ is

$$p = \pi t + \pi_0, \quad x = \xi t + \xi_0. \tag{78}$$

Here (π, ξ) is his velocity and (π_0, ξ_0) is his starting point. In his observation, v, P, Q behave as polynomials of t. Write $Q = q + \mathrm{i}s$, where q and s are real. Then $|g|^2 = |P|^2 e^{2q}$. Along l,

$$q - wt = C_0 t^m + C_1 t^{m-1} + \cdots + C_{m-1} t + C_m \tag{79}$$

where the coefficients $C_0, C_1, \cdots, C_m$ are polynomials in π, ξ and π_0, ξ_0. If $|A_i| \to 0$ as $t \to \pm\infty$ along l, then l is called a vanishing line of the soliton. If at least one of $C_0, C_1, \cdots, C_{m-1}$ is not zero and $|P| \neq 0$ along l, then the world line l is a vanishing line. If along l, $P = 0$ and e^{-q+wt} approaches infinity, l is also a vanishing line. The asymptotic behavior of the single soliton is seen by the observer, who is going along the non-vanishing world line.

Example 1. Let

$$v = a = \text{real constant}, \quad g = P(\mu p + x).$$

Then $w = 0$, $q = 0$. From (76) we have

$$\alpha_{i11} = 2\tau\mathrm{i}\frac{1}{(1+|P|^2)^2}\frac{\partial |P|^2}{\partial x_i},$$

$$\alpha_{i12} = -2\tau\mathrm{i}\left(\frac{1}{1+|P|^2}\frac{\partial \bar{P}}{\partial x_i} - \frac{\bar{P}}{(1+|P|^2)^2}\frac{\partial |P|^2}{\partial x_i}\right)e^{2a\mathrm{i}t}.$$

The soliton is periodic with time t and $x_i = x_i^0$, $p_i = p_i^0$, $t = t$ are non-vanishing world lines if $|P| \neq 0$ or $\dfrac{\partial \bar{P}}{\partial x_i} \neq 0$.

Example 2. Let

$$v = v(\mu), \quad g = \exp\{Q(x)\}.$$

Then $w = -2\mathrm{i}\,\mathrm{Im}\, v$, $u = 2\,\mathrm{Re}\, v$, $P = 1$, $q = \mathrm{Re}\, Q$, $s = \mathrm{Im}\, Q$. From (76), we have

$$a_{i11} = \frac{-4\tau\mathrm{i}}{(e^{-q+wt} + e^{q-wt})^2} \frac{\partial w}{\partial x_i} t,$$

$$a_{i12} = -2\tau\mathrm{i} \frac{\frac{\partial}{\partial x_i}(q - \mathrm{i}s)}{e^{-q+wt} + e^{q-wt}} e^{\mathrm{i}(u-s)}$$

$$+2\tau\mathrm{i} \frac{(e^{-q+wt} - e^{q-wt}) \frac{\partial w}{\partial x_i} t - 2e^{q-wt} \frac{\partial q}{\partial x_i}}{(e^{-q+wt} + e^{q-wt})^2} e^{\mathrm{i}(u-s)t}.$$

The vanishing world lines are defined by the expansion of $q - wt$, since $|P| = 1$.

§6. Multi-solitons and the interaction of solitons

Let (Ψ, A) be a single soliton obtained from the trivial solution with "parameters" (μ_1, g_1), with v given. It will be denoted by $\Sigma_{(\mu_1, g_1)}$. Applying the Darboux transformation with parameters (μ_2, g_2) to (Ψ, A) we obtain a solution $\Sigma_{(\mu_1, g_1, \mu_2, g_2)}$ which is called a double soliton. The Lax representation of the double soliton is

$$\Psi_2 = (\lambda I + \alpha_2')(\lambda I + \alpha_1)\Psi. \tag{80}$$

Here Ψ is the Lax representation of the trivial solution, $\lambda I + \alpha_1$ is the Darboux matrix based on Ψ and defined by (μ_1, g_1), $\lambda I + \alpha_2'$ is the Darboux matrix based on $\Psi_1 = (\lambda I + \alpha_1)\Psi$ and defined by (μ_2, g_2).

Let $\mathcal{O}_{(\mu_1, g_1)}$, $\mathcal{O}_{(\mu_2, g_2)}$ be the sets of vanishing world lines of $\Sigma_{(\mu_1, g_1)}$ and $\Sigma_{(\mu_2, g_2)}$, respectively. Consider the behavior of $\Sigma_{(\mu_1, g_1, \mu_2, g_2)}$ along a line $l_1 \in \mathcal{O}_{(\mu_1, g_1)}$. From the definition it is seen that as $t \to \infty$, α_1 behaves as a constant matrix α_1^0. Then

$$\Psi_2 \sim (\lambda I + \alpha_2'')(\lambda I + \alpha_1^0)\Psi. \tag{81}$$

Hence Ψ_2 behaves asymptotically as a Darboux transformation with parameters (μ_2, g_2), based on $\Psi_1' = (\lambda I + \alpha_1^0)$ which is a Lax representation of the trivial solution. Consequently, $\Sigma_{(\mu_1, g_1, \mu_2, g_2)}$ behaves asymptotically as a single soliton $\Sigma'_{(\mu_2, g_2)}$ along l_1. By the theorem of permutability, $\Psi_2 = (\lambda I + \alpha_1')(\lambda I + \alpha_2)\Psi$. Similarly, it can be seen that, along a line $l_2 \in \mathcal{O}_{(\mu_2, g_2)}$, $\Sigma_{(\mu_1, g_1, \mu_2, g_2)}$ behaves as a single soliton $\Sigma'_{(\mu_1, g_1)}$. Thus we have:

Theorem 6. *If $l \in \mathcal{O}_{(\mu_1,g_1)}$ (respectively, $\mathcal{O}_{(\mu_2,g_2)}$), then along l the double soliton $\Sigma_{(\mu_1,g_1,\mu_2,g_2)}$ behaves asymptotically as a single soliton $\Sigma'_{(\mu_2,g_2)}$ (respectively, $\Sigma'_{(\mu_1,g_1)}$).*

In particular, if $l \in \mathcal{O}_{(\mu_1,g_1)} \cap \mathcal{O}_{(\mu_2,g_2)}$, then $\Sigma_{(\mu_1,g_1,\mu_2,g_2)}$ approaches the trivial solution. Let $S_{(\mu_1,g_1)}$ (respectively, $S_{(\mu_2,g_2)}$) be the set of non-vanishing world lines of $\Sigma_{(\mu_1,g_1)}$ (respectively, $\Sigma_{(\mu_2,g_2)}$). If $S_{(\mu_1,g_1)} \cap S_{(\mu_2,g_2)} = \emptyset$, then $\Sigma_{(\mu_1,g_1,\mu_2,g_2)}$ splits into two single solitons $\Sigma'_{(\mu_1,g_1)}$ and $\Sigma'_{(\mu_2,g_2)}$ as $t \to \infty$. If $S_{(\mu_1,g_1)} \cap S_{(\mu_2,g_2)} \neq \emptyset$ and $l \in S_{(\mu_1,g_1)} \cap S_{(\mu_2,g_2)}$, then along l, the behavior of $\Sigma_{(\mu_1,g_1,\mu_2,g_2)}$ can be regarded as confinement of solitons.

The situation for p-solitons is similar. The most interesting phenomena of the interaction of KdV solitons and many other solitons are thus extended to the case of GSYMF. Furthermore, confinement of solitons seems to be an interesting phenomenon in soliton theory.

§7. Reduction to AKNS systems in $\mathbf{R}^{n,1}$

As has been pointed out by many authors, almost all known soliton equations can be obtained as reductions of self-dual Yang-Mills equations. Of course, each of these equations at most has four independent variables. In recent years, integrable systems with more independent variables have beeen considered by many authors [4, 10, 15]. Almost all of these integrable systems with many independent variables belong to the framework of the AKNS system [11]. However, we have:

Theorem 7. *The AKNS system in the space $\mathbf{R}^{n,1}$ is a reduction of the generalized self-dual Yang-Mills flows.*

Proof. Let $\Psi = \Psi(x,p,t)$ satisfy the Lax equation (16), (28), and let J_i $(i = 1, 2, \cdots, n)$ be constant diagonal matrices. Define

$$
\begin{aligned}
&\Phi(p,t) = \exp(-\textstyle\sum x_j J_j)\Psi, \\
&P_i(p,t) = -\exp(-\textstyle\sum x_j J_j) A_i \exp(\textstyle\sum x_j J_j), \qquad (82)\\
&U_a(p,t) = \exp(-\textstyle\sum x_j J_j) V_a \exp(\textstyle\sum x_j J_j).
\end{aligned}
$$

Assume that the right-hand sides are independent of x. Substituting them into the Lax equations (16), (28), we obtain

$$
\begin{aligned}
\frac{\partial \Phi}{\partial p_i} &= (\lambda J_i + P_i)\Phi, \\
\frac{\partial \Phi}{\partial t} &= \sum_{a=0}^{m+q} U_a \lambda^{m-a}\Phi.
\end{aligned}
\tag{83}
$$

This is just the AKNS system in $\mathbf{R}^{n,1}$.

From this theorem, one is lead to believe that the GSYMF is the most general known integrable system.

§ Acknowledgments

The work is supported by the Chinese National Research Project “Nonlinear Science”, the Scientific Foundation of the National Education Department of China for Doctor Program and the Research Foundation of the Education Commission of Shanghai.

References

[1] M. J. Ablowitz, S. Chakrayarty and L. A. Takhtajan, A self-dual Yang-Mills hierarchy and its reduction to integrable systems in 1+1 and 2+1 dimensions, Commun. Math. Phys., **159** (1993), 289–314.

[2] M. F. Atiyah, Instantons in two and four dimensions, Commun. Math. Phys., **93** (1984), 437–451.

[3] M. F. Atiyah, Geometry of Yang-Mills Fields, Lezioni Fermiane, Accademia Nazionale dei Lincei Scuola Normaie Superione, Pisa, 1979.

[4] R. Beals and K. Tenenblat, An intrinsic generalization for the wave and sine-Gordon equation, In: Differential Geometry, Pitman Monograph, **52**, Longman, Harlow, 1991, pp. 25–46.

[5] D. S. Freed and K. K. Uhlenbeck, Instantons and Four-Manifolds, Math. Sciences Res. Inst. Publ. I, Springer-Verlag, 1984.

[6] C. H. Gu, Bäcklund transformations and Darboux transformations, In: Soliton theory and its applications, (ed. C. H. Gu), Springer-Verlag and Zhejiang Science and Technology Publishing House, 1995, pp. 122–151.

[7] C. H. Gu, Integrable evolution systems based on generalized self-dual Yang-Mills equations and their soliton-like solutions, Lett. Math. Phys., **35** (1995), 61–74.

[8] C. H. Gu, Generalized self-dual Yang-Mills flows: explicit solutions and reductions, Proc. of KdV'95, Acta Appl. Math., **39** (1995), 349–360.

[9] C. H. Gu, On the interaction of solitons for a class of integrable systems in the space-time $\mathbf{R}^{n+1}$, Lett. Math. Phys., **26** (1992), 199–209.

[10] C. H. Gu and H. S. Hu, Explicit solutions to the intrinsic generalization for the wave and sine-Gordon equations, Lett. Math. Phys., **29** (1993), 1–11.

[11] C. H. Gu and Z. X. Zhou, On Darboux transformations for soliton equations in high-dimensional space-time, Lett. Math. Phys., **32** (1994), 1–10.

[12] L. J. Mason, Integrable and solvable systems, In: Further advances in twistor theory, (eds. L. J. Mason, L. P. Hughston and P. Z. Kobak), Longman Scientific Technical, 1995, pp. 1–73.

[13] K. Takasaki, A new approach to the self-dual Yang-Mills equations, Commun. Math. Phys., **94** (1984), 35–59.

[14] K. Takasaki, Integrable systems in gauge theory, Kähler geometry and super KP hierarchy — symmetries and algebraic points of view, In: Proc. International Congr. Math., Kyoto, 1990, Vol. II, pp. 1205–1214.

[15] C. L. Terng and K. Uhlenbeck, Bäcklund transformations and loop group actions, Commun. Pure Appl. Math., **53** (2000), 1–75.

[16] R. S. Ward, Integrable systems and twistors, In: Integrable systems, twistors, loop groups, and Riemann surfaces, (eds. N. J. Hitchin, G. B. Segal and R. S. Ward), Oxford, Clarendon Press, 1999, pp. 121–134.

[17] R. S. Ward, Integrable and solvable systems and relations among them, Phil. Trans. Roy. Soc. London Ser. A, **315** (1985), 451–457.

[18] R. S. Ward, Completely solvable gauge-field equations in dimension greater than four, Nuclear Phys. B, **236** (1984), 381–396.

[19] C. N. Yang, Condition of self-duality for SU_2 gauge fields on Euclidean 4-dimensional spaces, Phys. Rev. Lett., **38** (1977), 1377–1379.

Institute of Mathematics
Fudan University
Shanghai 200433
P. R. CHINA

E-mail address:
`guch@fudan.ac.cn`

Advanced Studies in Pure Mathematics 51, 2008
Surveys on Geometry and Integrable Systems
pp. 163–187

From CMC surfaces to Hamiltonian stationary Lagrangian surfaces

Frédéric Hélein and Pascal Romon

§1. Introduction

Minimal surfaces and surfaces with constant mean curvature (CMC) have fascinated differential geometers for over two centuries. Indeed these surfaces are solutions to variational problems whose formulation is elegant, modelling physical situations involving soap films and bubbles; however their richness has not been exhausted yet. Advances in the understanding of these surfaces draw on complex analysis, theory of Riemann surfaces, topology, nonlinear elliptic PDE theory and geometric measure theory. Furthermore, one of the most spectacular developments in the past twenty years has been the discovery that many problems in differential geometry – including those of minimal and CMC surfaces – are actually integrable systems.

The theory of integrable systems developed in the 1960's, beginning essentially with the study of a now famous example: the Korteweg-de Vries equation, $u_t + 6uu_x + u_{xxx} = 0$, modelling waves in a shallow flat channel[1]. In the 1960's mathematicians noticed the exceptional properties of the KdV equation: existence of solitary waves that "superpose" almost linearly, and an infinite number of conserved quantities. From these observations, C. Gardner, J. Greene, M. Kruskal and R. Miura [12] showed in 1967 that this equation could be solved completely by

Received December 27, 2000.
Revised February 8, 2002.
2000 *Mathematics Subject Classification.* Primary 49Q05; Secondary 53C42.

[1]Earlier already, in 1955, Fermi, Pasta and Ulam had unexpectedly discovered the soliton phenomenon (to their great surprise) while simulating a one-dimensional model in statistical mechanics on the Los Alamos computer.

reducing it to a linear one, thanks to a rather sophisticated nonlinear transformation called *inverse scattering*. In 1968, P. Lax [23] gave an interpretation of their method by noticing that the KdV equation can be rewritten as the evolutionary equation for an operator L (in fact $L = -\frac{d^2}{dx^2} + u$), namely $L_t = [A, L]$. In particular the spectrum of L is time independent, explaining the infinite number of conserved quantities. That was the starting point for a series of increasingly deeper observations on KdV, formulated notably by the Japanese and Russian schools, aiming at interpreting its miraculous properties: Hamiltonian structure, explicit formulation of some of its solutions (called *finite gap solutions*) through techniques from algebraic geometry, relation with dynamical systems on Lie algebras, and so on... For all these properties the KdV equation deserves the name of *completely integrable system*. Later on other equations coming from physics were identified as completely integrable systems: the one dimensional nonlinear Schrödinger, Toda field equation, and so on... Today however differential geometry seems to be the field where most of the completely integrable systems are discovered (see [33] for instance). We will see in the following some examples stemming from problems in Riemannian geometry. Other very interesting examples are linked to geometrical problems (but inspired by physics) in four dimensions: self-dual Yang-Mills connections, and self-dual Einstein metrics (the starting point being Penrose's theory of twistors).

The intrusion of integrable systems in differential geometry may seem a recent discovery, however it is not so, since a "prehistory" of completely integrable systems in geometry occurred one century ago. A number of clues had been spotted by nineteenth century geometers: conjugate families of CMC surfaces discussed by O. Bonnet in 1853 [5], the Enneper-Weierstrass representation for minimal surfaces (see below), construction of constant mean or constant Gaussian curvature surfaces by A. Enneper and his students in 1868 and 1880 [10, 8, 22, 24], construction of soliton-like solutions through iteration of Bäcklund transformations [3, 25]. A synthesis can be found in the book of G. Darboux [7]. Of course the geometers of that period did not have the point of view we have today, which is based on concepts from Lie algebra theory.

In this survey we wish to present this theory in an accessible way, through the examples of minimal surfaces, CMC surfaces and harmonic maps or Willmore surfaces. We will then present a more recent example, discovered by both authors [17, 18, 19].

§2. Minimal surfaces and CMC surfaces

Let m be a point of an oriented surface $\Sigma \subset \mathbb{R}^3$. We define the principal curvatures at m as follows. Consider the one-parameter family of affine planes P through m, perpendicular to the tangent plane to Σ at m, $T_m\Sigma$. Any such plane P cuts Σ along a (planar) curve and we let $k(P)$ be the (oriented) curvature of that curve at m, obviously dependent on P. As P revolves around the normal line at m, $k(P)$ oscillates; its minimal and maximal values are called *principal curvatures* and denoted by $k_1 \leq k_2$. If $k(P)$ is independent of P, i.e. $k_1 = k_2$, the point m is called *umbilic*. The quantity $H := \frac{1}{2}(k_1 + k_2)$ is the *mean curvature* of Σ at m and the product $K = k_1 k_2$ is the *Gaussian curvature.*

A minimal surface satisfies by definition $H = 0$ at all points. The reason behind the name is that such surfaces are stationary with respect to the area functional $\mathcal{A}(\Sigma) = \int_\Sigma da$ (not necessarily minimizing[2]). A very simple experimental procedure for producing minimal surfaces consists of dipping a closed metal wire in soap and water: when taken out of the liquid, the wire bounds a soap film, materializing a minimal surface.

Constant mean curvature surfaces are by convention those surfaces with constant non-zero mean curvature H, as opposed to minimal surfaces.

Using conformal coordinates turns out to be a very convenient way of studying the properties of these surfaces. Indeed, for any simply connected immersed surface Σ there exists a conformal parametrization $X : \Omega \to \mathbb{R}^3$, where Ω is an open subset of $\mathbb{C}$, i.e.

$$|\partial_x X|^2 - |\partial_y X|^2 - 2i\langle \partial_x X, \partial_y X\rangle = 0\,,$$

and $X(\Omega) = \Sigma$. Let u map $z = x + iy \in \Omega$ to the oriented unit normal vector at $X(z)$:

$$u(z) := \frac{\partial_x X \times \partial_y X}{|\partial_x X \times \partial_y X|}\,,$$

where $\times$ denotes the cross product in $\mathbb{R}^3$. The map u is called the Gauss map. Then the conformality assumption implies that the first and second fundamental forms take the following form:

$$(1) \qquad I := \begin{pmatrix} |\partial_x X|^2 & \langle \partial_x X, \partial_y X\rangle \\ \langle \partial_x X, \partial_y X\rangle & |\partial_y X|^2 \end{pmatrix} = e^{2\omega}\begin{pmatrix} 1 & 0 \\ 0 & 1 \end{pmatrix}$$

[2]However stationary surfaces will always be minimizing for small perturbations in a small enough compact subset.

and

$$\text{(2)}\quad II := \begin{pmatrix} \langle (\partial_x)^2 X, u\rangle & \langle \partial_x\partial_y X, u\rangle \\ \langle \partial_x\partial_y X, u\rangle & \langle (\partial_y)^2 X, u\rangle \end{pmatrix} = e^{2\omega}\begin{pmatrix} H+a & b \\ b & H-a \end{pmatrix}$$

with ω, H, a, b are real-valued functions. H is exactly the mean curvature. In particular

$$\Delta X = 2H\partial_x X \times \partial_y X.$$

Thus all minimal surfaces (respectively, CMC surfaces) are locally described by the conformal immersions X such that $\Delta X = 0$ (respectively, $\Delta X = 2H_0\partial_x X \times \partial_y X$ with H_0 a non-zero constant). As we will see, this way of stating the geometric problem is quite productive.

2.1. Minimal surfaces

The set of equations

$$\text{(3)}\qquad |\partial_x X|^2 - |\partial_y X|^2 - 2i\langle \partial_x X, \partial_y X\rangle = 0 \quad \text{and} \quad \Delta X = 0$$

may be solved locally by introducing a complex-valued function

$$f := \partial_z X = \frac{1}{2}(\partial_x X - i\partial_y X)$$

then (3) becomes

$$f^2 = 0 \quad \text{and} \quad \partial_{\bar z} f = 0\,,$$

since $\partial_{\bar z}\partial_z = \frac{1}{4}\Delta$. ($f^2$ denotes the dot product $(f^1)^2 + (f^2)^2 + (f^3)^2$.) The solution to these two equations is given by

$$\text{(4)}\qquad f = \begin{pmatrix} \frac{i}{2}(w^2-1) \\ \frac{1}{2}(w^2+1) \\ iw \end{pmatrix} h\,,$$

where w (respectively, h) is some meromorphic (respectively, holomorphic) function from Ω to $\mathbb{C}$. Thus

$$\text{(5)}\qquad X(z) = C + \operatorname{Re}\left[\int_{z_0}^{z} f(\zeta)d\zeta\right],$$

with C a constant in $\mathbb{R}^3$ and z_0 a fixed base point in Ω. This is the Enneper-Weierstrass representation formula.

The reader may observe the following fact: locally, any minimal surface is part of (continuous) one-parameter family of minimal surfaces given by

$$(6) \qquad X_\lambda(z) = C + \mathrm{Re}\left[\lambda^{-2}\int_{z_0}^{z} f(\zeta)d\zeta\right], \quad \lambda \in S^1 \subset \mathbb{C}^\star.$$

This family is called the conjugate family[3] of the surface Σ.

2.2. Constant mean curvature surfaces

Even before Enneper and Weierstrass discovered their representation formula, O. Bonnet [5] found that minimal surfaces and CMC surfaces could be deformed, giving birth to conjugate families. This relies on the fact that the immersion X can be reconstructed with the mere knowledge of the first and second fundamental forms of X. More specifically, given four real-valued functions ω, H, a, b, there exists a conformal immersion X whose first and second fundamental forms are given by (1) and (2), *under the condition that* ω, H, a, b satisfy some compatibility relations: the Gauss-Codazzi equations[4]. The crucial observation is that (ω, H, a, b) satisfies the Gauss-Codazzi equations if and only if the deformed data $(\omega, H, a_\lambda, b_\lambda)$ does, where

$$a_\lambda - ib_\lambda := \lambda^{-2}(a - ib) \text{ for all } \lambda \in \mathbb{C}^\star .$$

Hence starting with a conformal immersion X with constant or zero mean curvature, and deforming the first and second fundamental forms I and II by substituting $(\omega, H, a_\lambda, b_\lambda)$ for (ω, H, a, b), one gets tensors I_λ and II_λ on Ω that are the first and second fundamental forms of a new conformal immersion X_λ of constant or zero mean curvature. One constructs that way the conjugate family of X, which coincides with the Enneper-Weierstrass formula (6) in the case of minimal surfaces.

2.3. Changing viewpoint: the moving frame

We revisit here the construction of the conjugate family of a constant or zero mean curvature surface. Before doing this, let us make a detour into the theory of moving frames. Consider the projection p from the

---

[3]It depends only on the surface, not on the choice of the parametrization X.

[4]If H = constant, the Gauss-Codazzi equations are: $\frac{\partial}{\partial \bar{z}}\left(e^{2\omega}(a - ib)\right) = 0$ and $\Delta\omega + e^{2\omega}(H^2 - a^2 - b^2) = 0$.

group of affine isometries of $\mathbb{R}^3$, $SO(3)\ltimes\mathbb{R}^3$, to $\mathbb{R}^3$, defined by $(R,t)\mapsto t$. Recall that $SO(3)\ltimes\mathbb{R}^3$ can be identified with the group of four by four real matrices of type $\begin{pmatrix} R & t \\ 0 & 1 \end{pmatrix}$ with $R\in SO(3)$ and $t\in\mathbb{R}^3$. We say that F lifts X if $p\circ F = X$:

$$\begin{array}{ccc} & & SO(3)\ltimes\mathbb{R}^3 \\ & \overset{F}{\nearrow} & \downarrow p \\ \Omega & \overset{X}{\longrightarrow} & \mathbb{R}^3 \end{array}$$

If Ω is a simply connected domain in $\mathbb{C}$ we can always lift X. A moving frame (or *Darboux* frame) is a lift $F:\Omega\longrightarrow SO(3)\ltimes\mathbb{R}^3$ of the following form:

$$F(z) := \begin{pmatrix} e_1(z) & e_2(z) & u(z) & X(z) \\ 0 & 0 & 0 & 1 \end{pmatrix}$$

where, for $z\in\Omega$, $(e_1(z),e_2(z))$ is any positively oriented orthonormal basis of $T_{u(z)}S^2 \simeq T_{X(z)}\Sigma$, smoothly varying with z. The simplest (but not the only) way to define $(e_1(z),e_2(z))$ is through coordinates: $e_1 = e^{-\omega}\partial_x X$, $e_2 = e^{-\omega}\partial_y X$, where $e^{2\omega} := |\partial_x X|^2 = |\partial_y X|^2$. Notice that the triple (e_1,e_2,u) is a positively oriented orthonormal basis of $\mathbb{R}^3$ and as such can be identified with an element of $SO(3)$.

As seen above, X is tied to its conjugate family, hence u is tied to the family of Gauss maps u_λ of X_λ. We set

$$e_{1,\lambda} - ie_{2,\lambda} := 2\lambda e^{-\omega}\frac{\partial X_\lambda}{\partial z}.$$

obtaining thus a deformation F_λ of F defined by

$$F_\lambda(z) = \begin{pmatrix} \phi_\lambda(z) & X_\lambda(z) \\ 0 & 1 \end{pmatrix}, \text{ where } \phi_\lambda = \begin{pmatrix} e_{1,\lambda} & e_{2,\lambda} & u_\lambda \end{pmatrix}.$$

Since F_λ is group-valued, we may compute its Maurer-Cartan form $A_\lambda := F_\lambda^{-1}dF_\lambda$ and find

$$\begin{aligned}
A_\lambda &= \lambda^{-1}\begin{pmatrix} 0 & 0 & \langle\partial_z u, e_1\rangle & \langle\partial_z X, e_1\rangle \\ 0 & 0 & \langle\partial_z u, e_2\rangle & \langle\partial_z X, e_2\rangle \\ -\langle\partial_z u, e_1\rangle & -\langle\partial_z u, e_2\rangle & 0 & 0 \\ 0 & 0 & 0 & 0 \end{pmatrix} dz \\
&\quad + \star\, d\omega \begin{pmatrix} 0 & -1 & 0 & 0 \\ 1 & 0 & 0 & 0 \\ 0 & 0 & 0 & 0 \\ 0 & 0 & 0 & 0 \end{pmatrix} \\
&\quad + \lambda \begin{pmatrix} 0 & 0 & \langle\partial_{\bar z} u, e_1\rangle & \langle\partial_{\bar z} X, e_1\rangle \\ 0 & 0 & \langle\partial_{\bar z} u, e_2\rangle & \langle\partial_{\bar z} X, e_2\rangle \\ -\langle\partial_{\bar z} u, e_1\rangle & -\langle\partial_{\bar z} u, e_2\rangle & 0 & 0 \\ 0 & 0 & 0 & 0 \end{pmatrix} d\bar z.
\end{aligned}$$

Remarkably enough, A_λ splits into three pieces $\lambda^{-1}A_1' + A_0 + \lambda A_1''$ which can be easily read off from the original Maurer-Cartan form $A = F^{-1}dF$ of F. Indeed

$$F^{-1}dF = \begin{pmatrix} 0 & -\star\, d\omega & \langle du, e_1\rangle & \langle dX, e_1\rangle \\ \star d\omega & 0 & \langle du, e_2\rangle & \langle dX, e_2\rangle \\ -\langle du, e_1\rangle & -\langle du, e_2\rangle & 0 & 0 \\ 0 & 0 & 0 & 0 \end{pmatrix}.$$

We deduce a more direct way of constructing the conjugate family of X, namely start by lifting X to $F : \Omega \longrightarrow SO(3) \ltimes \mathbb{R}^3$, split its Maurer-Cartan form A into two pieces

$$A_0 = \star d\omega \begin{pmatrix} 0 & -1 & 0 & 0 \\ 1 & 0 & 0 & 0 \\ 0 & 0 & 0 & 0 \\ 0 & 0 & 0 & 0 \end{pmatrix}$$

and $A_1 := A - A_0$, then split again $A_1 = A_1' + A_1''$ where $A_1' := A_1(\frac{\partial}{\partial z})dz$ and $A_1'' = \overline{A_1'} := A_1(\frac{\partial}{\partial \bar z})d\bar z$. Then the Maurer-Cartan form of F_λ is just $A_\lambda = \lambda^{-1}A_1' + A_0 + \lambda A_1''$. Now given such an A_λ, one may ask whether a corresponding map $F_\lambda : \Omega \longrightarrow SO(3) \ltimes \mathbb{R}^3$ exists, such that $A_\lambda = F_\lambda^{-1}dF_\lambda$. The answer for a simply connected domain Ω is positive if and only if

$$dA_\lambda + A_\lambda \wedge A_\lambda = 0. \tag{7}$$

However this equation turns out to be exactly equivalent to these two conditions: $A = F^{-1}dF$ and X is a constant or zero mean curvature immersion.

2.4. Another point of view using the Gauss map

We have previously defined the Gauss map $u : \Omega \longrightarrow S^2$ of an immersion X, allowing us later on to characterize the second fundamental form and the lift F. There is more to this than one might expect, thanks to the following result.

Theorem 1 (Ruh-Vilms). *Let X be a conformal immersion. Then X has constant or zero mean curvature if and only if its Gauss map u is a harmonic map into the unit sphere, i.e. u is a solution of the equation*

$$\Delta u + u|\nabla u|^2 = 0. \tag{8}$$

Among harmonic maps into the sphere, one should single out two particular subclasses, namely the holomorphic and antiholomorphic maps (solutions to $i\partial_z u = u \times \partial_z u$ and $i\partial_z u = -u \times \partial_z u$, respectively). In the case where u is holomorphic, X is proportional to u, and its image Σ is a piece of a sphere. If u is antiholomorphic, then X is minimal. In the remaining cases – the most interesting ones for us – X is a CMC surface whose image does not lie in a sphere. In that latter case, the Ruh-Vilms theorem can be improved on, in the sense that X can be reconstructed from its Gauss map u (up to translations of course). This is achieved through various formulae (Kenmotsu [20], Sym-Bobenko), or using the procedure below. Rewrite equation (8) as a "closure" condition

$$\partial_x(u \times \partial_x u) + \partial_y(u \times \partial_y u) = 0\ .$$

This implies the existence of a map $B : \Omega \longrightarrow \mathbb{R}^3$ such that

$$\partial_x B = u \times \partial_y u, \quad \partial_y B = -u \times \partial_x u.$$

Then $X = B + u$.

Therefore, in order to build conformal CMC immersions, it suffices to construct harmonic maps (which are neither holomorphic nor antiholomorphic), and this turns out to be easier.

We may analyze harmonic maps into the 2-sphere in an analogous way to that in the previous section. Indeed let $\phi : \Omega \longrightarrow SO(3)$ lift u in the following sense: $\phi(z) = (e_1(z), e_2(z), u(z))$ where $(e_1(z), e_2(z))$

is any oriented orthonormal basis of $T_{u(z)}S^2$. Its Maurer-Cartan form $\alpha := \phi^{-1}d\phi$ splits into two pieces $\alpha = \alpha_0 + \alpha_1$ where

$$\alpha_0 := \langle de_1, e_2 \rangle \begin{pmatrix} 0 & -1 & 0 \\ 1 & 0 & 0 \\ 0 & 0 & 0 \end{pmatrix}.$$

Further, we decompose as above α_1 in its (1,0) and (0,1) parts $\alpha_1' + \alpha_1''$, and set $\alpha_\lambda = \lambda^{-1}\alpha_1' + \alpha_0 + \lambda\alpha_1''$ for any $\lambda \in \mathbb{C}^\star$. Then u is harmonic if and only if

$$d\alpha_\lambda + \alpha_\lambda \wedge \alpha_\lambda = 0 \quad \forall \lambda \in \mathbb{C}^\star. \tag{9}$$

Reciprocally, the condition (9) on a simply connected domain Ω ensures the existence for any $\lambda \in \mathbb{C}^\star$ of a map $\phi_\lambda : \Omega \longrightarrow SO(3)$ such that $d\phi_\lambda = \phi_\lambda \alpha_\lambda$, and consequently of a map u_λ.

This characterization of harmonic maps was obtained by K. Pohlmeyer [29], V. E. Zhakarov, A. B. Shabat [37] and K. Uhlenbeck [35], following a different approach, inspired by the theory of integrable systems.

§3. Harmonic maps: a completely integrable system

3.1. Introducing loop groups

The most efficient way to deal with harmonic maps $u : \Omega \longrightarrow S^2$ is to work with the family of maps ϕ_λ constructed above. For the sake of simplicity let us assume (without loss of generality) that $0 \in \Omega$ and $\phi(0) = \mathbb{1}$. From now on λ will be a non-zero complex number, unless specified otherwise. Recall from equation (9) the existence of a unique map $\phi_\lambda : \Omega \longrightarrow SO(3)$ (for fixed λ) such that

$$d\phi_\lambda = \phi_\lambda \alpha_\lambda \quad \text{and} \quad \phi_\lambda(0) = \mathbb{1}. \tag{10}$$

Setting

$$P := \begin{pmatrix} 1 & 0 & 0 \\ 0 & 1 & 0 \\ 0 & 0 & -1 \end{pmatrix},$$

we define a Lie algebra involution by $\tau : \xi \mapsto P\xi P^{-1}$, and observe that

$$\tau(\alpha_\lambda) = \alpha_{-\lambda} \quad \forall \lambda.$$

Hence[5]

$$\tau(\phi_\lambda) := P\phi_\lambda P^{-1} = \phi_{-\lambda}. \tag{11}$$

To verify this we set $\tilde{\phi}_\lambda := \tau(\phi_{-\lambda})$. Then $\tilde{\phi}_\lambda^{-1} d\tilde{\phi}_\lambda = \tau(\phi_{-\lambda}^{-1} d\phi_{-\lambda}) = \tau(\alpha_{-\lambda}) = \alpha_\lambda$; we conclude the proof by a uniqueness argument.

For this reason we consider the loop group

$$LSO(3) := \{S^1 \ni \lambda \mapsto g_\lambda \in SO(3)\},$$

with group law being pointwise multiplication, and its *twisted* subgroup

$$LSO(3)_\tau := \{S^1 \ni \lambda \mapsto g_\lambda \in SO(3) \,|\, \tau(g_\lambda) = g_{-\lambda}\}.$$

The family of maps $(\phi_\lambda)_{\lambda \in S^2}$ may therefore be considered as a single map $\phi_\lambda : \Omega \longrightarrow LSO(3)_\tau$. The construction of harmonic maps from Ω to S^2 amounts to finding maps $\phi_\lambda : \Omega \longrightarrow LSO(3)_\tau$ such that $\phi_\lambda^{-1} d\phi_\lambda = \lambda^{-1}\alpha_1' + \alpha_0 + \lambda\alpha_1''$.

Remark. The involution τ gives us a clear explanation for the splitting $\alpha = \alpha_0 + \alpha_1$. Indeed τ is a Lie algebra automorphism ($\tau([\xi,\eta]) = [\tau(\xi),\tau(\eta)]$), which squares to the identity. We may hence decompose the Lie algebra $\mathfrak{so}(3)$ as the sum of two eigenspaces $\mathfrak{so}(3)_0 \oplus \mathfrak{so}(3)_1$ associated respectively to the eigenvalues $(-1)^0$ and $(-1)^1$. Thus α_n is just the projection of α on the $\mathfrak{so}(3)_n$ term.

3.2. Weierstrass-type representation à *la* Dorfmeister, Pedit and Wu

As an application of the formalism introduced above, we describe here an algebraic algorithm for constructing all harmonic maps $\Omega \longrightarrow S^2$ (where Ω is simply connected) starting with holomorphic data. This construction is due to Dorfmeister, Pedit and Wu [9].

Step 1: choosing a potential. Let $a, b : \Omega \longrightarrow \mathbb{C}$ be holomorphic maps, and define a matrix-valued (actually loop algebra-valued) holomorphic 1-form

$$\mu_\lambda = \lambda^{-1} \begin{pmatrix} 0 & 0 & a \\ 0 & 0 & b \\ -a & -b & 0 \end{pmatrix} dz$$

which we call the *potential*.

[5] We use the same letter τ to denote a group automorphism and its differential at the identity, acting on the Lie algebra.

Step 2: integrating μ_λ. The potential trivially satisfies $d\mu_\lambda + \mu_\lambda \wedge \mu_\lambda = 0$, which is the necessary and sufficient condition for the existence of $g_\lambda : \Omega \longrightarrow LSO(3)^{\mathbb{C}}_\tau$ such that[6] $dg_\lambda = g_\lambda \mu_\lambda$.

Step 3: splitting. We write g_λ as a product $\phi_\lambda b_\lambda$, with $\phi_\lambda : \Omega \rightarrow LSO(3)_\tau$ and $b_\lambda : \Omega \longrightarrow L^+SO(3)^{\mathbb{C}}_\tau$, where $L^+SO(3)^{\mathbb{C}}_\tau$ is the subgroup consisting of loops $b_\lambda \in LSO(3)^{\mathbb{C}}_\tau$ that are the restriction of holomorphic maps (in λ) from the closed unit disk to $SO(3)^{\mathbb{C}}$.

The map ϕ_λ produced in this way is a lift (in the sense of the previous section) of a harmonic map into the sphere.

A few comments are necessary.

a) The least obvious and most complex operation in the algorithm is the decomposition $g_\lambda = \phi_\lambda b_\lambda$. It rests upon a difficult theorem from A. Pressley and G. Segal [30], stating more precisely that any loop $g_\lambda \in LSO(3)^{\mathbb{C}}$ can be written uniquely as the product of $\phi_\lambda \in LSO(3)$ and $b_\lambda \in L^+SO(3)^{\mathbb{C}}$ (hence the decomposition of maps is done pointwise in z). J. Dorfmeister, F. Pedit and H.-Y. Wu call this decomposition the Iwasawa decomposition, for it is an infinite dimensional analog of the classical Iwasawa decomposition (whose prototype is the Gram-Schmidt theorem: any real matrix is the product of an orthogonal matrix by an upper triangular one).

b) This algorithm accounts for the construction of *almost* all harmonic maps. Actually J. Dorfmeister, F. Pedit and H.-Y. Wu show how to associate to any harmonic map a unique such potential μ_λ where the data (a, b) is meromorphic, albeit with non accumulating poles. There are other constructions along the same lines which avoid using meromorphic data (at some cost though).

c) The algorithm parallels the Enneper-Weierstrass representation formula (hence its name). Indeed μ_λ is the analog of

$$f_\lambda dz := \lambda^{-2} \begin{pmatrix} \frac{i}{2}(w^2 - 1) \\ \\ \frac{1}{2}(w^2 + 1) \\ \\ iw \end{pmatrix} hdz,$$

---

[6] $LSO(3)^{\mathbb{C}}_\tau$ is the complexification of $LSO(3)_\tau$. As a matter of fact $LSO(3)^{\mathbb{C}}_\tau := \{S^1 \ni \lambda \mapsto g_\lambda \in SO(3)^{\mathbb{C}} \mid \tau(g_\lambda) = g_{-\lambda}\}$ where $SO(3)^{\mathbb{C}} := \{M \in M(3, \mathbb{C}) | \, {}^tMM = \mathbb{1} \text{ and } \det M = 1\}$.

The map g_λ obtained from μ_λ corresponds to the (standard) integral $\int^z f_\lambda dz$ (actually both integrals have the same expression, but the latter takes place in an abelian group). Finally the Iwasawa decomposition reduces to taking the real part. Notice that the analogy is not only in spirit, but that under some conditions, the DPW algorithm reduces to the Enneper-Weierstrass representation formula.

3.3. Generalizations

The result of J. Dorfmeister, F. Pedit and H.-Y. Wu applies to any harmonic map from a simply connected $\Omega \subset \mathbb{C}$ to a symmetric space G/H, with G compact semisimple. In [15] the first author has built a similar theory for Willmore surfaces. Recall the definition: for any surface Σ immersed in $\mathbb{R}^3$, we define the *Willmore functional* as

$$\mathcal{W}(\Sigma) := \int_\Sigma H^2 da,$$

where H denotes as usual the mean curvature. The critical points of this functional satisfy the fourth order PDE

$$\Delta_\Sigma H + 2H(H^2 - K) = 0,$$

where K is the Gaussian curvature, and they are called Willmore surfaces. Actually this problem had been tackled ten years earlier by K. Voss. Later on it was realized that G. Thomsen had also studied it in 1923 [34], and many important results on these surfaces can be found in W. Blaschke's book [4] (however in 1929, Willmore was only ten years old and Blaschke could not possibly guess that his "conformal minimal surfaces" would become famous under the name of Willmore surfaces). Let us add that S. Germain had already considered the same problem early in the 19th century.

The crucial property of Willmore surfaces is the invariance under conformal transformations of $\mathbb{R}^3 \cup \{\infty\}$ (also known as the Möbius group of S^3, which is isomorphic to the connected component of the identity in $SO(4,1)$). This group has dimension 10, and is generated by the translations, dilations and inversions of $\mathbb{R}^3$. Invariance means that for any surface Σ without boundary and any Möbius transformation T, $\mathcal{W}(T(\Sigma)) = \mathcal{W}(\Sigma)$. Consequently the Willmore problem does not rely on the Euclidean structure of $\mathbb{R}^3$ but rather on its conformal structure. Since $\mathbb{R}^3$ is (locally) conformally equivalent to the sphere S^3 or the hyperbolic space H^3, the (local) theory of Willmore surfaces is identical in all three spaces.

Another important property (rediscovered by R. Bryant [6], though already known to Blaschke) is the existence of the *conformal Gauss map*, a notion akin to the classical Gauss map of Euclidean geometry, but adapted to conformal geometry. This map takes values in the set $S^{3,1}$ of oriented spheres and planes of $\mathbb{R}^3$ (the notation $S^{3,1}$ will be explained below). If Σ is an oriented surface in $\mathbb{R}^3$, the conformal Gauss map γ maps any point m to the unique sphere or plane tangent to Σ at m such that their mean curvature vectors coincide — or equivalently with the same orientation and mean curvature. Note that the mean curvature of a sphere is the inverse of its radius; hence the conformal Gauss map is a plane if and only if H vanishes.

A study of conformal geometry shows that the set of oriented spheres and planes of $\mathbb{R}^3$ is canonically isomorphic to

$$\{y \in \mathbb{R}^{4,1} \mid |y|^2 = -(y^0)^2 + (y^1)^2 + (y^2)^2 + (y^3)^2 + (y^4)^2 = 1\}$$

which is indeed the unit sphere in Minkowski space $\mathbb{R}^{4,1}$, hence the notation. The set $S^{3,1}$ has a natural pseudo-riemannian structure, and is also a symmetric space, being written as $SO(4,1)/SO(3,1)$.

All this construction is motivated by the following analog of the Ruh-Vilms theorem.

Theorem 2 ([4], [6]). *Let $X : \Omega \to \Sigma$ be a conformal parametrization of a surface in $\mathbb{R}^3$ or S^3, and $\gamma : \Omega \to S^{3,1}$ its conformal Gauss map. Then*

- *γ is weakly conformal, namely for any $z \in \Omega$, γ is conformal at z or $d\gamma_z = 0$,*
- *Σ is a Willmore surface if and only if γ is harmonic.*

This suggests immediately a strategy for analyzing and constructing Willmore surfaces, by analogy to the theory of CMC surfaces. One should study harmonic maps from Ω to $S^{3,1}$ and deduce from them conformal Willmore immersions. However it is trickier to reconstruct the Willmore immersion from the conformal Gauss map than the CMC surface from the classical Gauss map. Although feasible in principle, singularities may arise at umbilic points, hence the difficulties. See [15] for details.

A slightly different strategy proposed in [15] allows us to avoid the problem with the umbilic points. It relies on another kind of Gauss map: for that purpose we choose $\hat{X} : \Omega \longrightarrow S^3$ *any* map such that $X(z) \neq \hat{X}(z)$, $\forall z \in \Omega$. Now there is a canonical way to associate to each pair of disjoint points in S^3 a 3-dimensional spacelike subspace in $\mathbb{R}^{4,1}$. Thus the pair $(X, \hat{X}) : \Omega \longrightarrow S^3 \times S^3$ induces a map $Z : \Omega \longrightarrow Gr_3(\mathbb{R}^{4,1})$, where $Gr_3(\mathbb{R}^{4,1})$ is the Grassmannian of 3-dimensional spacelike subspaces contained in $\mathbb{R}^{4,1}$. Of course another choice $\hat{X}'$ in place of $\hat{X}$ would lead to another map Z' and one goes from Z to Z' through a gauge transformation parametrized by a map in $\mathcal{C}^\infty(\Omega, \mathbb{R}^2)$. The key point here is that $Gr_3(\mathbb{R}^{4,1}) = SO_0(4,1)/SO(3) \times SO(1,1)$, so that any map Z can be lifted to a map $\phi : \Omega \longrightarrow SO_0(4,1)$; then the Maurer-Cartan form $\omega := \phi^{-1}d\phi$ can be deformed into a form $\omega_\lambda = \lambda^{-1}\omega_1' + \omega_0 + \lambda\omega_1''$, which is curvature–free if and only if X is a conformal Willmore immersion. Therefore we see a way to rely on the results of J. Dorfmeister, F. Pedit and H.-Y. Wu. Note that here ω_1' is *not* a $(1,0)$-form in general, except for special gauge choices of Z (which can be achieved only locally in general).

§4. Another example of an integrable system in geometry: Hamiltonian stationary Lagrangian surfaces

We will describe now another integrable system which also corresponds to a variational problem in differential geometry, and shares many traits with the problems seen above. Nevertheless this new system cannot be reduced to the study of a harmonic map via some Gauss map, as was the case for CMC or in spirit for Willmore surfaces. Details of this work can be found in [17, 18, 19].

4.1. Hamiltonian stationary Lagrangian submanifolds

A symplectic manifold is a (real) even-dimensional manifold $\mathcal{M}^{2n}$ endowed with a closed 2-form ω that is non-degenerate, i.e. for any point $m \in M$, and any covector $\alpha \in T_m^*\mathcal{M}$ there exists a unique vector $V \in T_m\mathcal{M}$ such that $\omega(V, .) = \alpha$. The form ω is called a *symplectic form*. Hereafter we will consider manifolds that are also endowed with a Riemannian metric g, compatible with ω in the sense that the tensor field J defined by $g(JV, W) = \omega(V, W)$ for any $V, W \in T_m\mathcal{M}$ is an almost-complex structure, i.e. $J^2 = -\mathbb{1}$.

The simplest example is $\mathbb{R}^{2n}$ with the standard scalar product $\langle\, .\, ,\, .\, \rangle$ and the symplectic form

$$\omega := dx^1 \wedge dy^1 + \cdots + dx^n \wedge dy^n.$$

The corresponding almost-complex structure J is actually complex (meaning that $\mathbb{R}^{2n} \simeq \mathbb{C}^n$ is indeed a complex manifold); it is given in the

$(x^1, y^1, \dots, x^n, y^n)$ coordinates by the matrix

$$\begin{pmatrix} 0 & -1 & \dots & 0 & 0 \\ 1 & 0 & \dots & 0 & 0 \\ \vdots & \vdots & & \vdots & \vdots \\ 0 & 0 & \dots & 0 & -1 \\ 0 & 0 & \dots & 1 & 0 \end{pmatrix}.$$

The metric and symplectic structures merge into the standard Hermitian product $\langle\, .\, , .\, \rangle_H := \langle\, .\, , .\, \rangle - i\omega(\, .\, , .\,)$. As a consequence, the group of linear transformations of $\mathbb{R}^{2n}$ preserving both the scalar product and the symplectic form is $U(n)$, the group which preserves the Hermitian product.

We now mix both structures to set up a variational problem. Using the metric g, one defines the volume of any immersed submanifold Σ of $\mathcal{M}$, $\mathcal{A}(\Sigma) := \int_\Sigma d\,Vol$. We study critical points of this functional but restricted to (i) *Lagrangian* submanifolds Σ and (ii) Hamiltonian variations (which in particular preserve the condition of being Lagrangian). A *Lagrangian* submanifold of $(\mathcal{M}, \omega)$ is a submanifold Σ of dimension n such that the restriction of ω to Σ vanishes: $\omega_{|\Sigma} = 0$. Equivalently, using the almost-complex structures, $JT_m\mathcal{M}$ is orthogonal to $T_m\mathcal{M}$ for any $m \in \mathcal{M}$. A vector field V on $\mathcal{M}$ is said to be Hamiltonian if there exists $h \in C^\infty(\mathcal{M}, \mathbb{R})$ such that $dh(.) = \omega(V, .)$ (in other words $V = -J\nabla h$). We write then $V = \xi_h$ and call V the *symplectic gradient* of h. It is easy to show that the flow of a Hamiltonian vector field preserves Lagrangian submanifolds, i.e. the image of a Lagrangian submanifold by the flow is again a Lagrangian submanifold.

A *Hamiltonian stationary* Lagrangian submanifold is a Lagrangian submanifold such that the volume is constant up to first order for any Hamiltonian variation: for any $h \in \mathcal{C}^\infty(\mathcal{M}, \mathbb{R})$, $\delta\mathcal{A}(\xi_h) = 0$. In general this notion is weaker than minimality (for a Lagrangian submanifold). Indeed let $\mathcal{H}$ be the mean curvature vector. By definition it is the unique normal vector field along Σ such that for any infinitesimal variation V of Σ,

$$\delta\mathcal{A}(\Sigma)(V) = -\int_\Sigma g(\mathcal{H}, V)d\,Vol\,.$$

Then a Hamiltonian stationary Lagrangian submanifold is a solution to the following system

$$(12) \qquad \begin{cases} \omega_{|\Sigma} & = \quad 0 \\ \displaystyle\int_\Sigma g(J\mathcal{H}, \nabla h)d\,Vol & = \quad 0, \quad \text{for any } h \in \mathcal{C}^\infty(\mathcal{M}, \mathbb{R}) \end{cases}$$

Examples

a) In $\mathbb{R}^2$ the symplectic form is just the volume form $dx \wedge dy$. For dimensional reasons, any curve is Lagrangian. The minimal curves are the straight lines, but Hamiltonian stationary curves are the critical points of length for area-preserving variations. Thus these are straight lines *and* circles. In this case the problem of Hamiltonian stationary Lagrangian submanifolds is akin to the isoperimetric problem.

b) In $\mathbb{R}^4$ an example of Hamiltonian stationary Lagrangian surface is the flat torus

$$S^1 \times S^1 = \{(x^1, y^1, x^2, y^2) \in \mathbb{R}^4 \mid (x^1)^2 + (y^1)^2 = (x^2)^2 + (y^2)^2 = \tfrac{1}{2}\}.$$

In [27], Y. G. Oh has conjectured that this torus minimizes area among all tori obtained from it through Hamiltonian deformations. Under regularity assumptions, a proof of that conjecture has been obtained by H. Anciaux [2].

4.2. Characterizing Hamiltonian stationary submanifolds

In order to understand further the geometry of these submanifolds, we need a better formulation than (12) for the variational problem, and in particular for the mean curvature vector.

Considering first the case $\mathcal{M} = \mathbb{R}^{2n} \simeq \mathbb{C}^n$, let us tackle the problem from the Ruh-Vilms theorem angle and connect it to the Gauss map of a Lagrangian immersion. Obviously this map takes values in the Grassmannian of oriented Lagrangian n-planes

$$Gr_{Lag}(\mathbb{C}^n) \simeq U(n)/SO(n) \simeq U(1) \times (SU(n)/SO(n)).$$

The Gauss map we will consider is the map $\gamma : \Sigma \to Gr_{Lag}(\mathbb{C}^n)$ sending each $m \in \Sigma$ to its oriented tangent space $\gamma(m) = T_m\Sigma$. As mentioned above, Gr_{Lag} splits into two parts and we can write $\gamma = (e^{i\beta}, \tilde{\gamma})$ where $e^{i\beta}$ (respectively, $\tilde{\gamma}$) takes values in $U(1)$ (respectively, $SU(n)/SO(n)$). The $\mathbb{R}/2\pi\mathbb{Z}$-valued map β is called the *Lagrangian angle* and we claim it is the only relevant part of the Gauss map, as far as Hamiltonian stationary submanifolds are concerned.

Lemma 1. *Let Σ be a Lagrangian submanifold. Then the mean curvature vector of Σ is*

$$\mathcal{H} = \tfrac{1}{2} J\nabla\beta,$$

where $e^{i\beta}$ is the $U(1)$ component of the Gauss map.

Note that, even if β is only defined modulo 2π, its gradient is well-defined everywhere. Finally we apply lemma 1 to equation (12) to conclude that

$$\int_\Sigma g(JH, \nabla h) d\,Vol = -\int_\Sigma g(\nabla\beta, \nabla h) d\,Vol = \int_\Sigma h\Delta_\Sigma \beta\, d\,Vol\,.$$

Hence:

Theorem 3. *A Lagrangian submanifold Σ of $\mathbb{C}^n$ is Hamiltonian stationary if and only if*

$$\Delta_\Sigma \beta = 0 \qquad (13)$$

Despite its compact form, this equation is a third order nonlinear PDE. Indeed the operator Δ_Σ depends (nonlinearly) on the immersion X, and β depends on the first order derivatives of X.

One may wonder how to extract painlessly the angle β. Simply pick any oriented orthonormal frame $(e_1, \ldots, e_n)$ of $T_m\Sigma$; then

$$e^{i\beta} = \theta(e_1, \ldots, e_n) \quad \text{where} \quad \theta = dz^1 \wedge \ldots \wedge dz^n\,.$$

This characterization offers a seamless generalization to Calabi-Aubin-Yau manifolds. These manifolds are complex Kähler manifolds (i.e. the almost-complex structure is complex and parallel) with flat Ricci tensor. As a crucial consequence, the canonical bundle is flat. Recall that the canonical bundle of a complex manifold is the (complex) one-dimensional bundle K of $(n,0)$-forms (locally generated by $dz^1\wedge\ldots\wedge dz^n$). The metric induces a connection on K whose curvature is a multiple of the Ricci form $\rho = Ric(J\,.\,,\,.)$; hence the flatness of K. Consequently there exists a (local) nonzero parallel section θ of K, which generalizes $dz^1\wedge\ldots\wedge dz^n$, and we define the Lagrangian angle in the same fashion.

The definition of Lagrangian angle along a Lagrangian submanifold can also be extended to Kähler-Einstein manifolds (see [36]). In that case the Ricci tensor is a multiple of the metric, and consequently the Ricci form is proportional to the symplectic form ω. Thus the curvature of the canonical bundle vanishes along a Lagrangian submanifold. We can define a parallel section θ along Σ only (not on the whole manifold) and give meaning to the Lagrangian angle. Then again $\mathcal{H} = \frac{1}{2}J\nabla\beta$.

4.3. Hamiltonian stationary Lagrangian surfaces in $\mathbb{R}^4$

We specialize now to Lagrangian surfaces in $\mathbb{R}^4 \simeq \mathbb{C}^2$. In this case we may assume without loss of generality that the surface is given by a conformal immersion $X : \Omega \to \mathbb{R}^4$. As usual we take Ω to be a simply connected domain in the plane.

Inspired by the previous constructions for CMC surfaces, we build a moving frame (e_1, e_2) in $\mathbb{R}^4$ such that

$$(14) \qquad \left\{ \begin{array}{rcl} \langle e_a(z), e_b(z)\rangle & = & \delta_{ab} \\ \omega(e_1(z), e_2(z)) & = & 0 \end{array} \right. \iff \langle e_a(z), e_b(z)\rangle_H = \delta_{ab}$$

for any $z \in \Omega$. In other words (e_1, e_2) is an unitary basis of $\mathbb{C}^2$ over $\mathbb{C}$, or (e_1, e_2) are the columns of a matrix in $U(2)$. That way we have constructed a map $F : \Omega \to U(2) \ltimes \mathbb{C}^2$ lifting X, defined by

$$F(z) := \begin{pmatrix} e_1(z) & e_2(z) & X(z) \\ 0 & 0 & 1 \end{pmatrix}.$$

Before going further we need to understand what we are looking for. If we lift X into $U(2) \ltimes \mathbb{C}^2$ (using Darboux frames for instance), the lift F will include first order information on X and the Euler-Lagrange equation (13) will be reduced to second order (as in the harmonic map case). If we hope to go one step further, it is judicious to lift the pair (X, γ) – which takes values in $(U(2) \ltimes \mathbb{C}^2)/SO(2)$ – where γ is the Gauss map defined above. Since (X, γ) contains first order data (on X), the lift should contain second order derivatives. Such a solution is feasible but hardly optimal; indeed the only relevant part of the Gauss map is the Lagrangian angle. So we want to lift the pair $(X, e^{i\beta})$, which is a $U(2) \ltimes \mathbb{C}^2/SU(2)$ map.

To achieve this, we require from the lift F (defined as above) that

$$\theta(e_1, e_2) = e^{i\beta} \text{ where } \beta \text{ is the Lagrangian angle along } \Sigma.$$

From now on, the lift will always be such (and we call it a *Lagrangian* lift). Let $\alpha := F^{-1}dF$ be the Maurer-Cartan form, taking values in the Lie algebra of $U(2) \ltimes \mathbb{C}^2$, $\mathfrak{g} = \mathfrak{u}(2) \oplus \mathbb{C}^2$. To have a better understanding of the geometry of this Lie algebra, we introduce the matrix

$$K := \begin{pmatrix} 0 & 0 & -1 & 0 & 0 \\ 0 & 0 & 0 & 1 & 0 \\ 1 & 0 & 0 & 0 & 0 \\ 0 & -1 & 0 & 0 & 0 \\ 0 & 0 & 0 & 0 & 1 \end{pmatrix},$$

(recall that although we write $U(2)$ we always think of real matrices), and define an automorphism τ on $U(2) \ltimes \mathbb{C}^2$ by $g \mapsto KgK^{-1}$. Abusing notation we write τ again for its differential at identity, acting linearly on $\mathfrak{g}$. Since $\tau^4 = id$, its action diagonalizes on $\mathfrak{g}^{\mathbb{C}} = (\mathfrak{u}(2) \oplus \mathbb{C}^2)^{\mathbb{C}}$ which splits into four eigenspaces $\mathfrak{g}_\ell^{\mathbb{C}}$ for each eigenvalue i^ℓ, where $\ell \in \{-1, 0, 1, 2\}$. So we may write

$$\alpha = \alpha_{-1} + \alpha_0 + \alpha_1 + \alpha_2 \ .$$

As a direct consequence of the choice of lift, each component has a geometrical meaning: $\alpha_{\pm 1}$ contain $\langle dX, e_a \rangle$, α_0 contains the $\mathfrak{su}(2)$ part, and α_2 is more or less $d\beta$, indeed this is the second order information we aimed at. As in the harmonic map case we can split into dz and $d\bar{z}$ parts, writing $\alpha_\ell = \alpha'_\ell + \alpha''_\ell$. We read off the splitting all the geometrical properties we need.

Theorem 4. ***a)*** *The Lagrangian immersion X is conformal if and only if $\alpha'_1 = \alpha''_{-1} = 0$.*

b) *Let X be a conformal Lagrangian immersion, then X is Hamiltonian stationary if and only if*

$$d\alpha_\lambda + \alpha_\lambda \wedge \alpha_\lambda = 0, \quad \textit{for any } \lambda \in S^1 \subset \mathbb{C}, \tag{15}$$

where

$$\alpha_\lambda := \lambda^{-2}\alpha'_2 + \lambda^{-1}\alpha_{-1} + \alpha_0 + \lambda\alpha_1 + \lambda^2\alpha''_2.$$

Equation (15) puts forward a family of deformations of $\alpha = F^{-1}dF$ in a manner similar to what happens in the CMC or harmonic case. In particular, (15) is the necessary and sufficient condition for the existence of $F_\lambda : \Omega \to U(2) \ltimes \mathbb{C}^2$ such that $dF_\lambda = F_\lambda\alpha_\lambda$; and each F_λ yields by projection on $\mathbb{C}^2$ a conformal Hamiltonian stationary Lagrangian immersion X_λ.

This result shows that the Hamiltonian stationary problem is an integrable system. We may at that point use appropriate tools like the twisted loop groups

$$L(U(2) \ltimes \mathbb{C}^2)_\tau := \{S^1 \ni \lambda \mapsto g_\lambda \in U(2) \ltimes \mathbb{C}^2 \,|\, \tau(g_\lambda) = g_{i\lambda}\},$$

and obtain essentially the same results as for CMC surfaces: a Weierstrass-type formula and a classification of all tori (which are of finite type again).

4.4. Generalization to other symmetric spaces

Switching from flat $\mathbb{R}^4$ to other Hermitian symmetric spaces of dimension 2 ($\mathbb{C}P^2$, $\mathbb{C}P^1 \times \mathbb{C}P^1$ and their non-compact duals), it remains true that Hamiltonian stationary Lagrangian surfaces are the solutions of an integrable system, even if the Lagrangian angle is not anymore globally defined (since these manifolds are not Calabi-Aubin-Yau) [19]. Nevertheless the Lagrangian angle can be defined along the surface only, because the manifolds are Kähler-Einstein. Furthermore, the integrable system retains the same structure; the only change lies in the Lie algebra $\mathfrak{g}$ in which the Maurer-Cartan form takes values. For instance, we write the complex projective plane as $\mathbb{C}P^2 = SU(3)/U(2)$ and $\mathfrak{g} = \mathfrak{su}(3)$. As above, we have a notion of an "appropriate" lift of $(X, e^{i\beta})$, and also an order four automorphism with the same geometric properties. All results hold, up to the loop group splittings which depend on the compactness of $\mathfrak{g}$; at that point we have either global or local results, according to the space under consideration.

The case of Lagrangian surfaces in $\mathbb{C}P^2$ deserves special attention due to its relation with conical singularities of Lagrangian surfaces in complex 3-manifolds. Indeed such singularities have a limit cone $C \subset \mathbb{C}^3$ whose intersection $\hat{\Sigma}$ with the unit sphere is a *Legendrian* surface, namely the tangent plane at any $m \in \hat{\Sigma} \subset S^5$ is orthogonal to both m and im. This is a *contact* condition: $\hat{\Sigma}$ is tangent to a four dimensional non integrable distribution in S^5. The Hopf fibration $\pi : S^5 \to \mathbb{C}P^2$ projects Legendrian surfaces down to Lagrangian surfaces. Conversely, any (simply connected) Lagrangian surface admits a unique Legendrian lift through π, up to a multiplicative constant, giving rise to a Lagrangian cone in $\mathbb{C}^3$. Moreover Hamiltonian stationarity is preserved from the cone to the projected surface in $\mathbb{C}P^2$. So is minimality, so that special-Lagrangian cones in $\mathbb{C}^3$ correspond to minimal Lagrangian surfaces in $\mathbb{C}P^2$, a subcase of our integrable system. Notice that an intermediate integrable system can be written for Legendrian surfaces in S^5.

4.5. Another approach for Lagrangian surfaces in $\mathbb{R}^4$

Finally let us inspect more carefully Hamiltonian stationary Lagrangian surfaces in $\mathbb{R}^4$. Using the splitting $\mathfrak{g}^{\mathbb{C}} = \mathfrak{g}_{-1}^{\mathbb{C}} + \mathfrak{g}_0^{\mathbb{C}} + \mathfrak{g}_1^{\mathbb{C}} + \mathfrak{g}_2^{\mathbb{C}}$, equation (15) uncouples to yield (almost) linear systems. It turns out that constructing such surfaces amounts to solving consecutively two linear PDEs, and integrating, a procedure much simpler than the infinite dimensional methods described previously. Even more, this process applies actually to Lagrangian surfaces themselves, not only the stationary ones. Before we proceed with the construction, let us point out its limitations: since the decoupling relies upon commutation properties, it will never apply to other Hermitian symmetric spaces like $\mathbb{C}P^2$, nor to the CMC case.

The simplest way to obtain these results consists in taking a particular lift F among the Lagrangian lifts of $(X, e^{i\beta})$ (X is as usual a conformal Lagrangian immersion, not necessarily stationary). Indeed, choose $F = (e^{i\beta}\epsilon_1, e^{i\beta}\epsilon_2, X)$ or $(-e^{i\beta}\epsilon_1, -e^{i\beta}\epsilon_2, X)$, where (ϵ_1, ϵ_2) is any unitary basis. Clearly F satisfies our axiom from section 4.3. We may actually write

$$e^{-J\beta/2}dX = \mathrm{Re}\left[\begin{pmatrix} s_1 \\ -is_1 \end{pmatrix} dz\right] + i\,\mathrm{Re}\left[\begin{pmatrix} s_2 \\ is_2 \end{pmatrix} dz\right],$$

where s_1 and s_2 are complex-valued functions. If we set

$$\phi := \begin{pmatrix} s_1 \\ \bar{s}_2 \end{pmatrix},$$

then ϕ satisfies the Dirac-type equation

$$\mathcal{D}\phi = M\phi, \tag{16}$$

where

$$\mathcal{D} := \begin{pmatrix} 0 & \partial/\partial z \\ -\partial/\partial\bar{z} & 0 \end{pmatrix} \quad \text{and} \quad M := \frac{1}{2}\begin{pmatrix} \partial\beta/\partial\bar{z} & 0 \\ 0 & \partial\beta/\partial z \end{pmatrix}.$$

Reciprocally, given any map $\beta : \Omega \to \mathbb{R}$ and any ϕ solution of (16), then the $\mathbb{C}^2$-valued one-form

$$\xi := e^{J\beta/2}\,\mathrm{Re}\left(\begin{pmatrix} s_1 \\ -is_1 \end{pmatrix} dz\right) + ie^{J\beta/2}\,\mathrm{Re}\left(\begin{pmatrix} s_2 \\ is_2 \end{pmatrix} dz\right)$$

is closed: $d\xi = 0$. Thus there is a map $X : \Omega \to \mathbb{C}^2 \simeq \mathbb{R}^4$ such that $dX = \xi$ and whenever ϕ does not vanish, X is a conformal Lagrangian immersion. Furthermore, X is Hamiltonian stationary if and only if β is harmonic.

An alternative approach was developed by R. Aiyama [1]. We explain here the connection with our framework. Write

$$X =: e^{i\beta/2}\begin{pmatrix} a + b \\ i(\bar{a} - \bar{b}) \end{pmatrix}, \text{ and } \Phi := \begin{pmatrix} a + \bar{b} \\ i(\bar{a} - b) \end{pmatrix}.$$

Clearly the components of the immersion X can be expressed in terms of Φ and β without integration. But Φ satisfies also a Dirac equation:

$\mathcal{D}\Phi = -\overline{M}\Phi$. The variables Φ and ϕ are related by

$$\phi = (\mathcal{D} + \overline{M}) \begin{pmatrix} 0 & -1 \\ 1 & 0 \end{pmatrix} \Phi.$$

We obtain in this way a Weierstrass representation of sorts for Lagrangian surfaces in $\mathbb{R}^4$. Analogous formulations have already been suggested for surfaces in $\mathbb{R}^3$, extending the Enneper-Weierstrass formula to the non-minimal case; see for instance the work of B. G. Konopelchenko [21]. It is quite instructive to compare his representation formula with the one for Lagrangian surfaces we have just obtained. The representation formula of B. G. Konopelchenko consists in choosing a real-valued function p on Ω, called the potential, and a map $\phi : \Omega \to \mathbb{C}^2$ (actually a spinor) such that $\mathcal{D}\phi = p\phi$ (with the same notation as before). Then the $\mathbb{R}^3$-valued one-form

$$\eta := \mathrm{Re} \begin{pmatrix} s_1^2 dz - \bar{s_2}^2 d\bar{z} \\ -i(s_1^2 dz - \bar{s_2}^2) d\bar{z} \\ s_1 s_2 dz + \overline{s_1 s_2} d\bar{z} \end{pmatrix}$$

satisfies $d\eta = 0$. Hence the existence of a map $Y : \Omega \to \mathbb{R}^3$ such that $dY = \eta$; furthermore Y is a weakly-conformal immersion with first fundamental form

$$I = (|s_1|^2 + |s_2|^2)^2 \begin{pmatrix} 1 & 0 \\ 0 & 1 \end{pmatrix}$$

and mean curvature given by $2p = H(|s_1|^2 + |s_2|^2)$.

However both formulae differ deep inside in that the quantities involved (ϕ, β, p) are of a different nature in each problem (e.g. p is a density while β is a function), and also because one is quadratic while the other is essentially linear. Still there are interesting analogies and in some particular cases, both Dirac equations coincide. Thus some Lagrangian surfaces in $\mathbb{R}^4$ correspond to some surfaces in $\mathbb{R}^3$. This Dirac-like approach yields quite simple expressions in the Hamiltonian stationary case. Notably it leads to a complete description of all such tori in $\mathbb{R}^4$. Using this representation, H. Anciaux [2] has obtained precise isoperimetric estimates in relation to Oh's conjecture.

In this survey, we have exhibited only some aspects of integrable systems theory, and skipped interesting developments such as finite type solutions (e.g. all CMC tori have finite type, as shown by U. Pinkall and I. Sterling [28] (see also [16])), the dressing action, and so on. We have also given few examples. For more details of the general theory see [13, 33, 16]; for Hamiltonian stationary Lagrangian surfaces, see [17, 18, 19].

References

[1] R. Aiyama, Totally real surfaces in the complex 2-space, In: Steps in Differential Geometry, Proceedings of the Colloquium on Differential Geometry, Debrecen, Hungary, 25-30 July, 2000, `http://pc121.math.klke.hu/diffgeo/`.
Lagrangian surfaces in the complex 2-space, In: Proceedings of the 5th International Workshops on Differential Geometry, Kyungpook National University, Korea, 2000.

[2] H. Anciaux, An isoperimetric inequality for Hamiltonian stationary Lagrangian tori in $\mathbb{C}^2$ related to Oh's conjecture, Math. Z., **241** (2002), 639–664.

[3] L. Bianchi, Lezioni di geometria differenziale, Spoerri, Pisa, 1902.

[4] W. Blaschke, Vorlesungen über Differentialgeometrie III, Springer-Verlag, 1929.

[5] O. Bonnet, Note sur une propriété de maximum relative à la sphère, Nouvelles Annales de Mathématiques, **t. XII** (1853), p. 433.

[6] R. Bryant, A duality theorem for Willmore surfaces, J. Differential Geom., **20** (1984), 23–53.
Surfaces in conformal geometry, Proc. Symp. Pure Math. Amer. Math. Soc., **48** (1988), 227–240.

[7] G. Darboux, Leçons sur la théorie générale des surfaces, livre VII, Chapitre X, Gauthiers-Villars, 1884.

[8] H. Dobriner, Über die Flächen mit einem System sphärischer Krümmungslinien, J. Reine Angew. Math., **94** (1883), 116–161.
Die Flächen Constanter Krümmung mit einem System Sphärischer Krümmungslinien dargestellt mit Hilfe von Theta Functionen Zweier Variabeln, Acta Math., **9** (1886), 73–104.

[9] J. Dorfmeister, F. Pedit and H. Y. Wu, Weierstrass type representation of harmonic maps into symmetric spaces, Comm. Anal. Geom., **6** (1998), 633–668.

[10] A. Enneper, Analytisch-geometrische Untersuchungen, Göttingen Nachr., 1868, 258–277 and 421–443.
Untersuchungen über die Flächen mit planen und sphärischen Krümmungslinien, Abh. Königl. Ges. Wissensch. Göttingen, **23** (1878) and **26** (1880).
Über die Flächen mit einem System sphärischen Krümmungslinien, J. Reine Angew. Math., **94** (1883), 329–341.

[11] A. Fordy, J. C. Wood, Harmonic maps and integrable systems, Aspects Math., **E23**, Vieweg, 1994, (see contents at `www.amsta.leeds.ac.uk /Pure/staff/wood/FordyWood/contents.html`).

[12] C. Gardner, J. Greene, M. Kruskal and R. Miura, Method for solving the Korteweg-de Vries equation, Phys. Rev. Lett., **19** (1967), 1095–1097.

[13] M. Guest, Harmonic maps, loop groups and integrable systems, Cambridge Univ. Press, Cambridge, 1997.

[14] F. Hélein, Applications harmoniques, lois de conservation et repères mobiles, Diderot éditeur, 1996, or Harmonic maps , integrable systems and moving frames, Diderot éditeur, 1997.

[15] F. Hélein, Willmore immersions and loop groups, J. Differential Geom., **50** (1998), 331–385. See also Weierstrass representation for Willmore surfaces, In: Harmonic morphisms, harmonic maps and related topics, (eds. C. K. Anand, P. Baird, E. Loubeau and J. C. Wood), Chapman & Hall/CRC Res. Notes Math., **413**.

[16] F. Hélein, Constant mean curvature surfaces, harmonic maps and integrable systems, Lectures Math. ETH Zürich, Birkhäuser, 2001.

[17] F. Hélein and P. Romon, Hamiltonian stationary Lagrangian surfaces in $\mathbb{C}^2$, Commun. Anal. Geom., **10** (2002), 79–126.

[18] F. Hélein and P. Romon, Weierstrass representation of Lagrangian surfaces in four-dimensional space using spinors and quaternions, Comment. Math. Helv., **75** (2000), 668–680.

[19] F. Hélein and P. Romon, Hamiltonian stationary Lagrangian surfaces in Hermitian symmetric spaces, In: Differential geometry and integrable systems, Contemp. Math., **308**, 2002, pp. 161–178.

[20] K. Kenmotsu, Weierstrass formula for surfaces of prescribed mean curvature, Math. Ann., **245** (1979), 89–99.

[21] B. G. Konopelchenko, Induced surfaces and their integrable dynamics, Stud. Appl. Math., **96** (1996), 9–51. See also: B. G. Konopelchenko and I. Taimanov, Constant curvature surfaces via integrable dynamical system, J. Phys. A: Math. Gen., **29** (1996), 1261–2165. and B. G. Konopelchenko and I. Taimanov, Generalized Weierstrass formulae, soliton equations and Willmore surfaces I. Tori of revolution and the mKdV equation, preprint.

[22] T. Kuen, Flächen von constanten negativem Krümmungsmass nach L. Bianchi, Sitzungsber. der Bayer. Akad., 1884.

[23] P. D. Lax, Integrals of nonlinear equations of evolution and solitary waves, Comm. Pure. Appl. Math., **21** (1968), 467–490.

[24] E. Lenz, Über Enneperschen Flächen konstanten negativen Krümmungsmasses, Dissertation, Univ. Göttingen, 1879.

[25] S. Lie, Zur Theorie des Flächen konstanter Krümmung, Archiv for Mathematik og Naturvidenskab (I u. II) (1879) and (III) (1880), Kristiania.

[26] M. Melko and I. Sterling, Integrable systems, harmonic maps and the classical theory of solitons, in [11].

[27] Y. G. Oh, Second variation and stabilities of minimal Lagrangian submanifolds in Kähler manifolds, Invent. Math., **101** (1990), 501–519.
Volume minimization of Lagrangian submanifolds under Hamiltonian deformations, Math. Z., **212** (1993), 175–192.

[28] U. Pinkall and I. Sterling, On the classification of constant mean curvature tori, Ann. of Math. (2), **130** (1989), 407–451.
[29] K. Pohlmeyer, Integrable Hamiltonian systems and interactions through constraints, Comm. Math. Phys., **46** (1976), 207–221.
[30] A. Pressley and G. Segal, Loop groups, Oxford Math. Monogr., Clarendon Press, Oxford, 1986.
[31] E. Ruh and J. Vilms, The tension field of the Gauss map, Trans. Amer. Math. Soc., **149** (1970), 569–573.
[32] W. Symes, Systems of Toda type, inverse spectral problems and representation theory, Invent. Math., **159** (1980), 13–51.
[33] C. L. Terng and K. Uhlenbeck, Integrable systems, J. Differential Geometry Surveys, **4**, International Press, Cambridge, MA, 1999.
[34] G. Thomsen, Über konforme Geometrie I: Grundlagen der konformen Flächentheoremrie, Abh. Math. Sem. Hamburg, 1923, 31–56.
[35] K. Uhlenbeck, Harmonic maps into Lie groups, J. Differential Geom., **30** (1989), 1–50.
[36] J. Wolfson, Minimal Lagrangian diffeomorphisms and the Monge-Ampère equations, J. Differential Geom., **46** (1997), 335–373.
[37] V. E. Zhakarov and A. B. Shabat, Integration of nonlinear equations of mathematical physics by the method of inverse scattering, Funktsional'nyi Analiz i Ego Prilozheniya, **13** (July-September 1979), 13–22.

Frédéric Hélein
CMLA, ENS de Cachan
61 avenue du Président Wilson
94235 Cachan Cedex
FRANCE

Current address:
Institut de Mathématiques de Jussieu, UMR 7586
Université Denis Diderot—Paris 7, Site de Chevaleret
175 rue du Chevaleret
75013 Paris
FRANCE

E-mail address:
helein@math.jussieu.fr

Pascal Romon
Université de Marne-la-Vallée
5 bd Descartes
Champs-sur-Marne
77454 Marne-la-Vallée Cedex 2
FRANCE

E-mail address:
pascal.romon@univ-paris-est.fr

Advanced Studies in Pure Mathematics 51, 2008
Surveys on Geometry and Integrable Systems
pp. 189–233

Special Lagrangian 3-folds and integrable systems

Dominic Joyce

§1. Introduction

This is the sixth in a series of papers [17, 18, 19, 20, 21] constructing explicit examples of special Lagrangian submanifolds (SL m-folds) in $\mathbb{C}^m$. The principal motivation for the series is to study the singularities of SL m-folds, especially when $m = 3$. This paper also has a second objective, which is to connect SL m-folds with the theory of integrable systems, and to arouse interest in special Lagrangian geometry within the integrable systems community.

We begin in §2 with a brief introduction to *special Lagrangian submanifolds* in $\mathbb{C}^m$, which are a class of real m-dimensional minimal submanifolds in $\mathbb{C}^m$, defined using calibrated geometry. Section 3 then gives a rather longer introduction to *harmonic maps* $\psi : S \to \mathbb{CP}^{m-1}$, where S is a Riemann surface. Such maps form an *integrable system*, and have a complex and highly-developed theory involving the Toda lattice equations, loop groups, and classification using spectral curves.

Section 4 explains the connection of this with special Lagrangian geometry. Let N be a *special Lagrangian cone* in $\mathbb{C}^3$, and set $\Sigma = N \cap \mathcal{S}^5$. Then Σ is a *minimal Legendrian surface* in $\mathcal{S}^5$, and so the image of a *conformal harmonic map* $\phi : S \to \mathcal{S}^5$ from a Riemann surface S. The projection $\psi = \pi \circ \phi$ of ϕ from $\mathcal{S}^5$ to $\mathbb{CP}^2$ is also conformal and harmonic, with Lagrangian image.

Thus, ψ can be analyzed in the integrable systems framework of §3. As the image of ψ is Lagrangian there is a simplification, in which the SU(3) Toda lattice equation reduces to the *Tzitzéica equation*, and the spectral curve acquires an extra symmetry. We use the integrable

Received January 30, 2001.
Revised October 1, 2001; August 22, 2007 (updated).

systems theory to give *parameter counts* for the expected families of SL T^2-cones in $\mathbb{C}^3$.

In §5 we give an explicit construction of special Lagrangian cones N in $\mathbb{C}^3$, involving two commuting o.d.e.s, and reducing to constructions given in [17, 19] in special cases. Taking the intersection with $\mathcal{S}^5$, we obtain families of explicit conformal harmonic maps $\phi : \mathbb{R}^2 \to \mathcal{S}^5$ and $\psi : \mathbb{R}^2 \to \mathbb{CP}^2$. Under some circumstances we can solve the conditions for these maps to be *doubly-periodic* in $\mathbb{R}^2$, and so to push down to harmonic maps $T^2 \to \mathcal{S}^5$ and $T^2 \to \mathbb{CP}^2$.

Section 6 analyzes this family of harmonic maps $\psi : \mathbb{R}^2 \to \mathbb{CP}^2$ from the point of view of integrable systems. We find that for generic initial data ψ is *superconformal*, and explicitly determine its harmonic sequence, Toda and Tzitzéica solutions, algebra of polynomial Killing fields, and spectral curve. In §7 we generalize the ideas of §5 to give a new construction of special Lagrangian 3-folds in $\mathbb{C}^3$, which involves *three* commuting o.d.e.s, and reduces to the construction of §5 in a special case.

We end with an open problem. The SL 3-folds of §7 look very similar to those of §5, and share many of the hallmarks of integrable systems – commuting o.d.e.s, elliptic functions, conserved quantities. The author wonders whether these examples can also be explained in terms of some higher-dimensional integrable system.

Acknowledgements. I would like to thank Ian McIntosh, Karen Uhlenbeck, Robert Bryant and Mark Haskins for helpful conversations. I would also like to thank the organizers of the 'Integrable Systems in Differential Geometry' conference, Tokyo, 2000, where I began to understand the material of §3 and §6.

Update added, August 2007: Since this paper was written in 2001, McIntosh [29], Carberry and McIntosh [7], Haskins [15], and others, have studied special Lagrangian cones in $\mathbb{C}^3$ as an integrable system, and proved some very interesting results. Also, Haskins and Kapouleas [16] have constructed examples of special Lagrangian cones in $\mathbb{C}^3$ with higher genus links by a gluing method, as suggested in §5.5 below.

§2. Special Lagrangian submanifolds in $\mathbb{C}^m$

Special Lagrangian submanifolds were introduced by Harvey and Lawson [13]. A recent introductory textbook on special Lagrangian and related geometry is [22]. We begin by defining *calibrations* and *calibrated submanifolds*, following Harvey and Lawson [13].

Definition 2.1 Let (M, g) be a Riemannian manifold. An *oriented tangent k-plane* V on M is a vector subspace V of some tangent space T_xM to M with $\dim V = k$, equipped with an orientation. If V is an oriented tangent k-plane on M then $g|_V$ is a Euclidean metric on V, so combining $g|_V$ with the orientation on V gives a natural *volume form* vol_V on V, which is a k-form on V.

Now let φ be a closed k-form on M. We say that φ is a *calibration* on M if for every oriented k-plane V on M we have $\varphi|_V \leqslant \mathrm{vol}_V$. Here $\varphi|_V = \alpha \cdot \mathrm{vol}_V$ for some $\alpha \in \mathbb{R}$, and $\varphi|_V \leqslant \mathrm{vol}_V$ if $\alpha \leqslant 1$. Let N be an oriented submanifold of M with dimension k. Then each tangent space T_xN for $x \in N$ is an oriented tangent k-plane. We say that N is a *calibrated submanifold* if $\varphi|_{T_xN} = \mathrm{vol}_{T_xN}$ for all $x \in N$.

It is easy to show that calibrated submanifolds are automatically *minimal submanifolds* [13, Th. II.4.2]. Here is the definition of special Lagrangian submanifolds in $\mathbb{C}^m$, taken from [13, §III].

Definition 2.2 Let $\mathbb{C}^m$ have complex coordinates $(z_1, \ldots, z_m)$, and define a metric g, a real 2-form ω and a complex m-form Ω on $\mathbb{C}^m$ by

$$\begin{aligned} g = |\mathrm{d}z_1|^2 + \cdots + |\mathrm{d}z_m|^2, \quad \omega &= \frac{i}{2}(\mathrm{d}z_1 \wedge \mathrm{d}\bar{z}_1 + \cdots + \mathrm{d}z_m \wedge \mathrm{d}\bar{z}_m), \\ \text{and} \quad \Omega &= \mathrm{d}z_1 \wedge \cdots \wedge \mathrm{d}z_m. \end{aligned} \tag{1}$$

Then $\mathrm{Re}\,\Omega$ and $\mathrm{Im}\,\Omega$ are real m-forms on $\mathbb{C}^m$. Let L be an oriented real submanifold of $\mathbb{C}^m$ of real dimension m, and let $\theta \in [0, 2\pi)$. We say that L is a *special Lagrangian submanifold* of $\mathbb{C}^m$ if L is calibrated with respect to $\mathrm{Re}\,\Omega$, in the sense of Definition 2.1. We will often abbreviate 'special Lagrangian' by 'SL', and 'm-dimensional submanifold' by 'm-fold', so that we shall talk about SL m-folds in $\mathbb{C}^m$.

As in [19] there is also a more general definition of special Lagrangian submanifolds involving a *phase* $\mathrm{e}^{i\theta}$, but we will not use it in this paper. Harvey and Lawson [13, Cor. III.1.11] give the following alternative characterization of special Lagrangian submanifolds.

Proposition 2.3. *Let L be a real m-dimensional submanifold of $\mathbb{C}^m$. Then L admits an orientation making it into an SL submanifold of $\mathbb{C}^m$ if and only if $\omega|_L \equiv 0$ and $\mathrm{Im}\,\Omega|_L \equiv 0$.*

Note that an m-dimensional submanifold L in $\mathbb{C}^m$ is called *Lagrangian* if $\omega|_L \equiv 0$. Thus special Lagrangian submanifolds are Lagrangian submanifolds satisfying the extra condition that $\operatorname{Im}\Omega|_L \equiv 0$, which is how they get their name.

§3. Harmonic maps and integrable systems

A map $\phi : S \to M$ of Riemannian manifolds is *harmonic* if it extremizes the energy functional $\int_S |d\phi|^2 dV$. When S is 2-dimensional, the energy is conformally invariant, so that we may take S to be a Riemann surface. In this case, if ϕ is conformal, then ϕ is harmonic if and only if $\phi(S)$ is *minimal* in M. Thus, harmonic maps are closely connected to minimal surfaces.

We shall describe a relationship, due to Bolton, Pedit and Woodward [2], between a special class of harmonic maps $\psi : S \to \mathbb{CP}^{m-1}$ called *superconformal* harmonic maps, and solutions of the *Toda lattice equations* for $\mathrm{SU}(m)$. Then we will explain how superconformal maps can be studied using *loop groups* and *loop algebras*.

This leads to the definition of *polynomial Killing fields* and a special class of superconformal maps called *finite type*, which include all maps from T^2. Finally we explain how to associate a Riemann surface called the *spectral curve* to each finite type superconformal map, and that finite type harmonic maps can be classified in terms of algebro-geometric data including the spectral curve.

This is a deep and complex subject, and we cannot do it justice in a few pages. A good general reference on the following material is Fordy and Wood [11], in particular, the articles by Bolton and Woodward [11, p. 59–82], McIntosh [11, p. 205–220] and Burstall and Pedit [11, p. 221–272].

3.1. The harmonic sequence and superconformal maps

Suppose S is a connected Riemann surface and $\psi : S \to \mathbb{CP}^{m-1}$ a harmonic map. Then the *harmonic sequence* (ψ_k) of ψ is a sequence of harmonic maps $\psi_k : S \to \mathbb{CP}^{m-1}$ with $\psi_0 = \psi$, defined in Bolton and Woodward [3, §1]. Each $\psi_k : S \to \mathbb{CP}^{m-1}$ defines a holomorphic line subbundle L_k of the trivial vector bundle $S \times \mathbb{C}^m$, where a section s of L_k is defined to be holomorphic if $\partial s/\partial\bar{z}$ is orthogonal to L_k.

The ψ_k and L_k are characterized by the following property. If U is an open subset of S and z a holomorphic coordinate on U, then any nonzero holomorphic section ϕ_0 of L_0 over U may be extended uniquely to a sequence of nonzero holomorphic sections ϕ_k of L_k over U satisfying

$$\begin{aligned}(2)\qquad \langle \phi_k, \phi_{k+1}\rangle = 0, \quad &\frac{\partial \phi_k}{\partial z} = \phi_{k+1} + \frac{\partial}{\partial z}\big(\log|\phi_k|^2\big)\phi_k \\ \text{and} \quad &\frac{\partial \phi_k}{\partial \bar z} = -\frac{|\phi_k|^2}{|\phi_{k-1}|^2}\phi_{k-1} \quad \text{for all } k,\end{aligned}$$

where $\langle\,,\rangle$ is the standard Hermitian product on $\mathbb{C}^m$. (Actually one should allow the ϕ_k to be meromorphic, but we will ignore this point.) If (ϕ_k), (ϕ'_k) both satisfy (2) then $\phi'_k = f\phi_k$ for some holomorphic $f : U \to \mathbb{C}^*$ and all k, where $\mathbb{C}^* = \mathbb{C}\setminus\{0\}$. Thus $\psi_k = [\phi_k]$ is independent of the choice of ϕ_0.

Note that ψ_k may not be defined for all $k \in \mathbb{Z}$. For if $\psi_k : S \to \mathbb{CP}^{m-1}$ is holomorphic then $\frac{\partial \phi_k}{\partial z} = 0$, so that $\phi_{k+1} = 0$ and ψ_{k+1} is undefined, and the sequence terminates above at ψ_k. Similarly, if ψ_k is antiholomorphic then ψ_{k-1} is undefined, and the sequence terminates below at ψ_k.

If ψ_k exists for all $k \in \mathbb{Z}$ then ψ is called *non-isotropic*. Otherwise ψ is called *isotropic*. Isotropic maps $\psi : S \to \mathbb{CP}^{m-1}$ were studied by Eells and Wood [9]. They all arise by projection from certain holomorphic maps into a complex flag manifold, and so are fairly easy to understand and construct.

Harmonic sequences have strong orthogonality properties. Points in $\mathbb{CP}^{m-1}$ are called *orthogonal* if the corresponding lines in $\mathbb{C}^m$ are orthogonal at all points, and two maps $\psi_j, \psi_k : S \to \mathbb{CP}^{m-1}$ are called orthogonal if $\psi_j(s)$ and $\psi_k(s)$ are orthogonal in $\mathbb{CP}^{m-1}$ for all $s \in S$.

Bolton and Woodward [3, Prop. 2.4] show that if some set of l consecutive terms in a harmonic sequence (ψ_k) are mutually orthogonal, then *every* set of l consecutive terms of (ψ_k) are mutually orthogonal. A harmonic map $\psi : S \to \mathbb{CP}^{m-1}$ and its harmonic sequence (ψ_k) are both called *l-orthogonal* if every set of l consecutive terms are mutually orthogonal.

Clearly, every harmonic sequence is 2-orthogonal. It is easy to show that $\psi = \psi_0$ is conformal if and only if ψ_1 and ψ_{-1} are orthogonal. Therefore, ψ is 3-orthogonal if and only if it is conformal, and then all the elements ψ_k of the harmonic sequence are also conformal.

The maximum number of mutually orthogonal elements of $\mathbb{CP}^{m-1}$ is m, and so a harmonic map $\psi : S \to \mathbb{CP}^{m-1}$ is at most m-orthogonal. A nonisotropic, m-orthogonal harmonic map $\psi : S \to \mathbb{CP}^{m-1}$ is called *superconformal*. The harmonic sequence of a superconformal map ψ is *periodic*, with period m, so that $\psi_{k+m} = \psi_k$ for all k.

A nonisotropic, conformal harmonic map $\psi : S \to \mathbb{CP}^2$ is superconformal, as ψ is 3-orthogonal because it is conformal, from above. Thus, every conformal harmonic map $\psi : S \to \mathbb{CP}^2$ is either isotropic or superconformal.

3.2. The Toda lattice equations

The *Toda lattice equations* for $\mathrm{SU}(m)$ may be written as follows. For all $k \in \mathbb{Z}$, let $\chi_k : \mathbb{C} \to (0, \infty)$ be differentiable functions satisfying

$$\chi_{k+m} = \chi_k \quad \text{for all } k \in \mathbb{Z}, \quad \chi_0\chi_1 \cdots \chi_{m-1} \equiv 1, \quad \text{and} \tag{3}$$

$$\frac{\partial^2}{\partial z \partial \bar{z}}(\log \chi_k) = \chi_{k+1}\chi_k^{-1} - \chi_k\chi_{k-1}^{-1} \quad \text{for all } k \in \mathbb{Z}. \tag{4}$$

They are important integrable equations in mathematical physics, and large classes of solutions to them may be constructed using loop algebra methods.

We shall show how to construct a solution of (3)–(4) from a superconformal map $\psi : S \to \mathbb{CP}^{m-1}$. Use the notation of §3.1, and suppose ψ is superconformal. Define functions $\chi_k : U \to (0, \infty)$ by $\chi_k = |\phi_k|^2$. Using the fact that $\partial^2\phi_k/\partial z\partial\bar{z} = \partial^2\phi_k/\partial\bar{z}\partial z$, one can show using (2) that

$$\frac{\partial^2}{\partial z \partial \bar{z}} \log |\phi_k|^2 = \frac{|\phi_{k+1}|^2}{|\phi_k|^2} - \frac{|\phi_k|^2}{|\phi_{k-1}|^2}.$$

Thus the χ_k satisfy (4).

To make the χ_k satisfy (3) as well, we need to choose the coordinate z and lifts ϕ_k more carefully. As ψ is superconformal, the ψ_k are periodic with period m. We shall arrange for the lifts ϕ_k also to be periodic with period m. Then $\chi_{k+m} = \chi_k$ for all k, the first equation of (3). This can be done by a suitable choice of holomorphic coordinate z.

It is not difficult to show that the ϕ_k satisfy $\phi_{k+m} = \xi\phi_k$ for all k, where ξ is a nonzero holomorphic function on U. If we change to a new holomorphic coordinate z' on U, then ξ is replaced by

$$\xi' = \left(\frac{\partial z'}{\partial z}\right)^{-m} \xi. \tag{5}$$

Thus, by changing coordinates we can arrange that $\xi \equiv 1$, so that $\phi_{k+m} = \phi_k$ for all k, as we want. A holomorphic coordinate z on an open subset U of S with this property is called *special* [11, p. 65]. Such coordinates are unique up to addition of a constant, and multiplication by an mth root of unity.

Suppose from now on that z is special, so that $\phi_{k+m} = \phi_k$ for all k. It remains to show that we can choose the ϕ_k such that the second equation of (3) holds. Regard the ϕ_k as complex column vectors, so that $(\phi_0\,\phi_1\cdots\phi_{m-1})$ is a complex $m\times m$ matrix. Then the determinant $\det(\phi_0\,\phi_1\cdots\phi_{m-1})$ is a nonzero holomorphic function on U.

From above, the ϕ_k are defined uniquely up to multiplication by some holomorphic function $f : U \to \mathbb{C}^*$. By multiplying the ϕ_k by a suitable f we can arrange that

$$\det(\phi_0\,\phi_1\cdots\phi_{m-1}) \equiv 1. \tag{6}$$

This fixes the ϕ_k uniquely up to multiplication by an mth root of unity. As ψ is superconformal, $\phi_0,\ldots,\phi_{m-1}$ are complex orthogonal in $\mathbb{C}^m$. It follows that

$$\chi_0\chi_1\cdots\chi_{m-1} \equiv |\phi_0|^2|\phi_1|^2\cdots|\phi_{m-1}|^2 \equiv \left|\det(\phi_0\,\phi_1\cdots\phi_{m-1})\right|^2 \equiv 1,$$

so that the second equation of (3) holds. Thus equations (3)–(4) hold, and the χ_k satisfy the Toda lattice equations for $\mathrm{SU}(m)$.

When S is a torus T^2 and $\psi : S \to \mathbb{CP}^{m-1}$ a superconformal harmonic map, from [2, Cor. 2.7] and [11, p. 67-8] there exists a global special holomorphic coordinate z on the universal cover $\mathbb{C}$ of T^2, which then yields a solution (χ_k) of the Toda lattice equations (3)–(4) on the whole of $\mathbb{C}$. In particular, there are no 'higher order singularities', and the ϕ_k and ξ do not have zeros or poles.

3.3. Toda frames and the reconstruction of ψ

Above we saw how to construct a solution (χ_k) of the Toda lattice equations for $\mathrm{SU}(m)$ out of a superconformal harmonic map $\psi : S \to \mathbb{CP}^{m-1}$. We shall now explain how to go the other way, and reconstruct ψ from (χ_k). We continue to use the same notation.

Define $F : U \to \mathrm{GL}(m,\mathbb{C})$ by $F = (f_0\,f_1\cdots f_{m-1})$, where $f_j = |\phi_j|^{-1}\phi_j$. Then $F : U \to \mathrm{SU}(m)$, as $\phi_0,\ldots,\phi_{m-1}$ are complex orthogonal and $\det(\phi_0\,\phi_1\cdots\phi_{m-1}) = 1$. We call F a *Toda frame* for ψ on U, [2, p. 126]. Define α to be the matrix-valued 1-form $F^{-1}\mathrm{d}F$ on U. Then by (2) we find that α is given by

$$
(7)\qquad \begin{pmatrix}
-\frac{i}{2}J\mathrm{d}\log\chi_0 & -\chi_1^{1/2}\chi_0^{-1/2}\mathrm{d}\bar z & & & \chi_0^{1/2}\chi_{m-1}^{-1/2}\mathrm{d}z \\
\chi_1^{1/2}\chi_0^{-1/2}\mathrm{d}z & -\frac{i}{2}J\mathrm{d}\log\chi_1 & \ddots & & \\
 & \chi_2^{1/2}\chi_1^{-1/2}\mathrm{d}z & \ddots & -\chi_{m-2}^{1/2}\chi_{m-3}^{-1/2}\mathrm{d}\bar z & \\
 & & \ddots & -\frac{i}{2}J\mathrm{d}\log\chi_{m-2} & -\chi_{m-1}^{1/2}\chi_{m-2}^{-1/2}\mathrm{d}\bar z \\
-\chi_0^{1/2}\chi_{m-1}^{-1/2}\mathrm{d}\bar z & & & \chi_{m-1}^{1/2}\chi_{m-2}^{-1/2}\mathrm{d}z & -\frac{i}{2}J\mathrm{d}\log\chi_{m-1}
\end{pmatrix}
$$

where J is the complex structure on U.

Now α is a 1-form on U with values in $\mathfrak{su}(m)$, so we may regard it as a *connection 1-form* upon the trivial $\mathrm{SU}(m)$-bundle over U. The connection $\mathrm{d}+\alpha$ is automatically *flat*, as α is of the form $F^{-1}\mathrm{d}F$, so that α satisfies

$$
(8)\qquad \mathrm{d}\alpha + \tfrac{1}{2}[\alpha\wedge\alpha] = 0.
$$

Furthermore, α depends only on the solution (χ_k) of the Toda equations.

Thus, to reconstruct ψ from (χ_k), we proceed as follows. Given (χ_k), we can write down the flat $\mathfrak{su}(m)$-connection $\mathrm{d}+\alpha$ on U. Then we retrieve the Toda frame $F: U \to \mathrm{SU}(m)$ by solving the equation $\mathrm{d}F = F\alpha$, which is in effect two commuting first-order linear o.d.e.s. If U is simply-connected there exists a solution F, which is unique up to multiplication $F \mapsto AF$ by some $A \in \mathrm{SU}(m)$.

Finally we define $\psi = [f_0]$, where f_0 is the first column of F. In this way, any solution (χ_k) of the Toda lattice equations on a simply-connected open set U in $\mathbb{C}$ generates a superconformal map $\psi: U \to \mathbb{CP}^{m-1}$, which is unique up to multiplication by $A \in \mathrm{SU}(m)$, that is, up to automorphisms of $\mathbb{CP}^{m-1}$.

3.4. Loop groups and loops of flat connections

A large part of the integrable systems literature on harmonic maps is formulated in terms of infinite-dimensional Lie groups known as *loop groups*. If G is a finite-dimensional Lie group, the *loop group* LG is the group of smooth maps $\mathcal{S}^1 \to G$, under pointwise multiplication and inverses, and the corresponding *loop algebra* $L\mathfrak{g}$ is the Lie algebra of smooth maps $\mathcal{S}^1 \to \mathfrak{g}$, where $\mathfrak{g}$ is the Lie algebra of G.

In the situation of §3.3, for each $\lambda \in \mathbb{C}$ with $|\lambda| = 1$, define an $\mathfrak{su}(m)$-valued 1-form α_λ on U to be

$$\begin{pmatrix} -\frac{i}{2}J\mathrm{d}\log\chi_0 & -\lambda\chi_1^{1/2}\chi_0^{-1/2}\mathrm{d}\bar{z} & & & \lambda^{-1}\chi_0^{1/2}\chi_{m-1}^{-1/2}\mathrm{d}z \\ \lambda^{-1}\chi_1^{1/2}\chi_0^{-1/2}\mathrm{d}z & -\frac{i}{2}J\mathrm{d}\log\chi_1 & \ddots & & \\ & \lambda^{-1}\chi_2^{1/2}\chi_1^{-1/2}\mathrm{d}z & \ddots & -\lambda\chi_{m-2}^{1/2}\chi_{m-3}^{-1/2}\mathrm{d}\bar{z} & \\ & & \ddots & -\frac{i}{2}J\mathrm{d}\log\chi_{m-2} & -\lambda\chi_{m-1}^{1/2}\chi_{m-2}^{-1/2}\mathrm{d}\bar{z} \\ -\lambda\chi_0^{1/2}\chi_{m-1}^{-1/2}\mathrm{d}\bar{z} & & & \lambda^{-1}\chi_{m-1}^{1/2}\chi_{m-2}^{-1/2}\mathrm{d}z & -\frac{i}{2}J\mathrm{d}\log\chi_{m-1} \end{pmatrix}. \tag{9}$$

When $\lambda = 1$ this coincides with the 1-form α of (7). Using the Toda lattice equations one can show that $\mathrm{d} + \alpha_\lambda$ is also a flat $\mathrm{SU}(m)$-connection, so that

$$\mathrm{d}\alpha_\lambda + \tfrac{1}{2}[\alpha_\lambda \wedge \alpha_\lambda] = 0. \tag{10}$$

Thus the family $\{\alpha_\lambda\}$ gives a *loop of flat connections.* We can interpret this in loop group terms as follows. We defined the α_λ as an $\mathcal{S}^1$ family of 1-forms on $U \subseteq \mathbb{C}$ with values in $\mathfrak{su}(m)$, but we can instead regard it as a single 1-form on U with values in the loop algebra $L\mathfrak{su}(m)$. So the α_λ give an $L\,\mathrm{SU}(m)$-connection on U, which turns out to be flat.

If U is simply-connected, there exists a smooth 1-parameter family of maps $F_\lambda : U \to \mathrm{SU}(m)$ with $F_\lambda^{-1}\mathrm{d}F_\lambda = \alpha_\lambda$, which are unique up to multiplication $F_\lambda \mapsto A_\lambda F_\lambda$ by elements $A_\lambda \in \mathrm{SU}(m)$. The family $\{F_\lambda\}$ is called an *extended Toda frame* for ψ. In loop group terms, we may interpret the F_λ as a map $U \to L\,\mathrm{SU}(m)$. It turns out that each F_λ is the Toda frame of a superconformal harmonic map $\psi_\lambda : U \to \mathbb{CP}^{m-1}$, where the special holomorphic coordinate on U is $\lambda^{-1}z$ rather than z.

Now if $\Phi : U \to L\,\mathrm{SU}(m)$ is any smooth map, then $\mathrm{d} + \Phi^{-1}\mathrm{d}\Phi$ is a flat $L\,\mathrm{SU}(m)$-connection on U, or equivalently a loop of flat $\mathrm{SU}(m)$-connections on U. This gives an enormous family of loops of flat $\mathrm{SU}(m)$-connections on U, most of which have nothing to do with harmonic maps into $\mathbb{CP}^{m-1}$. The important thing about the family $\{\alpha_\lambda\}$ is that it has two special algebraic properties.

The first property is that we may write α_λ in the form

$$\alpha_\lambda = (\alpha_1'\lambda + \alpha_0')\mathrm{d}z + (\alpha_{-1}''\lambda^{-1} + \alpha_0'')\mathrm{d}\bar{z}, \tag{11}$$

where $\alpha_1', \alpha_0', \alpha_{-1}''$ and α_0'' map $U \to \mathfrak{su}(m)^{\mathbb{C}} = \mathfrak{sl}(m, \mathbb{C})$. This equation says two things. Firstly, as a Laurent series in λ we have $\alpha_\lambda = \alpha_1\lambda +$

$\alpha_0 + \alpha_{-1}\lambda^{-1}$. Secondly, if we decompose α_λ into $(1,0)$ and $(0,1)$ parts as $\alpha_\lambda = \alpha'_\lambda \mathrm{d}z + \alpha''_\lambda \mathrm{d}\bar{z}$, then $\alpha'_\lambda = \alpha'_1\lambda + \alpha'_0$, so that α'_λ has no λ^{-1} component, and $\alpha''_\lambda = \alpha''_{-1}\lambda^{-1} + \alpha''_0$, so that α''_λ has no λ component.

The second property is this. Define $\zeta = \mathrm{e}^{2\pi i/m}$, so that $\zeta^m = 1$, and let Υ be the diagonal $m \times m$ matrix with entries $1, \zeta^{-1}, \zeta^{-2}, \ldots, \zeta^{1-m}$. Then

$$\alpha_{\zeta\lambda} = \Upsilon \alpha_\lambda \Upsilon^{-1} \quad \text{for all } \lambda \in \mathbb{C} \text{ with } |\lambda| = 1. \tag{12}$$

That is, α_λ is equivariant under $\mathbb{Z}_m$-actions on $\mathcal{S}^1$ and $\mathfrak{su}(m)$.

It follows from Bolton, Pedit and Woodward [2, §2] that an $\mathcal{S}^1$-family $\mathrm{d} + \alpha_\lambda$ of flat $\mathrm{SU}(m)$-connections on a simply-connected open subset $U \subseteq \mathbb{C}$ come from a solution of the Toda lattice equations, and hence from a superconformal map $\psi : U \to \mathbb{CP}^{m-1}$, if and only if the α_λ satisfy (11)–(12) and the additional condition that $\det(\alpha'_1)$ is nonzero except at isolated points in U.

3.5. Polynomial Killing fields

Polynomial Killing fields were introduced by Ferus et al. [10, §2] and used extensively by Burstall et al. [6], but in a somewhat different situation to us. Our treatment is based on McIntosh [28, App. A] and Bolton, Pedit and Woodward [2, §3].

We shall work with the Lie algebra $\mathfrak{u}(m)$ and its complexification $\mathfrak{gl}(m,\mathbb{C})$ rather than $\mathfrak{su}(m)$ and $\mathfrak{sl}(m,\mathbb{C})$. Let ζ and Υ be as in §3.4. For each $d \in \mathbb{N}$, let $\Lambda_d\mathfrak{gl}(m,\mathbb{C})$ be the vector space of maps $\eta : \mathbb{C}^* \to \mathfrak{gl}(m,\mathbb{C})$ of the form $\eta(\lambda) = \sum_{n=-d}^{d} \eta_n \lambda^n$, where $\eta_n \in \mathfrak{gl}(m,\mathbb{C})$, which satisfy

$$\eta(\zeta\lambda) = \Upsilon\eta(\lambda)\Upsilon^{-1} \quad \text{for all } \lambda \in \mathbb{C}^*. \tag{13}$$

That is, η is equivariant under the same $\mathbb{Z}_m$-actions as α_λ in (12).

Let $\Lambda_d\mathfrak{u}(m)$ be the real vector subspace of $\eta \in \Lambda_d\mathfrak{gl}(m,\mathbb{C})$ such that $\eta(\lambda)$ lies in $\mathfrak{u}(m)$ for all $\lambda \in \mathbb{C}$ with $|\lambda| = 1$. Then $\Lambda_d\mathfrak{gl}(m,\mathbb{C}) = \Lambda_d\mathfrak{u}(m) \otimes_{\mathbb{R}} \mathbb{C}$. Note that by restricting η to $\mathcal{S}^1$ in $\mathbb{C}^*$, we can regard $\Lambda_d\mathfrak{gl}(m,\mathbb{C})$ and $\Lambda_d\mathfrak{u}(m)$ as finite-dimensional vector subspaces of the loop algebras $L\mathfrak{gl}(m,\mathbb{C})$ and $L\mathfrak{u}(m)$.

We define a *polynomial Killing field* on U to be a map $\eta : U \to \Lambda_d\mathfrak{gl}(m,\mathbb{C})$ for some $d \in \mathbb{N}$ satisfying

$$\mathrm{d}\eta = [\eta, \alpha_\lambda]. \tag{14}$$

We call η *real* if it maps to $\Lambda_d\mathfrak{u}(m)$ in $\Lambda_d\mathfrak{gl}(m,\mathbb{C})$. We may write $\eta = \eta(\lambda, z)$ for $\eta \in \mathbb{C}^*$ and $z \in U$, and decompose η as

$$(15) \qquad \eta(\lambda, z) = \sum_{n=-d}^{d} \eta_n(z)\lambda^n, \quad \text{where } \eta_n \text{ maps } U \to \mathfrak{gl}(n, \mathbb{C}).$$

Using the decompositions (11) and (15) of α_λ and η, it is easy to show that (14) is equivalent to the equations

$$(16) \qquad \frac{\partial \eta_n}{\partial z} = [\eta_n, \alpha_0'] + [\eta_{n-1}, \alpha_1'] \quad \text{and}$$

$$(17) \qquad \frac{\partial \eta_n}{\partial \bar{z}} = [\eta_n, \alpha_0''] + [\eta_{n+1}, \alpha_{-1}''] \quad \text{for all } n,$$

where we set $\eta_n \equiv 0$ if $|n| > d$.

Define $\mathcal{A}$ to be the vector space of polynomial Killing fields. It is easy to see that the polynomial Killing fields form a Lie algebra under the obvious Lie bracket. In our case, where ψ is superconformal, this Lie algebra is *abelian*, and the polynomial Killing fields form a *commutative algebra* under matrix multiplication [28, p. 240-1]. The reason why we work with $\mathfrak{gl}(m,\mathbb{C})$ rather than $\mathfrak{sl}(m,\mathbb{C})$ is because $\mathfrak{sl}(m,\mathbb{C})$ is not closed under matrix multiplication.

Following [2, p. 133], we say that ψ is of *finite type* if there exists a real polynomial Killing field $\eta : U \to \Lambda_d\mathfrak{u}(m)$ for some $d \equiv 1 \bmod m$ with $\eta_d = \alpha_1'$ and $\eta_{d-1} = 2\alpha_0'$. All finite type solutions may be obtained by integrating commuting Hamiltonian o.d.e.s on the finite-dimensional manifold $\Lambda_d\mathfrak{u}(m)$, and so are fairly well understood. By [2, Cor. 3.7], every superconformal map corresponding to a doubly-periodic Toda solution on $\mathbb{C}$, and hence every superconformal T^2 in $\mathbb{CP}^{m-1}$, is of finite type.

3.6. Spectral curves

To each superconformal map $\psi : U \to \mathbb{CP}^{m-1}$ of finite type one can associate a Riemann surface known as a *spectral curve.* There are two different definitions of spectral curve in use in the literature. Here is the first. Fix $z \in U$, and define

(18)
$$Y = \{(\lambda, [v]) \in \mathbb{C}^* \times \mathbb{CP}^{m-1} : \forall \eta \in \mathcal{A}, \exists \mu \in \mathbb{C} \text{ with } \eta(\lambda, z)v = \mu v\}.$$

The *spectral curve* as defined by Ferus, Pedit, Pinkall and Sterling [10, §5] is the compactification $\tilde{Y}$ of Y in $\mathbb{CP}^1 \times \mathbb{CP}^{m-1}$. In the generic case, $\tilde{Y}$ is a compact, nonsingular Riemann surface.

However, McIntosh [28, App. A] uses a different definition. As each $\eta \in \mathcal{A}$ satisfies (13), if $(\lambda, [v]) \in Y$ then $(\zeta\lambda, [\Upsilon v]) \in Y$. So define $\nu : Y \to Y$ by

$$\nu : (\lambda, [v]) \mapsto (\zeta\lambda, [\Upsilon v]). \tag{19}$$

Then $\nu^m = 1$, and $\langle \nu \rangle$ is a free $\mathbb{Z}_m$-action on Y, which lifts to a free $\mathbb{Z}_m$-action on $\tilde{Y}$. The spectral curve used by McIntosh is equivalent to $\tilde{X} = \tilde{Y}/\langle \nu \rangle$. It is also generically a compact, nonsingular Riemann surface.

The point of (18) is that as $\mathcal{A}$ is a commutative algebra under matrix multiplication, the matrices $\eta(\lambda, z)$ can be simultaneously diagonalized for all $\eta \in \mathcal{A}$. Assume that all the common eigenspaces are 1-dimensional. Then $(\lambda, [v]) \in Y$ if $[v]$ is an eigenspace of $\eta(\lambda, z)$ for all $\eta \in \mathcal{A}$.

This suggests an alternative description of Y, using eigenvalues rather than eigenvectors. Pick $\eta \in \mathcal{A}$, and define

$$Y' = \{(\lambda, \mu) \in \mathbb{C}^* \times \mathbb{C} : \det\bigl(\mu I - \eta(\lambda, z)\bigr) = 0\}. \tag{20}$$

Define $\pi : Y \to Y'$ by $\pi : (\lambda, [v]) \mapsto (\lambda, \mu)$ when $\eta(\lambda, z)v = \mu v$. Then π is birational for generic η, and biholomorphic if η generates $\mathcal{A}$ over $\mathbb{C}[\lambda^m I, \lambda^{-m} I]$.

As a Riemann surface $\tilde{Y}$ is independent of base-point $z \in U$, but its embedding in $\mathbb{CP}^1 \times \mathbb{CP}^{m-1}$ does depend on z. However, Y' is independent of z. This is because (14) is equivalent to $\mathrm{d}(F_\lambda \eta F_\lambda^{-1}) = 0$, as $\alpha_\lambda = F_\lambda^{-1}\mathrm{d}F_\lambda$. Thus $F_\lambda \eta F_\lambda^{-1}$ is independent of z, so the eigenvalues of $\eta(\lambda, z)$ are independent of z.

Spectral curves are used by McIntosh [26, 27, 28] to give a construction of all finite type harmonic maps $\psi : \mathbb{C} \to \mathbb{CP}^{m-1}$, and hence of all nonisotropic harmonic maps $\psi : T^2 \to \mathbb{CP}^{m-1}$. Above we have only considered superconformal ψ, which are dealt with in [28]; the general case is studied in [26, 27], and is rather more complicated.

By an explicit construction, McIntosh establishes a 1-1 correspondence between finite type nonisotropic harmonic maps $\psi : \mathbb{C} \to \mathbb{CP}^{m-1}$ up to isometry, and quadruples of *spectral data* $(X, \sigma, \pi, \mathcal{L})$, where X is a Riemann surface, $\sigma : X \to X$ a real structure, $\pi : X \to \mathbb{CP}^1$ a branched cover, and $\mathcal{L}$ a holomorphic line bundle over X, all satisfying certain conditions. Understanding the set of such $(X, \sigma, \pi, \mathcal{L})$ is a fairly straightforward problem in algebraic geometry.

When ψ is doubly-periodic it pushes down to T^2, and all nonisotropic harmonic maps $\psi : T^2 \to \mathbb{CP}^{m-1}$ arise in this way. This reduces the classification of nonisotropic harmonic tori in $\mathbb{CP}^{m-1}$ to the problem of understanding the double-periodicity conditions upon $(X, \sigma, \pi, \mathcal{L})$, which will be discussed in §4.3.

§4. SL cones in $\mathbb{C}^3$ and the Tzitzéica equation

Suppose N is a *special Lagrangian cone* in $\mathbb{C}^3$. Then $\Sigma = N \cap \mathcal{S}^5$ is a *minimal Legendrian surface* in $\mathcal{S}^5$, and its projection $\pi(\Sigma)$ to $\mathbb{CP}^2$ is a *minimal Lagrangian surface* in $\mathbb{CP}^2$. Thus, Σ and $\pi(\Sigma)$ are the images of *conformal harmonic maps* $\phi : S \to \mathcal{S}^5$ and $\psi : S \to \mathbb{CP}^2$, where S is a Riemann surface.

Therefore, we can apply the theory of §3 to ψ. It turns out that as $\psi(S)$ is Lagrangian, the corresponding solutions $\chi_k : U \to (0, \infty)$ of the SU(3) Toda lattice equations simplify, coming from a single function $f : U \to \mathbb{R}$ satisfying the *Tzitzéica equation.* This will be explained in §4.1.

The programme of §3.4–§3.6 can then be applied, to interpret finite type solutions of the Tzitzéica equation in terms of *spectral data.* The difference with the SU(3) Toda lattice case is that the spectral curve X acquires an extra symmetry, a holomorphic involution ρ satisfying some compatibility conditions with the other data $\sigma, \pi, \mathcal{L}$. The details can be found in Sharipov [30] and Ma and Ma [25], on which this section is based.

4.1. Derivation of the Tzitzéica equation

Suppose N is a special Lagrangian cone in $\mathbb{C}^3$. Define $\Sigma = N \cap \mathcal{S}^5$. Then Σ is a *minimal Legendrian surface* in $\mathcal{S}^5$, as N is a minimal Lagrangian 3-fold in $\mathbb{C}^3$. It has a natural metric and orientation, and so inherits the structure of a Riemann surface. Let S be Σ, regarded as an abstract Riemann surface, and $\phi : S \to \Sigma$ the inclusion map.

Then ϕ is conformal by definition, and has minimal image, so it is harmonic. Let $\pi : \mathcal{S}^5 \to \mathbb{CP}^2$ be the Hopf projection, and define $\psi : S \to \mathbb{CP}^2$ by $\psi = \pi \circ \phi$. As ϕ has Legendrian image it follows that

ψ is also a conformal harmonic map, with *Lagrangian image*. Therefore we can consider the harmonic sequence (ψ_k) of ψ, as in §3.1.

Let $z = x+iy$ be a holomorphic coordinate on an open subset $U \subset S$. Then $g\big(\phi, \frac{\partial\phi}{\partial x}\big) = g\big(\phi, \frac{\partial\phi}{\partial y}\big) = 0$ as $|\phi| \equiv 1$, and $\omega\big(\phi, \frac{\partial\phi}{\partial x}\big) = \omega\big(\phi, \frac{\partial\phi}{\partial y}\big) = 0$ as N is Lagrangian. Hence, ϕ is complex orthogonal to $\frac{\partial\phi}{\partial\bar z}$, so ϕ is a *holomorphic* section of L_0.

Thus, by §3.1, there is a unique sequence (ϕ_k) with $\phi_0 = \phi$ satisfying equation (2). As $|\phi_0| \equiv 1$, we see that

$$\phi_0 = \phi, \quad \phi_1 = \frac{\partial\phi}{\partial z} \quad \text{and} \quad \phi_{-1} = -\left|\frac{\partial\phi}{\partial\bar z}\right|^{-2} \cdot \frac{\partial\phi}{\partial\bar z}. \tag{21}$$

Suppose now that $\psi : S \to \mathbb{CP}^2$ is *superconformal*, and that z is a *special* holomorphic coordinate. Then ϕ_k exists for all $k \in \mathbb{Z}$ and $\phi_{k+3} = \phi_k$, so that (21) determines ϕ_k for all k. Also, the functions $\chi_k = |\phi_k|^2$ satisfy the *Toda lattice equations* (3)–(4) for $m = 3$.

Now in this case the Toda lattice equations simplify. For $\chi_0 \equiv 1$ as $|\phi_0| \equiv 1$, and so $\chi_2 \equiv \chi_1^{-1}$ as $\chi_0\chi_1\chi_2 \equiv 1$. Defining $f = \log\chi_1$ gives $\chi_{3k} = 1$, $\chi_{3k+1} = \mathrm{e}^f$ and $\chi_{3k-1} = \mathrm{e}^{-f}$ for $f : U \to \mathbb{R}$, and then (4) reduces to the single equation

$$\frac{\partial^2 f}{\partial z \partial\bar z} = \mathrm{e}^{-2f} - \mathrm{e}^f. \tag{22}$$

This is the elliptic version of the *Tzitzéica equation.* The corresponding hyperbolic equation first arose in 1910 in a study by Georges Tzitzéica [31] of a class of surfaces in $\mathbb{RP}^3$ now known as *affine spheres.* The equation was rediscovered in a solitonic context by Bullough and Dodd [4], and amongst mathematical physicists is often known as the Bullough–Dodd equation.

We have shown that superconformal harmonic maps $\psi : S \to \mathbb{CP}^2$ coming from special Lagrangian cones in $\mathbb{C}^3$ are related to solutions of the Tzitzéica equation (22), in the same way that general superconformal harmonic maps $\psi : S \to \mathbb{CP}^{m-1}$ are related to solutions of the $\mathrm{SU}(m)$ Toda lattice equations. The converse also applies, in that starting with a solution of (22) one can reconstruct a special Lagrangian cone in $\mathbb{C}^3$ using the Toda frame method of §3.3.

4.2. Spectral data for the Tzitzéica equation

Suppose f is a solution of the Tzitzéica equation. Then as in §4.1 we get a solution (χ_k) of the SU(3) Toda lattice equations. If (χ_k) is of finite type then as in §3.6 it has a set of *spectral data* $(X, \sigma, \pi, \mathcal{L})$. But because the (χ_k) come from a solution of the Tzitzéica equation and so have a simplified structure, the spectral data $(X, \sigma, \pi, \mathcal{L})$ has an extra symmetry.

It turns out that this symmetry is a *holomorphic involution* $\rho : X \to X$. It commutes with σ, has the property that if $\pi(x) = [1, \lambda]$ then $\pi \circ \rho(x) = [1, -\lambda]$ for $x \in X$ and $\lambda \in \mathbb{C}$, and lifts to a holomorphic involution of the line bundle $\mathcal{L}$.

Thus, just as there is a correspondence between finite type solutions of the SU(m) Toda lattice equations and quadruples of spectral data $(X, \sigma, \pi, \mathcal{L})$ satisfying certain conditions, there is also a correspondence between solutions of the Tzitzéica equation and quintuples of spectral data $(X, \rho, \sigma, \pi, \mathcal{L})$ satisfying certain conditions.

This construction is used in two papers by Sharipov [30] and Ma and Ma [25]. Sharipov considers 'complex normal' surfaces in $\mathcal{S}^5$, which in our terminology are just Legendrian surfaces. He shows that minimal Legendrian tori correspond to solutions of the Tzitzéica equation (22), gives the spectral data for finite type solutions, and sketches how to write the maps $\phi : \mathbb{R}^2 \to \mathcal{S}^5$ in terms of Prym theta functions.

The paper by Ma and Ma is very similar. They consider 'totally real' surfaces in $\mathbb{CP}^2$, which in our terminology are just Lagrangian surfaces. They show that minimal Lagrangian tori correspond to solutions of (22), give the spectral data, and write the maps $\psi : \mathbb{R}^2 \to \mathbb{CP}^2$ in terms of Prym theta functions. The principal difference is that Ma and Ma give more proofs, more detail, and more explicit formulae.

4.3. Parameter counts for minimal tori in $\mathbb{CP}^2$

We shall now use the integrable systems set-up above to give parameter counts for the families of minimal tori in $\mathbb{CP}^2$ and of minimal Lagrangian tori in $\mathbb{CP}^2$ (equivalently, of special Lagrangian T^2-cones in $\mathbb{C}^3$ up to isomorphism).

Similar parameter counts for case of minimal tori in $\mathbb{CP}^{m-1}$ are given by McIntosh in [27, p. 516] and [28, Th. 5], and we follow his method, modifying it in the obvious way for the minimal Lagrangian case by requiring invariance under the holomorphic involution ρ. Each set of spectral data corresponds up to isometries of $\mathbb{CP}^2$ with a unique finite type harmonic map $\psi : \mathbb{R}^2 \to \mathbb{CP}^2$. We shall count the number of free parameters in the spectral data, and the number of restrictions for ψ to be doubly-periodic.

First consider general minimal tori in $\mathbb{CP}^2$. The spectral data for this is a quadruple $(X, \sigma, \pi, \mathcal{L})$. We suppose that X is a nonsingular Riemann surface of genus $p \geqslant 2$. (The case $p \leqslant 1$ is dealt with in [28, §5].) Now $\pi : X \to \mathbb{CP}^1$ is a 3-fold branched cover, which we can regard as a *meromorphic function*, identifying $\mathbb{CP}^1$ with $\mathbb{C} \cup \{\infty\}$. It is required that $\pi^{-1}(0)$ and $\pi^{-1}(\infty)$ should be single points, and thus triple branch points.

When $(X, \sigma, \pi, \mathcal{L})$ is generic, all other branch points of π will be double branch points, and there will be $2p$ of them by elementary topology. Let $\lambda_1, \ldots, \lambda_{2p}$ be the images of these branch points in $\mathbb{C} \setminus \{0\}$. Then the λ_k are distinct, and it is required [27, Prop. 1] that no λ_k lies on the unit circle.

Now σ acts on X, and if $\pi(x) = \lambda$ then $\pi \circ \sigma(x) = 1/\bar{\lambda}$. Clearly σ must take double branch points to double branch points, and so the set $\{\lambda_1, \ldots, \lambda_{2p}\}$ is closed under $\lambda \mapsto 1/\bar{\lambda}$. As no λ_k lies on the unit circle this swaps the λ_k in pairs. Order the λ_k so that $|\lambda_k| < 1$ and $\lambda_{k+p} = 1/\bar{\lambda}_k$ for $k = 1, \ldots, p$.

The triple (X, σ, π) depends on $\lambda_1, \ldots, \lambda_p$ and discrete data. Thus there are $2p$ real parameters in (X, σ, π). The set of σ-invariant line bundles $\mathcal{L}$ has dimension p, but $\mathcal{L}$ depends on a choice of base point in $\mathbb{R}^2$, so factoring out by translations in $\mathbb{R}^2$ shows that the choice of $\mathcal{L}$ really represents $p - 2$ degrees of freedom.

From [28, §5], the double-periodicity conditions for $\psi : \mathbb{R}^2 \to \mathbb{CP}^2$ depend only on (X, σ, π), being independent of $\mathcal{L}$. The condition for the Toda solution to be doubly-periodic is $2p - 4$ rationality conditions, and for ψ to be doubly periodic in $\mathbb{CP}^2$ is another 4 rationality conditions.

Effectively, this means that the moduli space $\mathcal{M}_p$ of data (X, σ, π) has dimension $2p$, and to be doubly-periodic requires that $2p$ real functions $f_1, \ldots, f_{2p}$ on this moduli space have rational values. If $f_1, \ldots, f_{2p}$ are locally transverse, then the set of (X, σ, π) giving doubly-periodic ψ will be countable and dense in $\mathcal{M}_p$. Having fixed (X, σ, π) there are $p - 2$ degrees of freedom to choose $\mathcal{L}$, all of which give doubly-periodic ψ.

Next we do a similar parameter count for minimal Lagrangian tori. To do this we have to include the holomorphic involution $\rho : X \to X$ as in §4.2. This has the property that if $\pi(x) = \lambda$ then $\pi \circ \rho(x) = -\lambda$. Clearly, ρ takes branch points to branch points, so the set $\{\lambda_1, \ldots, \lambda_{2p}\}$

is invariant under $\lambda \mapsto -\lambda$. It follows that p is even, say $p = 2d$, and we can order the λ_k such that $\lambda_{d+k} = -\lambda_k$ for $k = 1, \ldots, d$.

Thus, $\lambda_1, \ldots, \lambda_{4d}$ are determined by $\lambda_1, \ldots, \lambda_d$, so there are $2d$ real parameters in (X, ρ, σ, π). Suppose that $d \geqslant 2$, so that $p \geqslant 4$. (If $d = 1$ the solutions have an $\mathbb{R}$ symmetry group, and the parameter count is slightly different.) The condition for the Tzitzéica solution to be doubly-periodic is $2d - 4$ rationality conditions, and for ψ to be doubly-periodic in $\mathbb{CP}^2$ is another 4 rationality conditions. As L must be invariant under ρ and σ, it has $d - 2$ degrees of freedom.

Effectively, this means that the moduli space $\mathcal{M}'_{2d}$ of data (X, ρ, σ, π) has dimension $2d$, and to be doubly-periodic requires that $2d$ real functions $f_1, \ldots, f_{2d}$ on this moduli space have rational values. Having fixed (X, σ, π) there are $d - 2$ degrees of freedom to choose $\mathcal{L}$, all of which give doubly-periodic ψ.

Here are our conclusions in brief:

—Up to isometries of $\mathbb{CP}^2$, we expect the family of minimal tori in $\mathbb{CP}^2$ with spectral curve of genus $p \geqslant 2$ to depend on $2p$ rational numbers and $p - 2$ real numbers.

—Up to isometries of $\mathbb{CP}^2$, we expect the family of minimal Lagrangian tori in $\mathbb{CP}^2$ with spectral curve of genus $2d \geqslant 4$ to depend on $2d$ rational numbers and $d - 2$ real numbers.

As the double-periodicity conditions for Lagrangian $\psi : \mathbb{R}^2 \to \mathbb{CP}^2$ are equivalent to those for its Legendrian lift $\phi : \mathbb{R}^2 \to \mathcal{S}^5$, the second parameter count also gives the answer for the families of minimal Legendrian tori in $\mathcal{S}^5$ up to transformations in $\mathrm{U}(3)$, and of special Lagrangian T^2-cones in $\mathbb{C}^3$ up to transformations in $\mathrm{SU}(3)$.

The moral for special Lagrangian geometry is that one should expect very large numbers of SL T^2-cones in $\mathbb{C}^3$, which can even exist in continuous families up to isomorphisms. These provide many local models for singularities of SL 3-folds in Calabi–Yau 3-folds.

§5. A family of special Lagrangian cones in $\mathbb{C}^3$

In [19, §8] and [17, §6] the author gave two constructions of countable families of special Lagrangian T^2-cones in $\mathbb{C}^3$, the first using $\mathrm{U}(1)$-invariance, and the second by evolving a 1-parameter family of quadric cones in Lagrangian planes. Both constructions are related to work of other authors.

In particular, the section on U(1)-invariant SL cones in [19] essentially repeats the work of Castro and Urbano [8] on U(1)-invariant minimal tori in $\mathbb{CP}^2$, and was also discovered independently by Haskins [14]. The 'evolving quadrics' construction of [17] generalizes examples of Lawlor [24], and some of the examples it produces were also studied by Bryant [5, §3.5] from a different point of view. For more details, see [17, 19].

Motivated by these we shall now construct a more general family of special Lagrangian cones in $\mathbb{C}^3$ which includes those of [17, 19] as special cases. These some from a family of explicit conformal harmonic maps $\phi : \mathbb{R}^2 \to \mathcal{S}^5$ with Legendrian image, which will be analyzed from the integrable systems point of view in §6.

5.1. Constructing the family

Here is our main result.

Theorem 5.1. *Let $\beta_1, \beta_2, \beta_3$ and $\gamma_1, \gamma_2, \gamma_3$ be real numbers with not all β_j and not all γ_j zero, such that*

$$\beta_1+\beta_2+\beta_3=0, \quad \gamma_1+\gamma_2+\gamma_3=0 \quad \textit{and} \quad \beta_1\gamma_1+\beta_2\gamma_2+\beta_3\gamma_3=0. \tag{23}$$

Suppose $y_1, y_2, y_3 : \mathbb{R} \to \mathbb{C}$ and $v : \mathbb{R} \to \mathbb{R}$ are functions of s, and $z_1, z_2, z_3 : \mathbb{R} \to \mathbb{C}$ and $w : \mathbb{R} \to \mathbb{R}$ functions of t, satisfying

$$\frac{\mathrm{d}y_1}{\mathrm{d}s} = \beta_1\,\overline{y_2y_3}, \qquad \frac{\mathrm{d}y_2}{\mathrm{d}s} = \beta_2\,\overline{y_3y_1}, \qquad \frac{\mathrm{d}y_3}{\mathrm{d}s} = \beta_3\,\overline{y_1y_2}, \tag{24}$$

$$\frac{\mathrm{d}z_1}{\mathrm{d}t} = \gamma_1\,\overline{z_2z_3}, \qquad \frac{\mathrm{d}z_2}{\mathrm{d}t} = \gamma_2\,\overline{z_3z_1}, \qquad \frac{\mathrm{d}z_3}{\mathrm{d}t} = \gamma_3\,\overline{z_1z_2}, \tag{25}$$

$$|y_1|^2 = \beta_1 v + 1, \qquad |y_2|^2 = \beta_2 v + 1, \qquad |y_3|^2 = \beta_3 v + 1, \tag{26}$$

$$|z_1|^2 = \gamma_1 w + 1, \qquad |z_2|^2 = \gamma_2 w + 1, \qquad |z_3|^2 = \gamma_3 w + 1. \tag{27}$$

If (24)–(25) hold for all s, t and (26)–(27) hold for $s = t = 0$, then (26)–(27) hold for all s, t, for some functions v, w. Define $\Phi : \mathbb{R}^3 \to \mathbb{C}^3$ by

$$\Phi : (r,s,t) \mapsto \tfrac{1}{\sqrt{3}}\big(ry_1(s)z_1(t), ry_2(s)z_2(t), ry_3(s)z_3(t)\big). \tag{28}$$

Define a subset N of $\mathbb{C}^3$ by

$$N = \big\{\Phi(r,s,t) : r,s,t \in \mathbb{R}\big\}. \tag{29}$$

Then N is a special Lagrangian cone in $\mathbb{C}^3$.

Proof. Suppose (24) holds for all s, and (26) for $s=0$. From (24) we deduce that $\frac{\mathrm{d}}{\mathrm{d}s}\bigl(|y_j|^2\bigr)=2\beta_j\operatorname{Re}(y_1y_2y_3)$ for $j=1,2,3$. Comparing this with (26) shows that $v(s)$ should satisfy $\frac{\mathrm{d}v}{\mathrm{d}s}=2\operatorname{Re}(y_1y_2y_3)$. Therefore, setting $v(s)=v(0)+2\int_0^s\operatorname{Re}\bigl(y_1(u)y_2(u)y_3(u)\bigr)\mathrm{d}u$ shows that (26) holds for all s. Similarly, if (25) holds for all t and (27) holds for $t=0$, then it holds for all t. This proves the first part of the theorem.

For the second part, we must show that N is special Lagrangian wherever it is nonsingular, that is, wherever Φ is an immersion. Now Φ is an immersion at (r,s,t) when $\frac{\partial\Phi}{\partial r},\frac{\partial\Phi}{\partial s},\frac{\partial\Phi}{\partial t}$ are linearly independent, and then $T_{\Phi(r,s,t)}N=\bigl\langle\frac{\partial\Phi}{\partial r},\frac{\partial\Phi}{\partial s},\frac{\partial\Phi}{\partial t}\bigr\rangle_{\mathbb{R}}$. Thus we must show that $T_{\Phi(r,s,t)}N$ is an SL 3-plane $\mathbb{R}^3$ in $\mathbb{C}^3$ for all (r,s,t) for which $\frac{\partial\Phi}{\partial r},\frac{\partial\Phi}{\partial s},\frac{\partial\Phi}{\partial t}$ are linearly independent. By Proposition 2.3, this holds if and only if

$$\omega\Bigl(\frac{\partial\Phi}{\partial r},\frac{\partial\Phi}{\partial s}\Bigr)\equiv\omega\Bigl(\frac{\partial\Phi}{\partial r},\frac{\partial\Phi}{\partial t}\Bigr)\equiv\omega\Bigl(\frac{\partial\Phi}{\partial s},\frac{\partial\Phi}{\partial t}\Bigr)\equiv 0 \tag{30}$$

$$\text{and}\qquad \operatorname{Im}\Omega\Bigl(\frac{\partial\Phi}{\partial r},\frac{\partial\Phi}{\partial s},\frac{\partial\Phi}{\partial t}\Bigr)\equiv 0. \tag{31}$$

Using equations (24), (25) and (28) we find that

$$\frac{\partial\Phi}{\partial r}=\tfrac{1}{\sqrt{3}}\bigl(y_1z_1,y_2z_2,y_3z_3\bigr), \tag{32}$$

$$\frac{\partial\Phi}{\partial s}=\tfrac{1}{\sqrt{3}}\bigl(r\beta_1\,\overline{y_2y_3}z_1,r\beta_2\,\overline{y_3y_1}z_2,r\beta_3\,\overline{y_1y_2}z_3\bigr), \tag{33}$$

$$\frac{\partial\Phi}{\partial t}=\tfrac{1}{\sqrt{3}}\bigl(r\gamma_1y_1\,\overline{z_2z_3},r\gamma_2y_2\,\overline{z_3z_1},r\gamma_3y_3\,\overline{z_1z_2}\bigr). \tag{34}$$

From (1) we deduce that $\omega\bigl((a_1,a_2,a_3),(b_1,b_2,b_3)\bigr)=\operatorname{Im}(a_1\bar b_1+a_2\bar b_2+a_3\bar b_3)$. Thus from (32) and (33) we have

$$\begin{aligned}\omega\bigl(\tfrac{\partial\Phi}{\partial r},\tfrac{\partial\Phi}{\partial s}\bigr)&=\tfrac13 r\operatorname{Im}(y_1y_2y_3)\bigl(\beta_1|z_1|^2+\beta_2|z_2|^2+\beta_3|z_3|^2\bigr)\\&=\tfrac13 r\operatorname{Im}(y_1y_2y_3)\bigl(\beta_1(\gamma_1w+1)+\beta_2(\gamma_2w+1)+\beta_3(\gamma_3w+1)\bigr)\\&=\tfrac13 r\operatorname{Im}(y_1y_2y_3)\bigl(\beta_1+\beta_2+\beta_3+w(\beta_1\gamma_1+\beta_2\gamma_2+\beta_3\gamma_3)\bigr)=0,\end{aligned}$$

using (27) in the second line and (23) in the third. This proves the first equation of (30). The second follows in the same way, and the third from

$$\omega\bigl(\tfrac{\partial\Phi}{\partial s},\tfrac{\partial\Phi}{\partial t}\bigr)=\tfrac13 r^2\operatorname{Im}(\overline{y_1y_2y_3}z_1z_2z_3)(\beta_1\gamma_1+\beta_2\gamma_2+\beta_3\gamma_3)=0,$$

using (23), (33) and (34).

To prove (31), observe that

$$\Omega\Bigl(\frac{\partial\Phi}{\partial r},\frac{\partial\Phi}{\partial s},\frac{\partial\Phi}{\partial t}\Bigr)=\Bigl|\frac{\partial\Phi}{\partial r}\;\frac{\partial\Phi}{\partial s}\;\frac{\partial\Phi}{\partial t}\Bigr|-\frac{1}{3\sqrt{3}}\begin{vmatrix} y_1z_1 & r\beta_1\,\overline{y_2y_3}z_1 & r\gamma_1y_1\,\overline{z_2z_3} \\ y_2z_2 & r\beta_2\,\overline{y_3y_1}z_2 & r\gamma_2y_2\,\overline{z_3z_1} \\ y_3z_3 & r\beta_3\,\overline{y_1y_2}z_3 & r\gamma_3y_3\,\overline{z_1z_2}\end{vmatrix}$$

$$=\tfrac{1}{3\sqrt{3}}r^2\bigl(\beta_2|y_3y_1|^2\gamma_3|z_1z_2|^2+\beta_3|y_1y_2|^2\gamma_1|z_2z_3|^2+\beta_1|y_2y_3|^2\gamma_2|z_3z_1|^2$$
$$-\beta_3|y_1y_2|^2\gamma_2|z_3z_1|^2-\beta_1|y_2y_3|^2\gamma_3|z_1z_2|^2-\beta_2|y_3y_1|^2\gamma_1|z_2z_3|^2\bigr),$$

where in the first line the terms $|\ldots|$ are determinants of complex 3×3 matrices, and $\frac{\partial\Phi}{\partial r},\frac{\partial\Phi}{\partial s},\frac{\partial\Phi}{\partial t}$ are regarded are complex column matrices. As every term in the second line is real, $\Omega\bigl(\frac{\partial\Phi}{\partial r},\frac{\partial\Phi}{\partial s},\frac{\partial\Phi}{\partial t}\bigr)$ is real. Thus $\operatorname{Im}\Omega\bigl(\frac{\partial\Phi}{\partial r},\frac{\partial\Phi}{\partial s},\frac{\partial\Phi}{\partial t}\bigr)=0$, proving (31). Q.E.D.

5.2. Explicit solution of the o.d.e.s (24) and (25)

As in [19, §8] and [17, §6], we can simplify the solutions of (24) and (25). To do this, note first that (24) implies that

$$\tfrac{\mathrm{d}}{\mathrm{d}s}(y_1y_2y_3)=\beta_1|y_2y_3|^2+\beta_2|y_3y_1|^2+\beta_3|y_1y_2|^2.$$

As the right hand side is real, we have $\operatorname{Im}(y_1y_2y_3)\equiv B$ for some $B\in\mathbb{R}$.

Now $|y_1y_2y_3|^2=(\beta_1v+1)(\beta_2v+1)(\beta_3v+1)$, so this gives

$$\bigl|\operatorname{Re}(y_1y_2y_3)\bigr|^2=(\beta_1v+1)(\beta_2v+1)(\beta_3v+1)-B^2.$$

However, $\frac{\mathrm{d}v}{\mathrm{d}s}=2\operatorname{Re}(y_1y_2y_3)$, as in the proof of Theorem 5.1, and so v satisfies the o.d.e.

$$\Bigl(\frac{\mathrm{d}v}{\mathrm{d}s}\Bigr)^2=4\bigl((\beta_1v+1)(\beta_2v+1)(\beta_3v+1)-B^2\bigr).$$

Since $|y_j|^2=\beta_jv+1$ by (26) we may write $y_j(s)=\mathrm{e}^{i\delta_j(s)}\sqrt{\beta_jv(s)+1}$ for $j=1,2,3$, for real functions $\delta_1,\delta_2,\delta_3$. In this way we prove:

Proposition 5.2. *In the situation of Theorem 5.1 the functions y_1,y_2,y_3 may be written $y_j(s)=\mathrm{e}^{i\delta_j(s)}\sqrt{\beta_jv(s)+1}$, for $v,\delta_1,\delta_2,\delta_3:\mathbb{R}\rightarrow\mathbb{R}$. Define $Q(v)=(\beta_1v+1)(\beta_2v+1)(\beta_3v+1)$ and $\delta=\delta_1+\delta_2+\delta_3$. Then (26) holds automatically, and (24) is equivalent to*

$$\Bigl(\frac{\mathrm{d}v}{\mathrm{d}s}\Bigr)^2=4\bigl(Q(v)-B^2\bigr)\quad\text{and}\quad\frac{\mathrm{d}\delta_j}{\mathrm{d}s}=-\frac{\beta_jB}{\beta_jv+1}\tag{35}$$

for $j=1,2,3$, where $\operatorname{Im}(y_1y_2y_3)\equiv Q(v)^{1/2}\sin\delta\equiv B$ for some $B\in[-1,1]$.

Here is the corresponding result for z_1, z_2, z_3.

Proposition 5.3. *In the situation of Theorem 5.1 the functions z_1, z_2, z_3 may be written $z_j(t) = \mathrm{e}^{i\epsilon_j(t)}\sqrt{\gamma_j w(t)+1}$, for $w, \epsilon_1, \epsilon_2, \epsilon_3 : \mathbb{R} \to \mathbb{R}$. Define $R(w) = (\gamma_1 w + 1)(\gamma_2 w + 1)(\gamma_3 w + 1)$ and $\epsilon = \epsilon_1 + \epsilon_2 + \epsilon_3$. Then* (27) *holds automatically, and* (25) *is equivalent to*

$$\left(\frac{\mathrm{d}w}{\mathrm{d}t}\right)^2 = 4\big(R(w) - C^2\big) \quad \text{and} \quad \frac{\mathrm{d}\epsilon_j}{\mathrm{d}t} = -\frac{\gamma_j C}{\gamma_j w + 1} \tag{36}$$

for $j = 1, 2, 3$, where $\operatorname{Im}(z_1 z_2 z_3) \equiv R(w)^{1/2} \sin\epsilon \equiv C$ for some $C \in [-1, 1]$.

As in [19, §8.2], the o.d.e.s for v and w in (35) and (36) can be solved entirely explicitly in terms of the *Jacobi elliptic functions*. Then δ_j and ϵ_j can also be given explicitly, in terms of integrals involving the Jacobi elliptic functions, and so the solutions y_j, z_j of (24) and (25) are known explicitly in terms of elliptic functions. We shall not give these solutions here.

5.3. Conformal harmonic maps to $\mathcal{S}^5$ and $\mathbb{CP}^2$

As in §4.1, a special Lagrangian cone in $\mathbb{C}^3$ induces conformal harmonic maps $\phi : S \to \mathcal{S}^5$ and $\psi : S \to \mathbb{CP}^2$ from a Riemann surface S. We shall now write these out explicitly for the SL cones of Theorem 5.1. For convenience, we begin by choosing a normalization for the constants $\beta_1, \beta_2, \beta_3$ and $\gamma_1, \gamma_2, \gamma_3$.

Regarding $\boldsymbol{\beta} = (\beta_1, \beta_2, \beta_3)$ and $\boldsymbol{\gamma} = (\gamma_1, \gamma_2, \gamma_3)$ as vectors in $\mathbb{R}^3$, the conditions on $\boldsymbol{\beta}, \boldsymbol{\gamma}$ in Theorem 5.1 are that $\boldsymbol{\beta}$ and $\boldsymbol{\gamma}$ should be nonzero, and that $\boldsymbol{\beta}, \boldsymbol{\gamma}$ and $(1, 1, 1)$ should be orthogonal. However, multiplying $\boldsymbol{\beta}$ or $\boldsymbol{\gamma}$ by a nonzero constant has no effect on the set of special Lagrangian cones constructed in Theorem 5.1.

To see this, let $\beta_j, \gamma_j, y_j, z_j, v$ and w satisfy the conditions of the theorem, let $\sigma, \tau \in \mathbb{R}$ be nonzero, and define

$$\begin{aligned} \beta_j' &= \sigma\beta_j, \; y_j'(s) = y_j(\sigma s) \;\text{ for } j = 1, 2, 3, \text{ and } \; v'(s) = \sigma^{-1} v(\sigma s), \\ \gamma_j' &= \tau\gamma_j, \; z_j'(t) = z_j(\tau t) \;\text{ for } j = 1, 2, 3, \text{ and } \; w'(t) = \tau^{-1} w(\tau t). \end{aligned} \tag{37}$$

Then it is easy to show that $\beta_j', \gamma_j', y_j', z_j', v'$ and w' also satisfy the conditions of the theorem, yielding $\Phi' : \mathbb{R}^3 \to \mathbb{C}^3$ with $\Phi'(r, s, t) = \Phi(r, \sigma s, \tau t)$, so that the images of Φ' and Φ are the same special Lagrangian cone.

Therefore, we are free to rescale $\boldsymbol{\beta}$ and $\boldsymbol{\gamma}$ without changing the resulting set of SL cones. Fix $|\boldsymbol{\beta}| = |\boldsymbol{\gamma}| = 1$. Then $\boldsymbol{\beta}$ and $\boldsymbol{\gamma}$ lie on the unit circle in the plane $x_1 + x_2 + x_3 = 0$ in $\mathbb{R}^3$ and are orthogonal, so we may write

$$\begin{aligned} \boldsymbol{\beta} &= \cos\theta \cdot \tfrac{1}{\sqrt{2}}(1,-1,0) + \sin\theta \cdot \tfrac{1}{\sqrt{6}}(-1,-1,2) \\ \text{and} \quad \boldsymbol{\gamma} &= \cos\theta \cdot \tfrac{1}{\sqrt{6}}(-1,-1,2) - \sin\theta \cdot \tfrac{1}{\sqrt{2}}(1,-1,0) \end{aligned}$$

for some $\theta \in [0, 2\pi)$. (Here $\boldsymbol{\beta}$ determines $\boldsymbol{\gamma}$ up to sign, which we have chosen arbitrarily.) We shall show that with these choices, the map $(s,t) \mapsto \Phi(1,s,t)$ is conformal.

Theorem 5.4. *Fix $\theta \in [0, 2\pi)$, and define*

(38)
$$\beta_1 = \tfrac{1}{\sqrt{2}}\cos\theta - \tfrac{1}{\sqrt{6}}\sin\theta, \qquad \beta_2 = -\tfrac{1}{\sqrt{2}}\cos\theta - \tfrac{1}{\sqrt{6}}\sin\theta, \qquad \beta_3 = \tfrac{2}{\sqrt{6}}\sin\theta,$$

(39)
$$\gamma_1 = -\tfrac{1}{\sqrt{6}}\cos\theta - \tfrac{1}{\sqrt{2}}\sin\theta, \qquad \gamma_2 = -\tfrac{1}{\sqrt{6}}\cos\theta + \tfrac{1}{\sqrt{2}}\sin\theta, \qquad \gamma_3 = \tfrac{2}{\sqrt{6}}\cos\theta.$$

In the situation of Theorem 5.1, with these values of β_j and γ_j, we have $|\Phi(r,s,t)|^2 = r^2$ and $\frac{\partial\Phi}{\partial r}$, $\frac{\partial\Phi}{\partial s}$ and $\frac{\partial\Phi}{\partial t}$ are orthogonal with $|\frac{\partial\Phi}{\partial r}|^2 = 1$ and

$$\text{(40)} \qquad \left|\tfrac{\partial\Phi}{\partial s}\right|^2 = \left|\tfrac{\partial\Phi}{\partial t}\right|^2 = 2r^2\big(a + bv(s) + cw(t)\big),$$

(41)
$$\begin{aligned} \textit{where} \quad a &= \tfrac{1}{6}(\beta_1^2+\beta_2^2+\beta_3^2) = \tfrac{1}{6}(\gamma_1^2+\gamma_2^2+\gamma_3^2) = \tfrac{1}{6}, \\ \text{(42)} \qquad b &= -\tfrac{1}{6}(\beta_1^3+\beta_2^3+\beta_3^3) = \tfrac{1}{6}(\beta_1\gamma_1^2+\beta_2\gamma_2^2+\beta_3\gamma_3^2) = -\tfrac{1}{2}\beta_1\beta_2\beta_3, \\ \text{(43)} \quad \textit{and} \quad c &= \tfrac{1}{6}(\beta_1^2\gamma_1+\beta_2^2\gamma_2+\beta_3^2\gamma_3) = -\tfrac{1}{6}(\gamma_1^3+\gamma_2^3+\gamma_3^3) = -\tfrac{1}{2}\gamma_1\gamma_2\gamma_3. \end{aligned}$$

The maps $\phi : \mathbb{R}^2 \to \mathcal{S}^5$ and $\psi : \mathbb{R}^2 \to \mathbb{CP}^2$ defined by $\phi : (s,t) \mapsto \Phi(1,s,t)$ and $\psi : (s,t) \mapsto [\Phi(1,s,t)]$ are both conformal harmonic maps.

Proof. For the first part, by (23) and (26)–(28) we have

$$\begin{aligned} \big|\Phi(r,s,t)\big|^2 &= \tfrac{1}{3}r^2\big(|y_1|^2|z_1|^2 + |y_2|^2|z_2|^2 + |y_3|^2|z_3|^2\big) \\ &= \tfrac{1}{3}r^2\big((\beta_1 v+1)(\gamma_1 w+1)+(\beta_2 v+1)(\gamma_2 w+1)+(\beta_3 v+1)(\gamma_3 w+1)\big) \\ &= \tfrac{1}{3}r^2\big(3+(\beta_1+\beta_2+\beta_3)v+(\gamma_1+\gamma_2+\gamma_3)w+(\beta_1\gamma_1+\beta_2\gamma_2+\beta_3\gamma_3)vw\big) \\ &= r^2. \end{aligned}$$

The equation $|\frac{\partial\Phi}{\partial r}|^2 = 1$ follows in the same way.

To prove $\frac{\partial\Phi}{\partial r}, \frac{\partial\Phi}{\partial s}, \frac{\partial\Phi}{\partial t}$ are orthogonal we use (32)–(34) and the formula
$g\big((a_1,a_2,a_3),(b_1,b_2,b_3)\big) = \operatorname{Re}(a_1\bar b_1 + a_2\bar b_2 + a_3\bar b_3)$. Thus we have

$$\begin{aligned} g\big(\tfrac{\partial\Phi}{\partial r}, \tfrac{\partial\Phi}{\partial s}\big) &= \tfrac13 r \operatorname{Re}(y_1y_2y_3)\big(\beta_1|z_1|^2 + \beta_2|z_2|^2 + \beta_3|z_3|^2\big) \\ &= \tfrac13 r \operatorname{Re}(y_1y_2y_3)\big(\beta_1(\gamma_1 w + 1) + \beta_2(\gamma_2 w + 1) + \beta_3(\gamma_3 w + 1)\big) \\ &= \tfrac13 r \operatorname{Re}(y_1y_2y_3)\big(\beta_1 + \beta_2 + \beta_3 + w(\beta_1\gamma_1 + \beta_2\gamma_2 + \beta_3\gamma_3)\big) = 0, \end{aligned}$$

using (27) in the second line and (23) in the third. In the same way we show that $g\big(\frac{\partial\Phi}{\partial r}, \frac{\partial\Phi}{\partial t}\big) = g\big(\frac{\partial\Phi}{\partial s}, \frac{\partial\Phi}{\partial t}\big) = 0$, and so $\frac{\partial\Phi}{\partial r}, \frac{\partial\Phi}{\partial s}, \frac{\partial\Phi}{\partial t}$ are orthogonal.

Using equations (26), (27) and (33) we obtain

$$\begin{aligned} \big|\tfrac{\partial\Phi}{\partial s}\big|^2 &= \tfrac13 r^2\big[\beta_1^2|y_2|^2|y_3|^2|z_1|^2 + \beta_2^2|y_3|^2|y_1|^2|z_2|^2 + \beta_3^2|y_1|^2|y_2|^2|z_3|^2\big] \\ &= \tfrac13 r^2\big[\beta_1^2(\beta_2 v+1)(\beta_3 v+1)(\gamma_1 w+1) + \beta_2^2(\beta_3 v+1)(\beta_1 v+1)(\gamma_2 w+1) \\ &\qquad + \beta_3^2(\beta_1 v+1)(\beta_2 v+1)(\gamma_3 w+1)\big] \\ &= \tfrac13 r^2\big[(\beta_1^2+\beta_2^2+\beta_3^2) + v\big(\beta_1^2(\beta_2+\beta_3) + \beta_2^2(\beta_3+\beta_1) + \beta_3^2(\beta_1+\beta_2)\big) \\ &\quad + w(\beta_1^2\gamma_1 + \beta_2^2\gamma_2 + \beta_3^2\gamma_3) + vw\big(\beta_1^2(\beta_2+\beta_3)\gamma_1 + \beta_2^2(\beta_3+\beta_1)\gamma_2 + \beta_3^2(\beta_1+\beta_2)\gamma_3\big) \\ &\qquad + v^2\beta_1\beta_2\beta_3(\beta_1+\beta_2+\beta_3) + v^2 w\beta_1\beta_2\beta_3(\beta_1\gamma_1+\beta_2\gamma_2+\beta_3\gamma_3)\big]. \end{aligned}$$

By (23), the terms in v^2 and v^2w vanish. Also, using (38) and (39) we have

$$\begin{aligned} &\beta_1^2(\beta_2 + \beta_3)\gamma_1 + \beta_2^2(\beta_3 + \beta_1)\gamma_2 + \beta_3^2(\beta_1 + \beta_2)\gamma_3 \\ &\qquad = \tfrac{1}{\sqrt3}\big(\beta_1^2(\beta_2^2 - \beta_3^2) + \beta_2^2(\beta_3^2 - \beta_1^2) + \beta_3^2(\beta_1^2 - \beta_2^2)\big) = 0, \end{aligned}$$

so the term in vw vanishes. Thus, replacing $(\beta_2 + \beta_3)$ by $-\beta_1$, etc., we get

$$\big|\tfrac{\partial\Phi}{\partial s}\big|^2 = \tfrac13 r^2\big[(\beta_1^2 + \beta_2^2 + \beta_3^2) - v(\beta_1^3 + \beta_2^3 + \beta_3^3) + w(\beta_1^2\gamma_1 + \beta_2^2\gamma_2 + \beta_3^2\gamma_3)\big].$$

In the same way, we find that

$$\big|\tfrac{\partial\Phi}{\partial t}\big|^2 = \tfrac13 r^2\big[(\gamma_1^2 + \gamma_2^2 + \gamma_3^2) + v(\beta_1\gamma_1^2 + \beta_2\gamma_2^2 + \beta_3\gamma_3^2) - w(\gamma_1^3 + \gamma_2^3 + \gamma_3^3)\big].$$

Now using (38) and (39) one can show that

$$\begin{gathered} \beta_1^2+\beta_2^2+\beta_3^2 = \gamma_1^2+\gamma_2^2+\gamma_3^2, \quad -(\beta_1^3+\beta_2^3+\beta_3^3) = \beta_1\gamma_1^2+\beta_2\gamma_2^2+\beta_3\gamma_3^2 = -3\beta_1\beta_2\beta_3 \\ \text{and} \quad \beta_1^2\gamma_1+\beta_2^2\gamma_2+\beta_3^2\gamma_3 = -(\gamma_1^3+\gamma_2^3+\gamma_3^3) = -3\gamma_1\gamma_2\gamma_3. \end{gathered}$$

The last five equations prove (40)–(43), as we want. Finally, it follows from what we have proved so far that $\phi : \mathbb{R}^2 \to \mathcal{S}^5$ is a conformal map, and as its image is minimal, it is also harmonic. As ϕ has Legendrian image, ψ is also conformal and harmonic, in the usual way. Q.E.D.

As from §5.2 the functions y_j, z_j defining Φ are known explicitly in terms of integrals involving the Jacobi elliptic functions, we have constructed families of *explicit conformal harmonic maps* $\phi : \mathbb{R}^2 \to \mathcal{S}^5$ and $\psi : \mathbb{R}^2 \to \mathbb{CP}^2$.

5.4. Interesting special cases, and double periodicity

We now consider some special cases in which the y_j or z_j assume a simple form, and so explain how to recover the constructions of [19, §8] and [17, §6] from the more general construction above.

(a) Let $\kappa_1, \kappa_2, \kappa_3 \in \mathbb{R}$ with $\kappa_1 + \kappa_2 + \kappa_3 = -\pi/2$, and define $y_j = \mathrm{e}^{i(\beta_j s + \kappa_j)}$ for $j = 1, 2, 3$. Then it is easy to see that y_1, y_2, y_3 satisfy (24) and (26), with $v \equiv 0$ and $B = -1$. The corresponding special Lagrangian cones in Theorem 5.1 are invariant under the group action

$$(z_1, z_2, z_3) \mapsto (\mathrm{e}^{i\beta_1 s} z_1, \mathrm{e}^{i\beta_2 s} z_2, \mathrm{e}^{i\beta_3 s} z_3)$$

for $s \in \mathbb{R}$, which is a U(1) subgroup of SU(3) if $\beta_1, \beta_2, \beta_3$ are relatively rational, and an $\mathbb{R}$ subgroup otherwise. In this case, Theorem 5.1 reduces to the construction of U(1)-invariant SL cones in $\mathbb{C}^3$ given in [19, §8].

In the same way, we can take $z_j = \mathrm{e}^{i(\gamma_j t + \kappa_j)}$ for $j = 1, 2, 3$, with $w \equiv 0$ and $C = -1$, and two similar cases with $B = 1$ and $C = 1$, all of which give U(1)-invariant or $\mathbb{R}$-invariant SL cones in $\mathbb{C}^3$ coming from the construction of [19, §8]. (See also Castro and Urbano [8], and Haskins [14].)

(b) Take $B = 0$ in Proposition 5.2. Then (35) shows that the phases $\delta_1, \delta_2, \delta_3$ are constant, so we may as well fix them to be 0 or π, and take y_1, y_2, y_3 to be *real*. As in [17, §6.1] the y_j are given by simple formulae involving Jacobi elliptic functions (rather than integrals of Jacobi elliptic functions).

As the point (y_1, y_2, y_3) moves in $\mathbb{R}^3$ it sweeps out one of the two connected components of the curve

$$\{(x_1, x_2, x_3) \in \mathbb{R}^3 : x_1^2 + x_2^2 + x_3^2 = 3, \quad \gamma_1 x_1^2 + \gamma_2 x_2^2 + \gamma_3 x_3^2 = 0\}.$$

From this it follows that for fixed t, as r, s vary $\Phi(r, s, t)$ sweeps out a quadric cone in a Lagrangian $\mathbb{R}^3$ in $\mathbb{C}^3$. So the special Lagrangian cone

N of (29) is the total space of a 1-parameter family of such quadrics, and we recover the 'evolving quadrics' construction of [17].

In the same way, if $C = 0$ in Proposition 5.3 a similar thing happens, with s and t exchanged.

(c) Set $\theta = 0$ in Theorem 5.4. Then $\beta_3 = 0$, so y_3 is constant with $|y_3| = 1$ by (24) and (26), and y_1, y_2 are linear combinations of $\mathrm{e}^{\pm is/\sqrt{2}}$. Also $\gamma_1 = \gamma_2$, so $z_2 \equiv \mathrm{e}^{i\kappa} z_1$ for some $\kappa \in \mathbb{R}$.

The corresponding SL cones in $\mathbb{C}^3$ turn out to be invariant under a U(1) subgroup of SU(3) which fixes the third coordinate in $\mathbb{C}^3$, corresponding to translation in the s variable. Thus, after a linear coordinate change in $\mathbb{C}^3$, this case reduces to a special case of the U(1)-invariant cones in part (a), but with a different parametrization.

In the same way, for each of the five other values of $\theta \in [0, 2\pi)$ for which one of $\beta_2, \beta_3, \gamma_1, \gamma_2$ and γ_3 is zero, a similar thing happens.

Next we consider when the maps $\phi : \mathbb{R}^2 \to \mathcal{S}^5$ and $\psi : \mathbb{R}^2 \to \mathbb{CP}^2$ of Theorem 5.4 are *doubly-periodic* in $\mathbb{R}^2$. Then ϕ and ψ push down to conformal harmonic maps $T^2 \to \mathcal{S}^5$ and $T^2 \to \mathbb{CP}^2$ whose images are *minimal tori* in $\mathcal{S}^5$ and $\mathbb{CP}^2$, and the special Lagrangian cone N of Theorem 5.1 is a cone on T^2. We suppose for simplicity that β_j, γ_j are normalized as in equations (38)–(39).

It turns out that in cases (a)–(c) above the double-periodicity conditions are soluble:

(a) In case (a), suppose $\beta_1, \beta_2, \beta_3$ are *relatively rational*. This happens for a countable dense set of $\theta \in [0, 2\pi)$. Then $\beta_j = n_j/S$ for $S > 0$ and n_1, n_2, n_3 coprime integers. It follows that $y_j(s + S) = y_j(s)$ for $j = 1, 2, 3$ and $s \in \mathbb{R}$, so that the y_j are periodic in s.

For double periodicity in s, t, the z_j have only to be periodic up to multiplication by $\mathrm{e}^{i\beta_j s}$ for some $s \in \mathbb{R}$. Now θ and B are already fixed, but we are free to vary the constant C in Proposition 5.3. It is shown in [19, Th. 8.5] that double periodicity holds for a countable dense subset of $C \in [-1, 1]$.

(b) In case (b) with $B = 0$, y_1, y_2, y_3 are automatically periodic in s. We then need to vary the remaining data θ, C to make z_1, z_2, z_3 periodic in t.

Now w is always periodic in t, with period T, say, and the z_j transform as $z_j(t + T) = \mathrm{e}^{i\zeta_j} z_j(t)$, where $\zeta_1, \zeta_2, \zeta_3 \in \mathbb{R}$ with $\zeta_1 + \zeta_2 + \zeta_3 = 0$.

If $\zeta_j \in \pi\mathbb{Q}$ for $j = 1, 2, 3$ then $n\zeta_j \in 2\pi\mathbb{Z}$ for some positive integer n, and then z_1, z_2, z_3 are periodic with period nT. So, for double periodicity we need 2 functions of θ and C to be rational. In [17, Th.s 5.9, 6.3 & 6.4] it is shown that the z_j are periodic for a countable dense set of values of (θ, C).

(c) In case (c), y_1, y_2, y_3 are automatically periodic with period $2\sqrt{2}\pi$. Also, as $z_2 \equiv \mathrm{e}^{i\kappa} z_1$, the periodicity conditions for z_1, z_2, z_3 reduce to one rationality condition, rather than two. As in case (a), z_1, z_2, z_3 are periodic in t for a countable dense subset of $C \in [-1, 1]$.

What about double periodicity conditions in the general case? If $|B| = 1$ then v is constant and we are in case (a) above, so suppose $|B| < 1$, and similarly $|C| < 1$. Then v, w are automatically nonconstant and periodic in s, t, with periods S, T say, and the y_j and z_j transform as

$$y_j(s + S) = \mathrm{e}^{i\eta_j} y_j(s) \quad \text{and} \quad z_j(t + T) = \mathrm{e}^{i\zeta_j} z_j(t)$$

for some constants $\eta_j, \zeta_j \in \mathbb{R}$ with $\eta_1 + \eta_2 + \eta_3 = \zeta_1 + \zeta_2 + \zeta_3 = 0$. The conditions for the y_j and z_j to be periodic in s and t are that $\eta_j \in \pi\mathbb{Q}$ and $\zeta_j \in \pi\mathbb{Q}$ for $j = 1, 2, 3$ respectively.

Thus, for ϕ and ψ to be doubly-periodic we need the four functions η_1/π, η_2/π, ζ_1/π, ζ_2/π of the three variables θ, B, C to be rational. This is an *overdetermined* problem, so it seems likely that in the general case, the double periodicity conditions will have few solutions, or none. Other than parts (a)–(c) above, the author knows of no cases in which ϕ, ψ are doubly-periodic.

We can use (40) to give a formula for the area of the minimal tori in $\mathcal{S}^5$ or $\mathbb{CP}^2$ arising from the construction above.

Proposition 5.5. *Suppose that the map $\phi : \mathbb{R}^2 \to \mathcal{S}^5$ defined in Theorem 5.4 is doubly-periodic in (s, t), with image Σ, so that Σ is a minimal torus in $\mathcal{S}^5$. Let S, T be the periods of v, w in s and t, as above, and let the period lattice of $(s, t) \mapsto \Phi(1, s, t)$ in $\mathbb{R}^2$ be generated by $(a_{11}S, a_{12}T)$ and $(a_{21}S, a_{22}T)$ for integers a_{ij}. Let $N = |a_{11}a_{22} - a_{12}a_{21}|$. Then the area of Σ is*

$$\text{Area}(\Sigma) = 2N\Big(aST + bT\int_0^S v(s)\mathrm{d}s + cS\int_0^T w(t)\mathrm{d}t\Big). \tag{44}$$

Proof. As $\frac{\partial\Phi}{\partial s}$ and $\frac{\partial\Phi}{\partial t}$ are orthogonal, (40) implies that the area form on Σ is $2\big(a + bv(s) + cw(t)\big)\mathrm{d}s \wedge \mathrm{d}t$. Also, as the period lattice is generated by $(a_{11}S, a_{12}T)$ and $(a_{21}S, a_{22}T)$, we can divide Σ into $N = |a_{11}a_{22} - a_{12}a_{21}|$ copies of the basic rectangle $[0, S] \times [0, T]$, each of which has area $\int_0^T \int_0^S 2\big(a + bv(s) + cw(t)\big)\mathrm{d}s\mathrm{d}t$. Equation (44) follows immediately. Q.E.D.

Observe that v, w can be written explicitly using Jacobi elliptic functions as in §5.2, and so (44) could easily be evaluated numerically in examples using a computer. This may be valuable in studying singularities of special Lagrangian 3-folds, since the area of Σ is a crude measure of how nongeneric singularities modelled on the cone on Σ are in the family of all special Lagrangian 3-folds. Also, note that the area of Σ in $\mathcal{S}^5$ is the same as the area of its image in $\mathbb{CP}^2$, as the two are isometric.

5.5. Comparison with constant mean curvature tori in $\mathbb{R}^3$

There is a strong analogy between the minimal Lagrangian tori in $\mathbb{CP}^2$ constructed above, and the examples of *constant mean curvature (CMC) tori* in $\mathbb{R}^3$ constructed by Wente [32] and Abresch [1], known as *Wente tori.* Wente proved [32] using analysis that there exist immersed CMC tori in $\mathbb{R}^3$, and so provided the first counterexamples to a conjecture of Hopf that the only compact surfaces in $\mathbb{R}^3$ with constant mean curvature are round spheres.

Motivated by Wente's construction, Abresch [1] gave explicit formulae for the Wente tori in terms of elliptic integrals. Abresch's solutions are very similar in structure to those above. In particular, they have a 'separated variable' form, being given in terms of single-variable functions $f(s), g(t)$ rather than two-variable functions, and f and g may be written explicitly using Jacobi elliptic functions.

We can also exploit the analogy in another way. Kapouleas [23] used analytic methods to construct examples of compact CMC surfaces Σ in $\mathbb{R}^3$ for any genus $g \geqslant 3$. It seems very likely that one could use Kapouleas' method to construct examples of higher genus (immersed) minimal Lagrangian surfaces in $\mathbb{CP}^2$, and minimal Legendrian surfaces in $\mathcal{S}^5$. (*Note added in proof:* this has now been done by Haskins and Kapouleas [16].)

Kapouleas makes his examples by gluing together long segments of *Delaunay surfaces*, which are SO(2)-invariant CMC surfaces resembling a string of round 2-spheres joined by narrow, catenoid-like 'necks'.

The appropriate analogues of Delaunay surfaces in our problem are Legendrian surfaces in $\mathcal{S}^5$ invariant under the U(1)-action $(z_1, z_2, z_3) \mapsto (e^{is}z_1, e^{-is}z_2, z_3)$, for $s \in \mathbb{R}$.

In the notation of §5.1–§5.3, these have $\theta = 0$ and $B = -1$. When the remaining parameter $C \in [-1, 1]$ is nonzero and small, the corresponding minimal Legendrian surfaces resemble chains of round Legendrian $\mathcal{S}^2$'s in $\mathcal{S}^5$ joined by small necks.

§6. Interpretation using integrable systems

In Theorem 5.4 we constructed families of conformal harmonic maps $\phi : \mathbb{R}^2 \to \mathcal{S}^5$ and $\psi : \mathbb{R}^2 \to \mathbb{CP}^2$. We shall now analyze these in the integrable systems framework described in §3 and §4. We will show that they are generically superconformal, and explicitly determine their harmonic sequences, Toda and Tzitzéica solutions, loops of flat connections, polynomial Killing fields, and spectral curves. This goes some way towards redressing the 'dearth of examples' of superconformal harmonic tori referred to by Bolton and Woodward [11, p. 76]. We shall use the notation of §5.1–§5.3 throughout.

6.1. The harmonic sequence of ψ

In the situation of §3.1, take U to be $\mathbb{R}^2$ with complex coordinate $z = s + it$. Then $\frac{\partial}{\partial z} = \frac{1}{2}\frac{\partial}{\partial s} - \frac{i}{2}\frac{\partial}{\partial t}$ and $\frac{\partial}{\partial \bar z} = \frac{1}{2}\frac{\partial}{\partial s} + \frac{i}{2}\frac{\partial}{\partial t}$. Thus by (24), (25), (28) and the definition $\phi(s,t) = \Phi(1, s, t)$ we have

$$\begin{aligned}\frac{\partial \phi}{\partial z} &= \frac{1}{2\sqrt{3}}\big(\beta_1 \overline{y_2 y_3} z_1 - i\gamma_1 y_1 \overline{z_2 z_3}, \beta_2 \overline{y_3 y_1} z_2 - i\gamma_2 y_2 \overline{z_3 z_1}, \beta_3 \overline{y_1 y_2} z_3 - i\gamma_3 y_3 \overline{z_1 z_2}\big),\\ \frac{\partial \phi}{\partial \bar z} &= \frac{1}{2\sqrt{3}}\big(\beta_1 \overline{y_2 y_3} z_1 + i\gamma_1 y_1 \overline{z_2 z_3}, \beta_2 \overline{y_3 y_1} z_2 + i\gamma_2 y_2 \overline{z_3 z_1}, \beta_3 \overline{y_1 y_2} z_3 + i\gamma_3 y_3 \overline{z_1 z_2}\big).\end{aligned}$$

Calculation using (26) and (27) shows that $\langle \frac{\partial \phi}{\partial z}, \phi \rangle = \langle \frac{\partial \phi}{\partial \bar z}, \phi \rangle = 0$. Also, using (40) we find that $\left|\frac{\partial \phi}{\partial z}\right|^2 = \left|\frac{\partial \phi}{\partial \bar z}\right|^2 = a + bv(s) + cw(t)$.

As $\langle \frac{\partial \phi}{\partial \bar z}, \phi \rangle = 0$, by definition ϕ is a holomorphic section of the holomorphic line bundle L_0 over $\mathbb{C}$ associated to $\psi_0 = \psi : \mathbb{C} \to \mathbb{CP}^2$. Therefore, from §3.1, there exists a unique sequence of maps $\phi_k : \mathbb{C} \to \mathbb{C}^3$ with $\phi_0 = \phi$, which satisfy (2), and the harmonic sequence (ψ_k) of ψ is given by $\psi_k = [\phi_k]$.

From (2) we see that $\phi_{-1} = -|\phi_0|^2 |\frac{\partial \phi_0}{\partial \bar z}|^{-2} \frac{\partial \phi_0}{\partial \bar z}$ and $\phi_1 = \frac{\partial \phi_0}{\partial z}$, since $|\phi_0| \equiv 1$. Thus the equations above give

(45)
$$\phi_{-1} = -\frac{1}{2\sqrt{3}(a+bv+cw)}\bigl(\beta_1\overline{y_2y_3}z_1+i\gamma_1y_1\overline{z_2z_3}, \beta_2\overline{y_3y_1}z_2+i\gamma_2y_2\overline{z_3z_1}, \beta_3\overline{y_1y_2}z_3+i\gamma_3y_3\overline{z_1z_2}\bigr),$$

(46)
$$\phi_0 = \tfrac{1}{\sqrt{3}}\bigl(y_1z_1, y_2z_2, y_3z_3\bigr),$$

(47)
$$\phi_1 = \tfrac{1}{2\sqrt{3}}\bigl(\beta_1\overline{y_2y_3}z_1-i\gamma_1y_1\overline{z_2z_3}, \beta_2\overline{y_3y_1}z_2-i\gamma_2y_2\overline{z_3z_1}, \beta_3\overline{y_1y_2}z_3-i\gamma_3y_3\overline{z_1z_2}\bigr).$$

These satisfy

(48) $$|\phi_{-1}|^2 = (a+bv+cw)^{-1}, \quad |\phi_0|^2 = 1 \quad \text{and} \quad |\phi_1|^2 = a+bv+cw.$$

From (2) and the equation $|\phi_1|^2 = a+bv+cw$ we see that

$$\phi_2 = \frac{\partial \phi_1}{\partial z} - \frac{\partial}{\partial z}\bigl(\log(a+bv+cw)\bigr)\phi_1.$$

Substituting in for ϕ_1 from (47) gives a long and complicated expression for ϕ_2. After much calculation using equations (24)–(27), (35)–(36), (41)–(43) and other identities satisfied by β_j, γ_j and a, b, c, one can prove that

(49) $$\phi_2 = \xi\phi_{-1}, \quad \text{where} \quad \xi = cC + ibB.$$

We can now identify the harmonic sequence of ψ.

Proposition 6.1. *If bB and cC are not both zero then $\psi : \mathbb{R}^2 \to \mathbb{CP}^2$ is superconformal, and has harmonic sequence (ψ_k) given by*

(50) $$\psi_{3k-1}(s,t) = \bigl[\beta_1\overline{y_2y_3}z_1+i\gamma_1y_1\overline{z_2z_3}, \beta_2\overline{y_3y_1}z_2+i\gamma_2y_2\overline{z_3z_1}, \beta_3\overline{y_1y_2}z_3+i\gamma_3y_3\overline{z_1z_2}\bigr],$$

(51) $$\psi_{3k}(s,t) = \bigl[y_1z_1, y_2z_2, y_3z_3\bigr],$$

(52) $$\psi_{3k+1}(s,t) = \bigl[\beta_1\overline{y_2y_3}z_1-i\gamma_1y_1\overline{z_2z_3}, \beta_2\overline{y_3y_1}z_2-i\gamma_2y_2\overline{z_3z_1}, \beta_3\overline{y_1y_2}z_3-i\gamma_3y_3\overline{z_1z_2}\bigr],$$

for all $k \in \mathbb{Z}$. If $bB = cC = 0$ then ψ is isotropic, with finite harmonic sequence $\psi_{-1}, \psi_0, \psi_1$ given by equations (50)–(52) with $k = 0$.

Proof. Since $\phi_2 = \xi\phi_{-1}$ where $\xi = cC + ibB$ by (49), if $\xi \neq 0$ then the sequence (ϕ_k) exists for all k and is given by

$$\phi_{3k-1} = \xi^k \phi_{-1}, \quad \phi_{3k} = \xi^k \phi_0 \quad \text{and} \quad \phi_{3k+1} = \xi^k \phi_1.$$

Since $\psi_k = [\phi_k]$, equations (50)–(52) follow from (45)–(47). Thus ψ is nonisotropic, as ψ_k exists for all k. But any conformal map $\psi : S \to \mathbb{CP}^2$ is isotropic or superconformal from §3.1, so ψ is superconformal.

If on the other hand $\xi = 0$ then $\phi_2 = 0$, so ψ_2 does not exist. Thus ψ is isotropic. By (45)–(47), ψ_{-1}, ψ_0 and ψ_1 exist and are given by equations (50)–(52) with $k = 0$. But the harmonic sequence of an isotropic map $\psi : S \to \mathbb{CP}^m$ has length at most $m + 1$, so this is the whole of the harmonic sequence. Q.E.D.

In the case when $\xi = 0$ and ψ is isotropic, ψ_{-1} is holomorphic and ψ_1 antiholomorphic. This is not obvious, but may be proved directly. For instance, when $B = C = 0$ we may take the y_j and z_j to be *real*. Then ψ maps to $\mathbb{RP}^2$ in $\mathbb{CP}^2$, and both ψ_1 and ψ_{-1} map to the conic $\{[w_0, w_1, w_2] \in \mathbb{CP}^2 : w_0^2 + w_1^2 + w_2^2 = 0\}$, with $\psi_{-1} = \overline{\psi}_1$.

6.2. Solutions of the Toda lattice and Tzitzéica equations

In the rest of the section we assume that $\xi = cC + ibB \neq 0$, so that ψ is superconformal. Following §3.2, we shall construct a solution of the Toda lattice equations for SU(3) out of ψ. The first thing to do is to find a *special* holomorphic coordinate z' on $\mathbb{C}$, that is, one in which $\xi' = 1$ and the ϕ'_k are periodic with period 3. By (5), $z' = z'(z)$ is special if

$$\xi' = \left(\tfrac{\partial z'}{\partial z}\right)^{-3} \xi = 1.$$

Thus we need $\frac{\partial z'}{\partial z} = \xi^{1/3}$ for some fixed complex cube root $\xi^{1/3}$ of ξ. So define $z' = \xi^{1/3}(s + it)$. Then z' is a special holomorphic coordinate on $\mathbb{C}$.

Working with respect to z' rather than z, we get a new sequence (ϕ'_k) rather than (ϕ_k), with $\phi'_k = -i\xi^{-k/3}\phi_k$. Thus (45)–(48) yield

$$\phi'_{3k-1} = \frac{i\xi^{1/3}}{2\sqrt{3}(a+bv+cw)}\bigl(\beta_1\overline{y_2y_3}z_1+i\gamma_1y_1\overline{z_2z_3}, \quad \beta_2\overline{y_3y_1}z_2+i\gamma_2y_2\overline{z_3z_1},\beta_3\overline{y_1y_2}z_3+i\gamma_3y_3\overline{z_1z_2}\bigr), \tag{53}$$

$$\phi'_{3k} = \frac{-i}{\sqrt{3}}\bigl(y_1z_1,y_2z_2,y_3z_3\bigr), \tag{54}$$

$$\phi'_{3k+1} = \frac{-i\xi^{-1/3}}{2\sqrt{3}}\bigl(\beta_1\overline{y_2y_3}z_1-i\gamma_1y_1\overline{z_2z_3}, \quad \beta_2\overline{y_3y_1}z_2-i\gamma_2y_2\overline{z_3z_1},\beta_3\overline{y_1y_2}z_3-i\gamma_3y_3\overline{z_1z_2}\bigr), \tag{55}$$

$$\begin{aligned} \text{with}\quad &|\phi'_{3k-1}|^2 = |\xi|^{2/3}(a+bv+cw)^{-1}, \quad |\phi'_{3k}|^2 = 1 \\ \text{and}\quad &|\phi'_{3k+1}|^2 = |\xi|^{-2/3}(a+bv+cw) \quad \text{for all } k\in\mathbb{Z}. \end{aligned} \tag{56}$$

Here we have multiplied by $-i$ because then $\det(\phi'_0\phi'_1\phi'_2)\equiv 1$, as in (6). Thus the ϕ'_k satisfy all the conditions on the ϕ_k in §3.1–§3.2. So from §3.2 if we define $\chi_k = |\phi'_k|^2$, then the χ_k satisfy the Toda lattice equations for SU(3) with respect to z'. Therefore by (56) we have proved:

Proposition 6.2. *In the situation above, define $\chi_k : \mathbb{C}\to(0,\infty)$ by*

$$\begin{aligned} &\chi_{3k-1} = |\xi|^{2/3}(a+bv+cw)^{-1}, \quad \chi_{3k} = 1 \quad \textit{and} \\ &\chi_{3k+1} = |\xi|^{-2/3}(a+bv+cw) \quad \textit{for all } k\in\mathbb{Z}. \end{aligned} \tag{57}$$

Then the χ_k satisfy the Toda lattice equations for SU(3) *with respect to $z' = \xi^{1/3}(s+it)$. In terms of s,t, this means that $\chi_0\chi_1\chi_2\equiv 1$, $\chi_{k+3}=\chi_k$ and*

$$\frac{1}{4|\xi|^{2/3}}\left(\frac{\partial^2}{\partial s^2}+\frac{\partial^2}{\partial t^2}\right)(\log\chi_k) = \chi_{k+1}\chi_k^{-1}-\chi_k\chi_{k-1}^{-1} \quad \textit{for all } k\in\mathbb{Z}. \tag{58}$$

Here (58) holds because $\frac{\partial^2}{\partial z'\partial\bar z'} = \frac{1}{4|\xi|^{2/3}}\bigl(\frac{\partial^2}{\partial s^2}+\frac{\partial^2}{\partial t^2}\bigr)$. One can verify (58) explicitly using equations (35)–(36), (41)–(43), (57) and various identities between the β_j,γ_j,B,C,a,b and c. The proposition defines a simple class of doubly-periodic solutions χ_k of the Toda lattice equations for SU(3). From §4.1 we deduce:

Corollary 6.3. *Define $f:\mathbb{C}\to(0,\infty)$ by $f=\log(a+bv+cw)-\frac{2}{3}\log|\xi|$. Then f satisfies the Tzitzéica equation* (22) *with respect to $z' = \xi^{1/3}(s+it)$.*

Note that the functions $v(s)$, $w(t)$ may be written in terms of Jacobi elliptic functions as in §5.2, and so the solutions in the last two results are entirely explicit. They have a 'separated variable' form, that is, they are written in terms of single-variable functions $v(s)$ and $w(t)$, rather than more general two-variable functions $u(s,t)$. The author is not sure whether these solutions are already known.

6.3. Loops of flat connections and polynomial Killing fields

For the rest of §6 we will work with the special coordinate $z = \xi^{1/3}(s+it)$, dropping the notation z'. From §3.3, the *Toda frame* $F : \mathbb{R}^2 \to \mathrm{SU}(3)$ of ψ is given by $F = (f_0 f_1 f_2)$, where $f_k = |\phi_k'|^{-1}\phi_k'$. Using equations (53)–(56) we may write F down explicitly, but we will not do so as the expression is complicated. Then $\alpha = F^{-1}\mathrm{d}F$ is a flat SU(3) connection matrix on $\mathbb{R}^2$.

As in §3.4, we may extend $\mathrm{d}+\alpha$ to a loop of flat SU(3)-connections $\mathrm{d}+\alpha_\lambda$ for $\lambda \in \mathbb{C}$ with $|\lambda| = 1$. We shall write α_λ out explicitly. Decompose α_λ as

$$\alpha_\lambda = (\alpha_1'\lambda + \alpha_0')\mathrm{d}z + (\alpha_{-1}''\lambda^{-1} + \alpha_0'')\mathrm{d}\bar{z}, \tag{59}$$

as in (11). Then from (9) and (57) we find that

$$\alpha_1' = r^{-1/3}\begin{pmatrix} 0 & 0 & f^{1/2} \\ f^{1/2} & 0 & 0 \\ 0 & rf^{-1} & 0 \end{pmatrix}, \quad \alpha_0' = \tfrac{1}{2}\begin{pmatrix} 0 & 0 & 0 \\ 0 & \frac{\partial}{\partial z}(\log f) & 0 \\ 0 & 0 & -\frac{\partial}{\partial z}(\log f) \end{pmatrix}, \tag{60}$$

$$\alpha_{-1}'' = -r^{-1/3}\begin{pmatrix} 0 & f^{1/2} & 0 \\ 0 & 0 & rf^{-1} \\ f^{1/2} & 0 & 0 \end{pmatrix}, \quad \alpha_0'' = \tfrac{1}{2}\begin{pmatrix} 0 & 0 & 0 \\ 0 & -\frac{\partial}{\partial \bar{z}}(\log f) & 0 \\ 0 & 0 & \frac{\partial}{\partial \bar{z}}(\log f) \end{pmatrix}, \tag{61}$$

where $f = a + bv + cw$ and $r = |\xi|$.

We shall now construct a *polynomial Killing field* τ for ψ, as in §3.5, which is in fact the nontrivial polynomial Killing field of lowest degree.

Theorem 6.4. *Write $\xi = r\mathrm{e}^{i\theta}$ for $r > 0$ and $\theta \in \mathbb{R}$. Define functions $f, h : \mathbb{R}^2 \to \mathbb{R}$ and $g : \mathbb{R}^2 \to \mathbb{C}$ by*

$$f = a+bv+cw, \quad g = \frac{1}{2f^{1/2}}\Big(-b\frac{\mathrm{d}v}{\mathrm{d}s} + ic\frac{\mathrm{d}w}{\mathrm{d}t}\Big), \quad h = \frac{1}{12f}\Big(-b\frac{\mathrm{d}^2 v}{\mathrm{d}s^2} + c\frac{\mathrm{d}^2 w}{\mathrm{d}t^2}\Big), \tag{62}$$

and let $\tau = \sum_{n=-2}^{2} \lambda^n \tau_n$, *where*

$$(63) \quad \tau_2 = ie^{2i\theta/3} \begin{pmatrix} 0 & rf^{-1/2} & 0 \\ 0 & 0 & f \\ rf^{-1/2} & 0 & 0 \end{pmatrix}, \quad \tau_1 = ie^{i\theta/3} \begin{pmatrix} 0 & 0 & g \\ -g & 0 & 0 \\ 0 & 0 & 0 \end{pmatrix},$$

$$(64) \quad \tau_0 = i \begin{pmatrix} 2h & 0 & 0 \\ 0 & -h & 0 \\ 0 & 0 & -h \end{pmatrix} \qquad \tau_{-1} = ie^{-i\theta/3} \begin{pmatrix} 0 & -\bar{g} & 0 \\ 0 & 0 & 0 \\ \bar{g} & 0 & 0 \end{pmatrix},$$

$$(65) \qquad \text{and} \quad \tau_{-2} = ie^{-2i\theta/3} \begin{pmatrix} 0 & 0 & rf^{-1/2} \\ rf^{-1/2} & 0 & 0 \\ 0 & f & 0 \end{pmatrix}.$$

Then τ *is a real polynomial Killing field.*

To prove the theorem one must show that the τ_n satisfy (16) and (17). This is a long but straightforward calculation, using equations (35), (36),

$$\frac{\partial}{\partial z} = \frac{1}{2r^{1/3}e^{i\theta/3}}\Big(\frac{\partial}{\partial s} - i\frac{\partial}{\partial t}\Big) \quad \text{and} \quad \frac{\partial}{\partial \bar{z}} = \frac{1}{2r^{1/3}e^{-i\theta/3}}\Big(\frac{\partial}{\partial s} + i\frac{\partial}{\partial t}\Big),$$

and identities satisfied by the β_j, γ_j, B, C and ξ, and we leave it to the reader.

Both α_λ and τ have an extra $\mathbb{Z}_2$-symmetry, which follows from the fact that $\chi_0 \equiv 1$. Define $\kappa : \mathfrak{gl}(3,\mathbb{C}) \to \mathfrak{gl}(3,\mathbb{C})$ by

$$(66) \qquad \kappa : \begin{pmatrix} A_{11} & A_{12} & A_{13} \\ A_{21} & A_{22} & A_{23} \\ A_{31} & A_{32} & A_{33} \end{pmatrix} \longmapsto - \begin{pmatrix} A_{11} & A_{31} & A_{21} \\ A_{13} & A_{33} & A_{23} \\ A_{12} & A_{32} & A_{22} \end{pmatrix}.$$

Then κ is a Lie algebra automorphism, and $\kappa^2 = 1$. It is easy to show from (60)–(61) and (63)–(65) that

$$(67) \qquad \kappa(\alpha_\lambda) = \alpha_{-\lambda} \quad \text{and} \quad \kappa(\tau(\lambda)) = -\tau(-\lambda) \quad \text{for all } \lambda \in \mathbb{C}^*.$$

The action of κ on the algebra of polynomial Killing fields will induce the holomorphic involution ρ on the spectral curve discussed in §4.2.

We can now determine the algebra of polynomial Killing fields $\mathcal{A}$.

Theorem 6.5. *In the situation above, the algebra of polynomial Killing fields* $\mathcal{A}$ *is generated by* τ, $\lambda^3 I$ *and* $\lambda^{-3} I$.

Proof. Let $\mathcal{A}'$ be the subalgebra of $\mathcal{A}$ generated by τ, $\lambda^3 I$ and $\lambda^{-3}I$, and suppose for a contradiction that $\mathcal{A}' \neq \mathcal{A}$. Let $\eta \in \mathcal{A} \setminus \mathcal{A}'$, and take η to be real, and of lowest degree d. That is, $\eta = \sum_{n=-d}^{d} \lambda^n \eta_n$ with $\eta_{-n} = -\bar{\eta}_n^T$ for $n = 0, \ldots, d$, and every polynomial Killing field of degree less than d lies in $\mathcal{A}'$.

As $\eta_{d+1} = 0$, equations (16) with $n = d+1$ and (17) with $n = d$ show that η_d satisfies

$$(68) \qquad [\eta_d, \alpha_1'] = 0 \quad \text{and} \quad \frac{\partial \eta_d}{\partial \bar{z}} = [\eta_d, \alpha_0''].$$

Divide into the three cases (a) $d = 3k$, (b) $d = 3k+1$, and (c) $d = 3k+2$ for some $k = 0, 1, 2, \ldots$. We will prove a contradiction in each case in turn.

In case (a), equation (13) implies that η_d is diagonal, and then as f is nonzero, the first equations of (60) and (68) show that η_d is a multiple of the identity. So write $\eta_d = \epsilon I$ for some $\epsilon : \mathbb{R}^2 \to \mathbb{C}$. Taking the trace of equations (16) and (17) for $n = d$ gives $\frac{\partial \epsilon}{\partial z} = \frac{\partial \epsilon}{\partial \bar{z}} = 0$, as the trace of any commutator is zero. Thus ϵ is constant, and $\eta_d = \epsilon I$, $\eta_{-d} = -\bar{\epsilon} I$.

For $k > 0$, consider $\eta' = \eta - \epsilon(\lambda^3 I)^k + \bar{\epsilon}(\lambda^{-3}I)^{-k}$. This is a polynomial Killing field of degree less than d, as we have cancelled the terms in $\lambda^{\pm d}$. Therefore $\eta' \in \mathcal{A}'$. But $\eta = \eta' + \epsilon(\lambda^3 I)^k - \bar{\epsilon}(\lambda^{-3}I)^{-k}$, so $\eta \in \mathcal{A}'$, a contradiction. Also, when $k = 0$ we have $\eta = \epsilon I \in \mathcal{A}'$. This eliminates case (a).

Similarly, in case (b), equation (13) and the first equations of (60) and (68) imply that

$$\eta_d = \epsilon r^{-1/3} \begin{pmatrix} 0 & 0 & f^{1/2} \\ f^{1/2} & 0 & 0 \\ 0 & r f^{-1} & 0 \end{pmatrix},$$

for some function $\epsilon : \mathbb{R}^2 \to \mathbb{C}$. The second equation of (56) is equivalent to $\frac{\partial \epsilon}{\partial \bar{z}} = 0$, so that ϵ is holomorphic. Using the fact that $F_\lambda \eta F_\lambda^{-1}$ is independent of z one can show that ϵ must be constant. This determines η_d and η_{-d}.

By (63), the leading term of τ^2 is

$$-\lambda^4 e^{4i\theta/3} \begin{pmatrix} 0 & 0 & r f^{1/2} \\ r f^{1/2} & 0 & 0 \\ 0 & r^2 f^{-1} & 0 \end{pmatrix}.$$

Suppose for the moment that $d \geqslant 7$, so that $k \geqslant 2$. Consider

$$\eta' = \eta + (\lambda^3 I)^{k-1}\xi^{-4/3}\epsilon\tau^2 - (\lambda^{-3}I)^{k-1}\bar{\xi}^{-4/3}\bar{\epsilon}\tau^2.$$

We have cancelled the terms in $\lambda^{\pm d}$, so η' is a polynomial Killing field of degree less than d, and lies in $\mathcal{A}'$. So η lies in $\mathcal{A}'$, a contradiction.

The cases $d = 1$ and $d = 4$ must be dealt with separately. By explicit calculation we prove that η is a multiple of I when $d = 1$, and a linear combination of $I, \lambda^{\pm 3}I$, τ and τ^2 when $d = 4$. So $\eta \in \mathcal{A}'$, finishing case (b).

In the same way, in case (c) we find that

$$\eta_d = \epsilon r^{-2/3}\begin{pmatrix} 0 & rf^{-1/2} & 0 \\ 0 & 0 & f \\ rf^{-1/2} & 0 & 0 \end{pmatrix}$$

for some constant $\epsilon \in \mathbb{C}$. When $d \geqslant 5$ we define

$$\eta' = \eta + (\lambda^3 I)^k i\xi^{-2/3}\epsilon\tau + (\lambda^{-3}I)^k i\bar{\xi}^{-2/3}\bar{\epsilon}\tau,$$

and deduce that $\eta' \in \mathcal{A}'$, so that $\eta \in \mathcal{A}'$. The case $d = 2$ we deal with separately, by showing that η is a linear combination of τ and I, and so lies in $\mathcal{A}'$. This completes the proof. Q.E.D.

We can use similar ideas to show that ψ is of *finite type*, as in §3.5. Define

$$\eta = (\xi^{-4/3}\lambda^3 - \bar{\xi}^{-4/3}\lambda^{-3})\tau^2. \tag{69}$$

Then η is a real polynomial Killing field of degree 7, and (60) and (63) imply that $\eta_7 = \alpha_1'$ and $\eta_6 = 2\alpha_0'$. So, by definition, ψ is of finite type.

Furthermore, the proof of the theorem actually implies that *every* polynomial Killing field is of the form $P_0 I + P_1\tau + P_2\tau^2$, where P_0, P_1, P_2 are Laurent polynomials in $\lambda^{\pm 3}$. Writing τ^3 in this way, and using the $\mathbb{Z}_2$-symmetry (67) to eliminate some of the terms, we find that τ must satisfy a cubic equation

$$\tau^3 + D\tau + i(\xi^2\lambda^6 + E + \bar{\xi}^2\lambda^{-6})I = 0 \tag{70}$$

for some $D, E \in \mathbb{R}$. Then $\mathcal{A}$ is the quotient of the free commutative algebra generated by $\lambda^{\pm 3}I$ and τ by the ideal generated by this equation.

6.4. The spectral curve

Now we can calculate the spectral curve of ψ, as in §3.6. Define

$$Y' = \{(\lambda, \mu) \in \mathbb{C}^* \times \mathbb{C} : \det(\mu I - \tau(\lambda, z)) = 0\},$$

as in (20). Since $\mathcal{A}$ is generated by τ and $\lambda^{\pm 3} I$, this is biholomorphic to the curve Y of (18), and so the spectral curve $\tilde{Y}$ as defined by Ferus et al. [10, §5] is the compactification $\tilde{Y}$ of Y'.

Calculating using (63)–(65), we find that

$$\det(\mu I - \tau) = \mu^3 + D\mu + iE + i\xi^2\lambda^6 + i\bar{\xi}^2\lambda^{-6}, \quad \text{where} \tag{71}$$

$$D = f^2 + 2r^2 f^{-1} + 2|g|^2 + 3h^2 \quad \text{and} \tag{72}$$

$$E = -fg^2 - f\bar{g}^2 - 2f^2 h + 2r^2 f^{-1} h + 2|g|^2 h + 2h^3. \tag{73}$$

As Y' is independent of $z \in \mathbb{C}$, the functions D, E are constant, which may be verified directly using (35)–(36), (62) and identities satisfied by the $\beta_j, \gamma_j, a, b, c, B, C$ and r.

We can find explicit expressions for these constants by putting $v = w = 0$, which by (35)–(36) gives

$$\left(\tfrac{\mathrm{d}v}{\mathrm{d}s}\right)^2 = 4(1 - B^2), \quad \left(\tfrac{\mathrm{d}w}{\mathrm{d}t}\right)^2 = 4(1 - C^2) \quad \text{and} \quad \tfrac{\mathrm{d}^2 v}{\mathrm{d}s^2} = \tfrac{\mathrm{d}^2 w}{\mathrm{d}t^2} = 0.$$

Equation (62) gives values for f, g and h, and substituting these into (72) and (73) yields

$$D = a^2 + 2a^{-1}(b^2 + c^2) \quad \text{and} \quad E = 2\big(b^2(1 - B^2) - c^2(1 - C^2)\big). \tag{74}$$

This proves that the spectral curve as defined by Ferus et al. [10, §5] is the compactification $\tilde{Y}$ of

$$Y' = \{(\lambda, \mu) \in \mathbb{C}^* \times \mathbb{C} : \mu^3 + D\mu + iE + i\xi^2\lambda^6 + i\bar{\xi}^2\lambda^{-6} = 0\}, \tag{75}$$

where D and E are given by (74). It can be shown using elementary algebraic geometry that $\tilde{Y}$ is nonsingular for generic D, E, with genus 10. Note that the equation satisfied by μ in (75) is the same as that satisfied by τ in (70).

However, McIntosh [26, 27, 28] uses a different definition of the spectral curve. To find it we replace λ^3 by λ in (75), giving

$$X' = \{(\lambda, \mu) \in \mathbb{C}^* \times \mathbb{C} : \mu^3 + D\mu + iE + i\xi^2\lambda^2 + i\bar{\xi}^2\lambda^{-2} = 0\}, \tag{76}$$

and McIntosh's spectral curve is the compactification $\tilde{X}$ of X'. For generic D, E it is nonsingular with genus 4. The involutions $\sigma : \tilde{X} \to \tilde{X}$ and $\rho : \tilde{X} \to \tilde{X}$ discussed in §3.6 and §4.2 act by

$$\rho : (\lambda, \mu) \mapsto (-\lambda, \mu) \quad \text{and} \quad \sigma : (\lambda, \mu) \mapsto (\bar{\lambda}^{-1}, -\bar{\mu}). \tag{77}$$

It would be interesting to understand what properties of the spectral curve $\tilde{X}$ correspond to the fact that ψ is written in terms of single-variable functions $y_k(s)$ and $z_k(t)$, rather than more general two-variable functions of (s, t). Ian McIntosh has an explanation of this, which may appear elsewhere.

6.5. Interpretation using the ideas of §4

Finally we relate the calculations above to the material of §4. From §6.4 the spectral curve X as defined by McIntosh has genus 4. Thus in §4.3 we have $p = 4$ and $d = 2$. The parameter counts there show that the moduli space of all finite type genus 4 solutions of the Tzitzéica equation, up to translations in $\mathbb{R}^2$, should have dimension 4. All of them are expected to be doubly-periodic. For the corresponding maps $\phi : \mathbb{R}^2 \to \mathcal{S}^5$ and $\psi : \mathbb{R}^2 \to \mathbb{CP}^2$ to be doubly-periodic is 4 rationality conditions.

Now the family of genus 4 solutions of the Tzitzéica equations constructed in Corollary 6.3 depends up to translations in $\mathbb{R}^2$ on the 3 parameters θ, B, C of §5. Thus, we have not constructed all the genus 4 Tzitzéica solutions, but only a codimension 1 subset of them. This agrees with the analysis of §5.4, where we were unable to solve the double-periodicity conditions in general, because they amounted to 4 rationality conditions on 3 variables.

Here are two ways of thinking about why the construction yields only a codimension 1 subset of the Tzitzéica solutions. Firstly, our solutions have a 'separated variable' form, being written in terms of functions $v(s), w(t)$. It follows that the period vectors of the doubly-periodic Tzitzéica solutions will point along the s and t axes, and so be perpendicular in $\mathbb{R}^2$. However, the general genus 4 Tzitzéica solution will have period vectors which are not orthogonal, and to require them to be orthogonal is a codimension 1 condition.

Secondly, although the moduli space of quadruples $(\tilde{X}, \rho, \sigma, \pi)$ with $\tilde{X}$ genus 4 is four-dimensional, the subset which can be defined by an equation of the form (76) is only 3-dimensional. In §6.3 we saw that our solutions admit a degree 2 polynomial Killing field τ, which satisfies a cubic equation over $\mathbb{C}[\lambda^3 I, \lambda^{-3} I]$. It is this cubic equation which gives X' the simple form (76).

So we conclude that although the family of genus 4 Tzitzéica solutions has dimension 4, only a 3-dimensional subfamily of these admit a degree 2 polynomial Killing field τ, and it is this which is responsible for the special form (76) of the spectral curve, and for the other nice behaviour of these examples. For generic genus 4 Tzitzéica solutions the first non-trivial polynomial Killing field will be of higher degree, and so the spectral curve will be given by a (singular) equation of higher-degree in $\lambda^{\pm 2}$.

§7. Extension to three variables

Next we generalize Theorem 5.1 to a construction of special Lagrangian 3-folds in $\mathbb{C}^3$ in which all three variables r, s, t enter in a non-trivial way. The proof is similar to that of Theorem 5.1, so we will be brief.

Theorem 7.1. *Let $\alpha_1, \alpha_2, \alpha_3$, $\beta_1, \beta_2, \beta_3$ and $\gamma_1, \gamma_2, \gamma_3$ be real numbers with not all α_j, not all β_j and not all γ_j zero, such that*

$$\begin{aligned} &\alpha_1\beta_1 + \alpha_2\beta_2 + \alpha_3\beta_3 = 0, && \alpha_1\gamma_1 + \alpha_2\gamma_2 + \alpha_3\gamma_3 = 0, \\ &\beta_1\gamma_1 + \beta_2\gamma_2 + \beta_3\gamma_3 = 0 \quad \text{and} && \alpha_1\beta_1\gamma_1 + \alpha_2\beta_2\gamma_2 + \alpha_3\beta_3\gamma_3 = 0. \end{aligned} \tag{78}$$

Let I, J, K be open intervals in $\mathbb{R}$. Suppose that $x_1, x_2, x_3 : I \to \mathbb{C}$ and $u : I \to \mathbb{R}$ are functions of r, that $y_1, y_2, y_3 : J \to \mathbb{C}$ and $v : J \to \mathbb{R}$ are functions of s, and $z_1, z_2, z_3 : K \to \mathbb{C}$ and $w : K \to \mathbb{R}$ functions of t, satisfying

$$\frac{dx_1}{dr} = \alpha_1\,\overline{x_2 x_3}, \qquad \frac{dx_2}{dr} = \alpha_2\,\overline{x_3 x_1}, \qquad \frac{dx_3}{dr} = \alpha_3\,\overline{x_1 x_2}, \tag{79}$$

$$\frac{dy_1}{ds} = \beta_1\,\overline{y_2 y_3}, \qquad \frac{dy_2}{ds} = \beta_2\,\overline{y_3 y_1}, \qquad \frac{dy_3}{ds} = \beta_3\,\overline{y_1 y_2}, \tag{80}$$

$$\frac{dz_1}{dt} = \gamma_1\,\overline{z_2 z_3}, \qquad \frac{dz_2}{dt} = \gamma_2\,\overline{z_3 z_1}, \qquad \frac{dz_3}{dt} = \gamma_3\,\overline{z_1 z_2}, \tag{81}$$

$$|x_1|^2 = \alpha_1 u + 1, \qquad |x_2|^2 = \alpha_2 u + 1, \qquad |x_3|^2 = \alpha_3 u + 1, \tag{82}$$

$$|y_1|^2 = \beta_1 v + 1, \qquad |y_2|^2 = \beta_2 v + 1, \qquad |y_3|^2 = \beta_3 v + 1, \tag{83}$$

$$|z_1|^2 = \gamma_1 w + 1, \qquad |z_2|^2 = \gamma_2 w + 1, \qquad |z_3|^2 = \gamma_3 w + 1. \tag{84}$$

If (79)–(81) *hold for all r, s, t and* (82)–(84) *hold for some r, s, t, then* (82)–(84) *hold for all r, s, t, for some functions u, v, w. Define a map*

$\Phi : I \times J \times K \to \mathbb{C}^3$ by

$$\Phi : (r,s,t) \mapsto \big(x_1(r)y_1(s)z_1(t), x_2(r)y_2(s)z_2(t), x_3(r)y_3(s)z_3(t)\big). \tag{85}$$

Define a subset N of $\mathbb{C}^3$ by

$$N = \big\{\Phi(r,s,t) : r \in I,\ s \in J,\ t \in K\big\}. \tag{86}$$

Then N is a special Lagrangian 3-fold in $\mathbb{C}^3$.

Proof. The first part of the theorem, that if (79)–(81) hold for all r,s,t and (82)–(84) for some r,s,t, then (82)–(84) hold for all r,s,t, follows as in Theorem 5.1. For the second part, we must prove that N is special Lagrangian wherever Φ is an immersion. As in Theorem 5.1, this holds if and only if

$$\omega\Big(\frac{\partial\Phi}{\partial r},\frac{\partial\Phi}{\partial s}\Big) \equiv \omega\Big(\frac{\partial\Phi}{\partial r},\frac{\partial\Phi}{\partial t}\Big) \equiv \omega\Big(\frac{\partial\Phi}{\partial s},\frac{\partial\Phi}{\partial t}\Big) \equiv 0 \tag{87}$$

$$\text{and} \qquad \operatorname{Im}\Omega\Big(\frac{\partial\Phi}{\partial r},\frac{\partial\Phi}{\partial s},\frac{\partial\Phi}{\partial t}\Big) \equiv 0. \tag{88}$$

Using equations (79)–(81) and (85) we find that

$$\frac{\partial\Phi}{\partial r} = \big(\alpha_1\,\overline{x_2x_3}y_1z_1, \alpha_2\,\overline{x_3x_1}y_2z_2, \alpha_3\,\overline{x_1x_2}y_3z_3\big), \tag{89}$$

$$\frac{\partial\Phi}{\partial s} = \big(\beta_1x_1\,\overline{y_2y_3}z_1, \beta_2x_2\,\overline{y_3y_1}z_2, \beta_3x_3\,\overline{y_1y_2}z_3\big), \tag{90}$$

$$\frac{\partial\Phi}{\partial t} = \big(\gamma_1x_1y_1\,\overline{z_2z_3}, \gamma_2x_2y_2\,\overline{z_3z_1}, \gamma_3x_3y_3\,\overline{z_1z_2}\big). \tag{91}$$

Equations (89) and (90) give

$$\begin{aligned}
\omega\big(\tfrac{\partial\Phi}{\partial r},\tfrac{\partial\Phi}{\partial s}\big) &= \operatorname{Im}(\overline{x_1x_2x_3}y_1y_2y_3)\big(\alpha_1\beta_1|z_1|^2 + \alpha_2\beta_2|z_2|^2 + \alpha_3\beta_3|z_3|^2\big)\\
&= \operatorname{Im}(\overline{x_1x_2x_3}y_1y_2y_3)\big(\alpha_1\beta_1(\gamma_1w+1)+\alpha_2\beta_2(\gamma_2w+1)+\alpha_3\beta_3(\gamma_3w+1)\big)\\
&= \operatorname{Im}(\overline{x_1x_2x_3}y_1y_2y_3)\big(\alpha_1\beta_1+\alpha_2\beta_2+\alpha_3\beta_3+w(\alpha_1\beta_1\gamma_1+\alpha_2\beta_2\gamma_2+\alpha_3\beta_3\gamma_3)\big)=0,
\end{aligned}$$

using (84) in the second line and (78) in the third. This proves the first equation of (87). The second and third follow in a similar way.

To prove (88), observe that

$$\Omega\Big(\frac{\partial\Phi}{\partial r},\frac{\partial\Phi}{\partial s},\frac{\partial\Phi}{\partial t}\Big)=\Big|\frac{\partial\Phi}{\partial r}\;\frac{\partial\Phi}{\partial s}\;\frac{\partial\Phi}{\partial t}\Big|=\begin{vmatrix}\alpha_1\,\overline{x_2x_3}y_1z_1 & \beta_1x_1\,\overline{y_2y_3}z_1 & \gamma_1x_1y_1\,\overline{z_2z_3}\\ \alpha_2\,\overline{x_3x_1}y_2z_2 & \beta_2x_2\,\overline{y_3y_1}z_2 & \gamma_2x_2y_2\,\overline{z_3z_1}\\ \alpha_3\,\overline{x_1x_2}y_3z_3 & \beta_3x_3\,\overline{y_1y_2}z_3 & \gamma_3x_3y_3\,\overline{z_1z_2}\end{vmatrix}$$

$$\begin{aligned}&=\big(\alpha_1|x_2x_3|^2\beta_2|y_3y_1|^2\gamma_3|z_1z_2|^2+\alpha_2|x_3x_1|^2\beta_3|y_1y_2|^2\gamma_1|z_2z_3|^2\\&\quad+\alpha_3|x_1x_2|^2\beta_1|y_2y_3|^2\gamma_2|z_3z_1|^2-\alpha_1|x_2x_3|^2\beta_3|y_1y_2|^2\gamma_2|z_3z_1|^2\\&\quad-\alpha_2|x_3x_1|^2\beta_1|y_2y_3|^2\gamma_3|z_1z_2|^2-\alpha_3|x_1x_2|^2\beta_2|y_3y_1|^2\gamma_1|z_2z_3|^2\big).\end{aligned}$$

Thus $\Omega\big(\frac{\partial\Phi}{\partial r},\frac{\partial\Phi}{\partial s},\frac{\partial\Phi}{\partial t}\big)$ is real, and so $\operatorname{Im}\Omega\big(\frac{\partial\Phi}{\partial r},\frac{\partial\Phi}{\partial s},\frac{\partial\Phi}{\partial t}\big)=0$. Q.E.D.

Here are a few comments on the theorem.

(a) In Theorem 5.1 we took the ranges of s,t to be $\mathbb{R}$, but here we take r,s,t in intervals I,J,K in $\mathbb{R}$. This is because, by an argument in [19, Prop. 7.11], the conditions $\beta_1+\beta_2+\beta_3=0$ and $\gamma_1+\gamma_2+\gamma_3=0$ imply that solutions of (24) and (25) in some open interval extend automatically to all of $\mathbb{R}$.

However, in Theorem 7.1 we do not assume that $\alpha_1+\alpha_2+\alpha_3=0$, and so it could happen that $\alpha_1,\alpha_2,\alpha_3$ all have the same sign. In this case, solutions x_j to (79) will in general exist in some open interval $I\subset\mathbb{R}$ with $|x_j|\to\infty$ at the endpoints of I, so that they do not extend to $\mathbb{R}$. The same applies to (80) and (81).

(b) As in §5.2 we can write the x_k,y_k and z_k entirely explicitly in terms of integrals involving the Jacobi elliptic functions.

(c) As in Theorem 5.4, in the situation of Theorem 7.1, $\frac{\partial\Phi}{\partial r}$, $\frac{\partial\Phi}{\partial s}$ and $\frac{\partial\Phi}{\partial t}$ are always complex orthogonal. But in general they are not of the same length, so Φ is not conformal.

(d) We may recover Theorem 5.1 from Theorem 7.1 as follows. Put $\alpha_1=\alpha_2=\alpha_3=1$, so that (78) becomes equivalent to (23). Define

$$I=(-\infty,0),\quad x_1(r)=x_2(r)=x_3(r)=-r^{-1}\quad\text{and}\quad u(r)=r^{-2}-1,$$

and $J=K=\mathbb{R}$. Then (79) and (82) hold, and Theorem 7.1 becomes equivalent to Theorem 5.1, but with a different parametrization for r.

7.1. Description of the family of SL 3-folds

We shall now describe the family of SL 3-folds resulting from Theorem 7.1. We begin by studying the set of solutions $\alpha_j,\beta_j,\gamma_j$ to (78).

Define vectors $\boldsymbol{\alpha} = (\alpha_1, \alpha_2, \alpha_3)$, $\boldsymbol{\beta} = (\beta_1, \beta_2, \beta_3)$, $\boldsymbol{\gamma} = (\gamma_1, \gamma_2, \gamma_3)$, and

$$\begin{aligned}\boldsymbol{\alpha\beta} &= (\alpha_1\beta_1, \alpha_2\beta_2, \alpha_3\beta_3),\\ \boldsymbol{\alpha\gamma} &= (\alpha_1\gamma_1, \alpha_2\gamma_2, \alpha_3\gamma_3),\\ \boldsymbol{\beta\gamma} &= (\beta_1\gamma_1, \beta_2\gamma_2, \beta_3\gamma_3)\end{aligned}$$

in $\mathbb{R}^3$. Rescaling $\boldsymbol{\alpha}, \boldsymbol{\beta}$ and $\boldsymbol{\gamma}$ has no effect on the SL 3-folds constructed in Theorem 7.1, so let us assume $\boldsymbol{\alpha}, \boldsymbol{\beta}, \boldsymbol{\gamma}$ are unit vectors. We will show that a generic choice of $\boldsymbol{\alpha}$ determines $\boldsymbol{\beta}, \boldsymbol{\gamma}$, essentially uniquely.

Proposition 7.2. *Let $\boldsymbol{\alpha}$ be a unit vector in $\mathbb{R}^3$, with $\alpha_1, \alpha_2, \alpha_3$ distinct and nonzero. Then there exist unit vectors $\boldsymbol{\beta}, \boldsymbol{\gamma}$ satisfying (78), which are unique up to sign and exchanging $\boldsymbol{\beta}, \boldsymbol{\gamma}$.*

Proof. Equation (78) implies that $\boldsymbol{\alpha}, \boldsymbol{\beta}$ and $\boldsymbol{\alpha\beta}$ are orthogonal to $\boldsymbol{\gamma}$. As $\boldsymbol{\gamma} \neq 0$, it follows that $\boldsymbol{\alpha}, \boldsymbol{\beta}$ and $\boldsymbol{\alpha\beta}$ are linearly dependent. Therefore $\det\bigl(\boldsymbol{\alpha}\ \boldsymbol{\beta}\ \boldsymbol{\alpha\beta}\bigr) = 0$. This may be rewritten in matrix form as

(92)
$$Q(\boldsymbol{\beta}) = \tfrac{1}{2}\begin{pmatrix}\beta_1\\ \beta_2\\ \beta_3\end{pmatrix}^T \begin{pmatrix} 0 & \alpha_3(\alpha_2-\alpha_1) & \alpha_2(\alpha_1-\alpha_3)\\ \alpha_3(\alpha_2-\alpha_1) & 0 & \alpha_1(\alpha_3-\alpha_2)\\ \alpha_2(\alpha_1-\alpha_3) & \alpha_1(\alpha_3-\alpha_2) & 0\end{pmatrix}\begin{pmatrix}\beta_1\\ \beta_2\\ \beta_3\end{pmatrix} = 0.$$

Similar equations hold between the α_j and γ_j, and between the β_j and γ_j. Now the 3×3 matrix appearing in (92) has trace zero and determinant $2\alpha_1\alpha_2\alpha_3(\alpha_1-\alpha_3)(\alpha_2-\alpha_1)(\alpha_3-\alpha_2)$. As by assumption $\alpha_1, \alpha_2, \alpha_3$ are distinct and nonzero, this determinant is nonzero. Hence Q is a trace-free, nondegenerate quadratic form on $\mathbb{R}^3$.

Therefore, $\boldsymbol{\beta}$ must be a unit vector in the intersection of the plane $\boldsymbol{\alpha} \cdot \boldsymbol{\beta} = 0$ and the quadric cone (92) in $\mathbb{R}^3$. Let $\boldsymbol{\alpha}^\perp$ be the plane perpendicular to $\boldsymbol{\alpha}$, and consider the restriction $Q|_{\boldsymbol{\alpha}^\perp}$ of Q to $\boldsymbol{\alpha}^\perp$. As $\boldsymbol{\alpha}$ is a unit vector, we have

$$0 = \mathrm{Tr}(Q) = \mathrm{Tr}\bigl(Q|_{\boldsymbol{\alpha}^\perp}\bigr) + Q(\boldsymbol{\alpha}).$$

But $Q(\boldsymbol{\alpha}) = 0$ by (92), so $Q|_{\boldsymbol{\alpha}^\perp}$ is trace-free.

Thus, by the classification of quadratic forms on $\mathbb{R}^2$, there exists an orthonormal basis $\boldsymbol{\beta}, \boldsymbol{\gamma}$ for $\boldsymbol{\alpha}^\perp$ such that $Q(x\boldsymbol{\beta} + y\boldsymbol{\gamma}) = cxy$ for some c and all x, y in $\mathbb{R}$. If $c = 0$ then $Q|_{\boldsymbol{\alpha}^\perp} = 0$, so Q is degenerate, a contradiction. So $c \neq 0$, and therefore $\boldsymbol{\beta}, \boldsymbol{\gamma}$ are unique up to sign and order, with $Q(\boldsymbol{\beta}) = Q(\boldsymbol{\gamma}) = 0$.

As $\boldsymbol{\alpha}, \boldsymbol{\beta}, \boldsymbol{\gamma}$ are orthonormal they automatically satisfy the first three equations of (78). But by construction we have arranged that $\boldsymbol{\alpha}, \boldsymbol{\beta}$ and $\boldsymbol{\alpha\beta}$ are linearly dependent, so $\boldsymbol{\alpha\beta} = x\boldsymbol{\alpha} + y\boldsymbol{\beta}$ for $x, y \in \mathbb{R}$. The fourth equation of (78) then follows from the second and third. Q.E.D.

The moral of the proposition is that a generic choice of $\boldsymbol{\alpha}$ determines $\boldsymbol{\beta}$ and $\boldsymbol{\gamma}$ up to obvious symmetries. However, for a nongeneric choice of $\boldsymbol{\alpha}$ there can be more freedom in $\boldsymbol{\beta}$ and $\boldsymbol{\gamma}$. For instance, if we put $\boldsymbol{\alpha} = 3^{-1/2}(1,1,1)$ then $Q \equiv 0$, and $\boldsymbol{\beta}, \boldsymbol{\gamma}$ can be arbitrary orthonormal vectors in $\boldsymbol{\alpha}^\perp$.

We can now do a parameter count for the family of SL 3-folds coming from Theorem 7.1. The proposition shows that up to symmetries, the data $\alpha_j, \beta_j, \gamma_j$ has two interesting degrees of freedom. Also, as in Propositions 5.2 and 5.3 there exist constants $A, B, C \in \mathbb{R}$ such that

$$\operatorname{Im}(x_1x_2x_3) \equiv A, \quad \operatorname{Im}(y_1y_2y_3) \equiv B \quad \text{and} \quad \operatorname{Im}(z_1z_2z_3) \equiv C.$$

Together the $\alpha_j, \beta_j, \gamma_j$ and A, B, C determine N up to automorphisms of $\mathbb{C}^3$. Thus the construction of Theorem 7.1 yields a 5-dimensional family of SL 3-folds, up to automorphisms of $\mathbb{C}^3$.

We can also discuss the possible *signs* of the $\alpha_k, \beta_k, \gamma_k$. Suppose for simplicity that $\alpha_k, \beta_k, \gamma_k$ are all nonzero. Then the four equations of (78) constrain the signs of $\alpha_k, \beta_k, \gamma_k$, as in each equation the three terms cannot have the same sign, since their sum is zero. Now permuting $\boldsymbol{\alpha}, \boldsymbol{\beta}, \boldsymbol{\gamma}$, and reversing any of their signs, does not change the set of SL 3-folds constructed in Theorem 7.1.

Considering the constraints on the signs of the $\alpha_k, \beta_k, \gamma_k$, it is not difficult to show that by permuting and changing signs of $\boldsymbol{\alpha}, \boldsymbol{\beta}, \boldsymbol{\gamma}$ we may can arrange that the α_k are all positive, two of the β_k are positive and one negative, and two of the γ_k positive and one negative.

With this choice of signs, the argument in [19, Prop. 7.11] shows that solutions y_k, z_k to (80)–(81) and (83)–(84) automatically extend to $\mathbb{R}$, so we may take $J = K = \mathbb{R}$. However, solutions x_k to (79) and (82) generally exist only on a proper subinterval I of $\mathbb{R}$. Let us take I to be as large as possible.

The discussion of §5.4 suggests that we should try to arrange that the y_k are periodic in s and the z_k periodic in t. When this happens, ψ pushes down to an immersion $I \times T^2 \to \mathbb{C}^3$, whose image is a *closed* SL 3-fold in $\mathbb{C}^3$. The double-periodicity conditions in s, t in this case turn out to be equivalent to those in §5.4, and there are analogues of parts (a)–(c) of §5.4 in which one can prove they are soluble, which yield countably many families of closed, immersed SL 3-folds in $\mathbb{C}^3$ diffeomorphic to $T^2 \times \mathbb{R}$.

7.2. Conclusion: an open problem

Theorems 5.1 and 7.1 are clearly very similar. But in §6 we saw that the special Lagrangian cones of Theorem 5.1 can be put into a much larger integrable systems framework. Is there also an 'integrable systems' explanation for the SL 3-folds of Theorem 7.1? Certainly the solutions of Theorem 7.1 have many of the hallmarks of integrable systems: commuting o.d.e.s, elliptic functions, conserved quantities.

Also, there exist many interesting families of SL m-folds in $\mathbb{C}^m$ which can be written down explicitly (and so are 'integrable' in a trivial sense), or have some other nice properties. For examples, see papers by the author [17, 18, 19, 20, 21], and others such as Harvey and Lawson [13, III.3], Haskins [14] and Bryant [5]. When the special Lagrangian equations are reduced to an o.d.e., it often turns out to be a completely integrable Hamiltonian system, as in [19, §7.6].

At present, as the author understands it, integrable systems methods are only really effective for o.d.e.s, or p.d.e.s in two variables (s, t), though perhaps for equations involving many unknowns $f_1(t), \ldots, f_k(t)$ or $f_1(s,t), \ldots, f_k(s,t)$ in these variables. However, the SL 3-folds of Theorem 7.1 involve three variables (r, s, t) in a nontrivial way.

It is probably much too optimistic to hope that the special Lagrangian equations themselves are integrable in any meaningful sense. Nonetheless, it seems plausible to the author that there may exist large families of examples of SL m-folds in $\mathbb{C}^m$ for $m \geqslant 3$ which admit some kind of m-variable 'integrable systems' type description, and that these would be an interesting thing to study. The author suggests this to the integrable systems community as a worthwhile problem.

References

[1] U. Abresch, Constant mean curvature tori in terms of elliptic functions, J. Reine Angew. Math., **374** (1987), 169–192.

[2] J. Bolton, F. Pedit and L. M. Woodward, Minimal surfaces and the affine Toda field model, J. Reine Angew. Math., **459** (1995), 119–150.

[3] J. Bolton and L. M. Woodward, Congruence theorems for harmonic maps from a Riemann surface into $\mathbb{CP}^n$ and $\mathcal{S}^n$, J. London Math. Soc., **45** (1992), 363–376.

[4] R.K. Bullough and R. K. Dodd, Polynomial conserved densities for the sine-Gordon equations, Proc. Roy. Soc. London Ser. A, **352** (1977), 481–503.

[5] R. L. Bryant, Second order families of special Lagrangian 3-folds, In: Perspectives in Riemannian geometry, (eds. V. Apostolov, A. Dancer, N, Hitchin and M. Wang), CRM Proc. Lecture Notes, 40, Amer. Math. Soc., Providence, RI, 2006, pp. 63–98, math.DG/0007128.

[6] F. E. Burstall, D. Ferus, F. Pedit and U. Pinkall, Harmonic tori in symmetric spaces and commuting Hamiltonian systems on loop algebras, Ann. of Math. (2), **138** (1993), 173–212.

[7] E. Carberry and I. McIntosh, Minimal Lagrangian 2-tori in $\mathbb{CP}^2$ come in real families of every dimension, J. London Math. Soc., **69** (2004), 531–544, math.DG/0308031.

[8] I. Castro and F. Urbano, New examples of minimal Lagrangian tori in the complex projective plane, Manuscripta Math., **85** (1994), 265–281.

[9] J. Eells and J. C. Wood, Harmonic maps from surfaces into projective spaces, Advances in Math., **49** (1983), 217–263.

[10] D. Ferus, F. Pedit, U. Pinkall and I. Sterling, Minimal tori in $\mathcal{S}^4$, J. Reine Angew. Math., **429** (1992), 1–47.

[11] A. P. Fordy and J.C. Wood, Harmonic Maps and Integrable Systems, Aspects of Math., **E23**, Vieweg, Wiesbaden, 1994.

[12] F. R. Harvey, Spinors and calibrations, Perspect. Math., **9**, Academic Press, San Diego, 1990.

[13] R. Harvey and H. B. Lawson, Calibrated geometries, Acta Math., **148** (1982), 47–157.

[14] M. Haskins, Special Lagrangian cones, Amer. J. Math., **126** (2004), 845–871, math.DG/0005164.

[15] M. Haskins, The geometric complexity of special Lagrangian T^2-cones, Invent. Math., **157** (2004), 11–70, math.DG/0307129.

[16] M. Haskins and N. Kapouleas, Special Lagrangian cones with higher genus links, Invent. Math., **167** (2007), 223–294, math.DG/0512178.

[17] D. D. Joyce, Constructing special Lagrangian m-folds in $\mathbb{C}^m$ by evolving quadrics, Math. Ann., **320** (2001), 757–797, math.DG/0008155.

[18] D. D. Joyce, Evolution equations for special Lagrangian 3-folds in $\mathbb{C}^3$, Ann. Global Anal. Geom., **20** (2001), 345–403, math.DG/0010036.

[19] D. D. Joyce, Special Lagrangian m-folds in $\mathbb{C}^m$ with symmetries, Duke Math. J., **115** (2002), 1–51, math.DG/0008021.

[20] D. D. Joyce, Ruled special Lagrangian 3-folds in $\mathbb{C}^3$, Proc. London Math. Soc., **85** (2002), 233–256, math.DG/0012060.

[21] D. D. Joyce, Singularities of special Lagrangian fibrations and the SYZ Conjecture, Comm. Anal. Geom., **11** (2003), 859–907, math.DG/0011179.

[22] D. D. Joyce, Riemannian Holonomy Groups and Calibrated Geometry, Oxf. Grad. Texts Math., **12**, Oxford Univ. Press, Oxford, 2007.

[23] N. Kapouleas, Compact constant mean curvature surfaces in Euclidean three-space, J. Diff. Geom., **33** (1991), 683–715.
[24] G. Lawlor, The angle criterion, Invent. Math., **95** (1989), 437–446.
[25] H. Ma and Y. Ma, Totally real minimal tori in $\mathbb{CP}^2$, Math. Z., **249** (2005), 241–267, math.DG/0106141.
[26] I. McIntosh, A construction of all non-isotropic harmonic tori in complex projective space, Internat. J. Math., **6** (1995), 831–879.
[27] I. McIntosh, Two remarks on the construction of harmonic tori in $\mathbb{CP}^n$, Internat. J. Math., **7** (1996), 515–520.
[28] I. McIntosh, On the existence of superconformal 2-tori and doubly periodic affine Toda fields, J. Geom. Phys., **24** (1998), 223–243.
[29] I. McIntosh, Special Lagrangian cones in $\mathbb{C}^3$ and primitive harmonic maps, J. London Math. Soc., **67** (2003), 769–789, math.DG/0201157.
[30] R. A. Sharipov, Minimal tori in the five-dimensional sphere in $\mathbb{C}^3$, Theoret. and Math. Phys., **87** (1991), 363–369.
[31] G. Tzitzéica, Sur une nouvelle classe de surfaces, C. R. Acad. Sci. Paris, **150** (1910), 955–956.
[32] H. Wente, Counterexample to a conjecture of H. Hopf, Pacific J. Math., **121** (1986), 193–243.

The Mathematical Institute
24-29 St Giles
Oxford OX1 3LB
UK

E-mail address:
`joyce@maths.ox.ac.uk`

Advanced Studies in Pure Mathematics 51, 2008
Surveys on Geometry and Integrable Systems
pp. 235–257

Quantum product, topological recursion relations, and the Virasoro conjecture

Xiaobo Liu

The Virasoro conjecture predicts that the generating function of the Gromov-Witten invariants is annihilated by infinitely many differential operators which form a half branch of the Virasoro algebra. This conjecture was proposed by Eguchi, Hori and Xiong [EHX] and S. Katz (cf. [CK] [EJX]). It is a natural generalization of a conjecture of Witten (cf. [W2] [Ko] [W2]) and provides a powerful tool in the computation of Gromov-Witten invariants. The genus-0 Virasoro conjecture was proved in [LT] (cf. [DZ] and [G3] for alternative proofs). The genus-1 Virasoro conjecture for manifolds with semisimple quantum cohomology was proved in [DZ]. Without assuming semisimplicity, the genus-1 Virasoro conjecture was reduced to the genus-1 L_1-constraint on the small phase space in [L1]. It was also proved in [L1] that the genus-1 Virasoro conjecture holds if the quantum cohomology is not too degenerate (a condition weaker than semisimplicity). The essential part of the results in [L1] was extended to the genus-2 Virasoro conjecture in [L2]. The study of the genus-2 Virasoro conjecture is important because this is the first case where we do not have a formula to reduce the problem to the small phase space. The behavior of the Virasoro conjecture in this case will provide much needed insight in what we should expect in the higher genera cases. The techniques developed in [L2] could be easily adapted to the study of the higher genera Virasoro conjecture.

In this expository article, we will explain how to apply the main ideas in [L2] to the study of the Virasoro conjecture in all genera. In particular, we will explain how to use the quantum product on the big

Received January 27, 2001.
Revised August 12, 2001; August 17, 2007 (updated).
2000 *Mathematics Subject Classification.* Primary 53D45; Secondary 14N35.
Research partially supported by an Alfred P. Sloan Research Fellowship and an NSF postdoctoral Fellowship.

phase space to interpret topological recursion relations and the Virasoro conjecture. The quantum product on the big phase space does not have an identity element. The string vector field is the closest vector field to an identity. In our view point, various topological recursion relations are just different ways to express how close the string vector field is to an identity for the quantum product on the big phase space. Using such topological recursion relations, we can rephrase the Virasoro conjecture in terms of quantum powers of the Euler vector fields. More precisely, for $g \geq 1$, the genus-g Virasoro constraints compute the derivatives of the genus-g generating function along a sequence of vector fields constructed from (twisted) quantum powers of the Euler vector field. When the quantum cohomology is semisimple, quantum powers of the Euler vector field span the space of the primary vector fields. Therefore in this case, the Virasoro conjecture is strong enough to determine the genus-g generating function in terms of data of genus less than g. Such an interpretation of the Virasoro conjecture is very useful both in resolving the conjecture and in applying it to solve other problems.

To keep the article brief, we will emphasize the main ideas and omit most proofs. Interested readers are referred to [L2] for more details. The major exception to this rule is a complete proof of the genus-0 Virasoro conjecture, which simplifies the arguments in [LT]. The argument presented here is new and has not appeared elsewhere.

The author would like to thank the organizers of Mathematical Society of Japan - 9th International Research Institute on "Integrable Systems in Differential Geometry" for their invitation. He would also like to thank G. Tian for helpful discussions.

§1. Quantum product on the big phase space and topological recursion relations

1.1. Gromov-Witten invariants

For simplicity, we assume that V is a smooth projective variety with $H^{\text{odd}}(V;\mathbb{C}) = 0$. All results in this paper should also be true for compact symplectic manifolds except those concerning Virasoro constraints of genus bigger than 0. Gromov-Witten invariants are defined via the intersection theory of moduli spaces of stable maps from Riemann surfaces to V. For any element $A \in H_2(V, \mathbb{Z})$ and non-negative integers g and k, the moduli space $\overline{\mathcal{M}}_{g,k}(V, A)$ is defined to be the collection of all data $(C; x_1, \ldots, x_k; f)$ where C is a genus-g projective connected curve over $\mathbb{C}$ whose only possible singularities are simple double points, where $x_1, \ldots, x_k$ are smooth points on C (called marked points), and when f is an algebraic map from C to V which is stable with respect to $(C; x_1, \ldots, x_k)$, (i.e. there is no infinitesimal deformation for this data). Each marked point x_i defines a map, called the i-th evaluation map,

$$ev_i : \begin{array}{ccc} \overline{\mathcal{M}}_{g,k}(V, A) & \longrightarrow & V \\ (C; x_1, \ldots, x_k; f) & \longmapsto & f(x_i). \end{array}$$

It also defines a line bundle over $\overline{\mathcal{M}}_{g,k}(V, A)$, denoted by E_i, whose fiber over $(C; x_1, \ldots, x_k; f)$ is $T^*_{x_i}C$. For any cohomology classes $\gamma_1, \ldots, \gamma_k \in H^*(V, \mathbb{C})$ and non-negative integers $n_1, \ldots, n_k$, the corresponding descendant Gromov-Witten invariants are defined by

$$\langle \tau_{n_1}(\gamma_1) \cdots \tau_{n_k}(\gamma_k) \rangle_g := \sum_A q^A \int_{\left[\overline{\mathcal{M}}_{g,k}(V,A)\right]^{\text{virt}}} c_1(E_1)^{n_1} \cup \text{ev}_1^*(\gamma_1) \cup \cdots \cup c_1(E_k)^{n_k} \cup \text{ev}_k^*(\gamma_k),$$

where $\left[\overline{\mathcal{M}}_{g,k}(V, A)\right]^{\text{virt}}$ is the virtual fundamental class of $\overline{\mathcal{M}}_{g,k}(V, A)$ (cf. [LiT]) and q^A belongs to the Novikov ring (i.e. the multiplicative ring spanned by monomials $q^A = q_1^{a_1} \cdots q_r^{a_r}$ over the ring of rational numbers, where $\{q_1, \cdots, q_r\}$ is a fixed basis of $H_2(V, \mathbb{Z})$ and $A = \sum_{i=1}^r a_i q_i$). When all the n_i are zero, the corresponding invariants are called primary Gromov-Witten invariants.

1.2. Notational conventions

We will use d to denote the complex dimension of V and N the dimension of the space of cohomology classes $H^*(V, \mathbb{C})$. To define the generating functions, we need to fix a basis $\{\gamma_1, \ldots, \gamma_N\}$ of $H^*(V, \mathbb{C})$ with γ_1 equal to the identity for the cohomology ring of V and $\gamma_\alpha \in H^{p_\alpha, q_\alpha}(V, \mathbb{C})$ for every α. We also arrange the basis in such a way that the dimension of γ_α is non-decreasing with respect to α, and, if two cohomology classes have the same dimension, we also require that the holomorphic dimension p_α is non-decreasing.

Lower case Greek characters will be used to index the cohomology classes. The range of these indices is from 1 to N. Lower case Roman characters will be used to index the level of descendants. Their range is the set of all non-negative integers, $\mathbb{Z}_+$. All summations are over the entire ranges of the indices unless otherwise indicated. Let

$$\eta_{\alpha\beta} = \int_V \gamma_\alpha \cup \gamma_\beta$$

be the intersection form on $H^*(V,\mathbb{C})$. We will use $\eta = (\eta_{\alpha\beta})$ and $\eta^{-1} = (\eta^{\alpha\beta})$ to lower and raise indices. For example, $\gamma^\alpha := \eta^{\alpha\beta}\gamma_\beta$. Here we are using the summation convention that repeated indices (in this formula, β) are summed over their entire ranges. Let $\mathcal{C} = (\mathcal{C}_\alpha^\beta)$ be the matrix of multiplication by the first Chern class $c_1(V)$ in the ordinary cohomology ring, i.e. $c_1(V) \cup \gamma_\alpha = \mathcal{C}_\alpha^\beta \gamma_\beta$. Since we are dealing only with even dimensional cohomology classes, the $\mathcal{C}^k\eta$ are symmetric matrices for all $k \geq 0$, where the entries of $\mathcal{C}^k\eta$ are given by

$$(\mathcal{C}^k)_{\alpha\beta} = \int_V c_1(V)^k \cup \gamma_\alpha \cup \gamma_\beta.$$

Let $b_\alpha = p_\alpha - \frac{1}{2}(d-1)$. The following simple observations will be used throughout the calculations without mention: If $\eta^{\alpha\beta} \neq 0$ or $\eta_{\alpha\beta} \neq 0$, then $b_\alpha = 1 - b_\beta$. $\mathcal{C}_\alpha^\beta \neq 0$ implies $b_\beta = 1 + b_\alpha$, and $\mathcal{C}_{\alpha\beta} \neq 0$ implies $b_\beta = -b_\alpha$.

1.3. Generating functions and correlation functions

The genus-g generating function is defined to be

$$F_g = \sum_{k\geq 0} \frac{1}{k!} \sum_{\substack{\alpha_1,\ldots,\alpha_k \\ n_1,\ldots,n_k}} t_{n_1}^{\alpha_1} \cdots t_{n_k}^{\alpha_k} \langle \tau_{n_1}(\gamma_{\alpha_1}) \tau_{n_2}(\gamma_{\alpha_2}) \ldots \tau_{n_k}(\gamma_{\alpha_k}) \rangle_g ,$$

where $\{t_n^\alpha \mid n \in \mathbb{Z}_+, \alpha = 1, \cdots, N\}$ is an infinite set of parameters. We can think of these parameters as coordinates on an infinite dimensional vector space, called the *big phase space.* The finite dimensional subspace defined by $\{t_n^\alpha = 0 \text{ if } n > 0\}$ is called the *small phase space.* The function F_g is to be understood as a formal power series of t_n^α. For convenience, we will always identify the symbol $\tau_n(\gamma_\alpha)$ with the tangent vector field $\frac{\partial}{\partial t_n^\alpha}$ on the big phase space. We also consider $\tau_n(\gamma_\alpha)$ with $n < 0$ as the zero operator. For each α, we will abbreviate $\tau_0(\gamma_\alpha)$ as γ_α. We call a (formal) vector field $\mathcal{W} = \sum_{m,\alpha} f_{m,\alpha}\tau_m(\gamma_\alpha)$ a *primary vector field* if $f_{m,\alpha} = 0$ whenever $m > 0$, a *descendant vector field* if $f_{m,\alpha} = 0$ whenever $m = 0$.

Instead of coordinates $\{t_m^\alpha \mid m \in \mathbb{Z}_+, \alpha = 1, \ldots, N\}$, it is very convenient to use the following shifted coordinates on the big phase space

$$\tilde{t}_m^\alpha = t_m^\alpha - \delta_{m,1}\delta_{\alpha,1} = \begin{cases} t_m^\alpha - 1, & \text{if } m = \alpha = 1, \\ t_m^\alpha, & \text{otherwise.} \end{cases}$$

As in [LT], it is convenient to introduce a k-tensor $\langle\langle \underbrace{\cdots\cdots}_{k} \rangle\rangle$ defined by $\langle\langle\, \mathcal{W}_1 \mathcal{W}_2 \cdots \mathcal{W}_k \,\rangle\rangle_g \; :=$

$$\sum_{m_1,\alpha_1,\ldots,m_k,\alpha_k} f^1_{m_1,\alpha_1} \cdots f^k_{m_k,\alpha_k} \; \frac{\partial^k}{\partial t^{\alpha_1}_{m_1} \partial t^{\alpha_2}_{m_k} \cdots \partial t^{\alpha_k}_{m_k}} F_g,$$

for (formal) vector fields $\mathcal{W}_i = \sum_{m,\alpha} f^i_{m,\alpha} \frac{\partial}{\partial t^\alpha_m}$ where the $f^i_{m,\alpha}$ are (formal) functions on the big phase space. We can also view this tensor as the k-th covariant derivative of F_g. This tensor is called the *k-point (correlation) function.*

1.4. Quantum product

For any vector fields $\mathcal{U}$ and $\mathcal{W}$ on the big phase, define the **quantum product** of $\mathcal{U}$ and $\mathcal{W}$ by

$$\mathcal{U} \bullet \mathcal{W} := \langle\langle \mathcal{U} \mathcal{W} \gamma^\alpha \rangle\rangle_0 \, \gamma_\alpha.$$

By definition, the quantum product of two vector fields is always a primary vector field. This product is evidently commutative. It is also associative due to the *generalized WDVV equation*

$$\langle\langle\, \mathcal{W}_1 \, \mathcal{W}_2 \, \gamma^\alpha \,\rangle\rangle_0 \, \langle\langle\, \gamma_\alpha \, \mathcal{W}_3 \, \mathcal{W}_4 \,\rangle\rangle_0 = \langle\langle\, \mathcal{W}_1 \, \mathcal{W}_3 \, \gamma^\alpha \,\rangle\rangle_0 \, \langle\langle\, \gamma_\alpha \, \mathcal{W}_2 \, \mathcal{W}_4 \,\rangle\rangle_0 \,,$$

which follows in turn from the genus-0 topological recursion relation

$$\langle\langle\, \tau_m(\gamma_\alpha) \; \tau_n(\gamma_\beta) \; \tau_k(\gamma_\mu) \,\rangle\rangle_0 = \langle\langle\, \tau_{m-1}(\gamma_\alpha) \; \gamma_\sigma \,\rangle\rangle_0 \, \langle\langle\, \gamma^\sigma \; \tau_n(\gamma_\beta) \; \tau_k(\gamma_\mu) \,\rangle\rangle_0$$

for $m > 0$ (cf. [W1]). When restricted to tangent vector fields on the small phase, this is precisely the product in the quantum cohomology of V (called the big quantum cohomology by some authors). For any vector field $\mathcal{W}$ on the big phase space, we define $\mathcal{W}^k$ to be the k-th quantum power of $\mathcal{W}$. i.e., $\mathcal{W}^k = \underbrace{\mathcal{W} \bullet \mathcal{W} \bullet \cdots \mathcal{W}}_{k}$, for $k > 0$.

For the quantum product on the small phase space, the constant vector field γ_1, which was chosen to be the identity for the ordinary cohomology ring, is also the identity for the quantum cohomology. However on the big phase space, there is no identity vector field for the quantum product. Instead, the *string vector field*

$$\mathcal{S} = -\sum_{m,\alpha} \tilde{t}^\alpha_m \tau_{m-1}(\gamma_\alpha)$$

can be considered as a sort of identity in the following sense: First, the *string equation* can be written as

$$\langle\!\langle \mathcal{S} \rangle\!\rangle_g = \frac{1}{2}\delta_{g,0}\eta_{\alpha\beta}t_0^\alpha t_0^\beta.$$

Taking second derivatives of the genus-0 string equation, we obtain

$$\langle\!\langle \gamma_\alpha \mathcal{S} \gamma^\beta \rangle\!\rangle_0 = \delta_\alpha^\beta$$

for any α and β. This equation can be interpreted as $\mathcal{S} \bullet \mathcal{W} = \mathcal{W}$ for any primary vector field $\mathcal{W}$. Second, since $\mathcal{U} \bullet \mathcal{W}$ is always a primary field, the associativity of the quantum product implies

$$\mathcal{S} \bullet \mathcal{U} \bullet \mathcal{W} = \mathcal{U} \bullet \mathcal{W} \tag{1}$$

for all vector fields $\mathcal{U}$ and $\mathcal{W}$. Define $\overline{\mathcal{W}} := \mathcal{S} \bullet \mathcal{W}$ for any vector field $\mathcal{W}$. Then

$$\mathcal{W} \bullet \mathcal{V} = \overline{\mathcal{W}} \bullet \mathcal{V} \tag{2}$$

for any vector field $\mathcal{V}$. If we restrict the quantum product to the space of primary vector fields, then it has an identity $\overline{\mathcal{S}}$.

1.5. Topological recursion relations

In some sense, the topological recursion relations tells us how far $\mathcal{S}$ is from being an identity for the quantum product on the big phase space. To describe this interpretation, we introduce the following linear transformations (that is, linear with respect to multiplication by functions) on the space of vector fields on the big phase space:

Definition 1.1. For any vector field $\mathcal{W} = \sum_{n,\alpha} f_{n,\alpha}\tau_n(\gamma_\alpha)$, define

$$\tau_+(\mathcal{W}) := \sum_{n,\alpha} f_{n,\alpha}\tau_{n+1}(\gamma_\alpha), \qquad \tau_-(\mathcal{W}) := \sum_{n,\alpha} f_{n,\alpha}\tau_{n-1}(\gamma_\alpha),$$

$$T(\mathcal{W}) := \tau_+(\mathcal{W}) - \langle\!\langle \mathcal{W}\gamma^\alpha \rangle\!\rangle_0\, \gamma_\alpha, \qquad \pi(\mathcal{W}) := \sum_\alpha f_{0,\alpha}\gamma_\alpha.$$

Then $\tau_+\tau_-(\mathcal{W}) = \mathcal{W} - \pi(\mathcal{W})$, $\tau_-\tau_+(\mathcal{W}) = \tau_- T(\mathcal{W}) = \mathcal{W}$. Moreover the second derivatives of the genus-0 string equation have the following form (cf. [LT, Lemma 1.1 (3)]):

$$\langle\!\langle \mathcal{W}\,\mathcal{S}\,\mathcal{V} \rangle\!\rangle_0 = \langle\!\langle \tau_-(\mathcal{W})\,\mathcal{V} \rangle\!\rangle_0 + \langle\!\langle \mathcal{W}\,\tau_-(\mathcal{V}) \rangle\!\rangle_0 + \frac{1}{2}\nabla^2_{\mathcal{W},\mathcal{V}}\eta_{\alpha\beta}t_0^\alpha t_0^\beta \tag{3}$$

where $\mathcal{W}$ and $\mathcal{V}$ are arbitrary vector fields and $\nabla^2_{\mathcal{W},\mathcal{V}} = \nabla_{\mathcal{W}}\nabla_{\mathcal{V}} - \nabla_{\nabla_{\mathcal{W}}\mathcal{V}}$ is the second covariant derivative. In particular, this implies that

$$\overline{\mathcal{W}} = \langle\!\langle\, \tau_-(\mathcal{W})\gamma^\alpha \,\rangle\!\rangle_0\, \gamma_\alpha + \pi(\mathcal{W}). \tag{4}$$

An immediate consequence of this formula is the following:

$$T(\mathcal{W}) = \tau_+(\mathcal{W}) - \mathcal{S} \bullet \tau_+(\mathcal{W}). \tag{5}$$

The genus-0 topological recursion relation implies

$$T(\mathcal{W}) \bullet \mathcal{V} = 0 \tag{6}$$

for any vector fields $\mathcal{W}$ and $\mathcal{V}$. The converse of this equation is also true (cf. [L2]), i.e.

$$\mathcal{W} \bullet \mathcal{V} = 0 \ \text{ for all } \mathcal{V} \iff \mathcal{W} = T(\mathcal{U}) \ \text{ for some } \mathcal{U}.$$

It follows from the first derivatives of the genus-0 string equation that $T(\mathcal{S}) = \mathcal{D}$ where $\mathcal{D}$ is the *dilaton vector field* $\mathcal{D} = -\sum_{m,\alpha} \tilde{t}^\alpha_m\, \tau_m(\gamma_\alpha)$. The genus-0 *dilaton equation* and its first two derivatives have the following form (cf. [LT, Lemma 1.2])

$$\langle\!\langle\, \mathcal{D} \,\rangle\!\rangle_0 = -2F_0, \qquad \langle\!\langle\, \mathcal{D}\mathcal{W} \,\rangle\!\rangle_0 = -\langle\!\langle\, \mathcal{W} \,\rangle\!\rangle_0, \qquad \langle\!\langle\, \mathcal{D}\mathcal{W}\mathcal{V} \,\rangle\!\rangle_0 = 0 \tag{7}$$

for all vector fields $\mathcal{W}$ and $\mathcal{V}$. Equation (6) in particular implies that $\mathcal{D} \bullet \mathcal{W} = 0$ for all $\mathcal{W}$, which also follows from (7). Moreover the equation $T(\mathcal{D}) \bullet \mathcal{W} = 0$ for all $\mathcal{W}$ is equivalent to [LT, Lemma 5.2 (2)], which is also equivalent to the genus-0 $\widetilde{\mathcal{L}}_1$ constraint. Similar reasoning also applies to the genus-0 $\widetilde{\mathcal{L}}_2$ constraint by considering the vector field $T(R(\mathcal{D}))$ where R is defined in Definition 2.2.

Equation (6) has the following generalization, which is obtained by taking derivatives of the equation $T(\mathcal{W}) \bullet \mathcal{V} = 0$,

$$\Big\langle\!\Big\langle\, T^k(\mathcal{W})\, \mathcal{V}_1 \cdots \mathcal{V}_{k+1} \,\Big\rangle\!\Big\rangle_0 = 0 \tag{8}$$

for any vector fields $\mathcal{W}$ and $\mathcal{V}_1, \ldots, \mathcal{V}_{k+1}$ with $k \geq 1$. Note that the power of T in this formula is sharp in the sense that if we replace T^k by T^i for $i < k$, the formula is no longer correct. The same is also true for (10). For example,

$$\langle\!\langle\, T(\mathcal{V})\, \mathcal{W}_1\, \mathcal{W}_2\, \gamma^\alpha \,\rangle\!\rangle_0\, \gamma_\alpha = \mathcal{W}_1 \bullet \mathcal{W}_2 \bullet \mathcal{V}$$

for any vector fields $\mathcal{W}_1$, $\mathcal{W}_2$, and $\mathcal{V}$ (cf. [L2, Corollary 1.6]).

If $\mathcal{S}$ were an identity for the quantum product on the big phase space, $T(\mathcal{W})$ would always be a zero vector field because of (5). However, from the definition of T, we know $T(\mathcal{W}) \neq 0$ unless $\mathcal{W} = 0$. On the other hand, equation (6) tells us that at the genus-0 level, the operator T always annihilates vector fields from the viewpoint of the quantum product. This equation can be thought of as another way to interpret the genus-0 topological recursion relation. Similar interpretations can also be given for the higher genus topological recursion relations.

The genus-1 topological recursion relation is the following (cf. [DW]):

$$\langle\!\langle \tau_{n+1}(\gamma_\alpha) \rangle\!\rangle_1 = \langle\!\langle \tau_n(\gamma_\alpha)\, \gamma^\mu \rangle\!\rangle_0 \langle\!\langle \gamma_\mu \rangle\!\rangle_1 + \frac{1}{24} \langle\!\langle \tau_n(\gamma_\alpha)\, \gamma^\mu\, \gamma_\mu \rangle\!\rangle_0 .$$

This formula is equivalent to

$$\langle\!\langle T(\mathcal{W}) \rangle\!\rangle_1 = \frac{1}{24} \langle\!\langle \mathcal{W}\, \gamma^\mu\, \gamma_\mu \rangle\!\rangle_0 \tag{9}$$

for any vector field $\mathcal{W}$. For $g > 0$, we call a vector field $\mathcal{W}$ *trivial at the genus-g level* if $\langle\!\langle \mathcal{W} \rangle\!\rangle_g$ can be represented by data of genera less than g. Then the genus-1 topological recursion relation just means that $T(\mathcal{W})$ is trivial at the genus-1 level for all $\mathcal{W}$.

Equation (9) and its derivatives imply the following

$$\langle\!\langle T^{k+2}(\mathcal{W})\, \mathcal{V}_1 \cdots \mathcal{V}_k \rangle\!\rangle_1 = 0 \tag{10}$$

for $k \geq 0$ and any vector fields $\mathcal{W}, \mathcal{V}_1, \ldots, \mathcal{V}_k$.

The genus-2 topological recursion relations are much more complicated than genus-0 and genus-1 topological recursion relations. Two genus-2 topological recursion relations were given in [G2]. The first one was derived by using a formula due to Mumford:

$$\begin{aligned}
\langle\!\langle \tau_{i+2}(x) \rangle\!\rangle_2 &= \langle\!\langle \tau_{i+1}(x)\, \gamma^\alpha \rangle\!\rangle_0 \langle\!\langle \gamma_\alpha \rangle\!\rangle_2 + \langle\!\langle \tau_i(x)\, \gamma^\alpha \rangle\!\rangle_0 \langle\!\langle \tau_1(\gamma_\alpha) \rangle\!\rangle_2 \\
&- \langle\!\langle \tau_i(x)\, \gamma^\alpha \rangle\!\rangle_0 \langle\!\langle \gamma_\alpha\, \gamma^\beta \rangle\!\rangle_0 \langle\!\langle \gamma_\beta \rangle\!\rangle_2 + \frac{7}{10} \langle\!\langle \tau_i(x)\, \gamma^\alpha\, \gamma^\beta \rangle\!\rangle_0 \langle\!\langle \gamma_\alpha \rangle\!\rangle_1 \langle\!\langle \gamma_\beta \rangle\!\rangle_1 \\
&+ \frac{1}{10} \langle\!\langle \tau_i(x)\, \gamma^\alpha\, \gamma^\beta \rangle\!\rangle_0 \langle\!\langle \gamma_\alpha\, \gamma_\beta \rangle\!\rangle_1 - \frac{1}{240} \langle\!\langle \tau_i(x)\, \gamma_\alpha \rangle\!\rangle_1 \langle\!\langle \gamma^\alpha\, \gamma_\beta\, \gamma^\beta \rangle\!\rangle_0 \\
&+ \frac{13}{240} \langle\!\langle \tau_i(x)\, \gamma_\alpha\, \gamma^\alpha\, \gamma^\beta \rangle\!\rangle_0 \langle\!\langle \gamma_\beta \rangle\!\rangle_1 + \frac{1}{960} \langle\!\langle \tau_i(x)\, \gamma^\alpha\, \gamma_\alpha\, \gamma^\beta\, \gamma_\beta \rangle\!\rangle_0 .
\end{aligned}$$

This formula can be written in the following form: For any vector field $\mathcal{W}$ on the big phase space,

$$\begin{aligned}\langle\langle T^2(\mathcal{W}) \rangle\rangle_2 = &\frac{7}{10} \langle\langle \gamma_\alpha \rangle\rangle_1 \langle\langle \{\gamma^\alpha \bullet \mathcal{W}\} \rangle\rangle_1 + \frac{1}{10} \langle\langle \gamma_\alpha \{\gamma^\alpha \bullet \mathcal{W}\} \rangle\rangle_1 \\ &- \frac{1}{240} \langle\langle \mathcal{W} \{\gamma_\alpha \bullet \gamma^\alpha\} \rangle\rangle_1 + \frac{13}{240} \langle\langle \mathcal{W} \gamma_\alpha \gamma^\alpha \gamma^\beta \rangle\rangle_0 \langle\langle \gamma_\beta \rangle\rangle_1 \\ &+ \frac{1}{960} \langle\langle \mathcal{W} \gamma^\alpha \gamma_\alpha \gamma^\beta \gamma_\beta \rangle\rangle_0, \end{aligned} \tag{11}$$

where $T^2(\mathcal{W}) := T(T(\mathcal{W}))$. Another topological recursion relation in [G2] is a formal consequence of this formula and a formula in [BP] (cf. [L2]).

Unlike in the genus-1 case, $T(\mathcal{W})$ is no longer trivial at the genus-2 level in general. In fact, the genus-2 dilaton equation implies $\langle\langle T(\mathcal{S}) \rangle\rangle_2 = \langle\langle \mathcal{D} \rangle\rangle_2 = 2F_2$. Unless F_2 can be expressed as a function of F_0 and F_1, $T(\mathcal{S})$ is not trivial at the genus-2 level. However, the topological recursion relation (11) tells us that $T^2(\mathcal{W})$ is trivial at the genus-2 level for all vector fields $\mathcal{W}$. The topological recursion relations for genus $g = 1, 2$ are derived by using a formula for expressing the tautological class ψ_1^g (i.e. $c_1(E_1)^g$ according to the notation in Section 1.1) on the moduli space of stable curves $\overline{\mathcal{M}}_{g,1}$ in terms of boundary classes. It was conjectured in [G2] that polynomials of degree g in the tautological classes ψ_i are boundary classes on $\overline{\mathcal{M}}_{g,n}$. This conjecture was proved in [Io]. A somewhat stronger version of a special case of this conjecture would be that ψ_1^g is equal to a boundary class in $\overline{\mathcal{M}}_{g,1}$ without genus-g components. This would imply the following:

(12) $T^g(\mathcal{W})$ is trivial at the genus$-g$ level for $g \geq 1$.

The following topological recursion relation for all $g \geq 1$ was derived in [EX] under the assumption that the genus-g generating function is a function of the derivatives of genus-0 generating function:

$$\langle\langle \tau_{n+3g-1}(\gamma_\alpha) \rangle\rangle_g = \sum_{j=0}^{3g-2} \langle\langle \tau_{n+3g-2-j}(\gamma_\alpha)\ \gamma_\beta \rangle\rangle_0\, A_j^\beta$$

where $A_0^\beta = \langle\langle \gamma^\beta \rangle\rangle_g$ and $A_j^\beta = \langle\langle \tau_j(\gamma^\beta) \rangle\rangle_g - \sum_{k=0}^{j-1} \langle\langle \tau_k(\gamma^\beta)\ \gamma_\mu \rangle\rangle_0\, A_{j-1-k}^\mu$. An easy induction on j shows that $\langle\langle T^j(\gamma^\beta) \rangle\rangle_g = A_j^\beta$ and a similar induction argument also shows that this topological recursion relation is precisely

$$\langle\langle T^{3g-1}(\mathcal{W}) \rangle\rangle_g = 0 \tag{13}$$

for all vector fields $\mathcal{W}$. For $g = 1$ and 2, this equation follows from equations (8), (10), and (11). For general g, it follows from $\psi_1^{3g-1} = 0$ on $\overline{\mathcal{M}}_{g,1}$ since the complex dimension of $\overline{\mathcal{M}}_{g,1}$ is $3g-2$. This was first observed by Getzler.

To apply these topological recursion relations, we observe that

$$\mathcal{W} = \overline{\mathcal{W}} + \sum_{i=1}^{k-1} T^i(\overline{\tau_-^i(\mathcal{W})}) + T^k(\tau_-^k(\mathcal{W})) \tag{14}$$

for any vector field $\mathcal{W}$ and $k \geq 1$ (cf. [L2]). Since $\overline{\tau_-^i(\mathcal{W})}$ is a primary vector field, the descendant level of $T^i(\overline{\tau_-^i(\mathcal{W})})$ is at most i. In applications, we can choose k large enough to apply suitable topological recursion relations to get rid of the last term on the right hand side of (14). For example, to apply (12), we would choose $k = g$. To apply (13), we would choose $k = 3g-1$.

§2. Virasoro conjecture

Let

$$Z := \exp \sum_{g \geq 0} \lambda^{2g-2} F_g.$$

This is the partition function in the topological sigma model coupled to gravity. In the case that the underlying manifold is a point, Witten conjectured that Z is a τ-function of the KdV hierarchy (cf. [W1]). Witten's conjecture was proved in [Ko] and [W2]. In [DVV] [FKN] [KS], it was proved that a function is a τ function of the KdV hierarchy and satisfies the string equation if and only if it is annihilated by a specific sequence of differential operators satisfying the Virasoro bracket relation. Eguchi, Hori, and Xiong [EHX] defined a sequence of differential operators L_n, $n \geq -1$, which satisfy the Virasoro type relation

$$[L_m, L_n] = (m-n)L_{m+n}$$

for a class of compact symplectic manifolds whose Chern numbers satisfy a certain non-trivial equation. For general projective varieties V, these operators were modified by S. Katz so that the Virasoro relation is always satisfied. The *Virasoro conjecture* predicts that

$$L_n Z = 0$$

for $n \geq -1$. The genus-g Virasoro conjecture predicts that the coefficient of λ^{2g-2} in the Laurent polynomial $Z^{-1}L_n Z$ in terms of λ vanishes.

The Virasoro conjecture is one natural generalization of Witten's conjecture. Another way to generalize Witten's conjecture is to construct an integrable system to govern the Gromov-Witten theory for each projective variety. In [EY], it was conjectured that the integrable system corresponding to $\mathbb{P}^1$ is the Toda hierarchy. For general projective varieties, the corresponding integrable systems are not clear at the time of writing. The study of the Virasoro conjecture should shed more light in this direction. In fact, it was conjectured in [DZ] that the Virasoro constraints can be used to construct KdV type integrable systems for semisimple Frobenius manifolds. If this program is successful, we can then obtain an integrable system for each projective variety with semisimple quantum cohomology provided that the Virasoro conjecture can be proved.

The Virasoro operators L_n defined in [EHX] are very complicated. A basic idea in [LT] is that we can study the Virasoro conjecture through the study of certain globally defined vector fields on the big phase space. For this reason, we will not give the explicit form of these operators here. Instead, we will give a recursive description of the relevant vector fields following [L2].

2.1. Virasoro vector fields

We first define a new product on the space of vector fields on the big phase space.

Definition 2.1. *For* $\mathcal{W} = \sum_{m,\alpha} f_{m,\alpha}\tau_m(\gamma_\alpha)$, $\mathcal{V} = \sum_{m,\alpha} g_{m,\alpha}\tau_m(\gamma_\alpha)$, *we define* $\mathcal{W} * \mathcal{V} = \sum_{m,\alpha} f_{m,\alpha}g_{m,\alpha}\tau_m(\gamma_\alpha)$.

We can think of "$*$" as a commutative associative product on the space of vector fields with the identity

$$\mathcal{Z} := \sum_{m,\alpha} \tau_m(\gamma_\alpha).$$

Another important vector field for this product is $\mathcal{G} := \sum_{m,\alpha}(m + b_\alpha)\tau_m(\gamma_\alpha)$. We also need the following linear (over the ring of functions) transformations on the space of vector fields:

Definition 2.2. *For any vector field* $\mathcal{W} = \sum_{m,\alpha} f_{m,\alpha}\tau_m(\gamma_\alpha)$, *define* $C(\mathcal{W}) = \sum_{m,\alpha,\beta} f_{m,\alpha}\mathcal{C}_\alpha^\beta\tau_m(\gamma_\beta)$ *and* $R(\mathcal{W}) = \mathcal{G} * T(\mathcal{W}) + C(\mathcal{W})$.

As pointed out in [LT], the most important vector field in studying the Virasoro conjecture is the *Euler vector field*, which is defined by

$$\mathcal{X} := -\sum_{m,\alpha} (m + b_\alpha - b_1 - 1)\, \tilde{t}_m^\alpha\, \tau_m(\gamma_\alpha) - \sum_{m,\alpha,\beta} \mathcal{C}_\alpha^\beta \tilde{t}_m^\alpha\, \tau_{m-1}(\gamma_\beta).$$

This vector field satisfies the following *quasi-homogeneity equation*

$$\langle\langle \mathcal{X} \rangle\rangle_g = (3-d)(1-g)F_g + \frac{1}{2}\delta_{g,0} \sum_{\alpha,\beta} \mathcal{C}_{\alpha\beta} t_0^\alpha t_0^\beta - \frac{1}{24}\delta_{g,1} \int_V c_1(V) \cup c_{d-1}(V).$$

The second derivatives of the genus-0 quasi-homogeneity equation have the following form (cf. [LT, lemma 1.4 (3)]):

$$\text{(15)} \qquad \langle\langle\, \mathcal{W}\, \mathcal{X}\, \mathcal{V} \,\rangle\rangle_0 = \langle\langle\, Q(\mathcal{W})\, \mathcal{V} \,\rangle\rangle_0 + \langle\langle\, \mathcal{W}\, Q(\mathcal{V}) \,\rangle\rangle_0 + \frac{1}{2} \nabla^2_{\mathcal{W},\mathcal{V}} \mathcal{C}_{\alpha\beta} t_0^\alpha t_0^\beta$$

where

$$\text{(16)} \qquad Q(\mathcal{W}) := \mathcal{G} * \mathcal{W} + C\tau_-(\mathcal{W})$$

for any vector field $\mathcal{W}$. This equation implies the following (see [L2])

Theorem 2.3. *For any vector field* $\mathcal{W}$, $\mathcal{W} \bullet \mathcal{X} = \mathcal{S} \bullet R(\mathcal{W}) = \overline{R(\mathcal{W})}$.

Proof: Since both sides of this equation are tensors with respect to $\mathcal{W}$, we only need to prove the theorem for $\mathcal{W} = \tau_m(\gamma_\alpha)$. By definition, $R(\tau_m(\gamma_\alpha)) = (m + 1 + b_\alpha)\tau_{m+1}(\gamma_\alpha) - b_\beta \left\langle\left\langle\, \tau_m(\gamma_\alpha)\, \gamma^\beta \,\right\rangle\right\rangle_0 \gamma_\beta + \mathcal{C}_\alpha^\beta \tau_m(\gamma_\beta)$. Equation (3) implies that $\langle\langle\, \mathcal{S}\, R(\tau_m(\gamma_\alpha))\, \gamma^\mu \,\rangle\rangle_0 = (m + 1 + b_\alpha - b_\mu) \langle\langle\, \tau_m(\gamma_\alpha)\, \gamma^\mu \,\rangle\rangle_0 + \mathcal{C}_\alpha^\beta \langle\langle\, \tau_{m-1}(\gamma_\beta)\, \gamma^\mu \,\rangle\rangle_0 + \delta_m^0 \mathcal{C}_\alpha^\mu$. The right hand side of this equation is precisely $\langle\langle\, \mathcal{X}\, \tau_m(\gamma_\alpha)\, \gamma^\mu \,\rangle\rangle_0$ by (15). The theorem is thus proved. □

Theorem 2.3 implies that $\overline{R(\mathcal{W})} = \overline{R(\mathcal{V})}$ if $\overline{\mathcal{W}} = \overline{\mathcal{V}}$. Another consequence of the quasi-homogeneity equation is that

$$\text{(17)} \qquad [\overline{\mathcal{X}}^{\,m}, \overline{\mathcal{X}}^{\,k}] = (k - m)\, \overline{\mathcal{X}}^{\,m+k-1}$$

for all $m, k \geq 0$. This can be proved by using an argument similar to the one on the small phase space given in [L1].

Define $\mathcal{L}_k := -R^{k+1}(\mathcal{S})$ for $k \geq -1$. We call $\mathcal{L}_k$ the k-th *Virasoro vector field* due to its intimate connection with the Virasoro conjecture. Since $\mathcal{D} = T(\mathcal{S})$, $\mathcal{L}_0 = -\mathcal{X} - (b_1 + 1)\mathcal{D}$. Therefore $\overline{\mathcal{L}}_0 = -\overline{\mathcal{X}}$ because $\overline{\mathcal{D}} = 0$. So Theorem 2.3 implies

$$\text{(18)} \qquad \overline{\mathcal{L}}_k = -\overline{\mathcal{X}}^{\,k+1}$$

for $k \geq -1$. Here and thereafter, we will understand $\mathcal{X}^0$ as the string vector field $\mathcal{S}$ and $\overline{\mathcal{X}}^0$ as $\overline{\mathcal{S}}$.

Let ∇ be the covariant derivative defined by

$$\nabla_{\mathcal{V}}\mathcal{W} = \sum_{m,\alpha} (\mathcal{V} f_{m,\alpha})\, \tau_m(\gamma_\alpha)$$

for any vector fields $\mathcal{V}$ and $\mathcal{W} = \sum_{m,\alpha} f_{m,\alpha}\, \tau_m(\gamma_\alpha)$. It is straightforward to check

$$R\tau_-(\mathcal{W}) = \tau_- R(\mathcal{W}) - \mathcal{G} * \overline{\mathcal{W}} - \mathcal{W} \tag{19}$$

and

$$\nabla_{\mathcal{V}}(R(\mathcal{W})) = R(\nabla_{\mathcal{V}}\mathcal{W}) - \mathcal{G} * (\mathcal{V} \bullet \mathcal{W}) \tag{20}$$

for any vector fields $\mathcal{V}$ and $\mathcal{W}$ (cf. [L2]). By induction on k and using Theorem 2.3, we obtain

$$R^k\tau_-(\mathcal{W}) = \tau_- R^k(\mathcal{W}) - k\, R^{k-1}(\mathcal{W}) - \sum_{i=0}^{k-1} R^i\left(\mathcal{G} * (\mathcal{X}^{k-1-i} \bullet \mathcal{W})\right) \tag{21}$$

and

$$\nabla_{\mathcal{V}}(R^k(\mathcal{W})) = R^k(\nabla_{\mathcal{V}}\mathcal{W}) - \sum_{i=0}^{k-1} R^i\left(\mathcal{G} * (\mathcal{X}^{k-1-i} \bullet \mathcal{V} \bullet \mathcal{W})\right) \tag{22}$$

for $k \geq 1$. We can also use (21) to simplify (22) and obtain

$$\nabla_{\mathcal{V}}(R^k(\mathcal{W})) = R^k(\nabla_{\mathcal{V}}\mathcal{W}) - \tau_- R^k(\mathcal{V} \bullet \mathcal{W}) + k\, R^{k-1}(\mathcal{V} \bullet \mathcal{W}) \tag{23}$$

since $\tau_-(\mathcal{V} \bullet \mathcal{W}) = 0$. Using these equations, we can prove the following

Lemma 2.4. $[\mathcal{L}_m, \mathcal{L}_k] = (m-k)\mathcal{L}_{m+k}$.

Proof: It is straightforward to check that $\nabla_{\mathcal{W}}\mathcal{S} = -\tau_-(\mathcal{W})$ for any vector field $\mathcal{W}$. Therefore, applying (22) to $\mathcal{L}_k = -R^{k+1}(\mathcal{S})$ first, then using (1) and (21), we have

$$\nabla_{\mathcal{W}}\, \mathcal{L}_k = \tau_- R^{k+1}(\mathcal{W}) - (k+1)R^k(\mathcal{W}) \tag{24}$$

for any vector field $\mathcal{W}$. In particular, if $\mathcal{W} = \mathcal{L}_m$, we have $\nabla_{\mathcal{L}_m} \mathcal{L}_k = \tau_-(\mathcal{L}_{m+k+1}) - (k+1)\mathcal{L}_{m+k}$. Therefore $[\mathcal{L}_m, \mathcal{L}_k] = \nabla_{\mathcal{L}_m} \mathcal{L}_k - \nabla_{\mathcal{L}_k} \mathcal{L}_m = (m-k)\mathcal{L}_{m+k}$. □

2.2. Quantum powers of the Euler vector field

From (18), we can see that the quantum powers of the Euler vector field are very important in the study of the Virasoro vector fields. In this section, we compute $\langle\langle\, \mathcal{W}\, \mathcal{X}^k\, \mathcal{V}\,\rangle\rangle_0$ for $k \geq 1$ and arbitrary vector fields $\mathcal{W}$ and $\mathcal{V}$.

For any $k \geq 2$, define

$$Q_k := Q(Q-1)\cdots(Q-k+1)\tau_+^{k-1}$$

where Q is defined by (16). For convenience, we also define

$$Q_0 := \tau_- \qquad \text{and} \qquad Q_1 := Q.$$

These operators have the following properties

Lemma 2.5.

$$\begin{aligned}
&(i)\ \tau_+ Q_k = (Q-1)\cdots(Q-k)\,\tau_+^k - \mathcal{C}^k\pi \quad \text{for } k \geq 1,\\
&(ii)\ \tau_+^k\, Q\, \tau_+ = (Q-k)\,\tau_+^{k+1},\\
&(iii) Q\, \tau_+\, Q_k = Q_{k+1} - Q\mathcal{C}^k\pi,\\
&(iv)\ Q_k\,(Q-1)\,\tau_+ = Q_{k+1}.
\end{aligned}$$

The first two formulae follow from a straightforward induction on k. The last two formulae are direct consequences of the first two.

The aim of this section is to prove the following

Proposition 2.6. *For any vector fields $\mathcal{W}$ and $\mathcal{V}$ and $k \geq 1$,*

$$\begin{aligned}
\langle\langle\, \mathcal{W}\, \mathcal{X}^k\, \mathcal{V}\,\rangle\rangle_0 = \langle\langle\, Q_k(\mathcal{W})\, \mathcal{V}\,\rangle\rangle_0 + \langle\langle\, \mathcal{W}\, Q_k(\mathcal{V})\,\rangle\rangle_0 + \frac{1}{2}\nabla^2_{\mathcal{W},\mathcal{V}}(\mathcal{C}^k)_{\alpha\beta}\, t_0^\alpha t_0^\beta \\
+ \sum_{i=1}^{k-1} \langle\langle\, \mathcal{W}\, Q_i(\gamma^\alpha)\,\rangle\rangle_0\, \langle\langle\, \{(Q_{k-i} - Q_{k-1-i}\mathcal{C})(\gamma_\alpha)\}\, \mathcal{V}\,\rangle\rangle_0 .
\end{aligned}$$

Note that this proposition also makes sense for $k = 0$ and it is precisely (3) if one takes $\mathcal{X}^0 = \mathcal{S}$ and $(\mathcal{C}^0)_{\alpha\beta} = \eta_{\alpha\beta}$. By the genus-0 topological recursion relation, $\langle\langle\, \mathcal{W}\, \gamma^\alpha\,\rangle\rangle_0\, \langle\langle\, \gamma_\alpha\, \mathcal{S}\, \tau_+(\mathcal{V})\,\rangle\rangle_0 = \langle\langle\, \tau_+(\mathcal{W})\, \mathcal{S}\, \tau_+(\mathcal{V})\,\rangle\rangle_0$. Applying (3) to both sides of this equation and observing that we have $\nabla^2_{\tau_+(\mathcal{W}),\mathcal{V}}\ t_0^\alpha\, t_0^\beta = 0$, we obtain

$$\langle\langle\, \mathcal{W}\, \gamma^\alpha\,\rangle\rangle_0\, \langle\langle\, \gamma_\alpha\, \mathcal{V}\,\rangle\rangle_0 = \langle\langle\, \tau_+(\mathcal{W})\ \mathcal{V}\,\rangle\rangle_0 + \langle\langle\, \mathcal{W}\ \tau_+(\mathcal{V})\,\rangle\rangle_0 . \tag{25}$$

This is precisely [LT, equation (10)].

For $k = 1$, Proposition 2.6 is precisely (15). By the genus-0 topological recursion relation, $\langle\!\langle\, \mathcal{W}\, \mathcal{X}\, \gamma^\alpha \,\rangle\!\rangle_0 \langle\!\langle\, \gamma_\alpha\, \mathcal{V} \,\rangle\!\rangle_0 = \langle\!\langle\, \mathcal{W}\, \mathcal{X}\, \tau_+(\mathcal{V}) \,\rangle\!\rangle_0$. Applying (15) to both sides of this equation, then using (25) and Lemma 2.5 (i) and observing that $\frac{1}{2}\left(\nabla^2_{\mathcal{W},\,\gamma^\mu} (\mathcal{C}^k)_{\alpha\beta}\, t_0^\alpha\, t_0^\beta\right) \gamma_\mu = C^k \pi(\mathcal{W})$, we obtain

(26)
$$\langle\!\langle\, \mathcal{W}\, Q(\gamma^\alpha) \,\rangle\!\rangle_0 \langle\!\langle\, \gamma_\alpha\, \mathcal{V} \,\rangle\!\rangle_0 = - \langle\!\langle\, \{(Q-1)\tau_+(\mathcal{W})\}\, \mathcal{V} \,\rangle\!\rangle_0 + \langle\!\langle\, \mathcal{W}\, \{Q\tau_+(\mathcal{V})\} \,\rangle\!\rangle_0 .$$

We remark here that [LT, Lemma 3.2 amd 3.3] are simple consequences of this formula and (25). Note that by using Proposition 2.6 for $k = 2$ in a similar way, we also obtain [LT, Lemma 4.4], which was needed in the proof of the genus-0 Virasoro conjecture in [LT]. But it will not be needed in this paper.

Proof of Proposition 2.6: We prove this proposition by induction on k. When $k = 1$, it is just (15). For $k \geq 2$, we have for any vector fields $\mathcal{W}$, $\mathcal{V}$,

$$\langle\!\langle\, \mathcal{W}\, \mathcal{X}^k\, \mathcal{V} \,\rangle\!\rangle_0 = \langle\!\langle\, \mathcal{W}\, \mathcal{X}\, \gamma^\alpha \,\rangle\!\rangle_0 \langle\!\langle\, \gamma_\alpha\, \mathcal{X}^{k-1}\, \mathcal{V} \,\rangle\!\rangle_0 . \tag{27}$$

If $\mathcal{V} = \gamma^\beta$ for some β, this formula is just the associativity of the quantum product. The general case follows from this special case and the genus-0 topological recursion relation.

Applying (15) to the first term on the right hand side of (27) and then using the genus-0 topological recursion relation, $\langle\!\langle\, \mathcal{W}\, \mathcal{X}^k\, \mathcal{V} \,\rangle\!\rangle_0 =$

$$\langle\!\langle\, \{(\tau_+ Q + C\pi)(\mathcal{W})\}\, \mathcal{X}^{k-1}\, \mathcal{V} \,\rangle\!\rangle_0 + \langle\!\langle\, \mathcal{W}\, Q(\gamma^\alpha) \,\rangle\!\rangle_0 \langle\!\langle\, \gamma_\alpha\, \mathcal{X}^{k-1}\, \mathcal{V} \,\rangle\!\rangle_0 .$$

The proposition then follows from applying the induction hypothesis to the right hand side of this equation and simplifying the resulting expression using (26), Lemma 2.5 and the following simple observation:

$$\langle\!\langle\, \cdots \{P_1 C(\gamma^\alpha)\} \,\rangle\!\rangle_0\, P_2(\gamma_\alpha) = \langle\!\langle\, \cdots \{P_1(\gamma^\alpha)\} \,\rangle\!\rangle_0\, P_2 C(\gamma_\alpha) \tag{28}$$

for any linear transformations P_1, P_2 on the space of vector fields. □

If we set $\mathcal{W} = \mathcal{V} = \mathcal{D}$ in Proposition 2.6 and use the genus-0 dilaton equation (7), we obtain

Corollary 2.7. $\langle\!\langle\, Q_k(\mathcal{D}) \,\rangle\!\rangle_0 =$

$$\frac{1}{2}\sum_{i=1}^{k-1} \langle\!\langle\, Q_i(\gamma^\alpha) \,\rangle\!\rangle_0 \langle\!\langle\, \{(Q_{k-i} - Q_{k-1-i}C)(\gamma_\alpha)\} \,\rangle\!\rangle_0 + \frac{1}{2}(\mathcal{C}^k)_{\alpha\beta}\, t_0^\alpha t_0^\beta .$$

2.3. Genus-0 Virasoro conjecture

In this section we compute $\langle\langle \mathcal{L}_k \rangle\rangle_0$ using Corollary 2.7 and then relate it to the genus-0 Virasoro conjecture. For this purpose, we need to study relations between the operators R, T, and Q.

First, replacing $\mathcal{W}$ by $RT(\mathcal{W})$ in (14) with $k=1$, then using Theorem 2.3 and (19), we obtain $RT = T(R+T)$ (cf. [L2, Lemma 4.6]). By induction on k, we have

$$T^k R = (R - kT)T^k \tag{29}$$

for $k \geq 0$. By the definitions of R and Q, $R = QT$. Writing $R^k = R^{k-1}R$, then by induction on k and using (29), we obtain

$$R^k = Q(Q-1)\cdots(Q-k+1)T^k \tag{30}$$

for $k \geq 1$. Therefore

$$\begin{aligned} \mathcal{L}_k &= -R^{k+1}(\mathcal{S}) = -Q(Q-1)\cdots(Q-k)T^k(\mathcal{D}) \\ &= -Q_{k+1}(\mathcal{D}) - Q(Q-1)\cdots(Q-k)(T^k - \tau_+^k)(\mathcal{D}). \end{aligned} \tag{31}$$

On the other hand, writing

$$T^k(\mathcal{W}) = T^{k-1}(T(\mathcal{W})) = T^{k-1}\left(\tau_+(\mathcal{W}) - \langle\langle \mathcal{W}\gamma^\alpha \rangle\rangle_0 \gamma_\alpha\right),$$

then by induction on k and using (25), we obtain

$$T^k(\mathcal{W}) = \tau_+^k(\mathcal{W}) + \sum_{i=0}^{k-1}(-1)^{i+1}\langle\langle \mathcal{W}\,\tau_+^i(\gamma^\alpha) \rangle\rangle_0\, \tau_+^{k-1-i}(\gamma_\alpha) \tag{32}$$

for any vector field $\mathcal{W}$. In particular, replacing $\mathcal{W}$ by $\mathcal{D}$ and using (7), we have

$$(T^k - \tau_+^k)(\mathcal{D}) = \sum_{i=0}^{k-1}(-1)^i \langle\langle \tau_+^i(\gamma^\alpha) \rangle\rangle_0\, \tau_+^{k-1-i}(\gamma_\alpha). \tag{33}$$

Together with (31) and Corollary 2.7, we can use this equation to compute $\langle\langle \mathcal{L}_k \rangle\rangle_0$. To simplify the result, we also need the following

Lemma 2.8. *For any linear (over the ring of functions) transformation P on the space of vector fields and $k \geq 1$,*

$$\begin{aligned} &Q(Q-1)\cdots(Q-k)\left(\sum_{i=0}^{k-1}(-1)^i \langle\langle \{P\tau_+^i(\gamma^\alpha)\} \rangle\rangle_0\, \tau_+^{k-1-i}(\gamma_\alpha)\right) \\ &= -\sum_{i=1}^{k}\langle\langle \{PQ_i(\gamma^\alpha)\} \rangle\rangle_0 (Q_{k+1-i} - Q_{k-i}C)(\gamma_\alpha). \end{aligned}$$

Proof: Let $\mathcal{W}_{P,k} := \sum_{i=0}^{k-1}(-1)^i \langle\!\langle \{P\tau_+^i(\gamma^\alpha)\} \rangle\!\rangle_0 \, \tau_+^{k-1-i}(\gamma_\alpha)$. We prove the lemma by induction on k. The lemma holds for $k=1$ because $Q(Q-1)(\mathcal{W}_{P,1}) = \langle\!\langle P(\gamma^\alpha) \rangle\!\rangle_0 \, (b_\alpha - 1)Q(\gamma_\alpha) = -\langle\!\langle \{PQ(\gamma^\alpha)\} \rangle\!\rangle_0 \, Q(\gamma_\alpha)$. By Lemma 2.5 (i) and (ii), $Q\,\tau_+^k = \tau_+^k\,(Q+k) + \tau_+^{k-1} C\pi$ for $k \geq 1$. Using this equation and (28), it is straightforward to check that

$$(Q-k)\left(\langle\!\langle \{P\tau_+^i(\gamma^\alpha)\} \rangle\!\rangle_0 \, \tau_+^{k-1-i}(\gamma_\alpha)\right) =$$
$$-\langle\!\langle \{P(Q\tau_+^i - \tau_+^{i-1}C)(\gamma^\alpha)\} \rangle\!\rangle_0 \, \tau_+^{k-1-i}(\gamma_\alpha) + \langle\!\langle \{P\tau_+^i C(\gamma^\alpha)\} \rangle\!\rangle_0 \tau_+^{k-2-i}(\gamma_\alpha).$$

Consequently, we have

$$(Q-k)(\mathcal{W}_{P,k}) = \sum_{i=0}^{k-1}(-1)^{i+1} \langle\!\langle \{PQ\tau_+^i(\gamma^\alpha)\} \rangle\!\rangle_0 \, \tau_+^{k-1-i}(\gamma_\alpha).$$

Therefore, for $k \geq 2$, $Q(Q-1)\cdots(Q-k)(\mathcal{W}_{P,k}) =$

$$Q(Q-1)\cdots(Q-k+1)\left(\mathcal{W}_{(PQ\tau_+),\,(k-1)}\right) - \langle\!\langle \{PQ(\gamma^\alpha)\} \rangle\!\rangle_0 \, Q_k(\gamma_\alpha).$$

The lemma then follows from induction on k and by using Lemma 2.5 (iii). □

Combining Corollary 2.7, equations (31) and (33), and lemma 2.8 for P equal to the identity transformation, we obtain

Theorem 2.9. $\langle\!\langle \mathcal{L}_k \rangle\!\rangle_0 =$

$$\frac{1}{2}\sum_{i=1}^{k} \langle\!\langle Q_i(\gamma^\alpha) \rangle\!\rangle_0 \, \langle\!\langle \{(Q_{k+1-i} - Q_{k-i}C)(\gamma_\alpha)\} \rangle\!\rangle_0 - \frac{1}{2}(\mathcal{C}^{k+1})_{\alpha\beta}\, t_0^\alpha t_0^\beta.$$

Remark: The right hand side of this equation is equal to $\rho_{0,k}$ as defined in [L2, Section 5.1]. One can show this by using the following properties of the operator R_+ defined in [L2]: $R_+^k = Q_k\,\tau_+$ for $k \geq 1$.

Now we relate Theorem 2.9 to the genus-0 Virasoro conjecture. Since

$$\langle\!\langle \mathcal{L}_k\,\gamma^\alpha \rangle\!\rangle_0 = \gamma^\alpha \, \langle\!\langle \mathcal{L}_k \rangle\!\rangle_0 - \langle\!\langle \{\nabla_{\gamma^\alpha}\mathcal{L}_k\} \rangle\!\rangle_0\,,$$

we can use Theorem 2.9 and (24) to compute $\langle\!\langle \mathcal{L}_k\,\gamma^\alpha \rangle\!\rangle_0$. We can then compute $\mathcal{L}_{k+1} = R(\mathcal{L}_k)$ from $\mathcal{L}_k$ by using these formulae. We already know that the vector fields $\mathcal{L}_{-1}$ and $\mathcal{L}_0$ are precisely the first derivative parts of the Virasoro operators L_{-1} and L_0 defined in [EHX]. Therefore using the above method, we obtain $\mathcal{L}_1 =$

$$\sum_{m,\alpha}(m+b_\alpha)(m+b_\alpha+1)\tilde{t}_m^\alpha\tau_{m+1}(\gamma_\alpha)+\sum_{m,\alpha,\beta}(2m+2b_\alpha+1)\mathcal{C}_\alpha^\beta\tilde{t}_m^\alpha\tau_m(\gamma_\beta)$$
$$+\sum_{m,\alpha,\beta}(\mathcal{C}^2)_\alpha^\beta\tilde{t}_m^\alpha\tau_{m-1}(\gamma_\beta)-\sum_\alpha b_\alpha(b_\alpha-1)\langle\!\langle\gamma^\alpha\rangle\!\rangle_0\gamma_\alpha$$

and

$$\begin{aligned}\mathcal{L}_2 = &\sum_{m,\alpha}(m+b_\alpha)(m+b_\alpha+1)(m+b_\alpha+2)\tilde{t}_m^\alpha\tau_{m+2}(\gamma_\alpha)\\ &+\sum_{m,\alpha,\beta}\left\{3(m+b_\alpha)^2+6(m+b_\alpha)+2\right\}\mathcal{C}_\alpha^\beta\tilde{t}_m^\alpha\tau_{m+1}(\gamma_\beta)\\ &+\sum_{m,\alpha,\beta}3(m+b_\alpha+1)(\mathcal{C}^2)_\alpha^\beta\tilde{t}_m^\alpha\tau_m(\gamma_\beta)+\sum_{m,\alpha,\beta}(\mathcal{C}^3)_\alpha^\beta\tilde{t}_m^\alpha\tau_{m-1}(\gamma_\beta)\\ &-\sum_\alpha b_\alpha(b_\alpha^2-1)\{\langle\!\langle\tau_1(\gamma_\alpha)\rangle\!\rangle_0\gamma^\alpha+\langle\!\langle\gamma^\alpha\rangle\!\rangle_0\tau_1(\gamma_\alpha)\}\\ &-\sum_{\alpha,\beta}(3b_\beta^2-1)\mathcal{C}_\beta^\alpha\langle\!\langle\gamma_\alpha\rangle\!\rangle_0\gamma^\beta.\end{aligned}$$

These vector fields are slightly different from the vector field $\mathcal{L}_1$ and $\mathcal{L}_2$ in [LT]. The linear parts of these vector fields are the corresponding vector fields in [LT], but $\mathcal{L}_1$ and $\mathcal{L}_2$ here have extra non-linear terms. These vector fields do agree with the corresponding vector field in [DZ] and [G3]. Therefore Theorem 2.9 is precisely the genus-0 Virasoro conjecture as proved in [LT]. So the arguments presented here give a new proof of the genus-0 Virasoro conjecture. We also notice that equation (18) for $k = 1, 2$ is equivalent to Lemma 3.1 and Lemma 4.2 of [LT] because of (3). We can interpret them as the second derivatives of the genus-0 L_1 and L_2 constraints. As explained in [LT], they are eventually also equivalent to the genus-0 L_1 and L_2 constraints because of the dilaton equation.

Moreover, as a consequence of Lemma 2.4, we obtain

$$\mathcal{L}_k = \lim_{\lambda\to 0} Z^{-1}\circ L_k\circ Z,$$

where L_k is the k-th Virasoro operator defined in [EHX], since the sequence of vector fields on the right hand side of this equality also satisfy the same Virasoro type bracket relation as in Lemma 2.4 (cf. [G3]).

2.4. Higher genus Virasoro conjecture

For convenience, we introduce the following notation: For two functions f_1 and f_2 on the big phase space, we say $f_1 \overset{g}{\approx} f_2$ if $f_1 - f_2$ can be expressed as a function of $F_0, \ldots, F_{g-1}$ and their derivatives. For $g \geq 1$, the genus-g Virasoro conjecture gives an explicit formula for computing $\langle\!\langle \mathcal{L}_n \rangle\!\rangle_g$ in terms of data with genus less than g for $n \geq -1$. The precise way to express $\langle\!\langle \mathcal{L}_n \rangle\!\rangle_g$ in terms of data of genus less than g can be explicitly written out using the Virasoro operators L_n defined in [EHX] (see, for example, [L2, Section 5.1] for a description using recursive operators). A weaker form of this conjecture

$$\langle\!\langle \mathcal{L}_n \rangle\!\rangle_g \overset{g}{\approx} 0. \tag{34}$$

In other words, this conjecture predicts that the sequence of vector fields $\mathcal{L}_k$ are trivial at the genus-g level for all $g \geq 1$. Since the genus-g Virasoro conjecture only differs from its weak version (34) at the lower genus level, (34) captures the major difficulty as well as the computational power of the original Virasoro conjecture. Moreover, conjecture (34) makes sense for all compact symplectic manifolds while the original Virasoro conjecture requires a non-trivial topological condition for the underlying manifolds.

To compute $\langle\!\langle \mathcal{L}_n \rangle\!\rangle_g$, we can use either (12) or (13) together with formula (14). Because of these formulae, the computation of $\langle\!\langle \mathcal{L}_n \rangle\!\rangle_g$ is equivalent to computing the derivative of F_g along the vector field $\sum_{m=0}^{k-1} T^m(\overline{\tau_-^m(\mathcal{L}_n)})$ where $k = g$ or $3g-1$ depending on whether we are using (12) or (13). For this purpose, we first need to compute $\overline{\tau_-^m(\mathcal{L}_k)}$. For $m = 0$, $\overline{\tau_-^m(\mathcal{L}_n)} = \overline{\mathcal{L}}_n = -\overline{\mathcal{X}}^{n+1}$ by (18). Using (19), we can prove inductively that $\tau_-^m(\mathcal{L}_{n+1}) = R(\tau_-^m(\mathcal{L}_n)) + m\,\tau_-^{m-1}(\mathcal{L}_n) + \mathcal{G} * \overline{\tau_-^{m-1}(\mathcal{L}_n)}$ for all $m \geq 1$ and $n \geq -1$. Therefore, by Theorem 2.3, we have the following (cf. [L2])

Theorem 2.10.

$$\overline{\tau_-^m(\mathcal{L}_{n+1})} = \mathcal{X} \bullet \overline{\tau_-^m(\mathcal{L}_n)} + m\,\overline{\tau_-^{m-1}(\mathcal{L}_n)} + \mathcal{G} * \overline{\tau_-^{m-1}(\mathcal{L}_n)}$$

for all $m \geq 1$ and $n \geq -1$.

Using this recursion formula, we can express $\overline{\tau_-^m(\mathcal{L}_n)}$ in terms of twisted quantum powers of the Euler vector field (here the twisting is given by the operation $\mathcal{G}*$) and vector fields $\overline{\tau_-^i(\mathcal{S})}$ where $i = 1, \ldots, m$. For example, when $m = 1$, $\overline{\tau_-(\mathcal{L}_n)} = -\overline{\mathcal{X}}^{n+1} \bullet \overline{\tau_-(\mathcal{S})} - (n+1)\,\overline{\mathcal{X}}^{n} -$

$\sum_{j=0}^{n} \overline{\mathcal{X}}^{j} \bullet (\mathcal{G} * \overline{\mathcal{X}}^{n-j})$ for all $n \geq 0$. By the genus-1 topological recursion relation (9) and (14),

$$\langle\!\langle \mathcal{L}_n \rangle\!\rangle_1 = \langle\!\langle \overline{\mathcal{L}}_n \rangle\!\rangle_1 + \langle\!\langle T(\tau_-(\mathcal{L}_n)) \rangle\!\rangle_1$$
$$= -\langle\!\langle \overline{\mathcal{X}}_{n+1} \rangle\!\rangle_1 + \frac{1}{24} \left\langle\!\!\left\langle \overline{\tau_-(\mathcal{L}_n)}\, \gamma_\alpha \gamma^\alpha \right\rangle\!\!\right\rangle_0 .$$

This immediately explains the rather mysterious formulae for the genus-1 Virasoro conjecture in [L1]. We would also like to point out that for $i \geq 1$, $\tau_-^i(\mathcal{S})$ is zero when restricted to the small phase space. In this way, we can get rid of such vector fields in applications.

From the point of view of applications, the computational power of the Virasoro conjecture depends on how large the space spanned by the vector fields $\sum_{m=0}^{k-1} T^m(\overline{\tau_-^m(\mathcal{L}_n)})$ (with $n \geq -1$) is. In general, this space could be very small. For example, for all algebraic curves, the space spanned by the quantum powers of the Euler vector fields always has dimension 2, while the dimension of the small phase space can be arbitrarily large if the genus of the curve is large enough. If the space spanned by the quantum powers of the Euler vector field is small compared to the small phase space, we do not expect that the Virasoro conjecture will determine the generating functions. In the other extreme case, when the quantum cohomology is semisimple, the quantum powers of the Euler vector fields span the entire tangent space of the small phase space. Therefore in this case, the generating functions can be computed by using the Virasoro conjecture.

For genus $g \geq 1$, the topological recursion relation is not sufficient to prove the Virasoro conjecture. For $g = 1$ and 2, we need at least the equations derived from the algebraic relations in $H^4(\overline{\mathcal{M}}_{1,4})$ and $H^4(\overline{\mathcal{M}}_{2,3})$ which were proved in [G1] and [BP] respectively. In [L1] and [L2], we wrote these equations as equations for certain global tensors on the big phase space. Studying these tensors for quantum powers of the Euler vector field, we proved that for all $n \geq 3$,

$$\left\langle\!\!\left\langle \overline{\mathcal{X}}^{n} \right\rangle\!\!\right\rangle_1 \overset{1}{\approx} \frac{n}{2} \overline{\mathcal{X}}^{n-1} \left\langle\!\!\left\langle \overline{\mathcal{X}}^{2} \right\rangle\!\!\right\rangle_1 \tag{35}$$

and

$$\psi_n \overset{2}{\approx} \frac{n}{2(n-2)} T(\overline{\mathcal{X}}) \psi_{n-1} \tag{36}$$

where $\psi_n := \left\langle\!\!\left\langle \overline{\mathcal{X}}^{n} \right\rangle\!\!\right\rangle_2 - \left\langle\!\!\left\langle T(\overline{\tau_-(\mathcal{L}_{n-1})}) \right\rangle\!\!\right\rangle_2$. The difference of the two sides of each equation can be written down explicitly in terms of the lower genus data. The topological recursion relations (9) and (11) imply that $\langle\!\langle \mathcal{L}_n \rangle\!\rangle_1 \stackrel{1}{\approx} -\left\langle\!\left\langle \overline{\mathcal{X}}^{n+1} \right\rangle\!\right\rangle_1$, $\langle\!\langle \mathcal{L}_n \rangle\!\rangle_2 \stackrel{2}{\approx} -\psi_{n+1}$ for all n. Therefore equations (35) and (36) imply that the conjecture (34) of genus 1 and 2 can be reduced to the corresponding L_1-constraint. The same is also true for the original Virasoro conjecture (cf. [L1] and [L2]). Moreover, since the space of primary vector fields is finite dimensional, the infinite sequence of vector fields $\{\overline{\mathcal{X}}^k \mid k \geq 0\}$ must be linearly dependent at each point. This adds extra information to equations (35) and (36). In the case that the quantum cohomology is not too degenerate (which is a condition weaker than semisimplicity), this enables us to solve for $\left\langle\!\left\langle \overline{\mathcal{X}}^{n} \right\rangle\!\right\rangle_1$ and ψ_n from these equations and thus obtain the corresponding Virasoro conjecture (the weak version only for the genus-2 case). The reader is referred to [L1] and [L2] for details. In [L2] we also explicitly found the genus-2 generating function F_2 in terms of F_1 and F_0 when the quantum cohomology is not too degenerate. For genus bigger than 2, relations analogous to the ones in [G1] and [BP] are missing. Provided with such equations, the structures described here combined with the techniques developed in [LT], [L1] and [L2] could lead to the final resolution of the Virasoro conjecture for all genera.

Update added, August 2007: The first version of this paper was submitted in January 2001. Much progress on the Virasoro conjecture has been made since then. In particular, the genus-2 Virasoro conjecture for manifolds with semisimple quantum cohomology was proved in [L3]. When the underlying manifold has a torus action with isolated fixed points and also has semisimple quantum cohomology, Givental gave a scheme to reduce the Virasoro conjecture to the so-called R-conjecture, which only depends on genus-0 equivariant Gromov-Witten invariants. An outline for the proof of the R-conjecture for projective spaces was given in [Gi]. The R-conjecture has been verified for flag manifolds in [JK] and for Grassmannians in [BCK]. Moreover, the Virasoro conjecture for algebraic curves was proved in [OP].

References

[BP] P. Belorousski and R. Pandharipande, A descendent relation in genus 2, Ann. Scuola Norm. Sup. Pisa Cl. Sci. (4), **29** (2000), 171–191.

[BCK] A. Bertram, I. Ciocan-Fontanine and B. Kim, Two Proofs of a conjecture of Hori and Vafa, Duke Math. J., **126** (2005), 101–136.

[CK] D. Cox and S. Katz, Mirror Symmetry and Algebraic Geometry, Amer. Math. Soc., Providence, R.I., 1999.

[DVV] R. Dijkgraaf, H. Verlinde and E. Verlinde, Loop equations and Virasoro constraints in nonperturbative two-dimensional quantum gravity, Nuclear Phys. B, **348** (1991), 435–456.

[DW] R. Dijkgraaf and E. Witten, Mean field theory, topological field theory, and multimatrix models, Nuclear Phys. B, **342** (1990), 486–522.

[DZ] B. Dubrovin and Y. Zhang, Frobenius manifolds and Virasoro constraints, Selecta Math. (N. S.), **5** (1999), 423–466.

[EHX] T. Eguchi, K. Hori and C. Xiong, Quantum cohomology and Virasoro algebra, Phys. Lett. B, **402** (1997), 71–80.

[EJX] T. Eguchi, M. Jinzinji and C. Xiong, Quantum Cohomology and Free Field Representation, Nuclear Phys. B, **510** (1998), 608–622.

[EX] T. Eguchi and C. Xiong, Quantum cohomology at higher genus: Topological recursion relations and Virasoro conditions, Adv. Theor. Math. Phys., **2** (1998), 219–229.

[EY] T. Eguchi and S.-K. Yang, The topological CP^1 model and the large-N matrix integral, Modern Phys. Lett. A, **9** (1994), 2893–2902.

[FKN] M. Fukuma, H. Kawai and R. Nakayama, Continuum Schwinger-Dyson equations and universal structures in two-dimensional quantum gravity, Internat. J. Modern Phys. A, **6** (1991), 1385–1406.

[G1] E. Getzler, Intersection theory on $\bar{M}_{1,4}$ and elliptic Gromov-Witten Invariants, J. Amer. Math. Soc., **10** (1997), 973–998.

[G2] E. Getzler, Topological recursion relations in genus 2, Integrable systems and algebraic geometry (Kobe/Kyoto, 1997), 73–106.

[G3] E. Getzler, The Virasoro conjecture for Gromov-Witten invariants, In: Algebraic geometry: Hirzebruch 70, Warsaw, 1998, Contemp. Math., **241**, Amer. Math. Soc., Providence, RI, 1999, pp. 147–176.

[Gi] A. Givental, Gromov-Witten invariants and quantization of quadratic hamiltonians, Mosc. Math. J., **1** (2001), 551–568.

[Io] E. Ionel, Topological recursive relations in $H^{2g}(\mathcal{M}_{g,n})$, Invent. Math., **148** (2002), 627–658.

[JK] D. Joe and B. Kim, Equivariant mirrors and the Virasoro conjecture for flag manifolds, Int. Math. Res. Not., **15** (2003), 859–882.

[KS] V. Kac and A. Schwarz, Geometric interpretation of the partition function of 2D gravity, Phys. Lett. B, **257** (1991), 329–334.

[Ko] M. Kontsevich, Intersection theory on the moduli space of curves and the matrix airy function, Comm. Math. Phys., **147** (1992), 1–23.

[LiT] J. Li and G. Tian, Virtual moduli cycles and Gromov-Witten invariants of general symplectic manifolds, In: Topics in symplectic 4-manifolds, Irvine, CA, 1996, pp. 47–83.

[L1] X. Liu, Elliptic Gromov-Witten invariants and Virasoro conjecture, Comm. Math. Phys., **216** (2001), 705–728.

[L2] X. Liu, Quantum product on the big phase space and Virasoro conjecture, Adv. Math., **169** (2002), 313–375.

[L3] X. Liu, Genus-2 Gromov-Witten invariants for manifolds with semisimple quantum cohomology, Amer. J. Math., **129** (2007), 463–498.

[LT] X. Liu and G. Tian, Virasoro constraints for quantum cohomology, J. Differential Geom., **50** (1998), 537–591.

[OP] A. Okounkov and R. Pandharipande, Virasoro constraints for target curves, Invent. Math., **163** (2006), 47–108.

[RT1] Y. Ruan and G. Tian, A mathematical theory of quantum cohomology, J. Differential Geom., **42** (1995), 259–367.

[RT2] Y. Ruan and G. Tian, Higher genus symplectic invariants and sigma models coupled with gravity, Invent. Math., **130** (1997), 455–516.

[W1] E. Witten, Two dimensional gravity and intersection theory on moduli space, Surveys in Diff. Geom., **1** (1991), 243–310.

[W2] E. Witten, On the Kontsevich model and other models of two dimensional gravity, In: Proceedings of the XXth international conference on differential geometric methods in theoretical physics, New York, 1991, World Sci. Publishing, River Edge, NJ, 1992, pp. 176–216.

Department of Mathematics
University of Notre Dame
Notre Dame, IN 46556
USA

E-mail address:
`Xiaobo.Liu.43@nd.edu`

Advanced Studies in Pure Mathematics 51, 2008
Surveys on Geometry and Integrable Systems
pp. 259–283

A generalized Weierstrass representation for a submanifold S in $\mathbb{E}^n$ arising from the submanifold Dirac operator

Shigeki Matsutani

§1. Introduction

Using the submanifold quantum mechanical scheme [dC, JK], the restricted Dirac operator for a k dimensional spin (k-spin) submanifold S immersed in Euclidean space $\mathbb{E}^n$ $(0 < k < n)$ was defined [BJ, Ma1-10]. We call it the submanifold Dirac operator. The zero modes of the Dirac operator express the local properties of the submanifold, such as the Frenet-Serret and generalized Weierstrass formulae. We shall give a survey of this method from the point of view of quantum physics.

As motivation, we recall three facts.

(1) Let us consider an element Q of a ring of operators $\mathbf{P}$ defined over a Riemannian manifold M. The concept of the adjoint of Q is very subtle, as we shall explain briefly, following the book of Björk (see [Remark 1.2.16 in Bj]). Assume that M is Riemannian. For smooth functions f_1 and f_2 whose support is compact, we consider the following integral as a bilinear form of f_1 and f_2:

$$\int_M d\,\mathrm{vol}\,(f_1 Q f_2). \tag{1-1}$$

What is the natural adjoint of Q? One might regard the action on f_1 obtained by integration by parts as defining the adjoint. However the measure here depends on the local coordinates.

Received January 3, 2001
Revised May 5, 2003 and April 30, 2008
2000 *Mathematics Subject Classification.* Primary 53C42.

(2) In a quantum mechanical problem, we sometimes encounter the situation that an eigenfunction ψ of a differential operator P,

$$P\psi = E\psi, \tag{1-2}$$

belongs to a representation space of a group G. Suppose that P is decomposed as $P = P_1 + P_2$. Let us consider the kernel of P_2, $\mathrm{Ker}P_2$, in a certain function space. If an element $\psi_1 \in \mathrm{Ker}P_2$ satisfies

$$P_1\psi_1 = E\psi_1, \tag{1-3}$$

we may also obtain a representation of the group G.

(3) In quantum mechanics over a manifold M, there are typically two pairings on functions: 1) the global pairing $< \ , \ >$ induced from the L^2-norm, such as (1-1), and 2) the pointwise pairing $\cdot$ which is connected with the probability density. (The pointwise pairing can be regarded as $< \circ, \delta(p)\times >$ in terms of the Dirac distribution δ at p in M.)

Even though the concept of adjoint operator is subtle, we can sometimes define a natural adjoint operator if the measure is fixed for some reason, e.g. Haar measure. We note that the ordinary Lebesgue measure in Euclidean space $\mathbb{E}^n$ is Haar measure for the translation group. Quantum mechanics in $\mathbb{E}^n$ is based on this measure and the concept of adjoint operator plays an essential role. In such a case, by fixing a measure and L^2-type pairing $< \ , \ >: \Omega^* \times \Omega \to \mathbb{C}$, for an operator $Q \in \mathbf{P}$ whose domain is a function space Ω, we can define a right-adjoint operator, $Ad(Q)$, with the domain Ω^*, by

$$< f, Qg >=< f Ad(Q), g >, \quad \text{for } (f, g) \in \Omega^* \times \Omega. \tag{1-4}$$

Assume that there is a linear isomorphism $\varphi : \Omega \to \Omega^*$. Then we can define the left-adjoint operator Q^* by $Q^* f = \varphi^{-1}(\varphi(f)Ad(Q))$. The triple $(\Omega^* \times \Omega, < \ , \ >, \varphi)$ becomes a pre-Hilbert space $\mathcal{H}$ by taking the inner product $(\ , \ )_\varphi : \Omega \times \Omega \to \mathbb{C}$ as $(f, g)_\varphi =< \varphi(f), g >$. (Then $(P^* f, g)_\varphi = (f, Pg)_\varphi$.)

Suppose that $Q \in \mathbf{P}$ is *self-adjoint*, i.e. $Q^* = Q$. Then we have the following properties:

(1) The kernel of Q is isomorphic to that of $Ad(Q)$:

$$(\mathrm{Ker}(Q))^* = \varphi(\mathrm{Ker}(Q)) = \mathrm{Ker}(Ad(Q)). \tag{1-5}$$

(2) $((\mathrm{Ker}Q)^* \times \mathrm{Ker}Q, < \ , \ >, \varphi)$ becomes a pre-Hilbert space.

The projection π from $\Omega^* \times \Omega$ to $(\mathrm{Ker}Q)^* \times \mathrm{Ker}Q$ commutes with φ, i.e.

$$\text{(1-6)} \qquad \varphi\pi|_\Omega = \pi|_{\Omega^*}\varphi, \quad (\varphi(\pi|_\Omega f) = \pi|_{\Omega^*}\varphi(f) = \varphi(f)Ad(\pi|_\Omega)).$$

For π satisfying (1-6), we will say that π *is compatible with the inner product.* In fact (1-6) means that $\pi_\Omega{}^* = \pi_\Omega$ due to the relation $\pi|_\Omega{}^* f = \varphi^{-1}(\varphi(f)Ad(\pi|_\Omega)) = \pi|_\Omega f$.

Next assume that P_2 is not necessarily self-adjoint on the pre-Hilbert space $\mathcal{H} = (\Omega^* \times \Omega, < \ , \ >, \varphi)$. For certain P_2, we may be able to construct a transformation η_{sa} of the pre-Hilbert space and its operators, and find a pre-Hilbert space $\mathcal{H}'$ satisfying the following conditions:

(1) There exists an isomorphism $\eta_{\mathrm{sa}} : \Omega^* \times \Omega \to \tilde{\Omega}^* \times \tilde{\Omega}$.

(2) By defining a pairing $< \circ, \times >_{P_2} = < \eta_{\mathrm{sa}}|_{\Omega^*}\circ, \eta_{\mathrm{sa}}|_\Omega \times >$, and $\tilde{\varphi} = \eta_{\mathrm{sa}}|_{\Omega^*}\varphi\eta_{\mathrm{sa}}^{-1}|_{\tilde{\Omega}}$, we obtain $\mathcal{H}' = (\tilde{\Omega}^* \times \tilde{\Omega}, (\ , \ ), \tilde{\varphi})$.

(3) Every operator P is transformed as $\eta_{\mathrm{sa}}|_{\Omega^*}P\eta_{\mathrm{sa}}^{-1}|_{\tilde{\Omega}}$.

(4) P_2 is self-adjoint.

We call η_{sa} a *self-adjointization*: $\eta_{\mathrm{sa}} : \mathcal{H} \to \mathcal{H}'$.

As mentioned above, $\mathrm{Ker}(P_2) \subset \tilde{\Omega}$ also becomes a pre-Hilbert space, denoted by $\mathcal{H}''$. Letting the projection $\mathcal{H}' \to \mathcal{H}''$ be denoted by π_{P_2}, we have a sequence,

$$\text{(1-7)} \qquad \mathcal{H} \xrightarrow{\eta_{\mathrm{sa}}} \mathcal{H}' \xrightarrow{\pi_{P_2}} \mathcal{H}''.$$

This sequence is the key to submanifold quantum mechanics. Instead of considering $P\psi = E\psi$ in $\mathcal{H}$, we search for a solution of $\eta_{\mathrm{sa}}(P)\psi_1 = E\psi_1$.

Let us explain the main idea. For a smooth k-submanifold S embedded in $\mathbb{E}^n$ $(0 < k < n)$, we can find a natural adjoint operator for a differential operator defined over S by using the induced metric of S from $\mathbb{E}^n$. For the Schrödinger equation in $\mathbb{E}^n$ with the L^2-type Hilbert space $\mathcal{H}$,

$$\text{(1-8)} \qquad -\Delta\psi = E\psi,$$

we regard the Laplace operator Δ as a Casimir operator for the translation group. By considering Δ over a tubular neighborhood of S, Δ includes the normal differential operator $\partial_\perp$. We regard $\partial_\perp$ as the above P_2. As $\partial_\perp$ is not self-adjoint in general, we use the above sequence. In the self-adjointization $\eta_{\rm sa}$, we obtain an extra potential in the differential equation. By considering the kernel of the normal component of the differential $\partial_\perp$ and restricting the domain of $\eta_{\rm sa}(\Delta)$ to S, we define a differential operator

$$\Delta_{S\hookrightarrow\mathbb{E}^n} = \eta_{\rm sa}(\Delta)|_{{\rm Ker}\partial_\perp}|_S. \tag{1-9}$$

Then it turns out that

$$\Delta_{S\hookrightarrow\mathbb{E}^n} = \Delta_S + U(\kappa_i), \tag{1-10}$$

where Δ_S is the Beltrami-Laplace operator on S and $U(\kappa_i)$ is an invariant functional of the principal curvature functions κ_i of S in $\mathbb{E}^n$.

Due to the self-adjointness of $\partial_\perp$ in $\mathcal{H}''$, we may naturally consider the Hilbert space $\mathcal{H}''$ for $\Delta_{S\hookrightarrow\mathbb{E}^n}$. The pointwise product is valid in ${\rm Ker}(\partial_\perp)|_S$. Thus we can consider the submanifold Schrödinger equation,

$$-\Delta_{S\hookrightarrow\mathbb{E}^n}\psi = E\psi,$$

as a quantum mechanical problem and a representation of the translation group.

For the case of a smooth surface S embedded in $\mathbb{E}^3$, where K and H denote the Gauss and mean curvature functions, we obtain

$$\Delta_{S\hookrightarrow\mathbb{E}^3} = \Delta_S + H^2 - K. \tag{1-11}$$

It is expected that $\Delta_{S\hookrightarrow\mathbb{E}^n}$ and its zero mode exhibit extrinsic properties, e.g., umbilical points, of the submanifold.

Submanifold quantum mechanics was started by Jensen and Koppe about thirty years ago, and rediscovered by da Costa [JK, dC]. In this paper we give an exposition of the Dirac operator version of the above quantum system from our point of view [Ma1-10].

We recall the fact that the solutions $\{\Psi\}$ of the Dirac equation

$$\not{D}_{\mathbb{E}^n}\Psi = 0$$

give a local representation of the spin group. By letting $\cdot$ denote pointwise pairing, $(\varphi_{pt}(\{\Psi\})\times\{\Psi\},\cdot,\varphi_{pt})$ for a certain map φ_{pt} becomes a pre-Hilbert space and for an appropriate γ-matrix and solution Ψ, $\varphi_{pt}(\Psi)\gamma\Psi$ exhibits a section of the SO(n) principal bundle ${\rm SO}_{\mathbb{E}^n}(T\mathbb{E}^n)$ over $\mathbb{E}^n$.

Thus we apply the submanifold quantum mechanical scheme to the Dirac operator over a k-spin submanifold S immersed in $\mathbb{E}^n$. We obtain a representation of $SO_{\mathbb{E}^n}(T\mathbb{E}^n)|_S$, which is the generalized Weierstrass representation. Our main theorem is Theorem 3.15.

The organization of this article is as follows. Section 2 gives preliminaries on the geometrical setting [E] and conventions regarding Clifford modules [BGV, Tas]. In §3, we give a construction algorithm for the submanifold Dirac equation and investigate its properties. Section 4 presents an example.

I thank the organizers of the conference, especially Prof. M. A. Guest and Prof. Y. Ohnita for giving me a chance to talk there. It is acknowledged that Prof. M. A. Guest gave me helpful suggestions to write this article. I am grateful to Prof. B. G. Konopelchenko, Prof. U. Pinkall and Prof. F. Pedit for their interest in this work, Prof. T. Kori and Dr. Y. Homma for inviting me to their seminar and for crucial discussions. I thank Prof. S. Saito, Prof. K. Tamano and H. Mitsuhashi for critical discussions and encouragement.

§2. Preliminaries

2.1 Conventions [Mal]. For a fiber bundle A over a differential manifold M and an open set $U \subset M$, let $\Gamma(U, A)$ denote the set of smooth sections of the fiber bundle A over U. For a point p in M, let $\Gamma(p, A)$ denote the stalk at p.

We use the Einstein summation convention and let $\mathbb{C}$ ($\mathbb{R}$) denote the complex (real) field. For brevity, we use the notation $\partial_{u^\mu} = \partial/\partial u^\mu$. For a real number $x \in [n, n+1)$ and an integer n, [x] denotes n. Let $\mathbb{C}_M$ ($\mathbb{R}_M$) denote a complex (real) line bundle over a manifold M.

In order to define the submanifold Dirac operator, let us consider a smooth spin k-submanifold S immersed in Euclidean space $\mathbb{E}^n$. As mentioned in §3, the Dirac operator can be constructed locally. We shall assume that S satisfies several properties:

2.2 Assumptions/Notation.

(1) Let S be diffeomorphic to $\mathbb{R}^k$ with coordinates $(s^1, \dots, s^k)$.

(2) Let T_S be a tubular neighborhood of S with projection map $\pi_{T_S} : T_S \to S$.

(3) Let T_S have the natural bundle structure $T_S \approx \mathbb{R}^k \times \mathbb{R}^{n-k}$ with coordinates $(s^1, \dots, s^k, q^{k+1}, \dots, q^n) = (u^1, \dots, u^n)$. (Let $\alpha, \beta, \dots$ run from 1 to k and $\dot{\alpha}, \dot{\beta}, \dots$ run from $k+1$ to n. Let $\mu, \nu, \dots$ run from 1

to n.)

(4) For a fixed $q \in \mathbb{R}^{n-k}$, let $S_q = u^{-1}(\mathbb{R}^k \times \{q\})$.

(5) Let $\mathfrak{g}_{T_S}$, $\mathfrak{g}_{S_q}$ and $\mathfrak{g}_S$ denote the Riemannian metrics of T_S, S_q and S induced from that of $\mathbb{E}^n$, respectively.

(6) Let $g_{T_S\ i,j} = \mathfrak{g}_{T_S}(\partial_{x^i}, \partial_{x^j})$, $g_{T_S\ \mu,\nu} = \mathfrak{g}_{T_S}(\partial_{u^\mu}, \partial_{u^\nu})$, $g_{T_S} = \det g_{T_S\ \mu,\nu}$ and similarly for T_S, S and $\mathbb{E}^n$.

(7) Let $(dq^{\dot\alpha})_{\dot\alpha=k+1,\ldots,n}$ be an orthonormal basis, $\mathfrak{g}_{T_S}(\partial_{q^{\dot\alpha}}, \partial_{q^{\dot\beta}}) = \delta_{\dot\alpha,\dot\beta}$, $\mathfrak{g}_{T_S}(\partial_{q^{\dot\alpha}}, \partial_{s^\alpha}) = 0$ for $\alpha = 1, \ldots, k$ and $\dot\alpha = k+1, \ldots, n$. In other words,

$$\mathfrak{g}_{T_S} = \mathfrak{g}_{S_q} \oplus \delta_{\dot\alpha\dot\beta} dq^{\dot\alpha} \otimes dq^{\dot\beta}. \tag{2-1}$$

(8) At a point p in T_S, let an orthonormal frame of the cotangent space T^*T_S be denoted by $d\xi = (d\xi^\mu) = (d\zeta^\alpha, dq^{\dot\alpha})$.

We call a parametrization q satisfying (1)-(8) a *canonical parametrization.*

2.3 Notation. For a point p of S, let the *Weingarten map* be denoted by $-\gamma_{\dot\beta} : T_pS \to T_p\mathbb{E}^n$; for bases $\mathbf{e}_\alpha$ of TS and $\tilde{\mathbf{e}}_{\dot\beta} \in TS^\perp$ ($T_p\mathbb{E}^n = T_pS \oplus T_pS^\perp$),

$$\gamma_{\dot\beta}(\mathbf{e}_\alpha) = \partial_\alpha \tilde{\mathbf{e}}_{\dot\beta} = \gamma^{\dot\alpha}_{\ \dot\beta\alpha} \tilde{\mathbf{e}}_{\dot\alpha} + \gamma^{\beta}_{\ \dot\beta\alpha} \mathbf{e}_\beta. \tag{2-2}$$

In general it is not obvious whether a submanifold in $\mathbb{E}^n$ has a canonical parametrization q. In particular property (7) requires some arguments on the Weingarten map but we can use the following Proposition [Ma10].

2.4 Proposition. For a basis $\mathbf{e}_\alpha$ of TS, there is an orthonormal frame $\mathbf{e}_{\dot\alpha} = \delta_{\dot\alpha\dot\beta} dq^{\dot\beta} \in TS^\perp$ satisfying

$$\partial_\alpha \mathbf{e}_{\dot\beta} = \gamma^{\beta}_{\ \dot\beta\alpha} \mathbf{e}_\beta. \tag{2-3}$$

In terms of the above properties, we have the moving frame and the metric as follows.

2.5 Lemma. *For the moving frame $\mathbf{e}_{\dot\alpha} = \delta_{\dot\alpha\dot\beta} dq^{\dot\beta}$ in Proposition 2.4, the moving frame $\mathbf{E}_\mu = E^i{}_\mu \partial_i$, $(E^i{}_\mu = \partial_\mu x^i)$ in S_q is expressed by*

$$E^i{}_\alpha = e^i{}_\alpha + q^{\dot\alpha}\gamma^\beta{}_{\dot\alpha\alpha} e^i{}_\beta, \quad E^i{}_{\dot\alpha} = e^i{}_{\dot\alpha}. \tag{2-4}$$

2.6 Corollary. *Let $g_{T_S} = \det_{n\times n}(g_{T_S,\mu.\nu})$, $g_{S_q} = \det_{k\times k}(g_{S_q,\alpha.\beta})$ and $\mathfrak{g}_S = \mathfrak{g}_{S_q}|_{q=0}$.*

(1) $g_{T_S} = g_{S_q}$.

(2) $\mathfrak{g}_{S_q} = \mathfrak{g}_S + \mathfrak{g}_S^{(1)} q^{\dot\alpha} + \mathfrak{g}_S^{(2)} (q^{\dot\alpha})^2$, and locally

(2-5)
$$g_{TS_q}(\partial_\alpha, \partial_\beta) = g_{S\alpha\beta} + [\gamma^\gamma{}_{\dot\alpha\alpha} g_{S\gamma\beta} + g_{S\alpha\gamma}\gamma^\gamma{}_{\dot\alpha\beta}] q^{\dot\alpha} + [\gamma^\delta{}_{\dot\alpha\alpha} g_{S\delta\gamma}\gamma^\gamma{}_{\dot\beta\beta}] q^{\dot\alpha} q^{\dot\beta}.$$

(3) When we factorize g_{S_q} as $g_{S_q} = g_S \cdot \rho_{S_q}$, the factor ρ_{S_q} is given by

(2-6)
$$\begin{aligned}\rho_{S_q} =& 1 + 2\mathrm{tr}_{k\times k}(\gamma^\alpha{}_{\dot\alpha\beta}) q^{\dot\alpha} \\ &+ \left[2\mathrm{tr}_{k\times k}(\gamma^\alpha{}_{\dot\alpha\beta})\mathrm{tr}_{k\times k}(\gamma^\alpha{}_{\dot\beta\beta}) - \mathrm{tr}_{k\times k}(\gamma^\delta{}_{\dot\alpha\beta}\gamma^\alpha{}_{\dot\beta\delta})\right] q^{\dot\alpha} q^{\dot\beta} + \mathcal{O}(q^{\dot\alpha} q^{\dot\beta} q^{\dot\gamma})\end{aligned}$$

where $\mathcal{O}$ is the Landau symbol.

2.7 Clifford algebras and spinor representations [BGV, Tas].

In this section we recall some properties of the Clifford algebra and related notation.

(1) Let $\mathrm{CLIFF}(\mathbb{R}^n)$ denote the Clifford algebra for the vector space $\mathbb{R}^n$ and let $\mathrm{CLIFF}^{\mathbb{C}}(\mathbb{R}^n) = \mathrm{CLIFF}(\mathbb{R}^n)\otimes\mathbb{C}$. Let $\mathrm{CLIFF}^{\mathbb{C}^{\mathrm{even}}}(\mathbb{R}^n)$ denote the subalgebra of $\mathrm{CLIFF}^{\mathbb{C}}(\mathbb{R}^n)$ generated by elements of even degree in $\mathrm{CLIFF}^{\mathbb{C}}(\mathbb{R}^n)$. Further let $\mathrm{SPIN}(\mathbb{R}^n)$ denote the spin group for $\mathbb{R}^n$.

(2) For the exterior algebra $\wedge\mathbb{R}^n = \oplus_{j=0}^n \wedge^j \mathbb{R}^n$, there is a vector space isomorphism (symbol map) $\mathrm{CLIFF}(\mathbb{R}^n) \to \wedge\mathbb{R}^n$. Let its inverse be denoted by γ (γ-matrix).

(3) For n even, let $\mathrm{Cliff}(\mathbb{R}^n)$ denote a $\mathrm{CLIFF}^{\mathbb{C}}(\mathbb{R}^n)$-module whose endomorphism ring is isomorphic to $\mathrm{CLIFF}^{\mathbb{C}}(\mathbb{R}^n)$, which is a $2^{[n/2]}$ dimensional $\mathbb{C}$-vector space. For n odd, let $\mathrm{Cliff}(\mathbb{R}^n)$ denote a $\mathrm{CLIFF}^{\mathbb{C}}(\mathbb{R}^n)$-module whose endomorphism ring is isomorphic to $\mathrm{CLIFF}^{\mathbb{C}}(\mathbb{R}^n)$ as a

$2^{[n/2]}$ dimensional $\mathbb{C}$-vector space and whose elements are invariant for the action of $\gamma(e_1)\gamma(e_2)\dots\gamma(e_n)$. Here $e_1, e_2, \dots, e_n$ are an orthonormal basis of $\mathbb{R}^n$.

2.8 Conventions. (1) We recall the fact $\mathrm{CLIFF}^{\mathbb{C}}(\mathbb{R}^{n+2}) \approx \mathrm{END}(\mathbb{C}^2)\otimes \mathrm{CLIFF}^{\mathbb{C}}(\mathbb{R}^n)$, where $\mathrm{END}(\mathbb{C}^2)$ is the ring of endomorphisms of $\mathbb{C}^2$; $\mathrm{END}(\mathbb{C}^2)$ can be generated by the Pauli matrices

(2-8)
$$\sigma_0 = \begin{pmatrix} 1 & 0 \\ 0 & 1 \end{pmatrix}, \sigma_1 = \begin{pmatrix} 0 & 1 \\ 1 & 0 \end{pmatrix}, \sigma_2 = \begin{pmatrix} 0 & -i \\ i & 0 \end{pmatrix}, \sigma_3 = \begin{pmatrix} 1 & 0 \\ 0 & -1 \end{pmatrix}.$$

(2) For even n, since $\mathrm{CLIFF}^{\mathbb{C}}(\mathbb{R}^n) \approx \mathrm{END}(\mathbb{C}^{[n/2]})$, we use the conventions

(2-9)
$$\begin{aligned}
\gamma(e_1) &= \sigma_1 \otimes \sigma_1 \otimes \cdots \otimes \sigma_1 \otimes \sigma_1 \otimes \sigma_1 \otimes \sigma_1, \\
\gamma(e_2) &= \sigma_1 \otimes \sigma_1 \otimes \cdots \otimes \sigma_1 \otimes \sigma_1 \otimes \sigma_1 \otimes \sigma_2, \\
\gamma(e_3) &= \sigma_1 \otimes \sigma_1 \otimes \cdots \otimes \sigma_1 \otimes \sigma_1 \otimes \sigma_1 \otimes \sigma_3, \\
\gamma(e_4) &= \sigma_1 \otimes \sigma_1 \otimes \cdots \otimes \sigma_1 \otimes \sigma_1 \otimes \sigma_2 \otimes \sigma_0 \\
\gamma(e_5) &= \sigma_1 \otimes \sigma_1 \otimes \cdots \otimes \sigma_1 \otimes \sigma_1 \otimes \sigma_3 \otimes \sigma_0 \\
&\cdots \\
\gamma(e_{n-1}) &= \sigma_1 \otimes \sigma_3 \otimes \cdots \otimes \sigma_0 \otimes \sigma_0 \otimes \sigma_0 \otimes \sigma_0, \\
\gamma(e_n) &= \sigma_2 \otimes \sigma_0 \otimes \cdots \otimes \sigma_0 \otimes \sigma_0 \otimes \sigma_0 \otimes \sigma_0.
\end{aligned}$$

(3) For odd n, since $\mathrm{CLIFF}^{\mathbb{C}}(\mathbb{R}^n) \approx \mathrm{END}(\mathbb{C}^{(n-1)/2}) \oplus \mathrm{END}(\mathbb{C}^{(n-1)/2})$, we use the conventions

(2-10)
$$\begin{aligned}
\gamma(e_1) &= \sigma_1 \otimes \sigma_1 \otimes \cdots \otimes \sigma_1 \otimes \sigma_1 \otimes \sigma_1 \otimes \mathfrak{e}, \\
\gamma(e_2) &= \sigma_1 \otimes \sigma_1 \otimes \cdots \otimes \sigma_1 \otimes \sigma_1 \otimes \sigma_2 \otimes 1, \\
\gamma(e_3) &= \sigma_1 \otimes \sigma_1 \otimes \cdots \otimes \sigma_1 \otimes \sigma_1 \otimes \sigma_3 \otimes 1, \\
&\cdots \\
\gamma(e_{n-2}) &= \sigma_1 \otimes \sigma_3 \otimes \cdots \otimes \sigma_0 \otimes \sigma_0 \otimes \sigma_0 \otimes 1, \\
\gamma(e_{n-1}) &= \sigma_2 \otimes \sigma_0 \otimes \cdots \otimes \sigma_0 \otimes \sigma_0 \otimes \sigma_0 \otimes 1. \\
\gamma(e_n) &= \sigma_3 \otimes \sigma_0 \otimes \cdots \otimes \sigma_0 \otimes \sigma_0 \otimes \sigma_0 \otimes 1,
\end{aligned}$$

where $\mathfrak{e}$ is the generator of $\mathrm{CLIFF}^{\mathbb{C}}(\mathbb{R}) = \mathbb{R}[\mathfrak{e}]/(1-\mathfrak{e}^2)$.

(4) Let $b_+ = \begin{pmatrix} 1 \\ 0 \end{pmatrix}$ and $b_- = \begin{pmatrix} 0 \\ 1 \end{pmatrix}$. Then $\mathrm{Cliff}(\mathbb{R}^n)$ is spanned by

(2-11)
$$\Xi_\epsilon = b_{\epsilon_1} \otimes b_{\epsilon_2} \otimes \cdots \otimes b_{\epsilon_{[n/2]-1}} \otimes b_{\epsilon_{[n/2]}},$$

where $\epsilon = (\epsilon_1, \epsilon_2, \dots, \epsilon_{[n/2]})$ and $\epsilon_a = \pm$ $(a = 1, \dots, [n/2])$. Similarly, $\mathrm{Cliff}(\mathbb{R}^n)^*$ is spanned by

$$\overline{\Xi}_\epsilon = \overline{b}_{\epsilon_1} \otimes \overline{b}_{\epsilon_2} \otimes \cdots \otimes \overline{b}_{\epsilon_{[n/2]-1}} \otimes \overline{b}_{\epsilon_{[n/2]}}, \tag{2-12}$$

where $\overline{b}_+ = (1,0)$ and $\overline{b}_- = (0,1)$. Renumbering, let $\epsilon^{[c]}$ $(c = 1, \dots, 2^{[n/2]})$ denotes the ϵ's. The isomorphism $\varphi : \mathrm{Cliff}(\mathbb{R}^n) \to \mathrm{Cliff}(\mathbb{R}^n)^*$ is given by

$$\varphi\Big(\sum_{c=1}^{2^{[n/2]}} a_c \Xi_{\epsilon^{[c]}}\Big) = \sum_{c=1}^{2^{[n/2]}} \overline{a}_c \overline{\Xi}_{\epsilon^{[c]}}, \tag{2-13}$$

for $a_c \in \mathbb{C}$ and its complex conjugate $\overline{a}_c$.

(5) Defining

$$b_1 = \frac{1}{\sqrt{2}}\begin{pmatrix}1\\1\end{pmatrix}, b_2 = \frac{1}{\sqrt{2}}\begin{pmatrix}1\\i\end{pmatrix}, b_3 = \begin{pmatrix}1\\0\end{pmatrix}$$

and $\overline{b}_1 = \frac{1}{\sqrt{2}}(1,1)$, $\overline{b}_2 = \frac{1}{\sqrt{2}}(1,-i)$, $\overline{b}_3 = (1,0)$, we have

$$\overline{b}_a \sigma_b b_a = \delta_{a,b}, \quad (a, b = 1,2,3), \text{ (not summed over } a\text{)}. \tag{2-14}$$

(6) For $j = 1, \dots, [n/2]$, define

$$\Psi^{(2j-1)} = b_1 \otimes b_1 \otimes \cdots \otimes b_1 \otimes b_3 \otimes b_1 \otimes \cdots \otimes b_1,$$

$$\Psi^{(2j)} = b_1 \otimes b_1 \otimes \cdots \otimes b_1 \otimes b_2 \otimes b_1 \otimes \cdots \otimes b_1. \tag{2-15}$$

For odd n, we define,

$$\Psi^{(n)} = b_3 \otimes b_1 \otimes \cdots \otimes b_1 \otimes b_1 \otimes \cdots \otimes b_1, \tag{2-16}$$

and assume that $\mathfrak{e}\Psi^{(i)} = \Psi^{(i)}$. We define $\overline{\Psi}^{(k)}$ as in (2-13). Then $\Psi^{(a)}$ is an element of $\mathrm{Cliff}(\mathbb{R}^n)$ $(a = 1, \dots, n)$ satisfying

$$\overline{\Psi}^{(a)}_{\{dx\}} \gamma(e^b) \Psi^{(a)}_{\{dx\}} = \delta^a_b, \quad \text{(not summed over } a\text{)}. \tag{2-17}$$

As we wish to consider a spin principal subbundle over S induced from a spin principal bundle over $\mathbb{E}^n$, we recall the following facts:

2.9 Lemma. *(1) For $k < n$, $\mathrm{CLIFF}^{\mathbb{C}}(\mathbb{R}^k)$ is a natural vector subspace of $\mathrm{CLIFF}^{\mathbb{C}}(\mathbb{R}^n)$ (from the generators in 2.8), and $\mathrm{CLIFF}^{\mathbb{C}\mathrm{even}}(\mathbb{R}^k)$ is a subring of $\mathrm{CLIFF}^{\mathbb{C}\mathrm{even}}(\mathbb{R}^n)$. (2) For $k < n$, $\mathrm{SPIN}(\mathbb{R}^k)$ is a natural subgroup of $\mathrm{SPIN}(\mathbb{R}^n)$.*

Proof. It will suffice to comment on the cases $k = 2l$ and $n = 2l+2$. As in 2.8, we have a natural inclusion defined on the basis elements e_i of $\mathbb{R}^k$ and E_i of $\mathbb{R}^n$ by $\tau_{k,n} : \gamma(e_i) \mapsto \gamma(E_i) := \sigma_1 \otimes \gamma(e_i)$, $i = 1, 2, \cdots, k$; then we define $\tau_{k,n}(\gamma(e_i)\gamma(e_j)) := \tau_{k,n}(\gamma(e_i))\tau_{k,n}(\gamma(e_j))$ and so on. On the other hand, for $c^i \in \mathrm{CLIFF}(\mathbb{R}^k)$, we have a ring homomorphism $\iota_{k,n} : \mathrm{CLIFF}(\mathbb{R}^k) \hookrightarrow \mathrm{CLIFF}(\mathbb{R}^n)$ by $C_i = \sigma_0 \otimes c_i$. For $1 \le i, j \le k$, $\gamma(E_i)\gamma(E_j) = \iota_{k,n}(\gamma(e_i)\gamma(e_j))$. In $\mathrm{CLIFF}^{\mathrm{even}}(\mathbb{R}^n)$, the images of τ and ι agree. Thus (1) is proved. Accordingly $\exp(\tau_{k,n}(\gamma(e_j)\gamma(e_i)))$ can be regarded as elements of $\mathrm{SPIN}(\mathbb{R}^n)$ and (2) is proved. ■

§3. Construction of the submanifold Dirac operator

An algorithm to construct the submanifold Dirac operator is presented in the following six steps.

Step 1: Consider the Dirac equation $\not{D}_{\mathbb{E}^n}\Psi_{\mathbb{E}^n} = 0$ in a Euclidean space $\mathbb{E}^n$ and embed the k-spin submanifold S into $\mathbb{E}^n$ $(0 < k < n)$

Here we explain our notation for the Dirac equation $\not{D}_{\mathbb{E}^n}\Psi_{\mathbb{E}^n} = 0$.

3.1 Clifford modules. (1) Let $\mathrm{CLIFF}^{\mathbb{C}}{}_{\mathbb{E}^n}(T^*\mathbb{E}^n)$ denote the Clifford bundle over $\mathbb{E}^n$ which has the $\mathrm{CLIFF}^{\mathbb{C}}(\mathbb{R}^n)$ structure associated with the cotangent bundle $T^*\mathbb{E}^n$ and the set of differential forms $\Omega(\mathbb{E}^n) = \sum_{a=1}^n \Omega^a(\mathbb{E}^n)$.

(2) Let $\mathrm{Cliff}_{\mathbb{E}^n}(T^*\mathbb{E}^n)$ denote the Clifford module over $\mathbb{E}^n$ associated with the cotangent space $T^*\mathbb{E}^n$, which is modelled on $\mathrm{Cliff}(\mathbb{R}^n)$.

(3) Let φ_{pt} denote the natural bijection of $\mathrm{CLIFF}^{\mathbb{C}}{}_{\mathbb{E}^n}(T^*\mathbb{E}^n)$-modules

$$\varphi_{pt} : \Gamma(p, \mathrm{Cliff}_{\mathbb{E}^n}(T^*\mathbb{E}^n)) \to \Gamma(p, \mathrm{Cliff}_{\mathbb{E}^n}(T^*\mathbb{E}^n)^*), \tag{3-1}$$

for p in $\mathbb{E}^n$.

(4) For an orthonormal frame $\mathbf{e}^i \in T^*\mathbb{E}^n$ at $p \in \mathbb{E}^n$, let $\gamma_{\{\mathbf{e}\}}$ denote the γ-matrix as the map from $\Omega(\mathbb{E}^n)$ to $\mathrm{CLIFF}^{\mathbb{C}}{}_{\mathbb{E}^n}(T^*\mathbb{E}^n)$ as in (2-7),

$$\gamma_{\{\mathbf{e}\}} : \mathbf{e}^i \mapsto \gamma_{\{\mathbf{e}\}}(\mathbf{e}^i) \in \mathrm{CLIFF}^{\mathbb{C}}{}_{\mathbb{E}^n}(T^*\mathbb{E}^n)|_p. \tag{3-2}$$

For later convenience, we also employ the following notation for the one-form $du^\mu = E^\mu_i \mathbf{e}^i \in \Gamma(p, \Omega^1(\mathbb{E}^n))$:

$$\gamma_{\{\mathbf{e}\}}(du^\mu) = E^\mu_i \gamma_{\{\mathbf{e}\}}(\mathbf{e}^i). \tag{3-3}$$

For a Cartesian coordinate system $(x^i) = (x^1, \dots x^n)$ in $\mathbb{E}^n$ we fix $\mathbf{e}^i = dx^i$ and in T_S let $\mathbf{e}^\mu = d\xi^\mu$ in 2.2 (8).

(5) Let $\mathrm{SPIN}_{\mathbb{E}^n}(T^*\mathbb{E}^n)$ denote a spin principal bundle over $\mathbb{E}^n$. Let the natural bundle map from $\mathrm{SPIN}_{\mathbb{E}^n}(T^*\mathbb{E}^n)$ to the $\mathrm{SO}(n)$-principal bundle $\mathrm{SO}_{\mathbb{E}^n}(T^*\mathbb{E}^n)$ be denoted by $\tau_{\mathbb{E}^n}$; for $p \in \mathbb{E}^n$, the orthonormal frame $\mathbf{e} \in T^*_p\mathbb{E}^n$ and $\mathrm{e}^\Omega \in \mathrm{SPIN}_p(T^*\mathbb{E}^n)$, the action of $\mathrm{SO}(n)$ is defined by

$$\tau_{\mathbb{E}^n}(\mathrm{e}^\Omega)\mathrm{e}^i = \gamma^{-1}_{\{\mathbf{e}\}}(\mathrm{e}^\Omega \gamma_{\{\mathbf{e}\}}(\mathrm{e}^i)\mathrm{e}^{-\Omega}). \tag{3-4}$$

(6) The Dirac operator $\not{D}_{\mathbb{E}^n}$ is a map

$$\not{D}_{\mathbb{E}^n} : \Gamma(U, \mathrm{Cliff}_{\mathbb{E}^n}(T^*\mathbb{E}^n)) \to \Gamma(U, \mathrm{Cliff}_{\mathbb{E}^n}(T^*\mathbb{E}^n)). \tag{3-5}$$

3.2 Proposition. *For a point $p \in \mathbb{E}^n$ and a pair $(\overline{\Psi}_{\{dx\}}, \Psi_{\{dx\}})$ in $\Gamma(p, \mathrm{Cliff}_{\mathbb{E}^n}(T^*\mathbb{E}^n)^*) \times \Gamma(p, \mathrm{Cliff}_{\mathbb{E}^n}(T^*\mathbb{E}^n))$, the following holds:*

(1) $\delta_{ij}\overline{\Psi}_{\{dx\}}(x)\gamma_{\{dx\}}(dx^i)\Psi_{\{dx\}}(x)dx^j$ is an element of $\Gamma(p, \Omega^1(T^*\mathbb{E}^n))$.

(2) $\overline{\Psi}_{\{dx\}}(x)\Psi_{\{dx\}}(x)$ is an element of $\Gamma(p, \mathbb{C}_{\mathbb{E}^n})$.

Using the conventions (2-11), we have:

3.3 Proposition. *For a point p in $\mathbb{E}^n$, $\{\Psi^{[a]}_{\{dx\}} = \Xi_{\epsilon^{[a]}}\}_{a=1,\dots,2^{[n/2]}}$ satisfy the Dirac equation*

$$\not{D}_{x,\{dx\}}\Psi_{\mathbb{E}^n}(x) = 0,$$

and are bases of $\Gamma(p, \mathrm{Cliff}_{\mathbb{E}^n}(T^*\mathbb{E}^n))$ as a $\Gamma(p, \mathbb{C}_{\mathbb{E}^n})$-vector space. Thus they are also bases of the fiber $\mathrm{Cliff}_p(T^*\mathbb{E}^n)$ of $\mathrm{Cliff}_{\mathbb{E}^n}(T^*\mathbb{E}^n)$ at p.

Using (2-17), we have the following important proposition.

3.4 Proposition. *There exist $\Psi^{(i)}_{\{dx\}}$ $(i = 1, \dots, n)$ and $\overline{\Psi}^{(i)}_{\{dx\}} = \varphi_{pt}(\Psi^{(i)}_{\{dx\}})$ satisfying $\sum_{j,k=1}^n \overline{\Psi}^{(i)}_{\{dx\}}\gamma_{\{dx\}}(dx^j)\Psi^{(i)}_{\{dx\}}dx^k = dx^i$.*

As mentioned in Proposition 3.2 (2), $\mathcal{W}_p = \Gamma(p, \mathrm{Cliff}_{\mathbb{E}^n}(T^*\mathbb{E}^n)^*) \times \Gamma(p, \mathrm{Cliff}_{\mathbb{E}^n}(T^*\mathbb{E}^n))$ has a pointwise-pairing $\cdot$ for any point $p \in \mathbb{E}^n$, so

$\mathcal{H}_p = (\mathcal{W}_p, \cdot, \varphi_{pt})$ becomes a pre-Hilbert space, and by Proposition 3.2 (1), $\mathcal{W}_p$ gives the stalk $\Gamma(p, \mathrm{SO}_{\mathbb{E}^n}(T^*\mathbb{E}^n))$.

Step 2: Express the Dirac operator using the canonical parametrization in the tubular neighborhood T_S of S.

Let $\Gamma_{vc}(T_S, \mathrm{Cliff}_{\mathbb{E}^n}(T^*\mathbb{E}^n))$ denote the set of sections of the Clifford module $\mathrm{Cliff}_{\mathbb{E}^n}(T^*\mathbb{E}^n)$ whose support is in T_S. We go on to express the stalk at $p \in T_S$ by $\Gamma_{vc}(p, \mathrm{Cliff}_{\mathbb{E}^n}(T^*\mathbb{E}^n))$. Let $\Psi_{\{d\xi\}}$ denote a germ of $\Gamma(p, \mathrm{Cliff}_{\mathbb{E}^n}(T^*\mathbb{E}^n))$ for a point $p \in T_S$. Recalling (3-3), $dx^i = E^i{}_\mu du^\mu$ and decomposing $E^i{}_\mu = G_\mu{}^\nu \Lambda^i{}_\nu$ as $dx^i = \Lambda^i{}_\mu d\xi^\mu$, we can find $\mathrm{e}^\Omega \in \Gamma(p, \mathrm{SPIN}_{\mathbb{E}^n}(T^*\mathbb{E}^n))$ satisfying the following relations, called a *gauge transformation*:

$$\Psi_{\{d\xi\}}(u) = \mathrm{e}^{-\Omega}\Psi_{\{dx\}}(x), \quad \overline{\Psi}_{\{d\xi\}}(u) = \overline{\Psi}_{\{dx\}}(x)\mathrm{e}^{\Omega},$$

$$\begin{aligned} \mathrm{e}^{-\Omega}\gamma_{\{dx\}}(dx^i)\mathrm{e}^{\Omega} &= \gamma_{\{d\xi\}}(\Lambda^i{}_\mu d\xi^\mu) \\ &= \Lambda^i{}_\rho G_\mu{}^\rho G_\nu{}^\mu \gamma_{\{d\xi\}}(d\xi^\nu) \\ &= E^i{}_\mu \gamma_{\{d\xi\}}(du^\mu). \end{aligned} \tag{3-6}$$

As $\overline{\Psi}_{\{d\xi\}} = \varphi_{pt}(\Psi_{\{d\xi\}})$, φ_{pt} does not depend upon the orthonormal frame. In terms of these expressions, we have (see Proposition 3.2)

$$\begin{aligned} &\delta_{ij}\overline{\Psi}_{\{dx\}}(x)\gamma_{\{dx\}}(dx^i)\Psi_{\{dx\}}(x)dx^j \\ &\qquad = g_{T_S,\mu\nu}\overline{\Psi}_{\{d\xi\}}(u)\gamma_{\{d\xi\}}(du^\mu)\Psi_{\{d\xi\}}(u)du^\nu, \end{aligned}$$

$$\overline{\Psi}_{\{dx\}}(x)\Psi_{\{dx\}}(x) = \overline{\Psi}_{\{d\xi\}}(u)\Psi_{\{d\xi\}}(u). \tag{3-7}$$

Further, the representation of the Dirac operator $\not{D}_{x,\{dx\}}$ is transformed to $\not{D}_{u,\{d\xi\}}$ by means of the gauge transformation:

$$\begin{aligned} \not{D}_{u,\{d\xi\}} &= \mathrm{e}^{-\Omega}\not{D}_{x,\{dx\}}\mathrm{e}^{\Omega} = \gamma_{\{d\xi\}}(du^\mu)\mathrm{e}^{-\Omega}\partial_\mu \mathrm{e}^{\Omega} \\ &= \gamma_{\{d\xi\}}(du^\mu)(\partial_\mu + \partial_\mu\Omega). \end{aligned} \tag{3-8}$$

Then we have the following lemma:

3.5 Lemma. *We can regard $\not{D}_{u,\{d\xi\}}$ as a representation of a map $\not{D}_{T_S}$,*

$$\not{D}_{T_S} : \Gamma_{vc}(p, \mathrm{Cliff}_{\mathbb{E}^n}(T^*\mathbb{E}^n)) \to \Gamma_{vc}(p, \mathrm{Cliff}_{\mathbb{E}^n}(T^*\mathbb{E}^n)),$$

for a point p in T_S.

By Lemma 3.5 and the fact that the zero-section is in $\mathrm{Cliff}_{\mathbb{E}^n}(T^*\mathbb{E}^n)$, it is not difficult to prove that the solution space of $\not{D}_{u,\{d\xi\}}\Psi = 0$ in $\Gamma(p, \mathbb{C}^{2^{[n/2]}}_{\mathbb{E}^n})$ belongs to $\Gamma(p, \mathrm{Cliff}_{\mathbb{E}^n}(T^*\mathbb{E}^n))$. Since $\not{D}_{u,\{d\xi\}}$ is a $2^{[n/2]} \times 2^{[n/2]}$-matrix type first order differential operator of rank $2^{[n/2]}$, the solution space gives an orthonormal frame in $\Gamma(p, \mathrm{Cliff}_{\mathbb{E}^n}(T^*\mathbb{E}^n))$. In fact, by Proposition 3.2, we can take $\mathrm{e}^{-\Omega}\Psi^{[a]}_{\{dx\}}$.

From Propositions 3.2-3.5 and (3-6)-(3-8), we have a key proposition:

3.6 Proposition. *(1) For a point $p \in T_S$, there exists some $(\Psi^{[a]}_{\{d\xi\}})_{a=1,\dots,2^{[n/2]}}$ in $\Gamma(p, \mathbb{C}^{2^{[n/2]}}_{\mathbb{E}^n})$ forming an orthonormal frame in $\Gamma(p, \mathrm{Cliff}_{\mathbb{E}^n}(T^*\mathbb{E}^n))$ and satisfying the Dirac equation*

$$\not{D}_{u,\{d\xi\}}\Psi_{\{d\xi\}}(u) = 0, \tag{3-9}$$

i.e.,

$$\overline{\Psi}^{[a]}_{\{d\xi\}}\Psi^{[b]}_{\{d\xi\}} = \delta_{a,b}, \tag{3-10}$$

for $\overline{\Psi}^{[a]}_{\{d\xi\}} = \varphi_{pt}(\Psi^{[a]}_{\{d\xi\}})$.

(2) For $\Psi^{[a]}_{\{dx\}} \in \Gamma(p, \mathrm{Cliff}_{\mathbb{E}^n}(T^*\mathbb{E}^n))$ as in (1) and $\Psi^{[a]}_{\{dx\}} \in \Gamma(p, \mathrm{Cliff}_{\mathbb{E}^n}(T^*\mathbb{E}^n))$ as in Proposition 3.3, there exists a $(2^{[n/2]} \times 2^{[n/2]})$-matrix $\mathrm{e}^{\Omega} \in \Gamma(p, \mathrm{SPIN}_{\mathbb{E}^n}(T^*\mathbb{E}^n))$ satisfying

$$\Psi^{[a]}_{\{dx\}} = \mathrm{e}^{\Omega}\Psi^{[a]}_{\{d\xi\}}, \quad (a = 1, \dots, 2^{[n/2]}).$$

(3) For $\mathrm{e}^{\Omega} \in \Gamma(p, \mathrm{SPIN}_{\mathbb{E}^n}(T^*\mathbb{E}^n))$ in (2), by letting $\Psi^{(i)}_{\{d\xi\}} = \mathrm{e}^{-\Omega}\Psi^{(i)}_{\{dx\}}$ for $(i = 1, \dots, n)$ in Proposition 3.4, we have the relation

$$g_{T_S,\mu\nu}\overline{\Psi}^{(i)}_{\{d\xi\}}(u)\gamma_{\{d\xi\}}(du^{\mu})\Psi^{(i)}_{\{d\xi\}}(u)du^{\nu} = dx^i \quad (i = 1, \dots, n). \tag{3-11}$$

Proof. It suffices to establish the well-definedness of (3-11). We should check the dependence on the orthonormal frame $\Psi^{[a]}_{\{d\xi\}}$. Let us take another one $\Psi'^{[a]}_{\{d\xi\}}$. Defining $\mathrm{e}^{\Omega'}$ by $\Psi^{[a]}_{\{dx\}} = \mathrm{e}^{\Omega'}\Psi'^{[a]}_{\{d\xi\}}$, we have $\Psi^{[a]}_{\{d\xi\}} = \mathrm{e}^{-\Omega}\mathrm{e}^{\Omega'}\Psi'^{[a]}_{\{d\xi\}}$. When one rewrites (3-11) in terms of the frame

$\Psi'^{[a]}_{\{d\xi\}}$, there appears $e^{-\Omega'}e^{\Omega}\gamma_{\{d\xi\}}(du^\mu)e^{-\Omega}e^{\Omega'}$. However as both are solutions of the Dirac equation (3-9), the gauge transformation for $e^{-\Omega}e^{\Omega'}$ must leave the Dirac operator invariant. Hence the du^μ component gives

$$e^{-\Omega'}e^{\Omega}\gamma_{\{d\xi\}}(du^\mu)e^{-\Omega}e^{\Omega'} = \gamma_{\{d\xi\}}(du^\mu).$$

Thus (3-11) does not depend on the choice of orthonormal frame. ■

In order to investigate the domain of the new Dirac operator, we apply Lemma 2.9 to the fiber bundles. Let $\mathrm{CLIFF}^{\mathbb{C}}{}_S(T^*S)$ (or $\mathrm{Cliff}_S(T^*S)$) be restricted to S as $\mathrm{CLIFF}^{\mathbb{C}}{}_{\mathbb{E}^n}(T^*\mathbb{E}^n)|_S$ (or $\mathrm{Cliff}_{\mathbb{E}^n}(T^*\mathbb{E}^n)|_S$) as a vector bundle over S. As the fiber of $\mathrm{SPIN}_p(T^*S)$ is a subgroup of $\mathrm{SPIN}_p(T^*\mathbb{E}^n)$, we also have $\mathrm{SPIN}_S(T^*\mathbb{E}^n) = \mathrm{SPIN}_{\mathbb{E}^n}(T^*\mathbb{E}^n)|_S$. The reader should note the difference between $\Gamma(p, \mathrm{Cliff}_S(T^*\mathbb{E}^n))$ and $\Gamma(p, \mathrm{Cliff}_{\mathbb{E}^n}(T^*\mathbb{E}^n))$; the former is $\{\psi(s)\}$ and the latter is $\{\psi(u)\}$.

3.7 Definition. *(1) The inclusion modelled by Lemma 2.9 is denoted by* $\tau_{S,\mathbb{E}^n} : \mathrm{CLIFF}^{\mathbb{C}}{}_S(T^*S) \to \mathrm{CLIFF}^{\mathbb{C}}{}_S(T^*\mathbb{E}^n)$, $\gamma_{S,\{d\zeta\}}(d\zeta) \mapsto \gamma_{\{d\xi\}}(d\zeta))$. *This induces* $\mathrm{CLIFF}^{\mathbb{C}\mathrm{even}}_S(T^*S) \to \mathrm{CLIFF}^{\mathbb{C}\mathrm{even}}_S(T^*\mathbb{E}^n)$ *(a ring homomorphism). Here* $\gamma_{S,\{d\zeta\}}$ *denotes the* γ*-matrix over* S *associated with* T^*S *and its orthonormal frame* $\{d\zeta\}$*. (2) The corresponding inclusion of groups is* $\iota_{S,\mathbb{E}^n} : \mathrm{SPIN}_S(T^*S) \to \mathrm{SPIN}_S(T^*\mathbb{E}^n)$.

Step 3: To construct a self-adjointization of the normal differential operator, we define a pairing in T_S

As q is a natural coordinate of T_S, let us introduce the Haar measure with respect to the local affine transformation along the normal direction,

$$e^{b^{\dot\alpha}\partial_{q^{\dot\alpha}}}\, g_S^{1/2} d^k s d^{n-k} q = g_S^{1/2} d^k s d^{n-k} q.$$

We call this measure the *normal affine invariant measure*. We have a generic normal differential operator $\partial_\perp = b^{\dot\alpha}\partial_{q^{\dot\alpha}}$ for generic real constants $b^{\dot\alpha}$. As mentioned in the Introduction, we apply the submanifold quantum mechanics scheme to this operator.

3.8 Definition. *For a point* $p \in S$ *and* $(\overline{\Psi}_{1,\{d\xi\}}, \Psi_{2,\{d\xi\}}) \in$

$$\Gamma_{vc}(\pi_{T_S}^{-1}(p), \mathrm{Cliff}_{\mathbb{E}^n}(T^*\mathbb{E}^n)^*) \times \Gamma_{vc}(\pi_{T_S}^{-1}(p), \mathrm{Cliff}_{\mathbb{E}^n}(T^*\mathbb{E}^n)),$$

we introduce the L^2-type pairing $<|>$ as the fiber integral $< \Psi_{1,\{dx\}} | \not{\mathbf{D}}_{u,\{d\xi\}} \Psi_{2,\{d\xi\}} >=$

$$\int_{\pi_{T_S}^{-1}(p)} g_{T_S}^{1/2} d^{n-k} q \overline{\Psi}_{1,\{d\xi\}}(u)\, \not{\mathbf{D}}_{u,\{d\xi\}} \Psi_{2,\{d\xi\}}(u) \in \Gamma(p, \mathbb{C}_S),$$

and a transformation η_{sa} whose measure is the normal affine invariant measure $< \Psi_{1,\{d\xi\}} | \not{\mathbf{D}}_{u,\{d\xi\}} \Psi_{2,\{d\xi\}} >= (\Phi_{1,\{d\xi\}} | \not{\mathbb{D}}_{u,\{d\xi\}} \Phi_{2,\{dx\}}) =$

$$\int_{\pi_{T_S}^{-1}(p)} g_S^{1/2} d^{n-k} q \overline{\Phi}_{1,\{dx\}}(u)\, \not{\mathbb{D}}_{u,\{d\xi\}}(u) \Phi_{2,\{dx\}}(u) \in \Gamma(p, \mathbb{C}_S),$$

where $\overline{\Phi}_{1,\{d\xi\}} = \eta_{\mathrm{sa}}(\overline{\Psi}_{1,\{d\xi\}}) = {\rho_{S_q}}^{1/4} \overline{\Psi}_{1,\{d\xi\}}$, $\Phi_{2,\{d\xi\}} = \eta_{\mathrm{sa}}(\Psi_{2,\{d\xi\}}) = {\rho_{S_q}}^{1/4} \Psi_{2,\{d\xi\}}$,

$$\not{\mathbb{D}}_{u,\{d\xi\}} = \eta_{\mathrm{sa}}(\not{\mathbf{D}}_{u,\{d\xi\}}) = {\rho_{S_q}}^{1/4}\, \not{\mathbf{D}}_{u,\{d\xi\}} {\rho_{S_q}}^{-1/4}. \tag{3-12}$$

Further let $\tilde{\varphi}_{pt}$ denote $\eta_{\mathrm{sa}} \varphi_{pt} \eta_{\mathrm{sa}}^{-1}$.

We note that η_{sa} can be defined for more general differential operators but for simplicity, we only define it for the Dirac operator. Further as the metric $\mathfrak{g}_{T_S}$ is not singular, η_{sa} gives a diffeomorphism.

3.9 Lemma. *For points $p \in S_q$ and $p' \in S$, the triples*

$$\mathcal{H}'_p = (\eta_{\mathrm{sa}}(\Gamma_{vc}(p, \mathrm{Cliff}_{\mathbb{E}^n}(T^*\mathbb{E}^n)^*) \times \Gamma_{vc}(p, \mathrm{Cliff}_{\mathbb{E}^n}(T^*\mathbb{E}^n)), \cdot, \tilde{\varphi}_{pt}),$$

and

$$\mathcal{H}' = (\eta_{\mathrm{sa}}(\Gamma_{vc}(\pi_{T_S}^{-1} p', \mathrm{Cliff}_{\mathbb{E}^n}(T^*\mathbb{E}^n)^*) \times \Gamma_{vc}(\pi_{T_S}^{-1} p', \mathrm{Cliff}_{\mathbb{E}^n}(T^*\mathbb{E}^n)), (|), \tilde{\varphi}_{pt}),$$

become pre-Hilbert spaces. In $\mathcal{H}'$, The operator $\partial_\perp$ is skew-Hermitian, i.e., $\partial_\perp^* = -\partial_\perp$ for each $b_{\dot\alpha}$.

Step 4: Decompose $\not{\mathbb{D}}_{T_S}$ into its normal and tangential parts

Using the orthonormal frame $(d\zeta^1, \dots, d\zeta^k, dq^{k+1}, \dots, dq^n)$, let us decompose $\not{\mathbb{D}}_{T_S}$ into

$$\not{\mathbb{D}}_{u,\{d\xi\}} = \not{\mathbb{D}}^{\parallel}_{u,\{d\xi\}} + \not{\mathbb{D}}^{\perp}_{u,\{d\xi\}}, \tag{3-13}$$

where $\not{\mathbb{D}}^{\perp}_{u,\{d\xi\}} = \gamma_{\{d\xi\}}(dq^{\dot\alpha}) \partial_{q^{\dot\alpha}}$.

3.10 Lemma. *(1)* $\not{\mathbb{D}}^{\|}_{u,\{d\xi\}}$ *does not contain the vertical differential operator* $\partial_{q^{\dot\alpha}}$.

(2) ${\not{\mathbb{D}}^{\perp}_{u,\{d\xi\}}}^* = -\not{\mathbb{D}}^{\perp}_{u,\{d\xi\}}$ *in* $\mathcal{H}'$.

(3) For a point $p \in T_S$, $\not{\mathbb{D}}^{\perp}_{u,\{d\xi\}}$ *is a homomorphism of Clifford modules,*

$$\not{\mathbb{D}}^{\perp}_{u,\{d\xi\}} : \eta_{\rm sa}(\Gamma_{vc}(p, \mathrm{Cliff}(T^*T_S))) \to \eta_{\rm sa}(\Gamma_{vc}(p, \mathrm{Cliff}(T^*T_S))).$$

For $p \in T_S$, let us define a projection $\pi : \eta_{\rm sa}((\Gamma_{vc}(p, \mathrm{Cliff}(T^*T_S)^*) \times \Gamma_{vc}(p, \mathrm{Cliff}(T^*T_S)))) \to (\mathrm{Ker}_p(Ad(\partial_\perp)) \times \mathrm{Ker}_p(\partial_\perp))$, where $Ad(P)$ denotes the right-adjoint of P. We note that $\mathrm{Ker}_p(\partial_\perp)(\subset \Gamma_{vc}(p, \mathrm{Cliff}(T^*T_S))$ is given as the intersection of the $\mathrm{Ker}_p(\partial_{\dot\alpha})$ for all $\dot\alpha = k+1, \dots, n$.

Using (2-6) and the fact that $\rho_{S_q} = 1$, we obtain the following relations.

3.11 Lemma. *(1)* $\tilde\varphi_{pt}|_S = \varphi_{pt}|_S$ *and* $\eta_{\rm sa}(\Psi_{\{d\xi\}})|_S = \Psi_{\{d\xi\}}|_S$ *for* $p \in S$ *and* $\Psi_{\{d\xi\}} \in \Gamma_{vc}(p, \mathrm{Cliff}_{\mathbb{E}^n}(T^*\mathbb{E}^n))$.

(2) For a point $p \in T_S$, $\tilde\varphi_{pt}|_{\mathrm{Ker}_p(\partial_\perp)}$ *is a bijection:*

$$(\mathrm{Ker}_p(\partial_\perp))^* = \tilde\varphi_{pt}\,(\mathrm{Ker}_p(\partial_\perp)) \approx \mathrm{Ker}_p(Ad(\partial_\perp)).$$

(3) $\pi^* = \pi$, $\pi\tilde\varphi_{pt} = \tilde\varphi_{pt}\pi$ *and* $\pi\varphi_{pt} = \varphi_{pt}\pi$ *on* S.

(4) For a point $p \in T_S$, *the triple* $\mathcal{H}''_p = ((\mathrm{Ker}_p(\partial_\perp))^* \times \mathrm{Ker}_p(\partial_\perp), \cdot, \tilde\varphi_{pt})$ *becomes a pre-Hilbert space. Let* $\mathcal{H}_p(S \hookrightarrow \mathbb{E}^n) = \mathcal{H}''_p$ *for* $p \in S$.

(5) $\mathrm{Ker}_p(\partial_\perp) \subset \mathrm{Ker}_p(\not{D}^{\perp}_{u,\{d\xi\}})$.

Step 5: Define the submanifold Dirac operator $\not{D}_{S\hookrightarrow\mathbb{E}^n}$ for $S \hookrightarrow \mathbb{E}^n$ by restricting to $\mathrm{Ker}_S(\partial_\perp)$

Let us define $\not{D}_{S\hookrightarrow\mathbb{E}^n}$ by restricting to $\coprod_{p\in S}\mathrm{Ker}_p(\partial_\perp)$, i.e.,

$$\not{D}_{S\hookrightarrow\mathbb{E}^n} = \not{\mathbb{D}}_{T_S}|_{\coprod_{p\in S}\mathrm{Ker}_p(\partial_\perp)}, \tag{3-14}$$

with the local expression $\not{D}_{s,\{d\xi\}} = \not{\mathbb{D}}_{u,\{d\xi\}}|_{q=0,\partial_q=0} = \not{\mathbb{D}}^{\|}_{u,\{d\xi\}}|_{q=0}$ whose explicit form is given by the following proposition.

3.12 Proposition. *Abbreviating $\tau_{S,\mathbb{E}^n}(\gamma_{S,d\zeta}(ds^\alpha))$ by its image $\gamma_{\{d\xi\}}(ds^\alpha)$, the explicit form of $\not{D}_{s,\{d\xi\}}$ is*

$$\not{D}_{s,\{d\xi\}} = \tau_{S,\mathbb{E}^n}(\not{\mathbf{D}}_{S,s,\{d\zeta\}}) + \gamma_{\{d\xi\}}(dq^{\dot\alpha})\mathrm{tr}_{k\times k}(\gamma^{\alpha}{}_{\dot\alpha\beta}), \tag{3-15}$$

where $\not{\mathbf{D}}_{S,s,\{d\zeta\}}$ is a proper (or intrinsic) Dirac operator of S,

$$\not{\mathbf{D}}_{S,s,\{d\zeta\}} = \gamma_{\{d\xi\}}(ds^\alpha)(\partial_{s^\alpha} + \partial_{s^\alpha}\Omega|_{q=0}). \tag{3-16}$$

Here we fix the coordinate s and the orthonormal frame $\{d\zeta\}$.

Proof. [BJ, Ma1-10, MT] From (2-6), ${\rho_{S_q}}^{1/4} = 1+\dfrac{1}{2}\mathrm{tr}_{k\times k}(\gamma^{\alpha}{}_{\dot\alpha\beta})q^{\dot\alpha}+\mathcal{O}(q^{\dot\alpha}q^{\dot\beta})$. Hence,

$${\rho_{S_q}}^{1/4}\partial_{\dot\alpha}{\rho_{S_q}}^{-1/4} = \partial_{\dot\alpha} - \frac{1}{2}\mathrm{tr}_{k\times k}(\gamma^{\alpha}{}_{\dot\alpha\beta}) + \mathcal{O}(q^{\dot\alpha}),$$

$${\rho_{S_q}}^{1/4}\partial_{\alpha}{\rho_{S_q}}^{-1/4} = \partial_{\alpha} + \mathcal{O}(q^{\dot\alpha}).$$

The second term in (3-15) is thus obtained. As $\tau_{S,\mathbb{E}^n}$ induces the group inclusion $\iota_{S,\mathbb{E}^n}$, this completes the proof. ■

3.13 Proposition. *For a point S, the stalks $\Gamma(p, \mathrm{Cliff}_{\mathbb{E}^n}(T^*\mathbb{E}^n))$ and $\Gamma_{vc}(p, \mathrm{Cliff}_{\mathbb{E}^n}(T^*\mathbb{E}^n))$ are equivalent.*

As we work at a point from now on, we identify $\Gamma(p, \mathrm{Cliff}_{\mathbb{E}^n}(T^*\mathbb{E}^n))$ and $\Gamma_{vc}(p, \mathrm{Cliff}_{\mathbb{E}^n}(T^*\mathbb{E}^n))$. We are considering the kernel of $\not{D}^{\perp}_{u,\{d\xi\}}$ at p as the domain of $\not{D}_{S\hookrightarrow\mathbb{E}^n}$. For $\psi_0(s) \in \Gamma(p, \mathrm{Cliff}_S(T^*\mathbb{E}^n))$, we assume that $\Phi(u) \in \Gamma(p, \mathrm{Cliff}_{\mathbb{E}^n}(T^*\mathbb{E}^n))$ satisfies $\not{D}^{\perp}_{u,\{d\xi\}}\Phi = 0$ with the boundary condition $\Phi(s,0) = \psi_0(s)$ at S. We regard $\Gamma(p, \mathrm{Cliff}_S(T^*\mathbb{E}^n))$ as a subset of $\mathrm{Ker}_p(\partial_\perp)$. Further we can evaluate $\Phi(u) \in \mathrm{Ker}_p(\partial_\perp)$ around the point p in S as

$$\Phi(u) = \psi_0(s)+\sum \psi_{\dot\alpha}(s)q^{\dot\alpha}+\sum \psi_{\dot\alpha\dot\beta}(s)q^{\dot\alpha}q^{\dot\beta}+\dots, \quad \psi_{\dot\alpha}(s) = 0 \tag{3-17}$$

where $\psi_0(s) \in \Gamma(p, \mathrm{Cliff}_S(T^*\mathbb{E}^n))$, $\psi_{\dot\beta}(s), \psi_{\dot\gamma\dot\delta}(s) \in \Gamma(p, \mathbb{C}^{2^{[n/2]}}_{\mathbb{E}^n})$ and so on. Accordingly we can identify $\mathrm{Ker}_p(\partial_\perp)|_{q=0}$ with $\Gamma(p, \mathrm{Cliff}_S(T^*\mathbb{E}^n))$.

Noting that $\not{D}_{s,\{d\xi\}}$ is expressed in coordinate-free fashion and does not include $\partial_{q^{\dot\alpha}}$, we see that Definition 3.7, Lemmas 2.9, 3.11 and Proposition 3.12 have the following consequence:

3.14 Proposition. *The Dirac operator* $\not{D}_{S\hookrightarrow\mathbb{E}^n}$ *is an endomorphism of* $\mathrm{Cliff}_S(T^*\mathbb{E}^n)$ *at a point* p *in* S*, more precisely*

$$\not{D}_{S\hookrightarrow\mathbb{E}^n} : \Gamma(p, \mathrm{Cliff}_S(T^*\mathbb{E}^n)) \to \Gamma(p, \mathrm{Cliff}_S(T^*\mathbb{E}^n)). \tag{3-18}$$

Step 6: Consider the solution of the submanifold Dirac equation $\not{D}_{S\hookrightarrow\mathbb{E}^n}\psi = 0$

Let us consider the solution of $\not{D}_{S\hookrightarrow\mathbb{E}^n}\psi = 0$ at a point $p \in S$. For non-vanishing $\psi \in \Gamma(p, \mathbb{C}^{2^{[n/2]}}_{\mathbb{E}^n}))$ such that

$$\not{D}_{s,\{d\xi\}}\psi(s) = 0, \tag{3-19}$$

$\psi(s)$ can be regarded as an element of $\Gamma(S, \mathrm{Cliff}_S(T^*\mathbb{E}^n))$. By solving a boundary problem for $\Phi(u) \in \Gamma(S, \mathbb{C}^{2^{[n/2]}}_{\mathbb{E}^n})$,

$$\not{\mathbb{D}}^{\perp}_{u,\{d\xi\}}\Phi(u) = 0, \tag{3-20}$$

with boundary condition

$$\Phi|_S = \psi, \text{ on } S, \tag{3-21}$$

we have $\Phi(u)$ satisfying $\not{\mathbb{D}}_{u,\{d\xi\}}\Phi(u) = 0$ and $\not{\mathbf{D}}_{u,\{d\xi\}}\eta_{\mathrm{sa}}^{-1}(\Phi(u)) = 0$ at $p \in S$.

Noting that $\not{\mathbb{D}}_{u,\{d\xi\}}|_{q=0} = \not{D}_{s,\{d\xi\}} + \not{\mathbb{D}}^{\perp}_{u,\{d\xi\}}$, and $\not{D}_{s,\{d\xi\}}$ does not include the parameter q, we can apply the separation of variables method to this system. Thus it can be expected that there exists an orthonormal frame $(\eta_{\mathrm{sa}}^{-1}\Phi^{[a]}_{\{d\xi\}})_{a=1,\dots,2^{[n/2]}}$ in $\Gamma(p, \mathrm{Cliff}_{\mathbb{E}^n}(T^*\mathbb{E}^n))$ such that each $\Phi^{[a]}_{\{d\xi\}}$ belongs to $\mathrm{Ker}_p(\partial_\perp)$ and satisfies the Dirac equation (3-20); each $(\eta_{\mathrm{sa}}^{-1}\Phi^{[a]}_{\{d\xi\}})$ is a solution of the Dirac equation (3-9). Noting Lemma 3.11 (1), we regard $(\Phi^{[a]}_{\{d\xi\}}|_{q=0})_{a=1,\dots,2^{[n/2]}}$ as an orthonormal frame of $\Gamma(p, \mathrm{Cliff}_S(T^*\mathbb{E}^n))$ and solutions of the submanifold Dirac equation (3-19).

Conversely, as the Dirac operator $\not{D}_{s,\{d\xi\}}$ is also a $2^{[n/2]} \times 2^{[n/2]}$-matrix type first order differential operator, let us assume that we can find an orthonormal frame of $\Gamma(p, \mathrm{Cliff}_S(\mathbb{E}^n))$ belonging to the solution space of the submanifold Dirac equation (3-19). Let it be denoted by $\psi^{[a]}_{\{d\xi\}}$ $(a = 1, \dots, 2^{[n/2]})$, i.e., $\varphi_{pt}(\psi^{[a]}_{\{d\xi\}})\psi^{[b]}_{\{d\xi\}} = \delta_{a,b}$. For each $\psi^{[a]}_{\{d\xi\}}(s)$, we find an element $\tilde{\Psi}^{[a]}_{\{d\xi\}}(u)$ in the solution space of $\not{\mathbb{D}}_{u,\{d\xi\}}\eta_{\mathrm{sa}}(\Psi(u)) =$

0 or $\boldsymbol{D}_{u,\{d\xi\}}\Psi(u) = 0$ with the boundary condition $\tilde{\Psi}^{[a]}_{\{d\xi\}}(u)|_{q=0} = \psi^{[a]}_{\{d\xi\}}(s)$ at p.

By Proposition 3.6 (2), we have an element of $e^{\Omega} \in \Gamma(p, \mathrm{SPIN}_{\mathbb{E}^n}(\mathbb{E}^n))$ such that

$$\Psi^{[a]}_{\{dx\}} = e^{\Omega}\tilde{\Psi}^{[a]}_{\{d\xi\}}, \quad (a = 1, \dots, 2^{[n/2]}).$$

Using this element, we define $\Psi^{(i)}_{\{d\xi\}} = e^{-\Omega}\Psi^{(i)}_{\{dx\}}$, $(i = 1, \dots, n)$. Then Proposition 3.6 (3) gives the relation,

$$(3\text{-}22) \qquad g_{T_S,\mu\nu}\overline{\Psi}^{(i)}_{\{d\xi\}}\gamma_{\{d\xi\}}(du^{\mu})\Psi^{(i)}_{\{d\xi\}}du^{\nu} = dx^i, \quad (i = 1, \dots, n).$$

Therefore by defining $\psi^{(i)}_{\{d\xi\}}(s) = \Psi^{(i)}_{\{d\xi\}}(u)|_{q=0}$ and extracting the ds^{α} component, we have
(3-23)

$$g_{S,\alpha\beta}\overline{\psi}^{(i)}_{\{d\xi\}}\gamma_{\{d\xi\}}(ds^{\alpha})e^{\Omega_i}\psi^{(i)}_{\{d\xi\}} = \frac{\partial x^i}{\partial s^{\alpha}}, \quad (i = 1, \dots, n, \alpha = 1, \dots, k).$$

This is a representation of the tangent space of S, which we regard as the generalized Weierstrass representation. We note that $\psi^{(i)}_{\{d\xi\}}$ is also a solution of the submanifold Dirac equation (3-19).

As mentioned above, our theory is locally constructed but we can extend it to the case of a general smooth spin k-submanifold immersed in $\mathbb{E}^n$. Since $\mathbb{E}^n$ is spin, we can define $\mathrm{Cliff}_{\mathbb{E}^n}(T^*\mathbb{E}^n)$, $\mathrm{SPIN}_{\mathbb{E}^n}(T^*\mathbb{E}^n)$ and their global sections. By restricting it to S, we have $\mathrm{Cliff}_S(T^*\mathbb{E}^n)$ and a $\mathrm{SPIN}_S(T^*\mathbb{E}^n)$ structure over S. On the other hand, S has its own $\mathrm{Cliff}_S(T^*S)$ and $\mathrm{SPIN}_S(T^*S)$ structure. By Proposition 3.12, S has the submanifold Dirac operator $\not{D}_{S\hookrightarrow\mathbb{E}^n}$.

3.15 Theorem. *A smooth spin k-submanifold S immersed in $\mathbb{E}^n$ has a Clifford module $\mathrm{Cliff}_S(T^*\mathbb{E}^n)$, constructed locally in terms of the solution space of the submanifold Dirac equation $\not{D}_{S\hookrightarrow\mathbb{E}^n}\psi = 0$ at each chart. The solutions give the data at each point as follows:*

Let $p \in S$ be expressed in terms of an affine coordinate (x^i). For $\{\psi\}$ in $\Gamma(p, \mathbb{C}_S^{2^{[n/2]}})$ satisfying $\not{D}_{s,\{d\xi\}}\psi = 0$, there exists an orthonormal frame $\{\psi^{[a]}_{\{d\xi\}}\}_{a=1,\dots,2^{[n/2]}}$ $(\subset \{\psi\})$ of $\Gamma(p, \mathrm{Cliff}_S(T^*\mathbb{E}^n))$ as a $\Gamma(p, \mathbb{C}_{\mathbb{E}^n})$-vector space. For these bases, there exists some $e^{\Omega} \in \Gamma(p, \mathrm{Spin}_{\mathbb{E}^n}(T^*\mathbb{E}^n))$ which can be transformed to $\{\Psi^{[a]}_{\{dx\}}|_S\}_{a=1,\dots,2^{[n/2]}}$ by $\psi^{[a]}_{\{d\xi\}} = e^{-\Omega}\Psi^{[a]}_{\{dx\}}|_S$,

$(a = 1, \dots, 2^{[n/2]})$. *We define* $\psi^{(i)} = \mathrm{e}^{-\Omega}\Psi^{(i)}_{\{dx\}}|_S$ *and* $\overline{\psi}^{(i)} = \overline{\Psi}^{(i)}_{\{dx\}}\mathrm{e}^{\Omega}|_S$ $(i = 1, \dots, n)$. *Then the following relation holds:*

$$g_{S,\alpha,\beta}\overline{\psi}^{(i)}[\tau_{S,\mathbb{E}^n}(\gamma_{S,\{d\xi\}}(ds^\beta))]\psi^{(i)} = \partial_{s^\alpha}x^i \tag{3-24}$$

$(i = 1, \dots, n,\ s = 1, \dots, k)$.

This is the generalized Weierstrass representation for a spin submanifold immersed in $\mathbb{E}^n$.

Even though we fix the coordinate system, for a point p in S, $(\partial x^i/\partial s^\alpha)_{\alpha=1,\dots,k,i=1,\dots,n}$ gives the data of an embedding $T^*_pS \approx \mathbb{R}^k$ in $T^*_p\mathbb{E}^n \approx \mathbb{R}^n$ up to action of $\mathrm{GL}(\mathbb{R}^k)$ and $\mathrm{GL}(\mathbb{R}^{n-k})$. In other words, the submanifold Dirac operator $\not{D}_{S\hookrightarrow\mathbb{E}^n}$ exhibits the data of a Grassmannian bundle over S.

3.16 Remark. The generalized Weierstrass representation for a conformal surface in $\mathbb{E}^3$ was discovered by Kenmotsu [Ke] as a generalization of the Weierstrass representation for a minimal surface in $\mathbb{E}^3$. A Dirac type representation was found by Konopelchenko in 1995 [Ko1] and the author showed that the submanifold Dirac equation can be identified with this representation [Ma8]. He computed the submanifold Dirac operators for the generalized Weierstrass representation for a conformal surface in $\mathbb{E}^4$; this was also discovered by Konopelchenko [KL, Ko1-2] and Pinkall and Pedit [PP]. Recently there has been much work on the relations between submanifolds and Clifford bundles [Bo, Ko1-2, KL, KT, Fr, Tai1-2, Tr]. However, as far as the author is aware, the transformation η_{sa} has not appeared before.

Over the past decade, the author has developed the relation between Dirac operators and submanifolds, especially for curves immersed in $\mathbb{E}^n$, where the submanifold Dirac equation is identified with the Frenet-Serret equation [Ma1-7]. After the present conference, he was lead to the generalized Weierstrass representation of Theorem 3.15.

Although not mentioned in this article so far, physical considerations suggest that the Dirac operator has the following properties:

(1) The index of the Dirac operator is related to the topological index for curves [Ma2, 5].

(2) The operator determinants are associated with energy functionals, such as the Euler-Bernoulli functional and the Willmore functional [Ma6, 10].

(3) Deformations preserving all eigenvalues of the submanifold Dirac operators give rise to soliton equations, such as the MKdV equation [Ma2, 6, 7, MT], complex MKdV equation, nonlinear Schrödinger equation [Ma1, 3], and modified Novikov-Veselov equations [Ko1, Ko2, KL, KT, Tai1].

As Dirac operators are generally related to characteristic classes of fiber bundles, it is expected that the submanifold Dirac operator might be related to the Thom class and/or a generalization of the Riemann-Roch Theorem [PP].

§4. Dirac operator on a conformal surface in $\mathbb{E}^4$

As an example, we will consider the case of a conformal surface immersed in $\mathbb{E}^4$. In [Ma10], we gave an explicit local form (4-1) of the submanifold Dirac operator in this case and conjectured that the submanifold Dirac operator represents the surface. The conjecture was proved by Konopelchenko [KL, Ko2] and Pedit and Pinkall [PP], but we will sketch another proof by means of the submanifold Dirac operator method.

First we will give the properties of the Dirac operators for a conformal surface S in $\mathbb{E}^n$ [P]. We take the metric $g_{S\alpha\beta} = \rho\delta_{\alpha\beta}$, and the orthonormal frame $\{d\zeta\}$ given by $d\zeta^\alpha = \rho^{-1/2}ds^\alpha$. Take a complex parametrization of S, $dz = ds^1 \pm i ds^2$. We introduce another transformation $\eta_{\rm conf}$ as follows. For a point $p \in S$ and $\psi_{\{d\xi\}} \in \Gamma(p, \mathrm{Cliff}_S(T^*\mathbb{E}^n))$, let $\varphi_{\{d\xi\}} = \eta_{\rm conf}(\psi_{\{d\xi\}}) = \rho^{1/2}\psi_{\{d\xi\}}$, $\overline{\varphi}_{\{d\xi\}} = \eta_{\rm conf}(\overline{\psi}_{\{d\xi\}}) = \varphi_{pt}(\psi_{\{d\xi\}})$. Then we have the following properties.

4.1 Lemma. *(1) The Dirac operator of S is given by*

$$\not{D}_{s,\{d\zeta\}} = \rho^{-1}\sigma^\alpha\partial_\alpha\rho^{1/2},$$

by letting $\gamma_{S,\{d\zeta\}}(d\zeta^a) = \sigma^a$.

(2) For $(\overline{\varphi}_{\{d\xi\}}, \varphi_{\{d\xi\}})$ in $\eta_{\rm conf}\mathcal{H}_p(S \hookrightarrow \mathbb{E}^n)$, $\varphi_{\{d\xi\}}\tau_{S,\mathbb{E}^n}[\gamma_{\{d\xi\}}(d\xi^\alpha)]\varphi_{\{d\xi\}}ds^\alpha$ is independent of choice of local parametrization of S.

Proof. (1) is obvious [P, Ma8-9]. Setting $\gamma_{\{d\xi\}}(ds^\alpha) = \rho^{-1/2}\gamma_{\{d\xi\}}(d\xi^\alpha)$, the relation in Proposition 3.2 (1) becomes

$$g_{S\alpha\beta}\overline{\psi}_{\{d\xi\}}\tau_{S,\mathbb{E}^n}[\gamma_{\{d\xi\}}(ds^\alpha)]\psi_{\{d\xi\}}ds^\alpha = \overline{\varphi}_{\{d\xi\}}\tau_{S,\mathbb{E}^n}[\gamma_{\{d\xi\}}(d\xi^\alpha)]\varphi_{\{d\xi\}}ds^\alpha,$$

and thus (2) is proved ■

4.2 Proposition. *(1) For the case of a conformal surface in $\mathbb{E}^4$, the submanifold Dirac operator is given by*

$$\not{D}_{s,\{d\xi\}} = 2 \begin{pmatrix} & & \overline{p}_c & \partial \\ & & \overline{\partial} & -p_c \\ p_c & \partial & & \\ \overline{\partial} & -\overline{p}_c & & \end{pmatrix}, \tag{4-1}$$

where $\partial = (\partial_{s^1} - i\partial_{s^2})/2$, $\overline{\partial} = (\partial_{s^1} + i\partial_{s^2})/2$ and p_c is

$$p_c = -\frac{1}{2}\rho^{1/2}\mathrm{tr}_{2\times 2}(\gamma^\alpha_{3\beta} + i\gamma^\alpha_{4\beta}).$$

(2) Let $\mathbb{E}^4 \approx \mathbb{C}\times\mathbb{C} \in (Z^1, Z^2) = (x^1 + ix^2, x^3 + ix^4)$. For the affine coordinate (Z^1, Z^2) of the surface, the relations

$$dZ^1 = fmdz - gnd\overline{z}, \quad dZ^2 = f\overline{n}dz + g\overline{m}d\overline{z}, \quad d\overline{Z}^1 = \overline{dZ^1}, \quad d\overline{Z}^2 = \overline{dZ^2},$$

hold if $\varphi_1 = \begin{pmatrix} f \\ g \\ 0 \\ 0 \end{pmatrix}$ $\varphi_2 = \begin{pmatrix} 0 \\ 0 \\ m \\ n \end{pmatrix}$ *are solutions of*

$$\not{D}_{s,\{d\xi\}}\varphi_a = 0,$$

and $(|f|^2 + |g|^2)(|m|^2 + |n|^2) = \rho^{1/2}$.

Direct computation leads to the next lemma.

4.3 Lemma. *For the four dimensional Euclidean space $\mathbb{E}^4$, we can fix the representation of the γ-matrices of $\{dx^i\}$ system as $\gamma_{\{dx\}}(dx^i) = \sigma^1 \otimes \sigma^i$ for $i = 1, 2, 3$ and $\gamma_{\{dx\}}(dx^4) = \sigma^2 \otimes 1$. By letting*

$$\Psi_1 = \begin{pmatrix} 1 \\ 0 \\ 1 \\ 0 \end{pmatrix}, \quad \Psi_2 = \begin{pmatrix} 0 \\ 1 \\ 0 \\ 1 \end{pmatrix}, \quad \Psi_3 = \begin{pmatrix} 1 \\ 0 \\ 0 \\ 1 \end{pmatrix}, \quad \Psi_4 = \begin{pmatrix} 0 \\ 1 \\ 1 \\ 0 \end{pmatrix},$$

$\overline{\Psi}_1 = (0, 1, 0, 1)$, $\overline{\Psi}_2 = (1, 0, 1, 0)$, $\overline{\Psi}_3 = (0, -1, 1, 0)$, $\overline{\Psi}_4 = (1, 0, 0, -1)$, we have the following relations:

$$\sum_i \overline{\Psi}_1\gamma_{\{dx\}}(dx^i)\Psi_1 dx^i = 2dZ_1, \quad \sum_i \overline{\Psi}_2\gamma_{\{dx\}}(dx^i)\Psi_2 dx^i = 2d\overline{Z}_1,$$

$$\sum_i \overline{\Psi}_3\gamma_{\{dx\}}(dx^i)\Psi_3 dx^i = 2dZ_2, \quad \sum_i \overline{\Psi}_4\gamma_{\{dx\}}(dx^i)\Psi_4 dx^i = 2d\overline{Z}_2.$$

Proof of Proposition 4.2. (1) follows from Theorem 3.15. We consider (2). For the φ_a's, their independent partner solutions are given by $\varphi_3 = \begin{pmatrix} -\overline{g} \\ \overline{f} \\ 0 \\ 0 \end{pmatrix}$ $\varphi_4 = \begin{pmatrix} 0 \\ 0 \\ -\overline{n} \\ \overline{m} \end{pmatrix}$. Since we have fixed $\gamma_{S,\{d\xi\}}(d\zeta^\beta) = \sigma^\beta$, the convention in 2.8 gives $\tau_{S,\mathbb{E}^2}(\gamma_{S,\{d\xi\}}(d\zeta^\beta)) = \sigma^1 \otimes \sigma^\beta$. We define $\tilde{\varphi}_1 = \varphi_1 + \varphi_2$, $\tilde{\varphi}_2 = \varphi_3 + \varphi_4$, $\tilde{\varphi}_3 = \varphi_1 + \varphi_4$, and $\tilde{\varphi}_4 = \varphi_3 + \varphi_2$. Let us assume that $\tilde{\varphi}_a = \rho^{1/2}e^{\Omega}\Psi_a|_{q=0}$ $(a = 1, 2, 3, 4)$ as in Lemma 4.3. Then we find the spin matrix as

$$\rho^{1/2}e^{\Omega} = \begin{pmatrix} f & -\overline{g} & 0 & 0 \\ g & \overline{f} & 0 & 0 \\ 0 & 0 & m & -\overline{n} \\ 0 & 0 & n & \overline{m} \end{pmatrix}.$$

We have these dual bases, $\overline{\tilde{\varphi}}_a = \overline{\Psi}_a e^{-\Omega}|_{q^\beta=0}$, and obtain the relation,

$$2dZ_1 = \overline{\tilde{\varphi}}_1\sigma_1 \otimes \sigma^\alpha\tilde{\varphi}_1 ds^\alpha, \quad 2d\overline{Z}_1 = \overline{\tilde{\varphi}}_2\sigma_1 \otimes \sigma^\alpha\tilde{\varphi}_2 ds^\alpha,$$

$$2dZ_2 = \overline{\tilde{\varphi}}_3\sigma_1 \otimes \sigma^\alpha\tilde{\varphi}_3 ds^\alpha, \quad 2d\overline{Z}_2 = \overline{\tilde{\varphi}}_4\sigma_1 \otimes \sigma^\alpha\tilde{\varphi}_4 ds^\alpha.$$

Explicit computation gives (2). ■

Update added, April 2008: After writing this article, we realised that the operator η_{sa} is closely related to the half-density form in Theorem 18.1.34 in [H] and the quasi-regular representation or L^2 induced representation as an infinite-dimensional representation of a non-compact subgroup [Kob]. These are alluded to in [Ma11, Ma12].

[Kob] T. Kobayashi, "Lie gun to Lie kan 2" (Lie groups and Lie algebras 2) Iwanami, 1999 (in Japanese).

[H] L. Hörmander, "The analysis of linear partial differential operators III", Springer-Verlag, 1985.

[Ma11] S. Matsutani, Submanifold Dirac operators with torsion, Balkan Journal of Geometry and its Applications, **9** (2004), 69–82.

[Ma12] S. Matsutani, Generalized Weierstrass Relations and Frobenius Reciprocity, Math. Phys. Anal. Geom., **9** (2006), 353–369.

References

[BGV] N. Berline, E. Getzler and M. Vergne Heat Kernels and Dirac Operators, Springer-Verlag, 1996.

[Bj] J.-E. Björk, Analytic $\mathcal{D}$-Modules and Applications, Kluwer, Dordrecht, 1992.

[Bo] A. I. Bobenko Surfaces in terms of 2 by 2 matrices: old and new integrable cases In: Harmonic Maps and Integrable Systems, (eds. A. P. Fordy and J. C. Wood), Vieweg, 1994, pp. 83–127.

[BJ] M. Burgess and B. Jensen Fermions near two-dimensional surfaces, Phys. Rev. A, **48** (1993), 1861–1866.

[dC] R. C. T. da Costa Quantum mechanics of a constrained particle, Phys. Rev A (3), **23** (1981), 1982–1987.

[E] L. P. Eisenhart A Treatise on Differential Geometry, Ellis Horwood, New York, 1909.

[Fr] T. Friedrich On the spinor representation of surfaces in Euclidean 3-space, J. Geom. Phys., **28** (1997), 143–157.

[JK] H. Jensen and H. Koppe Quantum Mechanics with constraints, Ann. Phys., **63** (1971), 586–591.

[Ke] K. Kenmotsu Weierstrass formula for surfaces of prescribed mean curvature, Math. Ann., **245** (1979), 89–99.

[Ko1] B. G. Konopelchenko Induced surfaces and their integrable dynamics, Stud. Appl. Math., **96** (1996), 9–51.

[Ko2] B. G. Konopelchenko Weierstrass representations for surfaces in 4D spaces and their integrable deformations via DS hierarchy, Ann. Global Anal. Geom., **18** (2000), 61–74.

[KL] B. G. Konopelchenko and G. Landolfi Induced surfaces and their integrable dynamics II. Generalized Weierstrass representations in 4D spaces and deformations via DS hierarchy, Stud. Appl. Math., **104** (2000), 129–169.

[KT] B. G. Konopelchenko and I. A. Taimanov Constant mean curvature surfaces via an integrable dynamical system, J. Phys. A, **29** (1996), 1261–1265.

[Mal] A. Mallios Geometry of Vector Sheaves, An Axiomatic Approach to Differential Geometry I: Vector Sheaves. General Theory, Kluwer, Netherlands, 1998..

[Ma1] S. Matsutani Berry phase of Dirac particle in thin rod, J. Phys. Soc. Jpn., **61** (1992), 3825–3826.

[Ma2] S. Matsutani The relation between the modified Korteweg-de Vries equation and anomaly of Dirac field on a thin elastic rod, Prog. Theor. Phys., **5** (1994), 1005–1037.

[Ma3] S. Matsutani On the physical relation between the Dirac equation and the generalized mKdV equation on a thin elastic rod, Phys. Lett. A, **189** (1994), 27–31.

[Ma4] S. Matsutani MKdV equation and submanifold quantum mechanics, Sorushiron-Kenkyu, **94** (1994), A72–A75.

[Ma5] S. Matsutani Anomaly on a submanifold system: new index theorem related to a submanifold system, J. Phys. A., **28** (1995), 1399–1412.
[Ma6] S. Matsutani The physical realization of the Jimbo-Miwa theory of the modified Korteweg-de Vries equation on a thin elastic rod: Fermionic theory, Int. J. Mod. Phys. A, **10** (1995), 3091–3107.
[Ma7] S. Matsutani On the Relation between Modified KdV Solitons and Dirac Fields on a Thin Elastic Rod, thesis, Tokyo Metroplitan Univ., 1996.
[Ma8] S. Matsutani Constant mean curvature surface and Dirac operator, J. Phys. A, **30** (1997), 4019–4029.
[Ma9] S. Matsutani Immersion anomaly of Dirac operator on surface in $\mathbb{R}^3$, Rev. Math. Phys., **2** (1999), 171–186.
[Ma10] S. Matsutani Dirac operator on a conformal surface immersed in R^4: a way to further generalized Weierstrass equation, Rev. Math. Phys., **12** (2000), 431–444.
[MT] S. Matsutani and H. Tsuru Physical relation between quantum mechanics and soliton on a thin elastic rod, Phys. Rev. A, **46** (1992), 1144–1147.
[P] A. M. Polyakov Gauge Fields and Strings, Harwood Academic Publ., London, 1987.
[PP] F. Pedit and U. Pinkall Quaternionic analysis on Riemann surfaces and differential geometry, Doc. Math. Extra, **II** (1998), 389–400.
[Tai1] I. A. Taimanov Modified Novikov-Veselov equation and differential geometry of surface, Amer. Math. Soc. Transl. Ser. 2, **179** (1997), 133–151.
[Tai2] I. A. Taimanov The Weierstrass representation of closed surfaces in $\mathbb{R}^3$, Funct. Anal. Appl., **32** (1998), 258–267.
[Tas] T. Tasaka NijiKeishiki II (Quadratic Forms II), Iwanami, Tokyo, 1970, (Japanese).
[Tr] A. Trautman Dirac operator on a hypersurface, Acta Phys. Polon. B, 1995, 1283–1310.

8-21-1 Higashi-Linkan
Sagamihara 228-0811
JAPAN

E-mail address:
`RXB01142@nifty.com`

Advanced Studies in Pure Mathematics 51, 2008
Surveys on Geometry and Integrable Systems
pp. 285–314

Harmonic tori and their spectral data

Ian McIntosh

One of the earliest applications of modern integrable systems theory (or "soliton theory") to differential geometry was the solution of the problem of finding all constant mean curvature (CMC) tori in $\mathbb{R}^3$ (and therefore, by taking the Gauss map, finding all non-conformal harmonic maps from a torus to S^2). At its simplest level this proceeds from the recognition that the Gauss-Codazzi equations of a CMC torus are the elliptic sinh-Gordon equations

$$u_{z\bar{z}} + \sinh(4u) = 0, \ z = x + iy. \tag{1}$$

It was shown in the late 1980's ([24, 1]) that each doubly periodic solution of this equation can be written down in terms of the Riemann θ-function for a compact Riemann surface X, called the spectral curve (this also follows from Hitchin's work [10] on harmonic tori in S^3, which used a distinctly different approach). That this is true relies on two observations. First, (1) has a zero-curvature (or Lax pair) representation: it is the condition that

$$[\frac{\partial}{\partial z} - U_\zeta, \frac{\partial}{\partial \bar{z}} + U^\dagger_{\bar{\zeta}^{-1}}] = 0, \ U_\zeta = \begin{pmatrix} u_z & e^{-2u}\zeta^{-1} \\ e^{2u}\zeta^{-1} & -u_z \end{pmatrix}, \ \forall \zeta \in \mathbb{C}^*,$$

where '$\dagger$' denotes the Hermitian transpose. As a result this equation belongs to a hierarchy of infinitely many commuting equations, so that solutions to (1) may belong to an infinite dimensional family of deformations through solutions. These deformations are called the "higher flows" of the sinh-Gordon hierarchy. Secondly, each independent higher flow contributes to the number of independent Jacobi fields which the CMC surface admits: these belong to the kernel of the elliptic operator

Received January 30, 2001.
Revised September 3, 2001.
2000 *Mathematics Subject Classification.* Primary 53C43.
Partially supported by EPSRC Grant GR/M97534.

$\triangle + 4\cosh(4u)$. Thus for a torus there can only be finitely many independent higher flows. It follows that there must be a higher flow with respect to which the solution $u(z,\bar{z})$ is stationary. In this context this means there is a solution to

$$(2) \qquad d\xi_\zeta = [\xi_\zeta, \alpha_\zeta], \; \alpha_\zeta = U_\zeta dz - U^\dagger_{\bar{\zeta}^{-1}} d\bar{z},$$

in which the matrix $\xi_\zeta(z,\bar{z})$ is a Laurent polynomial in ζ: it is called a polynomial Killing field. The spectral data of the CMC torus consists of the eigenvalues and eigenlines of ξ_ζ. In particular, equation (2) means ξ_ζ is isospectral i.e. its characteristic polynomial is independent of z. This provides us a with an invariant planar algebraic curve which is essentially the Riemann surface X. Altogether the spectral data consists of the Riemann surface X, which always possesses a real involution, a rational function λ on X of degree 2, and a line bundle $\mathcal{L}$ over X satisfying a certain reality condition. The CMC surface is determined, up to Euclidean motions, by its spectral data. However, the existence of a polynomial Killing field is only a necessary condition for a CMC plane to be doubly periodic. If we call CMC planes "of finite type" when they possess a polynomial Killing field then one must still work at distinguishing the tori amongst the planes of finite type: this is a problem of closing periods on the surface (see e.g. [1, 8, 11]). This is also true for the Gauss map: the space of non-conformal harmonic maps $\varphi : \mathbb{R}^2 \to S^2$ of finite type is substantially larger than the set of non-conformal harmonic tori.

Essentially the same line of argument shows that all non-isotropic harmonic tori in $\mathbb{CP}^n, S^n$ [4, 9] and all non-conformal harmonic tori in rank 1 compact symmetric spaces [5] are of (semisimple) finite type. Although the construction of the spectral data is more complicated the principle is the same [17, 18]. However, these complications have the effect of obscuring the geometry of the original map. In [20] I proposed a more direct geometric construction of the map from the spectral data, and showed how this produces pluri-harmonic maps $\mathbb{R}^{2k} \to Gr_k(\mathbb{C}^{n+1})$ as well.

My aim here is to use the example of non-conformal harmonic maps $\varphi : \mathbb{R}^2 \to S^2$ as a way of motivating the geometric construction of [20]. To this end sections 1.1-1.5 describe the construction and properties of the spectral data for a map of semisimple finite type into S^2. The approach is more concrete than that of [17] and owes much to [9, 22, 25]. Having obtained the spectral data we examine it closely, in sections 1.6 and 1.7, to see exactly what is needed to reproduce the map. In particular, we obtain a clear understanding of the periodicity conditions by introducing a "singularisation" X' of X. Section 1.8 ties the previous discussion in with two other methods of reconstruction: the Symes' formula of [6] and the dressing orbit of the vacuum solution [7]. I give explicit formulae for computing φ from its (hyperelliptic) spectral curve. This is illustrated with the example of the bubbletons: these are CMC surfaces in $\mathbb{R}^3$ whose Gauss maps have rational nodal spectral curve. They are the solitons of CMC theory, some of which were known to geometers of the 19th century (see [21]). The calculations in section 1.8 are particularly satisfying because they allow us to compute (using Nick Schmitt's CMCLab) explicit pictures of some CMC surfaces (see figures 1 and 2).

Section 2 describes the generalization presented in [20], which constructs pluri-harmonic maps of $\mathbb{R}^{2k}$ into $Gr_k(\mathbb{C}^{n+1})$. The key point is that a pluri-harmonic map $\varphi : \mathbb{R}^{2k} \to Gr_k(\mathbb{C}^{n+1})$ of semisimple finite type arises as a composition: $\varphi = \psi \circ \gamma$ where

$$\mathbb{R}^{2k} \xrightarrow{\gamma} J(X') \xrightarrow{\psi} Gr_k(\mathbb{C}^{n+1}).$$

The middle factor is the generalized Jacobian of a singularisation X' of the spectral curve X. The map γ is a homomorphism and the map ψ is algebraic, derived from a section of a trivial $Gr_k(\mathbb{C}^{n+1})$-bundle over $J(X')$. No proofs are given here, they can be found in [20]. Nevertheless, I give the details for the construction of totally equivariant maps, which are characterized as being those whose spectral curve is the Riemann sphere.

Acknowledgments. I am very grateful to Professors M. Guest, R. Miyaoka and Y. Ohnita for their generous invitation to participate in the 9th MSJ-IRI in Tokyo, which was the motivation for this article. I am also grateful to Martin Kilian and Nick Schmitt for their ideas concerning the explicit construction of bubbletons, which motivated the discussion in section 1.8.

Notation. If V is a vector space then V^t will denote its dual, while V^* will denote $V - \{0\}$.

§1. Maps into S^2.

1.1. Maps of semisimple finite type.

Let us start with a harmonic map $\varphi : \mathbb{R}^2 \to S^2$ of semisimple finite type. To recall what this means we fix a framing $F : \mathbb{R}^2 \to SU_2$ with

$F(0) = I$ i.e. $\varphi = F \cdot T$ if we view $S^2 \simeq SU_2/T$ where T is the maximal torus of diagonal matrices. The Lie algebra $\mathfrak{su}_2$ splits into the vector space sum $\mathfrak{t} + \mathfrak{m}$ where $\mathfrak{t}$ contains all diagonal matrices and $\mathfrak{m}$ contains all off-diagonal matrices. Now define the $\mathfrak{su}_2$-valued 1-form $\alpha = F^{-1}dF$: with respect to the splitting of $\mathfrak{su}_2$ this decomposes into $\alpha_{\mathfrak{t}} + \alpha_{\mathfrak{m}}$. From these components we construct a $\mathbb{C}^*$-family of $\mathfrak{gl}_2$- valued 1-forms

$$\alpha_\zeta = \zeta^{-1}\alpha_{\mathfrak{m}}^{(1,0)} + \alpha_{\mathfrak{t}} + \zeta\alpha_{\mathfrak{m}}^{(0,1)}$$

where $\zeta \in \mathbb{C}^*$. The condition that φ is harmonic is precisely the condition that α_ζ satisfies the Maurer-Cartan equations for all ζ. In addition, it has two symmetries:

(3) $$\alpha_{\bar{\zeta}^{-1}} = -\alpha_\zeta^\dagger, \qquad \alpha_{-\zeta} = \nu(\alpha_\zeta),$$

where '$\dagger$' denotes the Hermitian transpose and for $A \in \mathfrak{gl}_2$, $\nu(A) = \mathrm{Ad}\tau \cdot A$ where

$$\tau = \begin{pmatrix} 1 & 0 \\ 0 & -1 \end{pmatrix}.$$

We say that φ is of semisimple finite type when:

(1a) there exists a smooth function $a : \mathbb{R}^2 \to \mathbb{C}^*$ and a complex coordinate z on $\mathbb{R}^2$ such that

$$\alpha_{\mathfrak{m}}(\partial/\partial z) = \mathrm{Ad}\begin{pmatrix} a & 0 \\ 0 & a^{-1} \end{pmatrix} \cdot \begin{pmatrix} 0 & 1 \\ 1 & 0 \end{pmatrix};$$

(1b) there exists a smooth map $\xi_\zeta : \mathbb{R}^2 \to C^\omega(\mathbb{C}^*, \mathfrak{gl}_2)$ satisfying
(i) $d\xi_\zeta + [\alpha_\zeta, \xi_\zeta] = 0$,
(ii) ξ_ζ also possesses the two symmetries in (3),
(iii) for all $z \in \mathbb{R}^2$ there is a positive integer p such that ξ_ζ is a Laurent polynomial in ζ of order $2p + 1$.

These properties together imply

(4) $$\xi_\zeta = \zeta^{-2p-1}\alpha_{\mathfrak{m}}(\frac{\partial}{\partial z}) + \ldots + \zeta^{2p+1}\alpha_{\mathfrak{m}}(\frac{\partial}{\partial \bar{z}}).$$

1.2. The symmetric spectral curve Σ.

Define, for each $z \in \mathbb{R}^2$,

$$\Sigma_A(z) = \{(\zeta, [v]) \in \mathbb{C}^* \times \mathbb{P}^1 : \exists \mu \in \mathbb{C} \text{ such that } \xi_\zeta(z)v = \mu v\ \}.$$

To maximise the domain of definition here, whenever ξ_ζ is either singular or zero at ζ_0 we replace it by $(\zeta - \zeta_0)^m\xi_\zeta$ where m is chosen so that this is regular and non-zero at ζ_0. It is clear that this describes an algebraic curve birationally equivalent to the planar curve with equation $\mu^2 + \det(\xi_\zeta) = 0$. Moreover $\Sigma_A(z)$ will be smooth (and unramified over the ζ-plane) at all points for which ξ_ζ (or its renormalisation) is not nilpotent. In particular this is true over the unit ζ-circle (for the symmetry conditions imply ξ is skew-Hermitian there). Further, from (4) and 1a we see that $\Sigma_A(z)$ completes to a curve $\Sigma(z)$ in $\mathbb{P}^1 \times \mathbb{P}^1$ by adding two smooth points over each of $\zeta = 0$ and $\zeta = \infty$.

This curve admits a fixed point free involution arising from one of the symmetries of ξ_ζ. Define

$$\begin{array}{rccc} \tilde{\nu}: & \mathbb{P}^1 \times \mathbb{P}^1 & \to & \mathbb{P}^1 \times \mathbb{P}^1 \\ & (\zeta, [v]) & \mapsto & (-\zeta, [\tau v]) \end{array}$$

Then $\tilde{\nu}$ induces a fixed point free involution on $\Sigma(z)$ and the quotient curve $\Sigma(z)/\tilde{\nu}$ is smooth wherever $\Sigma(z)$ is.

1.3. The quotient spectral curve X.

Here we construct a model of the quotient curve $\Sigma(z)/\tilde{\nu}$. First, for any $\eta_\zeta \in C^\omega(\mathbb{C}^*, \mathfrak{gl}_2)$ satisfying $\nu(\eta_\zeta) = \eta_{-\zeta}$ define

$$\hat{\eta} = \mathrm{Ad}\kappa \cdot \eta_\zeta, \qquad \kappa = \begin{pmatrix} 1 & 0 \\ 0 & \zeta \end{pmatrix}.$$

It is easy to check that $\hat{\eta}(-\zeta) = \hat{\eta}(\zeta)$ so that it is a function of $\lambda = \zeta^2$. Therefore, with an abuse of notation, let us use the notation

$$\eta_\lambda = \mathrm{Ad}\kappa \cdot \eta_\zeta, \quad \lambda = \zeta^2. \tag{5}$$

Now define

$$X_A(z) = \{(\lambda, [w]) \in \mathbb{C}^* \times \mathbb{P}^1 : \xi_\lambda w = \mu w\}$$

with the same convention at singular points or zeroes of ξ_λ as earlier. An easy computation shows that

$$\xi_\lambda = \lambda^{-p-1} \begin{pmatrix} 0 & a^2 \\ 0 & 0 \end{pmatrix} + \ldots + \lambda^{p+1} \begin{pmatrix} 0 & 0 \\ -\bar{a}^2 & 0 \end{pmatrix}.$$

Therefore $X_A(z)$ is completed in $\mathbb{P}^1 \times \mathbb{P}^1$ by adding the points $P_0 = (0, [1, 0])$ and $P_\infty = (\infty, [0, 1])$. We will call this complete curve $X(z)$.

Lemma 1. *$X(z)$ is isomorphic to the quotient curve $\Sigma(z)/\tilde{\nu}$.*

Proof. Let $f : \mathbb{C}^* \times \mathbb{P}^1 \to \mathbb{C}^* \times \mathbb{P}^1$ be given by $f(\zeta, [v]) = (\zeta^2, [\kappa v])$. Since $\xi_\lambda = \mathrm{Ad}\kappa \cdot \xi_\zeta$ this maps $\Sigma_A(z)$ onto $X_A(z)$ and exhibits it as an unramified double cover. Further, it is easy to check that $f \circ \tilde{\nu} = f$ so that $\Sigma_A/\tilde{\nu} \simeq X_A$. Finally, one readily checks that the restriction of f to Σ_A extends to Σ with image X. Q.E.D.

We deduce from this that $X(z)$ is smooth at both P_0 and P_∞.

Lemma 2. *$X(0) \simeq X(z)$ for all $z \in \mathbb{R}^2$.*

Proof. By 1b we have $d(\mathrm{Ad}F_\lambda \cdot \xi_\lambda(z)) = 0$, where F_λ is given by $F_\lambda^{-1} dF_\lambda = \alpha_\lambda$ and $F_\lambda(0) = I$. Hence

$$\mathrm{Ad}F_\lambda \cdot \xi_\lambda(z) = \xi_\lambda(0). \tag{6}$$

It follows that the map

$$X_A(0) \to X_A(z); \quad (\lambda, [v]) \mapsto (\lambda, [F_\lambda^{-1} v]) \tag{7}$$

is an isomorphism. To see that this extends to the complete curves we follow [9].

Define

$$H_+ = \exp(-z\lambda^p \xi_\lambda(0)) F_\lambda; \qquad H_- = \exp(-\bar{z}\lambda^{-p} \xi_\lambda(0)) F_\lambda.$$

Then

$$\begin{aligned} H_+^{-1} dH_+ &= -\mathrm{Ad}F_\lambda^{-1} \cdot \lambda^p \xi_\lambda(0) dz + \alpha_\lambda \\ &= -\lambda^p \xi_\lambda(z) + \alpha_\lambda \end{aligned}$$

which is polynomial in λ. Therefore H_+ is holomorphic in λ. A similar computation shows that H_- is holomorphic in λ^{-1}. Whenever $\xi_\lambda(0)v = \mu v$ we see that

$$F_\lambda^{-1} v = H_+^{-1} \exp(z\lambda^p \xi_\lambda(0)) v = e^{z\lambda^p \mu} H_+^{-1} v$$

so that the line $[F_\lambda^{-1} v]$ equals $[H_+^{-1} v]$. Similarly we can show that $[F_\lambda^{-1} v] = [H_-^{-1} v]$. Now, we also have

$$\xi_\lambda(z) = \mathrm{Ad}F_\lambda^{-1} \cdot \xi_\lambda(0) = \mathrm{Ad}H_\pm^{-1} \cdot \xi_\lambda(0)$$

and it follows that the isomorphism (7) extends to give $X(0) \simeq X(z)$. Q.E.D.

Remark. Notice that this proof shows that $H_+|_{\lambda=0}$ is upper triangular since the isomorphism fixes the point $(0, [1, 0])$. Likewise, $H_-|_{\lambda=0}$ must be lower triangular.

1.4. The eigenline bundle $\mathcal{E}$ and its dual $\mathcal{L}$.

Let $\mathcal{E}_z$ denote the eigenline bundle of $\xi_\lambda(z)$: it is the pullback to $X(z)$ of the tautological bundle over $\mathbb{P}^1$, using the projection $(\lambda, [v]) \mapsto [v]$. We will denote its restriction to X_A by $\mathcal{E}_{z,A}$ and, to avoid too many superscripts, we will denote the dual bundle by $\mathcal{L}_z$. The inclusion $\mathcal{E}_z \hookrightarrow \mathbb{P}^1 \times \mathbb{P}^1 \times \mathbb{C}^2$ pulls back the canonical coordinates e_1, e_2 on $\mathbb{C}^2$ to give two independent globally holomorphic sections of $\mathcal{L}_z$, which we also denote e_1, e_2 (or e_1^z, e_2^z when necessary). Notice that $e_1((0,1)^t) = 0$ and $e_2((1,0)^t) = 0$ from which it follows that $e_1 \in \Gamma(\mathcal{L}(-P_\infty))$ and $e_2 \in \Gamma(\mathcal{L}(-P_0))$.

Our next aim is to show that these sections span the space of global holomorphic sections of $\mathcal{L}$ and this characterizes $\mathcal{L}$. Indeed, one reason for working over X is that on Σ we do not have such a straightforward relationship between points and the sections arising from the coordinates e_1, e_2. First let $A = \mathbb{C}[X_A]$, the coordinate ring of X_A, let $B = \mathbb{C}[\lambda, \lambda^{-1}] \subset A$ and for any ring R use $R\langle\cdot\rangle$ to denote an R-module presented by its generators.

Theorem 1. *(i)* $\Gamma(\mathcal{L}_A) = B\langle e_1, e_2\rangle$, *(ii)* $\Gamma(\mathcal{L}) = \mathbb{C}\langle e_1, e_2\rangle$, *(iii)* $\deg \mathcal{L} = g+1$ *(where g is the genus of X).*

Proof. Let Y be the completion of the planar curve with equation $\mu^2 + \det(\xi_\lambda) = 0$, with Y_A that part of the curve lying over $\lambda \neq 0, \infty$. Then Y_A has coordinate ring $A_Y = \mathbb{C}[\lambda, \lambda^{-1}, \mu] \subset A$ and there is a degree 1 morphism $\alpha : X \to Y$. Set $M = \Gamma(X_A, \mathcal{L}) = \Gamma(Y_A, \alpha_*\mathcal{L})$. First we will show that $B\langle e_1, e_2\rangle \subset M$ is an A_Y-submodule. For any $v \in \Gamma(Y_A, \alpha_*\mathcal{E}) = \Gamma(X_A, \mathcal{E})$ we have $v = (e_1(v), e_2(v))^t$ and $\xi_\lambda v = \mu v$ implies $\mu e_i(v) = a_i e_1(v) + b_i e_2(v)$ for some $a_i, b_i \in B$. It follows that $\mu e_i \in B\langle e_1, e_2\rangle$. So $B\langle e_1, e_2\rangle$ is an A_Y-module.

Now it can only be a proper A_Y-module if its localisation at every maximal ideal $\mathfrak{p}$ is also proper in the corresponding localisation $M_\mathfrak{p}$. But at any smooth point $P \in Y_A$, with maximal ideal $\mathfrak{p}$, $M_\mathfrak{p}$ is the stalk $\mathcal{L}_P$ of $\mathcal{L}$ at P, and all its proper submodules are contained in $\mathcal{L}_P(-P)$ (the local sections which vanish at P). But in that case every section in M must vanish at P. This means that for every $v \in \mathcal{E}_P$ both $e_1(v)$ and $e_2(v)$ vanish at P, which is absurd (there is always a non-zero eigenvector). Since Y_A must have at least one smooth point we deduce that $M = B\langle e_1, e_1\rangle$.

(ii) Given (i) it suffices to show that if $\lambda^n e_i \in \Gamma(\mathcal{L})$ for $n \in \mathbb{Z}$ then $n = 0$. Since e_1 does not vanish at P_0 and e_2 does not vanish at P_∞ it suffices to show that neither λe_1 nor $\lambda^{-1}e_2$ are globally holomorphic. Consider first λe_1: it is globally holomorphic if and only if $\lambda e_1(v)$ is holomorphic at P_∞ for all locally holomorphic sections v of $\mathcal{E}$ about P_∞. By definition,

$$[v] = \left[\begin{pmatrix} 1 & 0 \\ 0 & \pm\zeta \end{pmatrix} w\right]$$

where $\zeta^{-2p-1}\xi_\zeta w = \mu\zeta^{-2p-1}w$. We may assume, without loss of generality, that

$$w = \begin{pmatrix} \bar{a}^{-1} + O(\zeta^{-1}) \\ \pm\bar{a} + O(\zeta^{-1}) \end{pmatrix}.$$

Therefore

$$v = \begin{pmatrix} \zeta^{-1}\bar{a}^{-1} + O(\zeta^{-2}) \\ \bar{a} + O(\zeta^{-1}) \end{pmatrix}.$$

Hence $\lambda e_1(v)$ has a first order pole at P_∞. A similar calculation shows for v a locally holomorphic section about P_0 we can take

$$v = \begin{pmatrix} a + O(\zeta) \\ \zeta a^{-1} + O(\zeta^2) \end{pmatrix}$$

and therefore $\lambda^{-1}e_2(v)$ has a first order pole at P_0.

(iii) Since λ has divisor $2P_0 - 2P_\infty$, (i) and (ii) imply $\Gamma(\mathcal{L}(-2P_0)) = 0$. So applying the Riemann-Roch formula gives $\deg\mathcal{L} \leq g+1$. Now we show $\deg\mathcal{L} \geq g+1$. For n any suitably large positive integer $\mathcal{L}(2nP_\infty)$ must be non-special so that $\dim\Gamma(\mathcal{L}(2nP_\infty)) = \deg\mathcal{L} + 2n + 1 - g$. But

$$\mathbb{C}\langle e_1, e_2, \lambda e_1, \lambda e_2, \ldots, \lambda^n e_1, \lambda^n e_2\rangle \subset \Gamma(\mathcal{L}(2nP_\infty))$$

so $\deg\mathcal{L} + 2n + 1 - g \geq 2n+2$. Q.E.D.

1.5. The real structure of $\Gamma(\mathcal{L})$.

An important property of $\Gamma(\mathcal{L})$ is that it possesses a Hermitian inner product: this comes from a reality condition on $\mathcal{L}$ and is essential since we intend to identify $\mathbb{P}\Gamma(\mathcal{L})$ with $S^2 \cong \mathbb{CP}^1$ as a Hermitian symmetric space. This reality condition arises as follows.

The real symmetry $\xi_{\bar{\lambda}^{-1}} = -\xi_\lambda^\dagger$ induces a real involution ρ on X for which $\overline{\rho_*\lambda} = \lambda^{-1}$ and $\overline{\rho_*\mu} = -\mu$. Notice that, since ξ_λ is skew-Hermitian over $|\lambda| = 1$, μ is pure imaginary there so ρ fixes all points over $|\lambda| = 1$; this will prove to be important later. Consequently the eigenline bundle $\mathcal{E}$ also satisfies a reality condition.

Proposition 1. $\overline{\rho_*\mathcal{E}} \simeq \mathcal{L}(-R)$ *where* R *is the ramification divisor of* $\lambda : X \to \mathbb{P}^1$.

Proof. Since $\overline{\rho_*\xi_\lambda} = -\xi_\lambda^t$, sections of $\overline{\rho_*\mathcal{E}}$ correspond to solutions of $\xi_\lambda^t w = \mu w$. Take any proper open subset $U \subset X$ for which $U = \lambda^{-1} \circ \lambda(U)$, and let $v : U \to \mathcal{E}$ be a trivialising section. If σ denotes the hyperelliptic involution on X then $\sigma_*\mu = -\mu$ and clearly $v, \sigma_* v$ are linearly independent at $P \in X$ if and only if P is not a ramification point. Take V to be the matrix with columns $v, \sigma_* v$, then we have $\det(V)$ vanishing only at ramification points. Define $W = \det(V).V^{-1t}$, then W is holomorphic in U and

$$\xi_\lambda^t W = W \begin{pmatrix} \mu & 0 \\ 0 & -\mu \end{pmatrix}.$$

It is easy to check that the columns of W are given by $w, -\sigma_* w$ where $e_1(w) = e_2(\sigma_* v)$ and $e_2(w) = -e_1(\sigma_* v)$. Therefore w corresponds to a trivialising section for $\overline{\rho_*\mathcal{E}}$ over U. Now consider the injective homomorphism of $\mathcal{O}_U$-modules

$$\begin{array}{ccc} \mathcal{O}_U\langle w\rangle & \to & \mathrm{Hom}(\mathcal{O}_U\langle v\rangle, \mathcal{O}_U) \\ f.w & \mapsto & (h.v \mapsto fh.w^t v) \end{array}$$

for $f, h \in \mathcal{O}_U$. Since $w^t v = \det(V)$ we see that the induced sequence of sheaves is

$$0 \to \overline{\rho_*\mathcal{E}} \to \mathcal{L} \to \mathcal{O}_R \to 0$$

where $\mathcal{O}_R$ is the skyscraper sheaf for the divisor R. Therefore $\overline{\rho_*\mathcal{E}} \simeq \mathcal{L}(-R)$. Q.E.D.

Consequently we have $\mathcal{L} \otimes \overline{\rho_*\mathcal{L}} \simeq \mathcal{O}_X(R)$. The inner product on $\mathbb{C}^2$ corresponds (at least over the unit circle) to the section

$$e_1 \otimes \overline{\rho_* e_1} + e_2 \otimes \overline{\rho_* e_2} \tag{8}$$

(which maps (v, w) to $\overline{\rho_* v}^t w$). Up to scaling this corresponds to an inner product on $\Gamma(\mathcal{L})$ determined in the following manner.

We first take any $s \in \Gamma(\mathcal{L})$ to identify $\mathcal{L}$ with the divisor line bundle $\mathcal{O}_X(D)$ where D is the divisor of zeroes of s. Second we fix a rational function f with divisor $D + \rho_* D - R$ for which $\overline{\rho_* f} = f$ and f is positive over $|\lambda| = 1$ (this is always possible since ρ fixes all points over the unit circle). Now we define

$$\begin{array}{l} h : \Gamma(\mathcal{L}) \times \Gamma(\mathcal{L}) \to \mathbb{C}; \\ h(s_1, s_2) = \frac{1}{2} \sum_{j=1}^{2} f(O_j)(s_1/s)(O_j)\overline{\rho_*(s_2/s)}(O_j) \end{array} \tag{9}$$

where O_1, O_2 are the two points over $\lambda = 1$. The proper interpretation of the right hand side is in terms of the trace map $\mathrm{Tr} : \mathcal{O}_X(R) \to \mathbb{C}$ which I will not explain in detail here (see, for example, [26]). But it is worth noting for future reference that this inner product clearly makes the subspaces $\Gamma(\mathcal{L}(-O_1)), \Gamma(\mathcal{L}(-O_2)) \subset \Gamma(\mathcal{L})$ orthogonal.

1.6. What the frame does.

Let us introduce $\hat{\Gamma}(\cdot)$ for spaces of analytic sections and let $\hat{A}$ denote the ring of analytic functions on X_A while $\hat{B}$ denotes the analytic functions on the punctured λ-plane $\mathbb{P}^1_\lambda \setminus \{0, \infty\}$. The map

$$\hat{\Gamma}(\mathcal{E}_{z,A}) \to \hat{\Gamma}(\mathcal{E}_{0,A}); \quad v \mapsto F_\lambda v$$

is clearly an isomorphism of $\hat{A}$-modules. Therefore it corresponds to a family of trivialising sections

$$\theta_z \in \hat{\Gamma}(\mathcal{E}_{0,A} \otimes \mathcal{L}_{z,A}).$$

Let $J(X)$ denote the Jacobian of X — the abelian variety of isomorphism classes of line bundles of degree zero. If $J_R(X)$ denotes the real subgroup of degree zero line bundles L for which $\overline{\rho_* L} \simeq L^{-1}$ then we deduce from the previous section that $\mathcal{E}_0 \otimes \mathcal{L}_z$ belongs to $J_R(X)$ for all z.

Proposition 2. *(i) Define $L : \mathbb{R}^2 \to J_R(X)$ by $L_z = \mathcal{E}_0 \otimes \mathcal{L}_z$. Then L is $\mathbb{R}$-linear (i.e. a homomorphism of real abelian groups). (ii) The section $\theta_z \exp(z\lambda^p \mu)$ is holomorphic and non-vanishing over P_0, while $\theta_z \exp(\bar{z}\lambda^{-p}\mu)$ is holomorphic and non-vanishing over P_∞.*

Proof. Observe that (ii) implies (i) since we deduce from it that L_z corresponds to the transition functions $\exp(z\lambda^p\mu)$ and $\exp(\bar{z}\lambda^{-p}\mu)$ patching from X_A to U_0 and U_∞ respectively, where the latter are open neighbourhoods of P_0 and P_∞ respectively.

To prove (ii) we recall from the proof of lemma 2 that if v_0 is a holomorphic section of $\mathcal{E}_0$ about P_0 then $F_\lambda^{-1} v_0 = H_+^{-1} \exp(z\lambda^p\mu) v_0$ so that $\exp(-z\lambda^p\mu) F_\lambda^{-1} v_0$ is a holomorphic section of $\mathcal{E}_z$ about P_0. But $F_\lambda^{-1} v_0$ corresponds to $v_0 \otimes \theta_z^{-1}$ so tensoring with $\exp(-z\lambda^p\mu)\theta_z^{-1}$ preserves holomorphicity about P_0. A similar argument using H_- about P_∞ proves the second part of (ii). Q.E.D.

Let $L_{z,A}$ denote the restriction to X_A of L_z. We want to make explicit the representation

$$\mathcal{F} : \hat{\Gamma}(L_A) \to \hat{B} \otimes \mathfrak{gl}_2$$

which gives us $\mathcal{F}(\theta) = F_\lambda$. It arises from the composite isomorphism

$$\epsilon_z : \hat{\Gamma}(\mathcal{E}_{z,A}(R)) \to \hat{B} \otimes \Gamma(\mathcal{L}_z)^t \to \hat{B} \otimes \mathbb{C}^2. \tag{10}$$

The second arrow is just the identification $\Gamma(\mathcal{L}_z)^t \to \mathbb{C}^2$ determined by e_1^z, e_2^z. The first arrow is the $\hat{B}$-module isomorphism dual to $\hat{B} \otimes \Gamma(\mathcal{L}) \simeq \hat{\Gamma}(\mathcal{L}_A)$, from theorem 1. This uses the fact, implicit in the proof of proposition 1, that $\lambda_* \mathcal{L}$ is dual to $\lambda_* \mathcal{E}(R)$. It follows that to any $\phi \in \hat{\Gamma}(L_A)$ there is some $\mathcal{F}(\phi) \in \hat{B} \otimes \mathfrak{gl}_2$ so that the following diagram commutes:

$$\begin{array}{ccc} \hat{\Gamma}(\mathcal{E}_{z,A}(R)) & \xrightarrow{\phi} & \hat{\Gamma}(\mathcal{E}_{0,A}(R)) \\ \epsilon_z \downarrow & & \downarrow \epsilon_0 \\ \hat{B} \otimes \mathbb{C}^2 & \xrightarrow{\mathcal{F}(\phi)} & \hat{B} \otimes \mathbb{C}^2. \end{array} \tag{11}$$

Next we will show that θ is almost completely determined by its behaviour at the points P_0, P_∞. First observe that from (4) we have

$$\mu^2 = -\det \xi_\zeta = \zeta^{-4p-2} + \ldots + \zeta^{4p+2} = \lambda^{-2p-1} + \ldots + \lambda^{2p+1}.$$

Since $\overline{\rho_* \mu} = -\mu$ whereas $\overline{\rho_* \zeta} = \zeta^{-1}$ we find, with the right sign choice for ζ, $\mu = \zeta^{-2p-1} + \ldots - \zeta^{2p+1}$. Therefore $\lambda^p \mu - \zeta^{-1}$ is holomorphic about P_0 while $\lambda^{-p}\mu + \zeta$ is holomorphic about P_∞. Consequently, as a corollary of proposition 2 we have:

Corollary 1. *θ_z is determined up to sign, amongst trivialising sections of L_A, by the properties that: (a) $\theta_z \exp(z\zeta^{-1})$ is holomorphic and non-vanishing over P_0 while $\theta_z \exp(-\bar{z}\zeta)$ is holomorphic and non-vanishing over P_∞ and, (b) $\det(\mathcal{F}(\theta_z)) = 1$.*

Proof. If ϕ is any other trivialising section with these properties then $\phi\theta_z^{-1}$ is a globally holomorphic function and therefore a constant, k say. But clearly $\det(\mathcal{F}(k\theta_z)) = k^2$ so that $k = \pm 1$. Q.E.D.

Remark. The unitary nature of $\mathcal{F}(\theta_z)$ on the unit circle is a reflection of the fact that $\overline{\rho_* \theta} = \theta^{-1}$.

Finally, let us use this corollary to display a simple characterisation for the map L. Since it is linear it is completely determined by $dL_0(\partial/\partial z)$ which lies in $T_1^{1,0}J(X)$ (here 1 denotes the identity in $J(X)$). By the corollary above $L(z)$ corresponds to the cohomology class $[c(z)]$ in $H^1(X, \mathcal{O}^*)$ for the 1-cocycle

$$c(z) = \{(e^{z\zeta^{-1}}, X_A, U_0), (e^{-\bar{z}\zeta}, X_A, U_\infty)\} \tag{12}$$

for the open cover X_A, U_0, U_∞, where now U_0, U_∞ are (disjoint) parameter discs (i.e. domains for ζ, ζ^{-1}). Therefore

$$\frac{\partial[c]}{\partial z}|_{z=0} = [(\zeta^{-1}, X_A, U_0), (1, X_A, U_\infty)] \in H^1(X, \mathcal{O}).$$

Now recall the isomorphism $H^1(X,\mathcal{O}) \simeq \Gamma(\Omega_X)^t$: it identifies $\partial[c]/\partial z$ with the map $f : \omega \mapsto res_{P_0}\zeta^{-1}\omega$ for $\omega \in \Gamma(\Omega_X)$. But now observe that

$$res_{P_0}\zeta^{-1}\omega = (\omega/d\zeta)(P_0) = \frac{\partial}{\partial\zeta}\int_{P_0}^{\zeta}\omega.$$

Hence $f = d\mathcal{A}_{P_0}(\partial/\partial\zeta)$ where $\mathcal{A}_{P_0} : X \to J(X)$ is the Abel map with base point P_0. Thus we learn:

Lemma 3. *The linear map $L : \mathbb{R}^2 \to J_R(X)$ is uniquely determined by the property that $dL_0(\partial/\partial z) = d\mathcal{A}_{P_0}(\partial/\partial\zeta)$.*

1.7. Periodicity conditions.

We have seen that the non-conformal doubly periodic harmonic map $\varphi : \mathbb{R}^2 \to S^2$ yields us spectral data $(X, \lambda, \mathcal{L})$ and it is easy to see how to reverse this procedure to reconstruct the map from this data. We first construct the linear map $L : \mathbb{R}^2 \to J_R(X)$ given by lemma 3 and define $\mathcal{L}_z = \mathcal{L} \otimes L_z$. By theorem 1 $\Gamma(\mathcal{L}_z)$ comes equipped with a frame e_1^z, e_2^z determined by the points P_∞, P_0: this frame is chosen to be unitary according to the trace inner product described above. With the frame we recover the map $\mathcal{F}$ in (11). Now we equip L_z with the unique (up to sign) trivialising section θ_z over X_A given by corollary 1. Thus we obtain the extended frame $F_\lambda = \mathcal{F}(\theta_z)$ and the map φ is recovered as $F_1 \circ [1,0]$ where $[1,0] \in \mathbb{CP}^1$.

However, we do not need the frame itself to obtain φ: it is clear that each line $\varphi(z) \in \mathbb{CP}^1$ corresponds to the line $\Gamma(\mathcal{L}_z(-P_\infty)) \in \mathbb{P}\Gamma(\mathcal{L}_z)$ where $\mathbb{P}\Gamma(\mathcal{L}_z)$ is identified with $\mathbb{CP}^1$ using θ_z. There is an invariant way of describing this identification which avoids explicit reference to θ_z and this helps us understand the periodicity conditions. To obtain this let us first consider the expression for φ in homogeneous coordinates: it can be written as

$$\varphi = [(e_1^0 f_1^z \theta_z)|_{\lambda=1}, (e_2^0 f_1^z \theta_z)|_{\lambda=1}],$$

where $f_j^z \in \Gamma(\mathcal{E}_{z,A}(R))$ is the B-module generator dual to to e_j^z. Now it is clear that if we choose some other (unitary) basis v_1, v_2 of $\Gamma(\mathcal{L})$ we obtain, up to isometry of S^2, the same map. In particular, following the remarks made earlier, we could choose this new basis such that v_1 vanishes at O_2 and v_2 vanishes at O_1. In that case

$$\varphi = [(f_1^z\theta_z)|_{O_1}, \alpha(f_1^z\theta_z)|_{O_2}],$$

where $\alpha : \mathcal{L}|O_2 \to \mathcal{L}|O_1$ is the fibre identification induced by the choice of v_1, v_2 (i.e. $\alpha(v_2|_{O_2}) = v_1|_{O_1}$).

Recall that up to scaling f_1 is determined purely by the vanishing of e_1 at P_∞. It follows that φ has periodicity $\varphi(z+\tau) = \varphi(z)$ precisely when both equations $L(z+\tau) = L(z)$ and

$$(\theta_{z+\tau}\theta_z^{-1})|_{O_2} = (\theta_{z+\tau}\theta_z^{-1})|_{O_1}$$

are satisfied. The latter condition is more simply interpreted as saying that the fibre identification $L_z|O_2 \to L_z|O_1$ given by $\theta_z|_{O_2} \mapsto \theta_z|_{O_1}$ is τ-periodic. This identication determines at each z a line bundle L'_z over X', the singular curve obtained from X by identifying O_1 with O_2 to obtain a node. Thus we have a τ-periodic map L' from $\mathbb{R}^2$ to $J(X')$, the (generalized) Jacobi variety for X'. Recall that the pullback of line bundles along $X \to X'$ induces a surjective homomorphism $\pi : J(X') \to J(X)$ whose fibre at L is $L|O_2 \otimes L^{-1}|O_1 \cong \mathbb{C}^*$. In fact $\overline{\rho_* L'} \cong L'^{-1}$ so L' takes values in a real subgroup $J_R(X')$ of $J(X')$. It can be shown that, when X is smooth of genus g, this group is a real compact torus of dimension $g+1$.

Lemma 4. *The map $L' : \mathbb{R}^2 \to J_R(X')$ defined above is linear and is uniquely determined by the property that $dL'_0(\partial/\partial z) = d\mathcal{A}'_{P_0}(\partial/\partial\zeta)$, where $\mathcal{A}'_{P_0} : X' - \{O\} \to J(X')$ is the Abel map for X' based at P_0.*

Proof. Since θ_z arises from the 1-cocycle $c(z)$ in (12) L'_z has 1-cocycle

$$c'(z) = \{(e^{z\zeta^{-1}}, X'_A, U_0), (e^{-\bar{z}\zeta}, X'_A, U_\infty)\}.$$

Now recall (from e.g. [26]) that

$$J(X') \simeq \Gamma(\Omega'_X)^t / H_1(X - \{O_1, O_2\}, \mathbb{Z}),$$

where Ω'_X is the sheaf of regular differentials on X': each such differential can be identified with a meromorphic differential on X whose only poles are simple ones at O_1 and O_2 i.e. $\Omega'_X \cong \Omega_X(O_1 + O_2)$. The Abel map for X' is defined by

$$\mathcal{A}'_{P_0} : X' - \{O\} \to J(X'); \quad P \mapsto \int_{P_0}^{P},$$

where O is the nodal point lying under O_1, O_2. To compute dL'_0 we simply repeat the computation prior to lemma 3 using $c'(z)$. Q.E.D.

Corollary 2. *The harmonic map $\varphi : \mathbb{R}^2 \to S^2$ with spectral data $X, \lambda, \mathcal{L}$ has period τ if and only if the related map $L' : \mathbb{R}^2 \to J_R(X')$ has period τ. This depends only on the data X, λ.*

In particular, if X has genus $g \leq 1$ the harmonic map is necessarily doubly periodic since $J_R(X')$ is topologically S^1 or $S^1 \times S^1$. These examples yield the Gauss maps of all Delaunay surfaces in $\mathbb{R}^3$ (i.e. the constant mean curvature surfaces of revolution) with the case $g = 0$ corresponding to the Gauss map of the cylinder. A more interesting class of singly periodic examples are the Gauss maps of the "bubbletons". The bubbletons are periodic CMC surfaces whose ends are asymptotic to the standard cylinder (see figure 2). They get their name because they correspond to soliton solutions of the sinh-Gordon equation, which governs the behaviour of the metric. As with KdV solitons, these solutions have rational nodal spectral curves. Using the theory above we can characterize these spectral curves as follows.

Proposition 3. *Let X be the rational nodal curve of arithmetic genus $g = 2r$ with equation*

$$\mu^2 = \lambda \prod_{j=1}^{r} (\lambda - a_j)^2 (1 - a_j\lambda)^2, \; a_j \in \mathbb{R}, \; 0 < a_j < 1. \tag{13}$$

Then X, λ is the spectral data for a singly periodic non-conformal harmonic map $\varphi : \mathbb{R}^2 \to S^2$ if and only if there exist positive integers $p_0, p_1, \ldots, p_r$ for which

$$a_j = \left(\frac{p_j}{p_0} \pm \sqrt{\frac{p_j^2}{p_0^2} - 1} \right)^2, \; j = 1, \ldots, r. \tag{14}$$

Proof. Let us set $\zeta = \sqrt{\lambda}$: this is a rational coordinate on X. Thus we identify X with the singularization of the Riemann sphere $\mathbb{P}_\zeta$ with the points $\pm\zeta_j$ identified, where $\zeta_j^2 = a_j$ and $\zeta_{r+j}^2 = a_j^{-1}$ for $j = 1, \ldots, r$. Notice that $\zeta_j \in \mathbb{R}$ since $a_j > 0$. X' is the further singularization obtained by additionally identifying $\pm\zeta_0$, where $\zeta_0 = 1$. We may assume the real involution is $\rho_*\zeta = \bar{\zeta}^{-1}$. A basis for $\Gamma(\Omega'_X)$ is given by

$$\omega_j = \frac{1}{2\pi i}\left(\frac{1}{\zeta - \zeta_j} - \frac{1}{\zeta + \zeta_j}\right) d\zeta, \; j = 0, \ldots, 2r,$$

Now X' is obtained from its normalisation (a smooth curve of genus $2r+1$) by shrinking half the homology generators to zero, hence $H_1(X',\mathbb{Z})$ is generated by γ_j, $j=0,\ldots,2r$ where each of these is the boundary of a small positively oriented disc containing ζ_j. It follows that $\oint_{\gamma_j}\omega_k=\delta_{jk}$. The real group $J_R(X')$ is isomorphic to

$$\{\omega\in\Gamma(\Omega'_X):\rho_*\omega=-\bar{\omega}\}^t/\{\gamma\in H_1(X',\mathbb{Z}):\rho_*\gamma\sim-\gamma\}$$

which we will write more simply as V^t/Γ. It is not hard to check that a basis for V is given by

$$v_0=\omega_0, v_j=\frac{1}{2}(\omega_j+\omega_{r+j}), v_{r+j}=\frac{i}{2}(\omega_j-\omega_{r+j}),\ j=1,\ldots,r$$

and generators for $\Gamma\subset V^t$ can be given by

$$\oint_{\gamma_0},\oint_{\gamma_j+\gamma_{r+j}},\ j=1,\ldots,r.$$

With respect to this basis for V the dual isomorphism $V^t\cong\mathbb{R}^{2r+1}$ identifies the generators for Γ with the first $r+1$ standard basis vectors for $\mathbb{R}^{2r+1}$.

The map $L':\mathbb{R}^2\to J_R(X')$ described above is covered by

$$\ell:\mathbb{R}^2\to V^t;\ \ell(z,\bar{z})=z\,res_0\zeta^{-1}-\bar{z}\,res_\infty\zeta$$

where e.g. $res_0\zeta^{-1}:V\to\mathbb{C}$ takes the residue of $\zeta^{-1}\omega$ at $\zeta=0$. In terms of the dual basis for V^t this has coordinates

$$\ell:\mathbb{R}^2\to\mathbb{R}^{2r+1};\ \ell(z,\bar{z})=zU+\bar{z}\bar{U}$$

where $U\in\mathbb{C}^{2r+1}$ has coordinates

$$U=\frac{-1}{2\pi i}(2,\ldots,\alpha_j^{-1}+\alpha_j,\ldots,i(\alpha_j^{-1}-\alpha_j),\ldots)$$

where $\alpha_j^2=a_j$. The map is periodic precisely when there exists $z\in\mathbb{C}$ for which

$$zU+\bar{z}\bar{U}=(p_0,p_1,\ldots,p_r,0\ldots,0),\ p_j\in\mathbb{Z}.$$

If we write $z=x+iy$ then these $2r+1$ equations become

$$2y=-\pi p_0,\ y(\alpha_j+\alpha_j^{-1})=-\pi p_j,\ x(\alpha_j-\alpha_j^{-1})=0,\ j=1,\ldots,r.$$

These equations have a solution for $a_j<1$ if and only if $x=0$ and

$$\alpha_j^2-\frac{2p_j}{p_0}\alpha_j+1=0.$$

Q.E.D.

Remark. The reader may wonder why we only consider $a_j \in \mathbb{R}$. The more general case of complex nodes also leads to periodic maps $\varphi : \mathbb{R}^2 \to S^2$. The conditions are that, writing $a_j = r_j^2 e^{2\theta_j}$, there must be positive integers $p_0, \ldots, p_r$ for which

$$r_j^2 - \frac{2p_j}{p_0}\cos(\theta_j) r_j + 1 = 0.$$

However, these are not the Gauss maps of periodic CMC surfaces unless $\theta_j = 0$. Indeed, it is not obvious even then that we obtain periodic CMC surfaces since none of the discussion above accounts for the extra condition that the CMC surface must also have a period when its Gauss map does. That this happens when (14) is satisfied follows from an argument I learned from Martin Kilian and Nick Schmitt, which exploits the dressing construction. Unfortunately to describe this closing argument would take us too far afield, although I will say something about the dressing construction in the next section.

1.8. Two reconstructions of the harmonic map: Symes' method and dressing the vacuum.

I know of three approaches to reconstructing the harmonic map from its spectral data. The first of these, which I will describe in a more general context later, boils down to writing the map down in terms of the θ-functions for X' (cf. [1]). The other two methods use a loop group and require one to be able to perform a certain loop group factorization. Until recently this had only theoretical interest, but with the advent of Nick Schmitt's CMCLab software it is now possible to perform explicit calculations involving the (approximate) factorization, so I want to take this opportunity to explain how to reproduce the map $\varphi : \mathbb{R}^2 \to S^2$ (and hence its associated family of CMC surfaces) from its spectral data. Before I begin we must recall some fundamentals about the application of loop groups to the construction of harmonic maps.

First, set $G^{\mathbb{C}} = SL_2(\mathbb{C})$ and let G denote its compact real form SU_2. For $\epsilon \in \mathbb{R}^+$ with $0 < \epsilon < 1$ we let C be the union of circles $\{\zeta : |\zeta| = \epsilon \text{ or } |\zeta| = \epsilon^{-1}\}$ on the Riemann sphere $\mathbb{C} \cup \{\infty\}$ and consider it as the common boundary of the two open sets

$$E = \{\zeta : \epsilon < |\zeta| < \epsilon^{-1}\},\ I = \{\zeta : |\zeta| < \epsilon \text{ or } |\zeta| > \epsilon^{-1}\}.$$

We will work with the loop group (of "twisted loops")

$$\Lambda_C G = \{C^\omega \text{ maps } g : C \to G^C | g_{\bar\zeta^{-1}} = g_\zeta^\dagger,\ \nu(g_\zeta) = g_{-\zeta}\}.$$

This loop group contains in particular $\Lambda_E G$, the subgroup of those $g \in \Lambda_C G$ which extend holomorphically into E, and $\Lambda_I G$, the subgroup of those $g \in \Lambda_C G$ which extend holomorphically into I such that g_0 is upper triangular with positive real diagonal entries. It is well known (see [16]) that every $g \in \Lambda_C G$ factorizes uniquely into $g_E g_I$ where $g_E \in \Lambda_E G$ and $g_I \in \Lambda_I G$: this is sometimes called the Iwasawa decomposition for $\Lambda_C G$.

The relevance of these groups to our harmonic maps can be encapsulated in the following theorem. First, notice that the simplest non-conformal map $\varphi : \mathbb{R}^2 \to S^2$, which has been dubbed the "vacuum solution", maps onto a great circle and is framed by the homomorphism

$$F^{(0)} : \mathbb{R}^2 \to SU_2;\ F^{(0)} = \exp(zA - \bar{z}A);\ A = \begin{pmatrix} 0 & 1 \\ 1 & 0 \end{pmatrix}.$$

This has extended frame

$$F^{(0)}_\zeta : \mathbb{R}^2 \to \Lambda_E G;\ F^{(0)}_\zeta = \exp(\zeta^{-1} z A - \zeta \bar{z} A).$$

Theorem 2. [6, 7] *Let $\varphi : \mathbb{R}^2 \to S^2$ be a non-conformal harmonic map of finite type with polynomial Killing field $\xi_\zeta(z)$, in the form (4).*

(1) *φ has an extended frame given by*

$$F_\zeta = \exp(z\zeta^{2p}\xi_\zeta(0))_E.$$

This is "Symes' formula" [6].

(2) *For some $0 < \epsilon < 1$ there exists $g_\zeta \in \Lambda_I G$ so that φ has an extended frame given by*

$$F_\zeta = (g_\zeta F^{(0)}_\zeta)_E.$$

This is "dressing the vacuum solution" [7].

Since in both formulae the frame satisfies the same Maurer-Cartan equations with $F_\zeta(0) = I$, each method gives the same extended frame. Now I will describe how to compute the polynomial Killing field ξ_ζ and the dressing matrix g_ζ corresponding to the spectral data $X, \lambda, \mathcal{L}$ for a particularly amenable choice of $\mathcal{L}$.

Proposition 4. *Let X, λ correspond to the curve with affine equation*

$$y^2 = \lambda \prod_{j=1}^{g} (\lambda - a_j)(1 - \bar{a}_j \lambda);\ 0 < |a_j| < 1.$$

and let $\mathcal{L} = \mathcal{O}_X(R_+)$ *where* R_+ *is the divisor* $P_0 + \sum_{j=1}^{g} R_j$ *for* $\lambda(R_j) = a_j$. *Then the non-conformal map* $\varphi : \mathbb{R}^2 \to S^2$ *with spectral data* $X, \lambda, \mathcal{L}$ *arises from:*

(1) *Symes' formula using* $\xi_\zeta(0) = \eta_\zeta - \eta^\dagger_{\bar\zeta^{-1}}$ *where*

$$\eta_\zeta = \begin{pmatrix} 0 & \zeta\prod_{j=1}^{g}(1-\bar a_j\zeta^2) \\ \zeta\prod_{j=1}^{g}(\zeta^2 - a_j) & 0 \end{pmatrix}; \tag{15}$$

(2) *dressing the vacuum solution by*

$$g_\zeta = \begin{pmatrix} h^{-1/4} & 0 \\ 0 & h^{1/4} \end{pmatrix}; \quad h = \prod_{j=1}^{g}(\frac{\zeta^2 - a_j}{1-\bar a_j\zeta^2}). \tag{16}$$

Proof. 1. Given an orthonormal basis e_1, e_2 for $\Gamma(\mathcal{L})$ we obtain a B-module morphism

$$K : \{f \in \mathbb{C}[X_A] : \overline{\rho^* f} = -f\} \to \{\xi_\zeta(z) : d\xi = [\xi, \alpha],\ \xi_{\bar\zeta^{-1}} = -\xi^\dagger_\zeta\}$$

in which each ξ is algebraic (indeed, a Laurent polynomial) in λ. In fact this map is an isomorphism for real algebraic ξ [19]. It arises from the commutative diagram

$$\begin{array}{ccc} \hat\Gamma(\mathcal{E}_{z,A}(R)) & \stackrel{\times f}{\to} & \hat\Gamma(\mathcal{E}_{z,A}(R)) \\ \epsilon_z \downarrow & & \downarrow \epsilon_z \\ \hat B \otimes \mathbb{C}^2 & \stackrel{\xi(f)}{\to} & \hat B \otimes \mathbb{C}^2 \end{array}$$

This gives $\xi_\lambda(z) = K(f)$ for each z, where we recall from (5) that $\xi_\lambda = \mathrm{Ad}\kappa \cdot \xi_\zeta$. Since $\theta f \theta^{-1} = f$ it follows, by combining this diagram and the diagram (11), that $\xi(0) = F\xi(z)F^{-1}$ whence $d\xi = [\xi, \alpha]$. For the purposes of Symes' formula we want to compute $K(f)$ at $z = 0$ for $f = y - \overline{\rho_* y}$. Since

$$K(\overline{\rho_* y}) = \overline{\rho_* K(y)}^t$$

it suffices to compute $K(y)$ at $z = 0$. A simple computation shows that with respect to the trace inner product (9) $\Gamma(\mathcal{O}_X(R_+))$ has an orthonormal basis given by

$$e_1 = \frac{y}{\lambda\prod_{j=1}^{g}(\lambda - a_j)}, \quad e_2 = 1. \tag{17}$$

Here we are identifying holomorphic sections of $\mathcal{L}$ with rational functions on X whose divisor of poles is no worse than R_+. Notice that e_1

generates $\Gamma(\mathcal{L}(-P_\infty))$ while e_2 generates $\Gamma(\mathcal{L}(-P_0))$. Now $K(y)$ is the matrix

$$\begin{pmatrix} \alpha & \beta \\ \gamma & \delta \end{pmatrix}$$

where

$$ye_1 = \alpha e_1 + \beta e_2, \quad ye_2 = \gamma e_1 + \delta e_2$$

so that at $z = 0$

$$K(y) = \begin{pmatrix} 0 & \prod_{j=1}^{g}(1 - \bar{a}_j\lambda) \\ \lambda\prod_{j=1}^{g}(\lambda - a_j) & 0 \end{pmatrix}.$$

Finally, let η_ζ be the twisted loop $\mathrm{Ad}\kappa^{-1} \cdot K(y)$ to obtain the formula (15).

2. Let us consider the geometric meaning of the equation (16). If we write these loops in their untwisted form, then $g_\lambda F_\lambda^{(0)} = F_\lambda b_\lambda$ where F_λ extends holomorphically to an annulus on the λ-sphere (which we will call E despite the abuse of notation) and b_λ extends holomorphically to a pair of discs about $\lambda = 0, \infty$ (which we will call I) and is upper triangular at $\lambda = 0$. A little thought shows that the columns of F^{-1t} represent $e_1^z\theta^{-1}, e_2^z\theta^{-1}$, thought of as sections of the rank two vector bundle $\lambda_*\mathcal{L}$ over E, with respect to the global frame e_1^0, e_2^0. Let ϕ_E denote the trivialisation of $\lambda_*\mathcal{L}$ determined by this global frame, then the equation $g_\lambda^{-1t}b_{\lambda,z=0}^{-1t} = I$ expresses the fact that there is some local trivialisation ϕ_I for $\lambda_*\mathcal{L}$ over I for which the transition relation on $E \cap I$ is

$$g^{-1t}\phi_I = \phi_E.$$

Therefore g_λ^t is the matrix whose columns are $\phi_I(e_1^0), \phi_I(e_2^0)$. Now we recall from [15] that ϕ_I is obtained by direct image from a trivialisation of $\mathcal{L}$ over $\lambda^{-1}(I)$ in the following way. Let s_I be a non-vanishing holomorphic section of $\mathcal{L}$ over $\lambda^{-1}(I)$. By definition $\Gamma(I, \lambda_*\mathcal{L}) = \Gamma(\lambda^{-1}, \mathcal{L})$ and s_I induces the trivialisation

$$\phi_I : \Gamma(I, \lambda_*\mathcal{L}) \to \mathrm{Hol}(I, \mathbb{C}^2); \quad s \mapsto (s_1, s_2)$$

where $s/s_I = s_1(\zeta^2) + \zeta s_2(\zeta^2)$. Any trivialisation ϕ_I obtained this way and which gives $\det(g_\lambda) = 1$ will provide a suitable matrix g_λ (the freedom here is right multiplication of g_λ by any element of $\Lambda_I G$ which commutes with $F_\lambda^{(0)}$ for all z). To calculate g_λ we let e_1, e_2 be the basis (17) and initially take $s_I = e_1$: this is appropriate since as a function it has a simple pole at P_0 and does not vanish at P_∞, therefore it represents a non-vanishing section of $\mathcal{L}$ over I provided I is small enough.

Now we write

$$e_1/e_1 = 1 + \zeta.0, \quad e_2/e_1 = 0 + \zeta.\prod_{j=1}^{g}(\frac{\zeta^2 - a_j}{1 - \bar{a}_j\zeta^2})^{1/2}.$$

However, this choice of s_I does not give $\det(g_\lambda) = 1$, so it remains to rescale s_I by the appropriate non-vanishing function to obtain (16). Q.E.D.

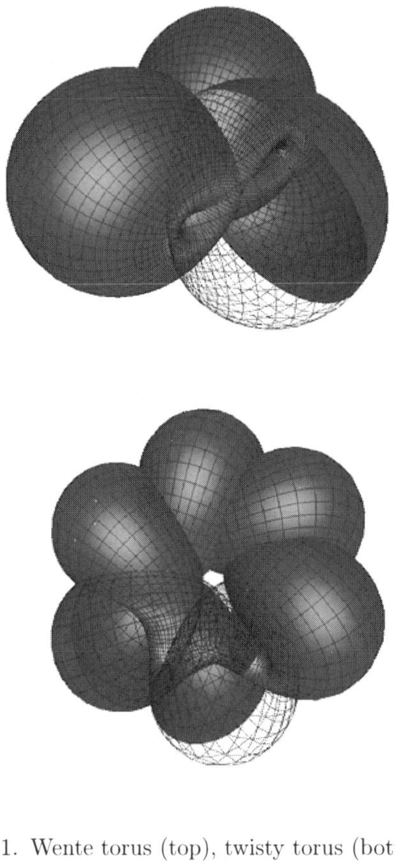

Fig. 1. Wente torus (top), twisty torus (bottom).

Remark 1. For simplicity define $\bar{\eta}_\zeta = -\eta^\dagger_{\bar{\zeta}^{-1}}$. It suffices to use $\bar{\eta}_\zeta$ instead of $\xi_\zeta(0)$ in Symes' formula, since $[\eta, \bar{\eta}] = 0$ and $\exp(z\zeta^{2g}\eta_\zeta)_E = I$ (since η_ζ is polynomial in ζ), therefore $\exp(z\zeta^{2g}\xi_\zeta(0))_E = \exp(z\zeta^{2g}\bar{\eta}_\zeta)_E$.

Moreover, by combining the extended frame with the Sym-Bobenko formula [2, 14] we can produce CMC tori once we know a choice of branch points for the spectral curve which satisfies the double periodicity condition (not just the periodicity condition above, which only makes the Gauss map periodic, but the full CMC periodicity condition described in [1]). The following examples for a genus two curve are due to Matthias Heil (private communication):

$$a_1 = 0.1413 + 0.1018i,\ a_2 = 0.1413 - 0.1018i,\ \text{(Wente torus)};$$
$$a_1 = 0.124 + 0.1485i,\ a_2 = 0.4387 - 0.071i\ \text{(twisty torus)}.$$

The corresponding CMC tori are drawn in figure 1.

Remark 2. In fact we can use the dressing construction to produce all harmonic maps with spectral data X, λ. For even though the dressing matrix (16) corresponds to the line bundle $\mathcal{O}_X(R_+)$ every other line bundle satisfying the reality condition is of the form $\mathcal{O}_X(R_+) \otimes L$ where $L \in J_R(X)$. It was shown in [17] that the full family of these is swept out by the "higher flows" described in [7]. That means an extended frame for the harmonic map with data $X, \lambda, \mathcal{O}_X(R_+) \otimes L$ is given by dressing the vacuum by

$$g_\zeta \exp\Big(\sum_{j=1}^{\infty}(t_j\zeta^j A^j - \bar{t}_j\zeta^{-j}A^{-j})\Big)$$

for some sequence $t_j \in \mathbb{C}$. Moreover, for a map of finite type only finitely many of the higher flows are independent, so there is no need for an infinite sum here. It can be shown that it suffices to have only $t_1, t_3, \ldots, t_{2g-1}$ taking any values and all other parameters zero: the first flow t_1 is just a z-translation of the surface domain.

Remark 3. By combining proposition 3 with proposition 4 we can compute the one and two bubbletons in figure 2. These have respectively $r = 1, g = 2$ and $r = 2, g = 4$. Using the previous remark we obtain a real g-parameter family of deformations of these surfaces. Each bubble can be moved relative to any other (or the cylinder) by a translation along the cylinder or a rotation about its circumference. Thus each bubble contributes two real parameters: there are r bubbles altogether. This demonstrates that $J_R(X') \cong (\mathbb{R} \times S^1)^r$. It is interesting to note that we can also think of the bubbletons as being constructed by dressing the vacuum by a rational loop on the ζ-sphere. We can re-scale the matrix g_ζ in (16) so that for a nodal curve (13) it becomes

$$\prod_{j=1}^{r}\begin{pmatrix} \bar{a}_j\zeta^2 - 1 & 0 \\ 0 & \zeta^2 - a_j \end{pmatrix}, \quad r = g/2. \tag{18}$$

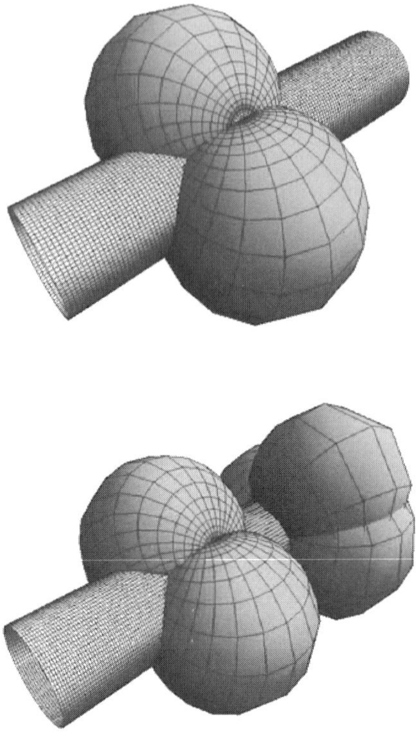

Fig. 2. One bubbleton and two bubbleton.

This dressing matrix produces the same surface and resembles a product of Bäcklund transforms in the sense of [28]. Although proposition 3 only proves that the Gauss map of the CMC surface is periodic (given the conditions (14) on each a_j) it turns out that the CMC surface itself is periodic. This can be shown by examining the effect on the monodromy matrix of $F_\zeta^{(0)}$ of dressing by any factor in the product (18). This approach was explained to me by Martin Kilian and Nick Schmitt. Their approach also explains the geometric significance of the positive integers $p_0, \ldots, p_n$ appearing in (14). The integer p_0 determines the number of times the cylindrical end of the bubbleton wraps around itself, while p_j is the number of "lobes" the j-th bubble possesses.

§2. Harmonic and pluri-harmonic maps into $Gr_k(\mathbb{C}^{n+1})$.

Let $k \leq (n+1)/2$. Here I will briefly recount the theory given in [20] for constructing pluri-harmonic maps $\varphi : \mathbb{R}^{2k} \to Gr_k(\mathbb{C}^{n+1})$ which generalizes the construction given above (recall that a map is pluri-harmonic if it is harmonic on any holomorphic curve: here $\mathbb{R}^{2k}$ is given the usual complex structure). At the end I will illustrate this with the example $X \cong \mathbb{P}^1$.

We assume that the spectral data here consists of a smooth compact Riemann surface X (of genus g) with real involution ρ together with a degree $n+1$ function λ on X and a line bundle $\mathcal{L}$ over X. We require: $\overline{\rho_*\lambda} = \lambda^{-1}$; the ramification divisor R of λ has no support over $|\lambda| = 1$; and ρ fixes every point over $|\lambda| = 1$. In that case $R = R_+ + \rho_* R_+$, where R_+ is the divisor of ramification over $|\lambda| > 1$. We can choose $\mathcal{L}$ to satisfy the reality condition $\overline{\rho_*\mathcal{L}} \cong \mathcal{L}^t(R)$ by taking any element of the compact real connected g-dimensional torus

$$\mathcal{N} = \{\mathcal{O}_X(R_+) \otimes L :\ L \in J_R\}$$

where J_R is the identity component of $\{L \in J(X) :\ L \cong \overline{\rho_* L}^{-1}\}$. It can be shown that for any such bundle $\lambda_*\mathcal{L}$ is a trivial rank $n+1$ bundle so $\dim(\Gamma(\mathcal{L})) = n+1$. Further, the trace pairing equips $\Gamma(\mathcal{L})$ with a Hermitian inner product.

As before, the geometry of the construction is best understood by working with the singularisation X' of X obtained by identifying the $n+1$ points $O_1, \ldots, O_{n+1}$ lying over $\lambda = 1$ together to obtain a nodal singularity O on X'. A line bundle $\mathcal{L}'$ over X' is best thought of as a line bundle $\mathcal{L}$ over X equipped with a linear identification of the fibres over $O_1, \ldots, O_{n+1}$: we can think of this as assigning a non-zero element to each stalk $\mathcal{L}|O_j$. In $Pic(X')$ (the algebraic group of all holomorphic line bundles over X') we distinguish the real variety

$$\mathcal{N}' = \{\mathcal{O}_{X'}(R_+) \otimes L' : L' \in J'_R\},$$

where $J'_R = \{L' \in J(X') : L' \cong \overline{\rho_* L'}^{-1}\}$. Let $\pi : Pic(X') \to Pic(X)$ be the natural epimorphism for which $\pi(\mathcal{L}') = \mathcal{L}$. An element of $\Gamma(\mathcal{L}')$ is a global section of $\mathcal{L}$ which "takes the same value" at each O_j using the fibre identification with which $\mathcal{L}'$ is equipped. For $\mathcal{L}' \in \mathcal{N}'$, since $\lambda_*\mathcal{L}$ is trivial, there is no non-zero global section of $\mathcal{L}$ which vanishes at every O_j, therefore $\dim(\Gamma(\mathcal{L}')) = 1$. Thus any non-zero global section of $\mathcal{L}'$ gives us a convenient representation for the fibre identification carried by $\mathcal{L}'$.

Over $\mathcal{N}'$ there exists a natural rank $n+1$ bundle E' whose fibre at $\mathcal{L}'$ is $\Gamma(\mathcal{L})$. For any k the Grassmann bundle $Gr_k(E')$ possesses a canonical trivialisation given pointwise as follows. Let

$$[e_1 \wedge \ldots \wedge e_k] \in Gr_k(E').$$

By taking any non-zero $s_{\mathcal{L}} \in \Gamma(\mathcal{L}')$ we can identify

$$e_j \mapsto v_j = (\frac{e_j|O_1}{s_{\mathcal{L}}|O_1}, \ldots, \frac{e_j|O_{n+1}}{s_{\mathcal{L}}|O_{n+1}}) \in \mathbb{C}^{n+1}$$

and this is projectively dependent only on $\mathcal{L}'$. Thus we have a natural map

$$[e_1 \wedge \ldots \wedge e_k] \mapsto [v_1 \wedge \ldots \wedge v_k] \in Gr_k(\mathbb{C}^{n+1}).$$

The relevance of this is that by taking a suitable section of $Gr_k(E')$ and applying this trivialisation we obtain a map $J'_R \cong \mathcal{N}' \to Gr_k(\mathbb{C}^{n+1})$ whose restriction to suitable subgroups of J'_R is (pluri)-harmonic. This result is true for any choice of isomorphism $J'_R \cong \mathcal{N}'$, so in fact we obtain not just one map but a family of them — these correspond to the deformations made available by the higher flows discussed earlier.

Now I must explain which section of $Gr_k(E')$ yields (pluri)-harmonic maps. Although we could discuss the construction of maps of any isotropy order we will stick with the simplest case of lowest isotropy order i.e. non-conformal maps. For this we take λ to have (at least) k double zeroes $P_1, \ldots, P_k$. Consequently the divisor of λ has the form

$$(\lambda) = 2P_1 + \ldots + 2P_k + E_0 - 2Q_1 - \ldots 2Q_k - E_\infty$$

where E_0, E_∞ are positive divisors of degree $n+1-2k$. Let D_∞ denote the positive divisor $Q_1 + \ldots Q_k + E_\infty$ of degree $n+1-k$, then D_∞ gives us a section of $Gr_k(E')$ by assigning to each $\mathcal{L}'$ the k-plane $\Gamma(\mathcal{L}(-D_\infty))$. Thus by our canonical trivialisation we have map

$$\psi : J'_R \cong \mathcal{N}' \to Gr_k(\mathbb{C}^{n+1}).$$

Now let $\gamma : \mathbb{R}^{2k} \to J'_R$ be the real homomorphism uniquely determined up to scalings by:

$$\partial\gamma/\partial z_j = \partial\mathcal{A}'_{P_j}/\partial\zeta_{P_j}$$

where $z_1, \ldots, z_k$ denote complex coordinates on $\mathbb{R}^{2k}$, ζ_{P_j} is a local coordinate about P_j and $\mathcal{A}'_{P_j}$ denotes the Abel map for X' with base point P_j.

Theorem 3. [20] *The map $\varphi = \psi \circ \gamma : \mathbb{R}^{2k} \to Gr_k(\mathbb{C}^{n+1})$ given above is pluri-harmonic. Indeed, the harmonic map obtained by restriction of φ to the complex line with tangent $\sum a_j \partial/\partial z_j$ is harmonic: it is also nowhere conformal iff $\sum a_j^2 \neq 0$.*

Remark. According to [17, 18] this theorem accounts for all non-conformal harmonic maps $\varphi : \mathbb{R}^2 \to \mathbb{CP}^n$ of semisimple finite type (and therefore all non-conformal tori). Indeed I believe it will account for all maps of semisimple finite type into $Gr_k(\mathbb{C}^{n+1})$ using a similar argument. The main unanswered question is to what extent the non-conformal (or more generally, non-isotropic) harmonic tori in $Gr_k(\mathbb{C}^{n+1})$ are accounted for by the tori of semisimple finite type. Some progress has been made in this direction (see [29]) but the problem is not yet settled.

2.1. Explicit formulae in terms of Riemann θ-functions.

In the construction above there is, up to scalings, a natural basis $e_1, \ldots, e_k$ for the k-plane $\Gamma(\mathcal{L}(-D_\infty)$. For each $j = 1, \ldots, k$ let D_j be the positive divisor $D_\infty + \sum_{k\neq j} Q_j$, which has degree n, and notice that for any j the divisor of poles of λ is $D_j + Q_j$. Since $\lambda_* \mathcal{L}$ is trivial the subspace $\Gamma(\mathcal{L}(-D_j) \subset \Gamma(\mathcal{L}(-D_\infty))$ is one dimensional and $\mathcal{L}(-D_j)$ is non-speciali of degree g. This means we can obtain a non-zero section of it using Riemann's θ-function. To obtain a formula for ψ we then have to understand the behaviour of the fibre identifications. It turns out that these can be incorporated by pulling back the θ-line bundle over $J(X)$ to $J(X')$ using π. An explicit formula for ψ is then obtained as follows. Throughout this discussion we take $\mathcal{L} = \mathcal{O}_X(R_+)$: any other choice of $\mathcal{L}$ simply amounts to a translation in the argument of the θ-function with no loss of generality.

We know that we can make identifications

$$J(X') \cong H^0(\Omega_X(\mathfrak{o}))^t / H_1(X \setminus \mathfrak{o}, \mathbb{Z}) \simeq \mathbb{C}^{g+n}/\Lambda', \tag{19}$$

where $\Omega_X(\mathfrak{o})$ is the sheaf of mermorphic differentials on X with divisor of poles no worse than $\mathfrak{o} = O_1 + \ldots + O_{n+1}$ and Λ' is a lattice on $2g+n$ generators. We choose coordinates so that $\pi : J(X') \to J(X)$ is covered by the map

$$\pi : \mathbb{C}^{g+n} \to \mathbb{C}^g; \quad \tilde{W} = (w_1, \ldots, w_{g+n}) \mapsto W = (w_1, \ldots, w_g).$$

Now let us define $\theta_0(\tilde{W}) = \theta(W)$ and for $j = 1, \ldots, n$ define

$$\theta_j(\tilde{W}) = \exp(2\pi i w_{g+j})\theta(W + \mathcal{A}(O_{j+1} - O_1)),$$

where θ is the classical Riemann θ-function on $\mathbb{C}^g$ corresponding to the induced isomorphism $J(X) \simeq \mathbb{C}^g/\pi(\Lambda')$. Each of $\theta_0, \ldots, \theta_n$ represents a global holomorphic section of the pullback by π of the θ-line bundle over $J(X)$ [20].

For $l = 1, \ldots, k$ let $\tilde{D}_l$ be the unique positive divisor (of degree g) in the linear system of $\mathcal{L}(-D_l)$ and let $\kappa_l \in \mathbb{C}^g$ be the appropriate translation for which $\theta(\mathcal{A}(P) + \kappa_l)$ has divisor of zeroes $\tilde{D}_l$. Finally, let f_l be a rational function on X with divisor $(f_l) = R_+ - D_l - \tilde{D}_l$ so that $f_l(P)\theta(\mathcal{A}(P) + \kappa_l)$ has divisor $R_+ - D_l$.

Proposition 5. [20] *Let* $v_l : \mathbb{C}^{g+n} \to \mathbb{C}^{n+1}$ *be defined by*

$$v_l(\tilde{W}) = (f_l(O_1)\theta_0(\tilde{W} + \kappa_l), \ldots, f_l(O_{n+1})\theta_n(\tilde{W} + \kappa_l))$$

Then, taking the base point $\mathcal{O}_{X'}(R_+)$ *on* $\mathcal{N}'$ *for the identification* $J'_R \cong \mathcal{N}'$, *the map* $\psi : J'_R \to Gr_k(\mathbb{C}^{n+1})$ *above is explicitly given by the* Λ'-*periodic map*

$$\psi(\tilde{W}) = [v_1(\tilde{W}) \wedge \ldots \wedge v_k(\tilde{W})].$$

An explicit formula for the function f can be obtained using Fay's prime form (see e.g. [23]). It remains to combine this with the real homomorphism $\gamma : \mathbb{R}^{2k} \to \mathbb{C}^{g+n}/\Lambda'$ which we have essentially computed earlier (cf. [20]). For illustration I will do these calculations explicitly for $X \cong \mathbb{P}^1$ in the next section.

2.2. Example: X is the Riemann sphere.

Let ζ be a rational parameter on $X \cong \mathbb{P}^1$ and define the real involution to be $\rho_*\zeta = \bar{\zeta}^{-1}$, then to satisfy all our conditions λ must be of the form

$$\lambda = \alpha \prod_{j=1}^{k} \frac{(\zeta - P_j)^2}{(\zeta - \bar{P}_j^{-1})^2} \prod_{i=1}^{n+1-2k} \frac{(\zeta - E_j)}{(\zeta - \bar{E}_j^{-1})}, \tag{20}$$

where the points $P_1, \ldots, P_k, E_1, \ldots, E_{n+1-2k}$ all lie inside $|\zeta| < 1$ (cf. [27]). The constant α is chosen so that $|\lambda| = 1$ over $|\zeta| = 1$.

First we construct the homomorphism $\gamma : \mathbb{R}^{2k} \to J'_R$. To fix the isomorphism (19) we choose the basis $\omega_1, \ldots, \omega_n$ of $H^0(\Omega(\mathfrak{o}))$ given by

$$\omega_m = \frac{1}{2\pi i}\left(\frac{1}{\zeta - O_{m+1}} - \frac{1}{\zeta - O_1}\right)d\zeta, \quad m = 1, \ldots, n.$$

Let $a_m \in H_1(X \setminus \mathfrak{o}, \mathbb{Z})$ be the class of a positively oriented cycle about O_{m+1} only, so that $\oint_{a_l} \omega_m = \delta_{lm}$. With these bases we have

$$J(X') \cong \mathbb{C}^n/\mathbb{Z}^n \overset{\exp(2\pi i \cdot)}{\to} (\mathbb{C}^*)^n.$$

Take $\zeta_{P_j} = \zeta - P_j$ for the local parameter about P_j and recall from earlier that as an element of $H^0(\Omega(\mathfrak{o}))^t \cong T_0 J(X')$

$$\frac{\partial \mathcal{A}'_{P_j}}{\partial \zeta_{P_j}} : \omega_m \mapsto res_{P_j} \zeta_{P_j}^{-1} \omega_m.$$

In our coordinates this is the vector $\frac{1}{2\pi i} U_j$ where $U_j \in \mathbb{C}^n$ has m-th coordinate

$$U_{jm} = \frac{1}{P_j - O_m} - \frac{1}{P_j - O_1}.$$

The map $\gamma : \mathbb{R}^{2k} \to (\mathbb{C}^*)^n$ is given by

$$\gamma(z_1, \ldots, z_k) = \exp(\sum_{j=1}^{k} (U_j z_j - \bar{U}_j \bar{z}_j)).$$

Now to apply proposition 5 we notice that since $J(X)$ is the trivial group we can take $\theta \equiv 1$. So for $\tilde{W} = (w_1, \ldots, w_n)$ we have simply

$$\theta_0(\tilde{W}) = 1,\ \theta_1(\tilde{W}) = \exp(2\pi i w_1), \ldots,\ \theta_n(\tilde{W}) = \exp(2\pi i w_n).$$

Finally, we need the divisors

$$D_l = 2Q_1 + \ldots + Q_l + \ldots 2Q_k + E_\infty, \quad l = 1, \ldots, k,$$

where $Q_j = \bar{P}_j^{-1}$ and $E_\infty = \bar{E}_1^{-1} + \ldots + \bar{E}_{n+1-2k}^{-1}$. Let f_l be any rational function with divisor $R_+ - D_l$ and define $v_l : \mathbb{C}^k \to \mathbb{C}^{n+1}$ by

$$v_l(z_1, \ldots, z_k) = (f_l(O_1), f_l(O_2)\gamma_1, \ldots, f_l(O_{n+1})\gamma_n) \tag{21}$$

where $\gamma_m = \exp(\sum_{j=1}^{k} (z_j U_{jm} - \bar{z}_j \bar{U}_{jm}))$.

Proposition 6. [20] *The pluri-harmonic map* $\varphi : \mathbb{R}^{2k} \to Gr_k(\mathbb{C}^{n+1})$ *with spectral data* $X \cong \mathbb{P}^1$ *and* λ *given by (20) is given by*

$$\varphi(z_1, \ldots, z_k) = [v_1 \wedge \ldots \wedge v_k].$$

This map is totally equivariant i.e. it can be framed by a homomorphism $\mathbb{R}^{2k} \to U_{n+1}$.

By a result of Kenmotsu [13] (see also [3]) the minimal (i.e. conformal harmonic) totally equivariant maps $\mathbb{R}^2 \to \mathbb{CP}^n$ include those minimal totally real maps which are isometric for the flat metric on $\mathbb{R}^2$. A study of their periodicity can be found in [12]. To pass from non-conformal to conformal maps in our construction (in the case $k = 1$ i.e. $\mathbb{CP}^n$) one insists that λ has a zero of degree 3 at P_1. In particular, this requires $n \geq 2$.

Remark. There is a geometric interpretation behind the form of v_l. Suppose $\varphi : \mathbb{R}^{2k} \to Gr_k(\mathbb{C}^{n+1})$ is totally equivariant with frame

$$F = \exp(z \cdot A - \bar{z} \cdot A^\dagger), \quad z \cdot A = \sum_{j=1}^{k} z_j A_j,$$

where $A_1, \ldots, A_k \in \mathfrak{gl}_{n+1}(\mathbb{C})$ are mutually commuting normal matrices. We will assume φ is based so that $\varphi(0) = [e_1 \wedge \ldots \wedge e_k]$ where the e_j are the standard basis vectors for $\mathbb{C}^{n+1}$. The matrices A_j and their Hermitian transposes may be simultaneously diagonalized by a unitary matrix: $MA_jM^{-1} = D_j$ where M is unitary and each D_j is diagonal. Therefore

$$\begin{aligned} M \circ \varphi &= MFM^{-1}M \circ [e_1 \wedge \ldots \wedge e_k] \\ &= \exp(z \cdot D - \bar{z} \cdot \bar{D}) \circ [u_1 \wedge \ldots \wedge u_k] \end{aligned}$$

where $u_1, \ldots, u_k$ are the first k columns of M. Thus $M \circ \varphi = [v_1 \wedge \ldots \wedge v_k]$ where $v_l = \exp(z \cdot D - \bar{z} \cdot \bar{D}) \circ u_l$. Notice that this is essentially the form of the map we derived above, using

$$D_j = \mathrm{diag}(1, U_{j1}, \ldots, U_{jn}), \ j = 1, \ldots, k.$$

The v_l appearing in (21) span the same k-plane as these but are not necessarily orthonormal.

References

[1] A. I. Bobenko, All constant mean curvature tori in R^3, S^3 and H^3 in terms of theta-functions, Math. Ann., **90** (1991), 209–245.

[2] A. I. Bobenko, Constant mean curvature surfaces and integrable equations, Uspekhi Mat. Nauk, **46** (1991), 3–42; Russian Math. Surv., **46** (1991), 1–45.

[3] J. Bolton and L. Woodward, Minimal surfaces in $\mathbb{CP}^n$ of constant curvature and Kähler angle, Math. Proc. Camb. Phil. Soc., **112** (1992), 287–296.

[4] F. E. Burstall, Harmonic tori in spheres and complex projective spaces, J. Reine. Angew. Math., **469** (1995), 149–177.

[5] F. E. Burstall, D. Ferus, F. Pedit and U. Pinkall, Harmonic tori in symmetric spaces and commuting Hamiltonian systems on loop algebras, Ann. of Math. (2), **138** (1993), 173–212.

[6] F. E. Burstall and F. Pedit, Harmonic maps via Adler-Kostant-Symes theory, In: Harmonic maps and integrable systems, (eds. A. P. Fordy and J. C. Wood), Aspects Math., **E23**, Vieweg, 1994.

[7] F. E. Burstall and F. Pedit, Dressing orbits of harmonic maps, Duke Math. J., **80** (1995), 353–382.

[8] N. Ercolani, H. Knörrer and E. Trubowitz, Hyperelliptic curves that generate constant mean curvature tori in $\mathbb{R}^3$, In: Integrable Systems, Luminy, 1991, Progr. Math., **115**, Birkhäuser, Boston, 1993, pp. 81–114.

[9] D. Ferus, F. Pedit, U. Pinkall and I. Sterling, Minimal tori in S^4, J. Reine Angew. Math., **429** (1992), 1–47.

[10] N. J. Hitchin, Harmonic maps from a 2-torus to the 3-sphere, J. Differential Geom., **31** (1990), 627–710.

[11] C. Jaggy, On the classification of constant mean curvature tori in $\mathbb{R}^3$, Comment. Math. Helv., **69** (1994), 640–658.

[12] G. Jensen and R. Liao, Families of flat minimal tori in $\mathbb{CP}^n$, J. Differential. Geom., **42** (1995), 113–132.

[13] K. Kenmotsu, On minimal immersions of $\mathbb{R}^2$ into $\mathbb{P}^n(\mathbb{C})$, J. Math. Soc. Japan, **37** (1985), 665–682.

[14] M. Kilian, I. McIntosh and N. Schmitt, New constant mean curvature surfaces, Experiment. Math., **9** (2000), 595–611.

[15] I. McIntosh, Global solutions of the elliptic 2D periodic Toda lattice, Nonlinearity, **7** (1994), 85–108.

[16] I. McIntosh, Infinite dimensional Lie groups and the two-dimensional Toda lattice, In: Harmonic maps and integrable systems, (eds. A. P. Fordy and J. C. Wood), Aspects Math., **E23**, Vieweg, 1994.

[17] I. McIntosh, A construction of all non-isotropic harmonic tori in complex projective space, Int. J. Math., **6** (1995), 831–879.

[18] I. McIntosh, Two remarks on the construction of harmonic tori in $\mathbf{CP}^n$, Int. J. Math., **7** (1996), 515–520.

[19] I. McIntosh, On the existence of superconformal 2-tori and doubly periodic affine Toda fields, J. Geom. Phys., **24** (1998), 223–243.

[20] I. McIntosh, Harmonic tori and generalised Jacobi varieties, Comm. Anal. Geom., **9** (2001), 423–449.

[21] M. Melko and I. Sterling, Integrable systems, harmonic maps and the classical theory of surfaces, In: Harmonic maps and integrable systems, (eds. A. P. Fordy and J. C. Wood), Aspects Math., **E23**, Vieweg, 1994.

[22] D. Mumford, An algebro-geometric construction of commuting operators and of solutions to the Toda lattice equation, Kortweg de Vries equation and related non-linear equations, Intl. Symp. on Algebraic Geometry, Kyoto, 1977, 115–153.

[23] D. Mumford, Tata lectures on theta II, Birkhäuser, Boston, 1983.
[24] U. Pinkall and I. Sterling, On the classification of constant mean curvature tori, Ann. of Math., **130** (1989), 407–451.
[25] A. Reimann and M. Semenov-Tian Shansky, Reduction of hamiltonian systems, affine Lie algebras and Lax equations, II, Invent. Math., **63** (1981), 423–432.
[26] J.-P. Serre, Algebraic groups and class fields, Grad. Texts in Math., **117**, Springer-Verlag, 1988.
[27] T. Taniguchi, Non-isotropic harmonic tori in complex projective spaces and configurations of points on Riemann surfaces, Tohoku Math. Publ., **14** (1999).
[28] C.-L. Terng and K. Uhlenbeck, Bäcklund transformations and loop group actions, Comm. Pure. Appl. Math., **53** (2000), 1–75.
[29] S. Udagawa, Harmonic maps from a two-torus into a complex Grassmann manifold, Int. J. Math., **6** (1995), 447–459.

Department of Mathematics
University of York
Heslington, York YO10 5DD
UK

E-mail address:
`im7@york.ac.uk`

Advanced Studies in Pure Mathematics 51, 2008
Surveys on Geometry and Integrable Systems
pp. 315–337

Isoparametric geometry and related fields

Reiko Miyaoka

§1. Introduction

The purpose of this paper is to give a perspective in the theory of isoparametric hypersurfaces and related topics. Starting with a brief introduction of the subject, we explain what is now going on, describing important results and remaining problems as well as new aspects.

The classification problem is now being solved in the cases $g = 4, 6$, where g is the number of principal curvatures. In §3, we give a simple proof of Cartan's theorem on the classification for $g = 3$, and generalize the strategy to any g. For this we use the relation between the curvature and the Lax equation, as well as Singer's strongly curvature-homogeneous theorem. On the other hand, the behavior of the kernel of the differential of the Gauss map of the focal submanifolds is important. In §4, we treat the degeneracy of the Gauss map, which seems related to the homogeneity.

Some isoparametric hypersurfaces and all the focal submanifolds provide us with many examples of austere submanifolds, whose twisted normal cones are special Lagrangian submanifolds in $\mathbb{C}^n$. In §5, we introduce this topic. We explain how to obtain explicit solutions of the special Lagrangian equation from isoparametric functions (§5.3). Special Lagrangian submanifolds are volume minimizing, and the topology of stable minimal submanifolds is restricted. We discuss the topology of austere submanifolds in §6, using Morse theory as in the proof of Lefschetz' theorem. In §7, we discuss hypersurface geometry from the viewpoint of Hamiltonian systems of hydrodynamic type.

Received February 8, 2002.
Revised March 3, 2007 (updated).
2000 *Mathematics Subject Classification.* Primary 53C42.

Except for some new results obtained mainly in §3, §5 and §6, we give complete references rather than proofs.

§2. Isoparametric hypersurfaces

First, we refer readers to the excellent survey on isoparametric geometry given by Thorbergsson [Th3], who proved that isoparametric submanifolds in S^n with codimension greater than one are homogeneous [Th2],[O].

By isoparametric hypersurfaces, we mean hypersurfaces with constant principal curvatures. A rich class of isoparametric hypersurfaces exists in the sphere, including all homogeneous hypersurfaces, which are classified as principal orbits of the linear isotropy representation of rank two symmetric spaces [HL]. Moreover, there are infinitely many non-homogeneous isoparametric hypersurfaces with four distinct principal curvatures, constructed by using Clifford algebras [OT],[FKM].

The fundamental facts on isoparametric hypersurfaces were obtained by E. Cartan and by E. Münzner [C1]~[C4],[Mu1],[Mu2]. Isoparametric hypersurfaces in S^n exist as families of parallel hypersurfaces; such a family includes two lower dimensional submanifolds called the focal submanifolds. All of these are level sets of the isoparametric function f, which is the restriction of the so-called Cartan-Münzner polynomial $F : \mathbb{R}^{n+1} \to \mathbb{R}$ to S^n. This polynomial is homogeneous and of degree equal to the number g of distinct principal curvatures $\lambda_1 > \lambda_2 > \cdots > \lambda_g$, and it satisfies

$$\begin{cases} |\nabla F|^2 = g^2 r^{2g-2}, & r(x) = |x|, \\ \triangle F = c r^{g-2}, & c = \dfrac{m_1 - m_2}{2} g^2, \end{cases}$$

where m_i is the multiplicity of λ_i [Mu1]. This implies that the hypersurfaces are algebraic, in particular, compact. Moreover, g is known to be limited to 1,2,3,4 and 6.

Isoparametric hypersurfaces with $g = 1$ are hyperspheres, $g = 2$ the Clifford hypersurfaces $S^k \times S^{n-1-k}$, and $g = 3$ the Cartan hypersurfaces given by tubes over standardly embedded Veronese surfaces corresponding to the four division algebras. The last result is known as Cartan's thoerem. Thus all isoparametric hypersurfaces with $g \leq 3$ are homogeneous. Two homogeneous examples for $g = 6$ are given as principal orbits of the isotropy representation of the symmetric spaces $G_2/SO(4)$

and $G_2 \times G_2/G_2$. Dorfmeister and Neher conjecture that these are all for $g = 6$. This has been confirmed in the case of principal curvatures of multiplicity one [DN],[M4], and is still open in the multiplicity two case.

For $g = 4$, H. Ozeki and M. Takeuchi [OT] were the first to construct non-homogeneous examples. In order to distinguish them from the homogeneous ones, they used the classification in [HL], but posed the question of how to distinguish them more naturally [OTT]. The number of principal curvatures and their multiplicities do not answer to this question, because there are homogeneous and non-homogeneous isoparametric families with $g = 4$ both having multiplicities $(m_1, m_2) = (4, 3)$.

Homogenous hypersurfaces in the sphere correspond to integrable Hamiltonian systems of hydrodynamic type in low dimensional cases [F2]. In fact, starting from the Hamiltonian density given by the isoparametric functions, we can deform the associated Hamiltonian system into the N-wave system. This argument uses the explicit forms of the isoparametric functions hence depends also on the classification of homogeneous hypersurfaces. The proof needs calculation for each case, which restricts the proof to lower dimensional cases, though it seems likely to be true for all homogeneous cases.

All this suggests the importance of characterizing homogeneity without using the classification. Ferus, Karcher and Münzner distinguished non-homogeneous hypersurfaces among examples of FKM-type, without using the classification, but using shape operators of the focal submanifolds. In particular, their kernel plays an important role. Incidentally, the author has characterized the homogeneity of isoparametric hypersurfaces with $g = 6$ by using the kernel structure of the focal submanifolds [M2],[M4].

While we were writing this article, a remarkable result on the classification in the case $g = 4$ was reported by Cecil, Chi and Jensen [CCJ]. They show that if the multiplicities satisfy $m_2 \geq 2m_1 - 1$, all isoparametric hypersurfaces are of FKM-type. Together with the known result by Takagi [T] and Ozeki-Takeuchi [OT], the remaining gaps in the classification are the following 4 cases: $(m_1, m_2) = (3, 4), (4, 5), (6, 9), (7, 8)$[1]. In the proof, they use special frame fields of the focal submanifolds, and using the structure equations and the symmetry satisfied by the structure coefficients, they construct Clifford algebras. Here an argument involving ideals of a polynomial ring plays an important role. The proof, therefore, depends on an algebraic argument, as in the proof of the

[1] Immervoll [I] obtained the same result.

Dorfmeister-Neher theorem for $(g, m) = (6, 1)$. For the latter, the author gives a simple geometric proof in [M4], and the case $(g, m) = (6, 2)$ is now in progress [M6]; we give some details in the next section.

§3. Homogeneity

In this section, we propose a strategy to prove the homogeneity of hypersurfaces.

3.1. Gauss equation and Cartan's theorem

Let M be an isoparametric hypersurface in S^n. Let $\lambda_1, \ldots, \lambda_g$ be the principal curvatures, and choose an orthonormal frame $\{e_i\}$ consisting of principal vectors $\{e_{a_1}, \ldots e_{a_m}\}$ of each λ_a, where m is the multiplicity (which depends on λ_a). With respect to this frame, we have the following fundamental formulae. Put

$$\tilde{\nabla}_{e_\alpha} e_\beta = \Lambda^\gamma_{\alpha\beta} e_\gamma + \lambda_\alpha \delta_{\alpha\beta} \xi, \quad \Lambda^\gamma_{\alpha\beta} = -\Lambda^\beta_{\alpha\gamma},$$

where $\tilde{\nabla}$ is the Levi-Civita connection of S^n, $1 \leq \alpha, \beta, k \leq n-1$, and we use the Einstein convention. The curvature tensor $R_{\alpha\beta\gamma\delta}$ of M is given by

$$(1) \qquad \begin{aligned} R_{\alpha\beta\gamma\delta} &= (1 + \lambda_\alpha \lambda_\beta)(\delta_{\beta\gamma}\delta_{\alpha\delta} - \delta_{\alpha\gamma}\delta_{\beta\delta}) \\ &= e_\alpha(\Lambda^\delta_{\beta\gamma}) - e_\beta(\Lambda^\delta_{\alpha\gamma}) + \Lambda^k_{\beta\gamma}\Lambda^\delta_{\alpha k} - \Lambda^k_{\alpha\gamma}\Lambda^\delta_{\beta k} \\ &\quad -\Lambda^k_{\alpha\beta}\Lambda^\delta_{k\gamma} + \Lambda^k_{\beta\alpha}\Lambda^\delta_{k\gamma}, \end{aligned}$$

which we call the Gauss equation. We have the following [M2]:

$$(2) \quad \Lambda^\gamma_{\alpha\beta}(\lambda_\beta - \lambda_\gamma) = \Lambda^\beta_{\gamma\alpha}(\lambda_\alpha - \lambda_\beta) = \Lambda^\alpha_{\beta\gamma}(\lambda_\gamma - \lambda_\alpha), \ (\text{distinct } \lambda_\alpha, \lambda_\beta, \lambda_\gamma)$$

$$(3) \qquad \Lambda^\gamma_{ab} = 0, \Lambda^\gamma_{aa} = \Lambda^\gamma_{bb}, \ (\lambda_a = \lambda_b \neq \lambda_\gamma \text{ and } a \neq b).$$

In particular for isoparametric hypersurfaces, we have

$$(4) \qquad \Lambda^\gamma_{\alpha\alpha} = 0 \quad \text{if} \quad \lambda_\gamma \neq \lambda_\alpha.$$

Using the Gauss equation, we obtain Cartan's theorem easily. This argument may be applied to the proof of the homogeneity for $g \geq 4$ (see Theorem 3.4).

Proposition 3.1. (Cartan's Theorem [C2]) *Isoparametric hypersurfaces with $g = 3$ are homogeneous.*

Proof: From the Gauss equation,

(5)

$$\begin{aligned} R_{aiaj} &= -(1+\lambda_a\lambda_i)\delta_{ij} \\ &= e_a(\Lambda^j_{ia}) - e_i(\Lambda^j_{aa}) + \Lambda^k_{ia}\Lambda^j_{ak} - \Lambda^k_{aa}\Lambda^j_{ik} - \Lambda^k_{ai}\Lambda^j_{ka} + \Lambda^k_{ia}\Lambda^j_{ka} \\ &= e_a(\Lambda^j_{ia}) + \Lambda^k_{ia}\Lambda^j_{ak} - \Lambda^k_{ai}\Lambda^j_{ka} + \Lambda^k_{ia}\Lambda^j_{ka}, \end{aligned}$$

we know that if λ_a, λ_i, λ_j are distinct, the left hand side vanishes. Moreover, the latter three terms in the last line vanish by (3), because in the summation in k, λ_k must be one of λ_a, λ_i, λ_j. This means that Λ^j_{ia} is constant along e_a. Since Λ^j_{ia} is non-zero only when $\lambda_a, \lambda_i, \lambda_j$ are distinct, and since we know $e_a(\Lambda^j_{ia}) = 0$, we conclude that Λ^j_{ia} is constant in all directions (see (2)). Now recall

Definition ([KN], p.357). A Riemannian manifold M is strongly curvature-homogeneous if, for any points x, y of M, there is a linear isomorphism of T_xM onto T_yM which maps g_x (the metric at x) and $(\nabla^k R)_x$ (higher covariant derivatives of the curvature tensor R), $k = 0, 1, 2, \ldots$ upon g_y and $(\nabla^k R)_y$, $k = 0, 1, 2, \ldots$.

Theorem 3.2. (Singer [S], Nomizu [N]) *If a connected Riemannian manifold M is strongly curvature-homogeneous, then it is locally homogeneous. If moreover, M is complete and simply connected, then it is homogeneous.*

Applying this, we know that M is locally intrinsically homogeneous. Then using the rigidity theorem on hypersurfaces with type number greater than 2 (which we apply to a non-minimal isoparametric hypersurface), we conclude that M is homogeneous. Q.E.D.

Cartan proved the theorem by determining all isoparametric functions explicitly, which made it possible to classify the hypersurfaces with $g = 3$.

3.2. Focal submanifolds.

In order to relate the Gauss equation to the Lax equation, we introduce the focal submanifolds of isoparametric hypersurfaces. As is well known, the principal curvatures are given by $\lambda_i = \cot\theta_i$, $\theta_i = \omega + \frac{(i-1)\pi}{g}$, $1 \le i \le g$, $0 < \omega < \frac{\pi}{g}$. Denote each curvature distribution by $D_i = D_{\lambda_i}$. Define a focal map $f_a \colon M \to S^n$ by

$$f_a(p) = \cos\theta_a p + \sin\theta_a \xi_p,$$

and let $\bar{p} = f_a(p)$. Then we have

$$df_a(e_{\hat{j}}) = \sin\theta_a(\lambda_a - \lambda_j)e_{\hat{j}},$$

where $e_{\hat{j}}$ is a principal vector with respect to λ_j, and the right hand side is considered as a vector in $T_{\bar{p}}S^n$ by a parallel translation in $\mathbb{R}^{n+1}$. We always use such identifications in what follows. The rank of f_a is constant and we obtain the focal submanifold M_a of M corresponding to $\lambda_a = \cot\theta_a$: $M_a = \{\cos\theta_a p + \sin\theta_a \xi_p \mid p \in M\}$. We have $T_{\bar{p}}M_a = \oplus_{j\neq a} D_j(q)$, for any $q \in {f_a}^{-1}(\bar{p})$. An orthonormal basis of the normal space of M_a at $\bar{p}$ is given by η_p and $e_{\hat{a}}(p)$, where

$$\eta_p = -\sin\theta_a p + \cos\theta_a \xi_p. \tag{6}$$

Lemma 3.3. [M2],[M4] *When we identify $T_{\bar{p}}M_a$ with $\oplus_{j=1}^{g-1} D_{a+j}(p)$ where the indices are modulo g, the second fundamental tensors B_{η_p} and B_{ζ_p} at $\bar{p}$, where $\zeta_p = e_a(p)$, are given respectively by the symmetric matrices*

$$B_{\eta_p} = \begin{pmatrix} \mu_1 I_{m_1} & 0 & 0 & 0 \\ 0 & \mu_2 I_{m_2} & 0 & 0 \\ 0 & 0 & \ddots & 0 \\ 0 & 0 & 0 & \mu_{g-1} I_{m_{g-1}} \end{pmatrix},$$

where

$$\mu_i = \frac{1+\lambda_i\lambda_a}{\lambda_a - \lambda_i}, \tag{7}$$

and

$$B_{\zeta_p} = \begin{pmatrix} 0 & B_{i+1\,i+2} & B_{i+1\,i+3} & \cdots & B_{i+1\,i+b} \\ B_{i+1\,i+2} & 0 & B_{i+2\,i+3} & \cdots & B_{i+2\,i+b} \\ B_{i+1\,i+3} & B_{i+2\,i+3} & 0 & & \vdots \\ \vdots & \vdots & & \ddots & B_{i+4\,i+b} \\ B_{i+1\,i+b} & B_{i+2\,i+b} & \cdots & B_{i+4\,i+b} & 0 \end{pmatrix},$$

where $b = g-1$ and

$$B_{jk} = \frac{1}{\sin\theta_a}\frac{\Lambda_{j_s a}^{k_t}}{\lambda_j - \lambda_a}, \qquad \lambda_{j_s} = \lambda_j, \lambda_{k_t} = \lambda_k,$$

is an $m_j \times m_k$ matrix.

In the following, we denote $\frac{\Lambda_{j_s a}^{k_t}}{\lambda_j - \lambda_a}$ by $\frac{\Lambda_{\hat{j}a}^{\hat{k}}}{\lambda_j - \lambda_a}$ for simplicity.

3.3. Isospectrality

Because any normal vector is expressed as η_q for some $q \in L_a$ as in (6), all the shape operators have eigenvalues μ_i, i.e., they are isospectral. For a unit normal vector $n_t = \cos t\eta + \sin t\zeta$, the shape operator $L(t) = B_{n_t}$ is given by

$$L(t) = \cos t B_\eta + \sin t B_\zeta. \tag{8}$$

Since $L(t)$ is isospectral and symmetric, we have orthogonal matrices $U(t) \in O(n-1-m_a)$ such that

$$B_{n_t} = L(t) = U(t)L(0)U(t)^{-1} = U(t)B_\eta U(t)^{-1}. \tag{9}$$

Define $H(t) \in \mathfrak{o}(n-1-m_a)$ by

$$H(t) = \frac{d}{dt}U(t)\,U(t)^{-1} \tag{10}$$

then L satisfies the Lax equation

$$\frac{d}{dt}L(t) = [H(t), L(t)], \tag{11}$$

i.e.

$$-\sin t B_\eta + \cos t B_\zeta = \cos t[H(t), B_\eta] + \sin t[H(t), B_\zeta]. \tag{12}$$

In particular, we obtain $B_\zeta = [H(0), B_\eta]$. Let $U(t)$ be the solution of (10) with $U(0) = I$. Now the eigenvector $e_i(t)$ of $L(t)$ with respect to the eigenvalue λ_i is given by $e_i(t) = U(t)e_i(0)$, where

$$\frac{d}{dt}e_i(t) = H(t)e_i(t).$$

If we use a moving frame $\{e_i(t)\}$ consisting of the principal vectors along the geodesic $p(t) \subset L_a(p)$ joining p and q such that $\eta = \eta_p$ and $\zeta = \eta_q$, we have by Lemma 3.3,

$$L(t) = \begin{pmatrix} \mu_1 I_{m_1} & 0 & 0 & 0 \\ 0 & \mu_2 I_{m_2} & 0 & 0 \\ 0 & 0 & \ddots & 0 \\ 0 & 0 & 0 & \mu_{g-1} I_{m_{g-1}} \end{pmatrix},$$

and

$$L(t+\pi/2) = \begin{pmatrix} 0 & & B_{ij} \\ & \ddots & \\ {}^tB_{ij} & & 0 \end{pmatrix},$$

where the $m_i \times m_j$ matrix B_{ij}, $i, j \neq a$ is given by

$$B_{ij} = (b_{\hat{i}\hat{j}}), \quad b_{\hat{i}\hat{j}} = \frac{\Lambda^{\hat{j}}_{\hat{i}a}(p(t))}{\sin\theta_a(\lambda_i - \lambda_a)}.$$

The following is crucial [M6].

Theorem 3.4. *An isoparametric hypersurface is homogeneous if and only if there exists a local frame so that* $e_a(\Lambda^j_{ia}) = 0$ *holds for all* a, i, j *such that* λ_a, λ_i *and* λ_j *are distinct.*

The philosophy of the proof is the same as the proof of Cartan's theorem in §3.1, though we need more arguments for larger g, because there are many choices of the frame when $m_i > 1$. We omit the details, but instead give the following lemma, which is rather easy to prove.

Lemma 3.5. $e_a(\Lambda^j_{ia}) = 0$ *holds for all* a, i, j *such that* λ_a, λ_i *and* λ_j *are distinct if and only if* $H(t) - H(0)$ *commutes with* $L(t)$.

Proof: Noting that $\dfrac{d}{dt} = \sin\theta_a \nabla_{e_a}$, we have

$$(13) \qquad \begin{aligned} \tfrac{d}{dt}L(t)_{ij} &= \sin\theta_a \nabla_{e_a}(\mu_i\delta_{ij}) \\ &= \sin\theta_a\{e_a(\mu_i\delta_{ij}) - \mu_k\delta_{kj}\Lambda^k_{ai}(t) - \mu_i\delta_{ik}\Lambda^k_{aj}(t)\} \\ &= [H(t), L(t)]_{ij}, \end{aligned}$$

where $H(t) = (\sin\theta_a\Lambda^i_{ak}(t))$, and i denotes the index of the rows. On the other hand, we have

(14)

$$\begin{aligned} \frac{d}{dt}L(t+\pi/2)_{ij} &= \sin\theta_a\nabla_{e_a}\Big(\frac{1}{\sin\theta_a}\frac{\Lambda^j_{ia}(t)}{\lambda_i - \lambda_a}\Big) \\ &= \sin\theta_a\{e_a(b_{ij}(t)) - b_{kj}(t)\Lambda^k_{ai}(t) - b_{ik}(t)\Lambda^k_{aj}(t)\} \\ &= e_a\Big(\frac{\Lambda^j_{ia}(t)}{\lambda_i - \lambda_a}\Big) + [H(t), L(t+\pi/2)]_{ij}. \end{aligned}$$

Because we also have $\frac{d}{dt}L(t+\pi/2)_{ij} = [H(t+\pi/2), L(t+\pi/2)]_{ij}$ by (13), $e_a(\Lambda^j_{ia}) = 0$ holds if and only if $H(t) - H(t+\pi/2)$ commutes with $L(t+\pi/2)$. In particular, when $H(0) - H(\pi/2)$ commutes with $L(\pi/2)$, noting that $\frac{d}{dt}L(t) = -\sin tL(0) + \cos tL(\pi/2) = L(t+\pi/2) = [H(t), L(t)]$, we obtain

$$\begin{aligned} \frac{d}{dt}L(t) &= -\sin tL(0) + \cos tL(\pi/2) = \sin tL(\pi) + \cos tL(\pi/2) \\ &= \sin t[H(\pi/2), L(\pi/2)] + \cos t[H(0), L(0)] \\ &= [H(0), L(t)], \end{aligned}$$

hence $H(t) - H(0)$ commutes with $L(t)$. Q.E.D.

These play an important role in the investigation of the case $g = 6$ in [M6].

Next, we demonstrate a relation between the Lax equation and the curvature.

Lemma 3.6. *The Lax equation satisfied by the shape operators of the focal submanifolds is a part of the Gauss equation of M.*

Proof: From (7), we have

$$\begin{aligned} \mu_i \delta_{ij} &= -\frac{(1+\lambda_a\lambda_i)\delta_{ij}}{\lambda_i - \lambda_a} = \frac{R_{aiaj}}{\lambda_i - \lambda_a} \\ &= \frac{e_a(\Lambda_{ia}^j)}{\lambda_i - \lambda_a} + \frac{\Lambda_{ia}^k}{\lambda_i - \lambda_a}\Lambda_{ak}^j - \frac{\Lambda_{ai}^k}{\lambda_i - \lambda_a}\Lambda_{ka}^j + \frac{\Lambda_{ia}^k}{\lambda_i - \lambda_a}\Lambda_{ka}^j \\ &= \sin\theta_a\{e_a(b_{ij}) - b_{ik}\Lambda_{aj}^k - b_{jk}\Lambda_{ai}^k\}, \end{aligned} \tag{15}$$

which coincides with (14), where we use (2) to obtain

$$\begin{aligned} -\frac{\Lambda_{ai}^k}{\lambda_i - \lambda_a}\Lambda_{ka}^j + \frac{\Lambda_{ia}^k}{\lambda_i - \lambda_a}\Lambda_{ka}^j &= -\frac{\Lambda_{ka}^j\Lambda_{ai}^k}{\lambda_i - \lambda_a}\left(1 - \frac{\lambda_i - \lambda_k}{\lambda_a - \lambda_k}\right) \\ &= \frac{\Lambda_{ka}^j\Lambda_{ai}^k}{\lambda_a - \lambda_k} = -\sin\theta_a b_{jk}\Lambda_{ai}^k. \end{aligned}$$

Q.E.D.

§4. Submanifolds with degenerate Gauss mapping

In [IKM], we discuss submanifolds in the sphere whose generalized Gauss map degenerates, in the sense that the rank of the Gauss map is less than the dimension of the submanifold. The Ferus number has the property that if the Gauss map has rank less than the Ferus number, then the submanifold must be totally geodesic. Here we consider submanifolds whose Gauss map degenerates with rank equal to the Ferus number. We find many examples by using isoparametric hypersurfaces.

For a submanifold M^p in S^n, by the Gauss map $\gamma : M \to \mathrm{Gr}(p+1, n+1)$, we mean the map assigning each point of M to the linear subspace spanned by the tangent space and the position vector. The Gauss map degenerates when the intersection of the kernels of the shape operators contains non-zero elements. In particular, for the focal submanifolds of isoparametric hypersurfaces, we can prove that if the kernel of the shape operator does not depend on the normal direction (hence the Gauss map degenerates), they are homogeneous in all cases but $g = 4$. Thus it is interesting to know what happens when $g = 4$, and how it relates with the homogeneity. In this case, the known examples consist of

(1) Homogeneous ones of Clifford type :
$(m_1, m_2) = (1, k)$, $(2, 2k-1)$, $(4, 4k-1)$, (9,6)

(2) Homogeneous ones of non-Clifford type :
$(m_1, m_2) = (2, 2), (4, 5)$

(3) Non-homogeneous ones of Clifford type :
$(m_1, m_2) = (3, 4k), (7, 8k)$.

Investigating all homogeneous examples (excluding, for a technical reason[2] the last case in (1)) we obtain the following :

Theorem 4.1. [IKM] *Let M be a homogeneous hypersurface with four principal curvatures. Let $M_\pm$ be the focal submanifolds of M. Then we have:*

(1) When $(m_1, m_2) = (1, k-2)$ and if $k \geq 3$, M_+ has degenerate Gauss mapping while M_- does not.

(2) When $(m_1, m_2) = (2, 2k-3), (4, 4k-5)$ and $k \geq 2$, M_- has degenerate Gauss mapping while M_+ does not.

(3) When $(m_1, m_2) = (2, 2)$, M_+ has degenerate Gauss mapping while M_- does not.

(4) When $(m_1, m_2) = (4, 5)$, M_- has degenerate Gauss mapping while M_+ does not.

For M_- in (2) with $(m_1, m_2) = (2, 2^p - 3)$, $p \geq 2$ and $(4, 2^q - 5)$, $q \geq 3$, and M_+ in (3), the degenerate Gauss map has rank equal to the Ferus number.

Here, $M_\pm$ are defined as in [FKM]. Note that M_+ is characterized in [W],[PT] as a Clifford-Stiefel submanifold. It is curious that neither all M_+ nor all M_- have degenerate Gauss map. Thus when $g = 4$, the homogeneity does not imply the invariance of the kernel of the Gauss map of both $M_\pm$. But it is still interesting because one of $M_\pm$ has this property. In fact, in [FKM], non-homogenous examples are found

[2]My student F. Kaneda investigated this case and obtained a similar result [K].

when they have kernels whose intersection has non-constant dimension. Conversely, it seems that if the intersection of the kernels of each focal submanifold has a constant dimension, then the hypersurfaces are homogeneous.

§5. Special Lagrangian submanifolds

Special Lagrangian geometry is the intersection of minimal submanifold theory with symplectic geometry. Special Lagrangian submanifolds in $\mathbb{C}^n$ consist of volume minimizing submanifolds among submanifolds with the same homology. They are necessarily non-compact, and may have singularities. Neverthress, they are interesting enough from the viewpoints of minimal submanifold theory and symplectic geometry, related with the stability of minimal graphs, and with integrability of finite and infinite dimensional Hamiltonian systems, as is shown below.

Rich families of examples of special Lagrangian submanifolds in $\mathbb{C}^n$ are obtained from isoparametric hypersurfaces and their focal submanifolds. They are nice examples because

—infinitely many homogeneous and non-homogeneous examples exist,

—they are given as a graph of explicit function, hence provide explicit solutions to the special Lagrangian equation,

—they have clear topology, and

—they are related with Hamiltonian system of hydrodynamic type.

In the following, we investigate these examples.

5.1. Definitions and known results

Let X be a Riemannian manifold with a closed p-form φ which satisfies

$$|\varphi(\zeta)| \leq \mathrm{vol}(\zeta)$$

for any oriented p-plane ζ of T_xX. Then (X, φ) is called a calibrated manifold, and φ a calibration. When a submanifold M of X satisfies $\varphi|_M = \mathrm{vol}$, M is called a φ manifold.

Fact 5.1. If M is a φ manifold of a calibrated manifold (X, φ), then M minimizes volume among manifolds in the same homology class.

Complex submanifolds of a Kähler manifold are examples of ω^p-manifolds, where ω is the Kähler form and p is the complex dimension of M.

When $X = \mathbb{C}^n$, using the coordinates $z_1, \ldots, z_n$, we can see that

$$\alpha^\theta = \Re\{e^{i\theta} dz\}, \quad dz = dz_1 \wedge \cdots \wedge dz_n$$

is a calibration, and $(\mathbb{C}^n, \alpha^\theta)$ is a calibrated manifold. Let J be the standard complex structure of $\mathbb{C}^n$, and let $\omega = \frac{i}{\sqrt{2}} \sum dz_j \wedge d\bar{z}_j$ be the Kähler form of $\mathbb{C}^n$. An oriented real n-plane ζ is called Lagrangian if $\omega|_\zeta = 0$, or equivalently, $Ju \perp \zeta$ holds for all $u \in \zeta$. Moreover, ζ is special Lagrangian if there exists $A \in SU(n)$ such that $\zeta = A\zeta_0$ where $\zeta_0 = \mathbb{R}^n = \{(x, 0) \in \mathbb{R}^n \oplus \mathbb{R}^n \cong \mathbb{C}^n\}$. A submanifold M in $\mathbb{C}^n$ is called *special Lagrangian* if all the tangent spaces of M are special Lagrangian.

Fact 5.2. [HaL] (1) M is special Lagrangian in $\mathbb{C}^n$ if and only if M is an α^θ-manifold for some θ.

(2) M is special Lagrangian in $\mathbb{C}^n$ if and only if M is minimal and Lagrangian.

5.2. Austere submanifolds and twisted normal cones

A submanifold of a Riemannian manifold is called *austere* [HaL] when all the shape operators have eigenvalues in pairs $\pm\lambda$ for each eigenvalue λ (including 0). All minimal surfaces are austere. A submanifold M of $S^{n-1} \subset \mathbb{R}^n$ is austere if and only if the cone CM is austere in $\mathbb{R}^n$. Austere 3-folds in $\mathbb{R}^n$ have been classified by Bryant [B].

Identifying $T_x^*\mathbb{R}^n = \mathbb{C}^n$, we have $J(X, Y) = (-Y, X)$ for $(X, Y) \in T\mathbb{C}^n$. Thus if M is a submanifold of $\mathbb{R}^n$, then the canonically embedded normal bundle

$$NM = \{(x, \nu(x)) \mid x \in M, \nu(x) \in NM\} \subset T\mathbb{R}^n \cong T^*\mathbb{R}^n$$

is Lagrangian.

Fact 5.3. [HaL] (1) When M is a submanifold of $\mathbb{R}^n$, the canonically embedded normal bundle NM is a special Lagrangian submanifold in $\mathbb{C}^n$ if and only if M is austere.

(2) When M is a submanifold of $S^{n-1} \subset \mathbb{R}^n$, the twisted normal cone

$$NCM = \{(tx, s\nu(x)) \mid (x, \nu(x)) \in T^\perp M, s, t \in \mathbb{R}\}$$

is a special Lagrangian submanifold in $\mathbb{C}^n$ if and only if M is austere.

Corollary 5.1. *Let M be a compact austere submanifold in S^{n-1}. Then the twisted normal cone is an n-dimensional cone of least mass in $\mathbb{R}^{2n}$.*

Examples of cones of least mass are given in [HaL] as twisted normal cones over compact minimal surfaces in S^3, Veronese surface in S^4 and Clifford hypersurfaces $S^k \times S^k$ in S^{2k+1}. The latter two are included in the class of submanifolds related to isoparametric hypersurfaces.

Recall that the principal curvatures $\lambda_i(t)$ of the level set $M_t = F^{-1}(t) \cap S^{n-1}$ of the isoparametric function $f = F|_{S^{n-1}}$ are given by

$$\lambda_i = \cot\left(\frac{\pi}{2g}(1-t) + \frac{(i-1)\pi}{g}\right), \quad i = 1, \ldots, g, \quad t \in (-1, 1),$$

where M_0 as well as $M_\pm = M_{\pm 1}$ are minimal.

Proposition 5.2. *Let M be a minimal isoparametric hypersurfaces in S^{n-1}. If each principal curvature has the same multiplicity, then M is austere. Thus the twisted normal cone over M is special Lagrangian, and volume minimizing in $\mathbb{C}^n$.*

Proof: The case $g = 1$ is trivial. When $g = 2$, we have two principal curvatures λ and μ, hence it is austere if and only if $\mu = -\lambda = 1$ and $m_1 = m_2$. When $g = 3$ and 6, the multiplicities are known to coincide, and austere means minimal with principal curvatures $\pm\sqrt{3}, 0$ for $g = 3$, and $\pm(2+\sqrt{3}), \pm 1, \pm(2-\sqrt{3})$, for $g = 6$. When $g = 4$, the principal curvatures $\lambda_1 > \lambda_2 > \lambda_3 > \lambda_4$ have multiplicities m_1, m_2, m_1, m_2, respectively. Thus it is austere if and only if $\lambda_4 = -\lambda_1 = \sqrt{2}+1, \lambda_3 = -\lambda_2 = \sqrt{2}-1$ and $m_1 = m_2 = 1$, or 2 [St]. Q.E.D.

All known examples of isoparametric hypersurfaces with principal curvatures having the same multiplicities are homogeneous. That is, all the cases for $g \leq 3$, and $m = m_1 = m_2 = 1, 2$ for $g = 4, 6$. Thus the twisted normal cones obtained in Proposition 5.2 are all homogeneous up to Dorfmeister-Neher's conjecture. However, the non-homogeneous case occurs as well.

Proposition 5.3. *The focal submanifolds of any isoparametric hypersurface in S^{n-1} are austere. The twisted normal cones over them are special Lagrangian, and volume minimizing in $\mathbb{C}^n$.*

Proof: When $g = 1$, each focal submanifold is a point manifold, and the twisted normal cone is $\mathbb{R}^n$. When $g = 2$, each focal submanifold is a great subsphere of S^{n-1} and the twisted normal cone is also $\mathbb{R}^n$. The eigenvalues of the shape operators $M_\pm$ are given by (7) as $\mu_i = \frac{1+\lambda\lambda_i}{\lambda_i-\lambda}$ where λ is the maximal or minimal principal curvature and $\lambda \neq \lambda_i$. When $g=3$, they are given by $\pm\frac{1}{\sqrt{3}}$ with the same multiplicity. When $g=4$, they are ± 1 with a common multiplicity, and 0. When $g=6$, they are $\pm\sqrt{3}, \pm\frac{1}{\sqrt{3}}$ and 0 with the same multiplicity. Hence $M_\pm$ are austere. Q.E.D.

Corollary 5.4. *There are infinitely many non-homogeneous compact austere submanifolds of dimension greater than two in the sphere. Thus there are infinitely many non-homogeneous special Lagrangian cones in* $\mathbb{C}^n$.

We have many other compact austere submanifolds in the sphere [IKM].

5.3. Expression as a graph and the SL equation

A Lagrangian submanifold of $\mathbb{C}^n$ is known to be expressable locally as a graph of the gradient of a function [HaL]. The converse is evident: let $G : U \to \mathbb{R}$ be a smooth function on an open domain U of $\mathbb{R}^n$. Then $\Gamma_G = \{(x, \nabla G) \in \mathbb{R}^n \oplus \mathbb{R}^n\}$ is Lagrangian. Moreover this is special Lagrangian with respect to α^θ if and only if Hess G satisfies

$$\Im\{e^{i\theta}\det_{\mathbb{C}}(I + i\text{Hess } G)\} = 0. \tag{16}$$

Though the Dirichlet problem for this equation is known to have solutions [HaL], not many explicit solutions are known. Here, we express the twisted normal cone obtained in Proposition 5.2 as a graph, and give the corresponding solutions of (16) explicitly by using the isoparametric functions.

Let $F : \mathbb{R}^n \to \mathbb{R}$ be a Cartan-Münzner function. The minimal isoparametric hypersurface in the unit sphere S^{n-1} of $\mathbb{R}^n$ is given by $M = F^{-1}(0) \cap S^{n-1}$, and its cone by $CM = F^{-1}(0)$, because of the homogeneity of F. This means that we have a local function

$$x_n = f(x_1, \ldots, x_{n-1})$$

such that $F((x_1, \ldots, x_{n-1}, f(x_1, \ldots, x_{n-1})) = 0$.

More generally, consider a hypersurface M given by a graph

$$(x_1, \ldots, x_{n-1}, f(x_1, \ldots, x_{n-1})) \in \mathbb{R}^n.$$

of a certain function f on $\mathbb{R}^{n-1}$. The canonically embedded normal bundle of M is given by

$$(17)\quad NM = \{((x_1, \dots, x_{n-1}, f), (-tf_1, \dots, -tf_{n-1}, t)) \in \mathbb{R}^n \oplus \mathbb{R}^n\},$$

where $f_i = \dfrac{\partial f}{\partial x_i}$, since the tangent vectors of M are given by $v_i = (0, \dots, 1, \dots, f_i)$. Now define a function $G : \mathbb{R}^n \to \mathbb{R}$ by

$$G(x_1, \dots, x_{n-1}, t) = -tf(x_1, \dots, x_{n-1}).$$

Then we have $\nabla G = (-tf_1, \dots, -tf_{n-1}, -f)$, and

$$(18)\quad \Gamma_G = \{((x_1, \dots, x_{n-1}, t), (-tf_1, \dots, -tf_{n-1}, -f))\} \in \mathbb{R}^n \oplus \mathbb{R}^n$$

is a Lagrangian submanifold of $\mathbb{C}^n$. Let us identify $\mathbb{C}^n$ with $\mathbb{R}^n \oplus \mathbb{R}^n$ by $(z_1, \dots, z_n) = (x_1, \dots, x_n, y_1, \dots, y_n)$, $z_j = x_j + iy_j$. Comparing (17) with (18), we see that $f + it$ in (17) is replaced by $t - if$ in (18). This means that Γ_G is the image of NM under $U = \operatorname{diag}\begin{pmatrix} 1 & \cdots & 1 & -i \end{pmatrix} \in U(n)$ acting on $\mathbb{C}^n$. Thus NM is special Lagrangian with respect to a calibration α^θ if and only if Γ_G is so with respect to $\alpha^{\theta - n\pi/2}$.

Any hypersurface can be expressed locally as a graph. Thus by the above procedure, we obtain a solution of the special Lagrangian equation from an austere hypersurface given as a graph. For instance, consider austere isoparametric hypersurfaces in S^{n-1}, i.e. isoparametric hypersurfaces satisfying the condition of Proposition 5.2. We give f and G explicitly in the first non-trivial case, $g = 3$. ([OT], pp.23-4). For $x \in \mathbb{F}$ where $\mathbb{F} = \mathbb{R}, \mathbb{C}, \mathbb{H}$ or the real Cayley algebra $\mathcal{C}$, define

$$t(x) = x + \bar{x}, \quad n(x) = x\bar{x}.$$

Then the isoparametric functions on $\mathbb{F}^3 \times \mathbb{R}^2$ are given by $F(u) =$

$$\tfrac{3\sqrt{3}}{2}\{-\xi_1\xi_2(\xi_1 + \xi_2) - \xi_1(n(x_1) - n(x_3)) - \xi_2(n(x_2) - n(x_3)) + t(x_1x_2x_3)\},$$

for $u = (x_1, x_2, x_3, \xi_1, \xi_2) \in \mathbb{F}^3 \times \mathbb{R}^2$. Solving $F(u) = 0$, we get $f = \xi_2$ as a function of x_1, x_2, x_3, ξ_1 on a suitable domain in $\mathbb{F}^3 \times \mathbb{R}$:

$$\begin{aligned} f &= \frac{-(\xi_1^2 + (n(x_2) - n(x_3))}{2\xi_1} \\ &\quad \pm \frac{\sqrt{(\xi_1^2 + (n(x_2) - n(x_3))^2 - 4\xi_1(\xi_1(n(x_1) - n(x_3)) - t(x_1x_2x_3))}}{2\xi_1}, \end{aligned}$$

and $G = -tf$.

For $g = 4$ and $m_1 = m_2 = 1, 2$, the polynomial is of degree 4 and we can solve it with respect to some variables [OT], pp.27. For $g = 6$, see [OT], pp.27–29.

§6. Topological aspects

The canonically embedded normal bundle over an austere submanifold M in $\mathbb{R}^n$ is volume minimizing, hence is stable. The stability of minimal submanifolds implies some topological restrictions [Pa],[M1]. We give a topological characterization of complete austere submanifolds in $\mathbb{R}^n$, as well as their normal bundles.

Theorem 6.1. *A complete proper austere submanifold M in $\mathbb{R}^n$ has the homotopy type of a CW complex of dimension not greater than $\frac{1}{2}\dim M$.*

Proof: This can be shown by using Morse theory in analogy with the proof of Lefschetz' theorem on algebraic varieties in [AF],[Mi]. First, since M is real analytic and proper, we can apply Morse theory to the squared distance function L_p^2, $p \in \mathbb{R}^n$. On each normal line of M, focal points are located symmetrically on both sides, with the same index at the corresponding focal points. Since the shape operator in this normal direction has at most $\dim M$ eigenvalues, which correspond to the focal points in a one to one way (counting multiplicities), there are at most $\frac{1}{2}\dim M$ focal points on each side. If M has a cycle of dimension greater than $\frac{1}{2}\dim M$, a function L_p^2 for some $p \in \mathbb{R}^n$ has a critical point of index greater than $\frac{1}{2}\dim M$, a contradiction. Q.E.D.

Corollary 6.2. *The normal bundle over a complete proper austere submanifold M in $\mathbb{R}^n$ has the homotopy type of a CW complex of dimension not greater than $\frac{1}{2}\dim M$*

More generally, it should be interesting to investigate the topology of special Lagrangian submanifolds, from the viewpoint of stability.

The cone over a minimal submanifold in S^{n-1} has a singularity at the origin, hence Theorem 6.1 does not apply in this case. Moreover, compact minimal submanifolds in S^{n-1} are not stable [S], and we cannot expect a topological restriction. In fact, the topology of an isoparametric hypersurface M can be described as follows: The cases $g = 1, 2$ are obvious. When $g = 3, 4, 6$, the non-trivial $\mathbb{Z}_2$ coefficient homology groups are $H_0(M) = H_{n-1}(M) = \mathbb{Z}_2$ and

$$\begin{array}{ll} g=3 & H_{mi}(M) = 2\mathbb{Z}_2, \quad i=1,2\,(m=1,2,4,8) \\ g=4 & H_{m_i}(M) = H_{n-1-m_i}(M) = \mathbb{Z}_2,\ H_{m_1+m_2}(M) = 2\mathbb{Z}_2,\ i=1,2 \\ g=6 & H_{mi}(M) = 2\mathbb{Z}_2, \quad i=1,2,3,4,5\,(m=1,2) \end{array}$$

hence the sum of the Betti numbers is $2g$. The non-trivial homology groups of the focal submanifolds $M_\pm$ are

$$\begin{array}{ll} g=3 & H_{mi}(\bar{M}) = \mathbb{Z}_2, \quad i=0,1,2\,(m=1,2,4,8) \\ g=4 & H_0(\bar{M}) = H_{m_i}(\bar{M}) = H_{m_1+m_2}(\bar{M}) = H_{n-1-m_i}(\bar{M}) = \mathbb{Z}_2,\ i=1,2 \\ g=6 & H_{mi}(\bar{M}) = \mathbb{Z}_2, \quad i=0,1,2,3,4,5\,(m=1,2) \end{array}$$

hence the sum of the Betti numbers is g. In relation to Morse theory, isoparametric hypersurfaces, and more generally, compact embedded Dupin hypersurfaces, are taut [Th1]. For more details, see [Ce],[CR],[M5].

§7. Relations to integrable systems

D. Joyce has proposed the problem of clarifying the relation between special Lagrangian submanifolds and some integrable systems [J1],[J2]. For instance, minimal surfaces in S^n, $\mathbb{C}P^n$ are given by conformal harmonic maps, whose equations are an integrable system known as the Toda equations [BPW],[M3]. Therefore the twisted normal cones over them are related to integrable systems. Limiting the discussion to examples obtained from isoparametric hypersurfaces, we consider this problem in higher dimensions.

Hamiltonian systems of hydrodynamic type [DuN1],[DuN2], correspond hypersurfaces in space forms [F1]. Roughly speaking, a certain Hamiltonian system on the space of curves (an infinite dimensional space with a certain symplectic structure) corresponds to the Weingarten equation satisfied by the shape operators of a hypersurface in $\mathbb{R}^n$ or in S^n. When these hypersurfaces are homogeneous, the Hamiltonian systems seem to be integrable [F2]. In view of Proposition 5.2, the twisted normal cone over an isoparametric hypersurface is special Lagrangian if and only if the hypersurface is homogeneous up to Dorfmeister-Neher's conjecture, hence corresponds to an integrable Hamiltonian system of hydrodynamic type. However, as for twisted normal cones over focal submanifolds, we have both homogeneous and non-homogeneous examples, and so we can say nothing about relations between the homogeneity and integrability in an easy way. Still we feel there should exist some relation between isoparametric geometry and integrable systems.

We shall give a brief explanation of how hypersurfaces correspond to Hamiltonian systems. Let $\mathcal{C} = \{u = u(x) : \mathbb{R} \to \mathbb{R}^n\}$ be a space of curves in $\mathbb{R}^n$ with a suitable decay condition so that $\mathcal{C}$ becomes a topological linear space. For functionals

$$F(u) = \int f(u, u_x, \dots)dx, \quad G(u) = \int g(u, u_x, \dots)dx$$

where f and g are polynomials in u and its derivatives, we consider a bracket

$$\{F, G\} = \int \frac{\delta F}{\delta u^i}\omega^{ij}\frac{\delta G}{\delta u^j}dx, \tag{19}$$

where δ is the variational derivative [P]

$$\frac{\delta F}{\delta u^i} = \frac{\partial f}{\partial u^i} - (\frac{\partial f}{\partial u^i_x})_x + (\frac{\partial f}{\partial u^i_{xx}})_{xx} - \cdots .$$

In order for $\{\,,\,\}$ to be a Poisson bracket, ω^{ij} should satisfy certain conditions. When this is of hydrodynamic type i.e., $\omega^{ij} = g^{ij}(u)d + b^{ij}_k(u)u^k_x$, we have:

Fact 7.1 [DuN1] (19) defines a Poisson bracket if and only if

(1) $\det g^{ij} \neq 0$

(2) $g^{ij} = g^{ji}$

(3) $\Gamma^j_{sk}(u)$ defined by $b^{ij}_k = -g^{is}(u)\Gamma^j_{sk}(u)$ is a torsion free flat connection ∇.

(4) $\nabla g = 0$.

When $h(u)$ is independent of the derivatives of u, the Hamiltonian equation with respect to the Hamiltonian function $H = \int h(u)dx$, namely

$$u^i_t = \omega^{ij}(u)\frac{\delta H}{\delta u^j} = v^i_j(u)u^j_x, \quad v^i_j(u) = \nabla^i\nabla_j h, \tag{20}$$

is called a Hamiltonian system of hydrodynamic type [DuN1]. Conversely, when there exists a flat metric g_{ij} such that

$$g_{ik}v^k_j = g_{jk}v^k_i, \quad \nabla_k v^i_j = \nabla_j v^i_k, \tag{21}$$

the system $u^i_t = v^i_j(u)u^j_x$ is of hydrodynamic type [Ts], and v is called local. With respect to flat coordinates, we can express it as

$$u^i_t = h_{ij}u^j_x, \quad h_{ij} = \frac{\partial^2 h}{\partial u^i \partial u^j} \tag{22}$$

Similarly for $\omega^{ij} = g^{ij}d - g^{is}\Gamma^j_{sk}u^k_x + cu^i_x d^{-1}u^j_x$, where c is a constant and d^{-1} is an integral operator, $\{\,,\}$ is a Poisson bracket if and only if $g^{ij} = g^{ji}$, $\nabla g = 0$ where the connection ∇ is defined as above and satisfies $\Gamma^j_{sk} = \Gamma^j_{ks}$, $R^{ij}_{kl} = c(\delta^i_k\delta^j_l - \delta^i_l\delta^j_k)$ for a constant c. Then (20) can be written as

$$u^i_t = (g^{is}\nabla_s\nabla_j h + c\delta^i_j h)u^j_x.$$

Conversely, in the system $u^i_t = v^i_j(u)u^j_x$, v^i_j is of the form $\nabla^i\nabla_j h + c\delta^i_j h$ if there exists a non-degenerate curvature 1 metric g_{ij}, for which v^i_j satisfies (21). In this case, v is called non-local.

Fact 7.2. [FM] The quadratic forms

$$\mathrm{I} = v^k_i g_{kl} v^l_j du^i du^j, \quad \mathrm{II} = v^k_i g_{kj} du^i du^j, \quad \mathrm{III} = g_{ij}du^i du^j,$$

satisfy the Gauss-Coddazi equations if and only if g_{ij} is of curvature 1 and v^i_j satisfies (21).

Therefore, a non-local hydrodynamic system corresponds to a hypersurface in $\mathbb{R}^{n+1}$, and through a certain transformation, a hypersurface in S^{n+1}. In fact when (20) is given, we can change the parameters t, x by the so-called reciprocal transformation [F1]

$$\begin{cases} dX = Bdx + Adt = \frac{1}{2}((u^i)^2 + 1)dx + (h_j u^j - h)dt \\ dT = Ndx + Mdt = dt, \end{cases}$$

then (20) changes into

$$u^i_t = V^i_j u^j_X = \{h_{ij}B - \delta^i_j A\}u^j_X,$$

which is non-local with respect to a new Hamiltonian given by $\tilde{h}(u) = h(u)/(\sum_k \frac{(u^k)^2}{2} + \frac{1}{2})$, hence corresponds to a hypersurface by Fact 7.2. In fact, putting

$$\begin{aligned} r &= (h_1 - u^1\frac{A}{B}, \ldots, h_n - u^n\frac{A}{B}, -\frac{A}{B}), \\ n &= (\frac{u^1}{B}, \ldots, \frac{u^n}{B}, \frac{1}{B} - 1), \end{aligned} \tag{23}$$

we obtain a hypersurface $r : \mathbb{R}^n \to \mathbb{R}^{n+1}$ with unit normal vector n satisfying the Weingarten equation

$$\frac{\partial n}{\partial u^j} = (V^{-1})^i_j(u)\frac{\partial r}{\partial u^i}.$$

Moreover, if we transform

$$\begin{cases} dX = Bdx + Adt = \frac{1}{2}(\sum(u^i)^2 + 1)dx + (h_j u^j - h)dt \\ dT = Ndx + Mdt = hdx + \frac{1}{2}(\sum(h_i)^2 + 1)dt, \end{cases}$$

then (20) changes into

$$u^i_T = \left\{\frac{1}{2}(\sum(u^k)^2 + 1)h_{ij} - (h_k u^k - h)\delta^i_j)\right\} u^j_X$$

and

$$(24) \qquad \begin{cases} r &= \dfrac{1}{AN - BM}(u^i A - h_i B, A, B + AN - BM) \\ n &= \dfrac{1}{BM - AN}(u^i M - h_i N, M - BM + AN, N) \end{cases}$$

defines a hypersurface in S^n, with unit normal vector n satisfying the Weingarten equation

$$(25) \qquad n_T = r_X,$$

which comes from the original Hamiltonian equation. Thus from a given Hamiltonian system of hydrodynamic type, we can construct locally a hypersurface in $\mathbb{R}^{n+1}$ and in S^{n+1}, whose shape operators satisfy (25) for r, n given by (23), and (24), respectively. Note that they correspond via stereographic projection. In particular when h is an isoparametric function on $\mathbb{R}^n$ for $n \leq 10$, (19) can be deformed into the N-wave system which is known to be integrable [F2]. Hence, homogeneous hypersurfaces seem to correspond to integrable systems. We are, however, not sure if the integrability of this system for homogeneous hypersurafaces can be shown directly.

There are many interesting geometrical objects such as Dupin hypersurfaces, tight and taut immersions, and Lie sphere geometry, which have a close relation to this Hamiltonian system. For more details, see [Ce],[CR],[F2],[M5].

References

[A] U. Abresch, Isoparametric hypersurfaces with four or six distinct principal curvatures. Necessary conditions on the multiplicities, Math. Ann., **264** (1983), 283–302.

[AF] A. Andreotti and T. Frankel, The Lefschetz theorem on hyperplane sections, Ann. of Math., **69** (1959), 713–717.

[B] R. L. Bryant, Some remarks on the geometry of austere submanifolds, Boletim da Sociedade Brasileira de Matematica (Nova Sie), **21** (1991), 133–157.

[BPW] J. Bolton, F. Pedit and L. Woodward, Minimal surfaces and the affine Toda field model, J. Reine Angew. Math., **459** (1995), 119–150.

[C1] E. Cartan, Familles de surfaces isoparamétriques dan les espaces à courbure constante, Ann. di Mat., **17** (1938), 177–191.

[C2] E. Cartan, Sur des familles remarkables d'hypersurfaces isoparamétriques dan les espaces sphérique, Math. Z., **45** (1939), 335–367.

[C3] E. Cartan, Sur quelques familles remarquables d'hypersurfaces, C. R. Congrès Math. Liège (1939), 30–41.

[C4] E. Cartan, Sur des familles d'hypersurfaces isoparamétriques des espaces sphèriques à 5 et à 9 dimensions, Revista Univ. Tucumán, **1** (1940), 5–22.

[Ce] T. Cecil, Lie sphere geometry, Universitext, Springer-Verlag, New York etc. (1992).

[CCJ] T. Cecil, Q. S. Chi and G. Jensen, Isoparametric hypersurfaces with four principal curvartures, to appear in Ann. of Math.

[CR] T. Cecil and P. Ryan, Tight and taut immersions of manifolds, Research Notes in Mathematics, **107**, Pitman, Boston, London, Melbourne (1985),

[DN] J. Dorfmeister and E. Neher, Isoparametric hypersufaces, case $g = 6$, $m = 1$, Comm. in Alg., **13** (1985), 2299–2368.

[DuN1] B. A. Dubrovin and S. P. Novikov, Hamiltonian formalism of one-dimensional systems of the hydrodynamic type and the Bogolyubov-Whitham averaging method, Dokl. Akad. Nauk SSSR, **270** (1983), 781–785.

[DuN2] B. A. Dubrovin and S. P. Novikov, Hydrodynamics of weakly deformed soliton lattices, Differential geometry and Hamiltonian theory, Russian Math. Surveys, **44** (1989), no. 6, 35–124.

[F1] E. V. Ferapontov, Dupin hypersurfaces and integrable hamiltonian systems of hydrodynamic type, which do not possess Riemann invariants, Diff. Geom. and its Appl., **5** (1995), 121–152.

[F2] E. V. Ferapontov, Isoparametric hypersurfaces in spheres, integrable nondiagonalizable systems of hydrodynamic type, and N-wave systems, Diff. Geom. and its Appl., **5** (1995), 335–369.

[FM] E. Ferapontov and O. Mokhov, Nonlocal Hamiltonian operators of hydrodynamic type that are connected with metrics of constant curvature, Russian Math. Surveys, **45** (1990), 218–219.

[FKM] D. Ferus, H. Karcher and H. F. Münzner, Cliffordalgebren und neue isoparametrische Hyperflächen, Math. Z., **177** (1981), 479–502.

[HaL] R. Harvey and H. B. Lawson, Calibrated geometries, Acta Math., **148** (1982), 47–157.

[HL] W. Y. Hsiang and H. B. Lawson, Minimal submanifolds of low cohomogeneity, J. Diff. Geom., **5** (1971), 1–38.

[I] S. Immervoll, The classification of isoparametric hypersurfaces with four distinct principal curvatures in spheres, to appear in Ann. of Math.

[IKM] G. Ishikawa, M. Kimura and R. Miyaoka, Submanifolds with degenerate Gauss mappings in the spheres, Advanced Stud. in Pure Math., **37** (2002), 115–149.

[J1] D. Joyce, Special Lagrangian 3-folds and integrable systems, this volume.

[J2] D. Joyce, Lectures on special Lagrangian geometry, math.DG/0111111 (2001).

[K] F. Kaneda, Isotropy orbits of the Hermitian symmetric spaces of EIII type, Master's thesis, Sophia Univ., 2003.

[KN] S. Kobayashi and K. Nomizu, Foundations of Differential Geometry II, Interscience Publ., 1963.

[Mi] J. W. Milnor, Morse theory, Ann. of Math. Studies, Princeton U.P., **51** (1963).

[M1] R. Miyaoka, L^2-harmonic 1-forms on a complete stable minimal hypersurface, In: Geometry and Global Analysis, Report of the First MSJ International Research Institute, (eds. by T. Kotake, S. Nishikawa and R. Schoen), 1993, pp. 289–294.

[M2] R. Miyaoka, The linear isotropy group of $G_2/SO(4)$, the Hopf fibering and isoparametric hypersurfaces, Osaka J. of Math., **30** (1993), 179–202.

[M3] R. Miyaoka, The family of isometric superconformal harmonic maps and the affine Toda equations, J. Reine Angew. Math., **481** (1996), 1–25.

[M4] R. Miyaoka, The Dorfmeister-Neher theorem on isoparametric hypersurfaces, math.DG/0602519.

[M5] R. Miyaoka, Isoparametric hypersurfaces, revisited, Sugaku Exposition, **17** (2004), 185–202,

[M6] R. Miyaoka, Geometry of G_2 orbits and isoparametric hypersurfaces, preprint, 2008.

[Mu1] H. F. Münzner, Isoparametrische Hyperflächen in Sphären I, Math. Ann., **251**, (1980), 57–71.

[Mu2] M. F. Münzner, Isoparametrische Hyperflächen in Sphären II, Math. Ann., **256** (1981), 215–232.

[N] K. Nomizu, Sur les algèbres de Lie de générateuer de Killing et l'homogéneité d'une variété riemannienne, Osaka J. Math., **14** (1962), 45–51.

[O] C. Olmos, Isoparametric submanifolds and their homogeneous structures, J. Diff. Geom., **38** (1993), 225–234.

[OT] H. Ozeki and M. Takeuchi, On some types of isoparametric hypersurfaces in spheres I, II, Tohoku Math. J., **27** (1975), 515–559, **28** (1976), 7–55.

[OTT] H. Ozeki, R. Takagi and M. Takeuchi, On isoparametric hypersurfaces (Japanese), Sugaku, **30** (1978), 23–32.

[P] R. Palais, Symmetries of solitons, Bull. Amer. Math. Soc., **34** (1997), 339–403.

[Pa] B. Palmer, Stability of minimal hypersurfaces, Comment Math. Helvetici, **66** (1991), 185–188.

[PT] U. Pinkall and G. Thorbergsson, Deformations of Dupin hypersurfaces, Proc. Amer. Math. Soc., **107** (1989), 1037–1043.

[S] J. Simons, Minimal varieties in Riemannian varieties, Ann. Math., **88** (1968), 62–105.

[Si] I. M. Singer, Infinitesimally homogeneous spaces, Comm. Pure Appl. Math., **13** (1960), 685–697.

[St] S. Stolz, Multiplicities of Dupin hypersurfaces, Invent. Math., **138** (1999), 253–279.

[T] R. Takagi, A class of hypersurfaces with constant principal curvatures in a shpere, J. Diff. Geom., **11** (1976), 225–233.

[TeT] C. L. Terng and G. Thorbergsson, Submanifold geometry in symmetric spaces, J. Diff. Geom., **42** (1995), 665–718.

[Th1] G. Thorbergsson, Dupin hypersurfaces, Bull. London Math. Soc., **15** (1983), 493–498.

[Th2] G. Thorbergsson, Isoparametric foliations and their buildings, Ann. of Math., **133** (1991), 429–446.

[Th3] G. Thorbergsson, A survey on isoparametric hypersurfaces and their generalizations, In: Handbook of Differential Geometry Vol. I, North-Holland, Amsterdam, 2000, pp. 963–995.

[Ts] S.P. Tsarëv, Poisson brackets and one-dimensional Hamiltonian system of hydrodynamic type, Dokl. Akad. Nauk SSSR, **282** (1985), 534–537.

[TT] R. Takagi and T. Takahashi, On the principal curvatures of homogeneous hypersurfaces in a sphere, In: Differential Geometry, in honor of K. Yano, Kinokuniya, 1972, 469–481.

[W] Q. M. Wang, On the topology of Clifford isoparametric hypersurfaces, J. Diff. Geom., **27** (1988), 55–66.

Mathematical Institute
Graduate School of Science
Tohoku University
Aramaki, Aoba-ku
Sendai 980-8578
JAPAN

E-mail address:
`r-miyaok@math.tohoku.ac.jp`

Advanced Studies in Pure Mathematics 51, 2008
Surveys on Geometry and Integrable Systems
pp. 339–345

Kähler Ricci solitons

Henrik Pedersen

§1. Introduction

We give a survey of Ricci solitons in a Kähler background. The emphasis is on joint work with Christina Tønnesen–Friedman and Galliano Valent [11].

Let (M, J) be a complex manifold. Consider pairs (g, V) consisting of a Kähler metric g and a real holomorphic vector field V on M, such that JV is an isometry of g and

$$\rho - \lambda\Omega = L_V\Omega, \tag{1}$$

where ρ is the Ricci form, Ω is the Kähler form and λ is a constant. Such structures are called *quasi-Einstein Kähler metrics* or *Kähler Ricci solitons* [4, 5, 7, 8, 12].

Remark 1. Quasi-Einstein metrics are solitons for the Hamilton flow [8]

$$\frac{d}{dt}g_t = -r_t + \frac{\overline{s_t}}{n}g_t, \tag{2}$$

where r_t is the Ricci curvature tensor and $\overline{s_t}$ is the average scalar curvature of g_t. Indeed, if g_0 is quasi-Einstein then $(\Phi_{-t})^*g_0$ solves (2), where $\Phi_t = \exp(tV)$. Thus if g_0 is quasi-Einstein but not Einstein, then g_t does not converge to an Einstein metric – it flows along V as a soliton.

Remark 2. Friedan [6] studied quasi-Einstein metrics in connection with bosonic σ-models. He showed that the one-loop renormalizability

Received January 3, 2001.
2000 *Mathematics Subject Classification.* Primary 53C25; Secondary 53C55.
Partially supported by European Contract HPRN-CT-2000-00101.

of the model is ensured if and only if the metric of the target space is quasi-Einstein.

Remark 3. If M is a compact manifold, then the equation

$$\rho - \lambda\Omega = L_V\Omega,$$

implies that $[\rho] = \lambda[\Omega] \in H^2(M,\mathbb{R})$. Thus a necessary condition for M to admit a quasi-Einstein Kähler metric is that $c_1(M) = [\frac{\rho}{2\pi}]$ has a sign. If $c_1 \leq 0$, then Calabi, Yau and Aubin showed that there exist Kähler-Einstein metrics. However, for $c_1 > 0$ we do not always have Kähler-Einstein metrics and the quasi-Einstein Kähler metrics serve as suitable generalizations.

Remark 4. Let M be compact and assume $c_1(M)$ is positive. Assume M is a Kähler-Ricci soliton with non-trivial V. Recall [2], that for any compact Kähler manifold, using Hodge theory and Kähler identities, we have

$$\rho - \rho_H = \sqrt{-1}\partial\bar{\partial}\varphi_\Omega = L_{\frac{1}{2}\nabla\varphi_\Omega}\Omega,$$

where ρ_H is the harmonic part of ρ and φ_Ω is called the Ricci potential. Indeed, we have $\varphi_\Omega = -Gs$, where G is the Green's operator of the Laplacian and s is the scalar curvature. Furthermore, the Futaki invariant of the Kähler class [2] associates to each holomorphic vector field X the integral

$$(m!)^{-1}\int_{M^{2m}} X(\varphi_\Omega)\Omega^m.$$

It follows easily that the Futaki invariant of the Kähler class on the vector field V is given as the L^2-norm of V. This observation tells us that the existence of quasi-Einstein Kähler metrics with non-trivial vector fields is an obstruction to the existence of Kähler-Einstein metrics.

Remark 5. In the compact case, Tian and Zhu [12] have proved uniqueness (modulo automorphisms) for Kähler-Ricci solitons with a fixed vector field.

The pair (g, V) is said to be *generalized quasi-Einstein* if

$$\rho - \rho_H = L_V\Omega, \tag{3}$$

where ρ_H is the harmonic part of the Ricci form ρ.

Remark 6. If (g, V) is quasi-Einstein Kähler then certainly

$$\rho_H = \lambda\Omega,$$

so we are indeed talking about a generalization.

Remark 7. The notion of generalized quasi-Einstein Kähler metrics is a generalization of constant scalar curvature Kähler metrics. Indeed, Guan [7] proved that a generalized quasi-Einstein Kähler metric has constant scalar curvature if and only if the Futaki invariant of the Kähler class vanishes. Thus, these metrics behave very much like extremal Kähler metrics [2].

§2. Constructions

In order to construct these solitons we look for Kähler metrics with at least one symmetry X. Therefore, the equations can be formulated in terms of complex coordinates on the Kähler quotient and a momentum map z. Recall for example the LeBrun Ansatz [9] (in real dimension four) for scalar-flat Kähler metrics

$$g = e^u w(dx^2 + dy^2) + wdz^2 + w^{-1}(dt + A)^2$$

$$w_{xx} + w_{yy} + (we^u)_{zz} = 0$$

$$u_{xx} + u_{yy} + (e^u)_{zz} = 0,$$

or the ansatz [10] for Kähler-Einstein metrics with g as above, where

$$w = \frac{u_z}{-2\lambda(z + B)}$$

$$u_{xx} + u_{yy} + z^2\left(z^{-2}(e^u)_z\right)_z = 0.$$

Here (x, y) are isothermal coordinates on the Kähler quotient, (u, w) are functions of (x, y, z), $X = \frac{\partial}{\partial t}$, A is a 1-form in (x, y, z) and (λ, B) are constants.

In higher dimensions a Kähler metric on M^{2m} with a symmetry X is given as

$$g = h + wdz^2 + w^{-1}\omega^2$$

where h is a Kähler metric on the Kähler quotient $B^{2(m-1)}$, ω is a connection 1-form and w is a function of the moment map coordinate z and of the complex coordinates ξ^μ on $B^{2(m-1)}$. The complex structure and the Kähler form are given by

$$J\omega = -wdz \quad \text{and} \quad \Omega = dz \wedge \omega + \Omega_h,$$

where Ω_h is the Kähler form on $B^{2(m-1)}$. Then the Kähler condition gives

$$\frac{\partial^2 h_{\mu\nu}}{\partial z^2} + 4\frac{\partial^2 w}{\partial \xi^\mu \partial \overline{\xi^\nu}} = 0. \tag{4}$$

To obtain the equations $\rho - \lambda\Omega = L_V\Omega$ for a soliton we furthermore assume that $V = -\frac{1}{2}cJX$ (for a constant c), and we have: let

$$u = \log(\frac{\det h}{w}) \quad \text{and} \quad w = \frac{c + u_z}{-2\lambda(z+B)};$$

then

$$4\frac{\partial^2 u}{\partial \xi^\mu \partial \overline{\xi^\nu}} = -2\lambda\left(h_{\mu\nu} - (z+B)\frac{\partial h_{\mu\nu}}{\partial z}\right). \tag{5}$$

In four dimensions this last equation is

$$u_{xx} + u_{yy} + z^2\left(z^{-2}(e^u)_z\right)_z + c\,z^2\left(z^{-2}e^u\right)_z = 0.$$

To solve these equations we proceed as follows. Let (B, g_B) be an $(m-1)$-dimensional compact Kähler manifold with scalar curvature s_B. Assume that the Kähler form Ω_B is such that the deRham class $[\frac{\Omega_B}{2\pi}]$ is contained in the image of $H^2(B,\mathbb{Z}) \to H^2(B,\mathbb{R})$. Let L be a holomorphic line bundle such that $c_1(L) = [\frac{-\Omega_B}{2\pi}]$. On the total space M of $(L-0) \xrightarrow{\pi} B$ we can form an S^1-symmetric Kähler metric

$$g = zg_B + wdz^2 + w^{-1}\omega^2$$

where z, being the coordinate of $(a,b) \subset (0,\infty]$, becomes the moment map of g with the obvious S^1 action on L, w is a positive function depending only on z, and ω is the connection one-form of the connection induced by g on the S^1-bundle

$$(L-0) \overset{(\pi,z)}{\longrightarrow} B \times (a,b).$$

That is $d\omega = \Omega_B$. Clearly, condition (4) is satisfied. From equation (5) we see that the base g_B must be *Kähler-Einstein* and

$$\left(\frac{z^{m-1}}{w}\right)_z + c\frac{z^{m-1}}{w} = -2\lambda z^m + \frac{s_B}{(m-1)}z^{m-1}.$$

§3. Global Metrics

If w^{-1} satisfies $w^{-1}(a) = 0$ and $(w^{-1})'(a) = 2$, then we can add a copy of B at $z = a$ and extend the Kähler metric g over the zero-section of the bundle $L \to B$. If moreover $b < \infty$, $w^{-1}(b) = 0$, and $(w^{-1})'(b) = -2$ then we can add another copy of B at $z = b$ and extend g to a Kähler metric on the total space of the $\mathbb{CP}_1$-bundle $\mathbb{P}(\mathcal{O} \oplus L)$ [11].

Theorem 1. *Let (B, g_B) be a non-positive compact Kähler-Einstein manifold of dimension $(m-1)$. Assume that $[\frac{\Omega_B}{2\pi}]$ is an integer cohomology class. Let L be a holomorphic line bundle on B such that $c_1(L) = [\frac{-\Omega_B}{2\pi}]$. Let X denote the Hamiltonian vector field generating the natural S^1 action on L and let J denote the complex structure on the total space of $L \to B$. Then, for a given $a > 0$ and a given $c > 0$, there exists a complete Kähler metric g on the total space of the bundle $L \to B$ such that the pair $(g, -\frac{1}{2}cJX)$ is quasi-Einstein, satisfying the equation*

$$\rho - \lambda\Omega = L_{-\frac{1}{2}cJX}\Omega,$$

where

$$\lambda = \frac{s_B - 2(m-1)}{2a(m-1)}.$$

Remark 8. Notice that there is no hope of producing quasi-Einstein metrics on the compact manifold $\mathbb{P}(\mathcal{O} \oplus L) \to B$. This follows from the fact that c_1 has no sign for $s_B \leq 0$: take any compact metric on M of the type described above. Since $\int_C \rho > 0$ and $\int_{E_0} \rho \wedge \Omega_B^{m-2} < 0$ when $s_B \leq 0$, we conclude that $c_1 = [\frac{\rho}{2\pi}]$ does not have a sign.

Remark 9. Koiso and Guan [8, 7] considered only *positive* s_B and obtained solutions on

$$M = \mathbb{P}(\mathcal{O} \oplus K^{\frac{p}{m}}) \to \mathbb{C}P_{m-1}; \qquad p = 1, \ldots, m-1,$$

where K is the canonical line bundle. For $p = m+1, m+2, \ldots$ they found complete non-compact quasi-Einstein Kähler metrics on the total space of $K^{\frac{p}{m}} \to B$. Note that $s_B = \frac{2m(m-1)}{p}$.

§4. Generalized Quasi-Einstein Manifolds

We use the same approach as above, but we stay in real dimension four. Therefore, B is a compact Riemann surface, and g_B is Kähler-Einstein ($\rho_B = \frac{s_B}{2}\Omega_B$). Finally, we assume that g can be extended to a smooth Kähler metric on the compact manifold $M = \mathbb{P}(\mathcal{O} \oplus L) \to B$. Now, taking traces, the equation $\rho - \rho_H = L_V\Omega$ implies

$$s - \overline{s} = -c\Delta z. \tag{6}$$

Conversely, if (6) is satisfied, then the Ricci potential is given as

$$\varphi_\Omega = -Gs = cz + \kappa,$$

where κ is some constant. Then $\nabla\varphi_\Omega = c\nabla z = 2V$, and since any compact Kähler metric satisfies

$$\rho - \rho_H - L_{\frac{1}{2}\nabla\varphi_\Omega}\Omega,$$

we conclude that (g, V) is generalized quasi-Einstein. Now, inserting

$$s = \frac{s_B}{z} - \frac{(\frac{z}{w})_{zz}}{z}$$

into (6), we see that there are in fact solutions w of the resulting equation satisfying the boundary conditions for compact metrics [11]:

Theorem 2. *Let $M = \mathbb{P}(\mathcal{O} \oplus L) \to B$, where L is a non-trivial holomorphic line bundle on a compact Riemann surface B. Then any Kähler class on M admits a generalized quasi-Einstein Kähler metric (g, V), where $V = \frac{1}{2}\nabla\varphi_\Omega$ and $W := V - iJV$ is a multiple of the holomorphic vector field generating the natural $\mathbb{C}^*$ action on L.*

Remark 10. If the genus of B is less than 2, then the metrics in the above theorem were constructed by Guan [7]. In particular, if the genus of B is equal to 0, $L = K^{\frac{1}{2}}$ and the Kähler class is a multiple of $c_1(M)$, we have the quasi-Einstein Kähler metric on $\mathbb{C}P_2\#\overline{\mathbb{C}P_2}$ constructed by Koiso [8].

Remark 11. If the genus of B is at least 2, we notice that, in contrast with the family of extremal Kähler metrics constructed by Tønnesen-Friedman [13], the generalized quasi-Einstein Kähler metrics above do exhaust the Kähler cone.

Acknowledgement. I am in great debt to Christina Tønnesen-Friedman and Galliano Valent. I would also like to take this opportunity to thank the organizers of the conference for their very successful work.

References

[1] T. Aubin, Equations du type de Monge-Ampère sur les variétés kähleriennes compactes, C. R. Acad. Sci. Paris, **283** (1976), 119–121.

[2] A. L. Besse, Einstein Manifolds, Springer-Verlag, 1987.

[3] E. Calabi, On Kähler manifolds with vanishing canonical class, In: Algebraic geometry and topology, a symposium in honor of S. Lefschetz, Princeton Univ. Press, 1955, pp. 78–89.

[4] H.-D. Cao, Existence of gradient Kähler Ricci solitons, In: Elliptic and Parabolic Methods in Geometry, Minneapolis, MN, 1994, (eds. B. Chow, R. Gulliver, S. Levy and J. Sullivan), AK Peters, 1996, pp. 1–16.

[5] T. Chave and G. Valent, On a class of compact and non-compact quasi-Einstein metrics and their renormalizability properties, Nuclear Phys. B, **478** (1996), 758–778.

[6] D. Friedan, Nonlinear models in $2+\varepsilon$ dimension, Phys. Rev. Lett., **45** (1980), 1057–1060.

[7] Z.-D. Guan, Quasi-Einstein metrics, Internat. J. Math., **6** (1995), 371–379.

[8] N. Koiso, On rotationally symmetric Hamilton's equation for Kähler-Einstein metrics, In: Recent Topics in Differential and Analytic Geometry, Adv. Stud. Pure Math., **18-I** (1990), 327–337.

[9] C. LeBrun, Scalar-flat Kähler metrics on blown-up ruled surfaces, J. Reine Angew. Math., **420** (1991), 161–177.

[10] H. Pedersen and Y. S. Poon, Hamiltonian construction of Kähler-Einstein metrics and Kähler metrics of constant scalar curvature, Comm. Math. Phys., **136** (1991), 309–326.

[11] H. Pedersen, C. Tønnesen-Friedman and G. Valent, Quasi-Einstein Kähler Metrics, Lett. Math. Phys., **50** (1999), 229–241.

[12] G. Tian and X. H. Zhu, Uniqueness of Kähler-Ricci solotions, Acta Math., **184** (2000), 271–305.

[13] C. Tønnesen-Friedman, Extremal Kähler Metrics on Minimal Ruled Surfaces, J. Reine Angew. Math., **502** (1998), 175–197.

[14] S. T. Yau, On the Ricci curvature of a compact Kähler manifold and the complex Monge-Ampère equation I, Comm. Pure Appl. Math., **31** (1978), 339–411.

University of Southern Denmark
Campusvej 55
5230 Odense M
DENMARK

E-mail address:
`henrik@sdu.dk`

Advanced Studies in Pure Mathematics 51, 2008
Surveys on Geometry and Integrable Systems
pp. 347–399

Period problems for mean curvature one surfaces in H^3 (with applications to surfaces of low total curvature)

Wayne Rossman, Masaaki Umehara and Kotaro Yamada

§1. Introduction

There is a wide body of knowledge about minimal surfaces in Euclidean 3-space $\boldsymbol{R}^3$, and there is a canonical local isometric correspondence (sometimes called the Lawson correspondence) between minimal surfaces in $\boldsymbol{R}^3$ and CMC-1 (constant mean curvature one) surfaces in hyperbolic 3-space H^3 (the complete simply-connected 3-manifold of constant sectional curvature -1). This has naturally led to the recent interest in and development of CMC-1 surfaces in H^3 in the last decade. There are now many known examples, and it is a natural next step to classify all such surfaces with low total absolute curvature.

By this canonical local isometric correspondence, minimal immersions in $\boldsymbol{R}^3$ are locally equivalent to CMC-1 immersions in H^3. But there are interesting differences between these two types of immersions on the global level. There are period problems on non-simply-connected domains of the immersions, which might be solved for one type of immersion but not the other. Solvability of the period problems is usually more likely in the H^3 case, leading to a wider variety of surfaces there. For example, a genus 1 surface with finite total curvature and two embedded ends cannot exist as a minimal surface in $\boldsymbol{R}^3$, but it does exist as a CMC-1 surface in H^3 [RS]. And a genus 0 surface with finite total curvature and two embedded ends exists as a minimal surface in $\boldsymbol{R}^3$ only if it is a surface of revolution, but it may exist as a CMC-1 surface in

Received February 24, 2001.
Revised January 8, 2002.
2000 *Mathematics Subject Classification.* Primary 53A10.

H^3 without being a surface of revolution (see Example 4.3). So there are many more possibilities for CMC-1 surfaces in H^3 than there are for minimal surfaces in $\boldsymbol{R}^3$. This means that it is more difficult to classify CMC-1 surfaces with low total curvature in H^3.

To find complete CMC-1 surfaces in H^3 with low total curvature, we must first determine the meromorphic data in the Bryant representation of the surfaces that can admit low total curvature, and then we must analyse when the parameters in the data can be adjusted to solve the period problems. Generally, finding the data is the easier step, and solving the period problems is the more difficult step. As the period problems are generally the crux of the problem, we have chosen the title of this paper to reflect this.

The total absolute curvature of a minimal surface in $\boldsymbol{R}^3$ is equal to the area of the image (counted with multiplicity) of the Gauss map of the surface, and complete minimal surfaces in $\boldsymbol{R}^3$ with total curvature at most 8π have been classified (see Lopez [Lop] and also Table 2). Furthermore, as the Gauss map of a complete conformally parametrized minimal surface is meromorphic, and has a well-defined limit at each end when the surface has finite total curvature, the area of the Gauss image must be an integer multiple of 4π.

However, unlike the case of minimal surfaces in $\boldsymbol{R}^3$, when searching for CMC-1 surfaces in H^3 with low total absolute curvature, we have a choice of two different Gauss maps: the *hyperbolic Gauss map* G and the *secondary Gauss map* g. So there are two ways to pose the question in H^3, with two very different answers. One way is to consider the true total absolute curvature, which is the area of the image of g, but since g might not be single-valued on the surface, the total curvature might not be an integer multiple of 4π, and this allows for many more possibilities. Furthermore, the Osserman inequality does not hold for the true total absolute curvature. The weaker Cohn-Vossen inequality is the best general lower bound for true absolute total curvature (with equality never holding [UY1]). So the true total absolute curvature is difficult to analyse, but it is important because of its clear geometric meaning.

The second way is to study the area of the image of G, which we call the *dual* total absolute curvature, as it is the true total curvature of the dual CMC-1 surface (which we define in Section 3) in H^3. This way has the advantage that G is single-valued on the surface, and so the dual total absolute curvature is always an integer multiple of 4π, like the case of minimal surfaces in $\boldsymbol{R}^3$. Furthermore, the dual total curvature satisfies not only the Cohn-Vossen inequality, but also the Osserman inequality [UY5, Yu2] (see also (3.13) in Section 3). So the dual total curvature shares more properties with the total curvature of minimal surfaces in $\boldsymbol{R}^3$, motivating our interest in it.

We shall refer to the true total absolute curvature of a CMC-1 immersion $f\colon M \to H^3$ of a Riemann surface as $\mathrm{TA}(f)$, and the dual total absolute curvature as $\mathrm{TA}(f^{\#})$.

We review the classification results for surfaces with $\mathrm{TA}(f) \leq 4\pi$ or $\mathrm{TA}(f^{\#}) \leq 4\pi$ in Section 2, which are results from [RUY4] and [RUY3]. An inequality for $\mathrm{TA}(f)$ stronger than the Cohn-Vossen inequality [RUY4] (for surfaces of genus zero with odd number of ends) is also introduced. In Section 3, we review basic notions and terminology. We introduce some important examples of CMC-1 surfaces in Section 4. Section 6 is devoted to describing the results in [RUY3], a partial classification of CMC-1 surfaces with $\mathrm{TA}(f^{\#}) \leq 8\pi$. In the final section 7, we introduce new results on partial classification of CMC-1 surfaces with $\mathrm{TA}(f) \leq 8\pi$. Since the proofs of these results are more technical and delicate than those of the results on $\mathrm{TA}(f^{\#})$, we include them in Appendix A.

§2. The cases $\mathrm{TA}(f)$ or $\mathrm{TA}(f^{\#}) \leq 4\pi$, and a natural extension

In [RUY4] the following theorem was proven:

Theorem 2.1. *Let $f\colon M \to H^3$ be a complete* CMC-1 *immersion of total absolute curvature* $\mathrm{TA}(f) \leq 4\pi$. *Then f is either*

- *a horosphere* (*Example* 4.1),
- *an Enneper cousin* (*Example* 4.2),
- *an embedded catenoid cousin* ($0 < l < 1$, $\delta = 1$ *and* $b = 0$ *in Example* 4.3),
- *a finite δ-fold covering of an embedded catenoid cousin* ($\delta \geq 2$, $0 < l \leq 1/\delta$ *and* $b = 0$ *in Example* 4.3), *or*
- *a warped catenoid cousin with injective secondary Gauss map* ($l = 1$, $\delta \in \boldsymbol{Z}^+$ *and* $b > 0$ *in Example* 4.3).

The horosphere is the only flat (and consequently totally umbilic) CMC-1 surface in H^3. The catenoid cousins are the only CMC-1 surfaces of revolution [Bry]. The Enneper cousins are isometric to minimal Enneper surfaces [Bry]. The warped catenoid cousins [UY1, RUY3] are less well known and are described more precisely in Section 4, as well as the other above three examples.

Although this theorem is simply stated, for the reasons given in the introduction the proof is more delicate than it would be if the condition $\mathrm{TA}(f) \leq 4\pi$ is replaced with $\mathrm{TA}(f^{\#}) \leq 4\pi$, or if minimal surfaces in $\boldsymbol{R}^3$ with $\mathrm{TA} \leq 4\pi$ are considered. CMC-1 surfaces f with $\mathrm{TA}(f^{\#}) \leq 4\pi$ are classified in Theorem 2.3 below. It is well-known that the only complete minimal surfaces in $\boldsymbol{R}^3$ with $\mathrm{TA} \leq 4\pi$ are the plane, the Enneper surface, and the catenoid (see Table 2).

We extend the above result in Section 7 to find an inclusive list of possibilities for CMC-1 surfaces with $\mathrm{TA}(f) \leq 8\pi$, and we consider which possibilities we can classify or find examples for, see Table 3. (Minimal surfaces in $\boldsymbol{R}^3$ with $\mathrm{TA} \leq 8\pi$ are classified by Lopez [Lop]. See Table 2.)

For a complete CMC-1 immersion f in H^3, equality in the Cohn-Vossen inequality never holds ([UY1, Theorem 4.3]). In particular, if f is of genus 0 with n ends, then

$$\mathrm{TA}(f) > 2\pi(n-2)\ . \tag{2.1}$$

When $n = 2$, the catenoid cousins show that (2.1) is sharp. However, we see from the above theorem that

$$\mathrm{TA}(f) > 4\pi \quad \text{for} \quad n = 3\ ,$$

which is stronger than the Cohn-Vossen inequality (2.1). The following theorem, which extends the above theorem and is proven in [RUY4], gives a sharper inequality than the Cohn-Vossen inequality when n is any odd integer:

Theorem 2.2. *Let* $f\colon \boldsymbol{C} \cup \{\infty\} \setminus \{p_1, \dots, p_{2m+1}\} \to H^3$ *be a complete conformal genus* 0 CMC-1 *immersion with* $2m+1$ *ends,* $m \in \boldsymbol{Z}^+$. *Then* $\mathrm{TA}(f) \geq 4\pi m$.

Remark. When $m = 1$, we know that the lower bound 4π in the theorem is sharp (see Example 4.4). However, we do not know if it is sharp for general m. For genus 0 CMC-1 surfaces with an even number $n \geq 4$ of ends, it is still an open question whether there exists any stronger lower bounds than that of the Cohn-Vossen inequality. It should be remarked that in Section 4 we have numerical examples with $n = 4$ whose total absolute curvature tends to 4π.

For the case of $\mathrm{TA}(f^{\#})$, the following theorem was proven in [RUY3]:

Theorem 2.3. *A complete* CMC-1 *immersion* f *with* $\mathrm{TA}(f^{\#}) \leq 4\pi$ *is congruent to one of the following:*

(1) a horosphere (Example 4.1*),*

(2) an Enneper cousin dual (Example 4.2*),*

(3) a catenoid cousin ($\delta = 1$*,* $l \neq 1$ *and* $b = 0$ *in Example* 4.3*), or*

(4) a warped catenoid cousin with embedded ends and injective hyperbolic Gauss map ($\delta = 1$*,* $l \in \boldsymbol{Z}$*,* $l \geq 2$ *and* $b > 0$ *in Example* 4.3*).*

§3. Preliminaries

Before we can state any results for the cases of higher $\mathrm{TA}(f)$ and higher $\mathrm{TA}(f^{\#})$, we must give some preliminaries here.

Let $f\colon M \to H^3$ be a conformal CMC-1 immersion of a Riemann surface M into H^3. Let ds^2, dA and K denote the induced metric, induced area element and Gaussian curvature, respectively. Then $K \leq 0$ and $d\sigma^2 := (-K)\,ds^2$ is a conformal pseudometric of constant curvature 1 on M. We call this pseudometric's developing map $g\colon \widetilde{M}(:=$ the universal cover of $M) \to \boldsymbol{CP}^1 = \boldsymbol{C} \cup \{\infty\}$ the *secondary Gauss map* of f. Namely, g is a conformal map so that its pull-back of the Fubini-Study metric of $\boldsymbol{CP}^1$ equals $d\sigma^2$:

$$d\sigma^2 = (-K)\,ds^2 = \frac{4\,dg\,d\bar{g}}{(1+g\bar{g})^2}\,. \tag{3.1}$$

Such a map g is determined by $d\sigma^2$ uniquely up to the change

$$g \mapsto a \star g := \frac{a_{11}g + a_{12}}{a_{21}g + a_{22}}\,, \qquad a = \begin{pmatrix} a_{11} & a_{12} \\ a_{21} & a_{22} \end{pmatrix} \in \mathrm{SU}(2)\,. \tag{3.2}$$

Since $d\sigma^2$ is invariant under the deck transformation group $\pi_1(M)$, there is a representation

(3.3)
$$\rho_g\colon \pi_1(M) \longrightarrow \mathrm{PSU}(2) \quad \text{such that} \quad g \circ \tau^{-1} = \rho_g(\tau) \star g \quad \bigl(\tau \in \pi_1(M)\bigr)\,,$$

where $\mathrm{PSU}(2) = \mathrm{SU}(2)/\{\pm\,\mathrm{id}\}$. The metric $d\sigma^2$ is called *reducible* if the image of ρ_g can be diagonalized simultaneously, and is called *irreducible* otherwise. In the case $d\sigma^2$ is reducible, we call it is $\mathcal{H}^3$*-reducible* if the image of ρ_g is the identity, and is called $\mathcal{H}^1$*-reducible* otherwise. We call a CMC-1 immersion $f\colon M \to H^3$ $\mathcal{H}^1$-reducible (respectively, $\mathcal{H}^3$-reducible) if the corresponding pseudometric $d\sigma^2$ is $\mathcal{H}^1$-reducible (respectively, $\mathcal{H}^3$-reducible). For details on reducibility, see Section 5.

In addition to g, two other holomorphic invariants G and Q are closely related to geometric properties of CMC-1 surfaces. The *hyperbolic Gauss map* $G\colon M \to \boldsymbol{CP}^1$ is holomorphic and is defined geometrically by identifying the ideal boundary of H^3 with $\boldsymbol{CP}^1$: $G(p)$ is the asymptotic class of the normal geodesic of $f(M)$ starting at $f(p)$ and oriented in the mean curvature vector's direction. The *Hopf differential* Q is a holomorphic symmetric 2-differential on M such that $-Q$ is the $(2,0)$-part of the complexified second fundamental form. The Gauss equation implies

$$ds^2 \cdot d\sigma^2 = 4\,Q \cdot \overline{Q}\,, \tag{3.4}$$

where $\cdot$ means the symmetric product. Moreover, these invariants are related by

$$S(g) - S(G) = 2Q\,, \tag{3.5}$$

where $S(\cdot)$ denotes the Schwarzian derivative:

$$S(h) := \left[\left(\frac{h''}{h'}\right)' - \frac{1}{2}\left(\frac{h''}{h'}\right)^2\right] dz^2 \qquad \left({}' = \frac{d}{dz}\right)$$

with respect to a local complex coordinate z on M.

In terms of g and Q, the induced metric ds^2 and complexification of the second fundamental form $I\!I$ are

$$ds^2 = (1+|g|^2)^2\left|\frac{Q}{dg}\right|^2, \qquad I\!I = -Q - \overline{Q} + ds^2\,.$$

Since $K \le 0$, we can define the *total absolute curvature* as

$$\mathrm{TA}(f) := \int_M (-K)\,dA \in [0,+\infty]\,.$$

Then $\mathrm{TA}(f)$ is the area of the image of M in $\boldsymbol{CP}^1$ of the secondary Gauss map g. $\mathrm{TA}(f)$ is generally not an integer multiple of 4π; for catenoid cousins [Bry, Example 2] and their δ-fold covers, $\mathrm{TA}(f)$ can be *any* positive real number.

For each conformal CMC-1 immersion $f\colon M \to H^3$, there is a holomorphic null immersion $F\colon \widetilde{M} \to \mathrm{SL}(2,\boldsymbol{C})$, the *lift* of f, satisfying the differential equation

$$dF = F\begin{pmatrix} g & -g^2 \\ 1 & -g \end{pmatrix}\omega\,, \qquad \omega = \frac{Q}{dg} \tag{3.6}$$

so that $f = FF^*$, where $F^* = {}^t\overline{F}$ [Bry, UY1]. Here we consider

$$H^3 = \mathrm{SL}(2,\boldsymbol{C})/\,\mathrm{SU}(2) = \{aa^* \mid a \in \mathrm{SL}(2,\boldsymbol{C})\}\,.$$

We call a pair (g,ω) the *Weierstrass data* of f. The lift F is said to be *null* because $\det F^{-1}dF$, the pull-back of the Killing form of $\mathrm{SL}(2,\boldsymbol{C})$ by F, vanishes identically on $\widetilde{M}$. Conversely, for a holomorphic null immersion $F\colon \widetilde{M} \to \mathrm{SL}(2,\boldsymbol{C})$, $f := FF^*$ is a conformal CMC-1 immersion of $\widetilde{M}$ into H^3. If $F = (F_{ij})$, equation (3.6) implies

$$g = -\frac{dF_{12}}{dF_{11}} = -\frac{dF_{22}}{dF_{21}}\,, \tag{3.7}$$

and it is shown in [Bry] that

$$G = \frac{dF_{11}}{dF_{21}} = \frac{dF_{12}}{dF_{22}}\,. \tag{3.8}$$

The inverse matrix F^{-1} is also a holomorphic null immersion, and produces a new CMC-1 immersion $f^\# = F^{-1}(F^{-1})^*\colon \widetilde{M} \to H^3$, called the *dual* of f [UY5]. The induced metric $ds^{2\#}$ and the Hopf differential $Q^\#$ of $f^\#$ are

$$ds^{2\#} = (1+|G|^2)^2\left|\frac{Q}{dG}\right|^2\,, \qquad Q^\# = -Q\,. \tag{3.9}$$

So $ds^{2\#}$ and $Q^\#$ are well-defined on M itself, even though $f^\#$ might be defined only on $\widetilde{M}$. This duality between f and $f^\#$ interchanges the roles of the hyperbolic Gauss map G and secondary Gauss map g. In particular, one has

$$dF\,F^{-1} = -(F^{-1})^{-1}d(F^{-1}) = \begin{pmatrix} G & -G^2 \\ 1 & -G \end{pmatrix}\frac{Q}{dG}\,. \tag{3.10}$$

Hence dFF^{-1} is single-valued on M, whereas $F^{-1}dF$ generally is not.

Since $ds^{2\#}$ is single-valued on M, we can define the *dual total absolute curvature*

$$\mathrm{TA}(f^\#) := \int_M (-K^\#)\,dA^\#,$$

where $K^\#$ (≤ 0) and $dA^\#$ are the Gaussian curvature and area element of $ds^{2\#}$, respectively. As

$$d\sigma^{2\#} := (-K^\#)ds^{2\#} = \frac{4\,dG\,d\overline{G}}{(1+|G|^2)^2} \tag{3.11}$$

is a pseudo-metric of constant curvature 1 with developing map G, $\mathrm{TA}(f^{\#})$ is the area of the image of G on $\boldsymbol{CP}^1 = S^2$. The following assertion is important for us:

Lemma 3.1 ([UY5, Yu2]). *The metric $ds^{2\#}$ is complete (respectively, nondegenerate) if and only if ds^2 is complete (respectively, nondegenerate).*

We now assume that the induced metric ds^2 (and consequently $ds^{2\#}$) on M is complete and that either $\mathrm{TA}(f) < \infty$ or $\mathrm{TA}(f^{\#}) < \infty$, hence there exists a compact Riemann surface $\overline{M}_\gamma$ of genus γ and a finite set of points $\{p_1, \ldots, p_n\} \subset \overline{M}_\gamma$ $(n \geq 1)$ so that M is biholomorphic to $\overline{M}_\gamma \setminus \{p_1, \ldots, p_n\}$ (see Theorem 9.1 of [Oss]). We call the points p_j the *ends* of f.

Unlike the Gauss map for minimal surface with $\mathrm{TA} < \infty$ in $\boldsymbol{R}^3$, the hyperbolic Gauss map G of the surface might not extend to a meromorphic function on $\overline{M}_\gamma$, as the Enneper cousin (Example 4.2) shows. However, the Hopf differential Q does extend to a meromorphic differential on $\overline{M}_\gamma$ [Bry]. We say an end p_j $(j = 1, \ldots, n)$ of a CMC-1 immersion is *regular* if G is meromorphic at p_j. When $\mathrm{TA}(f) < \infty$, an end p_j is regular precisely when the order of Q at p_j is at least -2, and otherwise G has an essential singularity at p_j [UY1]. Moreover, the pseudometric $d\sigma^2$ as in (3.1) has a *conical singularity* at each end p_j [Bry]. For a definition of conical singularity, see Section 5 (see also [UY3, UY7]).

Thus the orders of Q at the ends p_j are important for understanding the geometry of the surface, so we now introduce a notation that reflects this. We say a CMC-1 surface is of *type* $\mathbf{\Gamma}(d_1, \ldots, d_n)$ if it is given as a conformal immersion $f\colon \overline{M}_\gamma \setminus \{p_1, \ldots, p_n\} \to H^3$, where $\mathrm{ord}_{p_j} Q = d_j$ for $j = 1, \ldots, n$ (for example, if $Q = z^{-2}dz^2$ at $p_1 = 0$, then $d_1 = -2$). We use $\mathbf{\Gamma}$ because it is the capitalized form of γ, the genus of $\overline{M}_\gamma$. For instance, the class $\mathbf{I}(-4)$ means the class of surfaces of genus 1 with 1 end so that Q has a pole of order 4 at the end, and the class $\mathbf{O}(-2, -3)$ is the class of surfaces of genus 0 with two ends so that Q has a pole of order 2 at one end and a pole of order 3 at the other.

Analogue of the Osserman inequality. For a CMC-1 surface of genus γ with n ends, the second and third authors showed that the equality of the Cohn-Vossen inequality for the total absolute curvature never holds [UY1]:

$$\frac{1}{2\pi}\,\mathrm{TA}(f) > -\chi(M) = 2\gamma - 2 + n\,. \tag{3.12}$$

The catenoid cousins (Example 4.3) show that this inequality is the best possible.

On the other hand, the dual total absolute curvature satisfies an Osserman-type inequality [UY5]:

$$\frac{1}{2\pi}\,\mathrm{TA}(f^{\#}) \geq -\chi(M) + n = 2(\gamma + n - 1)\,. \tag{3.13}$$

Moreover, equality holds exactly when all the ends are embedded: This follows by noting that equality is equivalent to all ends being regular and embedded ([UY5]), and that any embedded end must be regular (proved recently by Collin, Hauswirth and Rosenberg [CHR1] and Yu [Yu3]).

Effects of transforming the lift $\boldsymbol{F}$. Here we consider the change $\hat{F} = aFb^{-1}$ of the lift F, where $a, b \in \mathrm{SL}(2, \boldsymbol{C})$. Then $\hat{F}$ is also a holomorphic null immersion, and the hyperbolic Gauss map $\hat{G}$, the secondary Gauss map $\hat{g}$ and the Hopf differential $\hat{Q}$ of $f = \hat{F}\hat{F}^*$ are given by (see [UY3])

$$\hat{G} = a \star G = \frac{a_{11}G + a_{12}}{a_{21}G + a_{22}}\,, \qquad \hat{g} = b \star g = \frac{b_{11}g + b_{12}}{b_{21}g + b_{22}}\,, \qquad \text{and} \qquad \hat{Q} = Q\,, \tag{3.14}$$

where $a = (a_{ij})$ and $b = (b_{ij})$. In particular, the change $\hat{F} = aF$ moves the surface by a rigid motion of H^3, and does not change g and Q.

SU(2)-monodromy conditions. Here we recall from [RUY1] the construction of CMC-1 surfaces with given hyperbolic Gauss map G and Hopf differential Q. Let $\overline{M}_\gamma$ be a compact Riemann surface and $M := \overline{M}_\gamma \setminus \{p_1, \dots, p_n\}$. Let G and Q be a meromorphic function and meromorphic 2-differential on $\overline{M}_\gamma$. We assume the pair (G, Q) satisfies the following two compatibility conditions:

(3.15)
For all $q \in M$, $\mathrm{ord}_q Q$ is equal to the branching order of G, and

(3.16)
for each end p_j, (branching order of G)$-d_j \geq 2$.

The first condition implies that the metric $ds^{2\#}$ as in (3.9) is non-degenerate at $q \in M$. The second condition implies that the metric $ds^{2\#}$ is complete at $p_j \in \overline{M}_\gamma$ $(j = 1, \dots, n)$. Our goal is to get a CMC-1 immersion $f \colon M \to H^3$ with hyperbolic Gauss map G and Hopf differential Q. If such an immersion exists, the induced metric ds^2 of f is non-degenerate and complete, by Lemma 3.1.

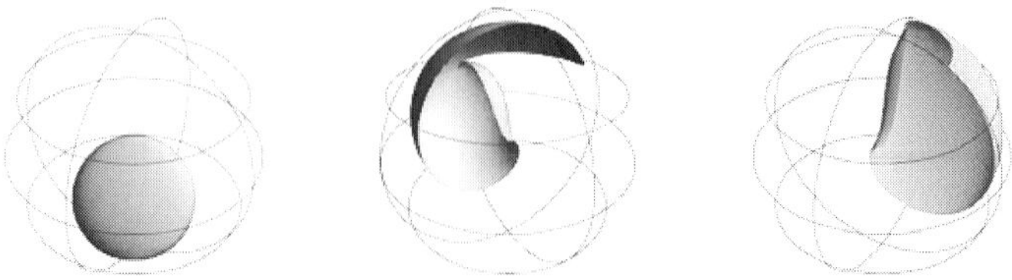

Fig. 1. A horosphere, and fundamental pieces (one-fourth of the surfaces with the ends cut away) of an Enneper cousin and the dual of an Enneper cousin. Figures are shown in the Poincaré model of H^3.

Since a pair (G,Q) satisfies (3.15) and (3.16), the differential equation (3.10) may have singularities at $\{p_1,\dots,p_n\}$, but is regular on M. Then there exists a solution $F\colon \widetilde{M}\to \mathrm{SL}(2,\boldsymbol{C})$, where $\widetilde{M}$ is the universal cover of M. Since the solution F of (3.10) is unique up to the change $F\mapsto Fa$ $(a\in\mathrm{SL}(2,\boldsymbol{C}))$, there exists a representation $\rho_F\colon \pi_1(M)\to\mathrm{SL}(2,\boldsymbol{C})$ such that

$$F\circ\tau = F\rho_F(\tau) \qquad \bigl(\tau\in\pi_1(M)\bigr)\,. \tag{3.17}$$

Here we consider an element τ of the fundamental group $\pi_1(M)$ as a deck transformation on $\widetilde{M}$. Thus:

Proposition 3.2. *If there exists a solution $F\colon\widetilde{M}\to\mathrm{SL}(2,\boldsymbol{C})$ of (3.10) for (G,Q) satisfying (3.15) and (3.16), then $f:=FF^*$ is a complete conformal CMC-1 immersion into H^3 which is well-defined on M if $\rho_F(\tau)\in\mathrm{SU}(2)$ for all $\tau\in\pi_1(M)$. Moreover, the hyperbolic Gauss map and the Hopf differential of f are G and Q, respectively.*

§4. Important Examples with $\mathrm{TA}(f)$ or $\mathrm{TA}(f^\#)\le 8\pi$

In this section, we shall introduce several important CMC-1 surfaces with $\mathrm{TA}(f)\le 8\pi$ or $\mathrm{TA}(f^\#)\le 8\pi$.

Example 4.1 (Horosphere). A horosphere (Figure 1) is the only surface of type $\mathbf{O}(0)$, with Weierstrass data given by

$$g=0,\qquad \omega = a\,dz \qquad (a\in\boldsymbol{C}\setminus\{0\})\,.$$

The holomorphic lift $F\colon \boldsymbol{C} \to \mathrm{SL}(2,\boldsymbol{C})$ of the surface with initial condition $F(0) = \mathrm{id}$ is given by

$$F = \begin{pmatrix} 1 & 0 \\ az & 1 \end{pmatrix} .$$

In particular the hyperbolic Gauss map is a constant function, as well as the secondary Gauss map $g = 0$. This surface is flat and totally umbilic. In particular, the total curvature and the dual total curvature of the surface are both equal to zero. Any flat or totally umbilic CMC-1 surfaces are parts of this surface. Planes in $\boldsymbol{R}^3$ are the corresponding minimal surfaces with the same Weierstrass data $(g,\omega) = (0, a\,dz)$.

Example 4.2 (Enneper cousin and dual of Enneper cousin)**.** The Enneper cousin is given in [Bry] (Figure 1), with the same Weierstrass data as the Enneper surface in $\boldsymbol{R}^3$:

$$g = z, \qquad \omega = a\,dz \qquad (a \in \boldsymbol{C} \setminus \{0\}) .$$

The holomorphic lift $F\colon \boldsymbol{C} \to \mathrm{SL}(2,\boldsymbol{C})$ of the surface with initial condition $F(0) = \mathrm{id}$ is given by

$$F = \begin{pmatrix} \cosh(az) & a^{-1}\sinh(az) - z\cosh(az) \\ a\sinh(az) & \cosh(az) - az\sinh(az) \end{pmatrix} .$$

In particular the hyperbolic Gauss map G is given by

$$G = a^{-1}\tanh(az) .$$

The Enneper cousin is in the class $\mathbf{O}(-4)$ and has a complete induced metric of total absolute curvature 4π. If one takes the inverse of F, one gets the dual of the Enneper cousin (Figure 1). Since

$$Fd(F^{-1}) = -dFF^{-1} = \begin{pmatrix} -a\cosh(az)\sinh(az) & \sinh^2(az) \\ -a^2\cosh^2(az) & a\cosh(az)\sinh(az) \end{pmatrix},$$

the Weierstrass data $(g^{\#},\omega^{\#})$ of the dual of the Enneper cousin given by

$$g^{\#} = a^{-1}\tanh(az), \qquad \omega^{\#} = a^2\cosh^2(az)\,dz .$$

This surface is also in the class $\mathbf{O}(-4)$ and has a complete induced metric of infinite total absolute curvature (see Lemma 3.1).

Example 4.3 (Catenoid cousins and warped catenoid cousins)**.** CMC-1 surfaces of type $\mathbf{O}(-2,-2)$ are classified in Theorem 6.2 in

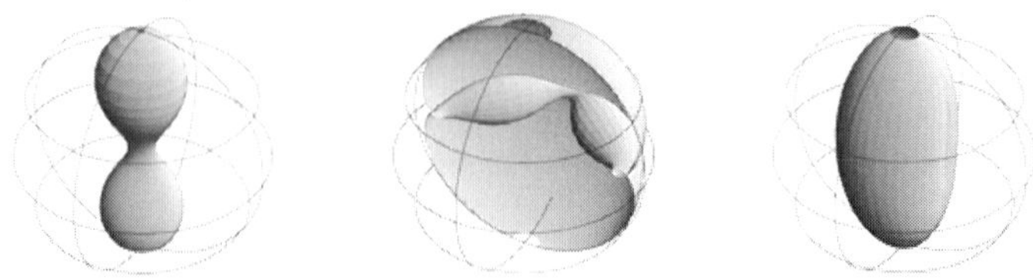

Fig. 2. A catenoid cousin with $l = 0.8$, and warped catenoid cousins with $(l, \delta, b) = (4, 1, 1/2)$ and $(1, 2, 1/2)$. The third surface has $\mathrm{TA}(f) = 4\pi$ because $l = 1$ even though its ends are not embedded.

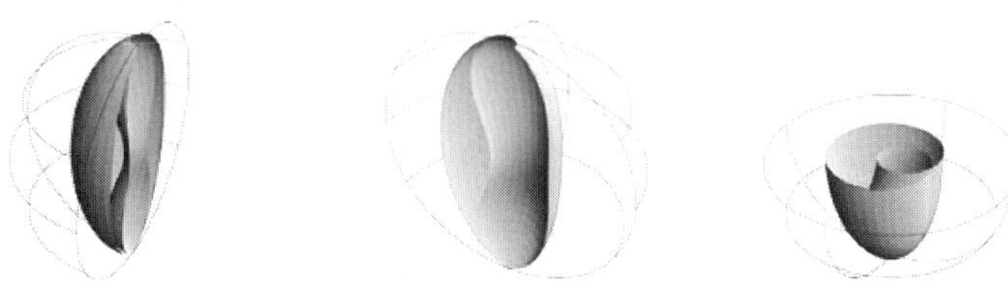

Fig. 3. Cut-away views of the third warped catenoid cousin in Figure 2.

[UY1]. Here we give a slightly refined version from [RUY4]: A complete conformal CMC-1 immersion $f\colon M = \boldsymbol{C} \setminus \{0\} \to H^3$ with regular ends have the following Weierstrass data

$$g = \frac{\delta^2 - l^2}{4l} z^l + b\,, \qquad \omega = \frac{Q}{dg} = z^{-l-1} dz\,, \tag{4.1}$$

with $l > 0$, $\delta \in \boldsymbol{Z}^+$, and $l \neq \delta$, and $b \geq 0$, where the case $b > 0$ occurs only when $l \in \boldsymbol{Z}^+$. When $b = 0$ and $\delta = 1$, the surface is called a *catenoid cousin*, which is rotationally symmetric. (The Weierstrass data of the catenoid cousin is often written as $g = z^\mu$ and $\omega = (1 - \mu^2) z^{-\mu-1} dz/(4\mu)$. This is equivalent to (4.1) for $b = 0$ and $\delta = 1$ and $l = \mu$ by a coordinate change $z \mapsto ((1 - \mu^2)/4\mu)^{(1/\mu)} z$.) Catenoid cousins are embedded when $0 < l < 1$ and have one curve of self-intersection when $l > 1$. When $b = 0$, f is a δ-fold cover of a catenoid cousin. When $b > 0$ (then automatically l is a positive integer), we call f a *warped catenoid cousin*, and its discrete symmetry group is the natural $\boldsymbol{Z}_2$ extension of the dihedral group D_l. Furthermore, the warped

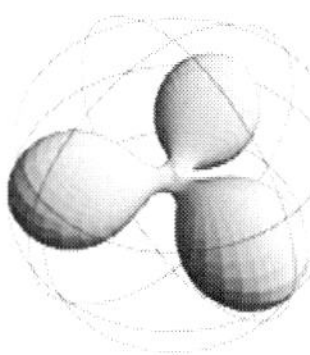

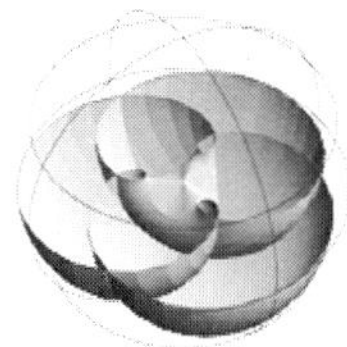

Fig. 4. Two different CMC-1 trinoids (proven to exist in [UY3]). Although these surfaces are proven to exist, and numerical experiments show that some of them are embedded (as one of the pictures here is), none have yet been proven to be embedded.

catenoid cousins can be written explicitly as $f = FF^*, F = F_0B$, where

$$F_0 = \sqrt{\frac{\delta^2 - l^2}{\delta}} \begin{pmatrix} \frac{1}{l-\delta} z^{(\delta-l)/2} & \frac{\delta - l}{4l} z^{(l+\delta)/2} \\ \frac{1}{l+\delta} z^{-(l+\delta)/2} & \frac{-(l+\delta)}{4l} z^{(l-\delta)/2} \end{pmatrix}, \quad B = \begin{pmatrix} 1 & -b \\ 0 & 1 \end{pmatrix}.$$

In particular, the hyperbolic Gauss map and Hopf differential are given by

$$G = z^\delta, \qquad Q = \frac{\delta^2 - l^2}{4z^2} dz^2,$$

which are equal to the Gauss map and Hopf differential of the catenoids in $\boldsymbol{R}^3$. The dual total curvature of a catenoid cousin is 4π, but its total curvature is $4\pi l$ $(l > 0)$, which can take any value in $(0, 4\pi) \cup (4\pi, \infty)$. On the other hand, the total absolute curvature and the dual total absolute curvature of warped catenoid cousins are always integer multiples of 4π. (See Figures 2 and 3).

Example 4.4 (Irreducible trinoids)**.** We take three real numbers $\mu_1, \mu_2, \mu_3 > -1$ such that

$$\cos^2 B_1 + \cos^2 B_2 + \cos^2 B_3 + 2\cos B_1 \cos B_2 \cos B_3 < 1, \tag{4.2}$$

where $B_j = \pi(\mu_j + 1)$ $(j = 1, 2, 3)$. We also assume

$$c_1^2 + c_2^2 + c_3^2 - 2(c_1c_2 + c_2c_3 + c_3c_1) \neq 0, \tag{4.3}$$

where $c_j = -\mu_j(\mu_j + 2)/2 \in \boldsymbol{R}$ $(j = 1, 2, 3)$. Then it is shown in [UY7] that there exists a unique CMC-1 surface $f_{\mu_1,\mu_2,\mu_3} \colon \boldsymbol{C} \setminus \{0, 1\} \to H^3$ of type $\mathbf{O}(-2, -2, -2)$ such that the pseudometric $d\sigma^2 = (-K)ds^2$ defined by (3.1) is irreducible and has conical singularities of orders μ_1, μ_2, μ_3

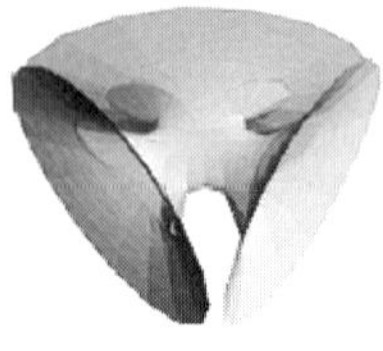

Type P

Type N

Fig. 5. Minimal trinoids of types P and N. The graphics were made by S. Tanaka of Hiroshima University.

at $z = 0, 1, \infty$, respectively. Moreover, any irreducible CMC-1 surface of type $\mathbf{O}(-2,-2,-2)$ whose ends are all embedded is congruent to some f_{μ_1,μ_2,μ_3}. All ends of these surfaces are asymptotic to catenoid cousin ends. The inequality (4.2) implies μ_1, μ_2 and μ_3 are all non-integers.

If we allow equality in (4.2), one of the μ_1, μ_2, μ_3 must be an integer. The corresponding CMC-1 surface might not exist for such μ_1, μ_2, μ_3 in general [UY7]. If it exists, its induced pseudometric $d\sigma^2$ must be reducible (see Lemmas A.1 and A.2).

The Hopf differential Q of f_{μ_1,μ_2,μ_3} is given by

$$Q = \frac{1}{2}\left(\frac{c_3z^2 + (c_2 - c_1 - c_3)z + c_1}{z^2(z-1)^2}\right) dz^2 . \tag{4.4}$$

Let q_1 and q_2 be zeros of Q, that is

$$c_3q_l^2 + (c_2 - c_1 - c_3)q_l + c_1 = 0 \qquad (l = 1, 2) . \tag{4.5}$$

By (4.3), $q_1 \neq q_2$ holds. The hyperbolic Gauss map is then given by

$$G = z + \frac{(q_1 - q_2)^2}{2\{2z - (q_1 + q_2)\}} . \tag{4.6}$$

In particular, all of these surfaces have dual total absolute curvature 8π. On the other hand, the total curvature is equal to $2\pi(4+\mu_1+\mu_2+\mu_3)$. If we set $\mu = \mu_1 = \mu_2 = \mu_3$, the condition (4.2) implies that $\mu > -2/3$, and then there exist $f_{\mu,\mu,\mu}$ for any μ arbitrarily close to $-2/3$, whose total curvatures tend to 4π. This implies Theorem 2.2 is sharp for $m = 1$.

It is interesting to compare these surfaces with minimal trinoids in $\boldsymbol{R}^3$. Minimal trinoids with three catenoid ends are classified in Barbanel [Bar], Lopez [Lop] and Kato [Kat]. Here, we adopt Kato's notation

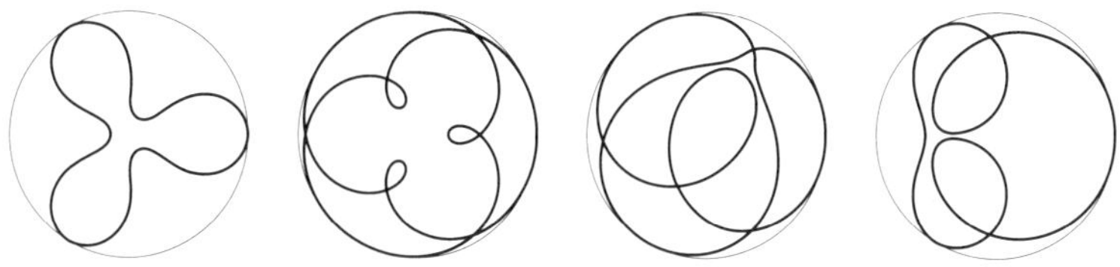

Type $(+,+,+)$ Type $(-,-,-)$ Type $(+,-,-)$ Type $(+,+,-)$

Fig. 6. Profile curves of trinoids f_{μ_1,μ_2,μ_3}.

[Kat]: The Weierstrass data of these trinoids $x_0 : \boldsymbol{C}\cup\{\infty\}\setminus\{0,p_1,p_2\} \to \boldsymbol{R}^3$ are given by

$$g = z - \frac{b(p_1^2{p_2}^2 + p_1^2 + {p_2}^2 - p_1p_2)}{f(z)}, \qquad \omega = -f(z)^2\,dz \qquad (b \in \boldsymbol{R})$$

where p_1 and p_2 are real numbers such that $(p_1-p_2)(1+p_1p_2) \neq 0$ and

$$f(z) := b\left(\frac{p_1(p_1-p_2)}{z-p_1} + \frac{p_2(p_2-p_1)}{z-p_2} + \frac{p_1p_2(p_1p_2+1)}{z}\right).$$

If the coefficients of $1/z^2$, $1/(z-p_1)^2$, $1/(z-p_2)^2$ in the Laurent expansion of the Hopf differential $Q = \omega\,dg$ at $z = 0, p_1, p_2$ are all the same signature, the surface is called of *type* P and otherwise it is called of *type* N. Type P surface are all Alexandrov-embedded. On the other hand, type N surfaces are not. (For a definition of Alexandrov embedded, see Cosín and Ros [CR].) These two classes consist of the two connected components of the set of minimal trinoids (Tanaka [Tan]; see Figure 5). In the case of CMC-1 trinoids in H^3, we would like to group the surfaces according to the signatures of c_1, c_2, c_3. For example, f_{μ_1,μ_2,μ_3} is called of type $(+,+,+)$ if c_1, c_2, c_3 are all positive, and it is called of type $(-,+,+)$ if one of c_1, c_2, c_3 is negative and the other two are positive. By numerical experiment, we see that these four types $(+,+,+)$, $(-,+,+)$, $(-,-,+)$ and $(-,-,-)$ are topologically distinct (see Figure 6). Surfaces of type $(+,+,+)$ have total curvature less than 8π, and it seems that only surfaces in this class can be embedded.

Example 4.5 (4-noids with $\mathrm{TA}(f) < 8\pi$)**.** A CMC-1 surface of genus 0 with 4 ends satisfies the Cohn-Vossen inequality $\mathrm{TA}(f) > 4\pi$ (see (3.12)). Though genus 0 surfaces with an odd number of ends satisfy a sharper inequality (Theorem 2.2), it seems that the Cohn-Vossen inequality is sharp for 4-noids, by numerical experiment: Let $a \in (0,1)$

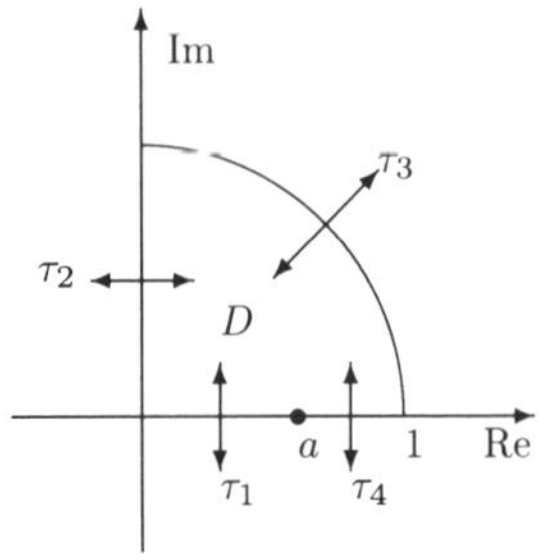

Fundamental region of a 4-noid

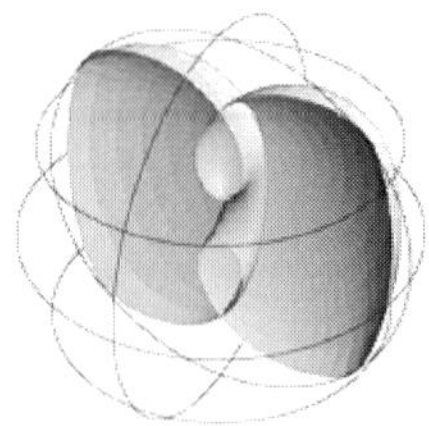

A 4-noid with $\mathrm{TA}(f) = 5\pi$

Fig. 7. 4-noid

be a real number and $M = \boldsymbol{C} \cup \{\infty\} \setminus \{a, -a, a^{-1}, -a^{-1}\}$. We set

$$G := \frac{pz^3 - z}{z^2 - p},$$
$$Q := -\frac{\mu(\mu+2)a^2(a^2 - a^{-2})^2}{(pa^4 - (3p^2-1)a^2 + p)} \frac{(pz^4 - (3p^2-1)z^2 + p)}{(z^2 - a^2)^2(z^2 - a^{-2})^2} dz^2,$$

where $\mu > -1$ and $p \in \boldsymbol{R} \setminus \{0, 1\}$ with $pa^4 - (3p^2-1)a^2 + p \neq 0$. If there exists a CMC-1 immersion $f\colon M \to H^3$ with hyperbolic Gauss map G and Hopf differential Q, then $\mathrm{TA}(f) = 4\pi(2\mu + 3)$. We shall solve the period problems using the method in [RUY1]: Let $D := \{z = re^{i\theta} \in \boldsymbol{C} \mid 0 < r < 1, 0 < \theta < \pi/2\}$. Then the Riemann surface M is obtained by reflection of D about ∂D. Let τ_1, τ_2, τ_3 and τ_4 be the reflections on the universal cover $\widetilde{M}$ of M, which are the lifts of the reflections on M about the segment $(0, a)$ on the real axis, the segment $(0, i)$ on the imaginary axis, the unit circle $|z| = 1$, and the segment $(a, 1)$ on the real axis, respectively (see Figure 7, left). Let $F\colon \widetilde{M} \to \mathrm{SL}(2, \boldsymbol{C})$ be a solution of (3.10). Since $\overline{G \circ \tau_j} = \sigma_j \star G$, $\overline{Q \circ \tau_j} = Q$, $j = 1, 2, 3, 4$, where

$$\sigma_1 = \sigma_4 = \mathrm{id}, \qquad \sigma_2 = \begin{pmatrix} i & 0 \\ 0 & -i \end{pmatrix}, \qquad \sigma_3 = \begin{pmatrix} 0 & i \\ i & 0 \end{pmatrix},$$

there exist matrices $\rho_F(\tau_j) \in \mathrm{SL}(2, \boldsymbol{C})$, $j = 1, 2, 3, 4$ such that $\overline{F \circ \tau_j} = \sigma_j F \rho_F(\tau_j)$, $j = 1, 2, 3, 4$. Moreover, by a similar argument to that in [RUY1, pp. 462–464], one can choose F such that

$$\rho_F(\tau_1) = \mathrm{id}, \quad \rho_F(\tau_2) = \sigma_2, \quad \rho_F(\tau_j) = \begin{pmatrix} q_j & i\gamma_j^1 \\ i\gamma_j^2 & \bar{q}_j \end{pmatrix} \quad (j = 3, 4),$$

where $\gamma_j^k \in \boldsymbol{R}$ and $q_j\bar{q}_j + \gamma_j^1\gamma_j^2 = 1$. Assume $\gamma_3^1\gamma_3^2 > 0$. Then there exists a unique solution F of (3.10) such that

$$\rho_F(\tau_1) = \mathrm{id},\ \rho_F(\tau_2) = \sigma_2,\ \rho_F(\tau_3) = \begin{pmatrix} q & i\gamma \\ i\gamma & \bar{q} \end{pmatrix},\ \rho_F(\tau_3) = \begin{pmatrix} q_4 & i\gamma_4^1 \\ i\gamma_4^2 & \bar{q}_4 \end{pmatrix}.$$

For given μ and a, if one can choose p so that $\gamma_4^1 = \gamma_4^2$, that is $\rho_F(\tau_j) \in$ SU(2), then there exists a CMC-1 immersion f of M into H^3 with hyperbolic Gauss map G and Hopf differential Q, by Proposition 4.7 in [RUY1].

By numerical calculation, for $\mu = -0.5$ and $a = 0.8$, there exists $p \simeq 1.4$ such that the period problem is solved. This surface thus has $\mathrm{TA}(f) = 8\pi$, and by continuity of the solvability of the period problems, clearly there exist surfaces with $\mathrm{TA}(f) < 8\pi$. Moreover, there exist such parameters a and p for $\mu \simeq -1$. So it seems that the Cohn-Vossen inequality for genus-zero 4-ended CMC-1 surfaces is sharp. Figure 7 shows the half cut of the surface with $\mathrm{TA}(f) = 5\pi$.

§5. Reducibility

To state the results for higher $\mathrm{TA}(f)$ or $\mathrm{TA}(f^\#)$, we review the notion of *reducibility*. For details, see [UY3, UY7, RUY1].

Metrics with conical singularities. Let $\overline{M}$ be a compact Riemann surface. A pseudometric $d\sigma^2$ on $\overline{M}$ is said to be an element of $\mathrm{Met}_1(\overline{M})$ if there exists a finite set of points $\{p_1, \dots, p_n\} \subset \overline{M}$ such that

(1) $d\sigma^2$ is a conformal metric of constant curvature 1 on $\overline{M} \setminus \{p_1, \dots, p_n\}$, and

(2) $\{p_1, \dots, p_n\}$ is the set of *conical singularities* of $d\sigma^2$, that is, for each $j = 1, \dots, n$, there exists a real number $\beta_j > -1$ so that $d\sigma^2$ is asymptotic to $c|z - p_j|^{2\beta_j}\, dz\, d\bar{z}$, where z is a complex coordinate of $\overline{M}$ around p_j and c is a positive constant.

We call the real number β_j the *order* of the conical singularity p_j, and write $\beta_j = \mathrm{ord}_{p_j} d\sigma^2$. The formal sum

$$\beta_1 p_1 + \dots + \beta_n p_n \tag{5.1}$$

is called the *divisor corresponding to* $d\sigma^2$.

Let $d\sigma^2 \in \mathrm{Met}_1(\overline{M})$ with divisor as in (5.1) and set $M := \overline{M} \setminus \{p_1, \dots, p_n\}$. Then there exists a holomorphic map $g\colon \widetilde{M} \longrightarrow \boldsymbol{C} \cup \{\infty\} =$

$\boldsymbol{CP}^1$ defined on the universal cover $\widetilde{M}$ of M such that

$$d\sigma^2 = \frac{4\,dg\,d\bar{g}}{(1+|g|^2)^2} = g^* ds_0^2 \,, \tag{5.2}$$

where ds_0^2 is the Fubini-Study metric of $\boldsymbol{CP}^1$. We call g the *developing map* of $d\sigma^2$. The developing map is unique up the change

$$g \longmapsto a \star g \qquad (a \in \mathrm{PSU}(2)) \,, \tag{5.3}$$

where $a \star g$ denotes the Möbius transformation of g with respect to a as in (3.2). Here we write $a \in \mathrm{PSU}(2)$ as a 2×2 matrix in $\mathrm{SU}(2)$ and identify a with $-a$.

For each deck transformation $\tau \in \pi_1(M)$ on $\widetilde{M}$, $d\sigma^2 = d\sigma^2 \circ \tau$ holds. So there exists a representation

(5.4)
$$\rho_g\colon \pi_1(M) \longrightarrow \mathrm{PSU}(2) \quad \text{such that} \quad \rho\circ\tau^{-1} = \rho_g(\tau)\star g \quad \text{for } \tau \in \pi_1(M) \,.$$

By a change of g as in (5.3), the corresponding representation changes by conjugation:

$$\rho_{a\star g} = a\rho_g a^{-1} \,. \tag{5.5}$$

Let τ_j be a deck transformation induced from a small loop on $\overline{M}$ surrounding a singularity p_j. Then by (5.5), one can choose the developing map g such that $\rho_g(\tau_j)$ is diagonal:

$$\rho_g(\tau_j) = \begin{pmatrix} e^{\pi i \nu_j} & 0 \\ 0 & e^{-\pi i \nu_j} \end{pmatrix} \qquad (\nu_j \in \boldsymbol{R}) \,,$$

namely, $g \circ \tau_j = e^{2\pi i \nu_j} g$. This implies that $(z-p_j)^{-\nu_j} g$ is single-valued on a neighborhood of p_j, where z is a complex coordinate around p_j. Then, replacing ν_j with $\nu_j + m$ ($m \in \boldsymbol{Z}$) if necessary, we can normalize

$$g = (z-p_j)^{\nu_j}\left(g_0 + g_1(z-p_j) + g_2(z-p_j)^2 + \dots\right) \qquad (g_0 \neq 0) \,. \tag{5.6}$$

By definition of the order and by equation (5.2), we have

$$\nu_j = \beta_j + 1 \qquad \text{or} \qquad -\beta_j - 1 \,.$$

Definition 5.1. A pseudometric $d\sigma^2 \in \mathrm{Met}_1(\overline{M})$ is called *reducible* if the representation ρ_g can be diagonalized simultaneously, where g is the developing map of $d\sigma^2$. More precisely, a reducible metric $d\sigma^2$ is called $\mathcal{H}^3$*-reducible* if the representation is trivial, and called $\mathcal{H}^1$*-reducible* otherwise. A pseudometric $d\sigma^2$ is called *irreducible* if it is not reducible.

By definition, a developing map g of an $\mathcal{H}^3$-reducible metric is a meromorphic function on $\overline{M}$ itself. Moreover, by (5.6), all conical singularities have integral orders, which coincide with the branching orders of the meromorphic function g. In this case, for any $a \in \mathrm{PSL}(2, \boldsymbol{C})$, $g_a := a \star g$ induces a new metric $d\sigma_a^2 := g_a^{\star} ds_0^2 \in \mathrm{Met}_1(\overline{M})$ with the same divisor as $d\sigma^2$. Since $d\sigma_a^2 = d\sigma^2$ if $a \in \mathrm{PSU}(2)$, we have a non-trivial deformation of $d\sigma^2$ preserving the divisor parametrized by a real 3-dimensional space $\mathcal{H}^3 = \mathrm{PSL}(2, \boldsymbol{C})/\mathrm{PSU}(2)$, which is the hyperbolic 3-space.

On the other hand, assume $d\sigma^2 \in \mathrm{Met}_1(\overline{M})$ is $\mathcal{H}^1$-reducible. Then there exists a developing map g such that the image of ρ_g consists of diagonal matrices. Let t be a positive real number and set

$$g_t := tg = \begin{pmatrix} t^{1/2} & 0 \\ 0 & t^{-1/2} \end{pmatrix} \star g \, .$$

Then by (5.5), $\rho_{g_t} = \rho_g$ holds. Thus, g_t induces a new metric $d\sigma_t^2 \in \mathrm{Met}_1(\overline{M})$. So we have one parameter family of pseudometrics $\{d\sigma_t^2\}$ preserving the corresponding divisor. This family is considered as a deformation of pseudometric parameterized by a geodesic line in $\mathcal{H}^3$. For details, see the Appendix in [RUY1].

We introduce a criterion for reducibility:

Lemma 5.2. *A metric $d\sigma^2 \in \mathrm{Met}_1(\overline{M})$ is reducible if and only if there exists a developing map such that $d \log g$ is a meromorphic 1-form on $\overline{M}$.*

Proof. Assume $d\sigma^2$ is reducible. Then one can choose the developing map g such that ρ_g is diagonal. Then for each deck transformation $\tau \in \pi_1(M)$,

$$g \circ \tau^{-1} = \begin{pmatrix} e^{\pi i \nu_\tau} & 0 \\ 0 & e^{-\pi i \nu_\tau} \end{pmatrix} \star g = e^{2\pi i \nu_\tau} g \qquad (\nu_\tau \in \boldsymbol{R})$$

holds. Hence we have $\log g \circ \tau = g + 2\pi i \nu_\tau$. Differentiating this, $d \log g \circ \tau = d \log g$ holds. Hence $d \log g$ is single-valued on $\overline{M}$.

Conversely, we assume $d \log g$ is well-defined on $\overline{M}$ for a developing map g. Then $\log g \circ \tau - \log g$ is a constant. Hence we have $g \circ \tau = \lambda_\tau g$ for some constant λ_τ. Then ρ_g is diagonal. Q.E.D.

Relationship with CMC-1 surfaces. Let $f\colon \overline{M}_\gamma \setminus \{p_1, \dots, p_n\} \to H^3$ be a complete conformal CMC-1 immersion, where $\overline{M}_\gamma$ is a compact

Riemann surface. If $\mathrm{TA}(f) < \infty$, then the pseudometric $d\sigma^2$ as in (3.1) is considered as an element of $\mathrm{Met}_1(\overline{M}_\gamma)$ (see [Bry]), and the secondary Gauss map g is the developing map of $d\sigma^2$. Let $\{q_1, \dots, q_m\}$ be the set of umbilic points of f, that is the zeros of Q and set $\xi_k := \mathrm{ord}_{q_k} Q$ $(k = 1, \dots, m)$. Then by (3.4), $d\sigma^2$ has a conical singularity of order ξ_k for each $k = 1, \dots, m$. Hence the divisor of $d\sigma^2$ is in the form

$$\mu_1 p_1 + \cdots + \mu_n p_n + \xi_1 q_1 + \cdots + \xi_m q_m \,, \tag{5.7}$$

where the μ_j $(j = 1, \dots, n)$ are the branch orders of g at each p_j.

Let F be a holomorphic lift of f as in (3.6). Then there exists a representation $\rho_F\colon \pi_1(M) \to \mathrm{SU}(2)$ as in (3.17). By (3.7), the secondary Gauss map g of F changes as $g \circ \tau^{-1} = \rho_F(\tau) \star g$ for each deck transformation $\tau \in \pi_1(M)$. Hence the representation ρ_g defined in (5.4) satisfies

$$\rho_g(\tau) = \pm \rho_F(\tau) \qquad \bigl(\tau \in \pi_1(M)\bigr) \,. \tag{5.8}$$

The immersion f is called $\mathcal{H}^3$-reducible (respectively, $\mathcal{H}^1$-reducible) if the corresponding pseudometric $d\sigma^2$ is $\mathcal{H}^3$-reducible (respectively, $\mathcal{H}^1$-reducible).

Lemma 5.3. *A* CMC-1 *immersion* $f\colon M \to H^3$ *is* $\mathcal{H}^3$*-reducible if and only if the dual immersion* $f^\#$ *is well-defined on* M.

Proof. Let F be a lift of f. Then $f^\# = F^{-1}(F^{-1})^*$ is well-defined on M if and only if $\rho_F = \pm\,\mathrm{id}$. This is equivalent to ρ_g being the trivial representation, by (5.8). Q.E.D.

§6. The case $\mathrm{TA}(f^\#) \le 8\pi$

We now have enough notation and facts to describe results on the case $\mathrm{TA}(f^\#) \le 8\pi$ [RUY3].

Let $f\colon \overline{M}_\gamma \setminus \{p_1, \dots, p_n\} \to H^3$ be a complete, conformal CMC-1 immersion, where $\overline{M}_\gamma$ is a Riemann surface of genus γ. Now we assume $\mathrm{TA}(f^\#) \le 8\pi$. If the hyperbolic Gauss map G has an essential singularity at any end p_j, then $\mathrm{TA}(f^\#) = +\infty$, since $\mathrm{TA}(f^\#)$ is the area of the image of G. So G is meromorphic on all of $\overline{M}_\gamma$. In particular, $\mathrm{TA}(f^\#) = 4\pi \deg G = 0$, 4π, or 8π.

Since $f^\#$ has finite total curvature, the Hopf differential $Q^\# = -Q$ can be extended to $\overline{M}_\gamma$ as a meromorphic 2-differential [Bry, Proposition

Type	TA($f^\#$)	Reducibility	Status	c.f.
$\mathbf{O}(0)$	0	$\mathcal{H}^3$-red.	classified^0	Horosphere
$\mathbf{O}(-4)$	4π	$\mathcal{H}^3$-red.	classified	Duals of Enneper cousins [RUY1, Example 5.4]
$\mathbf{O}(-2,-2)$	4π	reducible	classified	Catenoid cousins and warped catenoid cousins with embedded ends (i.e. $\delta = 1$) [Bry, Example 2], [UY1, RUY3, RUY4]
$\mathbf{O}(-5)$	8π	$\mathcal{H}^3$-red.	classified	[RUY3]
$\mathbf{O}(-6)$	8π	$\mathcal{H}^3$-red.	classified	[RUY3]
$\mathbf{O}(-2,-2)$	8π	red.	classified	Double covers of catenoid cousins and warped catenoid cousins with $\delta = 2$ [UY1, Theorem 6.2], [RUY3, RUY4]
$\mathbf{O}(-1,-4)$	8π	$\mathcal{H}^3$-red.	classified^0	[RUY3]
$\mathbf{O}(-2,-3)$	8π	$\mathcal{H}^1$-red.	classified	[RUY3]
$\mathbf{O}(-2,-4)$	8π	$\mathcal{H}^1$-red. $\mathcal{H}^3$-red.	classified classified	[RUY3] [RUY3]
$\mathbf{O}(-3,-3)$	8π	red.	existence	[RUY3]
$\mathbf{O}(-1,-1,-2)$	8π	$\mathcal{H}^3$-red.	classified^0	[RUY3]
$\mathbf{O}(-1,-2,-2)$	8π	$\mathcal{H}^1$-red. $\mathcal{H}^3$-red.	classified classified	[RUY3] [RUY3]
$\mathbf{O}(-2,-2,-2)$	8π	irred. $\mathcal{H}^1$-red. $\mathcal{H}^3$-red.	classified existence$^+$ existence$^+$	[UY6, Theorem 2.6] [RUY3] [RUY3]
$\mathbf{I}(-3)$	8π		unknown	
$\mathbf{I}(-4)$	8π		existence	Chen-Gackstatter cousins [RUY3]
$\mathbf{I}(-1,-1)$	8π		unknown$^+$	[RUY3]
$\mathbf{I}(-2,-2)$	8π		existence	Genus 1 catenoid cousins [RS]

Table 1. CMC-1 surfaces in H^3 with TA($f^\#$) $\leq 8\pi$ [RUY3].

5]. Hence $d_j = \mathrm{ord}_{p_j} Q$ is finite for each $j = 1, \ldots, n$. Our results from [RUY3] are shown in Table 1. In the table,

- *classified* means the complete list of the surfaces in such a class is known (and this means not only that we know all the possibilities for the form of the data (G, Q), but that we also know exactly for which (G, Q) the period problems of the immersions are solved).

Type	TA	The surface	c.f.
$\mathbf{O}(0)$	0	Plane	
$\mathbf{O}(-4)$	4π	Enneper's surface	
$\mathbf{O}(-5)$	8π		[Lop, Theorem 6]
$\mathbf{O}(-6)$	8π		[Lop, Theorem 6]
$\mathbf{O}(-2,-2)$	4π 8π	Catenoid Double cover of the catenoid	
$\mathbf{O}(-1,-3)$	8π		[Lop, Theorem 5]
$\mathbf{O}(-2,-3)$	8π		[Lop, Theorem 4, 5]
$\mathbf{O}(-2,-4)$	8π		[Lop, Theorem 5]
$\mathbf{O}(-3,-3)$	8π		[Lop, Theorem 4]
$\mathbf{O}(-1,-2,-2)$	8π		[Lop, Theorem 5]
$\mathbf{O}(-2,-2,-2)$	8π		[Lop, Theorem 5]
$\mathbf{I}(-4)$	8π	Chen-Gackstatter surface	[Lop, Theorem 5], [CG]

Table 2. The classification of complete minimal surfaces in $\boldsymbol{R}^3$ with TA $\leq 8\pi$ ([Lop]), for comparison with Table 1.

- *classified*^{0} means there exists a unique surface (up to isometries of H^3 and deformations that come from its reducibility).
- *existence* means that examples exist, but they are not yet classified.
- *existence*$^+$ means that all possibilities for the data (G, Q) are determined, but the period problems are solved only for special cases.
- *unknown* means that neither existence nor non-existence is known yet.
- *unknown*$^+$ means that all possibilities for the data (G, Q) are determined, but the period problems are still unsolved.

Any class and type of reducibility not listed in Table 1 cannot contain surfaces with $\mathrm{TA}(f^\#) \leq 8\pi$. For example, any irreducible or $\mathcal{H}^3$-reducible surface of type $\mathbf{O}(-2,-3)$ must have dual total absolute curvature at least 12π.

Table 2 shows the corresponding results for minimal surfaces in $\boldsymbol{R}^3$, the classification of complete minimal surfaces with TA $\leq 8\pi$ [Lop]. Comparing these two tables, one sees differences between the classes of minimal surfaces with TA $\leq 8\pi$ and the classes of CMC-1 surfaces with $\mathrm{TA}(f^\#) \leq 8\pi$. For example, there exist no mimimal surfaces of classes $\mathbf{O}(-1,-4)$ and $\mathbf{O}(-1,-1,-2)$ with TA $\leq 8\pi$, but CMC-1 surfaces of such types do exist.

§7. The case of $\mathrm{TA}(f) \leq 8\pi$

In the remainder of this paper, we shall give new results on the case of higher $\mathrm{TA}(f)$.

Preliminaries. First, we give further notation and facts that will be needed in our discussion.

Orders of the Gauss maps. Let $\overline{M}_\gamma$ be a compact Riemann surface of genus γ. For a complete conformal CMC-1 immersion $f\colon M = \overline{M}_\gamma \setminus \{p_1, \dots, p_n\} \to H^3$ with $\mathrm{TA}(f) < \infty$, we define μ_j and $\mu_j^\#$ to be the branching orders of the Gauss maps g and G, respectively, at an end p_j. Then the pseudometric $d\sigma^2$ as in (3.1) has a conical singularity of order $\mu_j > -1$ at each end p_j $(j = 1, \dots, n)$. Let $d_j = \operatorname{ord}_{p_j} Q$ $(j = 1, \dots, n)$. Then an end p_j is regular if and only if $d_j \geq -2$ (see Section 3, or [UY1]). If an end p_j is irregular, then $\mu_j^\# = \infty$. At a regular end p_j, the relation (3.5) implies that the Hopf differential Q expands as

$$Q = \left(\frac{1}{2} \frac{c_j - c_j^\#}{(z - p_j)^2} + \dots \right) dz^2 , \tag{7.1}$$

where

$$c_j = -\frac{1}{2}\mu_j(\mu_j + 2), \qquad c_j^\# = -\frac{1}{2}\mu_j^\#(\mu_j^\# + 2) \tag{7.2}$$

and z is a local complex coordinate around p_j.

Let $\{q_1, \dots, q_m\} \subset M$ be the m umbilic points of the surface, and let $\xi_k = \operatorname{ord}_{q_k} Q$. Since the total order of a holomorphic 2-differential is $-2\chi(\overline{M}_\gamma)$, we have

$$\sum_{j=1}^{n} d_j + \sum_{k=1}^{m} \xi_k = 4\gamma - 4, \qquad \text{in particular,} \quad \sum_{j=1}^{n} d_j \leq 4\gamma - 4 . \tag{7.3}$$

By (3.4) and (3.5), one has

(7.4)
$$\begin{aligned}\xi_k &= [\text{the branching order of } G \text{ at } q_k] = [\text{the branching order of } g \text{ at } q_k] \\ &= \operatorname{ord}_{q_k} d\sigma^2 = \operatorname{ord}_{q_k} Q .\end{aligned}$$

As in (2.4) of [RUY3], the Gauss-Bonnet theorem for $(\overline{M}_\gamma, d\sigma^2)$ implies

$$\frac{\mathrm{TA}(f)}{2\pi} = \chi(\overline{M}_\gamma) + \sum_{j=1}^{n} \mu_j + \sum_{k=1}^{m} \xi_k = (2\gamma - 2) + \sum_{j=1}^{n} \mu_j + \sum_{k=1}^{m} \xi_k \tag{7.5}$$

as well as

$$\frac{\mathrm{TA}(f^{\#})}{2\pi} = \chi(\overline{M}_\gamma) + \sum_{j=1}^{n} \mu_j^{\#} + \sum_{k=1}^{m} \xi_k \,, \tag{7.6}$$

which is obtained from the Gauss-Bonnet theorem for $d\sigma^{2\#} = (-K^{\#})ds^{2\#}$ [RUY3]. Combining this with (7.3), we have

$$\frac{\mathrm{TA}(f)}{2\pi} = 2\gamma - 2 + \sum_{j=1}^{n} (\mu_j - d_j) \,. \tag{7.7}$$

Proposition 4.1 in [UY1] implies that

$$\mu_j - d_j > 1, \qquad \text{in particular,} \qquad \mu_j - d_j \geq 2 \quad \text{if } \mu_j \in \boldsymbol{Z} \,. \tag{7.8}$$

An end p_j is regular if and only if $d_j \geq -2$, and then G is meromorphic at p_j. Thus

$$\mu_j^{\#} \text{ is a non-negative integer if } d_j \geq -2 \,. \tag{7.9}$$

In this case, one has (Lemma 3 of [UY5])

(7.10)

$\mu_j^{\#} - d_j \geq 2$ and the equality holds if and only if p_j is embedded.

By Proposition 4 of [Bry],

$$\mu_j > -1 \,, \tag{7.11}$$

hence equation (7.1) implies

$$\mu_j = \mu_j^{\#} \qquad \text{if} \quad d_j \geq -1 \,. \tag{7.12}$$

Finally, we note that

(7.13)

any meromorphic function on a Riemann surface $\overline{M}_\gamma$ of genus $\gamma \geq 1$ has at least three distinct branch points.

To prove this, let φ be a meromorphic function on $\overline{M}_\gamma$ with branch points $\{q_1, \ldots, q_N\}$ with branching order ν_k at q_k. Then the Riemann-Hurwitz relation implies

$$2 \deg \varphi = 2 - 2\gamma + \sum_{k=1}^{N} \nu_k \,.$$

On the other hand, since the multiplicity of φ at q_k is $\nu_k + 1$, $\deg \varphi \geq \nu_k + 1$ $(k = 1, \ldots, m)$. Thus $(N-2)\deg \varphi \geq 2(\gamma - 1) + N$. If $\gamma \geq 1$, then $\deg \varphi \geq 2$, and so $N \geq 3$.

Remark. Facts (7.7) and (7.8) imply that, for CMC-1 surfaces, equality never holds in the Cohn-Vossen inequality (see (3.12) and [UY1]).

Flux for CMC-1 surfaces. Let $f\colon \overline{M}_\gamma \setminus \{p_1, \dots, p_n\} \to H^3$ be a complete CMC-1 immersion. For each end p_j, the *flux* at p_j is defined as ([RUY2])

$$(7.14)\qquad \mathcal{F}l_j := \frac{1}{2\pi i}\int_{\tau_j}\begin{pmatrix} G & -G^2 \\ 1 & -G \end{pmatrix}\frac{Q}{dG} \in \mathfrak{sl}(2, \boldsymbol{C}) \qquad (j = 1, \dots, n)\,,$$

where G and Q are the hyperbolic Gauss map and the Hopf differential of f respectively, and τ_j is a loop surrounding the end p_j. Then the following balancing formula holds (Theorem 1 in [RUY2]):

$$(7.15)\qquad \sum_{j=1}^{n} \mathcal{F}l_j = 0\,.$$

Moreover, one has (Proposition 2 and Corollary 5 in [RUY2]):

Proposition 7.1. *Let $f\colon \overline{M}_\gamma \setminus \{p_1, \dots, p_n\} \to H^3$ be a complete* CMC-1 *immersion.*

(1) If an end p_j is regular and $\operatorname{ord}_{p_j} Q = -2$, then $\mathcal{F}l_j \neq 0$.

(2) If an end p_j is regular and embedded, $\mathcal{F}l_j = 0$ if and only if $\operatorname{ord}_{p_j} Q \geq 0$.

Then by the balancing formula (7.15), we have

Corollary 7.2. *There exists no complete* CMC-1 *surface of finite total curvature with only one end p that is regular, such that either one of the following holds:*

(1) $\operatorname{ord}_p Q = -2$.

(2) $\operatorname{ord}_p Q < 0$ *and the end is embedded.*

Results for TA(f) $\leq 8\pi$. First, we prepare the following lemma.

Lemma 7.3. *Let $f\colon M \to H^3$ be a complete* CMC-1 *immersion of genus γ and n ends with* $\mathrm{TA}(f) \leq 2\pi\rho$. *If f is not totally umbilic (not a horosphere), then the following hold:*

(1) $2\gamma < \rho + 1$ and $1 \leq n < \rho - 2\gamma + 2$.

(2) If $n = 1$, then $2\gamma - \rho - 3 < d_1 \leq 4\gamma - 4$ and $d_1 \neq -2$.

(3) If $\gamma = n = 1$, then $-\rho - 1 < d_1 \leq -3$.

(4) If $2 \leq n = \rho + 1 - 2\gamma$*, then* $d_j = -2$ *at all ends.*

(5) If $1 = n = \rho + 1 - 2\gamma$*, then* $d_1 \geq 0$ *and* $\mu_1 = 2 + d_1$.

Proof. The first item of the lemma is obtained from the Cohn-Vossen inequality (3.12). In particular, if $n = 1$, (7.7), (7.11) and (7.3) imply $\rho \geq 2\gamma - 2 + \mu_1 - d_1 > 2\gamma - 3 - d_1$ and $d_1 \leq 4\gamma - 4$. In this case, by the balancing formula (7.15), the flux $\mathcal{F}l_1$ must vanish. Hence by Corollary 7.2, $d_1 \neq -2$ holds, and the second item of the theorem follows. Even more particularly, if $\gamma = n = 1$, then $-\rho - 1 < d_1 \leq 0$ and $d_1 \neq -2$. Assume $d_1 \geq -1$. In this case, the end is regular, and then G is a meromorphic function on $\overline{M}_\gamma$. On the other hand, (7.3) implies that there is at most one umbilic point. Since a branch point of G is an umbilic point or an end, this implies that the number of branch points of G is at most 2, which contradicts (7.13). Hence the third item is proven.

Suppose $n = \rho - 2\gamma$. Then (7.3) implies

$$n + 1 \geq \sum_{j=1}^{n} (\mu_j - d_j) , \tag{7.16}$$

and we consider two cases:

Case 1 If $n \geq 2$, then (7.8) implies that $1 < \mu_j - d_j < 2$ for all j, so $\mu_j \notin \boldsymbol{Z}$ for all j, and hence (7.12) implies that $d_j \leq -2$ for all j. But by (7.16) and (7.11), we have $-2n \leq \sum_{j=1}^{n} d_j$, and so $d_j = -2$ for all j.

Case 2 If $n = 1$, then $1 < \mu_1 - d_1 \leq 2$ holds because of (7.16) and (7.8). Hence by (7.11), $d_1 \geq -2$. But Corollary 7.2 implies $d_1 \geq -1$. Then by (7.12), $\mu_1 \in \boldsymbol{Z}$ and $\mu_1 - d_1 = 2$ holds. Suppose $d_1 = -1$. Then $\mu_1^{\#} = \mu_1 = d_1 + 2 = 1$, and then by (7.10), the only end p_1 is regular and embedded. This contradicts (2) and (7.15). Hence $d_1 \geq 0$. Q.E.D.

Type	TA(f)	Reducibility	Status	cf.
$\mathbf{O}(0)$	0	$\mathcal{H}^3$-red.	classified	Horosphere
$\mathbf{O}(-4)$	4π	$\mathcal{H}^3$-red.	classified	Enneper cousins [Bry]
$\mathbf{O}(-5)$	8π	$\mathcal{H}^3$-red.	classified	Same as "dual" case
$\mathbf{O}(-6)$	8π	$\mathcal{H}^3$-red.	classified	Same as "dual" case
$\mathbf{O}(-2,-2)$	$(0, 8\pi]$	$\mathcal{H}^1$-red.	classified	Catenoid cousins and their δ-fold covers [Bry, Ex. 2],[UY1]
$\mathbf{O}(-2,-2)$	4π 8π	$\mathcal{H}^3$-red.	classified	Warped cat. cous. $l=1$ Warped cat. cous. $l=2$ [UY1, Thm 6.2], Exa. 4.3
$\mathbf{O}(-1,-4)$	8π	$\mathcal{H}^3$-red.	classified	Same as "dual" case
$\mathbf{O}(-2,-4)$	8π $(4\pi, 8\pi)$	$\mathcal{H}^3$-red. $\mathcal{H}^1$-red.	classified existence	Same as "dual" case Remark A.9
$\mathbf{O}(-2,-5)$	8π	$\mathcal{H}^1$-red.	existence	Remarks A.10, A.12
$\mathbf{O}(-3,-3)$		reducible	unknown	Remark A.11
$\mathbf{O}(-3,-4)$	8π	reducible	unknown	Remark A.12
$\mathbf{O}(0,-2,-2)$	$(4\pi, 8\pi)$	$\mathcal{H}^1$-red.	classified	Proposition A.15
$\mathbf{O}(-1,-2,-3)$	8π	$\mathcal{H}^1$-red.	unknown	
$\mathbf{O}(-1,-1,-2)$	8π	$\mathcal{H}^3$-red.	classified	Same as "dual" case
$\mathbf{O}(-1,-2,-2)$	8π $(4\pi, 8\pi)$ 8π	$\mathcal{H}^3$-red. $\mathcal{H}^1$-red. $\mathcal{H}^1$-red.	classified classified classified	Same as "dual" case Proposition A.16 Proposition A.17
$\mathbf{O}(-2,-2,-2)$	$(4\pi, 8\pi]$		existence	Classified for irred. embedded end case [UY6]
$\mathbf{O}(-2,-2,-3)$		irred./$\mathcal{H}^1$-red.	unknown	
$\mathbf{O}(-2,-2,-4)$	8π	irred./$\mathcal{H}^1$-red.	unknown	
$\mathbf{O}(-2,-3,-3)$	8π	irred./$\mathcal{H}^1$-red.	unknown	
$\mathbf{O}(-2,-2,-2,-2)$			existence	Example 4.5
$\mathbf{O}(-2,-2,-2,0)$	8π		existence	Remark A.19
$\mathbf{O}(-2,-2,-2,d)$ $d=-3,-2,-1,1$	8π when $d \geq -1$		unknown	

O$(-2,-2,-2,-2,-2)$	8π		unknown	Corollary 7.4
I(-3)			unknown	
I(-4)			unknown	
I$(-1,-1)$	8π		unknown	
I$(-2,-2)$			unknown	Remark A.20
I$(-2,-3)$			unknown	
I$(-2,-2,-2)$			unknown	Remark A.21

Table 3. Classification of CMC-1 surfaces in H^3 with $\mathrm{TA}(f) \leq 8\pi$

Lemma 7.3 gives the following corollary:

Corollary 7.4. *If* $f\colon M \to H^3$ *is a complete* CMC-1 *immersion with* $\mathrm{TA}(f) \leq 8\pi$*, then it is either*

(1) a surface of genus 0 *with at most* 5 *ends. (if it has* 5 *ends, then all* 5 *ends are regular with* $d_1 = d_2 = d_3 = d_4 = d_5 = -2$*), or*

(2) a surface of genus 1 *with at most* 3 *ends (if it has* 3 *ends, then all* 3 *ends are regular with* $d_1 = d_2 = d_3 = -2$*; if it has* 1 *end, then the end is irregular with* $d_1 = -3$ *or* $d_1 = -4$*).*

Proof. We only have to show that a CMC-1 surface with $\mathrm{TA}(f) \leq 8\pi$ of genus 2 and with 1 regular end satisfying $0 \leq d_1 \leq 4$ cannot exist. By (7.7), (7.12) and (7.6), such a surface would satisfy $\mathrm{TA}(f^\#) = \mathrm{TA}(f) \leq 8\pi$ and hence the hyperbolic Gauss map G is a meromorphic function on a compact Riemann surface $\overline{M}_2$ of genus 2 with $\deg G \leq 2$. Therefore $\mu_1^\#$ can be only 0 or 1, and so $d_1 \leq \mu_1^\# - 2 < 0$, a contradiction. Q.E.D.

Now we compile an *unfinished* classification of CMC-1 surfaces with $\mathrm{TA}(f) \leq 8\pi$ (see Table 3). In the "status" column of the table, *classified* means that the surfaces of such a class are completely classified (i.e. not only is the holomorphic data known, but the period problems are also completely solved), *existence* means that there exists such a surface, and *unknown* means that it is unknown if such a surface exists. Surfaces of any type not appearing in the table cannot exist with $\mathrm{TA}(f) \leq 8\pi$. The proofs of the existence and non-existence results are given in Appendix A.

§Appendix A. Detailed discussion of the case $\mathrm{TA}(f) \leq 8\pi$

In this appendix, we give a precise discussion of complete CMC-1 surfaces with $\mathrm{TA}(f) \leq 8\pi$.

Detailed Preliminaries. Here, we review facts which will be used to prove existence and non-existence for special cases.

Metrics in $\mathrm{Met}_1(\boldsymbol{C} \cup \{\infty\})$. In this subsection, we introduce special properties of pseudometrics in $\mathrm{Met}_1(\boldsymbol{C} \cup \{\infty\})$ (see Section 5).

Lemma A.1. *A pseudometric $d\sigma^2 \in \mathrm{Met}_1(\boldsymbol{C} \cup \{\infty\})$ with divisor as in* (5.1) *is $\mathcal{H}^3$-reducible if and only if all orders of conical singularities are integers.*

Proof. If $d\sigma^2$ is $\mathcal{H}^3$-reducible, then the developing map g is a meromorphic function on $\boldsymbol{C}\cup\{\infty\}$. So the branch orders must all be integers. Conversely, assume all conical singularities have integral orders. Then by (5.6), $\rho_g(\tau_j) = \pm\,\mathrm{id}$ for each j, where τ_j is the deck transformation on $M := \boldsymbol{C} \cup \{\infty\} \setminus \{p_1, \dots, p_n\}$ corresponding to the loop surrounding p_j. Since $\pi_1(M)$ is generated by $\tau_1, \dots, \tau_n$, ρ_g is the trivial representation. Q.E.D.

Lemma A.2. *Let $d\sigma^2 \in \mathrm{Met}_1(\boldsymbol{C} \cup \{\infty\})$ with divisor as in* (5.1)*. Assume the orders β_1 and β_2 are not integers, and β_j $(j \geq 3)$ are integers. Then $d\sigma^2$ is $\mathcal{H}^1$-reducible.*

Proof. Let g be a developing map such that $\rho_g(\tau_1)$ is diagonal. Here, as in the proof of the previous lemma, we have $\rho_g(\tau_j) = \pm\,\mathrm{id}$ $(j \geq 3)$. Then we have $\rho_g(\tau_1)\rho_g(\tau_2) = \pm\,\mathrm{id}$ because $\tau_1 \circ \cdots \circ \tau_n = \mathrm{id}$. Hence $\rho_g(\tau_2)$ is also a diagonal matrix. Q.E.D.

Lemma A.3 ([RUY4, Proposition A.1])**.** *There exists no metric $d\sigma^2 \in \mathrm{Met}_1(\boldsymbol{C} \cup \{\infty\})$ with divisor as in* (5.1) *such that only one β_j is a non-integer and all others are integers.*

A developing map of a reducible metric in $\mathrm{Met}_1(\boldsymbol{C}\cup\{\infty\})$ can be written explicitly as follows:

Lemma A.4 ([RUY4, Proposition B.1])**.** *Let $d\sigma^2 \in \mathrm{Met}_1(\boldsymbol{C}\cup\{\infty\})$ be reducible with divisor as in* (5.1)*. Assume*

$$p_n = \infty\,, \qquad \beta_1, \dots, \beta_m \notin \boldsymbol{Z}\,, \qquad \beta_{m+1}, \dots, \beta_{n-1} \in \boldsymbol{Z}\,.$$

Then there exists a developing map g of $d\sigma^2$ such that

$$g = (z - p_1)^{\nu_1} \dots (z - p_m)^{\nu_m}\, r(z) \qquad (\nu_1, \dots, \nu_m \in \boldsymbol{R} \setminus \boldsymbol{Z}) ,$$

where $r(z)$ is a rational function on $\boldsymbol{C} \cup \{\infty\}$.

Corollary A.5. *Let $d\sigma^2 \in \mathrm{Met}_1(\boldsymbol{C} \cup \{\infty\})$ be reducible with divisor as in* (5.1) *and $p_n = \infty$. Then there exists a developing map g such that*

(A.1)
$$dg = t\, \frac{(z - p_1)^{\alpha_1} \dots (z - p_{n-1})^{\alpha_{n-1}}}{\prod_{k=1}^{N} (z - a_k)^2}\, dz \qquad (\alpha_j = \beta_j \quad \mathit{or} \quad - \beta_j - 2) ,$$

where $a_1, \dots, a_N \in \boldsymbol{C} \setminus \{p_1, \dots, p_{n-1}\}$ are mutually distinct, t is a positive real number. Moreover, one has

(A.2)
$$-(\alpha_1 + \dots + \alpha_{n-1}) + 2N - 2 = \beta_n \quad \mathit{or} \quad - \beta_n - 2 .$$

Proof. If $d\sigma^2$ is $\mathcal{H}^3$-reducible, g is a meromorphic function on $\boldsymbol{C} \cup \{\infty\}$ which branches at $p_1, \dots, p_n$ with branch orders $\beta_j \in \boldsymbol{Z}^+$. Hence p_j is a zero of order β_j or a pole of order $\beta_j + 2$ of dg for each $j = 1, \dots, n$. Let $\{a_1, \dots, a_N\}$ be the simple poles of g on $\boldsymbol{C} \setminus \{p_1, \dots, p_{n-1}\}$, then each a_k is a pole of order 2 of dg. (The a_j are not branch points of g.) The zeros and poles of dg are the branch points and the simple poles of g. Hence we have (A.1) for $t \in \boldsymbol{C} \setminus \{0\}$. By a suitable change $g \mapsto e^{i\theta} g$ (which is a special form of the change (5.3)), we can choose g such that $t \in \boldsymbol{R}^+$. Since $\infty = p_n$ is a zero of order β_n or a pole of order $\beta_n + 2$ of dg, we have (A.2). Next we assume $d\sigma^2$ is $\mathcal{H}^1$-reducible. Without loss of generality, we may assume $\beta_1, \dots, \beta_m \notin \boldsymbol{Z}$ and $\beta_{m+1}, \dots, \beta_{n-1} \in \boldsymbol{Z}$. Then by Lemma A.4, we can choose the developing map g as $g = (z - p_1)^{\nu_1} \dots (z - p_m)^{\nu_m} r(z)$, where $r(z)$ is a rational function. By (5.6), we have $\nu_j = \beta_j + 1$ or $\nu_j = -\beta_j - 1$ $(j = 1, \dots, m)$. Differentiating this, we have

$$dg = (z - p_1)^{\alpha_1} \dots (z - p_m)^{\alpha_m} r_1(z)\, dz \qquad (\alpha_j = \beta_j \text{ or } \alpha_j = -\beta_j - 2)$$

where $r_1(z)$ is a rational function. Since each p_j $(j = m+1, \dots, n-1)$ is a branch point of g of order $\beta_j \in \boldsymbol{Z}$, we have (A.1) by an argument similar to the $\mathcal{H}^3$-reducible case. Moreover, since $\mathrm{ord}_\infty d\sigma^2 = \beta_n$, we have (A.2). Q.E.D.

Remark A.6. Let $M = \boldsymbol{C} \cup \{\infty\} \setminus \{p_1, \dots, p_n\}$, and $\widetilde{M}$ the universal cover. Then there exists a meromorphic function $g \colon \widetilde{M} \to \boldsymbol{C} \cup \{\infty\}$ satisfying (A.1) if and only if all of the residues of dg at the following points vanish:

(1) a_k $(k = 1, \ldots, n)$, and

(2) p_j such that α_j is a negative integer.

Construction of CMC-1 *surface from two Gauss maps.* In addition to the SU(2)-conditions for the period problem (Proposition 3.2), we introduce another method to construct CMC-1 surfaces [UY3]: Let M be a Riemann surface and $\widetilde{M}$ the universal cover of M.

Proposition A.7. *Let G and g be meromorphic functions defined on M and $\widetilde{M}$, respectively. Assume*

(1) $d\sigma^2 := 4\,dg\,d\bar{g}/(1+g\bar{g})^2$ is a pseudometric with conical singularities which is single-valued on M.

(2) The meromorphic differential $Q := (S(g) - S(G))/2$ is holomorphic on M.

(3) The metric $ds^2 := (1+|g|^2)^2|Q/dg|^2$ is a non-degenerate complete metric on M.

Then there exists a complete CMC-1 *immersion $f\colon M \to H^3$ with hyperbolic Gauss map G and secondary Gauss map g.*

Proof. By the second assumption , $\operatorname{ord}_p Q = \operatorname{ord}_p d\sigma^2$ for any point $p \in M$. Then, by Theorem 2.2 and Remark 2.3 in [UY3], there exists a CMC-1 immersion $f\colon M \to H^3$ whose hyperbolic Gauss map, secondary Gauss map and Hopf differential are G, g and Q, respectively. Moreover, by the third assumption, the induced metric is complete. Q.E.D.

Partial classification for $\mathrm{TA}(f) \le 8\pi$. By Corollary 7.4, a complete CMC-1 surface with $\mathrm{TA}(f) \le 8\pi$ is either a surface of genus 0 with at most 5 ends or a surface of genus 1 with at most 3 ends. We denote by γ and n the genus and the number of the ends, respectively.

The case $(\gamma, n) = (0, 1)$. In this case, we may assume $M = \boldsymbol{C}$ and the only end is $p_1 = \infty$. Since M is simply-connected, the representation ρ_g as in (3.3) is trivial, that is, such a surface is $\mathcal{H}^3$-reducible. Then by Lemma 5.3, the dual immersion $f^\#$ is also well-defined on M. And since the dual surface of $f^\#$ is f itself, the classification reduces to that for CMC-1 surfaces with dual absolute total curvature at most 8π, which is done in [RUY3].

The case $(\gamma, n) = (0, 2)$. In this case, the pseudometric $d\sigma^2$ as in (3.1) has the divisor $\mu_1 p_1 + \mu_2 p_2 + \xi_1 q_1 + \cdots + \xi_m q_m$ (see (5.7)), where p_1 and p_2 are the ends and $q_1, \ldots, q_m$ are umbilic points. Since ξ_k $(k = 1, \ldots, m)$

are integers, Lemma A.3 implies that a surface in this class satisfies either

(1) Both μ_1 and μ_2 are integers (the case of $\mathcal{H}^3$-reducible), or

(2) Both μ_1 and μ_2 are non-integral real numbers (the case of $\mathcal{H}^1$-reducible).

If both ends are regular, such surfaces are completely classified (see Example 4.3 or [UY1]), and the only possible case is $\mathbf{O}(-2,-2)$. So we may assume at least one end is irregular: $d_2 \le -3$. If $d_1 \ge -1$, then $\mu_1 \in \boldsymbol{Z}$ by (7.12), and hence we have the case (1). Hence g is a meromorphic function on the genus 0 Riemann surface $\boldsymbol{C} \cup \{\infty\}$, and $\mathrm{TA}(f) = 4\pi \deg g$. Thus $\deg g \le 2$, and hence μ_1 and μ_2 are 0 or 1. Then by (7.8), we have $d_1 \le -1$ and $\mu_1 = 1$. Moreover, by (7.7), we have $d_2 \ge -4$. On the other hand, if $d_1 \le -2$, by (7.3), (7.7) and (7.11), we have $-7 \le d_1 + d_2 \le -4$. Hence the possible cases are $(d_1, d_2) = (-1,-3), (-1,-4), (-2,-3), (-2,-4), (-2,-5), (-3,-3)$ and $(-3,-4)$. Throughout this subsection, we set $M = \boldsymbol{C} \setminus \{0\}$.

Proposition A.8. *There exists no complete* CMC-1 *immersion* $f\colon \boldsymbol{C}\setminus\{0\} \to H^3$ *with* $\mathrm{TA}(f) \le 8\pi$ *and of class* $\mathbf{O}(-1,-3)$ *or* $\mathbf{O}(-2,-3)$.

Proof. Assume f is of class $\mathbf{O}(-1,-3)$. In this case, $\mu_1 \in \boldsymbol{Z}$ by (7.12), and then f is $\mathcal{H}^3$-reducible (the case (1) above). Then the dual immersion $f^\#$ is also well-defined on M whose dual absolute total curvature is not greater than 8π. Such a surface cannot exist because of the results in [RUY3] (see Table 1).

Now suppose f is of class $\mathbf{O}(-2,-3)$. If $\mu_1 \in \boldsymbol{Z}$, then for the same reason as in the $\mathbf{O}(-1,-3)$ case, such a surface does not exist. Now assume $\mu_1 \notin \boldsymbol{Z}$. Then the surface is of type (2): $\mu_2 \notin \boldsymbol{Z}$. By the same argument as in the case $d_1 + d_2 = -5$ of $(\gamma, n) = (0,2)$ in the proof in [RUY4] of Theorem 2.1 (in this paper), such a surface cannot exist. Q.E.D.

By a similar argument to that in the proof of Proposition A.8, if $\mu_1 \in \boldsymbol{Z}$, the classification is the same as the dual case in [RUY3]. Hence the case $\mathbf{O}(-1,-4)$, and also the case $\mathbf{O}(-2,-4)$ with $\mu_1 \in \boldsymbol{Z}$ ($\mathcal{H}^3$-reducible), are classified. Furthermore, for the same reason, the $\mathbf{O}(-2,-5)$ case with $\mu_1 \in \boldsymbol{Z}$ and $\mathrm{TA}(f) \le 8\pi$ does not exist.

Remark A.9. In the case $\mathbf{O}(-2,-4)$ with (2) holding, we have the following examples: Let

$$dg = t\, z^{\mu} \frac{z^2 - a^2}{(z^2-1)^2}\, dz\,, \qquad Q = \theta \frac{z^2 - a^2}{z^2} dz^2\,,$$

where

$$a^2 = \frac{\mu+1}{\mu-1}\,, \qquad \theta = \frac{\mu(\mu+2)(\mu-1)}{4(\mu+1)}\,, \qquad -1 < \mu < 0\,.$$

Here t is a positive real number corresponding to the one parameter deformation coming from reducibility (see Section 5). Then the residues of dg at -1 and 1 vanish, and there exists the meromorphic function g defined on the universal cover of $\boldsymbol{C} \setminus \{0\}$ (see Remark A.6). We set $\omega = Q/dg$. Then by Theorem 2.4 in [UY1], one can check that there exists an immersion $f\colon \boldsymbol{C}\setminus\{0\} \to H^3$ with data (g, Q). For this example, $\mu_1 = \mu_2 = |\mu+1| - 1 = \mu$ because $-1 < \mu < 0$ (see Corollary A.5). Then, $\mathrm{TA}(f)/2\pi = 2(\mu+2) \in (4\pi, 8\pi)$.

Remark A.10. For the $\mathbf{O}(-2,-5)$ case, the following data gives examples: We set

$$dg = t\, z^{\mu} \frac{z^3 - a^3}{(z^3-1)^2}\, dz,\ Q = \theta \frac{z^3 - a^3}{z^2}\, dz^2,\ a^3 = \frac{\mu+1}{\mu-2},\ \theta = \frac{\mu(\mu^2-4)}{4(\mu+1)},$$

where $\mu \in \boldsymbol{R}\setminus\{0,-1,\pm 2\}$, $t \in \boldsymbol{R}^+$. Here t is a parameter corresponding to a deformation which comes from reducibility (see Section 5). The ends are $0, \infty$ and the umbilic points are $a, ae^{2\pi/3i}, ae^{4\pi/3i}$.

In this case, we have $\mu_1 = |\mu+1| - 1$ and $\mu_2 = |\mu| - 1$. Hence $\mu_1 + \mu_2 = |\mu+1| + |\mu| - 2 \geq -1$, where equality holds if and only if $-1 \leq \mu \leq 0$. Thus the total absolute curvature is $\mathrm{TA}(f) = 2\pi(-2 + \mu_1 + \mu_2 - d_1 - d_2) \geq 8\pi$ and equality holds if and only if $-1 < \mu < 0$.

Remark A.11. For the cases $\mathbf{O}(-3,-3)$ and $\mathbf{O}(-3,-4)$, all ends are irregular, and then one cannot solve the period problem immediately. In the dual total curvature case, a deformation procedure as in [RUY1] can be used to construct examples of type $\mathbf{O}(-3,-3)$ [RUY3]. Unfortunately, this procedure cannot be used here, because the hyperbolic Gauss map is not a rational function.

Remark A.12. In the cases of $\mathbf{O}(-3,-4)$ and $\mathbf{O}(-2,-5)$, it can be shown that $\mathrm{TA}(f) \geq 8\pi$. In fact, in these cases, the divisor corresponding the pseudometric $d\sigma^2$ is $\mu_1 p_1 + \mu_2 p_2 + \xi_1 q_1 + \cdots + \xi_m q_m$, and by (7.3), we have $\xi_1 + \cdots + \xi_m = 3$ is an odd integer. Then by Corollary 4.7 of [RUY4], we have $\mu_1 + \mu_2 \geq -1$. This shows that $\mathrm{TA}(f) \geq 8\pi$.

The case $(\gamma, n) = (0,3)$. If μ_1, μ_2 and μ_3 are integers, then by Lemma A.1 and Lemma 5.3, the surface is $\mathcal{H}^3$-reducible and its dual is also well-defined on M with dual total absolute curvature at most 8π. By [RUY3], these must be of type $\mathbf{O}(-1,-1,-2)$, $\mathbf{O}(-1,-2,-2)$, or $\mathbf{O}(-2,-2,-2)$, and the first two cases are classified. Also, examples exist in the third case as well [RUY3, Example 4.4]. Moreover, for any surface of type $\mathbf{O}(-1,-1,-2)$, μ_1 and μ_2 are integers, by (7.12). Then, by Lemma A.3, μ_3 is also an integer. Thus, surfaces of type $\mathbf{O}(-1,-1,-2)$ must be $\mathcal{H}^3$-reducible and are completely classified.

Next, we assume all $\mu_j \notin \boldsymbol{Z}$. Then (7.3), (7.7) and (7.11) imply that $-8 \leq d_1 + d_2 + d_3 \leq -4$, and (7.12) implies that $d_j \leq -2$ $(j = 1,2,3)$. Hence the possible cases are $\mathbf{O}(-2,-2,-2)$, $\mathbf{O}(-2,-2,-3)$, $\mathbf{O}(-2,-2,-4)$ and $\mathbf{O}(-2,-3,-3)$. For the case $\mathbf{O}(-2,-2,-2)$, that is, for surfaces with three regular ends, the second and third authors classified the irreducible ones with embedded ends ([UY7], see Example 4.4). For the cases $\mathbf{O}(-2,-3,-3)$ and $\mathbf{O}(-2,-2,-4)$, the sum of the orders of the umbilic points are an even integer, by (7.3). Then by Corollary 4.7 in [RUY4] and (7.7), we have $\mathrm{TA}(f) \geq 8\pi$, hence $\mathrm{TA}(f) = 8\pi$. By Lemma A.3, there exists no surface with only one non-integer μ_j. Then the remaining case is to assume that one μ_j, say μ_1, is an integer and $\mu_2, \mu_3 \notin \boldsymbol{Z}$. Then by (7.12), d_2, $d_3 \leq -2$. Also, by (7.7), (7.8) and (7.11), we have $-5 \leq d_2 + d_3$. Hence we have two possibilities: $(d_2, d_3) = (-2,-2)$ or $(d_2, d_3) = (-2,-3)$. When $(d_2, d_3) = (-2,-2)$, by (7.7) and (7.8), we have $\mu_1 - d_1 = 2$ or 3. And by (7.3), $d_1 \leq 0$. Hence we have the possibilities $\mathbf{O}(-3,-2,-2)$, $\mathbf{O}(-2,-2,-2)$, $\mathbf{O}(-1,-2,-2)$ and $\mathbf{O}(0,-2,-2)$. Similarly, when $(d_2, d_3) = (-2,-3)$, we have $\mu_1 - d_1 = 2$ and $d_1 \leq 1$. Hence the possibilities are $\mathbf{O}(-2,-2,-3)$, $\mathbf{O}(-1,-2,-3)$, $\mathbf{O}(0,-2,-3)$, $\mathbf{O}(1,-2,-3)$. In this case, the corresponding divisor of the pseudometric $d\sigma^2$ is

$$\mu_1 p_1 + \mu_2 p_2 + \mu_3 p_3 + \sum_{k=1}^{m} \xi_k q_k = \mu_2 p_2 + \mu_3 p_3 + (2 + d_1) p_1 + \sum_{k=1}^{m} \xi_k q_k \,,$$

where the q_k $(k = 1, \dots, m)$ are the umbilic points and ξ_k is the order of Q at q_k (see (5.7)). Here, by (7.3), $\xi_1 + \cdots + \xi_m = 1 - d_1$, so $\mu_1 + \sum_{k=1}^{m} \xi_k = d_1 + 2 + \sum_{k=1}^{m} \xi_k = 3$. Hence if $d_1 \geq -1$ (and so $\mu_1 \in \boldsymbol{Z}^+$), Corollary 4.7 of [RUY4] implies $\mu_2 + \mu_3 \geq -1$. This implies that $\mathrm{TA}(f) \geq 8\pi$, and so

(A.3) For a surface of type $\mathbf{O}(d,-2,-3)$ $(d \geq -1)$, $\qquad \mathrm{TA}(f) \geq 8\pi$.

Proposition A.13. *There exists no complete* CMC-1 *surface* f *with* $\mathrm{TA}(f) \leq 8\pi$ *and of type* $\mathbf{O}(0,-2,-3)$.

Proof. Assume such an immersion $f\colon \boldsymbol{C}\cup\{\infty\}\setminus\{p_1,p_2,p_3\}\to H^3$ exists. By (7.3), there is the only umbilic point q_1. We set the ends $(p_1,p_2,p_3)=(0,1,\infty)$ and the umbilic point $q_1=q\in\boldsymbol{C}\setminus\{0,1\}$. Then the Hopf differential Q has a zero only at q with order 1, and two poles at 1 and ∞ with orders 2 and 3, respectively. Thus, Q can be written as

$$Q:=\theta\frac{z-q}{(z-1)^2}\,dz^2 \qquad (\theta\in\boldsymbol{C}\setminus\{0\})\,.$$

On the other hand, by (7.7), (7.11), (7.8) and (7.12), we have $\mu_1=2$, and

$$\text{(A.4)}\qquad -2<\mu_2+\mu_3\le -1 \qquad \text{and} \qquad -1<\mu_j<0 \qquad (j=2,3)\,.$$

The secondary Gauss map branches at (p_1,p_2,p_3) and q with branch orders 2, μ_2, μ_3 and 1, respectively. Then by Corollary A.5, we can take the secondary Gauss map g such that

$$dg=t\,\frac{z^\alpha(z-1)^\nu(z-q)^\beta}{\prod_{j=1}^N(z-a_j)^2}\,dz \qquad (t\in\boldsymbol{R}\setminus\{0\})\,,$$

where

$$\nu=\mu_2 \quad\text{or}\quad -\mu_2-2\,, \qquad \alpha=2 \quad\text{or}\quad -4\,, \qquad \beta=1\quad\text{or}\quad -3\,,$$

and $a_j\in\boldsymbol{C}\setminus\{0,1,q\}$ $(j=1,\dots,N)$ are mutually distinct points.

Without loss of generality, we may assume $\nu=\mu_2$ (if not, we can take $1/g$ instead of g). Then by (A.2) in Corollary A.5, we have

$$-(\alpha+\beta)-\mu_2+2N-2=\mu_3 \quad\text{or}\quad -\mu_3-2\,,$$

so $\mu_2+\mu_3$ or $\mu_2-\mu_3$ is an odd integer. Then by (A.4), we have $\mu_2+\mu_3=-1$ and $(\alpha,\beta,N)=(2,1,2)$ or $(2,-3,0)$.

First, we assume $(\alpha,\beta,N)=(2,1,2)$, and we set $\mu_2=\mu$. Then

$$\text{(A.5)}\qquad dg=t\,\frac{z^2(z-1)^\mu(z-q)}{(z-a)^2(z-b)^2}\,dz \qquad (a,b\in\boldsymbol{C}\setminus\{0,1,q\},a\ne b)\,.$$

Such a g exists if and only if the residues of the right-hand side of (A.5) vanish:

$$\text{(A.6)}\qquad \frac{2}{a}+\frac{\mu}{a-1}+\frac{1}{a-q}-\frac{2}{a-b}=0\,,$$

$$\text{(A.7)}\qquad \frac{2}{b}+\frac{\mu}{b-1}+\frac{1}{b-q}-\frac{2}{b-a}=0\,.$$

Since $a \neq b$, these equations are equivalent to (A.6) – (A.7) and $b \times$ (A.6) – $a \times$ (A.7):

$$(\mu+1)(a^2+b^2)-(\mu q+1)(a+b)-2ab+2q=0\,,$$
$$(\mu+1)(a+b)ab-2(\mu q+q+2)ab+2q(a+b)=0\,. \tag{A.8}$$

On the other hand, let $\omega = Q/dg$, and consider the equation

$$X''-(\log\hat{\omega})'X'-\hat{Q}X=0 \qquad (\omega=\hat{\omega}\,dz,\quad Q=\hat{Q}\,dz^2)\,, \tag{A.9}$$

which is named (E.1) in [UY1]. The roots of the indicial equation of (A.9) at $z=0$ are 0 and -1. By Theorem 2.2 of [UY1], the log-term coefficient of (A.9) at $z=0$ must vanish if the surface exists:

$$\mu+2-\frac{2}{a}-\frac{2}{b}=0\,. \tag{A.10}$$

(See Appendix A of [RUY3] or Appendix A of [UY1].) Here, the solution of equations (A.8) and (A.10) is $a=b=q=4/(\mu+2)$, a contradiction. Hence the case $(\alpha,\beta,N)=(2,1,2)$ is impossible.

Next, we consider the case $(\alpha,\beta,N)=(2,-3,0)$. Then one has

$$dg=t\,\frac{z^2(z-1)^\mu}{(z-q)^3}\,dz \qquad (t\in \boldsymbol{C}\setminus\{0\})\,.$$

The residue at $z=q$ vanishes if and only if

$$(\mu+2)(\mu+1)q^2-4(\mu+1)q+2=0\,. \tag{A.11}$$

On the other hand, in the same way as the first case, the log-term coefficient of (A.9) at $z=0$ vanishes if and only if

$$\mu+2=\frac{4}{q}\,. \tag{A.12}$$

However, there is no pair (μ,q) satisfying (A.11) and (A.12) simultaneously. Hence this case is also impossible. Q.E.D.

Proposition A.14. *There exists no* CMC-1 *immersion of type* $\mathbf{O}(1,-2,-3)$ *with* $\mathrm{TA}(f)\leq 8\pi$.

Proof. Assume such an immersion $f\colon \boldsymbol{C}\cup\{\infty\}\setminus\{p_1,p_2,p_3\}\to H^3$ exists. Then we have $\mathrm{TA}(f)=8\pi$ because of (A.3), and by (7.7), (7.8) and (7.11), one has

$$\mu_1=3 \qquad\text{and}\qquad \mu_2+\mu_3=-1, \qquad -1<\mu_j<0 \quad (j=2,3)\,. \tag{A.13}$$

We set $(p_1, p_2, p_3) = (1, 0, \infty)$. By (7.3), there are no umbilic points. Then the Hopf differential Q can be written as

$$Q = \theta \, \frac{z-1}{z^2} \, dz^2 \qquad (\theta \in \boldsymbol{C} \setminus \{0\}) \, .$$

The secondary Gauss map g branches at 0, ∞ and 1 with orders μ_2, μ_3 and 3, respectively. Then by Corollary A.5, dg can be put in the following form:

$$dg = t \, z^{\mu_2} \frac{(z-1)^\alpha}{\prod_{j=1}^{N}(z-a_j)^2} \, dz \, ,$$

where $a_j \in \boldsymbol{C} \setminus \{0,1\}$ $(j = 1, \dots, N)$ are mutually distinct numbers, t is a positive real number, and $\alpha = 3$ or -5, and $-\mu_2 - \alpha + 2N - 2 = \mu_3$ or $\mu_3 - 2$. The second case is impossible because of (A.13). Hence $2N = \alpha + 1$, and then $\alpha = 3$ and $N = 2$. Thus we have the form

$$dg = t \, z^{\mu} \frac{(z-1)^3}{(z-a)^2(z-b)^2} \, dz \qquad (\mu = \mu_2) \, .$$

Such a g exists if the residues at $z = a$ and $z = b$ vanish:

$$\frac{\mu}{a} + \frac{3}{a-1} - \frac{2}{a-b} = 0 \quad \text{and} \quad \frac{\mu}{b} + \frac{3}{b-1} - \frac{2}{b-a} = 0 \, .$$

By direct calculation, we have

$$\begin{aligned} a &= \frac{-2 + \mu + \mu^2 + \sqrt{2}\sqrt{2 - \mu - \mu^2}}{(\mu+1)(\mu+2)} \, , \\ b &= \frac{-2 + \mu + \mu^2 - \sqrt{2}\sqrt{2 - \mu - \mu^2}}{(\mu+1)(\mu+2)} \, . \end{aligned}$$

Consider the equation (E.1) in [UY1] with $\omega = Q/dg$. Then, the indicial equation at $z = 1$ has the two roots 0 and -1 with difference 1. By direct calculation again, the log-term at $z = 1$ vanishes if and only if

$$\mu + 2 - \frac{2}{1-a} - \frac{2}{1-b} = -\frac{1}{3}(\mu + 2) = 0 \, ,$$

which is impossible because $\mu = \mu_2 > -1$. Q.E.D.

Proposition A.15. *For a non-zero real number μ $(-1 < \mu < 0)$ and positive integer m, set*

$$\text{(A.14)} \quad G = z^{m+1} \frac{mz - (m+2)}{(m+2)z - m} \qquad \text{and} \qquad g = z^{\mu+1} \frac{\mu z - (\mu+2)}{(\mu+2)z - \mu} \, .$$

Then there exists a one parameter family of conformal CMC-1 *immersions* $f\colon \boldsymbol{C}\setminus\{0,1\}\to H^3$ *of type* $\mathbf{O}(0,-2,-2)$ *with* $\mathrm{TA}(f)=4\pi(\mu+2)$, *whose hyperbolic Gauss map and secondary Gauss map are* G *and* tg $(t\in\boldsymbol{R}^+)$, *respectively.*

Conversely, any CMC-1 *surface of type* $\mathbf{O}(0,-2,-2)$ *with* $\mathrm{TA}(f)\le 8\pi$ *is obtained in such a manner. In particular,* $\mathrm{TA}(f)<8\pi$.

Proof. For g and G as in (A.14), set

$$Q:=\frac{1}{2}\left(S(g)-S(G)\right)=\frac{m(m+2)-\mu(\mu+2)}{4}\frac{dz^2}{z^2}\,.$$

Since $\mu\notin\boldsymbol{Z}$, the right-hand side is not identically zero. Moreover, one can easily check that the assumptions of Proposition A.7 hold. Hence there exists a complete CMC-1 immersion $f\colon \boldsymbol{C}\setminus\{0,1\}\to H^3$ with hyperbolic Gauss map G, secondary Gauss map g and Hopf differential Q. Conversely, suppose such a surface exists. Then without loss of generality, we set $M=\boldsymbol{C}\setminus\{0,1\}$ and $(p_1,p_2,p_3)=(1,0,\infty)$. By (7.3), there are no umbilic points. Then the Hopf differential Q has poles of order 2 at 0 and ∞, and has no zeros. Hence we have $Q=\theta\,z^{-2}\,dz^2$ $(\theta\in\boldsymbol{C}\setminus\{0\})$. This implies that the secondary and hyperbolic Gauss maps branch only at the ends. Hence both $S(g)$ and $S(G)$ have poles of order 2 at $z=0,1,\infty$ and holomorphic on M. More precisely, we have

$$\text{(A.15)}\qquad \begin{aligned} S(g)&=\frac{c_3z^2+(c_1-c_2-c_3)z+c_2}{z^2(z-1)^2}\,dz^2,\\ S(G)&=\frac{c_3^{\#}z^2+(c_1^{\#}-c_2^{\#}-c_3^{\#})z+c_2^{\#}}{z^2(z-1)^2}\,dz^2\,,\end{aligned}$$

where c_j and $c_j^{\#}$ are as in (7.2). Here, $\mu_1=\mu_1^{\#}$ because of (7.12). Hence we have

$$\begin{aligned}2Q&=S(g)-S(G)\\ &=\left(\frac{c_2-c_2^{\#}}{z^2}+\frac{(c_2-c_3)-(c_2^{\#}-c_3^{\#})}{z}-\frac{(c_2-c_3)-(c_2^{\#}-c_3^{\#})}{z-1}\right)dz^2\,.\end{aligned}$$

Thus we have

$$\text{(A.16)}\qquad (\mu_2-\mu_3)(\mu_2+\mu_3+2)=(\mu_2^{\#}-\mu_3^{\#})(\mu_2^{\#}+\mu_3^{\#}+2)\,.$$

On the other hand, (7.7), (7.8) and (7.11) imply that $\mu_1=\mu_1^{\#}=2$ or 3. If $\mu_1=3$, (7.7) implies that $\mu_2+\mu_3\le -1$. Then by (7.11),

$-1 < \mu_j < 0$ for $j = 2, 3$. Hence using (A.16), we have

$$1 > |\mu_2 - \mu_3| > |\mu_2^\# - \mu_3^\#| \,.$$

Here $\mu_2^\#$ and $\mu_3^\#$ are integers, hence $\mu_2^\# = \mu_3^\#$. The hyperbolic Gauss map G is a meromorphic function on $\boldsymbol{C} \cup \{\infty\}$. Then the Riemann-Hurwitz relation implies that $\boldsymbol{Z} \ni \deg G = \frac{1}{2}(2 + \mu_1^\# + \mu_2^\# + \mu_3^\#) = \mu_2^\# + 2 + \frac{1}{2}$. This is impossible.

When $\mu_1 = \mu_1^\# = 2$, by similar arguments, we have $-1 < \mu_j < 1$ $(j = 2, 3)$ and $\mu_2 + \mu_3 \le 0$. This implies that $|\mu_2 - \mu_3| < 2$. Thus by (A.16), we have $|\mu_2^\# - \mu_3^\#| = 0$ or 1. We may assume that $\mu_3^\# \ge \mu_2^\#$ (if not, exchange the ends 0 and ∞). Assume $\mu_3^\# - \mu_2^\# = 1$. In this case, $\boldsymbol{Z} \ni \deg G = \frac{1}{2}(2 + \mu_1^\# + \mu_2^\# + \mu_3^\#) = \mu_2^\# + 2 + \frac{1}{2}$, which is impossible. Hence, using also (A.16), we have $\mu_3^\# - \mu_2^\# = \mu_3 - \mu_2 = 0$. Moreover, $\mu_2 + \mu_3 = 2\mu_2 \le 0$, so $\mu_2 \le 0$. Putting all this together, we have

$$\mu_1 = \mu_1^\# = 2\,, \ -1 < \mu_2 = \mu_3 \le 0\,, \ \mu_2^\# = \mu_3^\#\,, \ \text{and } Q = \frac{c_2 - c_2^\#}{2z^2}\,dz^2\,.$$

If $\mu_2 = 0$, the secondary Gauss map g is a meromorphic function on $\boldsymbol{C} \cup \{\infty\}$ with only one branch point, which is impossible. Hence $\mu_2 < 0$. In this case, the pseudometric $d\sigma^2$ branches on the divisor $\mu_1 p_1 + \mu_2 p_2 + \mu_3 p_3$ because there are no umbilic points. Thus the secondary Gauss map g satisfies (A.15). One possibility of such a g is in the form (A.14) with $\mu = \mu_2$. On the other hand, since the surface is $\mathcal{H}^1$-reducible, g can be normalized as in (A.14) because of Corollary A.5. Since $S(G) = S(g) - 2Q$, the Schwarzian derivative of the hyperbolic Gauss map G is uniquely determined, and G is determined up to Möbius transformations. Then such a surface is unique, with given g and G. Q.E.D.

Proposition A.16. *Let $\mu \in (-1, 0)$, $m \ge 2$ an integer,*

$$a := -\frac{m + \mu + 2}{m - \mu - 2}, \qquad p := \frac{a\mu + a - a^2}{a\mu + a - 1}\,,$$

and $M = \boldsymbol{C} \cup \{\infty\} \setminus \{0, 1, p\}$. Then there exist a meromorphic function G on $\boldsymbol{C} \cup \{\infty\}$ and a meromorphic function g on the universal cover $\widetilde{M}$ of M such that
(A.17)

$$dG = z\frac{(z-p)^{m-2}}{(z-1)^{m+2}}\,dz \qquad \textit{and} \qquad dg = t\,z\frac{(z-1)^{\mu}(z-p)^{-\mu-2}}{(z-a)^2}\,dz$$

respectively, where $t \in \boldsymbol{R}^+$, and there exists a complete CMC-1 *immersion $f\colon M \to H^3$ whose hyperbolic Gauss map and secondary Gauss map* *are G and g, respectively. Moreover* $\mathrm{TA}(f) = 4\pi(\mu+2) \in (4\pi, 8\pi)$. *Conversely, an $\mathcal{H}^1$-reducible complete* CMC-1 *surface of class* $\mathbf{O}(-1,-2,-2)$ *with* $\mathrm{TA}(f) < 8\pi$ *is obtained in such a way.*

Proof. The residue of dG in (A.17) at $z = 1$ and the residue of dg in (A.17) at $z = a$ vanish. Thus there exist G and g such that (A.17) hold. Moreover, by direct calculation, we have

$$Q := \frac{1}{2}\bigl(S(g) - S(G)\bigr) = \frac{4m^2\bigl(m(m+2) - \mu(\mu+2)\bigr)}{(m+\mu)^2(2-m+\mu)^2} \frac{dz^2}{z(z-1)^2(z-p)^2} .$$

Then by Proposition A.7, there exists a CMC-1 immersion $f\colon M \to H^3$. One can easily check that f is complete and $\mathrm{TA}(f) = 4\pi(\mu+2)$. Conversely, assume such a surface exists. Then by (7.3), there is only one umbilic point q_1 of order one. We set $(p_1, p_2, p_3) = (0, 1, p)$ and $q_1 = \infty$, where $p \in \boldsymbol{C} \setminus \{0, 1\}$. By (7.7), (7.8), (7.12), (7.9) and (7.11), we have $\mu_1 = 1$ or 2. When $\mu_1 = 2$, by Corollary 4.7 of [RUY4], $\mu_2 + \mu_3 \geq -1$ holds, and then $\mathrm{TA}(f) \geq 8\pi$. Assume $\mu_1 = 1$ and $\mathrm{TA}(f) \leq 8\pi$. If one of μ_j $(j = 2, 3)$ is an integer, by Lemma A.3, the other is also an integer, and hence the surface is $\mathcal{H}^3$-reducible. Thus both μ_2 and μ_3 are non-integers. By (7.7) and (7.11), we have

$$\text{(A.18)} \qquad -2 < \mu_2 + \mu_3 \leq 0 , \qquad -1 < \mu_j < 1 \quad (j = 2, 3) .$$

Then the secondary Gauss map g branches at 0, 1, p and ∞ with orders μ_1, μ_2, μ_3 and 1, respectively. Then by Corollary A.5, g can be chosen in the form

$$dg = t\,\frac{(z-1)^{\nu_2}(z-p)^{\nu_3} z^{\alpha}}{\prod_{k=1}^{N}(z-a_k)^2}\,dz \qquad (t \in \boldsymbol{R}^+) ,$$

where $\nu_j = \mu_j$ or $-\mu_j - 2$, and $\alpha = 1$ or -3, and $\{a_1, \dots, a_N\} \subset \boldsymbol{C} \setminus \{0, 1, p\}$ are mutually distinct points. We may assume $\nu_2 = \mu_2$ (if not, take $1/g$ instead of g). Then by (A.2),

$$-\mu_2 - \nu_3 - \alpha + 2N - 2 = 1 \quad \text{or} \quad -3$$

holds. This implies that $\mu_2 + \mu_3$ or $\mu_2 - \mu_3$ is an even integer. Then by (A.18), we have

$$\mu := \mu_2 = \mu_3 \in (-1, 0), \qquad \nu_3 = -\mu - 2, \qquad \alpha = 1, \quad \text{and} \quad N = 1 .$$

Hence we have

$$\text{(A.19)} \qquad dg = t\,\frac{z(z-1)^{\mu}(z-p)^{-\mu-2}}{(z-a)^2}\,dz \qquad (t \in \boldsymbol{R}^+, \quad a \in \boldsymbol{C} \setminus \{0, 1, p\}) .$$

Such a map g exists on the universal cover of $\boldsymbol{C}\setminus\{0,1,p\}$ if and only if the residue at $z=a$ of the right-hand side of (A.19) vanishes, that is, if $p=(a\mu+a-a^2)/(a\mu+a-1)$. The Hopf differential of such a surface can be written in the form

$$Q=\frac{\theta\,dz^2}{z(z-1)^2(z-p)^2}\qquad(\theta\in\boldsymbol{C}\setminus\{0\}) \tag{A.20}$$

because it has poles of order 2 at $z=1$ and p, a pole of order 1 at $z=0$ and a zero of order 1 at $z=\infty$. Let $\mu_1^\#$, $\mu_2^\#$ and $\mu_3^\#$ be the branch orders of the hyperbolic Gauss map at p_1, p_2 and p_3, respectively. Then by (7.1), we have

$$c_2-c_2^\#=\frac{2\theta}{(1-p)^2},\qquad c_3-c_3^\#=\frac{2\theta}{(1-p)^2p}, \tag{A.21}$$

where c_j and $c_j^\#$ are as in (7.2). Then p and θ are positive real numbers. Without loss of generality, we may assume $\mu_2^\#\geq\mu_3^\#$. Then we have

$$0\geq c_2^\#-c_3^\#=\frac{2\theta}{p(1-p)}. \tag{A.22}$$

Hence $\mu_2^\#\neq\mu_3^\#$, that is, $\mu_2^\#>\mu_3^\#$. Since $\mu_1^\#=1$ by (7.12), the hyperbolic Gauss map branches at 0, 1, p and ∞ with branching order 1, $\mu_2^\#$, $\mu_3^\#$ and 1, respectively. Then the Riemann-Hurwitz relation implies that $\deg G=2+(\mu_2^\#+\mu_3^\#)/2<\mu_2^\#+2$. On the other hand, we have $\deg G\geq\mu_2^\#+1$. Hence we have $\deg G=\mu_2^\#+1$ and $\mu_3^\#=\mu_2^\#-2$. We set $m:=\mu_2^\#$. By a suitable Möbius transformation, we may set $G(p_2)=G(1)=\infty$. Since $z=1$ is a point of multiplicity $m+1$, G has no pole except $z=1$. Then dG can be written in the form

$$dG=c\,z\frac{(z-p)^{m-2}}{(z-1)^{m+2}}\,dz\qquad(c\in\boldsymbol{C}\setminus\{0\}),$$

and we can choose $c=1$ by a Möbius transformation again. Moreover, the Hopf differential $Q=(S(g)-S(G))/2$ is as in (A.20) if and only if $a=-(m+\mu+2)/(m-\mu-2)$. Q.E.D.

Proposition A.17. *Let m be a positive integer and $\mu\in(-1,0)$ a real number.*

(1) If $m\geq 3$, $p:=\frac{m(m+2)-\mu(\mu+2)}{(m-2)^2-\mu^2}$, $\theta:=\frac{(\mu-3m+2)^2(m(m+2)-\mu(\mu+2))}{((m-2)^2-\mu^2)^2}$, then there exists a complete CMC-1 *immersion $f\colon M:=\boldsymbol{C}\cup\{\infty\}\setminus\{0,1,p\}\to H^3$ with hyperbolic Gauss map G and Hopf differential Q so that $dG=z^2\frac{(z-p)^{m-3}}{(z-1)^{m+2}}\,dz$, $Q=\frac{\theta}{z(z-1)^2(z-p)^2}\,dz^2$.*

(2) If $m \geq 1$, $p := \frac{\mu+m+2}{\mu+m}$, $\theta := \frac{(m-\mu)(\mu+m+2)}{(m+\mu)^2}$, *then there exists a complete* CMC-1 *immersion* $f\colon M := \boldsymbol{C}\cup\{\infty\}\setminus\{0,1,p\} \to H^3$ *with hyperbolic Gauss map* G *and Hopf differential* Q *so that*

$$dG = z^2\frac{(z-p)^{m-1}}{(z-1)^{m+2}(z-a)^2}\,dz,\ Q = \frac{\theta}{z(z-1)^2(z-p)^2}\,dz^2,$$

where $a := \frac{m-\mu\pm\sqrt{9(m-\mu)^2+16m(\mu+1)+16\mu(m+1)}}{2(\mu+m)}$.

In each case, the immersion f *is complete, of type* $\mathbf{O}(-1,-2,-2)$, $\mathcal{H}^1$*-reducible and satisfies* $\mathrm{TA}(f) = 8\pi$. *Conversely, any* $\mathcal{H}^1$*-reducible* CMC-1 *immersion of class* $\mathbf{O}(-1,-2,-2)$ *with* $\mathrm{TA}(f) = 8\pi$ *is obtained in this way.*

Proof. Since the residue of dG as in (1) at $z=1$ vanishes, there exists a meromorphic function G. Since the metric $ds^{2\#}$ as in (3.9) is non-degenerate and complete on $M := \boldsymbol{C}\cup\{\infty\}\setminus\{0,1,p\}$, there exists a CMC-1 immersion $f\colon \widetilde{M} \to H^3$ with hyperbolic Gauss map G and Hopf differential Q as in (1), where $\widetilde{M}$ is the universal cover of M. (In fact, there exists a CMC-1 immersion $f^\#\colon \widetilde{M} \to H^3$ with Weierstrass data $(G, -Q/dG)$. Then taking the dual yields the desired immersion.) Let F be the lift of f. Then F is a solution of (3.10), and there exists a representation ρ_F as in (3.17).

The components F_{21} and F_{22} of F satisfy the equation $(\mathrm{E}.1)^\#$ in [RUY3]:
(A.23)

$$X'' - \left(\log(\hat{\omega}^\#)'\right)X' + \hat{Q}X = 0,\ \left(\omega^\# := \hat{\omega}^\#\,dz = \frac{Q}{dG},\ Q = \hat{Q}\,dz^2\right).$$

By a direct calculation, the roots of the indicial equation of (A.23) at $z=0$ are 0 and -2, and the log-term coefficient at $z=0$ vanishes (see Appendix A of [RUY3]). Hence F_{21} and F_{22} are meromorphic on a neighborhood of $z=0$, and then, the secondary Gauss map $g = -dF_{22}/dF_{21}$ is meromorphic at $z=0$. Hence, by (5.8), $\rho_F(\tau_1) = \pm\rho_g(\tau_1) = \pm\,\mathrm{id}$, where ρ_F is a representation corresponding to the secondary Gauss map g, and τ_1 is a deck transformation corresponding to a loop surrounding $z=0$. Moreover, the difference of the roots of the indicial equation at $z=1$ is $\mu+1 \notin \boldsymbol{Z}$. This implies that one can choose the secondary Gauss map g such that $g\circ\tau_2 = e^{2\pi i\mu}g$, where τ_2 is a deck transformation of $\widetilde{M}$ corresponding to a loop surrounding $z=1$. Then $\rho_F(\tau_2) = \pm\rho_g(\tau_2) = \mathrm{diag}\{e^{\pi i\mu}, e^{-\pi i\mu}\} \in \mathrm{SU}(2)$. Hence the representation ρ_F lies in $\mathrm{SU}(2)$, since τ_1 and τ_2 generate the fundamental group of M. Then by Proposition 3.2, the immersion f is well-defined on M, and by Lemma 3.1, f is a complete immersion. Using (7.1), we have $\mu_1 = 2, \quad \mu_2 = \mu, \quad \mu_3 = -\mu - 1$. Then by (7.7), we have $\mathrm{TA}(f) = 8\pi$.

In the case (2), we can prove the existence of f in a similar way.

Conversely, we assume a complete $\mathcal{H}^1$-reducible immersion $f\colon M \to H^3$ of type $\mathbf{O}(-1,-2,-2)$ with $\mathrm{TA}(f) = 8\pi$ exists. Without loss of generality, we may set $(p_1, p_2, p_3) = (0, 1, p)$ and the only umbilic point $q = \infty$. As shown in the proof of Proposition A.16, we have $\mu_1 = \mu_1^\# = 2$. Thus, by (7.7) and the assumption $\mathrm{TA}(f) = 8\pi$, we have $\mu_2 + \mu_3 = -1$. Hence by (7.11), we can set $\mu_2 = \mu$, $\mu_3 = -1 - \mu$, $-1 < \mu < 0$. Without loss of generality, we may assume $\mu_2^\# \geq \mu_3^\#$. Then by the Riemann-Hurwitz relation, we have

$$\text{(A.24)} \quad \deg G = \frac{1}{2}\left(2 + \mu_1^\# + \mu_2^\# + \mu_3^\# + 1\right) = \frac{5}{2} + \frac{\mu_2^\# + \mu_3^\#}{2} \leq \frac{5}{2} + \mu_2^\# .$$

On the other hand, $\deg G \geq \mu_2^\# + 1$. Thus we have $\deg G = \mu_2^\# + 1$ or $\mu_2^\# + 2$. We set $m := \mu_2^\#$. Assume $\deg G = \mu_2^\# + 1 = m + 1$. Then by (A.24), $\mu_3^\# = m - 3$. Hence, the hyperbolic Gauss map G branches at 0, 1, p and ∞ with branch orders 2, m, $m-3$ and 1, respectively. By a suitable Möbius transformation, we assume $G(1) = \infty$. The multiplicity of G at $z = 1$ is $m + 1 = \deg G$. Then G has no other poles on $\boldsymbol{C} \cup \{\infty\}$. Thus, dG can be written in the form

$$dG = c\, z^2 \frac{(z-p)^{m-3}}{(z-1)^{m+2}}\, dz ,$$

where $c \in \boldsymbol{C} \setminus \{0\}$. By a suitable Möbius transformation, we may set $c = 1$. On the other hand, the Hopf differential Q can be written in the form

$$Q = \frac{\theta}{z(z-1)^2(z-p)^2}\, dz^2$$

because f is type $\mathbf{O}(-1,-2,-2)$ and ∞ is the umbilic point of order 1. Thus, by (7.1), we have

$$\text{(A.25)} \qquad c_2 - c_2^\# = \frac{2\theta}{(1-p)^2} , \qquad c_3 - c_3^\# = \frac{2\theta}{(1-p)^2 p} ,$$

where c_j and $c_j^\#$ are as in (7.2). Thus we have the case (1).

Next, we assume $\deg G = m + 2$. Then by (A.24), we have $\mu_3^\# = m - 1$. If we set $G(1) = \infty$, then G has only one simple pole other than the pole $z=1$, since the multiplicity of G at $z=1$ is $m+1$. So dG can be written in the form

$$dG = c\,z^2\frac{(z-p)^{m-1}}{(z-1)^{m+2}(z-a)^2}\,dz \qquad (a\in \boldsymbol{C}\setminus\{0,1,p\})\,,$$

where $c\in \boldsymbol{C}\setminus\{0\}$, which can be set to $c=1$ by a suitable Möbius transformation. The residue of dG at $z=a$ vanishes if and only if $p=(a(m+1)+a^2)/(ma+2)$. On the other hand, the relation (A.25) also holds in this case. Thus we have the case (2). Q.E.D.

The case $(\gamma,n)=(0,4)$. If two of the μ_j are integers, (7.7) and (7.8) imply that $\mathrm{TA}(f)>8\pi$. So at most one μ_j is an integer. By (7.7), (7.3) and (7.11), we have $-9\le d_1+d_2+d_3+d_4\le -4$. When all $\mu_j\notin \boldsymbol{Z}$, all $d_j\le -2$. Hence the possible cases are $\mathbf{O}(-2,-2,-2,-3)$ and $\mathbf{O}(-2,-2,-2,-2)$ (see Example 4.5).

Assume $\mu_1\ge 0$ is an integer. Then $\mu_2,\mu_3,\mu_4\notin\boldsymbol{Z}$ and $d_2,d_3,d_4\le -2$. In this case, by (7.7) and (7.8), we have $-6\le d_2+d_3+d_4$. Hence $d_2=d_3=d_4=-2$ and $\mu_1-d_1=2$. This implies that $d_1\ge -2$. Moreover, by (7.3), we have $d_1\le 2$. Hence the possible cases are $\mathbf{O}(d,-2,-2,-2)$ with $-2\le d\le 2$. Moreover, $\mu_1=2+d$ holds and there are $2-d$ umbilic points. Then when $d\ge -1$, we have $\mu_1\in\boldsymbol{Z}^+$, and so by Corollary 4.7 in [RUY4], we have $\mathrm{TA}(f)\ge 8\pi$. So $\mathrm{TA}(f)=8\pi$.

Proposition A.18. *There exist no* CMC-1 *surfaces in* H^3 *with* $\mathrm{TA}(f)\le 8\pi$ *of class* $\mathbf{O}(2,-2,-2,-2)$.

Proof. Assume such an immersion $f\colon \boldsymbol{C}\cup\{\infty\}\setminus\{p_1,\dots,p_4\}\to H^3$ exists. Then there are no umbilic points, and by (7.7), (7.8) and (7.11), $\mu_1=\mu_1^\#=4$ holds. Let G be the hyperbolic Gauss map. Then $\deg G\ge 5$ because $\mu_1^\#=4$. Hence by the Riemann-Hurwitz relation, $10\le 2\deg G=\sum_{j=2}^4\mu_j^\#+\mu_1^\#+2=\sum_{j=2}^4\mu_j^\#+6$ holds. This implies that $\mu_2^\#+\mu_3^\#+\mu_4^\#$ is an even number not less than 4:

(A.26) $$\mu_2^\#+\mu_3^\#+\mu_4^\#=2l,\qquad (l\in\boldsymbol{Z}, l\ge 2)\,.$$

Since $\mathrm{TA}(f)=8\pi$, we have

(A.27) $$\mu_2+\mu_3+\mu_4=-2\,.$$

Hence by (7.11),

(A.28) $$-1<\mu_j<0\qquad (j=2,3,4)\,.$$

We set $(p_1, p_2, p_3, p_4) = (p, 0, 1, -1)$, where $p \in \boldsymbol{C} \cup \{\infty\}$. We may assume $p_1 \in \boldsymbol{C}$. In fact, if $p_1 = \infty$, the Möbius transformation

$$z \longmapsto \frac{z-1}{3z+1}, \tag{A.29}$$

maps the ends $(p_1, p_2, p_3, p_4) = (\infty, 0, 1, -1)$ to $(1/3, -1, 0, 1)$. The Hopf differential can be written as

$$Q = 2\theta^2 \frac{(z-p)^2}{z^2(z^2-1)^2}\, dz^2 \qquad (\theta \in \boldsymbol{C} \setminus \{0\}) .$$

By the relation (7.1), we have

$$c_2 - c_2^{\#} = 4\theta^2 p^2, \quad c_3 - c_3^{\#} = \theta^2(1-p)^2, \quad c_4 - c_4^{\#} = \theta^2(1+p)^2, \tag{A.30}$$

where c_j and $c_j^{\#}$ are as in (7.2). Since $-1 < \mu_j < 0$ and $\mu_j^{\#}$ is a non-negative integer, we have

$$0 < c_j < \frac{1}{2}, \qquad c_j^{\#} \leq 0, \tag{A.31}$$

and consequently, $c_j - c_j^{\#} > 0$ $(j = 2, 3, 4)$. Let

$$\alpha_2 = \theta p, \qquad \alpha_3 = \frac{1}{2}\theta(1-p), \qquad \alpha_4 = \frac{1}{2}\theta(1+p). \tag{A.32}$$

Then we have $4\alpha_j^2 = c_j - c_j^{\#}$, which implies that the α_j $(j = 2, 3, 4)$ are real numbers. And then $p = \alpha_2/(\alpha_3 + \alpha_4)$ and $\theta = \alpha_3 + \alpha_4$ are real numbers. Here, without loss of generality, we may set $0 < p < 1$. (In fact, if $p < 0$, applying the coordinate change $z \mapsto -z$, we have $p > 0$. Moreover, if $p > 1$, by the transformation (A.29), we have $0 < p < 1$.)

We choose the sign of θ as $\theta > 0$. Then, we have α_2, α_3 and α_4 are positive numbers. Moreover, by (A.32), we have

$$\alpha_2 + \alpha_3 = \alpha_4 . \tag{A.33}$$

Using this, we have

$$\mu_2^{\#}, \quad \mu_3^{\#} \leq \mu_4^{\#} . \tag{A.34}$$

In fact, by (A.33) we have $\alpha_j < \alpha_4$ for $j = 2, 3$. Then $c_j - c_j^{\#} < c_4 - c_4^{\#}$. Hence by (A.31), $-c_j^{\#} < \frac{1}{2} - c_4^{\#}$. By definition, this implies that $\mu_j^{\#}(\mu_j^{\#}+2) < 1 + \mu_4^{\#}(\mu_4^{\#}+2)$. Thus we have $(\mu_j^{\#} - \mu_4^{\#})(\mu_j^{\#} + \mu_4^{\#} + 2) < 1$

for $j = 2, 3$. As the $\mu_j^\#$ are non-negative integers, $\mu_j^\# - \mu_4^\# \leq 0$, which implies (A.34). By (A.30) and the definition of α_j, we have

$$\mu_j(\mu_j + 2) = \mu_j^\#(\mu_j^\# + 2) - 8\alpha_j^2 \qquad (j = 2, 3, 4) .$$

Since $\mu_j \in (-1, 0)$, this implies that
(A.35)

$$\mu_j + 1 = \sqrt{1 + \mu_j^\#(\mu_j^\# + 2) - 8\alpha_j^2} = \sqrt{(\mu_j^\# + 1)^2 - 8\alpha_j^2} \qquad (j = 2, 3, 4) .$$

Now, defining $m_j := \mu_j^\# + 1 \geq 1$ $(j = 2, 3, 4)$, (A.34) and (A.33) imply

(A.36) $$m_2\ ,\ m_3\ \leq\ m_4 \qquad \text{and} \qquad \alpha_2 + \alpha_3 = \alpha_4\ .$$

Moreover,

(A.37) $$m_2 + m_3 \neq m_4$$

holds. To prove this, if $m_2 + m_3 = m_4$, then $\mu_2^\# + \mu_3^\# + 2 = \mu_4^\# + 1$. This implies that $\mu_2^\# + \mu_3^\# + \mu_4^\#$ is an odd number, contradicting (A.26).

Using (A.35) and (A.36), the equality (A.27) can be written as

(A.38) $$\sqrt{m_2^2 - 8\alpha_2^2} + \sqrt{m_3^2 - 8\alpha_2^2} + \sqrt{m_4^2 - 8(\alpha_2 + \alpha_3)^2} = 1\ .$$

We shall prove that (A.38) cannot hold, making a contradiction. Let m_2, m_3, m_4 be positive integers which satisfy (A.36) and (A.37). Define

$$\varphi(\alpha_2, \alpha_3) := \sqrt{m_2^2 - 8\alpha_2^2} + \sqrt{m_3^2 - 8\alpha_3^2} + \sqrt{m_4^2 - 8(\alpha_2 + \alpha_3)^2}\ .$$

on the closure $\overline{D}$ of the open domain

$$D := \left\{(\alpha_2, \alpha_3)\ :\ 0 < \alpha_2 < \frac{m_2}{\sqrt{8}}, 0 < \alpha_3 < \frac{m_3}{\sqrt{8}}, 0 < \alpha_2 + \alpha_3 < \frac{m_4}{\sqrt{8}}\right\}$$

in the $\alpha_2\alpha_3$-plane. Then one has $\varphi(\alpha_2, \alpha_3) > 1$ if $(\alpha_2, \alpha_3) \in D$. To prove this, note that since φ is a continuous function on a compact set $\overline{D}$, it takes a minimum on $\overline{D}$. By a direct calculation, we have

$$\frac{\partial\varphi}{\partial\alpha_3} = \frac{-8\alpha_3}{\sqrt{m_3^2 - 8\alpha_3^2}} + \frac{-8\alpha_2 - 8\alpha_3}{\sqrt{m_4^2 - 8(\alpha_2 + \alpha_3)^2}} < 0 \qquad \text{on} \quad D\ .$$

So φ does not take its minimum in the interior D of $\overline{D}$, but rather on ∂D. Similarly, $\partial\varphi/\partial\alpha_2 < 0$ on D, so the minimum occurs at $(\alpha_2, \alpha_3) = (m_2/\sqrt{8}, m_3/\sqrt{8})$, where $\varphi(m_2/\sqrt{8}, m_3/\sqrt{8}) = \sqrt{m_4^2 - (m_2 + m_3)^2} \geq$

1, if the line $\alpha_2 + \alpha_3 = m_4/\sqrt{8}$ does not intersect ∂D. If the line $\alpha_2 + \alpha_3 = m_4/\sqrt{8}$ does intersect ∂D, then $m_2 + m_3 > m_4$ holds, and the minimum occurs somewhere on this line with α_2 in the interval $[(m_4 - m_3)/\sqrt{8}, m_2/\sqrt{8}]$. We have

$$\varphi(\alpha_2, m_4/\sqrt{8} - \alpha_2) = \sqrt{m_2^2 - 8\alpha_2^2} + \sqrt{m_3^2 - 8(m_4/\sqrt{8} - \alpha_2)^2}\,,$$

which takes its minimum values $\sqrt{(m_3 + m_2 - m_4)(m_3 + m_4 - m_2)}$ and $\sqrt{(m_3 + m_2 - m_4)(m_2 + m_4 - m_3)} \geq 1$ at the endpoints of the interval $[(m_4 - m_3)/\sqrt{8}, m_2/\sqrt{8}]$. Hence $\varphi > 1$ on D, contradicting (A.27) and proving the theorem. Q.E.D.

Remark A.19. There exist CMC-1 surfaces of class $\mathbf{O}(-2, -2, -2, 0)$ with $\mathrm{TA}(f) = 8\pi$: We set $(p_1, p_2, p_3, p_4) = (1, -1, \infty, 0)$ and $M := \boldsymbol{C} \cup \{\infty\} \setminus \{p_1, p_2, p_3, p_4\}$. Set
(A.39)

$$dg := \frac{(z^2-1)^{\mu}(z^2-q^2)z^2}{(z^2-a^2)^2}\,dz \qquad (a, q \in \boldsymbol{C} \setminus \{0, 1\},\ a \neq \pm q,\ \mu \in \boldsymbol{R})$$

and

$$Q := \frac{-\mu(\mu+2)}{q^2-1}\,\frac{z^2-q^2}{(z^2-1)^2}\,dz^2\,, \tag{A.40}$$

where $\frac{2\mu a}{a^2-1} + \frac{2a}{a^2-q^2} + \frac{1}{a} = 0$. Then the residues of dg at $z = \pm a$ vanish, and thus, there exists the secondary Gauss map g as in (A.39). We assume $-1 < \mu < -\frac{1}{2}$ and $a^2 = -\frac{1-\mu-q^2}{3+\mu-3q^2}$. Then by Theorem 2.4 of [UY1], there exists a CMC-1 immersion $f\colon M \to H^3$ with given g and Q. One can check that such an immersion is complete and has $\mathrm{TA}(f) = 8\pi$.

The case $(\gamma, n) = (0, 5)$. In this case, by Corollary 7.4, the only possible case is $\mathbf{O}(-2, -2, -2, -2, -2)$.

The case of $\gamma = 1$. By Corollary 7.4, a surface of this type has at most three ends, and if the surface has 3 ends, the only possible case is $\mathbf{I}(-2, -2, -2)$. If a surface has only one end, part (3) of Lemma 7.3 implies that it must be of type $\mathbf{I}(-3)$ or $\mathbf{I}(-4)$.

Now suppose there are two ends. By (7.3), (7.7) and (7.11), we have

$$-5 \leq d_1 + d_2 \leq 0\,. \tag{A.41}$$

Also, by (7.7) and (7.8), $\mathrm{TA}(f) = 8\pi$ if $d_1, d_2 \geq -1$. Suppose that both ends are regular (i.e. $d_1, d_2 \geq -2$). Then Theorem 7 of [RUY2] implies

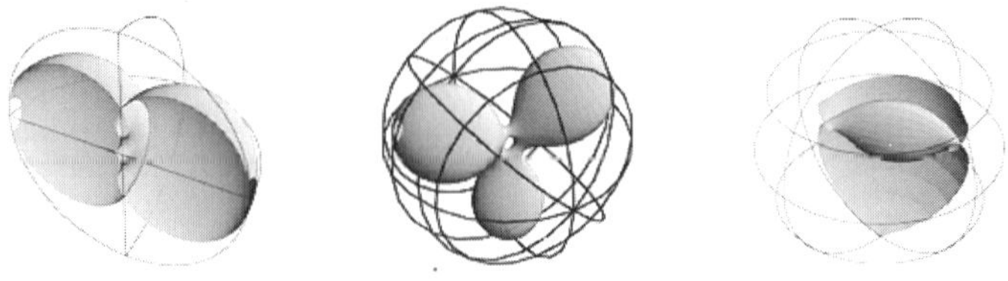

(a) (b) (c)

Fig. 8. Examples B.1, B.2 and B.3. The first two graphics were made by Katsunori Sato of Tokyo Institute of Technology.

that if $d_1 = -2$, then also $d_2 = -2$. Furthermore, by Lemma 3 of [UY5] combined with (7.7), (7.8) and (7.12), if $d_j \geq -1$, then the end at p_j is embedded. Therefore, when $d_j \geq -1$, Proposition 7.1 implies that the flux at the end p_j is zero if and only if $d_j \geq 0$. By the balancing formula (7.15) and Proposition 7.1, we conclude that the only possibilities are $\mathbf{I}(-2,-2)$, $\mathbf{I}(-1,-1)$, and $\mathbf{I}(0,0)$. But in fact the case $\mathbf{I}(0,0)$ cannot occur, because then (7.3) and (7.4) imply that the hyperbolic Gauss map G has at most two branch points, contradicting (7.13). If the end p_1 is irregular, $d_1 \leq -3$. Then by (A.41), we have $d_2 \geq -2$. In particular, the other end p_2 is regular. When $d_2 \geq -1$, then $\mu_1, \mu_2 \in \boldsymbol{Z}$, and (7.7) and (7.8) imply $\mu_1 - d_1 = \mu_2 - d_2 = 2$. In particular $d_1 = \mu_1 - 2 > -3$, a contradiction. Hence the only possible case is $\mathbf{I}(-2,-3)$.

Remark A.20. The genus one catenoid cousin in [RS] is of type $\mathbf{I}(-2,-2)$ (Figure 8 in Appendix B). However, the total absolute curvature seems to be strictly greater than 8π.

Remark A.21. There exists an example of CMC-1 surface of type $\mathbf{I}(-2,-2,-2)$, which is so-called the genus one trinoid ([RUY1], see Figure 8 in Appendix B), which is obtained by deforming minimal surface in $\boldsymbol{R}^3$. However, the absolute total curvature of the original minimal surface is 12π, so the obtained CMC-1 surface has $\mathrm{TA}(f)$ close to 12π. Thus, surfaces obtained by deformation are far from satisfying $\mathrm{TA}(f) \leq 8\pi$.

§Appendix B. Further examples

In this appendix, we introduce examples of interesting CMC-1 surfaces which do not appear in the classification table (Table 3).

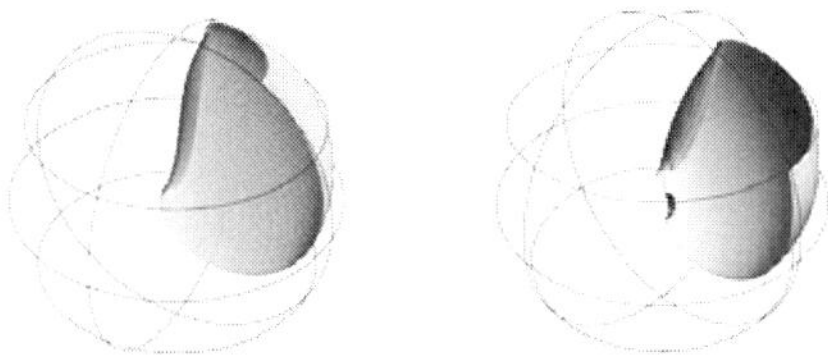

Fig. 9. Example B.4

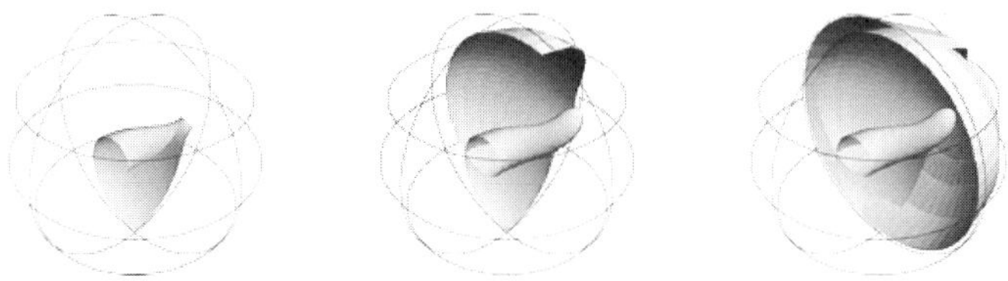

Fig. 10. Example B.5

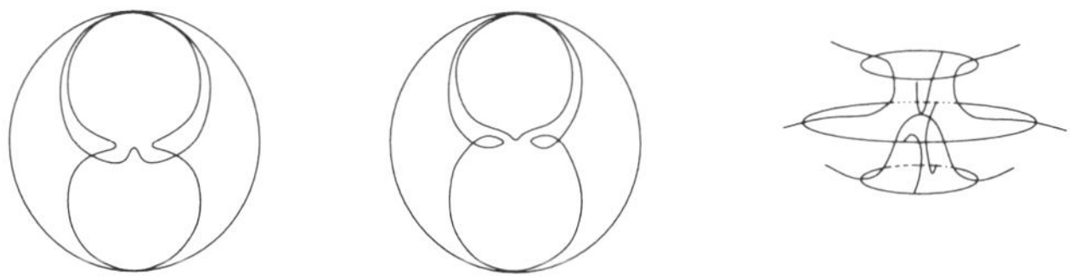

Fig. 11. Example B.6

Example B.1. There exists a genus 1 catenoid cousin in H^3 [RS] Figure 8 (a)). No corresponding minimal surface can exist, by Schoen's result [Sch]. Levitt and Rosenberg [LR] have proved that any complete properly embedded CMC-1 surface in H^3 with asymptotic boundary consisting of at most two points is a surface of revolution, which implies that this example and the last two examples in Figure 2 cannot be embedded, and we see that they are not. A CMC-1 genus 1 catenoid cousin in H^3 was proven to exist in [RS]. See Remark A.20 in Appendix A.

Example B.2. Figure 8 shows a genus 1 trinoid in H^3 proven to exist in [RUY1]. See Remark A.21 in Appendix A.

Example B.3. Figure 8 (c) shows 5 ended CMC-1 surface in H^3 found in [UY1]. Here we show only one of six congruent disks that form the surface. The full surface is constructed by reflections across planes containing boundary curves of the disk shown here.

Example B.4. Figure 9 shows genus 0 and genus 1 Enneper cousin duals. Each surface has a single end that triply wraps around its limiting point at the south pole of the sphere at infinity. These surfaces are of type $\mathbf{O}(-4)$ and $\mathbf{I}(-4)$, and have $\mathrm{TA}(f^{\#}) = 4\pi$ and $\mathrm{TA}(f^{\#}) = 8\pi$. In both cases only one of four congruent pieces (bounded by planar geodesics) of the surface is shown.

Example B.5. Figure 10 shows a CMC-1 surface in H^3, proven to exist in [UY1]. This example is interesting because the hyperbolic Gauss map has an essential singularity at one of its two ends, like the end of the Enneper cousin. And the geometric behavior of the end here is strikingly similar to that of the Enneper cousin's end (see the middle figure of Figure 1). Here we show three pictures consecutively including more of this end.

Example B.6. A CMC-1 "Costa cousin" in H^3 was proven to exist by Costa and Sousa Neto [CN]. In Figure 11, rather than showing graphics of this surface, we show two vertical cross sections by which the surface is reflectionally symmetric (including the "circles" at infinity), and a schematic of the central portion of the surface.

References

[Bar] E. L. Barbanel, Complete minimal surface in $\boldsymbol{R}^3$ of low total curvature, Ph. D. thesis, Univ. of Massachusetts, 1987.

[Bry] R. Bryant, Surfaces of mean curvature one in hyperbolic space, Astérisque, **154-155** (1987), 321–347.

[CR] C. P. Cosín and A. Ros, A plateau problem at infinity for properly immersed minimal surfaces with finite total curvature, Indiana Univ. Math. J., **50** (2001), 847–879.

[CG] C. C. Chen and F. Gackstatter, Elliptische und hyperelliptische Funktionen und vollständige Minimalflächen vom Enneperschen Typ, Math. Ann., **259** (1982), 359–369.

[CHR1] P. Collin, L. Hauswirth and H. Rosenberg, The geometry of finite topology Bryant surfaces, Ann. of Math. (2), **153** (2001), 623–659.

[CHR2] ———, The gaussian image of mean curvature one surfaces in H^3 of finite total curvature, In: Minimal Surfaces, Geometric Analysis and Symplectic Geometry, Adv. Stud. Pure Math., **34**, Math. Soc. Japan, Tokyo, 2002, pp. 9–14.

[CN] C. J. Costa and V. F. de Sousa Neto, Mean curvature 1 surfaces of Costa type in hyperbolic three-space, Tohoku Math. J. (2), **53** (2001), 617–628.

[ET1] R. Sa Earp and E. Toubiana, On the geometry of constant mean curvature one surfaces in hyperbolic space, Illinois J. Math., **45** (2001), 371–401.

[ET2] ———, Meromorphic data for mean curvature one surfaces in hyperbolic space, Tohoku Math. J. (2), **56** (2004), 27–64.

[GGN] C. C. Góes, M. E. E. L. Galvão and B. Nelli, A Weierstrass type representation for minimal surfaces in hyperbolic space with mean curvature equal to one, An. Acad. Brasil. Cienc., **70** (1998), 1–6.

[Kar] H. Karcher, Hyperbolic constant mean curvature one surfaces with compact fundamental domains, In: Global Theory of Minimal Surfaces, Clay Math. Proc., **2**, Amer. Math. Soc., Providence, RI, 2005, pp. 311–323.

[Kat] S. Kato, Construction of n-end catenoids with prescribed flux, Kodai Math. J., **18** (1995), 86–98.

[LR] G. Levitt and H. Rosenberg, Symmetry of constant mean curvature hypersurfaces in hyperbolic space, Duke Math. J., **52** (1985), 53–59.

[Lop] F. J. Lopez, The classification of complete minimal surfaces with total curvature greater than -12π, Trans. Amer. Math. Soc., **334** (1992), 49–74.

[MU] C. McCune and M. Umehara, An analogue of the UP-iteration for constant mean curvature one surfaces in hyperbolic 3-space, Differential Geom. Appl., **20** (2004), 197–207.

[Oss] R. Osserman, A Survey of Minimal Surfaces, 2nd ed., Dover, 1986.

[Ros] H. Rosenberg, In: Bryant surfaces, Lecture Notes in Math., **1775**, Springer-Verlag, 2002, pp. 67–111.

[RS] W. Rossman and K. Sato, Constant mean curvature surfaces with two ends in hyperbolic space, Experimental Math., **7** (1998), 101–119.

[RUY1] W. Rossman, M. Umehara and K. Yamada, Irreducible constant mean curvature 1 surfaces in hyperbolic space with positive genus, Tôhoku Math. J., **49** (1997), 449–484.

[RUY2] ———, A new flux for mean curvature 1 surfaces in hyperbolic 3-space, and applications, Proc. Amer. Math. Soc., **127** (1999), 2147–2154.

[RUY3] ———, Mean curvature 1 surfaces with low total curvature in hyperbolic 3-space I, Hiroshima Math. J., **34** (2004), 21–56.

[RUY4] ———, Mean curvature 1 surfaces with low total curvature in hyperbolic 3-space II, Tohoku Math. J. (2), **55** (2003), 375–395.

[RUY5] ———, Constructing mean curvature 1 surfaces in H^3 with irregular ends, In: Global theory of minimal surfaces, Clay Math. Proc., **2**, Amer. Math. Soc., Providence, RI, 2005, pp. 561–584.

[Sch] R. Schoen, Uniqueness, symmetry, and embeddedness of minimal surfaces, J. Diff. Geom., **18** (1982), 791–809.

[Sm] A. J. Small, Surfaces of constant mean curvature 1 in H^3 and algebraic curves on a quadric, Proc. Amer. Math. Soc., **122** (1994), 1211–1220.

[Tan] S. Tanaka, Minimal surfaces with three catenoidal ends, Master thesis, Hiroshima Univ., 2001.

[Tro1] M. Troyanov, Metric of constant curvature on a sphere with two conical singularities, In: Differential Geometry, Lect. Notes in Math., **1410**, Springer-Verlag, 1989, pp. 296–306.

[Tro2] ———, Prescribing curvature on compact surfaces with conical singularities, Trans. Amer. Math. Soc., **324** (1991), 793–821.

[UY1] M. Umehara and K. Yamada, Complete surfaces of constant mean curvature-1 in the hyperbolic 3-space, Ann. of Math. (2), **137** (1993), 611–638.

[UY2] ———, A parameterization of Weierstrass formulae and perturbation of some complete minimal surfaces of $\boldsymbol{R}^3$ into the hyperbolic 3-space, J. Reine Angew. Math., **432** (1992), 93–116.

[UY3] ———, Surfaces of constant mean curvature-c in $H^3(-c^2)$ with prescribed hyperbolic Gauss map, Math. Ann., **304** (1996), 203–224.

[UY4] ———, Another construction of a CMC 1 surface in H^3, Kyungpook Math. J., **35** (1996), 831–849.

[UY5] ———, A duality on CMC 1 surfaces in hyperbolic 3-space and a hyperbolic analogue of the Osserman Inequality, Tsukuba J. Math., **21** (1997), 229–237.

[UY6] ———, Geometry of surfaces of constant mean curvature 1 in the hyperbolic 3-space, Sugaku Expositions, **10** (1997), 41–55.

[UY7] ———, Metrics of constant curvature 1 with three conical singularities on the 2-sphere, Illinois J. Math., **44** (2000), 72–94.

[Yu1] Z. Yu, Value distribution of hyperbolic Gauss maps, Proc. Amer. Math. Soc., **125** (1997), 2997–3001.

[Yu2] ———, The inverse surface and the Osserman Inequality, Tsukuba J. Math., **22** (1998), 575–588.

[Yu3] ———, Surfaces of constant mean curvature one in the hyperbolic three-space with irregular ends, Tôhoku Math. J., **53** (2001), 305–318.

Wayne Rossman
Department of Mathematics
Faculty of Science
Kobe University
Rokko, Kobe 657-8501
JAPAN

E-mail address:
`wayne@math.kobe-u.ac.jp`

Masaaki Umehara
Department of Mathematics
Graduate School of Science
Osaka University
Osaka 560-0043
JAPAN

E-mail address:
`umehara@math.sci.osaka-u.ac.jp`

Kotaro Yamada
Faculty of Mathematics
Kyushu University
Hakozaki, Higashi-ku
Fukuoka 812-8581
JAPAN

E-mail address:
`kotaro@math.kyushu-u.ac.jp`

Advanced Studies in Pure Mathematics 51, 2008
Surveys on Geometry and Integrable Systems
pp. 401–488

Geometries and symmetries of soliton equations and integrable elliptic equations

Chuu-Lian Terng

§1. Introduction

In the classical literature, a differential equation is called "integrable" if it can be solved by quadratures. A Hamiltonian system in $2n$-dimensions is *completely integrable* if it has n independent commuting Hamiltonians. By the Arnold-Liouville Theorem, such systems have action-angle variables that linearize the flow, and these can be found by quadrature. This concept of integrability can be extended to PDEs, and one class consists of evolution equations on function spaces that have Hamiltonian structures and are completely integrable Hamiltonian systems in the sense of Liouville, i.e., there exist action angle variables. We call this class of equations soliton equations. The model examples are the Korteweg-de Vries equation, the non-linear Schrödinger equation (NLS equation), and the Sine-Gordon equation (SGE equation). For example, the action-angle variables are constructed for the KdV equation in [33], for the NLS equation in [34], and for flows in the $SL(n)$-hierarchy in [5]. Besides the Hamiltonian formulation and complete integrability, these soliton equations have many other remarkable properties including:

(1) infinite families of explicit solutions,

(2) a hierarchy of commuting flows described by partial differential equations,

---

Received March 18, 2002.
Revised January 24, 2003.
2000 *Mathematics Subject Classification.* Primary 37K25; Secondary 53A, 53C.
Research supported in part by NSF Grant DMS 9972172 and National Center for Theoretical Sciences, Taiwan.

(3) a Lax pair,

(4) an algebraic-geometric description of certain solutions,

(5) a scattering theory,

(6) an inverse scattering transform to solve the Cauchy problem,

(7) a construction of solutions using loop group factorizations (dressing actions).

We will give a brief review of properties (1), (2), (3), and (7), and refer the reader to [32, 46, 47, 53, 73, 74] for (4), and to [3, 4] for (5) and (6).

The existence of a Lax pair is one of the key properties of soliton equations. This was first constructed for the case of the KdV equation by Lax [49], who observed that the KdV equation can be written as the condition for an isospectral deformation of the Schrödinger operator on the line. Later, this was shown to be equivalent to the zero curvature condition of a family of connections ([2, 76]). Roughly speaking, a PDE for $q : \mathbb{R}^n \longrightarrow \mathbb{R}^m$ is said to have a *zero curvature formulation* if there is a family of connections θ_λ on $\mathbb{R}^n$, (defined by q and its derivatives, and a holomorphic parameter λ defined in some open subset of $\mathbb{C}$) so that the condition for θ_λ to be flat for all λ is that q solve the given PDE. The connection $\theta_\lambda = \sum_{i=1}^n A_i dx_i$ is flat if $d\theta_\lambda = -\theta_\lambda \wedge \theta_\lambda$ for all λ, or equivalently the n operators $\{\frac{\partial}{\partial x_i} + A_i \mid 1 \leq i \leq n\}$ commute, i.e.,

$$\left[\frac{\partial}{\partial x_i} + A_i, \frac{\partial}{\partial x_j} + A_j\right] = 0, \quad i \neq j.$$

We call θ_λ a *Lax pair* if $n = 2$, and a *Lax n-tuple* for general n. A Lax n-tuple naturally gives rise to a loop group factorization, which in turn provides a method for constructing explicit solutions and symmetries of the equations.

Another class of integrable PDEs are non-linear elliptic equations. Although these equations do not have Hamiltonian formulations, they do have zero curvature formulations that give rise to loop group factorizations, and hence the techniques developed for soliton equations can also be used to construct solutions and symmetries of these elliptic equations. In particular, we can find solutions of the equation by factorizations. One class of model examples are the equations for harmonic maps from $\mathbb{C}$ to a compact Lie group.

Some goals of this paper are to give a brief survey of the following:

—A systematic construction of integrable hierarchies associated to a complex semi-simple Lie algebra and finite order automorphisms.

—Some geometric integrable PDEs arising in differential geometry.

—Construction of solutions using loop group factorizations.

Another goal of this paper is to put some known results of evolution soliton equations and integrable elliptic systems together so that we can compare and see similarities and differences in these two theories.

The literature of integrable systems and their applications to differential geometry is huge. In this article, the author only covers a part of these research areas and the references are correspondingly restricted. We refer readers to the following survey books and articles [1, 25, 32, 34, 38, 46, 53, 54, 59, 58, 63], and articles in this volume for more complete references.

G-hierarchy

The ZS-AKNS construction of the $n \times n$-hierarchy of soliton flows works equally well when we replace $sl(n, \mathbb{C})$ by any complex, simple Lie algebra $\mathcal{G}$. In fact, let $a \in \mathcal{G}$, $\mathcal{G}_a = \{y \in \mathcal{G} \mid [a, y] = 0\}$ the centralizer of a, and $\mathcal{G}_a^{\perp} = \{\xi \in \mathcal{G} \mid (\xi, y) = 0$ for all $y \in \mathcal{G}_a\}$. Here $(\ ,\ )$ is a non-degenerate ad-invariant bilinear form of $\mathcal{G}$. It can be shown that there exists a sequence of polynomial differential operators on the space $C(\mathbb{R}, \mathcal{G}_a^{\perp})$ of smooth functions from $\mathbb{R}$ to $\mathcal{G}_a^{\perp}$,

$$\{Q_{b,j}(u) \mid b \in \mathcal{G}_a,\ \mathcal{G}_b = \mathcal{G}_a,\ j \geq 0 \text{ integer}\}.$$

These $Q_{b,j}(u)$ are determined uniquely from the following recursive formula

$$\begin{aligned}
&(Q_{b,j}(u))_x + [u, Q_{b,j}(u)] = [Q_{b,j+1}(u), a], \\
&Q_{b,0} = b, \quad Q_{a,1}(u) = u.
\end{aligned}$$

The (b, j)-flow is $u_t = (Q_{b,j}(u))_x + [u, Q_{b,j}(u)]$, which commutes with the (b', j')-flow. The hierarchy of these commuting flows is called the $\mathfrak{g}$AKNS-hierarchy in [75], and the G-hierarchy in [69].

It follows from the recursive formula that u is a solution of the (b, j)-th flow if and only if

$$\theta_\lambda = (a\lambda + u)dx + (b\lambda^j + Q_{b,1}(u)\lambda^{j-1} + \cdots + Q_{b,j}(u))dt$$

is flat for all $\lambda \in \mathbb{C}$. In other words, θ_λ is a Lax pair of the (b, j)-flow.

There are several ways to construct subhierarchies from the G-hierarchy by finding suitable invariant submanifolds (sometimes called *restrictions*). For example, there are many works concerning the KdV type equations (cf. [26, 30, 31, 62, 70]). In this paper, we explain several restrictions of the G-hierarchy using finite order automorphisms of $\mathcal{G}$ (cf. [41, 48, 63, 70]).

σ-twisted G-hierarchy

If σ is an order k automorphism of the complex Lie group G, then the $(b, nk+1)$-flow in the G-hierarchy leaves $C(\mathbb{R}, \mathcal{G}_a^\perp \cap \mathcal{G}_0)$ invariant, where $\mathcal{G}_0$ is the fixed point set of $d\sigma_e$ on $\mathcal{G}$. The hierarchy of the restriction of these flows to $C(\mathbb{R}, \mathcal{G}_a^\perp \cap \mathcal{G}_0)$ is called the *σ-twisted G-hierarchy*. The Kupershmidt-Wilson hierarchy is an example with $\mathcal{G} = sl(n, \mathbb{C})$ and $k = n$ ([48]).

U-hierarchy

Suppose τ is a conjugate linear Lie algebra involution of $\mathcal{G}$, and $\mathcal{U}$ is the fixed point set of τ, i.e., $\mathcal{U}$ is a real form of $\mathcal{G}$. Then the (b, j)-flow leaves $C(\mathbb{R}, \mathcal{G}_a^\perp \cap \mathcal{U})$ invariant. The hierarchy restricted to $C(\mathbb{R}, \mathcal{G}_a^\perp \cap \mathcal{U})$ is called the U-hierarchy. For example, the NLS equation occurs as the second flow in the $SU(2)$-hierarchy, and the 3-wave equation as the first flow in the $SU(3)$-hierarchy.

U/U_0-hierarchy

Suppose τ is a conjugate-linear involution, and σ is an order k complex linear, Lie algebra automorphism of $\mathcal{G}$ such that $\sigma\tau = \tau^{-1}\sigma^{-1}$. Then the $(b, nk+1)$-flow leaves $C(\mathbb{R}, \mathcal{G}_a^\perp \cap \mathcal{U}_0)$ invariant, where $\mathcal{U}_0$ is the Lie subalgebra of $\mathcal{G}$ that is fixed by both σ and τ. The hierarchy restricted to $C(\mathbb{R}, \mathcal{G}_a^\perp \cap \mathcal{U}_0)$ is called the U/U_0-hierarchy. For example, the 3rd flow in the $SU(2)/SO(2)$-hierarchy is the modified KdV equation with $k = 2$.

U/U_0-system

Let U/U_0 be the rank n symmetric space given by involutions τ, σ of G, $\mathcal{U} = \mathcal{U}_0 + \mathcal{U}_1$ the Cartan decomposition, and $\mathcal{A}$ a maximal abelian subspace of $\mathcal{U}_1$. Let $\{a_1, \cdots, a_n\}$ be a basis of $\mathcal{A}$. By putting the $(a_1, 1), \cdots, (a_n, 1)$-flows in the U/U_0-hierarchy together, we get the U/U_0 *-system* for maps $v : \mathbb{R}^n \to \mathcal{U}_\mathcal{A}^\perp \cap \mathcal{U}_1$:

$$[a_i, v_{x_j}] - [a_j, v_{x_i}] = [[a_i, v], [a_j, v]], \quad i \neq j,$$

where $\mathcal{U}_{\mathcal{A}} = \{y \in \mathcal{U} \mid [y, \xi] = 0 \ \forall \ \xi \in \mathcal{A}\}$. Note that v is a solution of the U/U_0-system if and only if

$$\theta_\lambda = \sum_{j=1}^{n} (a_i\lambda + [a_i, v]) \ dx_i$$

is flat for all $\lambda \in \mathbb{C}$, i.e., θ_λ is a Lax n-tuple of the U/U_0-system.

The -1-flow associated to U

There is also a sequence of negative flows associated to G (cf. [26, 31, 69]). We review the first one in this sequence (the -1-flow) below.

Let $a, b \in \mathcal{U}$ such that $[a, b] = 0$. The -1-flow associated to U is the following system for $g : \mathbb{R}^2 \to U$:

$$(g^{-1}g_x)_t = [a, g^{-1}bg],$$

with constraint $g^{-1}g_x \in \mathcal{U}_a^\perp$. The -1-flow has a Lax pair

$$(a\lambda + g^{-1}g_x)dx + \lambda^{-1}g^{-1}bg \ dt.$$

Elliptic (G, τ)-systems

The m-th elliptic (G, τ)-system is the equation for $(u_0, \cdots, u_m) : \mathbb{C} \to \oplus_{i=0}^m \mathcal{G}$ so that

$$\theta_\lambda = \sum_{j=0}^{m} \lambda^{-j}u_j dz + \lambda^j \tau(u_j) d\bar{z}$$

is flat for all $\lambda \in S^1$. The first (G, τ)-system is the equation for harmonic maps from $\mathbb{R}^2$ to U, where U is the fixed point set of τ.

Elliptic (G, τ, σ)-systems

Suppose σ is an order k automorphism of G such that $\sigma\tau = \tau\sigma$. Let $\mathcal{G}_j$ denote the eigenspace of σ_* on $\mathcal{G}$ with eigenvalue $e^{\frac{2\pi i j}{k}}$. We call the m-th elliptic (G, τ)-system with constraints $u_i \in \mathcal{G}_{-i}$ the m-th elliptic (G, τ, σ)-system. Solutions of the first (G, τ, σ)-system is the equation for primitive maps studied by Burstall and Pedit [15].

Dressing actions

To explain the symmetries and the construction of solutions of integrable systems, we need the dressing action of Zakharov and Shabat

[76]. Suppose G_+, G_- are subgroups of G and the multiplication map from $G_+ \times G_-$ to G is a bijection. Then every $g \in G$ can be factored uniquely as $g = g_+g_-$ with $g_+ \in G_+$ and $g_- \in G_-$. Moreover, the space of right cosets G/G_- can be identified with G_+, so the canonical action of G_- on G/G_- by left multiplication, $g_- \cdot (gG_-) = g_-gG_-$, induces an action $*$ of G_- on G_+. The action $*$ is called the *dressing action*. The dressing action can be computed by factorization. In fact, $g_- * g_+ = \tilde{g}_+$, where $g_-g_+ = \tilde{g}_+\tilde{g}_-$ with $\tilde{g}_+ \in G_+$ and $\tilde{g}_- \in G_-$. If the multiplication map from $G_+ \times G_-$ to G is injective but only onto an open, dense subset of G, then the dressing action $*$ is a local action, but the corresponding Lie algebra action is well-defined.

Iwasawa and Gauss factorizations

There are two well-known factorizations associated to a complex simple Lie group G. The *Iwasawa factorization* is $G = KAN$, where K is a maximal compact subgroup of G, A is abelian, and N is unipotent. We also refer to $G = KB$ as the Iwasawa factorization of G, where $B = AN$ is a Borel subgroup. Let $\mathcal{A}$ be a Cartan subalgebra of $\mathcal{G}$, $\mathcal{N}_+, \mathcal{N}_-$ the spaces spanned by all positive and negative roots respectively, and A, N_+, N_- the corresponding Lie subgroups of G. Then the multiplication map from $N_- \times A \times N_-$ to G is injective and onto an open dense subset of G. The set N_-AN_+ is called a *big cell* of G. The so-called *Gauss factorization* associated to G refers to the fact that any g in the big cell can be factorized uniquely as n_-an_+ with $n_\pm \in N_\pm$ and $a \in A$. For example, for $G = SL(n, \mathbb{C})$, let $K = SU(n)$, B_n the subgroup of upper triangular matrices with real diagonal, A_n the subgroup of diagonal matrices, and $N_+(n), N_-(n)$ the subgroups of strictly upper and lower triangular matrices. The Iwasawa factorization of $SL(n, \mathbb{C})$ is KB_n, which can be done using the Gram-Schmidt process. The Gauss factorization for the big cell is $N_-(n)A_nN_+(n)$, which can be carried out using the Gaussian elimination.

Loop group factorizations

We review three types of loop group factorizations that are needed for the study of symmetries of soliton equations and elliptic integrable systems. Let $L(G)$ denote the group of smooth maps $f : S^1 \to G$, $L_+(G)$ the subgroup of $f \in L(G)$ that are the boundary values of holomorphic maps defined on $|\lambda| < 1$, and $L_-(G)$ the subgroup of $f \in L(G)$ that can be extended holomorphically to $|\lambda| > 1$ in S^2 and $f(\infty) = e$. Let U be a maximal compact subgroup of G, and $L_e(U)$ the subgroup of $f \in L(G)$ such that the image of f lies in U and $f(1) = e$ (the identity of G).

—The *Gauss loop group factorization* (or the *Birkhoff factorization*) states that there is an open dense subset L' of $L(G)$ such that any $g \in L'$ can be factored uniquely as g_+g_- with $g_\pm \in L_\pm(G)$.

—The *Iwasawa loop group factorization*, proved in [57], states that the multiplication map from $L_e(U) \times L_+(G)$ to $L(G)$ is a bijection.

—Let $\epsilon > 0$, $\mathcal{O}_\epsilon = \{\lambda \in \mathbb{C} \mid |\lambda| < \epsilon\}$, and $\mathcal{O}_{\frac{1}{\epsilon}} = \{\lambda \in S^2 = \mathbb{C} \cup \{\infty\} \mid |\lambda| > 1/\epsilon\}$. Let $\mathbb{C}^* = \{\lambda \in \mathbb{C} \mid \lambda \neq 0\}$, and $\Omega^\tau(G)$ the group of holomorphic maps $f : (\mathcal{O}_\epsilon \cup \mathcal{O}_{1/\epsilon}) \cap \mathbb{C}^* \to G$ that satisfy the (G,τ)*-reality condition* $\tau(f(1/\bar{\lambda})) = f(\lambda)$, $\Omega^\tau_+(G)$ the subgroup of $f \in \Omega^\tau(G)$ that extend holomorphically to $\mathbb{C}^*$, and $\Omega^\tau_-(G)$ the subgroup of $f \in \Omega^\tau(G)$ that extend holomorphically to $\mathcal{O}_\epsilon \cup \mathcal{O}_{1/\epsilon}$ and $f(\infty) = e$. McIntosh proved ([51]) that the multiplication map from $\Omega^\tau_-(G) \times \Omega^\tau_+(G)$ to $\Omega^\tau(G)$ is a bijection.

These loop group factorizations play central roles in the study of integrable PDEs.

Solutions of soliton flows via loop group factorizations

Let $\mathcal{O}_{1/\epsilon} = \{\lambda \in \mathbb{C} \mid |\lambda| > 1/\epsilon\}$, and $\Lambda^\tau(G)$ the group of holomorphic maps $f : \mathcal{O}_{1/\epsilon} \to G$ that satisfies the U-reality condition

$$\tau(f(\bar{\lambda})) = f(\lambda),$$

where τ is the involution on $\mathcal{G}$ that defines the real form $\mathcal{U}$. Note that $f(r) \in U$ for real r. Let $\Lambda^\tau_+(G)$ denote the subgroup of $f \in \Lambda^\tau(G)$ that extend holomorphically to $\mathbb{C}$, and $\Lambda^\tau_-(G)$ the subgroup of $f \in \Lambda^\tau(G)$ that extend holomorphically to $1/\epsilon < |\lambda| \leq \infty$ and $f(\infty) = e$. The Gauss loop group factorization implies that the multiplication map from $\Lambda^\tau_+(G) \times \Lambda^\tau_-(G)$ to $\Lambda^\tau(G)$ is injective and its image is open and dense.

The Lax pair θ_λ of a soliton flow in the U-hierarchy is a flat $\mathcal{G}$-valued connection 1-form that satisfies the U-reality condition $\tau(\theta_{\bar{\lambda}}) = \theta_\lambda$. So $\theta_\lambda(x,t)$ can be viewed as a map from $(x,t) \in \mathbb{R}^2$ to the Lie algebra $\Lambda^\tau_+(\mathcal{G})$ of $\Lambda^\tau_+(G)$, or equivalently a $\Lambda^\tau_+(\mathcal{G})$-valued connection 1-form on $\mathbb{R}^2$. Therefore the trivialization $E_\lambda(x,t)$ of $\theta_\lambda(x,t)$ can be viewed as a map from $\mathbb{R}^2$ to $\Lambda^\tau_+(G)$. Given $g_- \in \Lambda^\tau_-(G)$, let $\tilde{E}(x,t)$ denote the dressing action of g_- on $E(x,t)$, i.e., $\tilde{E}(x,t)$ is obtained using the Gauss loop group factorization to factor $g_-E(x,t) = \tilde{E}(x,t)\tilde{g}(x,t)$ with $\tilde{E}(x,t) \in \Lambda^\tau_+(G)$ and $\tilde{g}(x,t) \in \Lambda^\tau_-(G)$ for each (x,t). It can be shown that $\tilde{E}(x,t)$ is again a trivialization of some solution of the soliton flow. This defines an action of $\Lambda^\tau_-(G)$ on the space of solutions, which we denote by $*$. Moreover, 0 is a solution. If $g_- \in \Lambda^{\tau}_{-}(G)$ is rational, then $g_- * 0$ can be computed explicitly and is a rational function of exponentials. These are the pure soliton solutions. For general $g_- \in \Lambda^{\tau}_{-}(G)$, $g_- * 0$ is a local analytic solution of the soliton flow. Algebraic geometric solutions are included in the orbit $\Lambda^{\tau}_{-}(G) * 0$. To construct general rapidly decaying solutions for the flows in the U-hierarchy, we need a new type of loop group factorization. Namely, factor fg as $\tilde{g}\tilde{f}$, where $f, \tilde{f} \in L_+(G)$ so that $f_b, \tilde{f}_b$ are the identity $e \in G$ at $\lambda = -1$ up to infinite order and $g, \tilde{g}$ are loops in U that have an essential singularity at $\lambda = -1$. Here $f_b(\lambda)$ and $\tilde{f}_b(\lambda)$ denote the B-component of $f(\lambda)$ and $\tilde{f}(\lambda)$ in the Iwasawa factorization $G = UB$ for each λ. For more details see section 5.3 and [69].

Solutions of elliptic systems via loop group factorizations

The Lax pair θ_λ of the m-th (G, τ)-system satisfies the (G, τ)-reality condition

$$\tau(g(1/\bar{\lambda})) = g(\lambda).$$

The trivialization E of θ_λ is a map from $\mathbb{C}$ to $\Omega^{\tau}_{+}(G)$. It follows from the McIntosh loop group factorization that the dressing action of $\Omega^{\tau}_{-}(G)$ induces an action on the space of solutions of the (G, τ)-systems. Since there are constant solutions for the (G, τ)-system, the $\Omega^{\tau}_{-}(G)$-orbits through these constant solutions give rise to a class of solutions. But these are not all the solutions. The (G, τ)-reality condition implies that the restriction of the trivialization E of a solution to the unit circle $|\lambda| = 1$ lies in U, i.e., E can be viewed as a map from $\mathbb{C}$ to $L(U)$. Dorfmeister, Pedit and Wu ([27]) use meromorphic maps and the Iwasawa loop group factorization $L(G) = L_e(U)L_+(G)$ to give a method of constructing all local solutions of the (G, τ)-systems. This is the so-called the *Weierstrass representation* or the *DPW method*.

Although methods of constructing solutions for both the U-hierarchy and the elliptic (G, τ)-systems are similar in spirit, initial data and techniques used are somewhat different. Moreover, while there is a canonical choice of initial data used in the factorization method to solve soliton flows, there is no clear canonical choice of meromorphic data for the (G, τ)-hierarchy. Since the (G, τ)-hierarchy contains the equation for harmonic maps from a domain of $\mathbb{R}^2$ to U, the main interest has been to understand the relation between the initial meromorphic data of the factorization method and the global geometry; for example, to find properties of meromorphic data which corresponds to a harmonic map from a complete surface M to U. This has been done when M is S^2 and more generally for harmonic maps of finite uniton numbers ([72, 14, 39]), and also when M is T^2 ([45, 56, 13]). For a detailed survey of results concerning harmonic maps, loop groups, and integrable systems, we refer the reader to [38].

When we study a geometric problem concerning maps f from a manifold M to a homogeneous space U/U_0, it is often useful to find a good lifting $\tilde{f} : M \to U$ and write down the geometric condition imposed on the map f in terms of the flat $\mathcal{U}$-valued 1-form $\tilde{f}^{-1}d\tilde{f}$. If there is a natural holomorphic deformation $F_\lambda : M \to G$ of such maps so that $F_0 = \tilde{f}$ and the flatness of $F_\lambda^{-1}dF_\lambda$ for all λ is equivalent to the flatness of $\tilde{f}^{-1}d\tilde{f}$ in some natural coordinate system on M, then the corresponding geometric PDE is often an integrable system with a zero curvature formulation.

Integrable systems in differential geometry

One of the main interests in classical differential geometry is to find natural geometric conditions for surfaces in $\mathbb{R}^3$ so that there are many explicit solutions and deformations. It is now known that the Gauss-Codazzi equations for surfaces with constant mean curvature, constant Gaussian curvature, and isothermic surfaces in $\mathbb{R}^3$ studied by classical differential geometers are integrable systems and Bäcklund and Ribaucour transformations can be constructed naturally using loop group factorizations (cf. [6, 10, 12, 17, 18, 37, 44, 56, 58, 67, 70, 71]).

In this paper, we give a brief review of the following subset of the known integrable geometric problems:

(i) The Gauss-Codazzi equations of n-submanifolds with constant sectional curvature in $\mathbb{R}^m$, S^m and hyperbolic space $\mathbb{H}^m$ are the U/U_0-system associated to certain real Grassmannian manifolds U/U_0 (cf. [10, 19, 36, 66, 68]).

(ii) The Gauss-Codazzi equations of flat Lagrangian submanifolds of $\mathbb{C}P^n$ is the $SU(n+1)/SO(n+1)$-system.

(iii) Indefinite affine spheres in $\mathbb{R}^3$ are given by solutions of the -1-flow in the $SL(3,\mathbb{R})/\mathbb{R}^+$-hierarchy ([7]).

(iv) Solutions of the -1-flow in the U/U_0-hierarchy give rise to harmonic maps from the Lorentz space $\mathbb{R}^{1,1}$ to U/U_0. (These are called sigma-models by physicists.)

(v) The first elliptic (G, τ)-system is the equation for harmonic maps from $\mathbb{R}^2$ to U ([72]). The first elliptic (G, τ, σ)-system is the equation for harmonic maps from $\mathbb{R}^2$ to the symmetric space U/U_0 if the order of σ is two ([13]), where U_0 is the fixed point set of σ in U.

(vi) The equation for minimal surfaces in $\mathbb{C}P^2$ is the first $(SL(3,\mathbb{C}), \tau, \sigma)$-system, where τ, σ gives the 3-symmetric space $SU(3)/T^2$ ([11, 9]).

(vii) Equations for minimal Lagrangian surfaces in $\mathbb{C}P^2$, minimal Legendrian surfaces in S^5, and minimal Lagrangian cones in $\mathbb{R}^6 = \mathbb{C}^3$ are given by the first $(SL(3,\mathbb{C}), \tau, \sigma)$-system, where τ, σ give the 6-symmetric space $SU(3)/SO(2)$ ([52]).

(viii) The equation for Hamiltonian stationary surfaces in $\mathbb{C}P^2$ is the second elliptic system associated to the 4-symmetric space $SU(3)/SU(2)$ ([43]).

Note that there may be several geometric problems associated to one integrable system. For example:

—The SGE equation is the equation for surfaces in $\mathbb{R}^3$ with Gaussian curvature $K = -1$, and is also the equation for harmonic maps from $\mathbb{R}^{1,1}$ to S^2. The reason here is that if M is a surface in $\mathbb{R}^3$ with $K = -1$, then the second fundamental form II of M is conformally equivalent to the flat Lorentzian metric and the Gauss map $\nu : M \to S^2$ is harmonic when M is equipped with metric II.

—The $U(n)/O(n)$-system is the equation for flat Lagrangian submanifolds in $\mathbb{R}^{2n}$ that lie in S^{2n-1}, the equation for flat Lagrangian submanifolds in $\mathbb{C}P^{n-1}$, and also the equation for flat Egoroff metrics. These three geometries are related as follows: the preimage of a flat Lagrangian submanifold in $\mathbb{C}P^{n-1}$ via the Hopf fibration $\pi : S^{2n-1} \to \mathbb{C}P^{n-1}$ is a flat Lagrangian submanifold in $\mathbb{R}^{2n}$ that lies in S^{2n-1}, and the induced metrics on these flat Lagrangian submanifolds are flat Egoroff metrics.

We now know that there is a very large collection of integrable geometric PDEs. In this paper, we only discuss a small subset of these examples. For a more extensive review of integrable systems in differential geometry, we refer the readers to the books [37, 58, 67], articles in this volume, and references therein.

Most of the integrable geometric PDEs mentioned in this paper are either the U/U_0-system, the -1-flow, or the (G, τ, σ)-systems. We would like to end this introduction by proposing a program: Find geometric problems whose equations are given by the U/U_0-system, the -1-flow, or the m-th (G, τ, σ)-system. Most examples given in this paper have $U = O(m)$ or $SU(m)$. We believe the success of this program for general compact Lie groups U should provide new natural classes of submanifolds in symmetric spaces and in homogeneous Riemannian manifolds with exceptional holonomy. This program also makes sense for any semi-simple, non-compact, Lie group U. We believe that interesting classes of submanifolds in pseudo-Riemannian symmetric spaces, in $SL(n, \mathbb{R})$-geometry (affine geometry), and in U-geometry will arise naturally from this program.

The author would like to thank Martin Guest for many helpful comments and suggestions.

§2. Soliton equations

We review the method of constructing a hierarchy of $n \times n$ soliton flows developed by Zakharov-Shabat [76] and Ablowitz-Kaup-Newell-Segur [2]. Their method works equally well if we replace the algebra of $n \times n$ matrices by a general semi-simple, complex Lie algebra $\mathcal{G}$ (cf. [41, 60, 69, 75]). We also review the construction of new hierarchies of flows by restricting the G-hierarchy to submanifolds naturally associated to finite order automorphisms of $\mathcal{G}$. Many interesting equations in differential geometry and mathematical physics are flows in these restricted hierarchies.

2.1. The G-hierarchy

Let $\langle\ ,\ \rangle$ be a non-degenerate, ad-invariant bilinear form on $\mathcal{G}$, $a \in \mathcal{G}$ regular, i.e., the centralizer $\mathcal{G}_a$ of a in $\mathcal{G}$ is abelian, and $\mathcal{G}_a^\perp = \{\xi \in \mathcal{G} \mid \langle \xi, \mathcal{G}_a \rangle = 0\}$. Let $\mathcal{S}(\mathbb{R}, \mathcal{G}_a^\perp)$ denote the space of rapidly decaying maps from $\mathbb{R}$ to $\mathcal{G}_a^\perp$.

There is a unique family of $\mathcal{G}$-valued maps $Q_{b,j}(u)$ parametrized by $\{b \in \mathcal{G} \mid \mathcal{G}_b = \mathcal{G}_a\}$ and positive integer j that satisfies the following conditions:

$$(2.1.1) \qquad (Q_{b,j}(u))_x + [u, Q_{b,j}(u)] = [Q_{b,j+1}(u), a], \quad Q_{b,0}(u) = b,$$

$$(2.1.2) \quad \sum_{j=0}^{\infty} Q_{b,j}(u)\lambda^{-j} \text{ is conjugate to } b \text{ as an asymptotic expansion.}$$

These conditions imply that $Q_{b,j}(u)$ is a polynomial in $u, \partial_x u, \ldots, \partial_x^{j-1} u$ (cf. [60, 69]). The *G-hierarchy* is a family of evolution equations on $\mathcal{S}(\mathbb{R}, \mathcal{G}_a^\perp)$ parametrized by (b, j), where $b \in \mathcal{G}_a$ such that $\mathcal{G}_b = \mathcal{G}_a$ and j is a positive integer. The (b, j)-flow is

$$u_t = (Q_{b,j}(u))_x + [u, Q_{b,j}(u)] = [Q_{b,j+1}(u), a]. \tag{2.1.3}$$

Recall that a $\mathcal{G}$-valued connection 1-form $\theta = \sum_{i=1}^n A_i(x) dx_i$ is *flat* if

$$d\theta = -\theta \wedge \theta,$$

i.e.,

$$-(A_i)_{x_j} + (A_j)_{x_i} + [A_i, A_j] = 0, \quad 1 \le i < j \le n.$$

The flatness of θ is equivalent to the solvability of the following linear system:

$$E_{x_i} = E A_i, \quad 1 \le i \le n. \tag{2.1.4}$$

Note that (2.1.4) can also be written as $E^{-1} dE = \theta$.

Definition 2.1.1. Let θ be a flat $\mathcal{G}$-valued connection 1-form on $\mathbb{R}^n$. A map $E : \mathbb{R}^n \to G$ is called *a trivialization of* θ if $E^{-1} dE = \theta$. A trivialization E of θ is called a *frame* of θ if E satisfies the initial condition $E(0) = e$, where e is the identity element of G.

The recursive formula (2.1.1) implies that u is a solution of the (b, j)-flow (2.1.3) if and only if

$$\theta_\lambda = (a\lambda + u)\, dx + (b\lambda^j + Q_{b,1}(u)\lambda^{j-1} + \cdots + Q_{b,j}(u))\, dt \tag{2.1.5}$$

is a flat $\mathcal{G}$-valued connection 1-form on the (x, t) plane for all $\lambda \in \mathbb{C}$. In other words, the (b, j)-flow has a Lax pair.

The Cauchy problem with rapidly decaying initial data for the (b, j)-flow (2.1.3) in the G-hierarchy is solved by the inverse scattering method (cf. [3]).

Theorem 2.1.2. *([3]). Suppose $a \in \mathcal{G}$ such that $\mathcal{G}_a$ is a maximal abelian subalgebra $\mathcal{A}$ of $\mathcal{G}$. Then there is an open dense subset $\mathcal{S}_0$ of $\mathcal{S}(\mathbb{R}, \mathcal{A}^\perp)$ such that if $u_0 \in \mathcal{S}_0$, then the Cauchy problem for the (b, j)-flow in the G-hierarchy,*

$$\begin{cases} u_t = (Q_{b,j}(u))_x + [u, Q_{b,j}(u)], \\ u(x, 0) = u_0(x), \end{cases}$$

has a unique solution u. Moreover, $u(x, t)$ is defined for all $(x, t) \in \mathbb{R}^2$ and $u(\cdot, t) \in \mathcal{S}(\mathbb{R}, \mathcal{A}^\perp)$.

The following is well-known, and the proof can be found in many places (cf. [1, 69]).

Theorem 2.1.3. *Let* $X_{b,j}$ *denote the vector field on* $\mathcal{S}(\mathbb{R}, \mathcal{A}^\perp)$ *defined by*

$$X_{b,j}(u) = (Q_{b,j}(u))_x + [u, Q_{b,j}(u)]. \tag{2.1.6}$$

Then $[X_{b,j}, X_{b',j'}] = 0$ *for all* $b, b' \in \mathcal{A}$ *and positive integers* j, j'. *In other words, the* (b, j)*-flow commutes with the* (b', j')*-flow.*

Example 2.1.4. *The* $SL(2,\mathbb{C})$*-hierarchy* (cf. [1, 54]).

Let $G = SL(2,\mathbb{C})$, $a = \mathrm{diag}(i, -i)$. Then $\mathcal{G}_a = \mathcal{A} = \mathbb{C}a$, and

$$\mathcal{G} \cap \mathcal{A}^\perp = \left\{ \begin{pmatrix} 0 & q \\ r & 0 \end{pmatrix} \,\middle|\, q, r \in \mathbb{C} \right\}.$$

Let $u = \begin{pmatrix} 0 & q \\ r & 0 \end{pmatrix}$. Use (2.1.1) and (2.1.2) and a direct computation to get the first three terms of $Q_{a,j}(u)$:

$$Q_{a,1}(u) = u, \quad Q_{a,2}(u) = \begin{pmatrix} \frac{iqr}{2} & \frac{iq_x}{2} \\ -\frac{ir_x}{2} & -\frac{iqr}{2} \end{pmatrix},$$

$$Q_{a,3}(u) = \begin{pmatrix} \frac{qr_x - q_x r}{4} & -\frac{q_{xx}}{4} + \frac{q^2 r}{2} \\ -\frac{r_{xx}}{4} + \frac{qr^2}{2} & -\frac{qr_x - q_x r}{4} \end{pmatrix}.$$

Then the (a, j)-flow, $j = 1, 2, 3$, in the $SL(2,\mathbb{C})$-hierarchy is the following evolution for q and r:

$$\begin{aligned} q_t &= q_x, \quad r_t = r_x, \\ q_t &= \frac{i}{2}(q_{xx} - 2q^2 r), \quad r_t = -\frac{i}{2}(r_{xx} - 2qr^2), \\ q_t &= -\frac{q_{xxx}}{4} + \frac{3}{2} qrq_x, \quad r_t = -\frac{r_{xxx}}{4} + \frac{3}{2} qrr_x. \end{aligned}$$

2.2. The U-hierarchy

Let τ be an involution of G such that its differential at the identity e (still denoted by τ) is a conjugate linear involution on the complex Lie algebra $\mathcal{G}$, and U the fixed point set of τ. The Lie algebra $\mathcal{U}$ of U is a real form of $\mathcal{G}$. Let $\mathcal{U}_a$ denote the centralizer of a in $\mathcal{U}$, and $\mathcal{U}_a^\perp$ the orthogonal complement of $\mathcal{U}_a$ in $\mathcal{U}$. Note that $\mathcal{U}_a^\perp = \mathcal{G}_a^\perp \cap \mathcal{U}$. It is known that the (b, j)-flow in the G-hierarchy leaves $\mathcal{S}(\mathbb{R}, \mathcal{U}_a^\perp)$-invariant (for more details see [70]). The restriction of the flow (2.1.3) to $\mathcal{S}(\mathbb{R}, \mathcal{U}_a^\perp)$ is the (b, j)-flow in the *U-hierarchy*. The Lax pair θ_λ defined by (2.1.5) is a $\mathcal{G}$-valued 1-form, and θ_λ satisfies the *U-reality condition*:

$$\tau(\theta_{\bar\lambda}) = \theta_\lambda. \tag{2.2.1}$$

Note that $\xi = \sum_j \xi_j \lambda^j$ satisfies the U-reality condition if and only if $\xi_j \in \mathcal{U}$ for all j.

Example 2.2.1. *The $SU(2)$-hierarchy.*

Let τ be the involution of $sl(2, \mathbb{C})$ defined by $\tau(\xi) = -\bar\xi^t$. Then the fixed point set of τ is the real form $\mathcal{U} = su(2)$. Let $a = \mathrm{diag}(i, -i) \in \mathcal{U}$. Then $\mathcal{U}_a = \mathcal{A} = \mathbb{R}a$, and

$$\mathcal{A}^\perp \cap \mathcal{U} = \left\{ \begin{pmatrix} 0 & q \\ -\bar q & 0 \end{pmatrix} \,\middle|\, q \in \mathbb{R} \right\}.$$

So $C(\mathbb{R}, \mathcal{A}^\perp \cap \mathcal{U})$ can be identified as $C(\mathbb{R}, \mathbb{C})$, and the $SU(2)$-hierarchy is the restriction of the $SL(2, \mathbb{C})$-hierarchy to the subspace $r = -\bar q$. The first three flows in the $SU(2)$-hierarchy are

$$\begin{aligned} q_t &= q_x, \\ q_t &= \frac{i}{2}(q_{xx} + 2|q|^2 q), \\ q_t &= -\frac{1}{4}(q_{xxx} + 6|q|^2 q). \end{aligned}$$

Note that the $(a, 2)$-flow in the $SU(2)$-hierarchy is the NLS equation.

2.3. The σ-twisted G-hierarchy

Let σ be an order k group automorphism of G such that its differential at the identity e (still denoted by σ) is a complex linear Lie algebra homomorphism of $\mathcal{G}$. Let

$$\mathcal{G} = \mathcal{G}_0 + \cdots + \mathcal{G}_{k-1},$$

where $\mathcal{G}_j$ is the eigenspace of σ with eigenvalue $e^{\frac{2j\pi i}{k}}$. Note that $\mathcal{G}_i = \mathcal{G}_j$ if $i \equiv j$ mod k, and

$$[\mathcal{G}_j, \mathcal{G}_r] \subset \mathcal{G}_{j+r}.$$

Let $\mathcal{A}$ be a maximal abelian subspace in $\mathcal{G}_1$, and $a \in \mathcal{A}$ *regular in* $\mathcal{G}_1$, i.e.,

$$\{x \in \mathcal{G}_1 \mid [x, a] = 0\} = \mathcal{A}.$$

It is known (cf. [70]) that if the image of u lies in $\mathcal{G}_0 \cap \mathcal{G}_a^\perp$, then

$$Q_{b,j}(u) \in \mathcal{G}_{1-j}. \tag{2.3.1}$$

Since $a \in \mathcal{G}_1$, the right hand side of the $(b, mk+1)$-flow satisfies

$$(Q_{b,mk+1}(u))_x + [u, Q_{b,mk+1}(u)] = [Q_{b,mk+2}, a] \in \mathcal{G}_{-mk} = \mathcal{G}_0.$$

In other words, the $(b, mk+1)$-flow leaves $\mathcal{S}(\mathbb{R}, \mathcal{G}_a^\perp \cap \mathcal{G}_0)$ invariant. *The σ-twisted G-hierarchy* is the restriction of the $(b, mk+1)$-flow in the G-hierarchy to $\mathcal{S}(\mathbb{R}, \mathcal{G}_a^\perp \cap \mathcal{G}_0)$ for $m = 1, 2, \cdots$.

It follows from (2.3.1) that the Lax pair of the $(b, mj+1)$-flow in the σ-twisted G-hierarchy satisfies the *(G, σ)-reality condition*:

$$\sigma(\theta_{e^{-\frac{2\pi i}{k}}\lambda}) = \theta_\lambda. \tag{2.3.2}$$

Note that $\xi = \sum_j \xi_j \lambda^j$ satisfies the (G, σ)-reality condition if and only if $\xi_j \in \mathcal{G}_j$ for all j.

Example 2.3.1. *Kupershmidt-Wilson hierarchy* ([48]).

Let $G = SL(n, \mathbb{C})$, and σ the order n automorphism of $SL(n, \mathbb{C})$ defined by $\sigma(g) = C^{-1} g C$, where $C = e_{21} + e_{32} + \cdots + e_{n,n-1} + e_{1n}$ is the permutation matrix $(12 \cdots n)$. Here e_{ij} denote the (i, j)-th elementary matrix in $gl(n)$. The eigenspace $\mathcal{G}_k$ of σ on $sl(n, \mathbb{C})$ with eigenvalue $\alpha = e^{\frac{2\pi i k}{n}}$ is the space of all $y = (y_{ij})$ such that $y_{i+1,j+1} = \alpha^k y_{ij}$ for all $1 \leq i, j \leq n$. Let $a = \mathrm{diag}(1, \alpha, \cdots, \alpha^{n-1}) \in \mathcal{G}_1$, and $\mathcal{A} = \mathbb{C}a$. Then $\mathcal{A}$ is a maximal abelian subalgebra of $\mathcal{G}_1$. The $(SL(n, \mathbb{C}), \sigma)$-hierarchy is the restriction of the $(jn+1)$-th flow in the $sl(n, \mathbb{C})$-hierarchy to $\mathcal{S}(R, \mathcal{G}_0 \cap \mathcal{G}_a^\perp)$. For example, for $n = 2$,

$$\mathcal{G}_0 \cap \mathcal{G}_a^\perp = \left\{ \begin{pmatrix} 0 & q \\ q & 0 \end{pmatrix} \middle| \; q \in \mathbb{C} \right\}.$$

The first flow is the translation $q_t = q_x$, and the third flow is the complex modified KdV equation

$$q_t = \frac{1}{4}(q_{xxx} - 6q^2 q_x).$$

2.4. The U/U_0-hierarchy

Let τ be a conjugate linear involution of $\mathcal{G}$, $\mathcal{U}$ its fixed point set, and σ a complex linear, order k automorphism of $\mathcal{G}$ such that

$$\tau\sigma = \sigma^{-1}\tau^{-1} = \sigma^{-1}\tau.$$

Let $\mathcal{G}_j$ denote the eigenspace of σ with eigenvalue $e^{\frac{2\pi i j}{k}}$. We claim that $\tau(\mathcal{G}_j) \subset \mathcal{G}_j$. To see this, let $\xi_j \in \mathcal{G}_j$. Then

$$\sigma(\tau(\xi_j)) = \tau(\sigma^{-1}(\xi_j)) = \tau(\alpha^{-j}\xi_j) = \overline{\alpha^{-j}}\tau(\xi_j) = \alpha^j\tau(\xi_j),$$

where $\alpha = e^{\frac{2\pi i}{k}}$, proving the claim. Let $\mathcal{U}_j = \mathcal{G}_j \cap \mathcal{U}$. Then we have

$$\mathcal{U} = \mathcal{U}_0 + \cdots + \mathcal{U}_{k-1}.$$

Let $\mathcal{A} \subset \mathcal{U}_1$ be a maximal abelian subspace in $\mathcal{U}_1$. An element $a \in \mathcal{A}$ is *regular in* $\mathcal{U}_1$ if

$$\{\xi \in \mathcal{U}_1 \mid [\xi, a] = 0\} = \mathcal{A}.$$

Let $b \in \mathcal{A}$, and $u \in \mathcal{G}_a^\perp \cap \mathcal{U}_0$. Then $Q_{b,j}(u) \in \mathcal{U}_{1-j}$ for all $j \geq 0$ ([70]). So the $(b, mk+1)$-flow in the σ-twisted G-hierarchy leaves $\mathcal{S}(\mathbb{R}, \mathcal{G}_a^\perp \cap \mathcal{U}_0)$ invariant. The restriction of these flows to $\mathcal{S}(\mathbb{R}, \mathcal{G}_a^\perp \cap \mathcal{U}_o)$ is called the *U/U_o-hierarchy*.

The Lax pair θ_λ of the $(b, mk+1)$-flow in the U/U_0-hierarchy satisfies the *U/U_o-reality condition*:

$$\tau(\theta_{\bar\lambda}) = \theta_\lambda, \quad \sigma(\theta_\lambda) = \theta_{e^{\frac{2\pi i}{k}}\lambda}. \tag{2.4.1}$$

Note that $\xi = \sum_j \xi_j \lambda^j$ satisfies the U/U_0-reality condition if and only if $\xi_j \in \mathcal{U}_j$ for all j.

When the order of σ is 2, the condition $\tau\sigma = \sigma^{-1}\tau^{-1}$ implies that τ and σ commute, U/U_0 is a symmetric space, and $\mathcal{U} = \mathcal{U}_0 + \mathcal{U}_1$ is a Cartan decomposition for the symmetric space U/U_0.

Example 2.4.1. *The $SU(2)/SO(2)$-hierarchy.*

Let $\tau(\xi) = -\bar\xi^t$ and $\sigma(\xi) = -\xi^t$ be involutions of $sl(2, \mathbb{C})$ that give the symmetric space $SU(2)/SO(2)$. Then

$$\mathcal{U}_0 = so(2), \quad \mathcal{U}_1 = \{i\xi \mid \xi \in sl(2, \mathbb{R}), \text{ symmetric}\}$$

with $SU(2)/SO(2)$ the corresponding symmetric space. Let $a = \mathrm{diag}(i, -i)$. Then $\mathcal{G}_a^\perp \cap \mathcal{U}_0 = so(2)$. So $\mathcal{S}(\mathbb{R}, \mathcal{G}_a^\perp \cap \mathcal{U}_0)$ can be identified as $\mathcal{S}(\mathbb{R}, \mathbb{R})$. The $(a,1)$- and $(a,3)$-flow in the $SU(2)/SO(2)$-hierarchy are

$$\begin{aligned} q_t &= q_x, \\ q_t &= -\frac{1}{4}(q_{xxx} + 6q^2 q). \end{aligned}$$

Note that the $(a,3)$-flow is *modified KdV equation* (mKdV equation).

Example 2.4.2. *The $SU(n)/SO(n)$-hierarchy.*

Let τ and ξ be involutions of $sl(n, \mathbb{C})$ defined by

$$\tau(\xi) = -\bar{\xi}^t, \quad \sigma(\xi) = -\xi^t.$$

Then $\tau\sigma = \sigma\tau$, $\mathcal{U} = su(n)$, $\mathcal{U}_0 = so(n)$, and $\mathcal{U}_1$ is the space of $iY \in su(n)$, where Y is real and symmetric with trace zero. The corresponding symmetric space is $\frac{U}{U_0} = \frac{SU(n)}{SO(n)}$. Let $\mathcal{A}$ denote the space of diagonal matrices in $su(n)$. Then $\mathcal{A}$ is a maximal abelian linear subspace in $\mathcal{U}_1$, and $\mathcal{A}^\perp \cap \mathcal{U}_0 = so(n)$. An element $a = i\mathrm{diag}(a_1, \cdots, a_n)$ is *regular in* $\mathcal{U}_1$ if $a_1, \cdots, a_n$ are distinct. Let $a \in \mathcal{A}$ be a regular element, and $b = i\mathrm{diag}(b_1, \cdots, b_n) \in \mathcal{A}$. The $(b, 1)$-flow in the $\frac{SU(n)}{SO(n)}$-hierarchy on $\mathcal{S}(\mathbb{R}, so(n))$ is the *reduced n-wave equation*

$$(u_{ij})_t = \frac{b_i - b_j}{a_i - a_j}(u_{ij})_x + \sum_k \left(\frac{b_k - b_j}{a_k - a_j} - \frac{b_i - b_k}{a_i - a_k}\right) u_{ik}u_{kj}, \quad i \neq j.$$

Example 2.4.3. Let U/U_0 be a symmetric space, $\mathcal{U} = \mathcal{U}_0 + \mathcal{U}_1$ a Cartan decomposition, $\mathcal{A}$ a maximal abelian subspace in $\mathcal{U}_1$, $a \in \mathcal{A}$ regular, and $b \in \mathcal{A}$. Note $\mathrm{ad}(a)^{-1}$ maps $\mathcal{U}_a^\perp \cap \mathcal{U}_0$ and $\mathcal{U}_a^\perp \cap \mathcal{U}_1$ isomorphically onto $\mathcal{U}_a^\perp \cap \mathcal{U}_1$ and $\mathcal{U}_a^\perp \cap \mathcal{U}_0$ respectively. So $\mathrm{ad}(b)\mathrm{ad}(a)^{-1}(\mathcal{U}_a^\perp \cap \mathcal{U}_0) \subset \mathcal{U}_a^\perp \cap \mathcal{U}_0$. The recursive formula (2.1.1) implies that

$$Q_{b,1}(u) = \mathrm{ad}(b)\mathrm{ad}(a)^{-1}(u). \tag{2.4.2}$$

So the $(b, 1)$-flow in the U/U_0-hierarchy is the equation for maps $u : \mathbb{R}^2 \to \mathcal{U}_a^\perp \cap \mathcal{U}_0$:

$$u_t = \mathrm{ad}(b)\mathrm{ad}(a)^{-1}(u_x) + [u, \mathrm{ad}(b)\mathrm{ad}(a)^{-1}(u)]. \tag{2.4.3}$$

This is the *reduced n-wave equation associated to U/U_0*, which has a Lax pair

$$\theta_\lambda = (a\lambda + u)dx + (b\lambda + \mathrm{ad}(b)\mathrm{ad}(a)^{-1}(u))dt. \tag{2.4.4}$$

2.5. The U/U_0-system

Let τ be a conjugate linear involution of $\mathcal{G}$, σ a complex linear involution of $\mathcal{G}$ such that $\tau\sigma = \sigma\tau$, $\mathcal{U}$ the fixed point set of τ, and U_0 the subgroup of U fixed by σ. Let $\mathcal{U} = \mathcal{U}_0 + \mathcal{U}_1$ denote the Cartan decomposition of the symmetric space U/U_0. Let $\mathcal{A}$ be a maximal abelian linear subspace of $\mathcal{U}_1$, and $a_1, \cdots, a_n$ a basis of $\mathcal{A}$. *The U/U_0-system* is the following system for $v : \mathbb{R}^n \to \mathcal{U}_{\mathcal{A}}^{\perp} \cap \mathcal{U}_1$:

$$[a_i, v_{x_j}] - [a_j, v_{x_i}] = [[a_i, v], [a_j, v]], \quad i \neq j. \tag{2.5.1}$$

It has a Lax n-tuple,

$$\theta_\lambda = \sum_{i=1}^{n} (a_i \lambda + [a_i, v]) dx_i, \tag{2.5.2}$$

which satisfies the U/U_0-reality condition (2.4.1). Moreover, the following statements are equivalent for smooth map $v : \mathbb{R}^n \to \mathcal{U}_{\mathcal{A}}^{\perp} \cap \mathcal{U}_1$:

(i) v is a solution of the U/U_0-system (2.5.1),
(ii) θ_λ defined by (2.5.2) is a flat $\mathcal{G}$-valued connection 1-form on $\mathbb{R}^n$ for all $\lambda \in C$,
(iii) θ_r is flat for some $r \in \mathbb{R}$.

We claim that the U/U_0-system is independent of the choice of basis of $\mathcal{A}$. If $b_1, \cdots, b_n$ is a basis of $\mathcal{A}$, then there exists a constant matrix (c_{ij}) such that $a_i = \sum_{j=1}^{n} c_{ij} b_j$. The U/U_0-system defined by the new base $b_1, \cdots, b_n$ is

$$[b_i, v_{y_j}] - [b_j, v_{y_i}] = [[b_i, v], [b_j, v]].$$

This is the same system as (2.5.1) if we make the coordinate transformation $y_i = \sum_{j=1}^{n} c_{ji} x_j$.

The U/U_0-system is given by the first commuting n-flows in the U/U_0-hierarchy, i.e.,

Proposition 2.5.1. *([68]). With the same notation as above, let $a_1, \cdots, a_n$ be a basis of $\mathcal{A}$ such that $a_1, \cdots, a_n$ are regular. Let $a = a_1$. Then $v : \mathbb{R}^n \to \mathcal{U}_{\mathcal{A}}^{\perp} \cap \mathcal{U}_1$ is a solution of the U/U_0-system (2.5.1) if and only if $u(x) = [a, v(x)]$ satisfies the $(a_j, 1)$-flow in the U/U_0-hierarchy,*

$$u_{x_j} = \mathrm{ad}(a_j)\mathrm{ad}(a)^{-1}(u_{x_1}) + [u, \mathrm{ad}(a_j)\mathrm{ad}(a)^{-1}(u)],$$

for all $1 \leq j \leq n$.

As a consequence of Theorem 2.1.2 and Proposition 2.5.1 we have

Corollary 2.5.2. *([68]). Suppose $a = a_1 \in \mathcal{A}$ is regular in $\mathcal{U}_1$. Then there exists an open dense subset $\mathcal{S}_0$ of $\mathcal{S}(\mathbb{R}, \mathcal{U}_a^{\perp} \cap \mathcal{U}_1)$ such that given any $v_0 \in \mathcal{S}_0$ there exists a unique solution v of (2.5.1) defined for all $x \in \mathbb{R}^n$ such that $v(x_1, 0, \cdots, 0) = v_0(x_1)$ and $v(\cdot, x_2, \cdots, x_n) \in \mathcal{S}(\mathbb{R}, \mathcal{U}_a^{\perp} \cap \mathcal{U}_1)$.*

Next we give some examples.

Example 2.5.3. The U-system.

Let τ be a conjugate linear involution of $\mathcal{G}$, and $\mathcal{U}$ the fixed point set of τ. Let $\tau_2(x,y) = (\tau(x), \tau(y))$ and $\sigma(x,y) = (y,x)$ be involutions of $\mathcal{G} \times \mathcal{G}$. Then $\tau_2\sigma = \sigma\tau_2$, and the corresponding symmetric space is $(U \times U)/\triangle(U) \simeq U$, where $\triangle(U)$ is the diagonal group $\{(g,g) \mid g \in U\}$. The $(U \times U)/\triangle(U)$-system is the U-system (2.5.1) for maps $v : \mathbb{R}^n \to \mathcal{A}^\perp \cap \mathcal{U}$, where $\mathcal{A}$ is a maximal abelian subalgebra of $\mathcal{U}$ and $\{a_1, \cdots, a_n\}$ is a basis of $\mathcal{A}$.

Example 2.5.4. The $\frac{O(2n)}{O(n)\times O(n)}$-system.

Here U/U_0 is the symmetric space $\frac{O(2n)}{O(n)\times O(n)}$, $G = O(2n, \mathbb{C})$, $\tau(g) = \bar{g}$, $\sigma(g) = \mathrm{I}_{n,n}\ g\ \mathrm{I}_{n,n}^{-1}$, where $I_{n,n}$ is the diagonal matrix with $a_{ii} = 1$ for $1 \le i \le n$ and $a_{ii} = -1$ for $n+1 \le i \le 2n$. So $\mathcal{U} = so(2n)$, $\mathcal{U}_0 = so(n) + so(n)$, and

$$\mathcal{U}_1 = \left\{ \begin{pmatrix} 0 & \xi \\ -\xi^t & 0 \end{pmatrix} \,\middle|\, \xi \in gl(n, \mathbb{R}) \right\}.$$

The linear subspace $\mathcal{A}$ spanned by

$$\{a_i = -e_{i,n+i} + e_{n+i,i} \mid 1 \le i \le n\}$$

is a maximal abelian subspace of $\mathcal{U}_1$, and

$$\mathcal{U}_1 \cap \mathcal{A}^\perp = \left\{ \begin{pmatrix} 0 & F \\ -F^t & 0 \end{pmatrix} \,\middle|\, F = (f_{ij}) \in gl(n, \mathbb{R}), f_{ii} = 0 \text{ for } 1 \le i \le n. \right\}.$$

The corresponding U/U_0-system (2.5.1) written in terms of $F = (f_{ij})$ is

$$\text{(2.5.3)} \qquad \begin{cases} (f_{ij})_{x_i} + (f_{ji})_{x_j} + \sum_k f_{ki} f_{kj} = 0, & \text{if } i \ne j, \\ (f_{ij})_{x_j} + (f_{ji})_{x_i} + \sum_k f_{ik} f_{jk} = 0, & \text{if } i \ne j, \\ (f_{ij})_{x_k} = f_{ik} f_{kj}, & \text{if } i,j,k \text{ are distinct.} \end{cases}$$

The Lax n-tuple θ_λ (2.5.2), written in matrix form is
(2.5.4)

$$\theta_\lambda = \begin{pmatrix} \delta F^t - F\delta & -\lambda\delta \\ \lambda\delta & -F^t\delta + \delta F \end{pmatrix}, \quad \text{where } \delta = \mathrm{diag}(dx_1, \cdots, dx_n).$$

Note that the first and the third equations of (2.5.3) imply that $\delta F^t - F\delta$ is flat, and the second and third equations of (2.5.3) imply that $-F^t\delta + \delta F$ is flat.

Example 2.5.5. The $\frac{U(n)}{O(n)}$-system.

Here $\mathcal{G} = gl(n, \mathbb{C})$, $\tau(\xi) = -\bar{\xi}^t$, and $\sigma(\xi) = -\xi^t$. Then $\mathcal{U} = u(n)$, $\mathcal{U}_0 = o(n)$, and

$$\mathcal{U}_1 = \{iF \mid F = (f_{ij}) \in gl(n, \mathbb{R}), f_{ij} = f_{ji}\}.$$

The linear subspace $\mathcal{A}$ spanned by

$$\{a_j = ie_{jj} \mid 1 \leq j \leq n\}$$

is a maximal abelian subspace of $\mathcal{U}_1$, and

$$\mathcal{U}_1 \cap \mathcal{A}^\perp = \{iF \mid F = (f_{ij}) \in gl(n, \mathbb{R}),\ f_{ij} = f_{ji},\ f_{ii} = 0 \text{ for } 1 \leq i, j \leq n\}.$$

The corresponding U/U_0-system written in terms of F is the restriction of system (2.5.3) to the linear subspace of $F = (f_{ij})$ such that $f_{ij} = f_{ji}$. So the $\frac{U(n)}{O(n)}$-system is the system for symmetric $F = (f_{ij})$:

$$\begin{cases} (f_{ij})_{x_i} + (f_{ij})_{x_j} + \sum_k f_{ik} f_{jk} = 0, & \text{if } i \neq j, \\ (f_{ij})_{x_k} = f_{ik} f_{kj}, & \text{if } i, j, k \text{ are distinct.} \end{cases} \tag{2.5.5}$$

Or, equivalently, $[\delta, F] = \delta F - F\delta$ is flat.

Example 2.5.6. The $\frac{SU(n)}{SO(n)}$-system.

Here $\mathcal{G} = sl(n, \mathbb{C})$, $\tau(\xi) = -\bar{\xi}^t$, and $\sigma(\xi) = -\xi^t$ for $\xi \in \mathcal{G}$. Then $\mathcal{U} = su(n)$, $\mathcal{U}_0 = so(n)$, and

$$\mathcal{U}_1 = \{iF \mid F = (f_{ij}) \in gl(n, \mathbb{R}), f_{ij} = f_{ji}, \sum_{i=1}^{n} f_{ii} = 0\}.$$

The linear subspace $\mathcal{A}$ spanned by

$$\{b_j = i(e_{jj} - e_{11}) \mid 2 \leq j \leq n\}$$

is a maximal abelian linear subspace of $\mathcal{U}_0$, and

$$\mathcal{A}^\perp \cap \mathcal{U}_1 = \{iF \mid F = (f_{ij}) \in gl(n, \mathbb{R}), f_{ij} = f_{ji}, f_{ii} = 0 \text{ for } 1 \leq i, j \leq n\}.$$

The $\frac{SU(n)}{SO(n)}$-system is

$$[b_i, F_{t_j}] - [b_j, F_{t_i}] = [b_i, F], [b_j, F]]. \quad 2 \leq i \neq j \leq n. \tag{2.5.6}$$

Example 2.5.7. The $U/U_0 = \frac{GL(2,\mathbb{H})}{(\mathbb{R}^+\times SU(2))^2}$-system ([8]).

Here $\mathcal{G} = gl(4,\mathbb{C})$. For $X \in gl(4,\mathbb{C})$, write

$$X = \begin{pmatrix} X_1 & X_2 \\ X_3 & X_4 \end{pmatrix}, \quad X_i \in gl(2,\mathbb{C}).$$

Let τ be the involution of $\mathcal{G}$ defined by

$$\tau(X) = \begin{pmatrix} J & 0 \\ 0 & J \end{pmatrix} \overline{X} \begin{pmatrix} J^{-1} & 0 \\ 0 & J^{-1} \end{pmatrix}, \quad \text{where } J = \begin{pmatrix} 0 & 1 \\ -1 & 0 \end{pmatrix}.$$

The fixed point set $\mathcal{U}$ of τ is the subalgebra of $X \in \mathcal{G}$ such that $J\bar{X}_iJ^{-1} = X_i$ for all $1 \leq i \leq 4$, i.e., X_i lies in the fixed point set of the involution of $gl(2,\mathbb{C})$ defined by $\tau_0(Y) = J\bar{Y}J^{-1}$. A direct computation implies that the fixed point set of τ_0 is

$$\left\{ \begin{pmatrix} z_1 & z_2 \\ -\bar{z}_2 & \bar{z}_1 \end{pmatrix} \;\middle|\; z_1, z_2 \in \mathbb{C} \right\} = \mathbb{R} \times su(2).$$

Note that $\mathbb{R} \times su(2)$ is isomorphic to the quaternions $\mathbb{H}$ as an associative algebra via the isomorphism

$$\mathbf{i} = \begin{pmatrix} i & 0 \\ 0 & -i \end{pmatrix}, \; \mathbf{j} = \begin{pmatrix} 0 & 1 \\ -1 & 0 \end{pmatrix}, \; \mathbf{k} = \begin{pmatrix} 0 & i \\ i & 0 \end{pmatrix}, \; \mathbf{1} = \begin{pmatrix} 1 & 0 \\ 0 & 1 \end{pmatrix}.$$

(It is easy to check that $\mathbf{ij} = \mathbf{k}$, $\mathbf{jk} = \mathbf{i}$, $\mathbf{ki} = \mathbf{j}$.) So we can view $\mathcal{U} = gl(2,\mathbb{H})$, i.e., the algebra of 2×2 matrices with entries in the quaternions $\mathbb{H}$.

Let σ be the involution on $gl(4,\mathbb{C})$ defined by

$$\sigma(Y) = \begin{pmatrix} \mathbf{1} & 0 \\ 0 & -\mathbf{1} \end{pmatrix} Y \begin{pmatrix} \mathbf{1} & 0 \\ 0 & -\mathbf{1} \end{pmatrix}^{-1}.$$

Then $\sigma\tau = \tau\sigma$, and

$$\mathcal{U}_0 = \left\{ \begin{pmatrix} P & 0 \\ 0 & Q \end{pmatrix} \;\middle|\; P, Q \in \mathbb{R} \times su(2) \right\},$$

$$\mathcal{U}_1 = \left\{ \begin{pmatrix} 0 & P \\ Q & 0 \end{pmatrix} \;\middle|\; P, Q \in \mathbb{R} \times su(2) \right\}.$$

Let

$$a_1 = \begin{pmatrix} 0 & \mathbf{k} \\ \mathbf{k} & 0 \end{pmatrix} \quad a_2 = \begin{pmatrix} 0 & \mathbf{j} \\ -\mathbf{j} & 0 \end{pmatrix}. \tag{2.5.7}$$

The space $\mathcal{A}$ spanned by a_1 and a_2 is a maximal abelian linear subspace of $\mathcal{U}_1$, and $\mathcal{A}^\perp \cap \mathcal{U}_1$ is the space of all matrices of the form

$$v = \begin{pmatrix} 0 & \begin{pmatrix} p_1 & p_2 \\ -\bar{p}_2 & \bar{p}_1 \end{pmatrix} \\ \begin{pmatrix} q_1 & \bar{p}_2 \\ -p_2 & \bar{q}_1 \end{pmatrix} & 0 \end{pmatrix}. \tag{2.5.8}$$

The $\frac{GL(2,\mathbb{H})}{(\mathbb{R}\times SU(2))^2}$-system is

$$\begin{cases} (p_2)_y + i(p_2)_x = -|p_1|^2 + |q_1|^2, \\ (\bar{q}_1 + p_1)_x = 2i\ (\bar{p}_2 - p_2)(\bar{q}_1 - p_1), \\ (\bar{q}_1 - p_1)_y = -2(p_2 + \bar{p}_2)(\bar{q}_1 + p_1). \end{cases} \tag{2.5.9}$$

Its Lax pair (2.5.2) is $\theta_\lambda =$

$$\begin{pmatrix} \begin{pmatrix} Z & W \\ -\bar{W} & \bar{Z} \end{pmatrix} & \mathbf{k}\lambda \\ \mathbf{k}\lambda & \begin{pmatrix} -\bar{Z} & \bar{W} \\ -W & -Z \end{pmatrix} \end{pmatrix} dx + \begin{pmatrix} \begin{pmatrix} -iZ & Y \\ -\bar{Y} & i\bar{Z} \end{pmatrix} & \mathbf{j}\lambda \\ -\mathbf{j}\lambda & \begin{pmatrix} -i\bar{Z} & -\bar{Y} \\ Y & iZ \end{pmatrix} \end{pmatrix} dy,$$

where

$$Z = -2ip_2, \quad W = i(\bar{q}_1 - p_1), \quad Y = (\bar{q}_1 + p_1). \tag{2.5.10}$$

Let $p_2 = \beta_1 + i\beta_2$, $\beta_1, \beta_2 \in \mathbb{R}$. Equate the imaginary part of the first equation in system (2.5.9) to get $(\beta_2)_y + (\beta_1)_x = 0$. So there exists u such that $\beta_1 = -\frac{u_y}{8}$ and $\beta_2 = \frac{u_x}{8}$, and hence $p_2 = \frac{-u_y + iu_x}{8}$. Substitute this into (2.5.9) to get

$$\begin{cases} u_{xx} + u_{yy} = 8(|p_1|^2 - |q_1|^2), \\ (\bar{q}_1 + p_1)_x = \frac{u_x}{2}(\bar{q}_1 - p_1), \\ (\bar{q}_1 - p_1)_y = \frac{u_y}{2}(\bar{q}_1 + p_1). \end{cases} \tag{2.5.11}$$

System (2.5.11) is gauge equivalent to system (2.5.9). To see this, we recall that v is a solution of the U/U_0-system (2.5.1) if and only if θ_0 is flat. So θ_0 is a $(\mathbb{R} \times su(2)) \times (\mathbb{R} \times su(2))$-valued flat connection 1-form. The $\mathbb{R} \times \mathbb{R}$-component of θ_0 is

$$\theta_0^0 = \begin{pmatrix} 2\beta_2\mathbf{1} & 0 \\ 0 & -2\beta_2\mathbf{1} \end{pmatrix} dx + \begin{pmatrix} -2\beta_1\mathbf{1} & 0 \\ 0 & 2\beta_1\mathbf{1} \end{pmatrix} dy = \begin{pmatrix} \frac{du}{4}\,\mathbf{1} & 0 \\ 0 & -\frac{du}{4}\,\mathbf{1} \end{pmatrix}.$$

Let

$$g = \begin{pmatrix} e^{u/4}\mathbf{1} & 0 \\ 0 & e^{-u/4}\mathbf{1} \end{pmatrix}.$$

The gauge transformation of θ_0 by g is

$$g * \theta_0 = \begin{pmatrix} \tau_1 & 0 \\ 0 & \tau_2 \end{pmatrix}, \quad \text{where}$$

$$\tau_1 = \begin{pmatrix} \frac{i(u_y dx - u_x dy)}{4} & i(\bar{q}_1 - p_1)dx + (\bar{q}_1 + p_1)dy \\ i(q_1 - \bar{p}_1)dx - (q_1 + \bar{p}_1)dy & -\frac{i(u_y dx - u_x dy)}{4} \end{pmatrix},$$

$$\tau_2 = \begin{pmatrix} \frac{i(u_y dx - u_x dy)}{4} & -i(q_1 - \bar{p}_1)dx - (q_1 + \bar{p}_1)dy \\ -i(\bar{q}_1 - p_1)dx + (\bar{q}_1 + p_1)dy & -\frac{i(u_y dx - u_x dy)}{4} \end{pmatrix}.$$

The connection $g * \theta_0$ is flat if and only if (u, p_1, q_1) is a solution of (2.5.11). So $g * \theta_\lambda$ is a Lax pair of (2.5.11). In other words, system (2.5.11) is gauge equivalent to the $\frac{GL(2,\mathbb{H})}{(\mathbb{R}^+ \times SU(2))^2}$-system.

Suppose (u, p_1, q_1) is a solution of (2.5.11) and q_1 is real. Let $p_1 = B_1 + iB_2$. Equate the imaginary part of the second and third equations of (2.5.11) to get

$$(B_2)_x = -u_x B_2/2, \quad (B_2)_y = -u_y B_2/2.$$

So $B_2 = ce^{-u/2}$ for some constant c, and (2.5.11) becomes the following system for real functions u, q_1, B_1.

$$\text{(2.5.12)} \qquad \begin{cases} u_{xx} + u_{yy} = 8(c^2 e^{-u} + B_1^2 - q_1^2), \\ (q_1 + B_1)_x = \frac{u_x}{2}(q_1 - B_1), \\ (q_1 - B_1)_y = \frac{u_y}{2}(q_1 + B_1). \end{cases}$$

If p_1 is also real, i.e., $c = 0$ in (2.5.12), then system (2.5.12) becomes the following system for real functions u, q_1, p_1:

$$\text{(2.5.13)} \qquad \begin{cases} u_{xx} + u_{yy} = 8(p_1^2 - q_1^2), \\ (q_1 + p_1)_x = \frac{u_x}{2}(q_1 - p_1), \\ (q_1 - p_1)_y = \frac{u_y}{2}(q_1 + p_1). \end{cases}$$

2.6. The −1-flow

Let $\mathcal{U}$ be the real form defined by the involution τ on $\mathcal{G}$, $a, b \in \mathcal{U}$ such that $[a, b] = 0$. The *−1-flow associated to U defined by a, b* is the following system for $u : \mathbb{R}^2 \to \mathcal{U} \cap \mathcal{U}_a^\perp$:

$$u_t = [a, g^{-1}bg], \quad \text{where } g^{-1}g_x = u. \tag{2.6.1}$$

This system has a Lax pair

$$\theta_\lambda = (a\lambda + u)dx + \lambda^{-1}g^{-1}bg\, dt, \quad \text{where } g : \mathbb{R}^2 \to U, \quad g^{-1}g_x = u.$$

Note that θ_λ satisfies the U-reality condition (2.2.1).

Theorem 2.6.1. *([68]). The -1-flow* (2.6.1) *commutes with all the flows in the U-hierarchy.*

Let σ be an order k automorphism of $\mathcal{G}$ such that $\tau\sigma = \sigma^{-1}\tau^{-1}$ and $\mathcal{U} = \mathcal{U}_0 + \cdots + \mathcal{U}_{k-1}$ as in section 2.4 Let $a \in \mathcal{U}_1$ a regular element, and $b \in \mathcal{U}_{k-1}$ such that $[a, b] = 0$, then the right hand side of (2.6.1) is a vector field on $C(\mathbb{R}, \mathcal{U}_0 \cap \mathcal{U}_a^\perp)$. The flow (2.6.1) restricted to the space $C(\mathbb{R}, \mathcal{U}_0 \cap \mathcal{U}_a^\perp)$ of smooth maps from $\mathbb{R}$ to $\mathcal{U}_0 \cap \mathcal{U}_a^\perp$ is *the -1-flow associated to U/U_0*, and θ_λ satisfies the U/U_0-reality condition (2.4.1).

We can also write the -1-flow (2.6.1) associated to U (or U/U_0) as an equation for $g : \mathbb{R}^2 \to U$ ($g : \mathbb{R}^2 \to U_0$ respectively):

$$(g^{-1}g_x)_t = [a, g^{-1}bg], \tag{2.6.2}$$

where $g^{-1}g_x \in \mathcal{U}_a^\perp$ ($\in \mathcal{U}_0 \cap \mathcal{U}_a^\perp$ respectively). Its Lax pair is

$$\theta_\lambda = (a\lambda + g^{-1}g_x)dx + \lambda^{-1}g^{-1}bgdt. \tag{2.6.3}$$

Example 2.6.2. The -1-flow associated to $SU(2)/SO(2)$.

Let $a = \mathrm{diag}(i, -i)$, and $b = -\frac{a}{4}$. Let $g = \begin{pmatrix} \cos\frac{q}{2} & \sin\frac{q}{2} \\ -\sin\frac{q}{2} & \cos\frac{q}{2} \end{pmatrix}$. Then $u = g^{-1}g_x = \frac{1}{2}\begin{pmatrix} 0 & q_x \\ -q_x & 0 \end{pmatrix}$, and the -1-flow (2.6.2) associated to $SU(2)/SO(2)$ is the sine-Gordon equation (SGE equation):

$$q_{xt} = \sin q,$$

and its Lax pair is

$$\theta_\lambda = \begin{pmatrix} i\lambda & \frac{q_x}{2} \\ -\frac{q_x}{2} & -i\lambda \end{pmatrix} dx - \frac{i\lambda}{4}\begin{pmatrix} \cos q & \sin q \\ \sin q & \cos q \end{pmatrix} dt.$$

Example 2.6.3. The -1-flow associated to $U/U_0 = SL(3, \mathbb{C})/\mathbb{R}^+$.

Here $\mathbb{R}^+$ is the subgroup

$$\mathbb{R}^+ = \{\mathrm{diag}(r, r^{-1}, 1) \mid r > 0\}$$

of $SL(3,\mathbb{R})$, $G = SL(3,\mathbb{C})$, $\tau(g) = \bar{g}$, and σ is the order 6 automorphism of $SL(3,\mathbb{C})$ defined by

$$\sigma(g) = C\ (g^t)^{-1}C^{-1}, \quad \text{where} \quad C = \begin{pmatrix} 0 & \alpha^2 & 0 \\ \alpha & 0 & 0 \\ 0 & 0 & 1 \end{pmatrix}, \quad \text{and} \quad \alpha = e^{\frac{2\pi i}{3}}.$$

The induced automorphism σ on $sl(3,\mathbb{C})$ is

$$\sigma(A) = -CA^tC^{-1}.$$

Note that the order of σ is 6, σ is complex linear on $sl(3,\mathbb{C})$, and $\sigma\tau = \tau^{-1}\sigma^{-1}$. Let $\beta = e^{\frac{2\pi i}{6}} = e^{\frac{\pi i}{3}}$. A direct computation implies that Y_j lies in the eigenspaces $\mathcal{G}_j$ of β^j if

$$Y_0 = \begin{pmatrix} s & 0 & 0 \\ 0 & -s & 0 \\ 0 & 0 & 0 \end{pmatrix}, \quad Y_1 = \begin{pmatrix} 0 & 0 & s_1 \\ s_2 & 0 & 0 \\ 0 & s_1 & 0 \end{pmatrix},$$

$$Y_2 = \begin{pmatrix} 0 & 0 & 0 \\ 0 & 0 & s \\ -s & 0 & 0 \end{pmatrix}, \quad Y_3 = \begin{pmatrix} s & 0 & 0 \\ 0 & s & 0 \\ 0 & 0 & -2s \end{pmatrix},$$

$$Y_4 = \begin{pmatrix} 0 & 0 & s \\ 0 & 0 & 0 \\ 0 & -s & 0 \end{pmatrix}, \quad Y_5 = \begin{pmatrix} 0 & s_1 & 0 \\ 0 & 0 & s_2 \\ s_2 & 0 & 0 \end{pmatrix}.$$

The fixed point set of τ is $\mathcal{U} = sl(3,\mathbb{R})$, and $\mathcal{U}_j = sl(3,\mathbb{R}) \cap \mathcal{G}_j$.

Let

$$a = \begin{pmatrix} 0 & 0 & 1 \\ 1 & 0 & 0 \\ 0 & 1 & 0 \end{pmatrix} \in \mathcal{U}_1, \quad b = \begin{pmatrix} 0 & 1 & 0 \\ 0 & 0 & 1 \\ 1 & 0 & 0 \end{pmatrix} \in \mathcal{U}_{-1} = \mathcal{U}_5.$$

Note that $[a, b] = 0$. The fixed point set U_0 of σ on U is the abelian group

$$U_0 = \{\mathrm{diag}(r, r^{-1}, 1) \mid r > 0\}.$$

A smooth map $g : \mathbb{R}^2 \to U_0$ is of the form

$$g = \begin{pmatrix} e^w & 0 & 0 \\ 0 & e^{-w} & 0 \\ 0 & 0 & 1 \end{pmatrix}$$

for some smooth function w. So $g^{-1}g_x = w_x \mathrm{diag}(1,-1,0)$, and

$$g^{-1}bg = \begin{pmatrix} 0 & e^{-2w} & 0 \\ 0 & 0 & e^w \\ e^w & 0 & 0 \end{pmatrix}.$$

Hence the -1-flow (2.6.2) associated to $SL(3,\mathbb{R})/U_0$ is

$$w_{xt}\mathrm{diag}(1,-1,0) = (e^w - e^{-2w})\mathrm{diag}(1,-1,0),$$

i.e., the *Tzitzeica equation*:

$$w_{xt} = e^w - e^{-2w}. \tag{2.6.4}$$

The corresponding Lax pair θ_λ (2.6.3) is

$$\theta_\lambda = \begin{pmatrix} w_x & 0 & \lambda \\ \lambda & -w_x & 0 \\ 0 & \lambda & 0 \end{pmatrix} dx + \lambda^{-1} \begin{pmatrix} 0 & e^{-2w} & 0 \\ 0 & 0 & e^w \\ e^w & 0 & 0 \end{pmatrix} dt. \tag{2.6.5}$$

Note that θ_λ satisfies the $\frac{SL(3,\mathbb{R})}{\mathbb{R}^+}$-reality condition:

$$\overline{\theta_{\bar\lambda}} = \theta_\lambda, \quad -C\theta_\lambda^t C^{-1} = \theta_{\beta\lambda}, \quad \text{where } \beta = e^{\frac{2\pi i}{6}} = e^{\frac{\pi i}{3}}. \tag{2.6.6}$$

Example 2.6.4. The -1-flow associated to $SL(n,\mathbb{R})/(\mathbb{R}^+)^{n-1}$.

Let $\mathcal{G} = sl(n,\mathbb{C})$, $\tau(\xi) = \bar\xi$, and $\sigma(\xi) = C\xi C^{-1}$, where

$$C = \mathrm{diag}(1, \alpha, \cdots, \alpha^{n-1})$$

and $\alpha = e^{\frac{2\pi i}{n}}$. Then the order of σ is n and $\tau\sigma = \sigma^{-1}\tau^{-1}$. The fixed point set $\mathcal{U}$ of τ is $sl(n,\mathbb{R})$. The eigenspace $\mathcal{G}_j$ of σ with eigenvalue $e^{\frac{2\pi j i}{n}}$ is spanned by $\{e_{i+j,i} \mid i = 1, \cdots, n\}$, where e_{ij} is the elementary matrix and $e_{ij} = e_{i'j'}$ if $i \equiv i'$ and $j \equiv j'$ mod n. Let $\mathcal{U}_j = \mathcal{G}_j \cap \mathcal{U}$, $a = e_{21} + e_{32} + \cdots + e_{n,n-1} + e_{1n}$, and $b = e_{12} + e_{23} + \cdots + e_{n-1,n} + e_{n1}$. Then $a \in \mathcal{U}_1$, $b \in \mathcal{U}_{-1}$, and $[a,b] = 0$. Let $g = \mathrm{diag}(e^{u_1}, \cdots, e^{u_n})$ with $\sum_i u_i = 0$. So $g^{-1}g_x = \mathrm{diag}((u_1)_x, \cdots, (u_n)_x)$. The -1-flow (2.6.2) associated to G, τ, σ is:

$$(u_i)_{xt} = e^{u_i - u_{i-1}} - e^{u_{i+1} - u_i}, \quad 1 \le i \le n,$$

where $u_{n+1} = u_1$ and $u_0 = u_n$.

2.7. The hyperbolic systems

Let $\mathcal{U}$ be the real form defined by the involution τ on $\mathcal{G}$. *The hyperbolic U-system* is the following system for $(u_0, u_1, v_0, v_1) : \mathbb{R}^2 \to \prod_{i=1}^4 \mathcal{U}$:

$$(2.7.1) \qquad \begin{cases} (u_1)_t = [u_1, v_0], \\ (u_0)_t = (v_0)_x + [u_1, v_1] + [u_0, v_0], \\ (v_1)_x = -[u_0, v_1]. \end{cases}$$

It has a Lax pair

$$(2.7.2) \qquad \Omega_\lambda = (u_1\lambda + u_0)dx + (\lambda^{-1}v_1 + v_0)dt,$$

which satisfies the U-reality condition (2.2.1).

Let σ be an order k automorphism of $\mathcal{G}$ such that $\tau\sigma = \sigma^{-1}\tau$, and $\mathcal{U} = \mathcal{U}_0 + \cdots + \mathcal{U}_{k-1}$, where $\mathcal{U}_j$ is the intersection of $\mathcal{U}$ and the $e^{\frac{2\pi i j}{k}}$-eigenspace of σ. *The hyperbolic U/U_0-system* is the restriction of the hyperbolic U-system (2.7.1) to $(u_0, u_1, v_0, v_1) \in \mathcal{U}_0 \times \mathcal{U}_1 \times \mathcal{U}_0 \times \mathcal{U}_{-1}$. The corresponding Lax pair (2.7.2) satisfies the U/U_0-reality condition (2.4.1).

§3. Geometries associated to soliton equations

We give geometric interpretations of certain soliton equations. For example, solutions of the $O(2n)/O(n) \times O(n)$-system give rise to orthogonal coordinates of $\mathbb{R}^n$ and flat submanifolds in $\mathbb{R}^{2n}$, solutions of the $U(n)/O(n)$-system give Egoroff flat metrics and flat Lagrangian submanifolds of $\mathbb{R}^{2n} = \mathbb{C}^n$, a subclass of solutions of the $GL(2,\mathbb{H})/(SU(2) \times \mathbb{R}^+)^2$-system give rise to Bonnet pairs in $\mathbb{R}^3$, and the -1-flow associated to $SL(3,\mathbb{R})/\mathbb{R}^+$ (the Tzitzeica equation) is the Gauss-Codazzi equation for affine spheres in the affine 3-space. If U/U_0 is a rank n symmetric space, then we can associate to each solution of the U/U_0-system a flat n-submanifold in U/U_0 and a flat n-submanifold in the tangent space of U/U_0. We also give a brief review of the relation between harmonic maps from $\mathbb{R}^{1,1}$ to U and solutions of the hyperbolic U-system.

3.1. The method of moving frames

Let (N, g) be an $(n+k)$-dimensional Riemannian manifold, ∇ the Levi-Civita connection of g, and $X : M^n \to N$ an immersion. We set up some notation next. The first fundamental form I is the induced metric.

Let ξ be a normal vector field on M, v a tangent vector field, $(\nabla_v\xi)^t$ and $(\nabla_v\xi)^\nu$ the tangential and normal components of $\nabla_v\xi$ respectively. The induced normal connection on the normal bundle $\nu(M)$ is defined by

$$\nabla^\nu_v\xi = (\nabla_v\xi)^\nu.$$

The second fundamental form II is a smooth section of $S^2(T^*M)\otimes\nu(M)$ defined by

$$\mathrm{II}_\xi(v_1, v_2) = -g(\nabla_{v_1}\xi, v_2).$$

Next we express $\mathrm{I}, \mathrm{II}, \nabla^\nu$ using a moving frame. Let $(e_1, \cdots, e_{n+k})$ be a local orthonormal frame on M such that $e_1, \cdots, e_n$ are tangent to M. We use the following index convention:

$$1 \leq A, B, C \leq n+k, \quad 1 \leq i, j, k \leq n, \quad n+1 \leq \alpha, \beta, \gamma \leq n+k.$$

Let w_A denote the dual coframe of e_A, and write

$$\nabla e_A = \sum_B w_{BA} e_B, \quad w_{AB} + w_{BA} = 0.$$

Then we have

$$\begin{aligned}
dX &= \sum_i w_i e_i,\\
dw_A &= -\sum_B w_{AB} \wedge w_B, \quad w_{AB} + w_{BA} = 0,\\
dw_{AB} &= -\sum_C w_{AC} \wedge w_{CB} + \sum_{CD} \tilde{R}_{ABCD}\, w_C \wedge w_D,
\end{aligned}$$

where $\tilde{R}_{ABCD}$ are the coefficients of the Riemann tensor of g. The first fundamental form of M is $\mathrm{I} = \sum_i w_i^2$. Let $w_{i\alpha} = \sum_j h^\alpha_{ij} w_j$. Since $w_\alpha = 0$, $dw_\alpha = -\sum_i w_{\alpha i} \wedge w_i = 0$. This implies that $h^\alpha_{ij} = h^\alpha_{ji}$. The second fundamental form and the normal connection are

$$\begin{aligned}
\mathrm{II} &= \sum_{i,\alpha} w_i w_{i\alpha} e_\alpha = \sum_{i,j,\alpha} h^\alpha_{ij} w_i w_j e_\alpha,\\
\nabla^\nu(e_\alpha) &= \sum_i w_{\beta\alpha} e_\beta
\end{aligned}$$

respectively. The normal curvature is the curvature of ∇^ν, i.e.,

$$\Omega^\nu_{\alpha\beta} = dw_{\alpha\beta} + \sum_\gamma w_{\alpha\gamma} \wedge w_{\gamma\beta}.$$

The normal bundle is *flat* if the normal connection is flat, i.e., $\Omega^\nu_{\alpha\beta} = 0$ for all α, β. The Levi-Civita connection of I is (w_{ij}), and the curvature is

$$\sum_{kl} R_{ijkl} w_k \wedge w_l = dw_{ij} + \sum_k w_{ik} \wedge w_{kj} = -\sum_\alpha w_{i\alpha} \wedge w_{\alpha j} + \tilde{R}_{ijkl}.$$

Note that given $\mathrm{I}, \mathrm{II}, \nabla^\perp$ is the same as given $w_i, w_{i\alpha}, w_{\alpha\beta}$. Moreover, the Levi-Civita connection of I can be obtained by solving

$$dw_i = -\sum_j w_{ij} \wedge w_j, \quad w_{ij} + w_{ji} = 0.$$

The Gauss-Codazzi equation is
(3.1.1)

$$\begin{cases} dw_{ij} + \sum_k w_{ik} \wedge w_{kj} = -\sum_\alpha w_{i\alpha} \wedge w_{\alpha j} + \sum_{kl} \tilde{R}_{ijkl} w_k \wedge w_l, \\ dw_{i\alpha} = -\sum_A w_{iA} \wedge w_{A\alpha} + \sum_{kl} \tilde{R}_{i\alpha kl} w_k \wedge w_l, \\ dw_{\alpha\beta} + \sum_\gamma w_{\alpha\gamma} \wedge w_{\gamma\beta} = -\sum_i w_{\alpha i} \wedge w_{i\beta} + \sum_{ij} \tilde{R}_{\alpha\beta ij} w_i \wedge w_j. \end{cases}$$

The Fundamental Theorem for Submanifolds states that I, II and ∇^ν together with the Gauss-Codazzi equation (3.1.1) determine the submanifold M up to isometries of N.

The mean curvature vector field is defined as the trace of II with respect to I, i.e.,

$$H = \mathrm{tr}_{\mathrm{I}} \mathrm{II} = \sum_{i\alpha} h^\alpha_{ii} e_\alpha.$$

The normal bundle $\nu(M)$ is said to be *non-degenerate* if the dimension of the space of all shape operators of M, $\{A_v \mid v \in \nu(M)_p\}$ is equal to $\mathrm{codim}(M)$ for all $p \in M$.

If $X : M \to N$ is a submanifold of a space form N^{n+k}, then the frame $F = (X, e_1, \cdots, e_{n+k})$ given above is a lift of X to $\mathrm{Iso}(N)$ and the Gauss-Codazzi equation for M is exactly the flatness of $F^{-1}dF$. When M satisfies certain geometric conditions, we often can find special coordinates and frames F on M so that $F^{-1}dF$ takes a special simple form. If moreover, such submanifolds admit a natural holomorphic deformation, then the Gauss-Codazzi equation for M is likely to be an integrable system. On the other hand, if the Lax n-tuple of the U/U_0-system can be interpreted as the connection 1-form of a submanifold, then we can read its geometry from the Lax n-tuple. This gives a natural method to find interesting submanifolds whose equations are integrable. We have had some success when U is an orthogonal group or a unitary group, but very little is known for other simple Lie groups U.

3.2. Orthogonal coordinates and the $\frac{O(2n)}{O(n)\times O(n)}$-system

A local coordinate system $(x_1, \cdots, x_n)$ of $\mathbb{R}^n$ is called an *orthogonal coordinate system* if the flat metric written in this coordinate system is diagonal, i.e., of the form $\sum_{i=1}^n g_{ii}(x)dx_i^2$. The theory of orthogonal coordinate systems of $\mathbb{R}^n$ was studied extensively by classical differential geometers (cf. Darboux [24]).

An elementary computation gives:

Proposition 3.2.1. *The Levi-Civita connection 1-form* (w_{ij}) *of the metric* $ds^2 = \sum_{i=1}^n b_i^2 dx_i^2$ *is*

$$w_{ij} = \frac{(b_i)_{x_j}}{b_j} dx_i - \frac{(b_j)_{x_i}}{b_i} dx_j.$$

So the Levi-Civita connection 1-form w of $ds^2 = \sum_{i=1}^n b_i^2 dx_i^2$ written in matrix form is

$$w = (w_{ij}) = \delta F - F^t \delta, \quad \text{where} \quad f_{ij} = \begin{cases} \frac{(b_i)_{x_j}}{b_j}, & \text{if } i \neq j \\ f_{ii} = 0, & \text{if } 1 \leq i \leq n. \end{cases}$$

Let $gl_*(n)$ denote the space of $\xi = (\xi_{ij}) \in gl(n, \mathbb{R})$ such that $\xi_{ii} = 0$ for $1 \leq i \leq n$. Recall that $F = (f_{ij}) : \mathbb{R}^n \to gl_*(n)$ is a solution of the $\frac{O(2n)}{O(n)\times O(n)}$-system (2.5.3) if and only if both $\delta F - F^t \delta$ and $\delta F^t - F\delta$ are flat connection 1-forms. Note that both connections have the same form as the Levi-Civita connections of orthogonal metrics. In the rest of the section, we try to answer the following question: Are there orthogonal coordinate systems of $\mathbb{R}^n$ whose Levi-Civita connections are $\delta F - F^t \delta$ and $\delta F^t - F\delta$?

Given $F : \mathbb{R}^n \to gl_*(n)$, there is a diagonal metric whose Levi-Civita connection 1-form is

$$w = \delta F - F^t \delta$$

if and only if there exist positive functions $b_1, \cdots, b_n$ so that

$$(b_i)_{x_j} = f_{ij} b_j, \quad i \neq j. \tag{3.2.1}$$

However, if system (3.2.1) is solvable for $b_1, \cdots, b_n$, then the mixed derivatives must be equal. This implies that

$$(f_{ij})_{x_k} = f_{ik} f_{kj}, \quad i, j, k \text{ distinct}. \tag{3.2.2}$$

It is a classical result that this condition is also sufficient for (3.2.1) to be solvable:

Theorem 3.2.2. *Given a smooth function $F = (f_{ij}) : \mathbb{R}^n \to gl_*(n)$, system (3.2.1) is solvable for $(b_1, \cdots, b_n)$ if and only if F satisfies (3.2.2). Moreover, given n smooth one variable functions $b_1^0, \cdots, b_n^0$, there exists a unique local solution $(b_1, \cdots, b_n)$ of (3.2.1) such that $b_i(0, \cdots, x_i, 0, \cdots, 0) = b_i^0(x_i)$.*

Corollary 3.2.3. *The space of local n-dimensional orthogonal metrics that have the same Levi-Civita connection 1-form is parametrized by n smooth positive functions of one variable.*

If F is a solution of the $\frac{O(2n)}{O(n)\times O(n)}$-system (2.5.3), then F is a solution of (3.2.2). So by Theorem 3.2.2 we can construct orthogonal coordinates of $\mathbb{R}^n$, whose Levi-Civita connections are $\delta F - F^t\delta$ and $\delta F^t - F\delta$. Therefore we have

Proposition 3.2.4. *Let $F = (f_{ij})$ be a solution of the $\frac{O(2n)}{O(n)\times O(n)}$-system (2.5.3), $\tau_1 = \delta F^t - F\delta$, $\tau_2 = \delta F - F^t\delta$, and $a_1^0, \cdots, a_n^0, b_1^0, \cdots, b_n^0$ smooth positive functions of one variable. Then there exist unique flat local orthogonal metrics $g_1 = \sum_{i=1}^n a_i^2(x)dx_i^2$ and $g_2 = \sum_{i=1}^n b_i(x)^2 dx_i^2$ such that*

(i) $a_i(0, \cdots, x_i, 0, \cdots) = a_i^0(x_i)$ and $b_i(0, \cdots, x_i, 0, \cdots) = b_i^0(x_i)$,

(ii) the Levi-Civita connection 1-form for g_1 and g_2 are τ_1 and τ_2 respectively,

(iii) there exist $O(n)$-valued maps $A = (\xi_1, \cdots, \xi_n)$ and $B = (\eta_1, \cdots, \eta_n)$ such that $A^{-1}dA = \tau_1$ and $B^{-1}dB = \tau_2$,

(iv) there exist ϕ and ψ defined on a neighborhood of the origin in $\mathbb{R}^n$ such that

$$d\phi = \sum_i a_i\eta_i dx_i, \quad d\psi = \sum_i b_i\xi_i dx_i,$$

(v) ϕ and ψ are local orthogonal coordinates on $\mathbb{R}^n$ with Levi-Civita connection τ_1 and τ_2 respectively.

The next theorem states that a subclass of orthogonal coordinate systems of $\mathbb{R}^n$ can be obtained using trivializations of τ_1 and τ_2.

Theorem 3.2.5. *Let $F = (f_{ij})$ be a solution of (2.5.3), and $A = (a_{ij})$, $B = (b_{ij})$ smooth $O(n)$-valued maps defined on an simply connected domain $\mathcal{O}$ of $\mathbb{R}^n$ satisfying*

$$A^{-1}dA = \delta F^t - F\delta, \quad B^{-1}dB = \delta F - F^t\delta. \tag{3.2.3}$$

If a_{mj}, b_{mj} never vanish on $\mathcal{O}$ for all $1 \le j, m \le n$, then:

(i) $ds_m^2 = a_{m1}^2 dx_1^2 + \cdots + a_{mn}^2 dx_n^2$ is a flat metric with $\delta F - F^t \delta$ as its Levi-Civita connection,

(ii) $d\tilde{s}_m^2 = b_{m1}^2 dx_1^2 + \cdots + b_{mn}^2 dx_n^2$ is a flat metric with $\delta F^t - F\delta$ as its Levi-Civita connection,

(iii) there exists a smooth map $X : \mathcal{O} \to gl(n, \mathbb{R})$ such that $dX = B\delta A^t$,

(iv) the m-th column X_m and the m-th row Y_m of X are local orthogonal coordinates for $\mathbb{R}^n$ such that the standard metric on $\mathbb{R}^n$ written in these coordinates are ds_m^2 and $d\tilde{s}_m^2$ respectively.

Proof. Let ξ_i denote the i-th column of A. W claim that

$$(\xi_j)_{x_k} = f_{jk}\xi_k, \quad j \neq k. \tag{3.2.4}$$

Note that (3.2.3) gives

$$\xi_i \cdot d\xi_j = f_{ji}dx_i - f_{ij}dx_j, \quad i \neq j. \tag{3.2.5}$$

This implies

$$(\xi_j)_{x_k} \cdot \xi_i = 0, \quad \text{if } i, j, k \text{ distinct.}$$

Since $\xi_j \cdot \xi_j = 1$, $(\xi_j)_{x_k} \cdot \xi_j = 0$. By (3.2.5), $\xi_k \cdot (\xi_j)_{x_k} = f_{jk}$. This proves (3.2.4). Equate each coordinate of (3.2.4) to get

$$(a_{mj})_{x_k} = f_{jk}a_{mk}, \quad 1 \le m \le n, \ j \neq k.$$

By Proposition 3.2.1, the Levi-Civita connection of ds_m^2 is

$$\frac{(a_{mj})_{x_k}}{a_{mk}} dx_j - \frac{(a_{mk})_{x_j}}{a_{mj}} dx_k = f_{jk}dx_j - f_{kj}dx_k,$$

i.e., the Levi-Civita connection 1-form for ds_m^2 is $\delta F - F^t\delta$. This proves (i). A similar argument gives (ii).

Since F is a solution of (2.5.3),

$$\theta_\lambda = \begin{pmatrix} \delta F^t - F\delta & -\delta\lambda \\ \delta\lambda & \delta F - F^t\delta \end{pmatrix}$$

is flat. Let $h = \begin{pmatrix} A & 0 \\ 0 & B \end{pmatrix}$. Then $h^{-1}dh = \theta_0 = \begin{pmatrix} \delta F^t - F\delta & 0 \\ 0 & \delta F - F^t\delta \end{pmatrix}$. The gauge transformation of θ_λ by h is

$$\Theta_\lambda = h\theta_\lambda h^{-1} - dhh^{-1} = \begin{pmatrix} 0 & -\lambda A\delta B^t \\ \lambda B\delta A^t & 0 \end{pmatrix}.$$

Since θ_λ is flat for all $\lambda \in \mathbb{C}$, $\Theta_\lambda = h * \theta_\lambda$ is flat for all λ, i.e., $d\Theta_\lambda = -\Theta_\lambda \wedge \Theta_\lambda$. This gives

$$\lambda \begin{pmatrix} 0 & -d\zeta^t \\ d\zeta & 0 \end{pmatrix} + \lambda^2 \begin{pmatrix} -\zeta^t \wedge \zeta & 0 \\ 0 & -\zeta \wedge \zeta^t \end{pmatrix} = 0,$$

where $\zeta = B\delta A^t$. Compare coefficients of λ to get $d\zeta = 0$. Since $\mathcal{O}$ is simply connected, there exists X such that

$$dX = B\delta A^t. \tag{3.2.6}$$

This proves (iii). Equate the m-th column and row of (3.2.6) to get

$$dX_m = B(a_{m1}dx_1, \cdots, a_{mn}dx_n)^t. \tag{3.2.7}$$

Recall that $A = (\xi_1, \cdots, \xi_n)$. Write equation (3.2.7) using columns of A and B to get $dX_m = \sum_{i=1}^n a_{mi}dx_i\eta_i$. Let η_i denote the i-th row of X, and $Y = X^t$. Then $dY = A\delta B^t$ and $dY_m = \sum_{i=1}^n b_{mi}dx_i\xi_i$. This proves (iv). Q.E.D.

3.3. Flat submanifolds and the $\frac{O(2n)}{O(n)\times O(n)}$-system

The $\frac{O(2n)}{O(n)\times O(n)}$-system can also be viewed as the Gauss-Codazzi equations for flat n-dimensional submanifolds in $\mathbb{R}^{2n}$ with flat and non-degenerate normal bundles. In fact, there is an isomorphism from the space of local n-dimensional flat submanifolds in $\mathbb{R}^{2n}$ with flat and non-degenerate normal bundle modulo rigid motions to the space of $(F, c_1, \cdots, c_n)$, where F is a local solution of the $\frac{O(2n)}{O(n)\times O(n)}$-system (2.5.3) and $c_1, \cdots, c_n$ are positive functions of one variable. We state this more precisely in the following two known theorems (cf. [68]):

Theorem 3.3.1. *Let M^n be an n-dimensional flat submanifold of $\mathbb{R}^{2n}$ with flat and non-degenerate normal bundle. Then there exist local coordinates $x_1, \cdots, x_n$, a parallel normal frame $e_{n+1}, \cdots, e_{2n}$, an $O(n)$-valued map $A = (a_{ij})$, and a map $b = (b_1, \cdots, b_n)$ such that the fundamental forms of M are*

$$\begin{cases} \mathrm{I} = \sum_{i=1}^n b_i^2 dx_i^2, \\ \mathrm{II} = \sum_{i,j=1}^n b_i a_{ji} dx_i e_{n+j}. \end{cases} \tag{3.3.1}$$

Moreover, let $f_{ij} = (b_i)_{x_j}/b_j$ for $1 \le i \ne j \le n$, $f_{ii} = 0$ for $1 \le i \le n$, and $F = (f_{ij})$. Then F is a solution of the $\frac{O(2n)}{O(n)\times O(n)}$-system (2.5.3).

Theorem 3.3.2. *Let F be a solution of the $\frac{O(2n)}{O(n)\times O(n)}$-system* (2.5.3), *and $b_1^0, \cdots, b_n^0$ be n smooth positive functions of one variable. Then there exist an open subset $\mathcal{O}$ of the origin in $\mathbb{R}^n$, smooth maps $A : \mathcal{O} \to O(n)$ and*

$$\phi = \begin{pmatrix} g & X \\ 0 & 1 \end{pmatrix} : \ \mathcal{O} \to GL(2n+1, \mathbb{R}) \tag{3.3.2}$$

with $g : \mathcal{O} \to O(2n)$, $X : \mathcal{O} \to \mathbb{R}^{2n}$, and $b_1, \cdots, b_n : \mathcal{O} \to \mathbb{R}$ such that

$$\begin{aligned} &A^{-1}dA = \delta F^t - F\delta, \\ &\phi^{-1}d\phi = \begin{pmatrix} 0 & -A\delta & 0 \\ \delta A^t & \delta F - F^t\delta & \varpi \\ 0 & 0 & 0 \end{pmatrix}, \\ &b_i(0, \cdots, x_i, 0, \cdots) = b_i^0(x_i), \quad 1 \le i \le n, \end{aligned}$$

where $\varpi = (b_1 dx_1, \cdots, b_n dx_n)^t$. Moreover,

(i) X is an immersion of a flat n-dimensional submanifolds of $\mathbb{R}^{2n}$ with flat and non-degenerate normal bundle,

(ii) $g = (e_{n+1}, \cdots, e_{2n}, e_1, \cdots, e_n)$ is a local orthonormal frame for X such that $e_{n+1}, \cdots, e_{2n}$ are parallel normal field,

(iii) $b_i(0, \cdots, x_i, 0, \cdots, 0) = b_i^0(x_i)$ for $1 \le i \le n$,

(iv) the fundamental forms of the immersion X are given as in (3.3.1),

(v) the Levi-Civita connection for the induced metric is $\delta F - F^t\delta$.

É Cartan proved that a flat n-dimensional submanifold cannot be locally isometrically immersed in S^{n+k} if $k < n-1$, but can be locally isometrically immersed into S^{2n-1}. Moreover, the normal bundle of a flat n-dimensional submanifold of S^{2n-1} is flat, and is non-degenerate viewed as a submanifold of $\mathbb{R}^{2n}$. By Theorem 3.3.1, flat n-dimensional submanifolds in S^{2n-1} give rise to solutions of the $\frac{O(2n)}{O(n)\times O(n)}$-system (2.5.3). This gives the following theorem of Tenenblat ([66]):

Theorem 3.3.3. *([66]). Let $X : M^n \to S^{2n-1}$ be an immersion of a flat submanifold. Then there exist local coordinates $x_1, \cdots, x_n$, a parallel normal frame $e_{n+2}, \cdots, e_{2n}$, and a smooth $O(n)$-valued map $A = (a_{ij})$ such that*

$$\mathrm{I} = \sum_{i=1}^{n} a_{1i}^2 dx_i^2, \quad \mathrm{II} = \sum_{i=1, j=2}^{n} a_{1i}a_{ji} dx_i e_{n+j}.$$

Set $f_{ij} = (a_{1i})_{x_j}/a_{1j}$ *for* $i \neq j$, $f_{ii} = 0$, *and* $F = (f_{ij})$. *Let* $e_i = X_{x_i}/a_{1i}$ *for* $1 \leq i \leq n$, *and* $g = (X, e_{n+2}, \cdots, e_{2n}, e_1, \cdots, e_n)$. *Then* F *is a solution of the* $\frac{O(2n)}{O(n)\times O(n)}$*-system* (2.5.3) *and*

$$g^{-1}dg = \begin{pmatrix} 0 & -A\delta \\ \delta A^t & -F^t\delta + \delta F \end{pmatrix}. \tag{3.3.3}$$

Conversely, let $F = (f_{ij})$ *be a solution of* (2.5.3), $A = (a_{ij})$ *an* $O(n)$*-valued map such that* $A^{-1}dA = \delta F^t - F\delta$ *and* $g \in O(2n)$ *a solution of* (3.3.3). *If* $a_{ij} > 0$ *for all* $1 \leq j \leq n$ *on an open subset* $\mathcal{O}$ *of* $\mathbb{R}^n$, *then the* i*-th column of* g *is an immersion of a flat submanifold in* S^{2n-1} *and the corresponding solution of* (2.5.3) *is* F.

Corollary 3.3.4. *Let* $M^n \subset S^{2n-1}$ *be a flat submanifold,* ξ *a parallel normal field such that the shape operator* A_ξ *is non-degenerate. Then* ξ *is an immersion of a flat* n*-submanifold in* S^{2n-1}. *Moreover, the solution of* (2.5.3) *corresponding to* ξ *is the same as the one corresponding to* M.

3.4. Egoroff metrics and the $\frac{U(n)}{O(n)}$-system

The $\frac{U(n)}{O(n)}$-system is the restriction of the $\frac{O(2n)}{O(n)\times O(n)}$-system to the subspace of symmetric real $n \times n$ matrices F. We have seen that each solution F of the $\frac{O(2n)}{O(n)\times O(n)}$-system gives rises to an $o(n)$-connection of some flat diagonal metric. In this section, we show that such a diagonal metric takes a special form:

Proposition 3.4.1. *Let* $ds^2 = \sum_{i=1}^n b_i^2 dx_i^2$ *be a metric,* $f_{ij} = (b_i)_{x_j}/b_j$ *for* $1 \leq i \neq j \leq n$, *and* $F = (f_{ij})$. *Then* $F = F^t$ *if and only if there exists a function* ϕ *such that* $b_i^2 = \phi_{x_i}$ *for all* $1 \leq i \leq n$.

Proof. Since $f_{ij} = \frac{(b_i)_{x_j}}{b_j}$, $F^t = F$ if and only if $\frac{(b_i)_{x_j}}{b_j} = \frac{(b_j)_{x_i}}{b_i}$, $i \neq j$. This is equivalent to $(b_i^2)_{x_j} = (b_j^2)_{x_i}$ for all $i \neq j$. Q.E.D.

Definition 3.4.2. An *Egoroff* metric is a flat metric of the form $\sum_{i=1}^n \phi_{x_i} dx_i^2$ for some smooth function ϕ.

It follows from Proposition 3.2.4, Theorem 3.2.5, and Proposition 3.4.1 that:

Theorem 3.4.3. *Let* F *be a solution of the* $\frac{U(n)}{O(n)}$*-system* (2.5.5), *and* $a_1, \cdots, a_n$ *smooth positive functions of one variable. Then there* *exists a smooth function ϕ defined on a simply connected open subset $\mathcal{O}$ of $\mathbb{R}^n$ such that*

$$\phi_{x_i}(0, \cdots, 0, x_i, 0, \cdots, 0) = a_i^2(x_i)$$

for $1 \leq i \leq n$ and the Levi-Civita connection for $\sum_{i=1}^n \phi_{x_i} dx_i^2$ is $[\delta, F]$. Moreover, let $A = (a_{ij})$ be an $O(n)$-valued map such that $A^{-1}dA = [\delta, F]$. Then:

(i) $ds_m^2 = \sum_{i=1}^n a_{mi}^2 dx_i^2$ is an Egoroff metric with $[\delta, F]$ as its Levi-Civita connection,

(ii) there exists a smooth map X from $\mathcal{O}$ to the space of symmetric matrices such that $dX = A\delta A^t$,

(iii) the m-th column X_m of X is a local orthogonal coordinate system for $\mathbb{R}^n$ and the flat metric of $\mathbb{R}^n$ written in this coordinate system is ds_m^2 as in (i).

3.5. Flat Lagrangian submanifolds and the $\frac{U(n)}{O(n)}$-system

In this section, we explain the relation between solutions of the $\frac{U(n)}{O(n)}$-system and the Gauss-Codazzi equations for flat, Lagrangian submanifolds of $\mathbb{R}^{2n}$. If these submanifolds also lie in S^{2n-1}, then they are invariant under the S^1-action of the Hopf fibration. Hence the projection of these submanifolds are flat Lagrangian submanifolds of $\mathbb{C}P^{n-1}$. For more details of the geometry of flat Lagrangian submanifolds of $\mathbb{C}P^{n-1}$ see [20].

Let $\langle\ ,\ \rangle$ and w be the standard inner product and symplectic form on $\mathbb{C}^n = \mathbb{R}^{2n}$ respectively, i.e.,

$$\langle X, Y\rangle = \mathrm{Re}(\bar{X}^t Y), \quad w(X, Y) = \mathrm{Im}(\bar{X}^t Y), \quad X, Y \in \mathbb{C}^n.$$

Write $Z \in \mathbb{C}^n$ as $Z = X + iY \in \mathbb{R}^n + i\mathbb{R}^n$, and $A \in gl(n, \mathbb{C})$ as $A = B + iC$ with $B, C \in gl(n, \mathbb{R})$. Then $A \in gl(n, \mathbb{C})$ is identified as $\begin{pmatrix} B & -C \\ C & B \end{pmatrix}$ in $gl(2n, \mathbb{R})$. This identifies $u(n)$ as the following subalgebra of $o(2n)$:

$$u(n) = \left\{ \begin{pmatrix} B & -C \\ C & B \end{pmatrix} \in o(2n) \,\middle|\, B \in o(n), C \in gl(n, \mathbb{R}) \text{ symmetric} \right\}.$$

The standard complex structure on $\mathbb{R}^{2n}$ is

$$J \begin{pmatrix} X \\ Y \end{pmatrix} = \begin{pmatrix} -Y \\ X \end{pmatrix}.$$

Definition 3.5.1. An n-dimensional submanifold M of $\mathbb{C}^n = \mathbb{R}^{2n}$ is *Lagrangian* if $w(v_1, v_2) = 0$ for all $v_1, v_2 \in TM$, or equivalently, $J(TM) = \nu(M)$.

The proposition below follows from the definition of Lagrangian submanifold:

Proposition 3.5.2. *Let $X : M^n \to \mathbb{R}^{2n}$ be a Lagrangian submanifold, and $(e_1, \cdots, e_n)$ a local orthonormal tangent frame. Then $(Je_1, \cdots, Je_n)$ is an orthonormal normal frame. Moreover, if*

$$g = (Je_1, \cdots, Je_n, e_1, \cdots, e_n)$$

then $g^{-1}dg$ is a $u(n)$-valued 1-form, i.e., it is of the form $\begin{pmatrix} \xi & -\eta \\ \eta & \xi \end{pmatrix}$, where ξ is an $o(n)$-valued 1-form and η is a 1-form with values in the space of symmetric matrices. Conversely, if M^n has a local orthonormal frame $g = (e_{n+1}, \cdots, e_{2n}, e_1, \cdots, e_n)$ such that $e_1, \cdots, e_n$ are tangent to M and $g^{-1}dg$ is $u(n)$-valued 1-form, then M is Lagrangian.

Proposition 3.5.3. *Let $F = (f_{ij})$ be the solution of the $\frac{O(2n)}{O(n)\times O(n)}$-system (2.5.3) corresponding to the flat n-submanifold M of $\mathbb{R}^{2n}$ with flat and non-degenerate normal bundle as in Theorem 3.3.1. Then the following statements are equivalent:*

(i) F is a solution of the $\frac{U(n)}{O(n)}$-system (2.5.5),

(ii) $F = F^t$,

(iii) M is Lagrangian.

Proof. It is obvious that (i) and (ii) are equivalent. Let $x_1, \cdots, x_n$, $b_1, \cdots, b_n$, $e_{n+1}, \cdots, e_{2n}$, and $A = (a_{ij})$ be as in Theorem 3.3.1. Let $e_i = \frac{X_{x_i}}{b_i}$, and $g = (e_{n+1}, \cdots, e_{2n}, e_1, \cdots, e_n)$. Then

$$g^{-1}dg = \begin{pmatrix} 0 & -A\delta \\ \delta A^t & [\delta, F] \end{pmatrix}.$$

To prove (ii) implies (iii), let

$$\phi = (\tilde{e}_{n+1}, \cdots, \tilde{e}_{2n}, e_1, \cdots, e_n) = g \begin{pmatrix} A & 0 \\ 0 & \mathrm{I} \end{pmatrix}.$$

Then $\tilde{e}_{n+1}, \cdots, \tilde{e}_{2n}$ are normal to X, and

$$\phi^{-1}d\phi = \begin{pmatrix} [\delta, F] & -\delta \\ \delta & [\delta, F] \end{pmatrix}. \tag{3.5.1}$$

Proposition 3.5.2 implies that M is Lagrangian.

(iii) implies (ii): Since M is Lagrangian and $(e_{n+1}, \cdots, e_{2n})$ is an orthonormal normal frame, $(Je_{n+1}, \cdots Je_{2n})$ is an orthonormal tangent frame for M. So there exists an $O(n)$-valued map h such that

$$\tilde{g} := (e_{n+1}, \cdots, e_{2n}, Je_{n+1}, \cdots, Je_{2n}) = g \begin{pmatrix} I & 0 \\ 0 & h^{-1} \end{pmatrix}.$$

Then

$$\tilde{g}^{-1} d\tilde{g} = \begin{pmatrix} 0 & -A\delta h^{-1} \\ h\delta A^t & h(\delta F - F^t\delta)h^{-1} - dhh^{-1} \end{pmatrix}.$$

But $\tilde{g}^{-1} d\tilde{g}$ is $u(n)$-valued. Hence

$$\begin{cases} h(\delta F - F^t\delta)h^{-1} - dhh^{-1} = 0, \\ A\delta h^t = h\delta A^t \end{cases} \tag{3.5.2}$$

The second equation of (3.5.2) gives $a_{ik}h_{jk} = h_{ik}a_{jk}$ for all i, j, k. This implies that the i-th rows (or i-th columns) of A and h are proportional. Since both A and h are in $O(n)$, we have $h = A$. So $h^{-1}dh = A^{-1}dA = \delta F - F^t\delta$. But $A^{-1}dA = \delta F^t - F\delta$. Hence

$$\delta F^t - F\delta = \delta F - F^t\delta.$$

Equate the ij-th entry of the above equation to get $f_{ji}dx_i - f_{ij}dx_j = f_{ij}dx_i - f_{ji}dx_j$. So F is symmetric. Q.E.D.

As a consequence of Proposition 3.5.3, Theorem 3.3.3, and Corollary 3.3.4, we have

Corollary 3.5.4. *Let F be a solution of the $\frac{U(n)}{O(n)}$-system (2.5.5), $A = (a_{ij})$ an $O(n)$-valued map satisfying $A^{-1}dA = [\delta, F]$, and $\tilde{g}$ an $U(n)$-valued map satisfying*

$$\tilde{g}^{-1} d\tilde{g} = \begin{pmatrix} 0 & -A\delta A^t \\ A\delta A^t & 0 \end{pmatrix}.$$

(Here $U(n)$ is embedded as a subgroup of $O(2n)$). Let e_{m+n} denote the m-th column of $\tilde{g}$ for $1 \leq m \leq n$. If $a_{m1}, \cdots, a_{mn}$ never vanishes in an open subset $\mathcal{O}$ of $\mathbb{R}^n$, then $e_{n+m} : \mathcal{O} \to S^{2n-1}$ is an n-dimensional immersed flat submanifold of S^{2n-1} that is Lagrangian in $\mathbb{R}^{2n}$. Conversely, if M^n is a flat submanifold of S^{2n-1} that is Lagrangian in $\mathbb{R}^{2n}$, then F defined in Theorem 3.3.3 is a solution of the $\frac{U(n)}{O(n)}$-system (2.5.5).

Proposition 3.5.5. *Let F be a solution of the $\frac{U(n)}{O(n)}$-system (2.5.5), M^n a flat submanifold of S^{2n-1} corresponding to F as in Corollary 3.5.4, and $\pi : S^{2n-1} \to \mathbb{C}P^{n-1}$ the Hopf fibration. Then $M = \pi^{-1}(\pi(M))$ and $\pi(M)$ is a flat Lagrangian submanifold of $\mathbb{C}P^{n-1}$.*

Proof. Let S^1 act on $\mathbb{R}^{2n} = \mathbb{C}^n$ by

$$e^{is} \cdot (z_1, \cdots, z_n) = (e^{is} z_1, e^{is} z_2, \cdots, e^{is} z_n).$$

This action leaves S^{2n-1} invariant, the orbit space S^{2n-1}/S^1 is $\mathbb{C}P^{n-1}$, and the projection $\pi : S^{2n-1} \to \mathbb{C}P^{n-1}$ is the Hopf fibration.

It suffices to show that M is invariant under the S^1-action on S^{2n-1}. Let X be the immersion, $(x_1, \cdots, x_n)$, $g = (X, e_{n+2}, \cdots, e_{2n}, e_1, \cdots, e_n)$, and $A = (a_{ij})$ as in Theorem 3.3.3. First we change coordinates from $x_1, \cdots, x_n$ to $t_1, \cdots, t_n$ such that

$$\begin{cases} x_1 = t_1 - t_2 - \cdots - t_n \\ x_j = t_j + t_1, \ 2 \leq j \leq n. \end{cases}$$

Then $\frac{\partial}{\partial t_1} = \frac{\partial}{\partial x_1} + \cdots \frac{\partial}{\partial x_n}$. Since

$$A^{-1} \frac{\partial A}{\partial t_1} = [\delta, F] \left(\frac{\partial}{\partial t_1} \right) = [\mathrm{I}_n, F] = 0,$$

we have $\frac{\partial A}{\partial t_1} = 0$. Here I_n is the identity $n \times n$ matrix. Let $\tilde{g} = g \begin{pmatrix} \mathrm{I}_n & 0 \\ 0 & A^t \end{pmatrix}$. Since $A^{-1} dA = [\delta, F]$,

$$\tilde{g}^{-1} d\tilde{g} = \begin{pmatrix} 0 & -A\delta A^t \\ A\delta A^t & 0 \end{pmatrix}.$$

So we have

$$\tilde{g}^{-1} \frac{\partial \tilde{g}}{\partial t_1} = \begin{pmatrix} 0 & -I \\ I & 0 \end{pmatrix} \tag{3.5.3}$$

This implies that

$$\tilde{g}(t_1, \cdots, t_n) = e^{it_1} \tilde{g}(1, t_2, \cdots, t_n).$$

But the first column of g and of $\tilde{g}$ are the immersion X. So X is invariant under the S^1-action on S^{2n-1}. Q.E.D.

It follows from elementary submanifold theory that $\tilde{M}$ is a flat Lagrangian submanifold of $\mathbb{C}P^{n-1}$ if and only if $\pi^{-1}(\tilde{M})$ is a flat submanifold of S^{2n-1} that is Lagrangian in $\mathbb{R}^{2n}$. Hence the Gauss-Codazzi equations for flat, Lagrangian submanifolds of $\mathbb{C}P^{n-1}$ is the $\frac{U(n)}{O(n)}$-system (2.5.5), or equivalently the $\frac{SU(n)}{SO(n)}$-system.

3.6. Bonnet pairs in $\mathbb{R}^3$ and the $\frac{GL(2,\mathbb{H})}{(SU(2)\times\mathbb{R}^+)^2}$-system

Let $X : M \to \mathbb{R}^3$ be an immersion. Locally, there exists a conformal coordinate system (x, y), i.e., the induced metric is of the form $\mathrm{I} = e^u(dx^2 + dy^2)$ for some smooth function u. Let H denote the mean curvature function of M. Since $\mathrm{II} - H\mathrm{I}$ is traceless, there is a smooth complex valued function $h = h_1 + ih_2$ such that

$$\mathrm{II} - H\,\mathrm{I} = h_1(dx^2 - dy^2) - 2h_2 dx dy = \mathrm{Re}(h dz^2),$$

where $z = x + iy$. The two fundamental forms of M are

$$(3.6.1)\qquad \begin{cases} \mathrm{I} = e^u(dx^2 + dy^2), \\ \mathrm{II} = H\,\mathrm{I} + \mathrm{Re}(h(dx + idy)^2) \\ \quad\;\; = (He^u + h_1)dx^2 - 2h_2 dx dy + (He^u - h_1)dy^2. \end{cases}$$

Let $e_1 = Xe^{-\frac{u}{2}}$, $e_2 = Xe^{-\frac{u}{2}}$, and $e_3 = e_1 \times e_2$, where $\times$ is the cross-product. Let w_1, w_2, w_3 be the dual coframe: $w_1 = e^{\frac{u}{2}}dx$, $w_2 = e^{\frac{u}{2}}dy$, $w_3 = 0$. Let $g = (e_1, e_2, e_3)$, and $(w_{ij}) = g^{-1}dg$, i.e., $de_i = \sum_{j=1}^3 w_{ji}e_j$, $1 \le i \le 3$. Then

$$(3.6.2)\qquad \begin{cases} w_{12} = \frac{1}{2}(u_y dx - u_x dy), \\ w_{13} = (He^{\frac{u}{2}} + h_1 e^{-\frac{u}{2}})dx - h_2 e^{-\frac{u}{2}}dy, \\ w_{23} = -h_2 e^{-\frac{u}{2}}dx + (He^{\frac{u}{2}} - h_1 e^{-\frac{u}{2}})dy, \end{cases}$$

The Gauss-Codazzi equations for M express the flatness of (w_{ij}), i.e.,

$$dw_{ij} = -\sum_{k=1}^{3} w_{ik} \wedge w_{kj}, \quad i \neq j.$$

Write this equation in terms of $u, H, h = h_1 + ih_2$ to get

(3.6.3)

$$\begin{cases} u_{xx} + u_{yy} = -2\left(H^2 e^u - (h_1^2 + h_2^2)e^{-u}\right), \\ (He^{\frac{u}{2}} + h_1 e^{-\frac{u}{2}})_y + (h_2 e^{-\frac{u}{2}})_x = \frac{1}{2}\left(u_y(He^{\frac{u}{2}} - h_1 e^{-\frac{u}{2}}) - u_x h_2 e^{-\frac{u}{2}}\right), \\ (He^{\frac{u}{2}} - h_1 e^{-\frac{u}{2}})_x + (h_2 e^{-\frac{u}{2}})_y = \frac{1}{2}\left(u_x(He^{\frac{u}{2}} + h_1 e^{-\frac{u}{2}}) - u_y h_2 e^{-\frac{u}{2}}\right). \end{cases}$$

A surface M in $\mathbb{R}^3$ is called *isothermic* if there exists a conformal line of curvature coordinate system, i.e., there is a coordinate system (x, y) such that both I and II are diagonalized, or equivalently, $h_2 = 0$ in (3.6.1). In this case, the Gauss-Codazzi equations (3.6.3) become

$$\begin{cases} u_{xx} + u_{yy} = -2\left(H^2 e^u - h_1^2 e^{-u}\right), \\ (He^{\frac{u}{2}} + h_1 e^{-\frac{u}{2}})_y = \frac{1}{2}u_y(He^{\frac{u}{2}} - h_1 e^{-\frac{u}{2}}), \\ (He^{\frac{u}{2}} - h_1 e^{-\frac{u}{2}})_x = \frac{1}{2}u_x(He^{\frac{u}{2}} + h_1 e^{-\frac{u}{2}}). \end{cases}$$

This implies that (u, p_1, q_1) is a solution of (2.5.13), where $p_1 = \frac{ih_1}{2}\, e^{-u/2}$ and $q_1 = \frac{H}{2}\, e^{u/2}$.

A pair of surfaces $(M, \tilde{M})$ in $\mathbb{R}^3$ is called a *Bonnet pair* if there is an isometry $f : M \to \tilde{M}$ so that $\tilde{H} = H \circ f$, where $\tilde{H}$ and H are the mean curvature functions of M and $\tilde{M}$ respectively and H is not a constant function. The following is a consequence of the Gauss-Codazzi equation (cf. [8]):

Proposition 3.6.1. *([8]). Let $(M, \tilde{M})$ be a Bonnet pair in $\mathbb{R}^3$. Then away from umbilic points there exist a conformal coordinate system (x, y), and smooth real functions u, h_1 and h_2 such that the two fundamental forms for $M, \tilde{M}$ are as follows:*

$$\text{(3.6.4a)} \qquad \begin{cases} \mathrm{I} = e^u(dx^2 + dy^2), \\ \mathrm{II} = H\ \mathrm{I} + \mathrm{Re}(h(dx + idy)^2) \\ \quad = (He^u + h_1)dx^2 - 2h_2 dxdy + (He^u - h_1)dy^2, \end{cases}$$

$$\text{(3.6.4b)} \qquad \begin{cases} \tilde{\mathrm{I}} = e^u(dx^2 + dy^2), \\ \tilde{\mathrm{II}} = H\ \tilde{\mathrm{I}} + \mathrm{Re}(\bar{h}(dx + idy)^2) \\ \quad = (He^u + h_1)dx^2 + 2h_2 dxdy + (He^u - h_1)dy^2. \end{cases}$$

Since both (u, H, h_1, h_2) and $(u, H, h_1, -h_2)$ are solutions of (3.6.3), we get

$$\text{(3.6.5)} \qquad \begin{cases} u_{xx} + u_{yy} = 2(-H^2 e^u + (h_1^2 + h_2^2)e^{-u}), \\ (he^{u/2} + h_1 e^{-u/2})_y = \frac{u_y}{2}(He^{u/2} - h_1 e^{-u/2}, \\ (h_2 e^{-u/2})_x = -\frac{u_x}{2} h_2 e^{-u/2}, \\ (He^{u/2} - h_1 e^{-u/2})_x = \frac{u_x}{2}(He^{u/2} + h_1 e^{-u/2}, \\ (h_2 e^{-u/2})_y = -\frac{u_y}{2}(h_2 e^{-u/2}. \end{cases}$$

Note that the third and the fifth equations of (3.6.5) imply $(h_2)_x = (h_2)_y = 0$. So h_2 is a constant, and we have

Theorem 3.6.2. *([8]). Let $(M, \tilde{M})$ be a Bonnet pair in $\mathbb{R}^3$, and (u, H, h_1, h_2) the corresponding solution of (3.6.5). Then h_2 is a constant. Moreover, set $p_1 = \frac{i}{2}(h_1 - ih_2)e^{-\frac{u}{2}}$, and $q_1 = \frac{1}{2}He^{\frac{u}{2}}$. Then (u, p_1, q_1) is a solution of the $\frac{GL(2,\mathbb{H})}{(\mathbb{R}^+ \times SU(2))^2}$-system (2.5.11). Conversely, if (u, p_1, q_1) is a solution of system (2.5.11) and q_1 is real, then there is a Bonnet pair with fundamental forms given by (3.6.4), where $H = 2q_1e^{-u/2}$ and $h_1 - ih_2 = -2i\, p_1e^{u/2}$.*

3.7. Curved flats in symmetric spaces

Let U/U_0 be a rank n Riemannian symmetric space, σ the corresponding involution on U, $\mathcal{U} = \mathcal{U}_0 + \mathcal{U}_1$ the eigenspace decomposition of $d\sigma_e$ on $\mathcal{U}$ corresponding to eigenvalues 1 and -1, $\mathcal{A}$ a maximal abelian linear subspace of $\mathcal{U}_1$, and $a_1, \cdots, a_n$ an orthonormal basis of $\mathcal{A}$. In this section, we associate to each solution of the U/U_0-system (2.5.1) a flat submanifold in $\mathcal{U}_1$. We also review the construction of curved flats in U/U_0 given by Ferus and Pedit [35].

Theorem 3.7.1. *Let $v : \mathbb{R}^n \to \mathcal{U}_1 \cap \mathcal{A}^\perp$ be a solution of the U/U_0-system (2.5.1), and $E(x, \lambda)$ the frame of the corresponding Lax n-tuple θ_λ (2.5.2), i.e.,*

$$E^{-1}dE = \theta_\lambda = \sum_{i=1}^{n}(a_i\lambda + [a_i, v])dx_i, \quad E(x, 0) = e,$$

Set $Y = \frac{\partial E}{\partial \lambda}E^{-1}\big|_{\lambda=0}$. Then Y is an immersed flat submanifold in $\mathcal{U}_1$ such that the tangent plane of Y is a maximal abelian subalgebra of $\mathcal{U}_1$ at every point. Conversely, locally all such flat submanifolds in $\mathcal{U}_1^0$ can be constructed this way, where $\mathcal{U}_1^0$ is the subset of regular points in $\mathcal{U}_1$.

Proof. Write $E_\lambda(x) = E(x,\lambda)$. Since $E^{-1}dE = \sum_i (a_i\lambda + [a_i, v])dx_i$, a direct computation gives

$$\begin{aligned}
dY &= \left(\frac{\partial}{\partial\lambda}(dE)\right) E^{-1} - \frac{\partial E}{\partial\lambda} E^{-1} dE E^{-1}\Big|_{\lambda=0} \\
&= \frac{\partial}{\partial\lambda}\left(E\left(\sum_i a_i\lambda + [a_i, v]\right) dx_i\right) E^{-1}\Big|_{\lambda=0} \\
&\quad - Y E_0 \left(\sum_i [a_i, v]\ dx_i\right) E_0^{-1} \\
&= \sum_i (Y E_0 [a_i, v] E_0^{-1} + E_0 a_i E_0^{-1} - Y E_0 [a_i, v] E_0^{-1})\ dx_i \\
&= \sum_{i=1}^n E_0 a_i E_0^{-1} dx_i.
\end{aligned}$$

Because θ_λ satisfies the U/U_0-reality condition, $\tau(E_\lambda) = E_{\bar\lambda}$ and $\sigma(E_\lambda) = E_{-\lambda}$. So $E_0 \in U_0$. Let $e_i = E_0\ a_i\ E_0^{-1}$. Since $a_1, \cdots, a_n$ are orthonormal and $E_0(x) \in U_0$, $\{e_i \mid 1 \le i \le n\}$ is an orthonormal tangent frame of Y. Hence Y is an immersion, and the induced metric is $\sum_{i=1}^n dx_i^2$.

Since $\mathrm{ad}(a_1)^2, \cdots, \mathrm{ad}(a_n)^2$ are commuting symmetric operators, there exist a set Λ of linear functionals of $\mathcal{A}$, an orthonormal common basis $\{p_\alpha \mid \alpha \in \Lambda\}$ for $\mathcal{A}^\perp \cap \mathcal{U}_1$, and an orthonormal basis $\{k_\alpha \mid \alpha \in \Lambda\}$ for $\mathcal{K} \cap \mathcal{K}_a^\perp$ such that $\mathrm{ad}(a)(p_\alpha) = \alpha(a) = k_\alpha$, $\mathrm{ad}(a)(k_\alpha) = -\alpha(a)p_\alpha$, for all $1 \le i \le n$ and $\alpha \in \Lambda$. Then $e_\alpha = E_0 p_\alpha E_0^{-1}$ is an orthonormal normal frame for the immersion Y in $\mathcal{U}_1$. Write the solution v of the U/U_0-system as $v(x) = \sum_\alpha v_\alpha(x) p_\alpha$ with respect to the decomposition $\mathcal{A}^\perp \cap \mathcal{U}_1 = \sum_\alpha \mathbb{R}\ p_\alpha$. Since $dE_0 = E_0 \sum_i [a_i, v] dx_i$, a direct computation gives

$$\begin{aligned}
de_i &= E_0 [E_0^{-1} dE_0, a_i] E_0^{-1} = \sum_{i,j} E_0 [[a_j, v], a_i] E_0^{-1} dx_j, \\
&= -\sum_{j,\alpha} v_\alpha \alpha(a_i) \alpha(a_j) dx_j e_\alpha.
\end{aligned}$$

Hence $w_{i\alpha} = v_\alpha \alpha(a_i) \sum_j \alpha(a_j) dx_j$. So the normal curvature $\sum_\alpha w_{i\alpha} \wedge w_{j\alpha}$ is zero.

To prove the converse, let M be a flat submanifold of $\mathcal{U}_1$ such that TM_p is a maximal abelian subalgebra of $\mathcal{U}_1$. Let x be a local flat, orthonormal coordinate of M, and $e_i = \frac{\partial}{\partial x_i}$ the orthonormal frame. Let $\mathcal{A}$ be a maximal abelian subspace of $\mathcal{U}_1$. Then every maximal abelian subspace of $\mathcal{U}_1$ is of the form $k\mathcal{A}k^{-1}$ for some $k \in U_0$. Since $M \subset \mathcal{U}_1^0$, we may assume that there exist $\mathcal{A}$-valued maps ξ_i and a U_0-valued map g such that $e_i = g\xi_i g^{-1}$. It follows from $(de_i, e_j) = 0$ and the fact that $(\, , \,)$ is ad-invariant that $d\xi_i = 0$ for all i. So $\xi_i = a_i$ is constant. In other words, $e_i = ga_i g^{-1}$. Note that if h is an A-valued map, then $e_i = ga_i g^{-1} = gha_i h^{-1} g^{-1}$. Choose an A-valued map h so that $(dh)h^{-1} = -\pi(g^{-1}dg)$, where π is the projection onto $\mathcal{U}_{\mathcal{A}}^{\perp} \cap \mathcal{U}_0$. Let $\tilde{g} = gh$. Then $e_i = \tilde{g}a_i\tilde{g}^{-1}$ and $\tilde{g}^{-1}d\tilde{g} \in \mathcal{U}_{\mathcal{A}}^{\perp} \cap \mathcal{U}_0$. Let X be the immersion of M into $\mathcal{U}_1$. Then $dX = \sum_{i=1}^n e_i dx_i = \sum_{i=1}^n \tilde{g}a_i\tilde{g}^{-1}dx_i$. Hence $(\tilde{g}a_i\tilde{g}^{-1})_{x_j} = (\tilde{g}a_j\tilde{g}^{-1})_{x_i}$ for all i, j. This implies $[\tilde{g}^{-1}\tilde{g}_{x_j}, a_i] = [\tilde{g}^{-1}\tilde{g}_{x_i}, a_j]$. So there exists an $\mathcal{A}^{\perp} \cap \mathcal{U}_1$-valued map v such that $\tilde{g}^{-1}\tilde{g}_{x_i} = [a_i, v]$. But this means $\sum_{i=1}^n [a_i, v]dx_i$ is flat and v is a solution of (2.5.1). Q.E.D.

Given an involution σ of U, there is a natural U-action on U defined by $g * x = gx\sigma(g)^{-1}$. The orbit at e is

$$M = \{g\sigma(g)^{-1} \mid g \in U\}.$$

Since the isotropy subgroup at e is U_0, the orbit M is diffeomorphic to U/U_0. Next we claim that M is totally geodesic. To see this, note that the map $f(g) = (\sigma(g))^{-1}$ is an isometry of U. So the fixed point set F of f is a totally geodesic submanifold of U. Note that $df_e = -d\sigma_e$. So $TF_e = \mathcal{U}_1$, and the dimension of F is equal to $\dim(\mathcal{U}_1)$. But M is fixed by f and $TM_e = \{x - d\sigma_e(x) \mid x \in \mathcal{U}\} = \mathcal{U}_1$. So M is an open subset of F. This proves the claim. This is the classical Cartan embedding of the symmetric space U/U_0 in U as a totally geodesic submanifold.

Note that U_0 acts on U/U_0 $(g\cdot(hU_0) = ghU_0)$. An element $x \in U/U_0$ is *regular* if the U_0-orbit at x is a principal orbit.

Theorem 3.7.2. ([35]). *With the same assumption as in Theorem 3.7.1, set $\psi(x) = E(x,1)E(x,-1)^{-1}$. Then ψ is an immersed flat submanifold of the symmetric space U/U_0 which is tangent to a flat of U/U_0 at every point. Conversely, locally all such flat submanifolds in N' can be constructed this way, where N' is the open dense subset of regular points in U/U_0.*

Proof. The reality condition implies that $E(x,1) \in U$ and

$$\psi(x) = E(x,1)E(x,-1)^{-1} = E(x,1)\sigma(E(x,1))^{-1}.$$

So the image of ψ lies in the symmetric space $U/U_0 = \{g\sigma(g)^{-1} \mid g \in U\}$. A direct computation gives

$$\psi^{-1}d\psi = 2\sum_{i=1}^{n} E_{-1}a_i E_{-1}^{-1}dx_i.$$

Thus ψ is a flat immersion into U/U_0 and $2(x_1, \cdots, x_n)$ is an orthonormal coordinate for the induced metric. The rest of the theorem can be proved in a similar manner as for Theorem 3.7.1. Q.E.D.

Ferus and Pedit called the flat submanifolds obtained in Theorem 3.7.2 *curved flats*.

3.8. Indefinite affine spheres in $\mathbb{R}^3$ and the -1-flow

Affine geometry (cf. [55]) studies the geometry of hypersurfaces in $\mathbb{R}^{n+1}$ invariant under the affine transformations $x \mapsto Ax + v$, where $A \in SL(n+1, \mathbb{R})$ and $v \in \mathbb{R}^{n+1}$. There are three local affine invariants, the affine metric, the Fubini cubic form, and the third fundamental form. These invariants satisfy certain integrability conditions, the Gauss-Codazzi equations. We first give a brief description of these invariants for affine surfaces in $\mathbb{R}^3$, then explain the relation between the Tzitzeica equation and indefinite affine spheres. Recall that the Tzitzeica equation is the -1-flow associated to $SL(3, \mathbb{R})/\mathbb{R}^+$ (see Example 2.6.3).

Let $X : M \to \mathbb{R}^3$ be a surface with non-degenerate second fundamental form, $g = (e_1, e_2, e_3)$ a local frame on M such that e_1, e_2 are tangent to M, e_3 is transversal to M, and $\det(e_1, e_2, e_3) = 1$. Let w^i denote the dual coframe of e_i, i.e., $dX = w^1 e_1 + w^2 e_2$. Let (w^i_j) denote the $sl(3, \mathbb{R})$-valued 1-form $g^{-1}dg$, i.e., $de_i = \sum_{j=1}^3 w_i^j e_j, \quad 1 \le i \le 3$. Then we have the structure equation:

$$(3.8.1) \qquad \begin{cases} dw^i = -\sum_{j=1}^3 w^i_j \wedge w^j = \sum_{j=1}^3 w^j \wedge w^i_j, & 1 \le i \le 2 \\ dw^j_i = -\sum_{k=1}^3 w^j_k \wedge w^k_i, & 1 \le i, j \le 3. \end{cases}$$

Since $w^3 = 0$ on M,

$$(3.8.2) \qquad w_i^3 = \sum_{j=1}^{2} h_{ij} w^j, \quad h_{ij} = h_{ji}.$$

A direct computation shows that the quadratic form

$$ds^2 = |\det(h_{ij})|^{-\frac{1}{4}} \sum_{ij} h_{ij} w^i w^j \tag{3.8.3}$$

is invariant under change of affine frames, and it is called the *affine metric* of M. An affine surface is called *definite* or *indefinite* if the affine metric is definite or indefinite respectively.

We can choose a vector field e_3 transversal to M so that

$$w_3^3 + \frac{1}{4} d(\log |\det(h_{ij})|) = 0. \tag{3.8.4}$$

Then

$$\nu = |\det(h_{ij})|^{\frac{1}{4}} e_3$$

is an affine invariant. The vector field ν is called the *affine normal* of M.

Take the exterior derivative of (3.8.2) to get

$$\sum_{j=1}^{2} dh_{ij} + h_{ij} w_3^3 + \sum_{k=1}^{2} (h_{ik} w_j^k + h_{kj} w_i^k) \wedge w^j = 0 \tag{3.8.5}$$

for $1 \leq j \leq 2$. Define h_{ijk} by

$$\sum_{k=1}^{2} h_{ijk} w^k = dh_{ij} + h_{ij} w_3^3 + \sum_{k=1}^{2} h_{ik} w_j^k + h_{kj} w_i^k. \tag{3.8.6}$$

Then (3.8.5) implies that $h_{ijk} = h_{ikj}$. But $h_{ij} = h_{ji}$. So h_{ijk} is symmetric in i, j, k. *The Fubini-Pick cubic form,*

$$J = \sum_{i,j,k} h_{ijk} w^i w^j w^k,$$

is an affine invariant.

Exterior differentiation of (3.8.4) gives $\sum_i w_3^i \wedge w_i^3 = 0$. Write $w_3^i = \sum_j \ell^{ij} w_j^3$. The *third fundamental form*,

$$\mathrm{III} = h^{\frac{1}{4}} w_3^i w_i^3,$$

is also an affine invariant. The trace of III with respect to the affine metric ds^2 is the *affine mean curvature*

$$L = \frac{1}{2} |\det(h_{ij})|^{\frac{1}{4}} \sum_{ij} h_{ij} \ell^{ij}.$$

The three affine invariants ds^2, J and III are completely determined by w^i and w_B^A, which satisfy the Gauss-Codazzi equations for affine surfaces. Conversely, suppose ds^2, J and III are given and satisfy the Gauss-Codazzi equations. Then h_{ij}, h_{ijk}, w^i, w_i^3, w_3^i, and w_3^3 can be computed from these three invariants. Moreover, we can find w_i^j by solving the linear system consisting of (3.8.6) and the first equation of (3.8.1). Then the Gauss-Codazzi equations, written in terms of w^i, w_B^A, are (3.8.1), i.e., the connection

$$\Omega = \begin{pmatrix} w_B^A & \tau \\ 0 & 0 \end{pmatrix}$$

is flat, where $\tau = (w^1, w^2, 0)^t$. Hence there exists

$$\psi = \begin{pmatrix} g & X \\ 0 & 1 \end{pmatrix}$$

such that $\psi^{-1} d\psi = \Omega$, where $g = (e_1, e_2, e_3) \in SL(3, \mathbb{R})$ and $X \in \mathbb{R}^3$. It follows that X is an immersion, e_1, e_2 are tangent to X, e_3 is the affine normal, and ds^2, J and III are the affine metric, Fubini-Pick form, and the third fundamental form for X respectively. This is the fundamental theorem for affine surfaces in $\mathbb{R}^3$.

A surface is called a *proper affine sphere* if there exists $p_0 \in \mathbb{R}^3$ such that the affine normal line $p + t\nu(p)$ passes through p_0 for all $p \in M$. We explain below the well-known fact (cf. [7]) that the equation for proper affine spheres with indefinite affine metric is the Tzitzeica equation (2.6.4).

Let w be a solution of the Tzitzeica equation (2.6.4), and θ_λ the corresponding Lax pair defined by (2.6.5), and $E(x, t, \lambda)$ the solution of

$$E^{-1} dE = \theta_\lambda, \quad E(0, 0, \lambda) = e.$$

(Here e is the identity matrix in $SL(3, \mathbb{R})$.) Fix a non-zero $r \in \mathbb{R}$, let $e_i(x, t)$ denote the i-th column of $E(x, t, r)$. We claim that $X = -e_3$ is an immersed indefinite affine sphere. To see this, we first note that

$$\theta_r = \begin{pmatrix} w_x dx & r^{-1} e^{-2w} dy & r dx \\ r dx & -w_x dx & r^{-1} e^w dy \\ r^{-1} e^w dy & r dx & 0 \end{pmatrix} = E(r)^{-1} dE(r).$$

Since θ_r is an $sl(3, \mathbb{R})$-valued flat 1-form, $E(r)$ is a map from $\mathbb{R}^2$ to $SL(3, \mathbb{R})$. Fix r, and let e_i denote $e_i(r)$. Equate each column of $dE(r) =$

$E(r)\theta_r$ to get

$$\begin{cases} dX = -de_3 = -rdx\ e_1 - r^{-1}e^w dy\ e_2, \\ de_1 = w_x dx\ e_1 + rdx\ e_2 + r^{-1}e^w dy\ e_3, \\ de_2 = r^{-1}e^{-2w}dy\ e_1 - w_x dx\ e_2 + rdx\ e_3. \end{cases}$$

This implies that e_1, e_2 are tangent to X,

$$\begin{aligned} &w^1 = -rdx, \quad w^2 = -r^{-1}e^w dy, \\ &w_1^3 = r^{-1}e^w dy = -w^2, \quad w_2^3 = rdx = -w^1, \\ &w_3^1 = -w^1, \quad w_3^2 = -w^2, \quad w_3^3 = 0. \end{aligned}$$

So $h_{11} = h_{22} = 0$, $h_{12} = -1$ and the affine metric is $2e^w dxdy$. Since $\det(h_{ij}) = -1$ and $w_3^3 = 0$, (3.8.4) is satisfied. Hence the affine normal is

$$\nu = |\det(h_{ij})|^{\frac{1}{4}} e_3 = e_3.$$

But $X = -e_3$ implies that all affine normal lines pass through the origin. In other words, X is an indefinite proper affine sphere.

Conversely, suppose X is an indefinite proper affine sphere in $\mathbb{R}^3$. We want to show that there exist a special coordinate system and a special affine frame so that the Gauss-Codazzi equation for X as an affine sphere is the Tzitzeica equation. First note that there exist a local asymptotic coordinate system (x, y) and a smooth function w such that the affine metric is

$$ds^2 = 2e^{2w} dxdy.$$

Let $e_1 = X_x$, $e_2 = X_y$, and e_3 parallel to the affine normal such that

$$\det(e_1, e_2, e_3) = 1.$$

Then

$$w^1 = dx, \quad w^2 = dy, \quad w_1^3 = e^{2w}dy, \quad w_2^3 = e^{2w}dx.$$

So $\det(h_{ij}) = -e^{4w}$. We may assume that all affine normal lines pass through the origin. So $X = fe_3$ for some function f. Exterior differentiation of $X = fe_3$ gives

$$w^1 e_1 + w^2 e_2 = df e_3 + f w_3^3 e_3 + f(w_3^1 e_1 + w_3^2 e_2).$$

Equate the coefficients of e_3 to get $df + fw_3^3 = 0$. Since e_3 is parallel to the affine normal, w_3^3 satisfies (3.8.4). Therefore $f = c \mid \det(h_{ij}) \mid ^{1/4} =$

ce^w for some constant c. By rescaling, we may assume $c = 1$. Equate coefficients of e_1 and e_2 to get

$$w_3^1 = e^{-w}dx, \quad w_3^2 = e^{-w}dy.$$

Therefore $\ell^{11} = \ell^{22} = 0$ and $\ell^{12} = \ell^{21} = e^{-3w}$. So the affine mean curvature is $L = 1$. Use (3.8.5) to get

$$w_1^2 = -\frac{h_{111}}{2}e^{-2w}dx, \quad w_2^1 = -\frac{h_{222}}{2}e^{-2w}dy.$$

Use $dw^i = -\sum_j w_j^i \wedge w^j$ to conclude that $w_1^1 = w_x dx, \quad w_2^2 = w_y dy$. Substitute w_A^B into $dw_j^i = -\sum_A w_A^i \wedge w_j^A$ for $i = 1, j = 2$ and $j = 1, i = 2$ to get

$$(h_{111}e^{-w})_y = 0, \quad (h_{222}e^{-w})_x = 0.$$

So $h_{111} = u_1(x)e^w$ and $h_{222} = u_2(y)e^w$ for some smooth function u_1, u_2 of one variable. By making a coordinate change to $(\tilde{x}, \tilde{y}) = (\tilde{x}(x), \tilde{y}(y))$, we may assume that

$$w_2^1 = e^{-w}dy, \quad w_1^2 = e^{-w}dx.$$

To summarize, we have shown that

$$g^{-1}dg = \begin{pmatrix} w_x dx & e^{-w}dy & e^{-w}dx \\ e^{-w}dx & w_y dy & e^{-w}dy \\ e^{2w}dy & e^{2w}dx & -dw \end{pmatrix},$$

where $g = (e_1, e_2, e_3)$. Change the frame g to $\tilde{g} = g\,\mathrm{diag}(1, e^{-w}, e^w)$. Then

$$\tilde{g}^{-1}d\tilde{g} = \begin{pmatrix} w_x dx & e^{-2w}dy & dx \\ dx & -w_x dx & e^w dy \\ e^w dy & dx & 0 \end{pmatrix}.$$

This is the Lax pair θ_λ (2.6.5) at $\lambda = 1$. So w is a solution of the Tzitzeica equation.

3.9. The -1 flow, hyperbolic system, and the sigma model

Let $\mathbb{R}^{1,1}$ denote the Lorentz space equipped with metric $2dxdt$. In this section, we discuss the relation between harmonic maps from $\mathbb{R}^{1,1}$ to a Lie group U and solutions of the -1-flow and the hyperbolic U-system.

First we recall a theorem of Uhlenbeck ([72]):

Theorem 3.9.1. *([72]). Let $s : \mathbb{R}^{1,1} \to U$ be a smooth map, $A = \frac{1}{2}s^{-1}s_x$, and $B = \frac{1}{2}s^{-1}s_y$. Then the following statements are equivalent:*

(i) *s is harmonic,*
(ii) $A_t = -B_x = [A, B]$,
(iii) $\Omega_\lambda = (1-\lambda)A\ dx + (1-\lambda^{-1})B\ dt$ *is flat for all* $\lambda \in \mathbb{C} \setminus 0$.

Corollary 3.9.2. *([72]). Suppose* $\theta_\lambda = (1-\lambda)Adx + (1-\lambda^{-1})Bdt$ *is flat for all* $\lambda \in \mathbb{C} \setminus 0$, *and* E_λ *satisfies* $E_\lambda^{-1}dE_\lambda = \theta_\lambda$. *Then* $s = E_{-1}$ *is a harmonic map from* $\mathbb{R}^{1,1}$ *into* U *such that* $s^{-1}ds = 2Adx + 2Bdt$.

The following Proposition is well-known:

Proposition 3.9.3. *Let* $i : N_0 \to N$ *be a totally geodesic submanifold of* N. *A smooth map* $s : M \to N_0$ *is a harmonic map if and only if* $i \circ s : M \to N$ *is a harmonic map.*

Proposition 3.9.4. *([68]). Let* $\mathcal{U}$ *be the real form of* $\mathcal{G}$ *defined by the involution* τ, $a, b \in \mathcal{U}$ *such that* $[a, b] = 0$, $g : \mathbb{R}^2 \to U$ *a solution of the* -1*-flow* (2.6.2) *associated to* U, *and* $E_\lambda(x,t) = E(x,t,\lambda)$ *the frame for the corresponding Lax pair* θ_λ (2.6.3), *i.e.,*

$$E^{-1}dE = (a\lambda + g^{-1}g_x)dx + \lambda^{-1}g^{-1}bg\ dt, \quad E(x,t,0) = e.$$

Then $s = E_{-1}E_1^{-1}$ *is a harmonic map from* $\mathbb{R}^{1,1}$ *to* U. *Moreover, if* σ *is an order* $k = 2m$ *automorphism such that* $\tau\sigma = \sigma^{-1}\tau$, $a \in \mathcal{U} \cap \mathcal{G}_1$, *and* $b \in \mathcal{U} \cap \mathcal{G}_{-1}$, *then* $s = E_{-1}E_1^{-1}$ *is a harmonic map from* $\mathbb{R}^{1,1}$ *to the symmetric space* U/H, *where* H *is the fixed point set of the involution* σ^m *and* $\mathcal{G}_j$ *is the eigenspace of* σ *on* $\mathcal{G}$.

Proof. Note that the gauge transformation of θ_λ by E_1 is $E_1 * \theta_\lambda = E_1\theta_\lambda E_1^{-1} - dE_1E_1^{-1} = (1-\lambda)E_1aE_1^{-1}\ dx + (1-\lambda^{-1})E_1g^{-1}bgE_1^{-1}\ dt$. Let $\psi_\lambda = E_\lambda E_1^{-1}$. Then $\psi_\lambda^{-1}d\psi_\lambda = E_1 * \theta_\lambda$. By Corollary 3.9.2, $s = \psi_{-1} = E_{-1}E_1^{-1}$ is a harmonic map from $\mathbb{R}^{1,1}$ to U and $s^{-1}s_x$, $s^{-1}s_t$ are conjugate to $2a$, $2b$ respectively.

If the order k of σ is 2, then $\tau\sigma = \sigma^{-1}\tau = \sigma\tau$ and σ leaves $\mathcal{U}$ invariant. Let K denote the fixed point set of σ in U, and $\mathcal{U} = \mathcal{K} + \mathcal{P}$ the decomposition of eigenspaces of σ on $\mathcal{U}$ with eigenvalues $1, -1$ respectively. Note that the reality condition is

$$\tau(\theta_{\bar{\lambda}}) = \theta_\lambda, \quad \sigma(\theta_\lambda) = \theta_{-\lambda}.$$

So E_λ satisfies the reality condition

$$\tau(E_{\bar{\lambda}}) = \theta_\lambda, \quad \sigma(E_\lambda) = E_{-\lambda}.$$

This implies that $s = E_{-1}E_1^{-1} = E_{-1}\sigma(E_{-1})^{-1}$. So the image of s lies in the totally geodesic submanifold $M = \{g\sigma(g)^{-1} \mid g \in U\}$ of U. But

M is the Cartan embedding of the symmetric space U/U_0 into U as a totally geodesic submanifold. It follows from Proposition 3.9.3 that s is a harmonic map to the symmetric space $M = U/K$.

If the order of σ is $k = 2m$, an even integer, then E_λ satisfies the (G, τ, σ)-reality condition

$$\tau(E_{\bar{\lambda}}) = E_\lambda, \quad \sigma(E_\lambda) = E_{e^{\frac{2\pi i}{2m}}\lambda}.$$

Hence we have $\sigma^m(E_\lambda) = E_{-\lambda}$. So $E_1 = \sigma^m(E_{-1})$, and $s = E_{-1}\sigma^m(E_{-1})^{-1}$. Therefore s is a harmonic map from $\mathbb{R}^{1,1}$ into the symmetric space U/H. Q.E.D.

Example 3.9.5. Let $G = SL(2, \mathbb{C})$, $\tau(\xi) = -\bar{\xi}^t$, and $\sigma(\xi) = -\xi^t$. Note that

$$SU(2) = \left\{ \begin{pmatrix} w & z \\ -\bar{z} & \bar{w} \end{pmatrix} \,\middle|\, z, w \in \mathbb{C}, |w|^2 + |z|^2 = 1 \right\} = S^3,$$

and the Cartan embedding of $SU(2)/SO(2)$ is the totally geodesic 2-sphere

$$\left\{ \begin{pmatrix} w & ir \\ ir & -\bar{w} \end{pmatrix} \,\middle|\, r \in \mathbb{R},\ z \in \mathbb{C},\ r^2 + |w|^2 = 1. \right\}$$

The -1-flow associated to $SU(2)/SO(2)$ is the SGE equation, so solutions of the SGE equation give rise to harmonic maps from $\mathbb{R}^{1,1}$ to S^2.

Example 3.9.6. Let $w : \mathbb{R}^2 \to \mathbb{R}$ be a solution of the Tzitzeica equation (2.6.4), the -1-flow associated to the homogeneous space $\frac{SL(3,\mathbb{R})}{\mathbb{R}^+}$ given in Example 2.6.3. For this example, the order of σ is 6. Let θ_λ be the corresponding Lax pair (2.6.5), and E_λ the frame of θ_λ. A direct computation shows that

$$\sigma^3(\xi) = -PA^tP, \quad \text{where } P = \begin{pmatrix} 0 & 1 & 0 \\ 1 & 0 & 0 \\ 0 & 0 & 1 \end{pmatrix}.$$

So the fixed point set of the involution σ^3 in $sl(3, \mathbb{R})$ is $so(2, 1)$, where $SO(2, 1)$ is the isometry group of the quadratic form $2x_1x_2 + x_3^2$ on $\mathbb{R}^3$. Hence $E_{-1}E_1^{-1}$ is a harmonic map from $\mathbb{R}^{1,1}$ to the symmetric space $\frac{SL(3,\mathbb{R})}{SO(2,1)}$.

The proof of Proposition 3.9.4 also implies

Proposition 3.9.7. *Let* $(u_0, u_1, v_0, v_1) : \mathbb{R}^2 \to \prod_{i=1}^4 \mathcal{U}$ *be a solution of the hyperbolic system* (2.7.1) *associated to* U*, and* $E(x,t,\lambda)$ *the frame of the corresponding Lax pair* (2.7.2). *Then* $s = E(\cdot,\cdot,-1)E(\cdot,\cdot,1)^{-1}$ *is a harmonic map from* $\mathbb{R}^{1,1}$ *to* U*. Moreover, if* σ *is an involution on* $\mathcal{U}$*,* $\mathcal{U} = \mathcal{U}_0 + \mathcal{U}_1$ *is the eigenspace decomposition, and* $u_0, v_0 \in \mathcal{U}_0$*,* $u_1, v_1 \in \mathcal{U}_1$*, then the image of* s *lies in the symmetric space* U/U_0 *(embedded in* U *via the Cartan embedding) and is a harmonic map from* $\mathbb{R}^{1,1}$ *to* U/U_0.

§4. Dressing actions and factorizations

Suppose that G is a Lie group, and that G_+, G_- are subgroups of G such that the multiplication maps $G_+ \times G_- \to G$ and $G_- \times G_+ \to G$ defined by $(g_+, g_-) \mapsto g_+g_-$ and $(g_-, g_+) \mapsto g_-g_+$ respectively are bijections. Thus given any $g \in G$, there exist unique $g_+ \in G_+$ and $g_- \in G_-$ so that $g = g_+g_-$, and unique $h_+ \in G_+$ and $h_- \in G_-$ such that $g = h_-h_+$. The dressing action of G_+ on G_- is defined as follows: Factor g_+g_- as $\tilde{g}_-\tilde{g}_+$ with $\tilde{g}_\pm \in G_\pm$. Then the dressing action of G_+ on G_- is $g_+ * g_- = \tilde{g}_-$. The dressing action of G_- on G_+ is defined similarly.

If the multiplication maps are injective and the images are open dense subsets of G, then the dressing actions are defined on an open neighborhood of the identity e in $G_\pm$. Moreover, the corresponding Lie algebra actions are well-defined.

There are two well known factorizations for a semi-simple Lie group, the Iwasawa and the Gauss factorizations. The analogous loop group factorizations are those given by Pressley and Segal in [57] and the Birkhoff factorization respectively. The dressing actions of these loop group factorizations play important roles in finding solutions and explaining the hidden symmetries of integrable systems. We will review these loop group factorizations.

4.1. Iwasawa and Gauss factorizations

Let G be a complex, semi-simple Lie group, U a maximal compact subgroup, and B a Borel subgroup. *The Iwasawa factorization* of G is $G = UB$, i.e., every $g \in G$ can be factored uniquely as ub, where $u \in U$ and $b \in B$. Let A be a maximal abelian subgroup of G, and N_+ and N_- the unipotent subgroups generated by the set of positive roots and negative roots with respect to a fixed simple root system of $\mathcal{A}$ respectively. The multiplication map from $N_- \times A \times N_+$ to G is injective and the image is an open dense subset of G, called a *big cell* of G. *The Gauss factorization* is the factorization of a big cell of G, i.e., every element g in the big cell can be factored uniquely as n_-an_+ with $n_\pm \in N_\pm$ and $a \in A$. Let $B_+ = AN_+$. We call the factorization of g in the big cell as n_-b_+ with $n_- \in N_-$ and $b_+ \in B_+$ again the Gauss factorization.

Example 4.1.1. Let $G = SL(n, \mathbb{C})$, $\triangle_+(n)$ the subgroup of upper triangular $g \in SL(n, \mathbb{C})$, and $\triangle_-(n)$ the subgroup of lower triangular matrix $g \in SL(n, \mathbb{C})$ with 1's on the the diagonal. The multiplication maps $\triangle_+(n) \times \triangle_-(n) \to SL(n, \mathbb{C})$ and $\triangle_-(n) \times \triangle_+(n) \to SL(n, \mathbb{C})$ are injective and the images are open and dense. Moreover, the factorization of $g \in SL(n, \mathbb{C})$ can be carried out using Gaussian elimination on rows and columns of g. This is the Gauss factorization of $SL(n, \mathbb{C})$.

Example 4.1.2. Let $G = SL(n, \mathbb{C})$, $U = SU(n)$, and $B_+(n)$ the subgroup of upper triangular matrices with real diagonal entries. The multiplication maps $B_+(n) \times U(n) \to SL(n, \mathbb{C})$ and $U(n) \times B_+(n) \to SL(n, \mathbb{C})$ are bijective. Moreover, the factorization can be carried out by applying the Gram-Schmidt process on rows and columns of g. This is the Iwasawa factorization of $SL(n, \mathbb{C})$.

4.2. Factorizations of loop groups

Let G be a complex, semi-simple Lie group, τ an involution of G that gives the compact real form U, and σ an order k automorphism of G. Let B be a Borel subgroup of G such that $G = UB$ is the Iwasawa factorization. Given an open subset $\mathcal{O}$ of S^2, let $\mathrm{Hol}(\mathcal{O}, G)$ denote the group all holomorphic maps $f : \mathcal{O} \to G$ with multiplication defined by $(fg)(\lambda) = f(\lambda)g(\lambda)$. Let $\epsilon > 0$,

$$
\begin{aligned}
&S^2 = \mathbb{C} \cup \{\infty\}, \quad \mathbb{C}^* = \{\lambda \in \mathbb{C} \mid \lambda \neq 0\}, \\
&\mathcal{O}_\epsilon = \{\lambda \in \mathbb{C} \mid |\lambda| < \epsilon\}, \quad \mathcal{O}_{1/\epsilon} = \{\lambda \in S^2 \mid |\lambda| > 1/\epsilon\}.
\end{aligned}
$$

To explain symmetries of soliton flows we need to consider the following groups:

$$
\begin{aligned}
\Lambda(G) &= \mathrm{Hol}(\mathbb{C} \cap \mathcal{O}_{1/\epsilon}, G), \\
\Lambda_+(G) &= \mathrm{Hol}(\mathbb{C}, G), \\
\Lambda_-(G) &= \{f \in \mathrm{Hol}(\mathcal{O}_{1/\epsilon}, G) \mid f(\infty) = e\}.
\end{aligned}
$$

The following is the Gauss loop group factorization (the Birkhoff factorization):

Theorem 4.2.1. The Gauss loop group factorization. *The multiplication maps from $\Lambda_+(G) \times \Lambda_-(G)$ and $\Lambda_-(G) \times \Lambda_+(G)$ to $\Lambda(G)$ are injective and the images are open and dense. In particular, there exists an open dense subset $\Lambda(G)_0$ of $\Lambda(G)$ such that given $g \in \Lambda(G)_0$, g can be factored uniquely as $g = g_+g_- = h_-h_+$ with $g_+, h_+ \in \Lambda_+(G)$ and $g_-, h_- \in \Lambda_-(G)$.*

Suppose $\tau\sigma = \sigma^{-1}\tau$. Let $\mathcal{G} = \mathcal{G}_0 + \cdots + \mathcal{G}_{k-1}$ be the eigenspace decomposition of σ, and $\mathcal{U}_j = \mathcal{U} \cap \mathcal{G}_j$. Let $\hat{\tau}$ and $\tilde{\sigma}$ be the automorphism of $\Lambda(G)$ defined by

$$\hat{\tau}(g)(\lambda) = \overline{\tau(g(\bar{\lambda}))}, \tag{4.2.1a}$$

$$\tilde{\sigma}(g)(\lambda) = \sigma(g(e^{-2\pi i/k}\,\lambda)). \tag{4.2.1b}$$

Let $\Lambda^\sigma(G)$ denote the fixed point set of $\tilde{\sigma}$ on $\Lambda(G)$. Since $\tau\sigma = \sigma^{-1}\tau$, a direct computation implies that $\tilde{\tau}$ leaves $\Lambda^\sigma(G)$ invariant. Let $\Lambda^{\tau,\sigma}(G)$ denote the subgroup of $g \in \Lambda(G)$ that is fixed by $\hat{\tau}$ and $\tilde{\sigma}$. Let $\Lambda^\tau(G)$ denote the subgroup of $\Lambda(G)$ fixed by $\hat{\tau}$. Then

$$\begin{aligned}
\Lambda^\tau(G) &= \{f \in \Lambda(G) \mid f \text{ satisfies } U\text{−reality condition (2.2.1)}\},\\
\Lambda^{\tau,\sigma}(G) &= \{f \in \Lambda(G) \mid f \text{ satisfies } U/U_0\text{−reality condition (2.4.1)}\},\\
\Lambda^\tau_\pm(G) &= \Lambda^\tau(G) \cap \Lambda_\pm(G),\\
\Lambda^{\tau,\sigma}_\pm(G) &= \Lambda^{\tau,\sigma}(G) \cap \Lambda_\pm(G).
\end{aligned}$$

Corollary 4.2.2. *Suppose $g \in \Lambda(G)$ is factored as $g = g_+g_-$ with $g_+ \in \Lambda_+(G)$ and $g_- \in \Lambda_-(G)$. If $\tau\sigma = \sigma^{-1}\tau$, then*

(i) *$g \in \Lambda^\tau(G)$ implies that $g_\pm \in \Lambda^\tau_\pm(G)$,*
(ii) *$g \in \Lambda^{\tau,\sigma}(G)$ implies that $g_\pm \in \Lambda^{\tau,\sigma}_\pm(G)$.*

To explain symmetries of the elliptic integrable systems, we need to consider the (G, τ)-reality condition

$$\tau(g(1/\bar{\lambda}) = g(\lambda), \tag{4.2.2}$$

and the following groups:

$$L(G) = C^\infty(S^1, G),$$
$$L_+(G) = \{f \in L(G) \mid f \text{ extends holomorphically to } \mid \lambda \mid\ < 1\},$$
$$L_e(U) = \{f \in C^\infty(S^1, U) \mid f(1) = e\},$$

$$\Omega^\tau(G) = \{f \in \mathrm{Hol}((\mathcal{O}_\epsilon \cup \mathcal{O}_{1/\epsilon}) \cap \mathbb{C}^*, G) \mid f \text{ satisfies the } (G,\tau)\text{−reality condition (4.2.2)}\},$$
$$\Omega^\tau_+(G) = \{f \in \Omega^\tau(G) \mid f \text{ extends holomorphically to } \mathbb{C}^*\},$$
$$\Omega^\tau_-(G) = \{f \in \Omega^\tau(G) \mid f \text{ extends holomorphically to } \mathcal{O}_\epsilon \cup \mathcal{O}_{1/\epsilon} \text{ and} f(\infty) = \mathrm{I}\}.$$

For the (G, τ, σ)-system, we also need the subgroup of $g \in \Omega^\tau(G)$ that satisfies the (G, τ, σ)-*reality condition*:

$$\tau(g(1/\bar{\lambda})) = g(\lambda), \quad \sigma(g(e^{-\frac{2\pi i}{k}}\lambda)) = g(\lambda). \tag{4.2.3}$$

Theorem 4.2.3. *([51]). The multiplication map from $\Omega^\tau_-(G) \times \Omega^\tau_+(G)$ to $\Omega^\tau(G)$ is a bijection.*

Corollary 4.2.4. *Given $g_+ \in \Omega^\tau_+(G)$ and $g_- \in \Omega^\tau_-(G)$, g_-g_+ can be factored uniquely as $\tilde{g}_+\tilde{g}_-$ with $\tilde{g}_+ \in \Omega^\tau_+(G)$ and $\tilde{g}_- \in \Omega^\tau_-(G)$. Moreover, if $\tau\sigma = \sigma\tau$ and g_+, g_- satisfy the (G, τ, σ)-reality condition* (4.2.3), *then $\tilde{g}_+ \in \Omega^{\tau,\sigma}_+(G)$ and $\tilde{g}_- \in \Omega^{\tau,\sigma}_-(G)$.*

Theorem 4.2.5. The Iwasawa loop group factorization *([57]). The multiplication maps $L_e(U) \times L_+(G) \to L(G)$ and $L_+(G) \times L_e(U) \to L(G)$ are bijections.*

§5. Symmetries of the U-hierarchy

Let τ be a conjugate linear involution of G, and U its fixed point set. In this Chapter, we assume U is compact. Let $\mathcal{A}$ be a maximal abelian subalgebra of $\mathcal{U}$. The U-hierarchy has three types of symmetries, i.e., three actions on the space of solutions of the (b, j)-flow in the U-hierarchy:

—An action of the infinite dimensional abelian algebra $\hat{\mathcal{A}}_+$ of polynomial maps from $\mathbb{C}$ to $\mathcal{A} \otimes \mathbb{C}$.

—An action of $\Lambda^\tau_-(G)$.

—An action of the subgroup of $f \in L_+^\tau(G)$ such that the infinite jet of $f_b - I$ at $\lambda = -1$ is 0, where $f = f_u f_b$ with $f_u \in U$, $f_b \in B$, and $G = UB$ is the Iwasawa factorization.

The first two symmetries arise naturally from the dressing actions of the factorization theorems given in section 4.2. The third action comes from the dressing action of a new factorization.

5.1. The action of $\Lambda_-^\tau(G)$

It follows from the Gauss loop group factorization 4.2.1 that the group $\Lambda_-^\tau(G)$ acts on $\Lambda_+^\tau(G)$ by local dressing action. This induces an action of $\Lambda_-^\tau(G)$ on the space of germs of solutions of the (b,j)-flow (2.1.3) in the U-hierarchy at the origin as follows: Let $u : \mathbb{R}^2 \to \mathcal{A}^\perp \cap \mathcal{U}$ be a solution of the (b,j)-th flow (2.1.3), θ_λ the corresponding Lax pair (2.1.5), and $E(x,t,\lambda)$ the frame of u, i.e., E is the solution of

$$\begin{cases} E^{-1}E_x = a\lambda + u, \\ E^{-1}E_t = \sum_{i=0}^{j} Q_{b,j-i}(u)\lambda^i, \\ E(0,\lambda) = e. \end{cases}$$

Then $E(x,t)(\lambda) = E(x,t,\lambda)$ is holomorphic in $\lambda \in \mathbb{C}$, i.e., $E(x,t) \in \Lambda_+(G)$ for all (x,t). Since θ_λ satisfies the U-reality condition $\tau(\theta_{\bar\lambda}) = \theta_\lambda$, $E(x,t)$ satisfies $\tau(E(x,t)(\bar\lambda)) = E(x,t)(\lambda)$. In other words, $E(x,t) \in \Lambda_+^\tau(G)$. Given $g \in \Lambda_-^\tau(G)$, by the Gauss loop group factorization 4.2.1 and Corollary 4.2.2 there is an open subset $\mathcal{O}$ of the origin in $\mathbb{R}^2$ such that the dressing action of g at $E(x,t)$ is defined for all $(x,t) \in \mathcal{O}$. Let $g * E(x,t)$ denote the dressing action of g at $E(x,t)$. This is obtained as follows: Factor $gE(x,t)$ as

$$gE(x,t) = \tilde{E}(x,t)\tilde{g}(x,t)$$

with $\tilde{E}(x,t) \in \Lambda_+^\tau(G)$, and $\tilde{g}(x,t) \in \Lambda_-^\tau(G)$. Then

$$g * E(x,t) = \tilde{E}(x,t).$$

Expanding $\tilde{g}(x)(\lambda)$ at $\lambda = \infty$ we have:

$$\tilde{g}(x,t)(\lambda) = I + g_1(x,t)\lambda^{-1} + g_2(x,t)\lambda^{-2} + \cdots.$$

The following results are known (cf. [70]):

(i) $\tilde{u} = u + [a, g_1]$ is again a solution of the (b,j)-flow.

(ii) $g * u = \tilde{u}$ defines an action of $\Lambda_-^\tau(G)$ on the space of local solutions of the (b, j)-flow.

(iii) $\tilde{E}(x, \lambda)$ is the frame of $\tilde{u}$.

(iv) Suppose $g \in \Lambda^\tau(G)$ is a rational map with only simple poles, i.e., g satisfies the U-reality condition and is of the form $g(\lambda) = I + \sum_{j=1}^k \frac{\xi_j}{\lambda - \alpha_j}$ for some $\xi_j \in \mathcal{G}$ and $\alpha_j \in \mathbb{C} \setminus \{0\}$. Then $g * u$ can be computed explicitly in terms of g, u, E.

(v) If U is compact, $g \in \Lambda^\tau(G)$ is rational, and u is a smooth solution defined on all $(x, t) \in \mathbb{R}^2$ that is rapidly decaying in x, then $g * u$ is also defined on all $\mathbb{R}^2$ and is rapidly decaying in x.

We claim that $\Lambda_-^{\tau,\sigma}(G)$ acts on the space of solutions of flows in the U/U_0-hierarchy. To see this, let E be the frame of a solution u of the $(b, mk+1)$-flow in the U/U_0-hierarchy. Then $E(x, t, \cdot) \in \Lambda_+^{\tau,\sigma}(G)$. By Corollary 4.2.2, $g * E$ satisfies the (G, τ, σ)-reality condition. Hence $g * u$ is a solution of the $(b, mk+1)$-flow in the U/U_0-hierarchy. We give an explicit example next.

Example 5.1.1. ([10]). Let $\tau(y) = \bar{y}$, and $\sigma(y) = \mathrm{I}_{n,n}\, y\, \mathrm{I}_{n,n}^{-1}$ be the involutions of $O(2n, \mathbb{C})$ that give the symmetric space $\frac{O(2n)}{O(n)\times O(n)}$ as in Example 2.5.4. Let $v = \begin{pmatrix} 0 & F \\ -F^t & 0 \end{pmatrix}$ be a solution of the $\frac{O(2n)}{O(n)\times O(n)}$-system (2.5.3), θ_λ the corresponding Lax n-tuple (2.5.4), and E the frame of v. We give an explicit construction of the action of certain rational map with two poles in $\Lambda_-^{\tau,\sigma}(G)$ on v below. Let W, Z be two unit vectors of $\mathbb{R}^n$, π the projection of $\mathbb{C}^{2n}$ onto the complex linear subspace spanned by $\begin{pmatrix} W \\ iZ \end{pmatrix}$, $s \in \mathbb{R}$ non-zero, and

$$h_{is,\pi}(\lambda) = \left(\pi + \frac{\lambda - is}{\lambda + is}(\mathrm{I} - \pi)\right)\left(\bar{\pi} + \frac{\lambda + is}{\lambda - is}(\mathrm{I} - \bar{\pi})\right). \tag{5.1.1}$$

A direct computation shows that $h_{is,\pi} \in \Lambda_-^{\tau,\sigma}(G)$. Then we have:

(1) $h_{is,\pi} E(x) = \tilde{E}(x) h_{is,\tilde{\pi}(x)}$, where $\tilde{\pi}(x)$ is the Hermitian projection onto the complex linear subspace spanned by

$$\begin{pmatrix} \tilde{W}(x) \\ i\tilde{Z}(x) \end{pmatrix} = E(x, -is)^{-1} \begin{pmatrix} W \\ iZ \end{pmatrix}.$$

(2) $(\tilde{W}, i\tilde{Z})^t$ is a solution of the following first order system:

$$\begin{pmatrix} \tilde{W} \\ i\tilde{Z} \end{pmatrix}_{x_j} = -(-is\, a_j + [a_j, v]) \begin{pmatrix} \tilde{W} \\ i\tilde{Z} \end{pmatrix}. \tag{5.1.2}$$

(3) Given v, system (5.1.2) is solvable for $\begin{pmatrix} \tilde{W} \\ i\tilde{Z} \end{pmatrix}$ if and only if v is a solution of (2.5.3).

(4) Let $\xi = (\xi_{ij})$, $\phi(\xi) = \xi - \sum_i \xi_{ii} e_{ii}$, and $\tilde{F} = F - 2s\, \phi(\tilde{\pi})$. Then

$$h_{is,\pi} * \begin{pmatrix} 0 & F \\ -F^t & 0 \end{pmatrix} = \begin{pmatrix} 0 & \tilde{F} \\ -\tilde{F}^t & 0 \end{pmatrix}.$$

$\tilde{E} = h_{is,\pi} * E$ is the frame of $\tilde{F} = h_{is,\pi} * F$.

5.2. The orbit $\Lambda_-^\tau(G) * 0$

Note that $u = 0$ is a trivial solution of the (b, j)-flow in the U-hierarchy and the corresponding Lax pair is $\theta_\lambda = a\lambda dx + b\lambda^j dt$. So the frame E^0 of $u = 0$ is $E^0(x, t, \lambda) = \exp(ax\lambda + b\lambda^j t)$. Given $g \in \Lambda_-^\tau(G)$, to compute $g * 0$, the first step is to factor

$$gE^0(x, t) = \tilde{E}(x, t)\tilde{g}(x, t), \tag{5.2.1}$$

with $\tilde{E}(x, t) \in \Lambda_+^\tau(G)$, $\tilde{g}(x, t) \in \Lambda_-^\tau(G)$. The second step is to expand $\tilde{g}(x, t)$ as

$$\tilde{g}(x, t)(\lambda) = \mathrm{I} + \tilde{g}_1(x, t)\lambda^{-1} + \tilde{g}_2(x, t)\lambda^{-2} + \cdots.$$

Then $g * 0 = [a, \tilde{g}_1]$ is a solution of the (b, j)-flow.

The orbit $\Lambda_-^\tau(G) * 0$ contains several interesting classes of solutions (cf. [69]):

(1) If $g \in \Lambda_-^\tau(G)$, then $g * 0$ is a local analytic solution of the (b, j)-flow.

(2) If $g \in \Lambda_-^\tau(G)$ is a rational map, then the factorization 5.2.1 can be carried out using residue calculus and linear algebra. In fact, $\tilde{g}$ can be given explicitly in terms of a rational function of exponentials. Hence $g * 0$ can be written explicitly as a rational function of exponentials. Moreover, $(g * 0)(x, t)$ is defined for all $(x, t) \in \mathbb{R}^2$, is rapidly decaying as $|x| \to \infty$ for each $t \in \mathbb{R}$, and is a pure soliton solution.

(3) If $g \in \Lambda_-^\tau(G)$ such that $g^{-1}(\lambda)ag(\lambda)$ is a polynomial in λ^{-1}, then $g * 0$ is a finite type solution, and $g * 0$ can be obtained either by solving a system of compatible first order differential equations or by algebraic geometric methods.

5.3. Rapidly decaying solutions

A global solution $u(x,t)$ of the (b,j)-flow in the U-hierarchy is called a *Schwartz class solution* if $u(x,t)$ is rapidly decaying as $|x| \to \infty$ for each $t \in \mathbb{R}$. The orbit $\Lambda_-^\tau(G) * 0$ contains soliton solutions, which are Schwartz class solutions. But most Schwartz class solutions of the (b,j)-flow do not belong to this orbit. We need to use a different loop group factorization than the ones given in section 4.2 to construct general Schwartz class solutions.

Since we assume U is the compact real form of G, there is a Borel subgroup B such that $G = UB$ (the Iwasawa factorization). Let SD^τ denote the group of meromorphic maps $g : \mathbb{C} \setminus \mathbb{R} \to G$ that satisfying the following conditions:

(a) g has an asymptotic expansion $g(\lambda) \sim I + g_1\lambda^{-1} + g_2\lambda^{-2} + \cdots$ at $\lambda = \infty$,

(b) g satisfies the U-reality condition (2.2.1),

(c) $\lim_{s\to 0^\pm} g(r+is) = g_\pm(r)$ is smooth,

(d) $h_+ - I$ is in the Schwartz class, where $g_+ = v_+h_+$ with $v_+ \in U$ and $h_+ \in B$. The U-reality condition implies that $g_- = \tau(g_+)$.

Let D_-^τ denote the subgroup of $g \in SD^\tau$ that is holomorphic in $\mathbb{C} \setminus \mathbb{R}$. Then D_-^τ is isomorphic to the subgroup of $f \in L_+(G)$ such that $f - \mathrm{I}$ vanishes up to infinite order at $\lambda = -1$. To see this, we consider the following linear fractional transformation

$$\lambda = \phi(z) = i(1-z)/(1+z). \tag{5.3.1}$$

Note that ϕ has the following properties:

(i) ϕ maps the unit circle $|z| = 1$ to the real axis,
(ii) $\phi(-1) = \infty$,
(iii) ϕ maps the unit disk $|z| < 1$ to the upper half plane.

Theorem 5.3.1. *([69]). Let $g \in D_-^\tau$, ϕ the linear fractional transformation defined by* (5.3.1), *and $\Phi(g)(z) = g(\phi(z))$ for $|z| \neq 1$. Then:*

(1) Φ is one to one,

(2) $\Phi(g_+)(e^{i\theta})$ *and* $\Phi(g_-)(e^{i\theta})$ *are the limit of* $\Phi(g)(z)$ *as* $z \to e^{i\theta}$ *with* $|z| < 1$ *and* $|z| > 1$ *respectively. Moreover,* $\Phi(g_+) \in L_+^\tau(G)$.

(3) $\Phi(g)$ *satisfies the* (G, τ)*-reality condition,* $\tau(\Phi(g)(1/\bar{\lambda})) = \Phi(g)(\lambda)$.

(4) Let $G = UB$ *be the Iwasawa factorization of* G*. Factor* $\Phi(g) = f_u f_b$ *with* $f_u \in U$ *and* $f_b \in B$*. Then the infinite jet of* $f_b - \mathrm{I}$ *at* $z = -1$ *is zero.*

The above Theorem identifies $g \in D_-^\tau$ with $\Phi(g) \in L_+^\tau(G)$, whose B-component is equal to the identity up to infinite order at $z = -1$. Recall that the frame of the trivial solution $u = 0$ of the (b, j)-flow in the U-hierarchy is $E^0(x, t)(\lambda) = e^{ax\lambda + b\lambda^j t}$. Note that $\Phi(E^0(x, t))(z)$ is smooth for all $z \in S^1$ except at $z = -1$, where it has an essential singularity. So we *cannot* use the dressing action from the Gauss factorization $L(G) = L_e(U)L_+(G)$ and the identification Φ to induce an action of D_-^τ at $u = 0$. However, we can still factor gE^0 as $E\tilde{g}$ with $E \in \Lambda_+^\tau(G)$ and $\tilde{g} \in D_-^\tau$. Intuitively speaking, the essential singularity at $z = -1$ is compensated by the infinite flatness of $\tau(\Phi(g_+))^{-1}\Phi(g_+)(z)$ at $z = -1$.

Theorem 5.3.2. *([69]). There is an open dense subset* $\mathcal{D}$ *of* SD^τ *such that if* $g \in \mathcal{D}$ *then for each* $(x, t) \in \mathbb{R}^2$*, We can factor*

$$g(\lambda)e^{ax\lambda + b\lambda^j t} = E(x, t, \lambda)g(x, t, \lambda)$$

uniquely such that $E(x, t, \cdot) \in \Lambda_+^\tau(G)$ *and* $g(x, t, \cdot) \in SD^\tau$*. Moreover,*

(i) $E^{-1}E_x$ *is of the form* $a\lambda + u(x, t)$*,*

(ii) $u(x, t)$ *is a Schwartz class solution of the* (b, j)*-flow,*

(iii) E *is the frame of* u.

Let $g\sharp 0$ denote the solution u constructed in Theorem 5.3.2. Then:

Theorem 5.3.3. *([69]).*

(1) $(D_-^\tau \sharp 0) \cap (\Lambda_-^\tau(G) * 0) = \{0\}$.
(2) $SD^\tau \sharp 0$ *is open and dense in the space of Schwartz class solutions of the* (b, j)*-flow in the* U*-hierarchy.*

5.4. The action of an infinite dimensional abelian group

Let $j > 0$ be an integer, $b \in \mathcal{A}$, $\xi_{b,j} \in \Lambda_+^\tau(\mathcal{G})$ defined by $\xi_{b,j}(\lambda) = b\lambda^j$, and $e_{b,j}(t)$ the one-parameter subgroup of $\Lambda_+^\tau(G)$ defined by $\xi_{b,j}$, i.e.,

$$e_{b,j}(t)(\lambda) = e^{b\lambda^j t}.$$

Let $\hat{A}_+$ be the subgroup of $\Lambda_+^\tau(G)$ generated by $\{e_{b,j}(t) \mid b \in \mathcal{A}, j \in \mathbb{N}, t \in \mathbb{R}\}$. The Lie algebra $\hat{\mathcal{A}}_+$ of $\hat{A}_+$ is the subalgebra of $\Lambda_+^\tau(\mathcal{G})$ generated by $\{\xi_{b,j} \mid b \in \mathcal{A}, j \in \mathbb{N}\}$. It follows from Theorem 5.3.2 that given $f \in \Lambda_-^\tau(G)$ we can factor $f^{-1}e_{a,1}(x) = E(x)m^{-1}(x)$ with $E(x) \in \Lambda_+^\tau(G)$ and $m(x) \in SD^\tau$. Expand $m(x)(\lambda)$ at $\lambda = \infty$ to get

$$m(x)(\lambda) = \mathrm{I} + m_1(x)\lambda^{-1} + m_2(x)\lambda^{-2} + \cdots .$$

Define $\mathcal{F}(f) = u^f := [a, m_1]$. Then $\mathcal{F}$ is a map from $SD^\tau(G)$ to $\mathcal{S}(\mathbb{R}, \mathcal{A}^\perp \cap \mathcal{U})$.

Given $b \in \mathcal{A}$, a positive integer j, and $f \in SD^\tau$, it follows from Theorem 5.3.2 that we can factor

$$f^{-1}e_{a,1}(x)e_{b,j}(t) = E(x,t)m(x,t)^{-1}$$

with $E(x,t) \in \Lambda_+^\tau(G)$ and $m(x,t) \in SD^\tau$. A straightforward direct computation implies that (cf. [69]):

(i) $E^{-1}E_x(x,t,\lambda)$ must be of the form $a\lambda + u(x,t)$ and $u = [a, m_1]$, where m_1 is the coefficient of λ^{-1} of the expansion of m as

$$m(x,t)(\lambda) = \mathrm{I} + m_1(x,t)\lambda^{-1} + m_2(x,t)\lambda^{-2} + \cdots .$$

(ii) u is a solution of the (b,j)-flow in the U-hierarchy.

By definition of the dressing action, $m(x,t) = (e_{a,1}(x)e_{b,j}(t)) * f$. Hence the (b,j)-flow arises naturally from the dressing action of $\hat{A}_+ \subset \Lambda_+^\tau(G)$ on SD^τ. Moreover,

$$\mathcal{F}(e_{b,j} * f) = u^{e_{b,j}(t)*f} = \phi_{b,j}(t)(u^f) = \phi_{b,j}(t)(\mathcal{F}(f)),$$

where $\phi_{b,j}(t)$ is the one-parameter subgroup generated by the vector field $X_{b,j}$ corresponding the (b,j)-flow in the U-hierarchy (i.e., $X_{b,j}$ defined by (2.1.6)). In other words, the infinitesimal dressing action of the abelian algebra $\hat{\mathcal{A}}_+$ on SD^τ gives the U-hierarchy of commuting flows. For more details of this discussion see [69].

5.5. Geometric transformations

Let $g \in \Lambda_-^{\tau,\sigma}(G)$, v a solution of some integrable system associated to U/U_0, and E the frame of v. Factor gE as $\tilde{E}\tilde{g}$ with $\tilde{E}(x,t) \in \Lambda_+^{\tau,\sigma}(G)$ and $\tilde{g} \in \Lambda_-^{\tau,\sigma}(G)$. Then $g * v = \tilde{v}$, and $g * E = \tilde{E}$ is the frame of $g * v$. We have seen in Chapter 3 that geometries associated to solutions of an integrable system can often be read from their frames at some special value $\lambda = \lambda_0$. So $E_{\lambda_0} \mapsto g * E_{\lambda_0}$ gives rise to a geometric transformation for the corresponding geometries. For example, it is known that solutions of the SGE equation correspond to surfaces in $\mathbb{R}^3$ with constant Gaussian curvature $K = -1$. The Lax pair of the SGE equation satisfies the $SU(2)/SO(2)$-reality condition, and the dressing action of $g_{is,\pi}(\lambda) = \pi + \frac{\lambda - is}{\lambda + is}(\mathrm{I} - \pi)$ on the space of solutions of the SGE equation corresponds to the classical Bäcklund transformation of surfaces in $\mathbb{R}^3$ with $K = -1$ (cf. [70]). In this section, we use another example to demonstrate this correspondence between the dressing action and geometric transformations. We describe the geometric transformation of flat submanifolds that corresponds to the action of the rational element $h_{is,\pi}$ (defined by (5.1.1)) on the solutions of the $\frac{O(2n)}{O(n)\times O(n)}$-system.

First, we need to recall the following definition given by Dajczer and Tojeiro in [21, 22].

Definition 5.5.1. Let M^n and $\tilde{M}^n$ be submanifolds of S^{2n-1} with flat normal bundle. A vector bundle isomorphism $P : \nu(M) \to \nu(\tilde{M})$, which covers a diffeomorphism $\ell : M \to \tilde{M}$, is called a *Ribaucour Transformation* if P satisfies the following properties:

(a) If ξ is a parallel normal vector field of M, then $P \circ \xi \circ \ell^{-1}$ is a parallel normal field of $\tilde{M}$.

(b) Let $\xi \in \nu_x(M)$, and $\gamma_{x,\xi}$ the normal geodesic with ξ as the tangent vector at $t = 0$. Then for each $\xi \in \nu(M)_x$, $\gamma_{x,\xi}$ and $\gamma_{\ell(x),P(\xi)}$ intersect at a point that is equidistant from x and $\ell(x)$ (the distance depends on x).

(c) If η is an eigenvector of the shape operator A_ξ of M, then $\ell_*(\eta)$ is an eigenvector of the shape operator $A_{P(\xi)}$ of $\tilde{M}$. Moreover, the geodesics $\gamma_{x,\eta}$ and $\gamma_{\ell(x),\ell_*(\eta)}$ intersect at a point equidistant to x and $\ell(x)$.

Dajczer and Tojeiro used geometric methods to prove the existence of Ribaucour transformations between flat n-submanifolds of S^{2n-1} in [21]. These Ribaucour transformations are exactly the ones obtained from dressing actions of $h_{is,\pi}$ given in Example 5.1.1, i.e.,

Theorem 5.5.2. *([10]). Let F be a solution of the $\frac{O(2n)}{O(n)\times O(n)}$-system (2.5.3), and E the frame of the corresponding Lax n-tuple (2.5.4). Let $h_{is,\pi} \in \Lambda_-^{\tau,\sigma}(G)$ defined by (5.1.1), $\tilde{F} = h_{is,\pi} * F$, and $\tilde{E} = h_{is,\pi} * E$ as in Example 5.1.1. Let M be a flat n-submanifold in S^{2n-1} associated to F as in Theorem 3.3.3. Then there exist a flat n-submanifold $\tilde{M}$ in S^{2n-1} and a Ribaucour transformation $P : \nu(M) \to \nu(\tilde{M})$ constructed from $\tilde{E} = h_{is,\pi} * E$ such that the solution of the $\frac{O(2n)}{O(n)\times O(n)}$-system for $\tilde{M}$ is $\tilde{F} = h_{z,\pi} * F$.*

5.6. The characteristic initial value problem for the -1-flow

Given $a, b \in \mathcal{U}$ such that $[a, b] = 0$, the -1-flow in the U-hierarchy defined by a, b is the following equation for $g : \mathbb{R}^2 \to U$

$$(g^{-1}g_x)_t = [a, g^{-1}bg] \tag{5.6.1}$$

with the constraint $g^{-1}g_x \in [a, \mathcal{U}]$. It has a Lax pair

$$\theta_\lambda = (a\lambda + g^{-1}g_x)dx + \lambda^{-1}g^{-1}bgdt.$$

Equation (5.6.1) is hyperbolic, and the x-, t-curves are the characteristics. The *characteristic initial value problem* (or the *degenerate Goursat problem*) is the initial value problem with initial data defined on two characteristic axes, i.e., given $h_1, h_2 : \mathbb{R} \to U$ satisfying $h_1^{-1}(h_1)_x \in [a, \mathcal{U}]$ and $h_1(0) = h_2(0)$, solve

$$\begin{cases} (g^{-1}g_x)_t = [a, g^{-1}bg], \\ g(x, 0) = h_1(x), \quad g(0, t) = h_2(t). \end{cases} \tag{5.6.2}$$

If we write $u = g^{-1}g_x, \quad v = g^{-1}bg$, then the -1-flow equation (5.6.1) becomes the following system for (u, v),

$$u_t = [a, v], \quad v_x = -[u, v], \tag{5.6.3}$$

The Lax pair is

$$\theta_\lambda = (a\lambda + u)dx + \lambda^{-1}vdt. \tag{5.6.4}$$

Let M_b denote the adjoint U-orbit in $\mathcal{U}$ at b. Since $u(x, 0) = h_1^{-1}h_1'(x)$ and $v(0, t) = h_2(t)^{-1}bh_2(t) \in M_b$, the characteristic initial value problem

(5.6.2) becomes the following initial value problem for (5.6.3): given $\xi : \mathbb{R} \to [a, \mathcal{U}]$ and $\eta : \mathbb{R} \to M_b$, find $(u, v) : \mathbb{R}^2 \to [a, \mathcal{U}] \times M_b$ so that

$$\begin{cases} u_t = [a, v], & v_x = -[u, v], \\ u(x, 0) = \xi(x), & v(0, t) = \eta(t). \end{cases} \tag{5.6.5}$$

In [28], Dorfmeister and Eitner use the Gauss loop group factorization to construct all local solutions of the Tzitzeica equation (2.6.4). Their construction in fact solves the characteristic initial value problem (5.6.5) for the -1 flow in the U-hierarchy:

Theorem 5.6.1. *([28]). Let $\xi, \eta : \mathbb{R} \to [a, \mathcal{U}] \times M_b$ be smooth maps, and $L_+(x, \lambda)$ and $L_-(t, \lambda)$ solutions of*

$$\begin{cases} (L_+)^{-1}(L_+)_x = a\lambda + \xi(x), \\ L_+(0, \lambda) = I, \end{cases} \qquad \begin{cases} (L_-)^{-1}(L_-)_t = \lambda^{-1}\eta(t), \\ L_-(0, \lambda) = I, \end{cases}$$

respectively. Factor

$$L_-^{-1}(t, \lambda)L_+(x, \lambda) = V_+(x, t, \lambda)V_-^{-1}(x, t, \lambda) \tag{5.6.6}$$

with $V_\pm(x, t, \cdot) \in L_\pm(G)$ via the Gauss loop group factorization. Set

$$\phi(x, t, \lambda) = L_-(t, \lambda)V_+(x, t, \lambda) = L_+(x, \lambda)V_-(x, t, \lambda).$$

Then $\phi^{-1}\phi_x = a\lambda + u(x, t)$ and $\phi^{-1}\phi_t = \lambda^{-1}v(x, t)$ for some u, v, and (u, v) solves the initial value problem of the -1-flow (5.6.5) in the U-hierarchy.

Proof. Differentiate $\phi = L_-V_+ = L_+V_-$ to get

$$\phi^{-1}\phi_x = V_-^{-1}(a\lambda + \xi(x))V_- + V_-^{-1}(V_-)_x = V_+^{-1}(V_+)_x.$$

So $\phi^{-1}\phi_x \in \mathcal{L}_+(\mathcal{G})$ and

$$\phi^{-1}\phi_x = \pi_+(V_-^{-1}(a\lambda + \xi(x))V_-),$$

where $\pi_\pm$ is the projection of $\mathcal{L}(\mathcal{G})$ onto $\mathcal{L}_\pm(\mathcal{G})$ with respect to the decomposition $\mathcal{L}(\mathcal{G}) = \mathcal{L}_+(\mathcal{G}) + \mathcal{L}_-(\mathcal{G})$. Expand

$$V_-(x, t, \lambda) = I + m_1(x, t)\lambda^{-1} + \cdots .$$

A direct computation shows that

$$\pi_+(V_-^{-1}(a\lambda + \xi(x))V_-) = a\lambda + \xi(x) + [a, m_1(x, t)].$$

Hence $\phi^{-1}\phi_x = a\lambda + u(x,t)$, where $u = \xi + [a, m_1]$. A similar argument implies that

$$\phi^{-1}\phi_t = \pi_-(V_+^{-1}\lambda^{-1}\eta(t)V_+) = \lambda^{-1}g_0(x,t)\eta(t)g_0^{-1}(x,t),$$

where $g_0(x,t)$ is the constant term in the expansion

$$V_+(x,t,\lambda) = \sum_{j=0}^{\infty} g_j(x,t)\lambda^j.$$

This implies that (u,v) is a solution of (5.6.3), where

$$u(x,t) = \xi(x) + [a, m_1(x,t)], \quad v(x,t) = g_0(x,t)\eta(t)g_0(x,t)^{-1}.$$

It remains to prove (u,v) satisfies the initial conditions. This can be seen from the factorization 5.6.6. Note that

$$L_-(0,\lambda)^{-1}L_+(x,\lambda) = V_+(x,0,\lambda)V_-(x,0,\lambda)^{-1}.$$

Since $L_-(0,\lambda) = I$, the right hand side lies in $L_+(G)$. Hence $V_-(x,0,\lambda) = I$, which proves that $m_1(x,0) = 0$. Therefore $u(x,0) = \xi(x)$. A similarl argument implies that $v(0,t) = \eta(t)$. Q.E.D.

They also show that every local solution of (5.6.3) can be constructed using suitable $\xi(x)$ and $\eta(t)$. To see this, let (u,v) be a solution of (5.6.3), and $\phi(x,t,\lambda)$ the trivialization of the corresponding Lax pair:

$$\phi^{-1}d\phi = (a\lambda + u)dx + \lambda^{-1}vdt, \quad \phi(0,0,\lambda) = I.$$

Use the Gauss loop group factorization to factor

$$\phi(x,t,\cdot) = L_-(x,t,\cdot)V_+(x,t,\cdot) = L_+(x,t,\cdot)V_-(x,t,\cdot).$$

Differentiate the above equation to get

$$\begin{aligned}
L_-^{-1}(L_-)_x &= \pi_-(V_+(a\lambda + u)V_+^{-1}) = 0,\\
L_-^{-1}(L_-)_t &= \pi_-(V_+\lambda^{-1}vV_+^{-1}) = \lambda^{-1}g_0vg_0^{-1},\\
L_+^{-1}(L_+)_x &= \pi_+(V_-(a\lambda + u)V_-^{-1}) = a\lambda + u + [m_1, a],\\
L_+^{-1}(L_+)_t &= \pi_-(V_-\lambda^{-1}vV_-^{-1}) = 0,
\end{aligned}$$

where g_0 is the constant term in the power series expansion of $V_+(x,t,\lambda)$ in λ and m_1 is the coefficient of λ^{-1} of V_-. This implies that $(L_-)_x = 0$, $(L_+)_t = 0$, $L_+^{-1}(L_+)_x = a\lambda + u(x,0)$ and $L_-^{-1}(L_-)_t = \lambda^{-1}v(0,t)$.

Let τ be the involution of G with the real form U as its fixed point set, and σ an order k automorphism of G such that $\sigma\tau = \tau^{-1}\sigma^{-1}$. Let $\mathcal{G}_j$ denote the eigenspace of σ_* of $\mathcal{G}$ with eigenvalue $e^{\frac{2\pi ij}{k}}$, and $\mathcal{U}_j = \mathcal{U} \cap \mathcal{G}_j$. Let $a \in \mathcal{G}_1$, $b \in \mathcal{G}_{-1}$ such that $[a, b] = 0$. The -1-flow in the U/U_0-hierarchy is the restriction of the -1-flow in the U-hierarchy (5.6.3) to the space of maps $(u, v) : \mathbb{R}^2 \to [a, \mathcal{U}_{-1}] \times \mathrm{Ad}(U_0)b$. It is easy to see that the solution constructed for initial data $\xi : \mathbb{R} \to [a, \mathcal{U}_{-1}]$ and $\eta : \mathbb{R} \to \mathrm{Ad}(U_0)(b)$ in Theorem 5.6.1 is a solution of the -1-flow in the U/U_0-hierarchy. In other words, the characteristic initial value problem for the -1 flow in the U/U_0-hierarchy can be solved by the algorithm given in Theorem 5.6.1.

§6. Elliptic systems associated to G, τ, σ

Let G be a complex Lie group, and τ a conjugate linear involution of $\mathcal{G}$, and σ an order k complex linear automorphism of $\mathcal{G}$ such that $\tau\sigma = \sigma\tau$. Let $\mathcal{G}_j$ be the eigenspace of σ with eigenvalue $e^{\frac{2\pi ji}{k}}$. So we have $\mathcal{G}_j = \mathcal{G}_m$ if $j \equiv m \pmod{k}$, and

$$\mathcal{G} = \mathcal{G}_0 + \mathcal{G}_1 + \cdots + \mathcal{G}_{k-1}, \quad [\mathcal{G}_j, \mathcal{G}_r] \subset \mathcal{G}_{j+r}.$$

We claim that $\tau(\mathcal{G}_j) \subset \mathcal{G}_{-j}$. To see this, let $\sigma(\xi_j) = \alpha^j \xi_j$, where $\alpha = e^{\frac{2\pi i}{k}}$. Then

$$\sigma(\tau(\xi_j)) = \tau(\sigma(\xi_j)) = \tau(\alpha^j \xi_j) = \bar{\alpha}^j \tau(\xi_j) = \alpha^{-j}\tau(\xi_j).$$

Let $\mathcal{U}$ be the fixed point set of τ, and U_σ denote the fixed point set of σ on U. Since $\sigma\tau = \tau\sigma$, $\sigma(U) \subset U$ and $\sigma \mid U$ is an order k automorphism of U. The quotient space U/U_σ is called a *k-symmetric space*.

We will construct the sequences of U- and U/U_σ- systems.

6.1. The m-th (G, τ)-system

The m-th (G, τ)-system (also called the *m-th elliptic U-system*) is the following system for $(u_0, \cdots, u_m) : \mathbb{C} \to \prod_{i=0}^m \mathcal{G}$:

$$\begin{cases} (u_j)_{\bar{z}} = \sum_{i=0}^{m-j} [u_{i+j}, \tau(u_i)], & \text{if } 1 \le j \le m, \\ (u_0)_{\bar{z}} - (\tau(u_0))_z = \sum_{i=0}^m [u_i, \tau(u_i)]. \end{cases} \tag{6.1.1}$$

It has a Lax pair:

$$\theta_\lambda = \sum_{j=0}^m u_j \lambda^{-j} dz + \tau(u_j)\lambda^j d\bar{z}. \tag{6.1.2}$$

Equation (6.1.1) is also referred to as the *m-th elliptic U-system*, where U is the fixed point set of τ.

The Lax pair (6.1.2) satisfies the (G,τ)-*reality condition* (4.2.2), i.e.,

$$\tau(\theta_{1/\bar{\lambda}}) = \theta_\lambda.$$

Note that $\xi = \sum_j \xi_j \lambda^j$ satisfies the (G,τ)-reality condition if and only if $\xi_{-j} = \tau(\xi_j)$ for all j.

6.2. The m-th (G,τ,σ)-system

The m-th (G,τ,σ)*-system* is the following equation for $(u_0,\cdots,u_m) : \mathbb{C} \to \oplus_{j=0}^m \mathcal{G}_{-j}$,

$$\begin{cases} (u_j)_{\bar{z}} = \sum_{i=0}^{m-j} [u_{i+j}, \tau(u_i)], \text{ if } 1 \leq j \leq m, \\ -(u_0)_{\bar{z}} + (\tau(u_0))_z + \sum_{j=0}^m [u_j, \tau(u_j)] = 0. \end{cases} \tag{6.2.1}$$

It has a Lax pair

$$\theta_\lambda = \sum_{i=0}^m u_i \lambda^{-i} dz + \tau(u_i)\lambda^i d\bar{z}. \tag{6.2.2}$$

Note that:

(i) The m-th (G,τ,σ)-system is the restriction of the m-th (G,τ)-system to the space of maps $(u_0,\cdots,u_m)$ with values in $\oplus_{j=0}^m \mathcal{G}_{-j}$.

(ii) The Lax pair of the m-th (G,τ,σ)-system satisfies the (G,τ,σ)-*reality condition* 4.2.3, i.e.,

$$\tau(\theta_{1/\bar{\lambda}}) = \theta_\lambda, \quad \sigma(\theta_\lambda) = \theta_{e^{\frac{2\pi i}{k}}\lambda}.$$

(iii) $\xi(\lambda) = \sum_j \xi_j \lambda^j$ satisfies the (G,τ,σ)-reality condition if and only if $\xi_j \in \mathcal{G}_j$ and $\xi_{-j} = \tau(\xi_j)$ for all j.

Let U be the fixed point set of τ, and U_σ the fixed point set of σ on U. System (6.2.1) will also referred to as the *m-th elliptic* U/U_σ*-system*.

A direct computation gives the following Proposition:

Proposition 6.2.1. *Let τ be an involution, and σ an order k automorphism of G such that $\sigma\tau = \tau\sigma$, and $1 \leq m < \frac{k}{2}$. If $\psi : \mathbb{C} \to U$ is a map such that*

$$\psi^{-1}\psi_z = u_o + \cdots + u_m \in \mathcal{G}_0 + \mathcal{G}_{-1} + \cdots + \mathcal{G}_{-m}, \tag{6.2.3}$$

then $(u_0, \cdots, u_m)$ *is a solution of the* m*-th* (G, τ, σ)*-system. Conversely, if*

$(u_0, \cdots, u_m)$ *is a solution of the* m*-th* (G, τ, σ)*-system* (6.2.1), *then there exists* $\psi : \mathbb{C} \to U$ *such that* $\psi^{-1}d\psi = \sum_{j=0}^{m} u_j dz + \tau(u_j)d\bar{z}$.

Definition 6.2.2. Let τ, σ, k, and U be as in Proposition 6.2.1, and $1 \leq m < \frac{k}{2}$. A map $\psi : \mathbb{C} \to U$ is called a (σ, m)-map if $\psi^{-1}\psi_z \in \oplus_{j=1}^{m}\mathcal{G}_{-j}$.

Definition 6.2.3. ([15]). Let U/U_σ denote the k-symmetric space (with $k \geq 3$) given by τ, σ, and $\pi : U \to U/U_\sigma$ the natural projection. A map $\phi : \mathbb{C} \to U/U_0$ is called *primitive* if there is a lift $\psi : \mathbb{C} \to U$ (i.e., $\pi \circ \psi = \phi$) so that $\psi^{-1}\psi_z \in \mathcal{G}_0 + \mathcal{G}_{-1}$. In other words, there is a lift ψ that is a $(\sigma, 1)$-map.

By Proposition 6.2.1, the equation for (σ, m)-maps is the m-th (G, τ, σ)-system, and the equation for primitive maps is the first (G, τ, σ)-system. We refer the reader to [15] for a more detailed study of primitive maps.

Example 6.2.4. Let $\mathcal{G} = sl(3, \mathbb{C})$, $\tau(\xi) = -\bar{\xi}^t$, and $D = \begin{pmatrix} 0 & 1 & 0 \\ -1 & 0 & 0 \\ 0 & 0 & 1 \end{pmatrix}$.

Let $\sigma(\xi) = -D\xi^t D^{-1}$. Note that $D^{-1} = D^t$, $\sigma^2(\xi) = \text{diag}(-1, -1, 1)\xi\text{diag}(-1, -1, 1)$, σ has order 4, and $\sigma\tau = \tau\sigma$. A direct computation shows that the eigenspaces $\mathcal{G}_j$ with eigenvalues $(\sqrt{-1}\ )^j$ are:

$$\mathcal{G}_0 = \left\{ \begin{pmatrix} \xi & 0 \\ 0 & 0 \end{pmatrix} \middle| \ \xi \in sl(2, \mathbb{C}) \right\},$$

$$\mathcal{G}_1 = \left\{ \begin{pmatrix} 0 & 0 & a \\ 0 & 0 & b \\ ib & -ia & 0 \end{pmatrix} \middle| \ a, b \in \mathbb{C} \right\},$$

$$\mathcal{G}_2 = \mathbb{C}\ \text{diag}(1, 1, -2),$$

$$\mathcal{G}_3 = \left\{ \begin{pmatrix} 0 & 0 & a \\ 0 & 0 & b \\ -ib & ia & 0 \end{pmatrix} \middle| \ a, b \in \mathbb{C} \right\}.$$

The 2nd (G, σ, τ)-system (or the second elliptic $SU(3)/SU(2)$-system) is the system for $(u_0, u_1, u_2) : \mathbb{C} \to \mathcal{G}_0 \times \mathcal{G}_{-1} \times \mathcal{G}_{-2}$:

$$\text{(6.2.4)} \qquad \begin{cases} (u_2)_{\bar{z}} = 0, \\ -(u_1)_{\bar{z}} + [u_1, \tau(u_0)] + [u_2, \tau(u_1)] = 0, \\ -(u_0)_{\bar{z}} + (\tau(u_0))_z + [u_0, \tau(u_0)] + [u_1, \tau(u_1)] = 0. \end{cases}$$

Example 6.2.5. ([38, 51]). Let $\mathcal{G} = sl(n, \mathbb{C})$, $\tau(\xi) = -\bar{\xi}^t$, and $\sigma(\xi) = C\xi C^{-1}$, where $C = \text{diag}(1, \alpha, \cdots, \alpha^{n-1})$ and $\alpha = e^{\frac{2\pi i}{n}}$. Then $\tau\sigma = \sigma\tau$, and the eigenspace $\mathcal{G}_j$ of σ is spanned by $\{e_{i,i+j} \mid 1 \le i \le n\}$. Here we use the notation that $e_{ij} = e_{i'j'}$ if $i \equiv i'$ and $j \equiv j'$ (mod n). The first (G, τ, σ)-system is the equation for $A_0 = \text{diag}(u_1, \cdots, u_n)$ and $A_1 = \sum_{i=1}^n v_i e_{i,i-1}$ so that

$$\theta_\lambda = (A_0 + A_1\lambda^{-1})\, dz - (\bar{A}_0^t + \lambda \bar{A}_1^t)\, d\bar{z}$$

is flat for all λ. If $v_i > 0$ for all $1 \le i \le n$ and $v_1 \cdots v_n = 1$, then flatness of θ_λ implies that we can write $u_i = (w_i)_z$ and $v_i = e^{w_i - w_{i-1}}$ for some $w_1, \cdots, w_n$. The first (G, τ, σ)-system written in terms of w_i's is the *2-dimensional elliptic periodic Toda lattice*:

$$2(w_i)_{z\bar{z}} = e^{2(w_{i+1} - w_i)} - e^{2(w_i - w_{i-1})}.$$

6.3. The normalized system

The normalized m-th (G, τ)-system or the *normalized m-th U-system* is the system for $v_1, \cdots, v_m : \mathbb{R}^2 \to \mathcal{G}$:

$$(6.3.1) \qquad (v_j)_{\bar{z}} = \sum_{i=1}^{m-j} [v_{i+j}, \tau(v_i)] - \sum_{i=1}^{m} [v_j, \tau(v_i)], \quad 1 \le j < m.$$

It has a Lax pair

$$(6.3.2) \qquad \Theta_\lambda = \sum_{j=1}^{m} (\lambda^{-j} - 1) v_j \, dz + (\lambda^j - 1)\tau(v_j)\, d\bar{z},$$

which satisfies the (G, τ)-reality condition (4.2.2).

We claim that the Lax pair (6.3.2) is gauge equivalent to the Lax pair (6.1.2) for the m-th (G, τ)-system (6.1.1). Hence system (6.3.1) and (6.1.1) are gauge equivalent. To see this, let $(u_0, \cdots, u_m)$ be a solution of (6.1.1),

$$\theta_\lambda = \sum_{j=0}^{m} u_j \lambda^{-j} dz + \tau(u_j)\lambda^j d\bar{z}$$

its Lax pair, and $E(z, \bar{z})(\lambda)$ the frame of θ_λ, i.e.,

$$E^{-1}E_z = \sum_{j=0}^{m} u_j \lambda^{-j}, \quad E^{-1}E_{\bar{z}} = \sum_{j=0}^{m} \tau(u_j)\lambda^j, \quad E(0,0)(\lambda) = e.$$

Let $g = E(\cdot,\cdot,1)$. Since θ_λ satisfies the (G,τ)-reality condition, E satisfies

$$\tau(E(\cdot,\cdot,1/\bar{\lambda})) = E(\cdot,\cdot,\lambda).$$

Hence $E(\cdot,\cdot,\lambda) \in U$ if $\mid \lambda \mid = 1$. In particular, $g \in U$. The gauge transformation of θ_λ by g is

$$\tilde{\theta}_\lambda = g\theta_\lambda g^{-1} - dgg^{-1} = \sum_{j=1}^{m}(\lambda^{-j} - 1)gu_jg^{-1}dz + (\lambda^j - 1)g\tau(u_j)g^{-1}d\bar{z}$$
$$= \sum_{j=1}^{m}(\lambda^{-j} - 1)gu_jg^{-1}dz + (\lambda^j - 1)\tau(gu_jg^{-1})d\bar{z}.$$

So $(v_1, \cdots, v_m)$ is a solution of (6.3.1), where $v_i = gu_jg^{-1}$, and

$$F(z,\bar{z})(\lambda) = E(z,\bar{z})(\lambda)(E(z,\bar{z})(1))^{-1}$$

is the frame of $\tilde{\theta}_\lambda$.

§7. Geometries associated to integrable elliptic systems

Let τ be the involution of G whose fixed point set is the maximal compact subgroup U of G, and σ an order k automorphism of G such that $\sigma\tau = \tau\sigma$. Let $\mathcal{G}_j$ denote the eigenspace of σ_* on $\mathcal{G}$ with eigenvalue $e^{\frac{2\pi j i}{k}}$. Since $\sigma\tau = \tau\sigma$, we have $\sigma(U) \subset U$, and $\sigma|_U$ is an order k automorphism of U. Let U_σ denote the fixed point set of σ in U. The quotient U/U_σ is a symmetric space if $k = 2$, is a *k-symmetric space* if $k > 2$.

It is known that the first (G,τ)-system is the equation for harmonic maps from $\mathbb{C}$ to U ([72]). The first (G,τ,σ)-system is the equation for harmonic maps from $\mathbb{C}$ to the symmetric space U/U_σ if the order of σ is 2, and is the equation for primitive maps if $k > 2$ ([15]). It is also known that a primitive map $\phi : \mathbb{C} \to U/U_\sigma$ is harmonic if U/U_σ is equipped with a U-invariant metric and $\mathcal{G}_1$ is isotropic ([15]).

The first (G,τ,σ)-system also arises naturally in the study of surfaces in symmetric spaces with certain geometric properties. For example, constant mean curvature surfaces of simply connected 3-dimensional space forms $N^3(c)$ ([56, 6]), minimal surfaces in $\mathbb{C}P^2$ ([11, 9]), minimal Lagrangian surfaces in $\mathbb{C}P^2$ ([50, 52]), minimal Legendrian surfaces in S^5 ([64, 42]), and special Lagrangian cone in $\mathbb{R}^6 = \mathbb{C}^3$ ([42, 52]).

The only known surface geometry associated to the m-th (G, τ, σ)-system for $m > 1$ was given by Hélein and Roman. They showed that the equations for Hamiltonian stationary surfaces in 4-dimension Hermitian symmetric spaces are the second elliptic system associated to certain 4-symmetric spaces (cf. [43]).

If the equation for surfaces with special geometric properties is the m-th (G, τ, σ)-system, then the techniques developed for integrable systems can be applied to study the corresponding surfaces. In particular, the finite type (or finite gap) solutions give rise to tori with given geometric properties. This has been done for constant mean curvature tori of $N^3(c)$ in [56] and [6], for minimal tori of $\mathbb{C}P^2$ in [11, 9], and for minimal Legendrian tori of S^5 in [64, 52].

We will give a very brief review of some of the results mentioned above. For more details, we refer the readers to [38, 39] for harmonic maps, to [56, 6] for constant mean curvature surfaces in 3-dimensional space forms, to [11, 9] for minimal surfaces in $\mathbb{C}P^2$, and to [43] for Hamiltonian stationary surfaces in four dimensional Hermitian symmetric spaces.

7.1. Harmonic maps from $\mathbb{R}^2$ to U and the first (G, τ)-system

First we state some results of Uhlenbeck ([72]) on harmonic maps from $\mathbb{C}$ or S^2 to $U(n)$.

Theorem 7.1.1. *([72]). Let G be a complex semi-simple Lie group, $\mathcal{U}$ the real form defined by the conjugate linear involution τ, $s : \mathbb{C} \to U$ a smooth map, and $A = -\frac{1}{2}s^{-1}s_z$. Then the following statements are equivalent:*

(i) s is harmonic,

(ii) $A_{\bar{z}} = -[A, \tau(A)]$,

(iii) $(\lambda^{-1} - 1)A\ dz + (\lambda - 1)\tau(A)\ d\bar{z}$ is flat for all $\lambda \in \mathbb{C} \setminus \{0\}$, i.e., A is a solution of the normalized 1st (G, τ)-system.

Corollary 7.1.2. *([72]). Suppose*

$$\theta_\lambda = (\lambda^{-1} - 1)A(z, \bar{z})dz + (\lambda - 1)\tau(A(z, \bar{z}))d\bar{z}$$

is a flat $\mathcal{G}$-valued 1-form for all $\lambda \in \mathbb{C} \setminus 0$, and E_λ the corresponding frame (i.e., $E_\lambda^{-1}dE_\lambda = \theta_\lambda$ and $E_\lambda(0) = e$). Then E_{-1} is a harmonic map from $\mathbb{C}$ to U.

Using the ellipticity of the harmonic map equation, Uhlenbeck proved that there are trivializations of the Lax pair of harmonic maps from S^2 to $U(n)$ that are polynomials in the spectral parameter:

Theorem 7.1.3. *([72]). Let $s : S^2 \to U(n)$ be a harmonic map, and E the frame of the corresponding Lax pair $\theta_\lambda = -\frac{\lambda^{-1}-1}{2}\, s^{-1}s_z\, dz - \frac{\lambda-1}{2}\, s^{-1}s_{\bar z}\, d\bar z$. Then there exist $\gamma \in L_e(U)$ and smooth maps $\pi_i : S^2 \to Gr(k_i, \mathbb{C}^n)$ such that $\gamma(\lambda)E(\cdot,\cdot,\lambda) = (\pi_1 + \lambda\pi_1^\perp)\cdots(\pi_r + \lambda\pi_r^\perp)$.*

A harmonic map from a domain of $\mathbb{C}$ to $U(n)$ is called a *finite uniton* if the corresponding Lax pair admits a trivialization that is polynomial in λ ([72]). The above theorem implies that all harmonic maps from S^2 to $U(n)$ are *finite unitons*.

A proof similar to that of Proposition 3.9.4 gives

Proposition 7.1.4. *Let τ be the involution of G that defines the real form U, $(u_0, u_1) : \mathbb{C} \to \mathcal{G} \times \mathcal{G}$ a solution of the first (G,τ)-system (6.1.1), and $E_\lambda(z,\bar z)$ the frame of the corresponding Lax pair (6.1.2). Then $s = E_{-1}E_1^{-1}$ is a harmonic map from $\mathbb{C}$ to U. Moreover, let σ be an involution of $\mathcal{G}$ that commutes with τ, and $\mathcal{G} = \mathcal{G}_0 + \mathcal{G}_1$ the eigenspace decomposition of σ. If $(u_0, u_1) \in \mathcal{G}_0 \times \mathcal{G}_1$, then $s = E_{-1}E_1^{-1}$ is a harmonic map from $\mathbb{C}$ to the symmetric space U/U_0.*

Corollary 7.1.5. *Let τ be a conjugate liner involution, and σ an order $k = 2m$ automorphism of $\mathcal{G}$ such that $\sigma\tau = \tau\sigma$. Let (u_1, u_0) be a solution of the first (G,τ,σ)-system (6.2.4), and E the frame of the corresponding Lax pair θ_λ (defined by (6.2.2)). Then $s = E(\cdot,\cdot,-1)E(\cdot,\cdot,1)^{-1}$ is a harmonic map from $\mathbb{C}$ to the symmetric space U/H, where H is the fixed point set of the involution σ^m on U.*

The following Theorem is proved by Burstall and Pedit.

Theorem 7.1.6. *([15]). Let $(u_0, u_1) : \mathbb{C} \to \mathcal{G}_0 \times \mathcal{G}_{-1}$ be a solution of the first (G,τ,σ)-system, U/U_σ the k-symmetric space corresponding to (τ,σ), and $k > 2$. Let $\pi : U \to U/U_\sigma$ be the natural fibration. If $E(z,\bar z,\lambda)$ is a trivialization of the Lax pair of the first (G,τ,σ)-system, then $\phi = \pi \circ E(\cdot,\cdot,1)$ is primitive. Moreover, if U/U_0 is equipped with an invariant metric and $\mathcal{G}_{-1}$ is isotropic, then ϕ is harmonic.*

7.2. The first (G,τ,σ)-system and surface geometry

Let $N^n(c)$ denote the simply connected space form of constant sectional curvature c, i.e., $N^n(0) = \mathbb{R}^n$, $N^n(1) = S^n$ the unit sphere in

$\mathbb{R}^{n+1}$, and $N^n(-1) = \mathbb{H}^n = \{x \in \mathbb{R}^{n,1} \mid \langle x, x \rangle = -1\} \subset \mathbb{R}^{n,1}$. The Gauss map ϕ of a k-dimensional submanifold M in $N^n(c)$ is the map from M to the symmetric space $Y(k, c)$, where

$$Y(n,0) = \mathrm{Gr}(k, \mathbb{R}^n), \quad Y(n,1) = \mathrm{Gr}(k, \mathbb{R}^{n+1}), \quad Y(n,-1) = \mathrm{Gr}(k, \mathbb{R}^{n,1}).$$

A theorem of Ruh and Vilms states that the Gauss map of a k-submanifold with parallel mean curvature vector in $N^n(c)$ is harmonic. Moreover, the Gauss-Codazzi equation for constant mean curvature (CMC) surfaces in $N^3(c)$ is the first (G, τ, σ)-system, where τ, σ are the involutions that define the symmetric space $Y(3, c)$. Since the equation for harmonic maps from $\mathbb{C}$ to $Y(3, c)$ defined by τ, σ is the first (G, τ, σ)-system, techniques developed for the first (G, τ, σ)-system (or harmonic maps) can be used to study CMC surfaces in $N^3(c)$ (cf. [56, 6]).

There are natural definitions of Gauss maps for surfaces and Lagrangian surfaces in $\mathbb{C}P^2$, for Legendrian surfaces in S^5, and Lagrangian cones in $\mathbb{R}^6 = \mathbb{C}^3$. The target manifolds of these Gauss maps are now k-symmetric spaces. The minimality of surfaces is equivalent to their Gauss maps being primitive. Hence equations of these surfaces are the corresponding first (G, τ, σ)-system.

Example 7.2.1. Minimal surfaces in $\mathbb{C}P^2$

Let $f : M \to \mathbb{C}P^2$ be an immersed surface, $L \to \mathbb{C}P^2$ the tautological complex line bundle, z a local conformal coordinate on M, and f_0 a local cross section of $f^*(L)$. Choose f_1, f_2 so that $(f_0, f_1, f_2) \in SU(3)$ and

$$\mathbb{C}f_0 + \mathbb{C}f_1 = \mathbb{C}f_0 + \mathbb{C}\frac{\partial f_0}{\partial z}.$$

The Gauss map of M is the map ϕ from M to the flag manifold $Fl(\mathbb{C}^3)$ of $\mathbb{C}^3$ defined by $\phi(f) =$ the flag $(\mathbb{C}f_0, \mathbb{C}f_0 + \mathbb{C}f_1, \mathbb{C}^3)$. Note that $Fl(\mathbb{C}^3) = SU(3)/T^2$ is a 3-symmetric space given by $\tau(g) = (g^t)^{-1}$ and $\sigma(g) = CgC^{-1}$, where $C = \mathrm{diag}(1, e^{\frac{2\pi i}{3}}, e^{\frac{4\pi i}{3}})$. It is proved in [11, 9] that M is minimal in $\mathbb{C}P^2$ if and only if the Gauss map $\phi : \mathbb{C} \to SU(3)/T^2$ is primitive.

Example 7.2.2. Minimal Legendrian surfaces in S^5

Let $v_1 = (1,0,0)^t$, and $\mathbb{R}^3$ the real part of $\mathbb{C}^3$. Let Fl_1 denote the $SU(3)$-orbit of $(v_1, \mathbb{R}^3)$, i.e.,

$$\begin{aligned} Fl_1 &= \{(gv_1, g(\mathbb{R}^3) \mid g \in SU(3)\} \\ &= \{(v, V) \mid v \in S^5, v \in V, V \text{ is Lagrangian linear subspace of } \mathbb{C}^3\} \\ &= \frac{SU(3)}{1 \times SO(2)} \end{aligned}$$

Note that Fl_1 is a 6-symmetric space corresponding to automorphisms τ, σ of $SL(3,\mathbb{C})$ defined by $\tau(g) = (\bar{g}^t)^{-1}$ and $\sigma(g) = R(g^t)^{-1}R^{-1}$, where R is the rotation that fixes the x-axis and rotate $\frac{\pi}{3}$ in the yz-plane.

Let α be the standard contact form on S^5. A surface M in S^5 is *Legendrian* if the restriction of α to M is zero. It is easy to see that M is Legendrian if and only if the cone

$$C(M) = \{tx \mid t > 0, x \in M\}$$

is Lagrangian in $\mathbb{C}^3$. If $M \subset S^5$ is Legendrian, then there is a natural map ϕ from M to Fl_1 defined by $\phi(x) = (x, V(x))$, where $V(x)$ is the real linear subspace $\mathbb{R}x + TM_x$. It is known that (cf. [42, 52, 50, 64]) that the following statements are equivalent:

(i) M is minimal Legendrian in S^5,

(ii) the cone $C(M)$ is minimal Lagrangian in $\mathbb{R}^6 = \mathbb{C}^5$,

(iii) the Gauss map $\phi : M \to Fl_1$ is primitive.

Let $\pi : S^5 \to \mathbb{C}P^2$ be the Hopf fibration, N a surface in $\mathbb{C}P^2$, and $\tilde{N}$ a horizontal lift of N in S^5 with respect to the connection α (the contact form). Then N is minimal Lagrangian in $\mathbb{C}P^2$ if and only $\tilde{N}$ is minimal Legendrian in S^5. Hence there are three surface geometries associated to the first (G, τ, σ)-system associated to the 6-symmetric space Fl_1: minimal Lagrangian surfaces in $\mathbb{C}P^2$, minimal Legendrian surfaces in S^5, and minimal Lagrangian cones in $\mathbb{R}^6$.

Example 7.2.3. Hamiltonian stationary surfaces in $\mathbb{C}P^2$

Let N be a Kähler manifold. Given a smooth function f on N, let X_f denote the Hamiltonian vector field associated to f. A Lagrangian submanifold M is called *Hamiltonian stationary* if it is a critical point of the area functional A with respect to any Hamiltonian deformation, i.e.,

$$\left.\frac{\partial}{\partial t}\right|_{t=0} A(\phi_t(M)) = 0$$

for all f, where ϕ_t is the one-parameter subgroup generated by X_f. This class of submanifolds was studied by Schoen and Wolfson in [61]. When N is a four dimensional Hermitian symmetric space U/H, Hélein and Romon proved that the Gauss-Codazzi equation for Hamiltonian stationary surfaces is the 2nd (G, τ, σ)-system, where τ is the involution that gives U and σ is an order four automorphism such that σ^2 gives rise to the natural complex structure of U/H. In particular, they proved that if M is a Hamiltonian stationary Lagrangian surface of $\mathbb{C}P^2$, then locally the Gauss-Codazzi equation for M is the 2nd (G, τ, σ)-system (6.2.4) given by Example 6.2.4. Conversely, if (u_0, u_1, u_2) is a solution of (6.2.4), then for each non-zero $r \in \mathbb{R}$, $E_r E_{-r}^{-1}$ is a Hamiltonian stationary Lagrangian surface of $\mathbb{C}P^2$, where E_λ is the frame of the Lax pair (6.2.2) corresponding to (u_0, u_1, u_2).

§8. Symmetries of the (G, τ)-systems

There have been extensive studies on harmonic maps from a Riemann surface to a compact Lie group U. For example, there are loop group actions, finite unitons, finite type solutions, and a method of constructing all local harmonic maps from meromorphic data. The equation for harmonic maps from $\mathbb{C}$ to U is the first (G, τ)-system. Most results for the first (G, τ)-system hold for the m-th (G, τ)-system as well. We will give a brief review here. For more details, see [27, 38, 39, 72].

8.1. The action of $\Omega_-^\tau(G)$

Let $u = (u_0, \cdots, u_m) : \mathbb{C} \to \prod_{i=0}^m \mathcal{G}$ be a solution of the m-th (G, τ)-system (6.1.1), θ_λ the corresponding Lax pair (6.1.2), and E the frame of θ_λ, i.e.,

$$\begin{cases} E^{-1}E_z = \sum_{i=0}^m u_i \lambda^{-i}, \\ E^{-1}E_{\bar z} = \sum_{i=0}^m \tau(u_i)\lambda^i, \\ E(0,0,\lambda) = \mathrm{I}. \end{cases}$$

Let $E(z, \bar z)(\lambda) = E(z, \bar z, \lambda)$. Since θ_λ satisfies the (G, τ)-reality condition,

$$\tau(E(z, \bar z)(1/\bar\lambda)) = E(z, \bar z)(\lambda),$$

i.e., $E(z, \bar z) \in \Omega_+^\tau(G)$. Given $g \in \Omega_-^\tau(G)$, we can use Theorem 4.2.3 to factor $gE(z, \bar z) = \tilde E(z, \bar z)\tilde g(z, \bar z)$ with $\tilde E(z, \bar z) \in \Omega_+^\tau(G)$ and $\tilde g(z, \bar z) \in \Omega_-^\tau(G)$ for z in an open subset of the origin, i.e., $\tilde E(z, \bar z) = g * E(z, \bar z)$

the dressing action. A direct computation gives

$$\tilde{E}^{-1}\tilde{E}_z = -\tilde{g}_z\tilde{g}^{-1} + \tilde{g}\left(\sum_{j=0}^{m} u_j\lambda^{-j}\right)\tilde{g}^{-1}. \tag{8.1.1}$$

Since $\tilde{g}(z,\bar{z})(\lambda)$ is holomorphic at $\lambda = 0$, the right hand side of (8.1.1) has a pole of order at most m at $\lambda = 0$. Hence there exist some $\tilde{u}_0, \cdots, \tilde{u}_m$ such that

$$\tilde{E}^{-1}\tilde{E}_z = \sum_{j=0}^{m} \tilde{u}_j\lambda^{-j}.$$

Since $\tilde{E}$ satisfies the (G,τ)-reality condition $\tau(\tilde{E}(1/\bar{\lambda})) = \tilde{E}(\lambda)$, $\tilde{E}^{-1}d\tilde{E}$ satisfies the (G,τ)-reality condition (4.2.2). Hence

$$\tilde{E}^{-1}d\tilde{E} = \sum_{j=0}^{m} \lambda^{-j}\tilde{u}_j dz + \lambda^j\tau(\tilde{u}_j)\, d\bar{z}.$$

In other words, $\tilde{u} = (\tilde{u}_0, \cdots, \tilde{u}_m)$ is a solution of the m-th (G,τ)-system. Moreover, $g * u = \tilde{u}$ defines an action of $\Omega_-^\tau(G)$ on the space of solutions of the m-th (G,τ)-system. This gives the following theorem of Uhlenbeck [72] (see also [38, 40]).

Theorem 8.1.1. *([38, 72]). Let E be the frame of a solution u of the m-th (G,τ)-system (6.1.1), and $g \in \Omega_-^\tau(G)$. Then the dressing action $\tilde{E}(z,\bar{z}) = g * E(z,\bar{z})$ is the frame of another solution $\tilde{u} = g * u$. Moreover, $(g,u) \mapsto g * u$ defines an action of $\Omega_-^\tau(G)$ on the space of solutions of the m-th (G,τ)-system.*

If $g \in \Omega_-^\tau(G)$ is a rational map with only simple poles, then the factorization of $gE(z,\bar{z}) = \tilde{E}(z,\bar{z})\tilde{g}(z,\bar{z})$ with $\tilde{E}(z,\bar{z}) \in \Omega_+^\tau(G)$ and $\tilde{g}(z,\bar{z}) \in \Omega_-^\tau(G)$ can be computed by an explicit formula in terms of g and E. In fact, if g has only one simple pole at $\alpha \in \mathbb{C} \setminus S^1$, then the factorization can be done by one of the following methods:

(i) Equate the residues of both sides of

$$g(\lambda)E(z,\bar{z},\lambda) = \tilde{E}(z,\bar{z},\lambda)\tilde{g}(z,\bar{z},\lambda)$$

at the pole $\lambda = \alpha$ to get an algebraic formula for $g * u$ in terms of g and E.

(ii) Let $\tilde{\theta}_\lambda = \tilde{E}^{-1}d\tilde{E}$. Equate the coefficient of λ^j in $\tilde{\theta}_\lambda\tilde{g} = d\tilde{g} + \tilde{g}\theta_\lambda$ for each j to get a system of compatible ordinary differential equations. Then $g*u$ can be obtained from the solution of this system of compatible ODEs.

Example 8.1.2. ([72]). Let $G = GL(n, \mathbb{C})$, and $\tau(g) = (\bar{g}^t)^{-1}$. The fixed point set U of τ is $U(n)$. Let V be a complex linear subspace of $\mathbb{C}^n$, π the Hermitian projection of $\mathbb{C}^n$ onto V, $\pi^\perp = \mathrm{I} - \pi$, $\alpha \in \mathbb{C} \setminus S^1$, and

$$f_{\alpha,\pi}(\lambda) = \pi + \zeta_\alpha(\lambda)\pi^\perp,$$

where $\zeta_\alpha(\lambda) = \frac{(\lambda-\alpha)(\bar{\alpha}-1)}{(\bar{\alpha}\lambda-1)(1-\alpha)}$. Note that $f_{\alpha,\pi}$ satisfies the (G,τ)-reality condition:

$$\overline{f(1/\bar{\lambda}))}^t f(\lambda) = \mathrm{I}.$$

If E is the frame of a solution u of the m-th (G,τ)-system (6.1.1), then for each $(z,\bar{z})$ the factorization $f_{\alpha,\pi}E(z,\bar{z})$ must be of the form

$$f_{\alpha,\pi}E(z,\bar{z}) = \tilde{E}(z,\bar{z})\tilde{f}_{\alpha,\tilde{\pi}(z,\bar{z})} \tag{8.1.2}$$

for some $\tilde{E}(z,\bar{z}) \in \Omega_+^\tau(G)$ and projection $\tilde{\pi}(z,\bar{z})$. Method (i) leads to the conclusion that the image $\tilde{V}(z,\bar{z})$ of $\tilde{\pi}(z,\bar{z})$ is

$$\tilde{V}(z,\bar{z}) = \overline{(E(z,\bar{z})(\alpha))}^t(V). \tag{8.1.3}$$

Moreover,

$$f_{\alpha,\pi} * E = f_{\alpha,\pi}Ef^{-1}_{\alpha,\tilde{\pi}(z,\bar{z})} = (\pi + \zeta_\alpha(\lambda)\pi^\perp)\ E\ (\tilde{\pi} + \zeta_\alpha(\lambda)^{-1}\tilde{\pi}^\perp) \tag{8.1.4}$$

is the frame of $f_{\alpha,\pi} * u$. For example, if $a \in \mathcal{U}$ is a constant, then a is a constant solution of the 1st normalized (G,τ)-system (6.1.1) with Lax pair $\theta_\lambda = a(\lambda^{-1}dz + \lambda d\bar{z})$ and frame $E_\lambda(z) = \exp(a\lambda^{-1}z + a\lambda\bar{z})$. The corresponding harmonic map is

$$s = E_{-1}(z)E_1^{-1}(z) = \exp(-2a(z+\bar{z})) = \exp(-4ax), \quad z = x + iy,$$

which is a geodesic. Since E is given explicitly for the constant solution, $f_{\alpha,\pi} * a$ is given explicitly and so is the harmonic map $f_{\alpha,\pi} * s$.

8.2. The DPW method and harmonic maps with finite uniton number

It is well-known that minimal surfaces in $\mathbb{R}^3$ have Weierstrass representations, i.e., they can be constructed from meromorphic functions. Dorfmeister, Pedit, and Wu gave a construction (the DPW method) of harmonic maps using meromorphic maps and the Iwasawa loop group factorization (Theorem 4.2.5). They call this construction of harmonic maps the *Weierstrass representation of harmonic maps*. The equation for harmonic maps from $\mathbb{C}$ to U is the first normalized (G,τ)-system. The DPW method works for the m-th normalized (G,τ)-system (6.3.1) as well. To explain the DPW method, we need the Iwasawa loop group factorization $L(G) = L_e(U) \times L_+(G)$, i.e., every $g \in L(G)$ can be factored uniquely as $g = g_1g_2$ with $g_1 \in L_e(U)$ and $g_2 \in L_+(G)$. Let U denote the fixed point set of τ. Recall that $L_e(U)$ is the subgroup of $g \in L(U)$ such that $g(1) = e$ and $L_+(G)$ the space of smooth loops $g : S^1 \to G$ that are boundary value of holomorphic maps defined in $|\lambda| < 1$. The following Theorem was proved in [27] for the first (G,τ)-system (the harmonic map equation), but their proof works for the m-th (G,τ)-system as well.

Theorem 8.2.1. *([27]). Let $\mathcal{O}$ be a simply connected, open subset of $\mathbb{C}$, and $\mu(z,\lambda) = \sum_{j\geq -m} h_j(z)\lambda^j$ holomorphic in $z \in \mathcal{O}$ and smooth in $\lambda \in S^1$. Let $H : \mathcal{O} \times S^1 \to G$ be a solution of*

$$\begin{cases} H^{-1}H_z = \sum_{j\geq -m} h_j(z)\lambda^j, \\ H^{-1}H_{\bar z} = 0. \end{cases}$$

Then:

(i) H can be factored as $H(z,\lambda) = F(z,\bar z,\lambda)\phi(z,\bar z,\lambda)$ such that $F(z,\bar z,\cdot) \in L_e(U)$ and $\phi(z,\bar z,\cdot) \in L_+(G)$.

(ii) $F^{-1}F_z$ is of the form $\sum_{i=1}^m(\lambda^{-i}-1)f_i$ and $f_\mu = (f_1,\cdots,f_m)$ is a solution of the normalized m-th (G,τ)-system.

(iii) Every solution of the m-th (G,τ)-system can be constructed from some μ.

Proof. We give a sketch of the proof (for more details see [27]). Statement (i) follows from the Iwasawa loop group factorization 4.2.5. The Iwasawa loop group factorization $L(G) = L_e(U)L_+(G)$ implies that there is a Lie algebra factorization

$$\mathcal{L}(\mathcal{G}) = \mathcal{L}_e(\mathcal{U}) + \mathcal{L}_+(\mathcal{G}). \tag{8.2.1}$$

In fact, we can use Fourier series to write down the Lie algebra factorization easily: Given $\xi = \sum_{j\in\mathbb{Z}}\xi_j\lambda^j$, then $\xi = \eta + \zeta$, where

$$\eta = \sum_{j=1}^{\infty}(\xi_{-j}(\lambda^{-j}-1) + \tau(\xi_{-j})(\lambda^j - 1)) \ \in \ \mathcal{L}_e(\mathcal{U}),$$

$$\zeta = b_0 + \sum_{j=1}^{\infty}(\xi_j - \tau(\xi_{-j}))\lambda^j \ \in \ \mathcal{L}_+(\mathcal{G}).$$

Since $F = H\phi^{-1}$, we have $F^{-1}dF = \phi H^{-1}dH\phi^{-1} - (d\phi)\phi^{-1}$. Let p_1, p_2 denote the projection of $\mathcal{L}(\mathcal{G})$ onto $\mathcal{L}_e(\mathcal{U})$ and $\mathcal{L}_+(\mathcal{G})$ with respect to (8.2.1). Since $(d\phi)\phi^{-1} \in \mathcal{L}_+(\mathcal{G})$ and $F^{-1}dF \in \mathcal{L}(\mathcal{U})$, we have $F^{-1}dF = p_1(\phi(H^{-1}dH)\phi^{-1})$. It follows from the fact that $\phi^{-1}d\phi \in \mathcal{L}_+(\mathcal{G})$ and $H^{-1}dH = \sum_{j\geq -m} h_j(z)\lambda^{-j}dz$ that we have $F^{-1}dF = \sum_{j=0}^{m} f_j(\lambda^{-j} - 1)dz + \tau(f_j)(\lambda^j - 1)d\bar{z}$ for some $f_0, \cdots, f_m$. This proves (ii).

Let $u = (u_1, \cdots, u_m)$ be a solution of the m-th (G, τ)-system, and E a trivialization of the corresponding Lax pair. To prove (iii), it suffices to find $h(z, \lambda)$ so that $g = Eh^{-1}$ is holomorphic in $z \in \mathcal{O}$. Since we want

$$g^{-1}dg = h\left(\sum_{j=1}^{m}(\lambda^{-j} - 1)u_j dz + (\lambda^j - 1)\tau(u_j)d\bar{z}\right)h^{-1} - dhh^{-1}$$

has no $d\bar{z}$ term, we must solve for h from $h^{-1}h_{\bar{z}} = \sum_{j=1}^{m}(\lambda^j - 1)\tau(u_j)$. Since the right hand side lies in $\mathcal{L}_+(\mathcal{G})$, $h(z, \cdot)$ lies in $L_+(G)$. Hence $g^{-1}dg = \sum_{j=1}^{m}(\lambda^{-j} - 1)hu_jh^{-1}dz$. Hence g is holomorphic in z and $g^{-1}g_z$ is of the form $\sum_{j\geq -m} h_j(z)\lambda^j$. Q.E.D.

It is proved in [27] that finite type solutions arise from constant normalized potentials. We give a brief explanation next. Let $\xi \in L(\mathcal{G})$, and $H = \exp(z\xi(\lambda))$. So $H^{-1}H_z = \xi(\lambda)$, $H^{-1}H_{\bar{z}} = 0$, and $H(0, \lambda) = e$. Factor

$$\exp(z\xi) = F(x, y)\phi(x, y), \tag{8.2.2}$$

with $F(x, y) \in L_e(U), \phi(x, y) \in L(G)$, where $z = x + iy$. Then

$$H\xi H^{-1} = \exp(z\xi)\xi\exp(-z\xi) = \xi = F\phi\xi\phi^{-1}F^{-1}.$$

This implies that

$$F^{-1}\xi F = \phi\xi\phi^{-1}. \tag{8.2.3}$$

Differentiate (8.2.2) to get $\xi dz = \phi^{-1}F^{-1}dF\phi + \phi^{-1}d\phi$. We obtain $\phi\xi\phi^{-1}dz = F^{-1}dF + d\phi\phi^{-1}$. Hence $F^{-1}dF = p_1(\phi\xi\phi^{-1}(dx + i\ dy))$, where p_1 is the projection of $\mathcal{L}(\mathcal{G})$ to $\mathcal{L}_e(\mathcal{U})$. By (8.2.3), we get $F^{-1}dF = p_1(F^{-1}\xi F(dx + i\ dy))$. But $d(F^{-1}\xi F) = [F^{-1}\xi F, F^{-1}dF]$. So we have

$$d(F^{-1}\xi F) = [F^{-1}\xi F, p_1(F^{-1}\xi F(dx + i\ dy))].$$

Equivalently,

$$\begin{cases} (F^{-1}\xi F)_x = [F^{-1}\xi F, p_1(F^{-1}\xi F)], \\ (F^{-1}\xi F)_y = [F^{-1}\xi F, p_1(\sqrt{-1}\ F^{-1}\xi F)]. \end{cases} \tag{8.2.4}$$

Let $\xi(\lambda) = \lambda^{d-m}V(\lambda)$. Then (8.2.4) becomes

$$\begin{cases} (F^{-1}VF)_x = [F^{-1}VF, p_1(\lambda^{d-m}F^{-1}VF)], \\ (F^{-1}VF)_y = [F^{-1}VF, p_1(\sqrt{-1}\ \lambda^{d-m}F^{-1}VF)]. \end{cases} \tag{8.2.5}$$

Let $\eta = F^{-1}VF$. Then 8.2.5 becomes

$$\begin{cases} \eta_x = [\eta, p_1(\lambda^{d-m}\eta)], \\ \eta_y = [\eta, p_1(i\ \lambda^{d-m}\eta)]. \end{cases} \tag{8.2.6}$$

Note that this equation leaves the following finite dimensional submanifold of $\mathcal{L}(\mathcal{G})$ invariant:

$$\mathcal{V}_d = \left\{ \eta \in \mathcal{L}(\mathcal{G}) \middle| \ \eta(\lambda) = \sum_{|j| \leq d} \eta_j \lambda^j \right\}.$$

Hence given V in $\mathcal{V}_d$, we can solve the ODE system ((8.2.6)) to get $\eta(x, y)$ such that $\eta(0, 0) = V$. System (8.2.6) is solvable if $p_1(\lambda^{d-m}\eta dz)$ is flat. So there exists $F(x, y) \in L_e(U)$ such that

$$F^{-1}dF = p_1(\lambda^{d-m}\eta dz),$$

i.e., F is a trivialization of the Lax pair of a solution of the normalized m-th (G, τ)-system. This is the method of constructing finite type solutions developed by Pinkall and Sterling in [56] and Burstall, Ferus, Pedit and Pinkall in [13].

All local solutions can also be constructed from meromorphic data μ that are polynomial in λ^{-1}. To explain this, we need

Theorem 8.2.2. *([27]). With the same notation as in Theorem 8.2.1, then there exists a discrete set $S \subset \mathcal{O}$ such that for $z \in \mathcal{O} \setminus S$, H can be factored as*

$$H(z, \lambda) = g_-(z, \lambda)g_+(z, \lambda)$$

with $g_-(z, \cdot) \in L_-^\tau(G)$ and $g_+(z, \cdot) \in L_+^\tau(G)$ via the Gauss loop group factorization. Moreover,

(i) $g_-(z, \lambda)$ is holomorphic in $z \in \mathcal{O} \setminus S$ and has poles at $z \in S$,

(ii) $g_-^{-1}dg_- = \sum_{j=1}^m \lambda^{-j}\eta_j(z)dz$ for some $\mathcal{G}$-valued meromorphic map η_j on $\mathcal{O}$.

Note that if we factor g_- via the Iwasawa loop group factorization (Theorem 4.2.5), then the $L_e(U)$ factor of g_- is the same E constructed in Theorem 8.2.1. This follows from $g_- = Hg_+^{-1} = E\phi g_+^{-1} = E(\phi g_+^{-1})$. The converse is also true. In fact, we have

Corollary 8.2.3. *Let* $\mu(\lambda, z) = \sum_{j=1}^m \lambda^{-j}\eta_j(z)$ *such that* η_j *are meromorphic. If there exists* $h(z,\lambda)$ *satisfying* $h^{-1}dh/dz = \mu$*, then the* $L_e(U)$*-factor* $E(z,\cdot)$ *of* $h(z,\cdot)$ *is a trivialization of some solution* $f_\mu = (f_1, \cdots, f_m)$ *of the* m*-th* (G,τ)*-system, i.e.,*

$$E^{-1}dE = \sum_{j=1}^{m}(\lambda^{-j} - 1)f_j(z)dz + (\lambda^j - 1)\tau(f_j)d\bar{z}.$$

Moreover, every local solution of the m*-th* (G,τ)*-system can be constructed this way.*

The 1-form $\mu(z,\lambda) = \sum_{j=1}^m \eta_j(z)\lambda^{-j}$ is called the *meromorphic potential* or the *normalized potential*. However, for a general normalized potential μ the solution f_μ might have singularities. An important problem is to identify meromorphic potentials μ so that the corresponding solution f_μ of the m-th (G,τ)-system can be extended to a complete surface. Burstall and Guest [14] have identified μ's that give rise to harmonic maps with finite uniton number. We explain some of their results next.

Burstall and Guest noted that if $\mu = \lambda^{-1}h(z)$ is nilpotent and $h(z)$ has no simple poles, then the equation

$$\begin{cases} H^{-1}H_z = \lambda^{-1}h(z), \\ H^{-1}H_{\bar{z}} = 0 \end{cases} \tag{8.2.7}$$

can be solved by integrations. We use $G = SL(n,\mathbb{C})$ to explain this. Let $\mathcal{N}$ denote the strictly upper triangular matrices in $sl(n,\mathbb{C})$, and $h : \mathcal{O} \to \mathcal{N}$ meromorphic. To solve (8.2.7), we may assume

$$H(z,\lambda) = I + b_1(z)\lambda^{-1} + b_2(z)\lambda^{-2} + \cdots$$

with meromorphic b_j's. Equate coefficients of λ^j to get

$$(b_1)_z = h, \ (b_2)_z = b_1 h, \ (b_3)_z = b_2 h, \ \cdots .$$

Since $\mathcal{N}^n = 0$, if we assume that $h(z)$ has no simple poles and the initial data $b_j(0) = 0$ then $b_1, \cdots, b_{n-1}$ can be solved by integration, $b_j = 0$ for all $j \geq n$, and $H(z,\lambda)$ is a polynomial of degree $\leq n-1$ in λ^{-1}.

Motivated by this computation and Uhlenbeck's finite uniton solutions, they make the following definition:

Definition 8.2.4. A harmonic map s from a Riemann surface M to U is said to have *finite uniton number* if there is a meromorphic $h : M \to \mathcal{G}$ such that (8.2.7) has a solution $H(z, \lambda)$ satisfying the following conditions:

(i) $H(z, \lambda)$ is meromorphic in $z \in M$ and a polynomial in λ and λ^{-1},

(ii) $s = E(\cdot, -1)$, where $E(z, \cdot)$ is the $L_e(U)$-component of the Iwasawa factorization of $H(z, \cdot)$.

In other words, s is the harmonic map constructed from the normalized potential $\mu = \lambda^{-1}h(z)$.

Theorem 8.2.5. *([14, 39]). If M is a Riemann surface and $s : M \to U$ is harmonic map with finite uniton number, then there exists a complex extended solution H (associated to s) of the form*

$$H(z, \lambda) = \exp(\lambda^{-1} b_1(z) + \cdots + \lambda^{-r} b_r(z)),$$

where $b_1, \cdots, b_r$ are meromorphic maps from M to the nilpotent subalgebra $\mathcal{N}$ of the Iwasawa decomposition $\mathcal{G} = \mathcal{K} + \mathcal{A} + \mathcal{N}$. Moreover,

(i) integer r can be computed in terms of root system of G,

(ii) the maps $b_2, \cdots, b_r$ satisfies a meromorphic ordinary differential equation, which can be solved by quadrature for any choice of b_1.

In fact, the normalized potential μ corresponding to the harmonic map constructed by Theorem 8.2.5 is $\mu = \lambda^{-1}(b_1)_z$.

8.3. Some comparisons

Let G be a complex, semi-simple Lie group, and U the maximal compact subgroup of G, and τ the corresponding involution with fixed point U. We have discussed the constructions of solutions of soliton equations in the U-hierarchy in Chapter 5 and of equations in the (G, τ)-hierarchy in section 8.2. Loop group factorizations are used in both cases. In this section, we give a summary and some comparisons of these constructions of solutions for the two hierarchies. To make the exposition easier to follow, we will not give references in this section (for references see the previous sections).

Let $\mathcal{A}$ be a maximal abelian subalgebra of $\mathcal{U}$, and $a \in \mathcal{A}$ a regular element. For the U-hierarchy defined by a, the data we use to construct solutions for the (b, j)-flow in the $\mathcal{U}$-hierarchy of soliton equations is one of the following types of maps:

(i) f is a holomorphic map from a neighborhood of $\lambda = \infty$ in $S^2 = \mathbb{C} \cup \{\infty\}$ to G that satisfies the U-reality condition, $\tau(f(\bar{\lambda})) = f(\lambda)$, and $f(\infty) = \mathrm{I}$,

(ii) $f : \mathbb{R} \to G$ is smooth, has an asymptotic expansion at ∞, $f(\infty) = \mathrm{I}$, f is the boundary value of a holomorphic map on the upper half plane, and f_b is rapidly decaying at infinity, where $f = f_u f_b$ is the pointwise Iwasawa factorization of $G = UB$, i.e., $f_u \in U$ and $f_b \in B$,

(iii) $f = f_1 f_2$, where $f_1 : S^2 = \mathbb{C} \cup \{\infty\} \to G$ is a rational map of type (i) and f_2 is of type (ii),

(iv) $f = f_1 f_2$, where f_1 is of type (i) and f_2 is of type (ii).

To construct solutions, we start with an f of type (i), (ii), (iii), or (iv), then factor $f^{-1} e_{a,1}(x) e_{b,j}(t)$ as $E(x,t) m(x,t)^{-1}$ with $E(x,t) \in \Lambda_+^\tau(G)$ and $m(x,t)$ of type (i), (ii), (iii) or (iv) accordingly, where $e_{\xi,j}(t) = e^{\xi \lambda^j t}$. Then

$$u^f(x,t) = [a, m_1(x,t)]$$

is a solution of the (b, j)-flow in the U-hierarchy, where $m_1(x,t)$ is the coefficient of λ^{-1} in the expansion of $m(x,t)(\lambda)$ at $\lambda = \infty$:

$$m(x,t,\lambda) \sim \mathrm{I} + m_1(x,t)\lambda^{-1} + \cdots .$$

Moreover, we know:

(1) $u^f = u^g$ if and only if $f = hg$ for some A-valued map h.

(2) u^f is a local real analytic solution if f is of type (i).

(3) If f is of type (iii), then $u^f(x,t)$ is a solution defined for all $(x,t) \in \mathbb{R}^2$ and is rapidly decaying in x for each fixed t. The space of such solutions u^f is open and dense in the space of all rapidly decaying solutions.

(4) If u is a finite gap solution (an algebraic geometric solution described by theta functions), then there exists an f of type (i) such that $f^{-1} a f$ is a polynomial in λ^{-1} and $u = u^f$.

(5) If f is of type (i) and is a rational map from S^2 to G, then u^f is a pure soliton solution.

For the normalized m-th (G, τ)-system, we start with meromorphic potential $\mu(z, \lambda) = \sum_{j=1}^m \eta_j(z) \lambda^{-j}\, dz$. There are two steps to construct a solution:

Step 1. Find a solution $H(z,\lambda)$ of $H^{-1}dH = \mu$ that is smooth for all $\lambda \in S^1$ and meromorphic in $z \in \mathcal{O} \subset \mathbb{C}$.

Step 2. Factor H as $F\phi$ with $F \in L_e(U)$ and $\phi \in L_+(G)$. Then $F^{-1}F_z$ is of the form $\sum_{j=1}^m (\lambda^{-j} - 1)v_i$ for some $v_1, \cdots, v_m$. Hence

$$v_\mu = (v_1, \cdots, v_m), \qquad s_\mu = F(\cdot, -1)$$

are a solution of the normalized m-th (G,τ)-system and a harmonic map from $\mathcal{O}$ to U respectively.

For the first normalized (G,τ)-system, to go beyond solutions with finite uniton numbers we note that:

—There is no simple condition on μ to guarantee that Step 1 can be done.

—Every local smooth solution can be constructed from some μ. However, in general, there is no canonical choice of μ.

—One of the main open problems is to identify the set of μ so that s_μ can be extended to a harmonic map on a closed surface.

References

[1] M. J. Ablowitz and P. A. Clarkson, Solitons, non-linear evolution equations and inverse scattering, Cambridge Univ. Press, 1991.

[2] M. J. Ablowitz, D. J. Kaup, A. C. Newell and H. Segur, The inverse scattering transform - Fourier analysis for nonlinear problems, Stud. Appl. Math., **53** (1974), 249–315.

[3] R. Beals and R. R. Coifman, Scattering and inverse scattering for first order systems, Commun. Pure Appl. Math., **37** (1984), 39–90.

[4] R. Beals, P. Deift and C. Tomei, Direct and inverse scattering on the line, Math. Surveys Monogr., **28**, Amer. Math. Soc., Providence, RI, 1988.

[5] R. Beals and D. H. Sattinger. On the complete integrability of complete integrable systems, Commun. Math. Phys., **138** (1991), 409–436.

[6] A. I. Bobenko, All constant mean curvature tori in R^3, S^3, H^3 in terms of theta functions, Math. Ann., **290** (1991), 1–45.

[7] A. I. Bobenko, Discrete indefinite affine spheres, Discrete integrable geometry and physics, Oxford Lecture Ser. Math. Appl., Oxford Univ. Press, New York, **16** (1999), 113–138.

[8] A. I. Bobenko, Exploring surfaces through methods from the theory of integrable systems: the Bonnet Problem, this volume, math.DG/0006216.

[9] J. Bolton, F. Pedit and L. Woodward, Minimal surfaces and the affine field model, J. Reine Angew. Math., **459** (1995), 119–150.
[10] M. Brück, X. Du, J. Park and C. L. Terng, Submanifold geometry of real Grassmannian systems, Mem. Amer. Math. Soc., **155**, no. 735, Amer. Math. Soc., 2002.
[11] F. Burstall, Harmonic tori in spheres and complex projective spaces, J. Reine Angew. Math., **469** (1995), 149–177.
[12] F. Burstall, Isothermic surfaces: conformal geometry, Clifford algebras and integrable systems, Integrable systems, geometry, and topology, AMS/IP Stud. Adv. Math., **36**, Amer. Math. Soc., Providence, RI, 2006.
[13] F. E. Burstall, D. Ferus, F. Pedit and U. Pinkall, Harmonic tori in symmetric spaces and commuting Hamiltonian systems on loop algebras, Ann. of Math. (2), **37** (1993), 173–212.
[14] F. E. Burstall and M. A. Guest, Harmonic two-spheres in compact symmetric spaces, Math. Ann., **309** (1997), 541–572.
[15] F. E. Burstall and F. Pedit, Harmonic maps via Adler-Kostant-Symes Theory, In: Harmonic maps and Integrable Systems, Aspects Math., **E23**, Vieweg, Braunschweig, 1994, pp. 221–272.
[16] F. E. Burstall and F. Pedit, Dressing orbits of harmonic maps, Duke Math. J., **80** (1995), 353–382.
[17] J. Cieśliński, P. Goldstein and A. Sym, Isothermic surfaces in E^3 as soliton surfaces, Phys. Lett. A, **205** (1995), 37–43.
[18] J. Cieśliński, The Darboux-Bianchi transformation for isothermic surfaces. Classical results versus the soliton approach, Differential Geom. Appl., **7** (1997), 1–28.
[19] M. Dajczer and R. Tojeiro, Isometric immersions and the generalized Laplace and elliptic sinh-Gordon equations, J. Reine Angew Math., **467** (1995), 109–147.
[20] M. Dajczer and R. Tojeiro, Flat totally real submanifolds of CP^n and the symmetric generalized wave equation, Tohoku Math. J., **47** (1995), 117–123.
[21] M. Dajczer and R. Tojeiro, The Ribaucour transformation for flat Lagrangian submanifolds, J. Geom. Anal., **10** (2000), 269–280.
[22] M. Dajczer and R. Tojeiro, An extension of the classical Ribaucour transformation, Proc. London Math. Soc., **85** (2002), 211–232.
[23] M. Dajczer and R. Tojeiro, Commuting Codazzi tensors and the Ribaucour transformation for submanifolds, Results Math., **44** (2003), 258–278.
[24] G. Darboux, Lecons sur les Systèmes Orthogonaux, Gauthier-Villars, 2nd ed., 1910.
[25] P. Defit, Integrable Hamiltonian systems, Dynamical systems and probabilitstic methods in PDE, Lectures in Appl. Math., **31** (1991), 103–140.
[26] J. Dorfmeister, H. Gradl and J. Szmigielski, Systems of PDEs obtained from factorization in loop groups, Acta Appl. Math., **53** (1998), 1–58.

[27] J. Dorfmeister, F. Pedit and H. Wu, Weierstrass type representation of harmonic maps into symmetric spaces, Comm. Anal. Geom., **6** (1998), 633–668.

[28] J. Dorfmeister and U. Eitner, Weierstrass type representation of affine spheres, Abh. Math. Sem. Univ. Hamburg, **71** (2001), 225–250.

[29] J. Dorfmeister, Generalized Weierstrass representations of surfaces, preprint.

[30] V. G. Drinfel'd and V. V. Sokolov, Equations of Korteweg-de Vries type and simple Lie algebras, Dokl. Akad. Nauk SSSR, **258** (1981), 11–16.

[31] V. G. Drinfel'd and V. V. Sokolov, Lie algebras and equations of Korteweg-de Vries type (Russian), Current problems in mathematics, **24** (1984), 81–180, Itogi Nauki i Tekhniki, Akad. Nauk SSSR, Vsesoyuz. Inst. Nauchn. i Tekhn. Inform., Moscow.

[32] B. A. Dubrovin, Theta-functions and nonlinear equations (Russian), With an appendix by I. M. Krichever, Uspekhi Mat. Nauk, **36** (1981), no. 2(218), 11–80; English translation: Russian Math. Surveys, 1981, 11–92.

[33] L. D. Faddeev and V. E. Zakharov, KdV equation as a completely integrable Hamiltonian system, Func. Anal. Appl., **5** (1972), 280–287 (English translation).

[34] L. D. Faddeev and L. A. Takhtajan, Hamiltonian methods in the theory of solitons, Springer-Verlag, 1987.

[35] D. Ferus and F. Pedit, Curved flats in symmetric spaces, Manuscripta Math., **91** (1996), 445–454.

[36] D. Ferus and F. Pedit, Isometric immersions of space forms and soliton theory, Math. Ann., **305** (1996), 329–342.

[37] C. Gu, H. Hu and Z. Zhou, Darboux transformations in soliton theory and its geometric applications (Chinese), Modern Mathematics Series, 1999, Shanghai Scientific and Technical Publishers.

[38] M. Guest, Harmonic maps, loop groups, and integrable systems, Cambridge Univ. Press, 1997.

[39] M. Guest, An update on harmonic maps of finite uniton numbers, via the zero curvature equation, In: Integrable systems, topology, and physics, Tokyo, 2000, Contemp. Math., **309**, Amer. Math. Soc., Providence, RI, 2002, pp. 85–113.

[40] J. Harnad, Y. Saint-Aubin and S. Shnider, Bäcklund transformations for nonlinear sigma models with values in Riemannian symmetric spaces, Comm. Math. Phys., **92** (1984), 329–367.

[41] J. Harnad, Y. Saint-Aubin and S. Shnider, The Soliton Correlation Matrix and the Reduction Problem for Integrable Systems, Commun. Math. Phys., **93** (1984), 33–56.

[42] M. Haskins, Special Lagrangian cones, Amer. J. Math., **126** (2004), 845–871.

[43] F. Hélein and P. Romon, Hamiltonian stationary Lagrangian surfaces in Hermitian symmetric spaces, In: Differential geometry and integrable

systems, Tokyo, 2000, Contemp. Math., **308**, Amer. Math. Soc., Providence, RI, 2002, pp. 161–178.

[44] U. Hertrich-Jeromin and F. Pedit, Remarks on the Darboux transform of isothermic surfaces, Doc. Math., **2** (1997), 313–333.

[45] N. J. Hitchin, Harmonic maps from a 2-torus to the 3-sphere, J. Differential Geom., **31** (1990), 627–710.

[46] I. M. Krichever, Methods of algebraic geometry in the theory of nonlinear equations (Russian), Uspehi Mat. Nauk, **32** (1977), no. 6(198), 183–208; English translation: Russian Math., Surveys, **36** (1981), 11–92.

[47] I. M. Krichever and S. P. Novikov, Holomorphic bundles over algebraic curves, and nonlinear equations (Russian), Uspekhi Mat. Nauk, **35** (1980), (216), 47–68; English translation: Russian Math., Surveys 35 (1980), 53–80.

[48] B. A. Kupershmidt and G. Wilson, Modifying Lax equations and the second Hamiltonian structure, Invent. Math., **62** (1981), 403–436.

[49] P. D. Lax, Integrals of nonlinear equations of evolution and solitary waves, Comm. Pure. Appl. Math., **21** (1968), 467–490.

[50] H. Ma and Y. Ma, Totally real minimal tori in CP^2, Math. Z., **249** (2005), 241–267.

[51] I. McIntosh, Global solutions of the elliptic 2D periodic Toda lattice, Nonlinearity, **7** (1994), 85–108.

[52] I. McIntosh, Special Lagrangian cones in $\mathbb{C}^3$ and primitive harmonic maps, J. London Math. Soc. (2), **67** (2003), 769–789.

[53] H. P. McKean, Integrable systems and algebraic curves, In: Global analysis, Proc. Biennial Sem. Canad. Math. Congr., Univ. Calgary, Calgary, Alta., 1978, Lecture Notes in Math., **755**, Springer-Verlag, 1979, pp. 83–200.

[54] A. C. Newell, Solitons in Mathematics and Physics, CBMS-NSF Regional Conf. Ser. in Appl. Math., **48**, Society for Industrial and Applied Mathematics (SIAM), Philadelphia, PA, 1985.

[55] K. Nomizu and T. Sasaki, Affine differential geometry, In: Geometry of affine immersions, Cambridge Tracts in Math., **111**, Cambridge Univ. Press, 1994.

[56] U. Pinkall and I. Sterling, On the classification of constant mean curvature tori, Ann. of Math. (2), **130** (1989), 407–451.

[57] A. Pressley and G. B. Segal, Loop Groups, Oxford Science Publ., Clarendon Press, Oxford, 1986.

[58] C. Rogers and W. K. Schief, Bäcklund and Darboux transformations, In: Geometry and modern applications in soliton theory, Cambridge Texts Appl. Math., Cambridge Univ. Press, 2002.

[59] R. S. Palais, The symmetries of solitons, Bull. Amer. Math. Soc. (N.S.), **34** (1997), 339–403.

[60] D. H. Sattinger, Hamiltonian hierarchies on semi-simple Lie algebras, Stud. Appl. Math., **72** (1984), 65–86.

[61] R. Schoen and J. Wolfson, Minimizing volume among Lagrangian submanifolds, Proc. Sympos. Pure Math., **65**, Amer. Math. Soc., Providence, RI, 1999.

[62] G. B. Segal and G. Wilson, Loop groups and equations of KdV type, Inst. Hautes Études Sci. Publ. Math., **61** (1985), 5–65.

[63] M. A. Semenov-Tian-Shansky, Integrable systems and factorization problems, In: Factorization and integrable systems, Faro, 2000, Oper. Theory Adv. Appl., **141**, Birkhauser, Basel, 2003, pp. 155–218.

[64] R. Sharipov, Minimal tori in the five dimensional sphere in $\mathbb{C}^3$, Theoret. Math. Phys., **87** (1991), 363–369.

[65] L. A. Takhtajan, Hamiltonian system connected with the Dirac equation in Differential Geometry, Lie group and Mechanics, J. Sov. Math., **8**, (1977), 219–228.

[66] K. Tenenblat, Bäcklund's theorem for submanifolds of space forms and a generalized wave equation, Bol. Soc. Brasil. Mat., **16** (1985), 69–94.

[67] K. Tenenblat, Transformations of manifolds and applications to differential equations, Pitman Monogr. Surveys Pure Appl. Math., **93**, Longman, Harlow, 1998.

[68] C. L. Terng, Soliton equations and differential geometry, J. Differential Geom., **45** (1997), 407–445.

[69] C. L. Terng and K. Uhlenbeck, Poisson actions and scattering theory for integrable systems, Surveys in Differential Geometry: Integrable systems; A supplement to J. Differential Geom., **4** (1998), 315–402.

[70] C. L. Terng and K. Uhlenbeck, Bäcklund transformations and loop group actions, Comm. Pure. Appl. Math., **53** (2000), 1–75.

[71] C. L. Terng and K. Uhlenbeck, Geometry of solitons, Notice Amer. Math. Soc., **47** (2000), 17–25.

[72] K. Uhlenbeck, Harmonic maps into Lie group (classical solutions of the Chiral model), J. Differential Geom., **30** (1989), 1–50.

[73] G. Wilson, Algebraic curves and soliton equations, In: Geometry today, Rome, 1984, Progr. Math., **60**, Birkhä user Boston, Boston, MA, 1985, pp. 303–329.

[74] G. Wilson, Infinite dimensional Lie groups and algebraic geometry in soliton theory, Philos. Trans. Roy. Soc. London Ser. A, **315** (1985), 393–404.

[75] G. Wilson, The τ-functions of the $\mathcal{G}$AKNS equations, Integrable systems, the Verdier Memorial, Progress in Math., **115** (1991), 147–162.

[76] V. E. Zakharov and A. B. Shabat, Integration of non-linear equations of mathematical physics by the inverse scattering method, II, Funct. Anal. Appl., **13** (1979), 166–174.

Department of Mathematics
University of California at Irvine
Irvine, CA 92697-3875
USA

E-mail address:
`cterng@math.uci.edu`

Advanced Studies in Pure Mathematics 51, 2008
Surveys on Geometry and Integrable Systems
pp. 489–510

A-integrability of geodesic flows and geodesic equivalence

Peter Topalov

§1. Introduction

A common property of the metrics that admit non-trivial geodesic equivalence is that their geodesic flows admit families of pairwise commuting integrals of a special type (see [14, 7, 16, 17, 13]). The metrics studied in [1] and [20] have integrals with analogous properties. We consider this property as a definition of a new class of (pseudo) Riemannian metrics that we call *A-integrable metrics.* We prove that these metrics inherit the main properties of the metrics that admits non-trivial geodesic equivalence (see [16]).

Let M^n be a smooth n-dimensional manifold. By $\Gamma(E)$ we denote the space of the smooth sections of the vector bundle $\pi : E \to M^n$. Let $P \in \Gamma(\mathrm{Hom}(TM, TM))$. We say that P is *diagonalizable over* $\mathbb{R}$ *on* M^n if for every point $x \in M^n$ there exists a (real) basis in T_xM^n such that the operator $P(x)$ has diagonal form $P(x) = \mathrm{diag}(\lambda_1, \ldots, \lambda_n)$.

Definition 1. *Let g be a (pseudo) Riemannian metric on the manifold M^n and let $A \in \Gamma(\mathrm{Hom}(TM, TM))$ be a self-adjoint operator which is diagonalizable over $\mathbb{R}$ on M^n with respect to g. The metric g is called* A-integrable *if the functions from the one-parameter family*

$$I_c(\xi) \stackrel{\mathrm{def}}{=} \det(A + c\mathbf{1})g((A + c\mathbf{1})^{-1}\xi, \xi) \tag{1}$$

are in involution with respect to the symplectic structure $\omega_g \stackrel{\mathrm{def}}{=} FL_g^*\omega$. Here $FL_g : TM \to T^*M$ is the Legendre transformation corresponding to the metric g and ω is the canonical symplectic structure on T^*M.

Received January 3, 2001.
Revised November 4, 2001.
2000 *Mathematics Subject Classification.* Primary 53D25.

Remark 1. *It seems that the assumption that A is diagonalizable over $\mathbb{R}$ is not essential and the theorems proved in the present paper are likely to be true without this restriction. We refer to [15] where the pseudo-Riemannian analogues of the results proved in [14, 7, 16, 17, 13] are given. The case of arbitrary A is extremely interesting because in this case we do not have orthogonal separation of the variables and the corresponding systems are not locally of Stäckel form.*

Remark 2. *If the functions I_c are in involution, then they are integrals of the geodesic flow of the metric g. Indeed, it is easily seen that $I_c(\xi) = S_{n-1}(\xi)c^{n-1} + S_{n-2}(\xi)c^{n-2} \cdots + S_0(\xi)$ and $S_{n-1}(\xi) = g(\xi, \xi)$.*

Remark 3. *It follows from the definition that every metric is $\mathbf{1}$-integrable, where $\mathbf{1}$ denotes the identity. Moreover, if g is A-integrable, then for any real constant α the metric g is $(A + \alpha\mathbf{1})$-integrable .*

We assign to any operator A on the manifold M^n a decomposition of the manifold $M^n = S(A) \bigsqcup_{i=1}^{n} M^i(A)$, where $M^i(A)$ denotes the set of *the stable points of type* i, i.e., the set of the points $x \in M^n$ such that the operator $A(y) \in \mathrm{Hom}(T_yM, T_yM)$ has exactly i distinct eigenvalues for every point y in an open neighborhood of the point x. Denote by $M(A)$ the set of the stable points $\bigsqcup_{i=1}^{n} M^i(A)$. The closed set $S(M)$ is called *set of the singular points.* It is easily seen that S is nowhere dense.

Definition 2. *The (pseudo) Riemannian metrics g and $\bar{g}$ are called* geodesically equivalent *iff they have the same geodesics (considered as unparametrized curves).*

If g is not a constant multiple of $\bar{g}$, then we say that the metrics are *non-trivially geodesically equivalent.* We say that the (pseudo) Riemannian metric g *admits non-trivial geodesical equivalence* if there exists a (pseudo) Riemannian metric $\bar{g}$ such that g and $\bar{g}$ are non-trivially geodesically equivalent. If, in addition, the metrics g and $\bar{g}$ are simultaneously diagonalizable over $\mathbb{R}$, then we say that the metric g *admits a non-trivial geodesical equivalence of Levi-Civita type (LC-type).*

Theorem 1. *The (pseudo) Riemannian metric g admits a non-trivial geodesical equivalence of LC-type if and only if g is A-integrable, the operator A is non-degenerate on M^n, $A \neq$ constant $\times\mathbf{1}$, and the eigenspaces of A are integrable on $M(A)$. If the metric g is A-integrable and the eigenspaces of A are integrable, then the metrics given by the formula*

$$g_c(\xi, \xi) \stackrel{\mathrm{def}}{=} \frac{1}{\det(A + c\mathbf{1})} g((A + c\mathbf{1})^{-1}\xi, \xi), \tag{2}$$

defined for every real parameter c such that the operator $(A+c\mathbf{1})$ is non-degenerate on M^n, are geodesically equivalent to g. Conversely, if there exists a metric $\bar{g}$ such that g and $\bar{g}$ are simultaneously diagonalizable over $\mathbb{R}$ and geodesically equivalent, then the metric g is A-integrable, where the operator A is given by the formula

$$A^i_j(g,\bar{g}) \overset{\text{def}}{=} \left|\frac{\det \bar{g}}{\det g}\right|^{\frac{1}{n+1}} \bar{g}^{i\alpha} g_{\alpha j}. \tag{3}$$

By definition, the operator A is *non-degenerate on* M^n iff $\det A \neq 0$ on M^n.

Remark 4. *Suppose that the operator $A = A(g,\bar{g})$ is derived from a pair of geodesically equivalent metrics g and $\bar{g}$ (see formula (3)); then the formula*

$$I_{\alpha,\beta}(g,\bar{g})(\xi) \overset{\text{def}}{=} \det(\alpha A + \beta\mathbf{1}) g((\alpha A + \beta\mathbf{1})^{-1}\xi, \xi), \tag{4}$$

gives a family of pairwise commuting integrals of the geodesic flow of the metric g.

Remark 5. *Suppose that g is a (pseudo) Riemannian metric and let A be a self-adjoint operator which is diagonalizable over $\mathbb{R}$ with respect to g. If the metrics given by formula (2) are geodesically equivalent, then the functions given by formula (1) are pairwise commuting integrals of the geodesic flow of the metric g. It is easily checked that $A(g,g_0) = A$, where $g_0 \overset{\text{def}}{=} g_c|_{c=0}$ and the operator $A(g,g_0)$ is derived from formula (3).*

The next theorem gives a description of the decomposition of the manifold corresponding to an A-integrable metric.

Theorem 2. *Suppose that the metric g is A-integrable. If the manifold M^n is connected, then $M^n = S(A) \bigsqcup M^m(A)$, where m is a natural number $m \leq n$. The set of singular points $S(A)$ is nowhere dense and coincides with the set of points on M^n where the number of the distinct eigenvalues of the operator A is less than m. There exist quadratic forms $B_1, \ldots, B_m$ on M^n such that any form from the family given by formula (1) is a linear combination (with constant coefficients) of the forms $B_1, \ldots, B_m$. Considered as functions on the tangent bundle TM^n the forms $B_1, \ldots, B_m$ are functionally independent almost everywhere.*

The smooth functions $F_1, \ldots, F_k$ given on the smooth manifold V^q are called *functionally independent* on V^q iff the set of the points $x \in V^q$ where the differentials $d_xF_1, \ldots, d_xF_k$ are linearly independent is dense in V^q.

Definition 3. *The natural number m that appears in Theorem 2 is called* rank of the A-integrable metric g.

It follows from Theorem 2 that if the operator A of some A-integrable metric g has n different eigenvalues at some point $x_0 \in M^n$, then the geodesic flow of the metric g is completely integrable.

The theorems that follow allow us to produce new A-integrable metrics from a given one.

Theorem 3. *Suppose that the (pseudo) Riemannian metric g is A-integrable; then for every real-analytic function $f(t)$ such that the operator $f(A)$ is non-degenerate and well defined on M^n the metric $g_f(\xi, \xi) \stackrel{\text{def}}{=} g(f(A)\xi, \xi)$ is also A-integrable.*

Theorem 4. *Suppose that the (pseudo) Riemannian metric g is A-integrable; then, provided the operator $(A + c\mathbf{1})$ is non-degenerate on M^n, the metric g_c given by formula (2) is $(\alpha A + \beta\mathbf{1})(A + c\mathbf{1})^{-1}$-integrable, where α and β are arbitrary real parameters.*

Therefore, from an A-integrable metric g we obtain a family (*hierarchy*) of A-integrable metrics

$$\mathbf{S}(g, A) \stackrel{\text{def}}{=} \{g_f | f \text{ satisfies the conditions of Theorem 3}\}. \tag{5}$$

Many classical examples of integrable geodesic flows and mechanical systems lie in hierarchies obtained from a fixed A-integrable metric (see [16] for details). Perhaps the most remarkable fact is that the Liouville integrability of the ellipsoid and the Poisson sphere follow from the Liouville integrability of the standard sphere (all these metrics lie in a fixed hierarchy). Analogous results are true also in the cases of the hyperbolic and the Euclidean plane. Let us describe briefly the corresponding constructions (for details see [16, 13] and [6]).

(a) **The standard sphere and the free motion of a rigid body in Euclidean space.** Let

$$S^n \stackrel{\text{def}}{=} \{\sum_{i=1}^{n+1} x_i^2 = 1\}$$

be the unit sphere embedded in the Euclidean space $\mathbb{R}^{n+1}$ supplied with the Euclidean metric $g_0 \stackrel{\text{def}}{=} dx_1^2 + \cdots + dx_{n+1}^2$. Denote by g the restriction of g_0 to the sphere S^n. Consider a non-degenerate linear transformation C of $\mathbb{R}^{n+1}$. The transformation C induces in a natural way a transformation μ_C of the sphere S^n ($\mu_C(x) \stackrel{\text{def}}{=} \frac{Cx}{||Cx||}$, where $||y||$ is the Euclidean norm of $y \in \mathbb{R}^{n+1}$). It is clear that μ_C preserves the geodesics on the sphere. Therefore, the metrics g and $\bar{g} \stackrel{\text{def}}{=} \mu^* g$ are geodesically equivalent. It follows from Theorem 1 that g is A-integrable where $A \stackrel{\text{def}}{=} A(g, \bar{g})$. Theorem 3 shows that the metrics $g^{(k)}(\xi, \xi) \stackrel{\text{def}}{=} g(A^k \xi, \xi)$ ($k = 1, 2, \dots$) are also A-integrable. It is proved in [16] that $g^{(1)}$ is isometric to the metric of a suitable ellipsoid and the metric $g^{(2)}$ is geodesically equivalent to the metric of a suitable Poisson sphere. Varying the matrix C we obtain all possible ellipsoids and Poisson spheres.

(b) **The hyperbolic plane and the free motion of a rigid body in Minkowski space.** Denote by

$$\mathbf{H}_0^n \stackrel{\text{def}}{=} \{-x_0^2 + x_1^2 + \cdots + x_n^2 = -1\}$$

and

$$\mathbf{S}_1^n \stackrel{\text{def}}{=} \{-x_0^2 + x_1^2 + \cdots + x_n^2 = 1\}$$

the hyperboloids of two sheets and one sheet respectively, embedded in the Minkowski space $\mathbb{R}^{1,n}$, $g_0 \stackrel{\text{def}}{=} -dx_0^2 + dx_1^2 + \cdots + dx_n^2$, and consider the submanifold $M^n \stackrel{\text{def}}{=} \mathbf{H}_0^n \bigsqcup \mathbf{S}_1^n$. Let g be the restriction of the metric g_0 on M^n. Recall that the hyperbolic plane $\mathbf{H}^n$ is isometric to the positive part ($x_0 > 0$) of the hyperboloid of two sheets. As in the previous case any non-degenerate linear transformation C of the plane $\mathbb{R}^{n+1}$ induces a partially defined mapping $\mu_C : M^n \to M^n$ that preserves the geodesics on M^n. It is clear that the metrics g and $\bar{g} \stackrel{\text{def}}{=} \mu_C^* g$ are geodesically equivalent. The metric $\bar{g}$ is not smooth at the points of M^n where the mapping μ_C is not defined. Nevertheless, the operator $A = A(g, \bar{g})$ derived from formula (3) is smooth. As in the previous case we construct the metrics $g^{(k)}$ ($k = 1, 2, \dots$). It is proved in [13] that the metric $g^{(2)}$ (restricted to the hyperbolic plane $\mathbf{H}^n$) is geodesically equivalent to the metric of the analogue of the Poisson sphere corresponding to the free motion of the rigid body in Minkowski space.

(c) **The Euclidean space and the Clebsch case of motion of a rigid body.** Let us consider the projective plane

$$\mathbb{R}P^n \stackrel{\text{def}}{=} \{(x_1 : \cdots : x_{n+1})\}$$

and fix the affine chart $\mathbb{R}^n \cong \{x_{n+1} = 1\} \hookrightarrow \mathbb{R}^{n+1}$. Any non-degenerate linear transformation L of $\mathbb{R}^{n+1}$ gives a projective transformation of $\mathbb{R}P^n$ that acts on the affine chart $\mathbb{R}^n$ as a linear-fractional transformation that we denote by μ_L. It is clear that the partially defined mapping $\mu_L : \mathbb{R}^n - \to \mathbb{R}^n$ preserves the geodesics of the Euclidean metric $g \stackrel{\text{def}}{=} dx_1^2 + \cdots + dx_n^2$ on $\mathbb{R}^n$. It allows us to construct a partially defined metric $\bar{g} \stackrel{\text{def}}{=} \mu_L^* g$ that is geodesically equivalent to the Euclidean metric g. The corresponding operator $A = A(g, \bar{g})$ given by formula (3) is smooth and the metric g is A-integrable. It is proved that the corresponding hierarchy $\mathbf{S}(g, A)$ (given by Theorem 3) contains a metric that is geodesically equivalent to the analogue of the Poisson sphere corresponding to the Clebsch case of motion of the rigid body (see [13]). We consider item c) in more detail in Appendix 1.

The next theorem allows us to construct A-integrable metrics on Cartesian products.

Suppose that the (pseudo) Riemannian metric g_1 given on the manifold M_1 ($\dim M_1 = n_1$) is A_1-integrable and the (pseudo) Riemannian metric g_2 given on the manifold M_2 ($\dim M_2 = n_2$) is A_2-integrable. Denote by M the Cartesian product $M_1 \times M_2$ and let $\pi_i : M \to M_i$ ($i = 1, 2$) be the projections. Let us consider the operators $\tilde{A}_1(\xi) \stackrel{\text{def}}{=} (d_u\pi_1|_{K_2})^{-1}(A_1(d_u\pi_1(\xi)))$, $\xi \in T_uM$, $K_{2u} \stackrel{\text{def}}{=} \ker d_u\pi_2$, and $\tilde{A}_2(\xi) \stackrel{\text{def}}{=} (d_u\pi_2|_{K_1})^{-1}(A_2(d_u\pi_2(\xi)))$, $\xi \in T_uM$, $K_{1u} \stackrel{\text{def}}{=} \ker d_u\pi_1$. Denote by $\sigma_1(A_i), \ldots, \sigma_{n_i}(A_i)$ the elementary symmetric polynomials of the operator A_i ($i = 1, 2$). Let L be a linear operator in the vector space V, $\dim V = n$. The elementary symmetric polynomials of the operator L are defined by the equality $\det(L + c\mathbf{1}) = c^n + \sigma_1(L)c^{n-1} + \cdots + \sigma_n(L)$. It is clear that $\sigma_k(A_i)$ are smooth functions on M_i. Let us consider the functions $\sigma_k^i \stackrel{\text{def}}{=} \pi_i^* \sigma_k(A_i)$. Define the operators $G_1 \stackrel{\text{def}}{=} \tilde{A}_1^{n_2} - \sigma_1^2 \tilde{A}_1^{n_2-1} + \cdots + (-1)^{n_2}\sigma_{n_2}^2 \mathbf{1}_1$ and $G_2 \stackrel{\text{def}}{=} \tilde{A}_2^{n_1} - \sigma_1^1 \tilde{A}_2^{n_1-1} + \cdots + (-1)^{n_1}\sigma_{n_1}^1 \mathbf{1}_2$, where $\mathbf{1}_1$ is the projection on K_{2u} with respect to K_{1u} and $\mathbf{1}_2$ is the projection on K_{1u} with respect to K_{2u}. Define also the forms $\tilde{g}_i(\xi, \eta) \stackrel{\text{def}}{=} g_i(d_u\pi_i(\xi), d_u\pi_i(\eta))$ ($i = 1, 2$), $\xi, \eta \in T_uM$. Let us consider the set $\Lambda_i \stackrel{\text{def}}{=} \{\lambda | \lambda \in \operatorname{Spect} A_i(x), x \in M_i\}$, where $\operatorname{Spect} A$ denotes the spectrum of the linear operator A.

Theorem 5. *Suppose that $\Lambda_1 \cap \Lambda_2 = \emptyset$; then the (pseudo) Riemannian metric*

$$g(\xi, \eta) \stackrel{\text{def}}{=} \tilde{g}_1(G_1\xi, \eta) + \tilde{g}_2(G_2\xi, \eta), \ \xi, \eta \in T_uM, \tag{6}$$

is A-integrable where $A \stackrel{\text{def}}{=} \tilde{A}_1 + \tilde{A}_2$.

§2. Proof of Theorem 1.

We follow the construction of the proof of Theorem 2 in [17]. Assume that the metric g is A-integrable. It follows from our assumption that any two functions from the 1-parameter family

$$I_c(\xi,\xi) \stackrel{\text{def}}{=} \det(A+c\mathbf{1})g((A+c\mathbf{1})^{-1}\xi,\xi) \tag{7}$$

are in involution. Suppose that the point $x_0 \in M^n$ is stable of type m $(0 \leq m \leq n)$, i.e., $x_0 \in M^m(A)$. Therefore, there is an open neighborhood $U(x_0)$ of the point such that the operator A has exactly $m \leq n$ distinct eigenvalues $\phi_1(x) < \phi_2(x) < \cdots < \phi_m(x)$. The corresponding eigenspaces give a natural splitting $TU \cong W_1 \oplus \cdots \oplus W_m$, $\dim W_k = l_k$. Let $X_{k\alpha}$ $(k = 1,\ldots,m; \alpha = 1,\ldots,l_k)$ be an orthonormal frame on U such that $A(X_{k\alpha}) = \phi_k X_{k\alpha}$, $W_k = \operatorname{Span}(X_{k1},\ldots,X_{kl_k})$. Making the Legendre transformation corresponding to the Riemannian metric g we obtain

$$I_c = \Phi(c)\left\{\frac{1}{\phi_1+c}A_1+\cdots+\frac{1}{\phi_m+c}A_m\right\}, \tag{8}$$

where $\Phi(c) \stackrel{\text{def}}{=} (\phi_1+c)^{l_1}\cdots(\phi_m+c)^{l_m}$, $A_k \stackrel{\text{def}}{=} \sum_{\alpha=1}^{l_k}\epsilon_{k\alpha}V_{k\alpha}^2$, $\epsilon_{k\alpha}=\pm 1$, and $V_{k\alpha} \stackrel{\text{def}}{=} \langle p, X_{k\alpha}\rangle$, $p \in T^*U(x_0)$.

First of all, let us consider the case $m = 1$. We obtain $A = \phi\,\mathbf{1}$. Hence, $I_c = (\phi+c)^{n-1}g$ is a 1-parameter family of first integrals of the geodesic flow of the metric g. This gives $0 \equiv \{g, I_c\} = (n-1)(\phi+c)^{n-2}\{g,\phi\}g$. Hence, $\{g,\phi\} \equiv 0$. This yields $\phi = $ constant and $A =$ constant $\times\mathbf{1}$. This contradicts the assumption that $A \neq$ constant $\times\mathbf{1}$.

We will suppose further that $m \geq 2$. Let us take m distinct real numbers $c_1,\ldots,c_m$ such that $c_i \neq -\phi_j$ $(i,j = 1,\ldots,m)$ on $U(x_0)$. Note that such constants exist if $U(x_0)$ is sufficiently small. If necessary, we shrink the neighborhood $U(x_0)$. We obtain m integrals $F_k \stackrel{\text{def}}{=} I_{c_k}$ in involution such that

$$\left\{\begin{array}{ccc} \frac{1}{\Phi_1}F_1 & = & \frac{1}{\phi_1+c_1}A_1+\cdots+\frac{1}{\phi_m+c_1}A_m, \\ \vdots & & \\ \frac{1}{\Phi_m}F_m & = & \frac{1}{\phi_1+c_m}A_1+\cdots+\frac{1}{\phi_m+c_m}A_m. \end{array}\right.$$

As in section 2 in [17], we prove that

$$A_l = \sum_k (-1)^{l+k} \frac{1}{\Pi_l} \frac{1}{C_k} \left(\frac{\prod_{ij} (\phi_i + c_j)}{\prod_{\substack{i \neq l \\ j \neq k}} (\phi_i + c_j)} \frac{1}{\Phi_k} \right) F_k$$

$$= \sum_k (-1)^{l+k} \frac{1}{\Pi_l} \frac{1}{C_k} \frac{(\phi_l + c_1) \cdots (\phi_l + c_{k-1})(\phi_l + c_{k+1}) \cdots (\phi_l + c_m)}{(\phi_1 + c_k)^{l_1 - 1} (\phi_2 + c_k)^{l_2 - 1} \cdots (\phi_m + c_k)^{l_m - 1}} F_k,$$

where $\Pi_l \stackrel{\text{def}}{=} (\phi_m - \phi_l) \cdots (\phi_{l+1} - \phi_l)(\phi_l - \phi_{l-1}) \cdots (\phi_l - \phi_1)$ and $C_k \stackrel{\text{def}}{=} (c_m - c_k) \cdots (c_{k+1} - c_k)(c_k - c_{k-1}) \cdots (c_k - c_1)$. Hence,

$$\mathcal{A}_p \stackrel{\text{def}}{=} \frac{A_p}{(\phi_p + c_1) \cdots (\phi_p + c_{m-1})} \Pi_p C_m [(\phi_1 + c_m)^{l_1 - 1} \cdots (\phi_m + c_m)^{l_m - 1}]$$

$$= \sum_k (-1)^{p+k} \frac{C_m}{C_k} \left(\frac{\phi_1 + c_m}{\phi_1 + c_k} \right)^{l_1 - 1} \cdots \left(\frac{\phi_p + c_m}{\phi_p + c_k} \right)^{l_p} \cdots \left(\frac{\phi_m + c_m}{\phi_m + c_k} \right)^{l_m - 1} F_k,$$

and we write this as

$$\mathcal{A}_p = \sum_k a_{pk} F_k. \tag{9}$$

We are going to prove that

1) $\phi_k =$ constant, if $\dim W_k > 1$;
2) $L_{X_{p\alpha}} \phi_q = 0$ $(p \neq q)$, if $\dim W_q = 1$.

It follows from (9) and the Stäckel-Painlevé theorem proved in [17] (see Proposition 1 in [17]) that $L_{X_{q\alpha}} a_{pk} = 0$ $(q \neq p)$. Without loss of generality we can suppose that $l_1 \geq l_2 \geq \cdots \geq l_m$. If $l_1 = 1$, then $a_{qk} = (-1)^{q+k} \frac{C_m}{C_k} \frac{\phi_q + c_m}{\phi_q + c_k}$. Hence, $L_{X_{p1}} \phi_q = 0$ $(p \neq q)$. Suppose that $l_1 > 1$. Denote by s the largest natural number $m \geq s \geq 1$ such that $l_s > 1$. Observing that $\frac{\phi_i + c_k}{\phi_i + c_m} = (c_k - c_m) \left(\frac{1}{c_k - c_m} + \frac{1}{\phi_i + c_m} \right)$ and applying Lemma 3 in [17] to the coefficients a_{pk} $(k = 1, \ldots, s)$ of the first two equations of (9) we obtain $\phi_k =$ constant $(k = 1, \ldots, s)$. Therefore, $a_{(s+l)1} =$ constant $\times \left(\frac{\phi_{s+l} + c_m}{\phi_{s+l} + c_1} \right)$ $(l = 1, \ldots, m - s)$. It follows that the functions ϕ_{s+l} $(l = 1, \ldots, m - s)$ satisfy condition 2).

Assume in addition that the eigenspaces of the operator A are integrable on $M(A)$. It follows from (9) and the Stäckel-Painlevé theorem that there is a chart $\{(\bar{x}_1, \ldots, \bar{x}_m)\}$, $\bar{x}_k \stackrel{\text{def}}{=} (x_{k1}, \ldots, x_{kl_k})$, given in a neighborhood of the point x_0, such that $A_k = A_k(\bar{x}_i, \bar{p}_k)$, where $p_{k\alpha}$ are the conjugate impulses. Therefore, $A_k = \frac{1}{\Pi_k} Q_k(\bar{x}_k, \bar{p}_k)$. By definition, $g = A_1 + \cdots + A_m$. Hence,

$$g = \frac{Q_1(\bar{x}_1, \bar{p}_1)}{\Pi_1} + \cdots + \frac{Q_m(\bar{x}_m, \bar{p}_m)}{\Pi_m}. \tag{10}$$

Therefore,

$$g = \Pi_1 G_1(\bar{x}_1, d\bar{x}_1) + \cdots + \Pi_m G_m(\bar{x}_m, d\bar{x}_m), \tag{11}$$

where G_k are quadratic forms. Let us consider the metric $\bar{g} \stackrel{\text{def}}{=} \frac{1}{\det A} g A^{-1}$. We have

$$\bar{g} = \rho^1 \Pi_1 G_1(\bar{x}_1, d\bar{x}_1) + \cdots + \rho^m \Pi_m G_m(\bar{x}_m, d\bar{x}_m), \tag{12}$$

where $\rho^i \stackrel{\text{def}}{=} \frac{1}{\phi_1 \dots \phi_m} \frac{1}{\phi_i}$. Now, it is easily checked (using the local coordinates) that the metrics g and $\bar{g}$ given by formulae (11) and (12) are geodesically equivalent. The same fact also follows from the Levi-Civita's theorem (see [5, 3, 11, 16]). Note that Levi-Civita proved his theorem when the metrics g and $\bar{g}$ are positive definite. Nevertheless, it is easily seen that his proof works also in the case of simultaneously diagonalizable pseudo-Riemannian metrics. Finally, the first statement of Theorem 1 follows from the fact that the metric $\bar{g}$ is globally defined on M^n and the set of stable points M is everywhere dense in M^n (see [7]). Formula (2) follows from Proposition 1 in [16].

Conversely, suppose that the (pseudo) Riemannian metric g admits non-trivial geodesical equivalence of LC-type, i.e., there exists a metric $\bar{g}$ such that the metrics g and $\bar{g}$ are non-trivially geodesically equivalent and simultaneously diagonalizable over $\mathbb{R}$. Let us consider the operator $A = A(g, \bar{g})$ given by formula (3). A is a self-adjoint operator with respect to the both metrics and $\bar{g}(\xi, \xi) = \frac{1}{\det A} g(A^{-1}\xi, \xi)$. We prove the involutivity of the functions I_c given by formula (1) in Appendix 2. Theorem 1 is proved.

§3. Properties of the A-integrable metrics.

In this section we prove theorems 2, 3, 4 and 5 that give some of the main properties of the A-integrable metrics.

3.1. Proof of Theorem 2.

Denote by $N(x)$ the number of distinct eigenvalues of the operator A at the point $x \in M^n$. Suppose that $m = \max_{x \in M^n} N(x)$. It is clear that if $N(x) = k$, then $N(x) \geq k$ in a neighborhood of the point x. Therefore, if $N(x) = m$, then x is stable point of type m. It is easily seen that the set of stable points is dense in M^n.

Consider the case when the geodesic flow of the metric g is complete. The general case is easily reduced to this one. Suppose that $M^{m_1}(A) \neq \emptyset$ where $m_1 < m$. Let us take $x_0 \in V_0 \subset M^m(A)$ and $x_1 \in V_1 \subset M^{m_1}(A)$ where V_0 and V_1 are open sets. Taking V_0 and V_1 sufficiently small we can find real constants $c_1, \dots, c_m$ that are not eigenvalues of the operator $(-A_x)$ if x belongs to $V_0 \sqcup V_1$. Let us consider the forms $B_i \stackrel{\text{def}}{=} I_{c_i}$ $(i = 1, \dots, m)$. It is easily shown that for any real number c there exist constants $a_1, \dots, a_m$ such that $I_c = \sum_{i=1}^m a_i B_i$ on V_0. Indeed, it follows from formula (8) and Lemma 2 in [17] that there are smooth functions $a_1, \dots, a_m$ defined in V_0 such that $I_c = \sum_{i=1}^m a_i B_i$. Differentiating the last equality with respect to the Hamiltonian vector field corresponding to the metric g we obtain $0 \equiv \{g, I_c\} = \sum_i \{g, a_i\} B_i$. Hence, $\{g, a_i\} \equiv 0$. This gives $a_i = \text{constant}$.

Let us take a geodesic line $\gamma(t)$ that connects the points x_0 and x_1 ($\gamma(0) = x_0$ and $\gamma(t_0) = x_1$). Without loss of generality we can suppose that the points x_0 and x_1 are not conjugate. It follows from formula (8) and Lemma 2 in [17] that the set Z_{x_0} of points $v \in T_{x_0}M^n$ where the differentials $d_v B_1, \dots, d_v B_m$ are linearly independent is dense in $T_{x_0}M^n$. Hence, there exists a point $v_0 \in Z_{x_0}$ such that $\exp v_0 \in V_1$. The geodesic flow of the metric g preserves the functions $B_1, \dots, B_m$. Therefore, the differentials $d_{v_1} B_1, \dots, d_{v_1} B_m$, where $v_1 \stackrel{\text{def}}{=} \frac{d}{dt}|_{t=1} \exp t v_0$, are also linearly independent. On the other hand, on V_1 we have $B_m = \sum_{j=1}^{m_1} \lambda_j B_j$, $\lambda_j = \text{constant}$. This is a contradiction. Therefore, $m_1 = m$. This proves that all stable points on M^n are of type m. Therefore, $M^n = S(A) \bigsqcup M^m(A)$. It is clear that if $N(x) = m$, then $x \in M^m(A)$. Hence, $S(A) = \{x \in M^n | N(x) < m\}$. Let us prove the last statement of the theorem. Suppose that there is an open set $U \in M^m$ such that for any $x \in U$ the forms $B_1(x), \dots, B_m(x)$ (defined above) are linearly dependent (as quadratic forms in the tangent space $T_x U$). Taking U sufficiently small we can choose constants $\alpha_1, \dots, \alpha_m$ such that $B_k = \sum_{j=1}^m \beta_{kj} I_{\alpha_j}$, where β_{kj} are constants and the forms $I_{\alpha_1}(x), \dots, I_{\alpha_m}(x)$ are linearly independent for any fixed $x \in U$. It is clear that the matrix (β_{kj}) is degenerate. Hence, the differentials $d_v B_1, \dots, d_v B_m$ are linearly dependent on TU. But the arguments used above show that we can find a point $v_0 \in TU$ where the differentials $d_{v_0} B_1, \dots, d_{v_0} B_m$ are linearly independent. This contradiction shows that the set K of points $x \in M^n$ where the forms $B_1(x), \dots, B_m(x)$ form a basis of the linear space generated by the forms $\{I_c(x)|c \in \mathbb{R}\}$ is dense in M^n. Therefore, for any real constant c we have $I_c = \sum_{j=1}^m \mu_j B_j$ where μ_j are functions on K that are locally constants. Let us prove that these functions are constant on K. Suppose that $I_c = \sum_j \mu_j^\alpha B_j$ in some open set $U_\alpha \subset K$ and $I_c = \sum_j \mu_j^\beta B_j$ in some open set $U_\beta \subset K$. Let us take $x_\alpha \in U_\alpha$. It is easily seen that the set of the points $\xi \in T_{x_\alpha} M^n$ where the differentials $d_\xi B_j$ $(j = 1, \dots, m)$ are linearly independent is dense in $T_{x_\alpha} M^n$. Therefore, we can find $\xi_\alpha \in T_{x_\alpha} M^n$ such that $\exp \xi_\alpha = x_\beta \in U_\beta$ and the differentials $d_{\xi_\alpha} B_j$ are linearly independent. Denote by $\zeta^t : TM^n \to TM^n$ the geodesic flow of the metric g and let $\xi_\beta \stackrel{\text{def}}{=} \frac{d}{dt}|_{t=1} \exp(t\xi_\alpha)$. We have that $d_{\xi_\alpha} I_c = \sum_j \mu_j^\alpha d_{\xi_\alpha} B_j$. On the other hand

$$\begin{aligned} d_{\xi_\alpha} I_c &= d_{\xi_\beta} I_c \circ \zeta_*^1 \\ &= \sum_j \mu_j^\beta d_{\xi_\beta} B_j \circ \zeta_*^1 \\ &= \sum_j \mu_j^\beta d_{\xi_\alpha} B_j. \end{aligned}$$

Hence, $\mu_j^\alpha = \mu_j^\beta \stackrel{\text{def}}{=} \mu_j$ $(j = 1, \dots, m)$. This gives $I_c = \sum_j \mu_j B_j$ on K. Therefore, $I_c = \sum_j \mu_j B_j$ on M^n. It is clear that the functions $B_1, \dots, B_m$ are functionally independent almost everywhere on TM^n. This proves the theorem.

3.2. Proof of Theorem 3.

Suppose that the metric g is A-integrable. It is sufficient to prove the theorem in a neighborhood of an arbitrary stable point. Suppose that $x_0 \in M^n$ is a stable point of type m and let the smooth functions $\phi_1(x) < \phi_2(x) < \cdots < \phi_m(x)$ be the eigenvalues of the operator $A(x)$ in a neighborhood $U(x_0)$ of the point x_0. As in the proof of Theorem 1 we see that the condition $m = 1$ implies that $A = \text{constant} \times \mathbf{1}$. In this case the statement of Theorem 3 is obvious. Consider the case $m \geq 2$. Let us consider the functions $F_k \stackrel{\text{def}}{=} I_{c_k}$, where $c_1, \dots, c_m$ are different real constants such that $c_i \neq -\phi_j$ $i, j = 1, \dots, m$ on $U(x_0)$. As in the proof of Theorem 1 we see that equality (9) holds and the functions $\phi_1, \dots, \phi_m$ satisfy conditions 1) and 2) of section 2 (following formula (3)). We have to prove that the 1-parameter family of functions $I_c^f \stackrel{\text{def}}{=} \det(A + c\mathbf{1}) g_f((A + c\mathbf{1})^{-1}\xi, \xi)$ are in involution. Let us consider the functions $F_k^f \stackrel{\text{def}}{=} I_{c_k}^f$. As in the proof of Theorem 1 we see that

$$\begin{aligned} A_p^f &\stackrel{\text{def}}{=} \frac{f(\phi_p)^{-1}A_p}{(\phi_p+c_1)\cdots(\phi_p+c_{m-1})}\Pi_p C_m[(\phi_1+c_m)^{l_1-1}\cdots(\phi_m+c_m)^{l_m-1}] \\ &= \sum_k (-1)^{p+k}\frac{C_m}{C_k}\left(\frac{\phi_1+c_m}{\phi_1+c_k}\right)^{l_1-1}\cdots\left(\frac{\phi_p+c_m}{\phi_p+c_k}\right)^{l_p}\cdots\left(\frac{\phi_m+c_m}{\phi_m+c_k}\right)^{l_m-1} F_k^f \\ &= \sum_k a_{pk}F_k^f. \end{aligned}$$

It is clear that the coefficients a_{pk} satisfy the Stäckel condition (see Proposition 1 in [17]). Moreover, $A_p^f = f(\phi_p)^{-1}A_p$, where the functions $A_1,\dots,A_m$ are in involution. Hence, the functions $A_1^f,\dots,A_m^f$ are in involution. Finally, a variant of the Stäckel-Painlevé theorem shows that the functions $F_1^f,\dots,F_m^f$ are in involution. Note that $g_f = \lim_{c\to\infty} c^{1-n}I_c^f$. Therefore, the functions $F_1^f,\dots,F_m^f$ are integrals of the geodesic flow of the metric g_f. Let us fix an arbitrary real constant α. It follows from the definition of the function I_α^f that $I_\alpha^f = \sum_{j=1}^m \alpha_j F_j^f$, where α_j are smooth functions on $U(x_0)$. The arguments used in the proof of Theorem 2 show that α_j are constants. Therefore, the functions from the family I_c^f are in involution on $M(A)$. Finally, the statement of the theorem follows from the fact that the set of stable points $M(A)$ is everywhere dense in M. Theorem 3 is proved.

3.3. Proof of Theorem 4.

It is sufficient to prove that the functions

$$I_s(\xi) \stackrel{\text{def}}{=} \det\left(\frac{\alpha A+\beta\mathbf{1}}{A+c\mathbf{1}}+s\mathbf{1}\right) g_c\left(\left(\frac{\alpha A+\beta\mathbf{1}}{A+c\mathbf{1}}+s\mathbf{1}\right)^{-1}\xi,\xi\right) \tag{13}$$

are in involution with respect to the symplectic structure $\omega_{g_c} \stackrel{\text{def}}{=} FL_{g_c}^*\omega$, where FL_{g_c} is the Legendre transformation corresponding to the metric g_c and ω is the canonical symplectic structure on T^*M^n. Applying the Legendre transformation and using matrix representation we obtain

$$\begin{aligned} I_s &= \frac{\det((c+\alpha)A+(cs+\beta)E)}{\det(A+cE)}((c+\alpha)A+(cs+\beta)E)^{-1}(A+cE)g_c^{-1} \\ &= \det((c+\alpha)A+(s+c\beta)E)((c+\alpha)A+(s+c\beta)E)^{-1}(A+cE)^2 g^{-1}, \end{aligned}$$

where E is the unit matrix. It follows from Theorem 3 that the metric $g((A+c\mathbf{1})^{-2}\xi,\xi)$ is A-integrable. This proves Theorem 4.

3.4. Proof of Theorem 5.

Suppose that the conditions of Theorem 5 are satisfied. Let us fix two points $x_0 \in M_1$ and $y_0 \in M_2$ and suppose that x_0 is stable of type $m_1 \leq n_1$ and y_0 is stable of type $m_2 \leq n_2$. Consider the point $u_0 = (x_0, y_0) \in M$. It is clear that the point u_0 is stable of type $m_1 + m_2$ with respect to the operator $A \stackrel{\mathrm{def}}{=} \tilde{A}_1 + \tilde{A}_2$. There is an open neighborhood $U(x_0) \subset M_1$ where the operator A_1 has exactly m_1 distinct eigenvalues $\phi_1 < \cdots < \phi_{m_1}$ and an open neighborhood $V(y_0) \subset M_2$ where the operator A_2 has exactly m_2 distinct eigenvalues $\phi_{m_1+1} < \cdots < \phi_{m_1+m_2}$. Denote by $\tilde{\phi}_k$ the functions $\pi_1^*\phi_k$ $(k = 1, \ldots, m_1)$ and $\pi_2^*\phi_k$ $(k = m_1+1, \ldots, m_1+m_2)$ on $U(x_0) \times V(y_0)$. There exist an orthonormal (with respect to the metric g_1) frame $X_{k\alpha_k}$ $(k = 1, \ldots, m_1,\ \alpha_k = 1, \ldots, l_k,\ \sum_{k=1}^{m_1} l_k = n_1)$ on $U(x_0)$, such that $A_1(X_{k\alpha_k}) = \phi_k X_{k\alpha_k}$, and an orthonormal (with respect to the metric g_2) frame $X_{k\alpha_k}$ $(k = m_1+1, \ldots, m_1+m_2,\ \alpha_k = 1, \ldots, l_k,\ \sum_{k=m_1+1}^{m_1+m_2} l_k = n_2)$ on $V(y_0)$, such that $A_2(X_{k\alpha_k}) = \phi_k X_{k\alpha_k}$. Using the natural identification $T_{(x,y)}M = K_2 \oplus K_1 \cong T_xM_1 \oplus T_yM_2$ we obtain a frame $\widetilde{X}_{k\alpha_k}$ $(k = 1, \ldots, m_1+m_2,\ \alpha_k = 1, \ldots, l_k)$ on $U(x_0) \times V(y_0)$. We have to prove that the one-parameter family of functions $I_c(\xi) \stackrel{\mathrm{def}}{=} \det(A+c)g((A+c)^{-1}\xi, \xi)$, $\xi \in TM$ are in involution with respect to the symplectic structure ω_g. Making the Legendre transformation corresponding to the metric g we obtain

$$I_c = \widetilde{\Phi}(c) \left\{ \frac{\widetilde{P}_1}{D_1(\tilde{\phi}_1 + c)} + \cdots + \frac{\widetilde{P}_{m_1+m_2}}{D_{m_1+m_2}(\tilde{\phi}_{m_1+m_2} + c)} \right\}, \tag{14}$$

where $\widetilde{\Phi}(c) \stackrel{\mathrm{def}}{=} (\tilde{\phi}_1 + c)^{l_1} \cdots (\tilde{\phi}_{m_1+m_2} + c)^{l_{m_1+m_2}}$, $\widetilde{P}_k \stackrel{\mathrm{def}}{=} \sum_{\alpha=1}^{l_k} \epsilon_{k\alpha} \widetilde{P}_{k\alpha}^2$, $\epsilon_{k\alpha} = \pm 1$, $\widetilde{P}_{k\alpha} \stackrel{\mathrm{def}}{=} \left\langle p, \widetilde{X}_{k\alpha} \right\rangle$, $p \in T^*M$, and

$$\widetilde{D}_k \stackrel{\mathrm{def}}{=} \begin{cases} (\tilde{\phi}_k - \tilde{\phi}_{m_1+1})^{l_{m_1+1}} \cdots (\tilde{\phi}_k - \tilde{\phi}_{m_1+m_2})^{l_{m_1+m_2}}, & 1 \leq k \leq m_1, \\ (\tilde{\phi}_k - \tilde{\phi}_1)^{l_1} \cdots (\tilde{\phi}_k - \tilde{\phi}_{m_1})^{l_{m_1}}, & m_1 + 1 \leq k \leq m_1 + m_2. \end{cases}$$

As in the proof of Theorem 1 we take $m = m_1 + m_2 \geq 2$ real constants $c_1, \ldots, c_m$ such that $c_i + \tilde{\phi}_j \neq 0$ $(i, j = 1, \ldots, m)$ on $U(x_0) \times V(y_0)$.

Denoting by F_k the functions I_{c_k} we obtain (see formula (9))

$$(15)\quad \widetilde{P}_p \stackrel{\text{def}}{=} \widetilde{P}_p \frac{\widetilde{\Pi}_p}{\widetilde{D}_p} \widetilde{C}_m \frac{(\tilde{\phi}_1 + c_m)^{l_1-1} \cdots (\tilde{\phi}_m + c_m)^{l_m-1}}{(\tilde{\phi}_p + c_1) \cdots (\tilde{\phi}_p + c_{m-1})}$$

$$= \sum_k (-1)^{p+k} \frac{\widetilde{C}_m}{\widetilde{C}_k} \left(\frac{\tilde{\phi}_1 + c_m}{\tilde{\phi}_1 + c_k} \right)^{l_1-1} \cdots \left(\frac{\tilde{\phi}_m + c_m}{\tilde{\phi}_m + c_k} \right)^{l_m-1} \frac{\tilde{\phi}_p + c_m}{\tilde{\phi}_p + c_k} F_k$$

$$= \sum_{k=1}^{m} q_{pk} F_k,$$

where $\widetilde{\Pi}_k \stackrel{\text{def}}{=} (\tilde{\phi}_m - \tilde{\phi}_k) \cdots (\tilde{\phi}_k - \tilde{\phi}_1)$ and $\widetilde{C}_k \stackrel{\text{def}}{=} (c_m - c_k) \cdots (c_k - c_1)$ $(k = 1, \ldots, m)$. By assumption, the metric g_1 is A_1-integrable and the metric g_2 is A_2-integrable. Hence, the functions $Q_c(\xi) \stackrel{\text{def}}{=} \det(A_1 + c) g_1((A_1 + c)^{-1}\xi, \xi)$, $\xi \in TM_1$ are in involution with respect to the symplectic structure ω_{g_1} on TM_1 and the functions $R_c(\xi) \stackrel{\text{def}}{=} \det(A_2 + c) g_2((A_2 + c)^{-1}\eta, \eta)$, $\eta \in TM_2$ are in involution with respect to the symplectic structure ω_{g_2} on TM_2. Using the constants $c_1, \ldots, c_{m_1}$ and formula (9) we obtain

$$P_s^{(1)} \stackrel{\text{def}}{=} P_s^{(1)} \Pi_s^{(1)} C_{m_1}^{(1)} \frac{(\phi_1 + c_{m_1})^{l_1-1} \cdots (\phi_{m_1} + c_{m_1})^{l_{m_1}-1}}{(\phi_s + c_1) \cdots (\phi_s + c_{m_1-1})}$$

$$= \sum_k (-1)^{s+k} \frac{C_{m_1}^{(1)}}{C_k^{(1)}} \left(\frac{\phi_1 + c_{m_1}}{\phi_1 + c_k} \right)^{l_1-1} \cdots \left(\frac{\phi_{m_1} + c_{m_1}}{\phi_{m_1} + c_k} \right)^{l_{m_1}-1} \frac{\phi_s + c_{m_1}}{\phi_s + c_k} Q_k,$$

which we write as

$$(16)\qquad P_s^{(1)} = \sum_{k=1}^{m_1} a_{sk} Q_k, \; Q_k \stackrel{\text{def}}{=} Q_{c_k}.$$

Here $P_s^{(1)} \stackrel{\text{def}}{=} \sum_{\alpha=1}^{l_s} \epsilon_{s\alpha} P_{s\alpha}^2$, $\epsilon_{s\alpha} = \pm 1$, $P_{s\alpha} \stackrel{\text{def}}{=} \langle p, X_{s\alpha} \rangle$, $p \in T^*M_1$, and $\Pi_s^{(1)} \stackrel{\text{def}}{=} (\phi_{m_1} - \phi_s) \cdots (\phi_s - \phi_1)$, $C_s^{(1)} \stackrel{\text{def}}{=} (c_{m_1} - c_s) \cdots (c_s - c_1)$, $s = 1, \ldots, m_1$. Analogously,

$$P_s^{(2)} \stackrel{\text{def}}{=} P_s^{(2)} \Pi_s^{(2)} C_m^{(2)} \frac{(\phi_{m_1+1} + c_m)^{l_{m_1+1}-1} \cdots (\phi_m + c_m)^{l_m-1}}{(\phi_s + c_{m_1+1}) \cdots (\phi_s + c_{m-1})}$$

$$= \sum_k (-1)^{s+k} \frac{C_m^{(2)}}{C_k^{(2)}} \left(\frac{\phi_{m_1+1} + c_m}{\phi_{m_1+1} + c_k} \right)^{l_{m_1+1}-1} \cdots \left(\frac{\phi_m + c_m}{\phi_m + c_k} \right)^{l_m-1} \frac{\phi_s + c_m}{\phi_s + c_k} R_k$$

which we write as

$$(17) \qquad P_s^{(2)} = \sum_{k=m_1+1}^{m} b_{sk} R_k,\ R_k \stackrel{\text{def}}{=} R_{c_k}.$$

Here $P_s^{(2)} \stackrel{\text{def}}{=} \sum_{\alpha=1}^{l_s} \epsilon_{s\alpha} P_{s\alpha}^2$, $\epsilon_{s\alpha} = \pm 1$, $P_{s\alpha} \stackrel{\text{def}}{=} \langle p, X_{s\alpha}\rangle$, $p \in T^*M_2$, and $\Pi_s^{(2)} \stackrel{\text{def}}{=} (\phi_m - \phi_s)\cdots(\phi_s - \phi_{m_1+1})$, $C_s^{(2)} \stackrel{\text{def}}{=} (c_m - c_s)\cdots(c_s - c_{m_1+1})$, $s = m_1+1,\ldots,m$. It follows from the Stäckel-Painlevé theorem that the functions $P_s^{(1)}$, $s = 1,\ldots,m_1$ are in involution with respect to the canonical symplectic structure on T^*M_1 and the functions $P_s^{(2)}$, $s = m_1+1,\ldots,m$ are in involution with respect to the canonical symplectic structure on T^*M_2. As in the proof of Theorem 1 we see that if $l_k > 1$, then $\phi_k = \text{constant}$, and if $k \neq q$, then $L_{X_{k\alpha}}\phi_q = 0$. It follows from formula (15), (16) and (17) that

$$\widetilde{P}_p = \begin{cases} \widetilde{P}_p^{(1)} B_p & , \quad 1 \le p \le m_1, \\ \widetilde{P}_p^{(2)} B_p & , \quad m_1+1 \le p \le m, \end{cases}$$

where B_p are smooth functions on $U(x_0) \times V(y_0)$ such that $L_{\widetilde{X}_{k\alpha}} B_p$, if $k \neq p$, and $\widetilde{P}_p^{(i)} \stackrel{\text{def}}{=} p_i^*(P_p^{(i)}) \in C^\infty(T^*M)$, where $p_i : T^*M \cong (T^*M_1) \times (T^*M_2) \to T^*M_i$ is the natural projection on T^*M_i ($i = 1,2$). Therefore, the functions $\widetilde{P}_p$ ($p = 1,\ldots,m$) are in involution with respect to the canonical symplectic structure on T^*M. Finally, applying a variant of the Stäckel-Painlevé theorem to equation (15) we obtain that the functions F_k are in involution. Theorem 5 is proved.

§4. **Appendix 1**

In this appendix we apply the theorems proved in the previous sections to the A-integrable metric described in item c) of section 1.

Let us consider the projective transformation of the plane $\mathbb{R}^n$ given by the formula

$$\mu : \begin{cases} x_1 & \mapsto & (\lambda_1 x_1)/x_n \\ & \vdots & \\ x_{n-1} & \mapsto & (\lambda_{n-1}x_{n-1})/x_n \\ x_n & \mapsto & \lambda_n - \lambda_n/x_n \end{cases}$$

where λ_i ($i = 1,\ldots,n$) are non-zero constants. Consider the Euclidean metric $g \stackrel{\text{def}}{=} dx_1^2 + \cdots + dx_n^2$ in $\mathbb{R}^n$ (see [15], where the pseudo-Riemannian analogue of the present construction is given). The geodesics of the metric g are straight lines. Denote by $\bar g$ the pull-back $\mu^* g$. Note that the metric $\bar g$ is not defined over the hyperplane $Z \overset{\text{def}}{=} \{x_n = 0\}$. It is clear that the metrics g and $\bar g$ are geodesically equivalent on $D \overset{\text{def}}{=} \mathbb{R}^n \setminus Z$. We have

$$\begin{aligned}\bar g &\overset{\text{def}}{=} \mu^*(g)\\ &= \lambda_1^2\left(\frac{x_n dx_1 - x_1 dx_n}{x_n^2}\right)^2 + \cdots + \lambda_{n-1}^2\left(\frac{x_n dx_{n-1} - x_{n-1}dx_n}{x_n^2}\right)^2\\ &\qquad +\lambda_n^2\left(\frac{dx_n}{x_n^2}\right)^2 .\end{aligned} \tag{18}$$

Let us compute the operator $A \overset{\text{def}}{=} A(g, \bar g)$. We need the following simple lemmas.

Lemma 1. *Consider the matrix*

$$C \overset{\text{def}}{=} \bar a \otimes \bar a + b, \tag{19}$$

where $b \overset{\text{def}}{=} \operatorname{diag}(b_1, \dots, b_n)$, $b_i \neq 0$ $(i = 1, \dots, n)$, $\bar a \overset{\text{def}}{=} (a_1, \dots, a_n)'$ (the symbol $'$ denotes the transposition of a matrix) and $\bar a \otimes \bar a$ stands for the matrix $\bar a \bar a'$. Then the inverse matrix C^{-1} is given by the formula

$$C^{-1} = b^{-1} - \frac{1}{D}(b^{-1}\bar a) \otimes (b^{-1}\bar a), \tag{20}$$

where $D \overset{\text{def}}{=} \langle b^{-1}\bar a, \bar a\rangle + 1$ and $\langle \bar x, \bar y\rangle \overset{\text{def}}{=} \sum_{i=1}^n x_i y_i$. Moreover, $\det C = (\prod_{k=1}^n b_k)D$.

Proof of Lemma 1. The proof is by direct calculation. We have $(\bar a\bar a' + b)(b^{-1} - (1/D)(b^{-1}\bar a)(b^{-1}\bar a)') = \bar a(\bar a' b^{-1}) + E - \langle \bar a, b^{-1}\bar a\rangle \bar a(b^{-1}\bar a)'/D$ $-(1/D)\bar a(b^{-1}\bar a)' = E$ where E denotes the unit matrix. Lemma 1 is proved.

Lemma 2. *Consider the matrix $C_0 \overset{\text{def}}{=} \bar a \otimes \bar a + \operatorname{diag}(b_1, \dots, b_{n-1}, 0)$, where $b_i \neq 0$ $(i = 1, \dots, n-1)$. Then the inverse matrix C_0^{-1} is given by the formula $C_0^{-1} =$*

$$\frac{1}{a_n^2}\left\{\sum_{i=1}^{n-1}\frac{a_n^2}{b_i}E_{ii} + \left(\sum_{i=1}^{n-1}\frac{a_i^2}{b_i} + 1\right)E_{nn} - \sum_{i=1}^{n-1}\frac{a_n a_i}{b_i}E_{in} - \sum_{i=1}^{n-1}\frac{a_n a_i}{b_i}E_{ni}\right\}$$

where the matrix E_{ij} has elements $e_{kl} \overset{\text{def}}{=} \delta_{ki}\delta_{lj}$ $(k, l = 1, \dots, n)$. We have $\det C_0 = (\prod_{k=1}^{n-1} b_k)a_n^2$.

Lemma 2 easily follows from Lemma 1 and the formula $C_0^{-1} = \lim_{\alpha\to+0}(C_0+\alpha)^{-1}$.

Denote by $\bar{G}$ the matrix corresponding to the metric $\bar{g}$, i.e. the Gramian of the metric g. Taking $\bar{a} = \bar{x} \overset{\text{def}}{=} (x_1,\dots,x_n)$, $b_i = \lambda_n^2/\lambda_i^2$ $(i = 1,\dots,n-1)$, and applying Lemma 2 we see that

$$\bar{G}^{-1} = \frac{x_n^2}{\lambda_n^2}\left(\bar{x}\otimes\bar{x} + \operatorname{diag}\left(\frac{\lambda_n^2}{\lambda_1^2},\dots,\frac{\lambda_n^2}{\lambda_{n-1}^2},0\right)\right). \tag{21}$$

As a corollary we obtain that $\det\bar{G} = (\prod_{i=1}^n \lambda_i^2)/x_n^{2n+2}$. Finally, combining (3) and (21) we obtain (up to multiplication by a constant) that

$$A(g,\bar{g}) = \bar{x}\otimes\bar{x} + \operatorname{diag}\left(\frac{\lambda_n^2}{\lambda_1^2},\dots,\frac{\lambda_n^2}{\lambda_{n-1}^2},0\right). \tag{22}$$

It follows from Theorem 1 formula (22) and Lemma 1 that the quadratic forms

$$\begin{aligned} I_c &= D_c\left(\frac{dx_1^2}{d_1+c}+\dots+\frac{dx_{n-1}^2}{d_{n-1}+c}+\frac{dx_n^2}{c}\right) \\ &- \left(\frac{x_1dx_1}{d_1+c}+\dots+\frac{x_{n-1}dx_{n-1}}{d_{n-1}+c}+\frac{x_ndx_n}{c}\right)^2, \end{aligned} \tag{23}$$

where $d_k \overset{\text{def}}{=} \lambda_n^2/\lambda_k^2$ $(k = 1,\dots,n-1)$, are pairwise commuting integrals of the geodesic flow of the metric g on all half-planes $\{x_n > 0\}$ and $\{x_n < 0\}$. The quadratic forms I_c are smoothly defined on the whole of $\mathbb{R}^n$. Therefore, they are pairwise commuting integrals of the geodesic flow the metric g on $\mathbb{R}^n$.

Suppose that $d_i \neq d_j$ $(i\neq j)$ and $d_i \neq 0$ $(i = 1,\dots,n-1)$. Consider the characteristic polynomial of the matrix $A \overset{\text{def}}{=} A(g,\bar{g})$

$$\begin{aligned} \chi_A(c) &\overset{\text{def}}{=} \det(A+cE) \\ &= \prod_{k=1}^n (d_k+c)\left\{\frac{x_1^2}{d_1+c}+\dots+\frac{x_{n-1}^2}{d_{n-1}+c}+\frac{x_n^2}{c}+1\right\}, \end{aligned}$$

where we put $d_n = 0$. It is clear that if $x_1\dots x_n \neq 0$, then the polynomial $\chi_A(c)$ has n different roots. We have proved the next theorem.

Theorem 6. *The metric $g \overset{\text{def}}{=} \sum_{i=1}^n dx_i^2$ is A-integrable. If $d_i \neq d_j$ $(i\neq j)$ and $d_i \neq 0$ $(i = 1,\dots,n-1)$, then the rank of the A-integrable metric g is n.*

It follows from Theorem 1 that the one-parameter family of metrics $g_c(g, \bar{g}) \stackrel{\text{def}}{=} \frac{1}{\det(A+c\mathbf{1})} g(A + c\mathbf{1})^{-1}$ are geodesically equivalent to g. We have

$$G_c \stackrel{\text{def}}{=} \frac{1}{\det(A + cE)}(A + cE)^{-1} \tag{24}$$

where G_c denotes the Gramian of the metric g_c. It follows from Lemma 1 that up to a multiplication on a constant

$$\begin{aligned} g_c &= \left\{ D_c \left(\frac{dx_1^2}{d_1 + c} + \cdots + \frac{dx_{n-1}^2}{d_{n-1} + c} + \frac{dx_n^2}{c} \right) \right. \\ &- \left. \left(\frac{x_1 dx_1}{d_1 + c} + \cdots + \frac{x_{n-1} dx_{n-1}}{d_{n-1} + c} + \frac{x_n dx_n}{c} \right)^2 \right\} / D_c^2, \end{aligned} \tag{25}$$

where

$$D_c \stackrel{\text{def}}{=} \frac{x_1^2}{d_1 + c} + \cdots + \frac{x_{n-1}^2}{d_{n-1} + c} + \frac{x_n^2}{c} + 1. \tag{26}$$

Remark 6. *The metric g_c is not defined on the quadric $Z_c \stackrel{\text{def}}{=} \{D_c = 0\}$. Nevertheless, from now on when we say that two metrics are geodesically equivalent we mean they are geodesically equivalent in the domain where both of them are defined.*

The metrics g_c given by formula (25) have the very special property given by the next theorem.

Theorem 7. *Let us assume that $d_i \neq d_j$ $(i \neq j)$ and $d_i \neq 0$ $(i = 1, \dots, n-1)$. Then the restrictions of the metrics g_c given by formula (25) on every quadric Q_α $(\alpha \neq c)$ from the confocal family $Q_\alpha \stackrel{\text{def}}{=} \{D_\alpha = 0\}$ are geodesically equivalent to the restriction of the standard Euclidean metric g on Q_α.*

As a corollary we obtain the theorem proved in [14] and independently in [12] that the standard ellipsoid admits a non-trivial geodesic equivalence. Theorem 7 shows that the same result is true for the hyperboloids.

Proof of Theorem 7. Suppose that $g = \sum_{i=1}^n dx_i^2$. Let us fix a point $\bar{x}_0 = (x_1^0, \dots, x_n^0) \in \mathbb{R}^n$ such that $x_1^0 \dots x_n^0 \neq 0$. It is easily seen that there exist n different real constants $\alpha_1, \dots, \alpha_n$ such that $Q_{\alpha_1}(\bar{x}_0) = 0, \dots, Q_{\alpha_n}(\bar{x}_0) = 0$. It follows from formula (23) that

$$I_{\alpha_l}|_{\bar{x}_0} = -(d_{\bar{x}_0} Q_{\alpha_l})^2/4. \tag{27}$$

The forms $d_{\bar{x}_0}Q_{\alpha_l}$ are linearly independent. It follows from Remark 2 and the non-degeneracy of the corresponding Vandermonde determinant that the forms $S_0|_{\bar{x}_0},\dots,S_{n-1}|_{\bar{x}_0}$ are simultaneously diagonalizable in the frame which is dual to the forms $d_{\bar{x}_0}Q_{\alpha_l}$ $(l=1,\dots,n)$. The form S_{n-1} is conformally equivalent to the metric g and the form S_0 is conformally equivalent to $\bar{g}$. Hence, the gradients (with respect to the metric g) $\nabla_{\bar{x}_0}Q_{\alpha_1},\dots,\nabla_{\bar{x}_0}Q_{\alpha_n}$ coincide with the principal directions of the metrics $g|_{\bar{x}_0}$ and $\bar{g}|_{\bar{x}_0}$. Finally, the statement of the theorem follows from Levi-Civita's theorem about the local form of the Riemannian metrics that permit geodesic equivalence (see [5, 16, 17]). Theorem 7 is proved.

Let $(x_1,\dots,x_n)$ be the coordinates in $\mathbb{R}^n$, and $(x_1,\dots,x_n;p_1,\dots,p_n)$ the corresponding coordinates in $T^*\mathbb{R}^n$. Applying Theorem 3 we obtain the next theorem.

Theorem 8. *For every fixed real κ and every fixed integer l, consider the one-parameter family of functions on $T^*\mathbb{R}^n$ given by the formula $I_\alpha^{(l)}(\bar{p}) \stackrel{\text{def}}{=} \langle I_\alpha^{(l)}\bar{p},\bar{p}\rangle$, where $\bar{p}\stackrel{\text{def}}{=}(p_1,\dots,p_n)$, $\langle\bar{\xi},\bar{\eta}\rangle \stackrel{\text{def}}{=} \sum_{i=1}^n \xi_i\eta_i$,*

$$\begin{aligned} I_\alpha^{(l)} &\stackrel{\text{def}}{=} \det(A+\alpha E)(A+\alpha E)^{-1}(A+\kappa E)^{-l} \\ &= S_{n-1}^{(l)}(\kappa)\alpha^{n-1}+S_{n-2}^{(l)}(\kappa)\alpha^{n-2}+\dots+S_0^{(l)}(\kappa) \end{aligned}$$

and the matrix A is given by the formula

$$A \stackrel{\text{def}}{=} \bar{x}\otimes\bar{x}+\operatorname{diag}(d_1,\dots,d_{n-1},0)\,. \tag{28}$$

These functions are in involution with respect to the canonical symplectic structure on $T^*\mathbb{R}^n$. If $d_i\neq d_j$ $(i\neq j)$ and $d_i\neq 0$ $(i=1,\dots,n-1)$, then the functions $S_0^{(l)},\dots,S_{n-1}^{(l)}$ are functionally independent on $T^*\mathbb{R}^n$.

Remark 7. *Taking $l=0$ we derive the integrals obtained by K. Uhlenbeck in [19]. Another way of obtaining these integrals was proposed by J. Moser in [8] and K. Kiyohara in [4]. The families of functions in involution given by Theorem 8 generalize all these results.*

Remark 8. *The case $l=-1$, $\kappa=0$, gives a family of pairwise commuting functions $S_0^{(-1)}(0),\dots,S_{n-1}^{(-1)}(0)$ on T^*M^n. The function $S_{n-1}^{(-1)}(0)$ coincides with the Hamiltonian of the metric*

$$g_{hyp} \stackrel{\text{def}}{=} \frac{1}{x_n^2}\sum_{i=1}^n dx_i^2. \tag{29}$$

Therefore, the functions $S_0^{(-1)}(0),\dots,S_{n-1}^{(-1)}(0)$ on T^*M^n are a complete family of integrals of the geodesic flow of the hyperbolic plane.

§5. Appendix 2

Suppose that the (pseudo) Riemannian metrics g and $\bar{g}$, given on the smooth manifold M^n, are geodesically equivalent. In this appendix we prove that the functions given by the formula $I_c(\xi) \overset{\text{def}}{=} \det(A+c\mathbf{1})g((A+c\mathbf{1})^{-1}\xi,\xi)$, $\xi \in TM^n$, where c is real parameter and the operator $A = A(g,\bar{g})$ is given by formula (3), are in involution with respect to the symplectic structure ω_g on TM^n (see [15]).

Let us fix a point $x_0 \in M^n$. It is clear that there exists an open neighborhood $U(x_0)$ of the point x_0 and an interval $(-\epsilon,\epsilon)$, $\epsilon > 0$, such that the operator $(A+c\mathbf{1})|_x$ is invertible if $x \in U(x_0)$ and $c \in (-\epsilon,\epsilon)$. Consider the one-parameter family of geodesically equivalent metrics on $U(x_0)$

$$g_c(g,\bar{g})(\xi,\eta) \overset{\text{def}}{=} \frac{1}{\det(A+c\mathbf{1})} g((A+c\mathbf{1})^{-1}\xi,\eta),\ \xi,\eta \in TM^n, \tag{30}$$

where $c \in (-\epsilon,\epsilon)$ (see Proposition 1 in [16]). Let us fix some $\kappa \in (-\epsilon,\epsilon)$. The metrics g and g_κ are geodesically equivalent on $U(x_0)$. Consider the following subsequence of geodesically equivalent metrics

$$\begin{array}{ccc} \downarrow & & \downarrow \\ g & \overset{g.e.}{\longleftrightarrow} & g_\kappa \\ \downarrow & & \downarrow \\ g^{(1)} & \overset{g.e.}{\longleftrightarrow} & g_\kappa^{(1)} \\ \downarrow & & \downarrow \\ g^{(2)} & \overset{g.e.}{\longleftrightarrow} & g_\kappa^{(2)} \\ \downarrow & & \downarrow \end{array}$$

where $g^{(l)} \overset{\text{def}}{=} gA(g,g_\kappa)^l$ and $g_\kappa^{(l)} \overset{\text{def}}{=} g_\kappa A(g,g_\kappa)^l$ (see [10, 16]). The metrics $g^{(2)}$ and $g_\kappa^{(2)}$ are geodesically equivalent. It is easily seen that $A(g,g_\kappa) = A+\kappa\mathbf{1}$. Hence, $g_\kappa^{(2)} \overset{\text{def}}{=} g_\kappa A(g,g_\kappa)^2 = \frac{1}{\det(A+\kappa\mathbf{1})}g(A+\kappa\mathbf{1})$. Let us consider the one-parameter family of integrals $I_\alpha(g_\kappa^{(2)},g^{(2)})$ of the geodesic flow of the metric $g_\kappa^{(2)}$. A direct computation show that $A(g_\kappa^{(2)},g^{(2)}) = (A+\kappa\mathbf{1})^{-1}$. Applying the Legendre transformation corresponding to the metric $g_\kappa^{(2)}$ we see that the functions

$${g_\kappa^{(2)}}^{-1} = \det(A+\kappa\mathbf{1})(A+\kappa\mathbf{1})^{-1}g^{-1} \tag{31}$$

and $I_\alpha(g^{(2)}_\kappa, g^{(2)}) \overset{\text{def}}{=}$

$$\det((A+\kappa\mathbf{1})^{-1}+\alpha\mathbf{1})((A+\kappa)^{-1}+\alpha\mathbf{1})^{-1}{g^{(2)}_\kappa}^{-1} \tag{32}$$

$$=\det(\alpha A+(\kappa\alpha+1)\mathbf{1})(\alpha A+(\kappa\alpha+1)\mathbf{1})^{-1}g^{-1} \tag{33}$$

are in involution with respect to the canonical symplectic structure on T^*M^n. Note that after applying the Legendre transformation corresponding to the metric g the family of integrals (1) takes the form $I_c(g,\bar{g}) = \det(A+c)(A+c)^{-1}g^{-1}$, where c is an arbitrary real parameter. Hence, the Poisson brackets $\{I_\kappa, I_{\kappa+\frac{1}{\alpha}}\}$ vanish for all $\alpha \neq 0$ and $\kappa \in (-\epsilon, \epsilon)$. Fixing n different real numbers $\kappa_i \in (\epsilon, \epsilon)$, $i = 1, \dots, n$, $\kappa_i \neq \kappa_j$ $(i \neq j)$, we obtain that $\{I_{\kappa_i}, I_c\} = 0$ for all real values of the parameter c. Finally, recall that $I_\mu = S_{n-1}\mu^{n-1} + S_{n-2}\mu^{n-2} + \cdots + S_0$. The last equality and the non-degeneracy of the corresponding Vandermonde determinant show that the functions S_l $(l = 1, \dots, n)$ are linear combinations with constant coefficients of the functions I_{κ_i} $(i = 1, \dots, n)$. Therefore, the integrals given by formula (1) are in involution on $T^*U(x_0)$. The point x_0 was arbitrary. This proves the statement.

Acknowledgments. The author takes the opportunity to thank V. Kozlov, K. Kiyohara, G. Popov and V. Matveev for useful discussions. The author is partially supported by MESC grant MM-810/98. The paper was written during the author's stay at the University of Nantes. He would like to thank CNRS for financial support. The author is also grateful to the 9th MSJ-IRI.

References

[1] S. Benenti, Inertia tensors and Stäckel systems in Euclidean space, Rend. Sem. Mat. Univ. di Torino, **50** (1992), 1–20.

[2] U. Dini, Sopra un problema che si presenta nella theoria generale delle rappresetazioni geografice di una superficie su un'altra, Ann. di Math. ser.2, **3** (1869), 269–293.

[3] L. Eisenhart, Riemannian Geometry, Princeton Univ. Press, 1960.

[4] K. Kiyohara, Two Classes of Riemannian Manifolds Whose Geodesic Flows Are Integrable, Mem. Amer. Math. Soc., **130**, no. 619, Amer. Math. Soc., 1997.

[5] T. Levi-Civita, Sulle trasformazioni delle equazioni dinamiche, Ann. di Mat. serie 2^a, **24** (1896), 255–300.

[6] V. Matveev and P. Topalov, Quantum integrability of Beltrami-Laplace operator as geodesic equivalence, Math. Z., **238** (2001), 833–866.

[7] V. S. Matveev and P. J. Topalov, Trajectory equivalence and corresponding integrals, Regular and Chaotic Dynamics, **2** (1998), 30–45.
[8] J. Moser, Integrable Hamiltonian systems and spectral theory, Lezioni Fermiante, Academia Nazionale Dei Lincei, Pisa, 1981.
[9] P. Painlevé, Sur les intégrale quadratiques des équations de la Dinamique, Compt. Rend., **124** (1897), 221–224.
[10] N. S. Sinyukov, An invariant transformation of Riemannian manifolds with common geodesics, Soviet Math. Docl., **137** (1961), 1312–1314.
[11] N. S. Sinyukov, Geodesic mappings of Riemannian spaces, Nauka, Moscow, 1979.
[12] S. Tabachnikov, Projectively equivalent metrics, exact transverse line field and the geodesic flow on the ellipsoid, Comment. Math. Helv., **74** (1999), 306–321.
[13] P. Topalov, Families of metrics geodesically equivalent to the analogues of the Poisson sphere, J. Math. Phys., **41** (2000), 7510–7520.
[14] P. J. Topalov and V. S. Matveev, Geodesic equivalence via integrability, Geom. Dedicata, **96** (2003), 91–115.
[15] P. Topalov, Geodesic hierarchies and involutivity, J. Math. Phys., **42** (2001), 3898–3914.
[16] P. Topalov, Hierarchy of integrable geodesic flows, Publ. Mat.,**44** (2000), 257–276.
[17] P. Topalov, Integrability criterion of geodesical equivalence. Hierarchies, Acta Appl. Math., **59** (1999), 271–298.
[18] P. Topalov, Tensor invariants of natural mechanical systems on compact surfaces, and the corresponding integrals, Sb. Mat., **188** (1997), 307–326.
[19] K. Uhlenbeck, Minimal 2-spheres and tori in S^k, preprint, 1975.
[20] K. Yamato, A class of Riemannian manifolds with integrable geodesic flows, J. Math. Soc. Japan, **47** (1995), 719–733.

Department of Mathematics
Northeastern University
Boston, MA 02115
USA

E-mail address:
`p.topalov@neu.edu`

Advanced Studies in Pure Mathematics

A SERIES OF UP-TO-DATE GUIDES OF LASTING INTEREST
TO ADVANCED MATHEMATICS

Volume 1 Algebraic Varieties and Analytic Varieties.
Edited by S. Iitaka. February, 1983

Volume 2 Galois Groups and their Representations.
Edited by Y. Ihara. December, 1983

Volume 3 Geometry of Geodesics and Related Topics.
Edited by K. Shiohama. June, 1984

Volume 4 Group Representations and Systems of Differential Equations.
Edited by K. Okamoto. March, 1985

Volume 5 Foliations.
Edited by I. Tamura. February, 1986

Volume 6 Algebraic Groups and Related Topics.
Edited by R. Hotta. March, 1985

Volume 7 Automorphic Forms and Number Theory.
Edited by I. Satake. February, 1986

Volume 8 Complex Analytic Singularities.
Edited by T. Suwa and P. Wagreich. February, 1987

Volume 9 Homotopy Theory and Related Topics.
Edited by H. Toda. February, 1987

Volume 10 Algebraic Geometry, Sendai, 1985.
Edited by T. Oda. July, 1987

Volume 11 Commutative Algebra and Combinatorics.
Edited by M. Nagata and H. Matsumura.
October, 1987

Volume 12 Galois Representations and Arithmetic Algebraic Geometry.
Edited by Y. Ihara. November, 1987

Volume 13 Investigations in Number Theory.
Edited by T. Kubota. March, 1988

Volume 14 Representations of Lie Groups, Kyoto, Hiroshima, 1986.
Edited by K. Okamoto and T. Oshima. February, 1989

Volume 15 Automorphic Forms and Geometry of Arithmetic Varieties.
Edited by K. Hashimoto and Y. Namikawa. July, 1989

Volume 16	Conformal Field Theory and Solvable Lattice Models. Edited by M. Jimbo, T. Miwa and A. Tsuchiya. June, 1988
Volume 17	Algebraic Number Theory—in honor of K. Iwasawa. Edited by J. Coates, R. Greenberg, B. Mazur and I. Satake. August, 1989
Volume 18	I—Recent Topics in Differential and Analytic Geometry. Edited by T. Ochiai. November, 1990 II—Kähler Metric and Moduli Spaces. Edited by Edited by T. Ochiai. November, 1990
Volume 19	Integrable Systems in Quantum Field Theory and Statistical Mechanics. Edited by M. Jimbo, T. Miwa and A. Tsuchiya. November, 1989
Volume 20	Aspects of Low Dimensional Manifolds. Edited by Y. Matsumoto and S. Morita. December, 1992
Volume 21	Zeta Functions in Geometry. Edited by N. Kurokawa and T. Sunada. December, 1992
Volume 22	Progress in Differential Geometry. Edited by K. Shiohama. September, 1993
Volume 23	Spectral and Scattering Theory and Applications. Edited by K. Yajima. February, 1994
Volume 24	Progress in Algebraic Combinatorics. Edited by E. Bannai and A. Munemasa. May, 1996
Volume 25	CR-Geometry and Overdetermined Systems. Edited by T. Akahori, G. Komatsu, K. Miyajima, M. Namba and K. Yamaguchi. July, 1997
Volume 26	Analysis on Homogeneous Spaces and Representation Theory of Lie Groups, Okayama-Kyoto. Edited by M. Kashiwara, T. Kobayashi, T. Matsuki, K. Nishiyama and T. Oshima. April, 2000
Volume 27	Arrangements – Tokyo 1998. Edited by M. Falk and H. Terao. August, 2000
Volume 28	Combinatorial Methods in Representation Theory. Edited by M. Kashiwara, K. Koike, S. Okada, I. Terada, and H.-F. Yamada. December, 2000

Volume 29 Singularities – Sapporo 1998.
Edited by J.-P. Brasselet and T. Suwa.
December, 2000

Volume 30 Class Field Theory – Its Centenary and Prospect.
Edited by K. Miyake. April, 2001

Volume 31 Taniguchi Conference on Mathematics Nara '98.
Edited by M. Maruyama and T. Sunada. May, 2001

Volume 32 Groups and Combinatorics — in memory of Michio Suzuki.
Edited by E. Bannai, H. Suzuki, H. Yamaki and T. Yoshida. October, 2001

Volume 33 Computational Commutative Algebra and Combinatorics.
Edited by T. Hibi. February, 2002

Volume 34 Minimal Surfaces, Geometric Analysis and Symplectic Geometry.
Edited by K. Fukaya, S. Nishikawa and J. Spruck.
June, 2002

Volume 35 Higher Dimensional Birational Geometry.
Edited by S. Mori and Y. Miyaoka. July, 2002

Volume 36 Algebraic Geometry 2000, Azumino.
Edited by S. Usui, M. Green, L. Illusie, K. Kato, E. Looijenga, S. Mukai and S. Saito. November, 2002

Volume 37 Lie Groups, Geometric Structures and Differential Equations — One Hundred Years after Sophus Lie —.
Edited by T. Morimoto, H. Sato and K. Yamaguchi.
December, 2002

Volume 38 Operator Algebras and Applications.
Edited by H. Kosaki. January, 2004

Volume 39 Stochastic Analysis on Large Scale Interacting Systems.
Edited by T. Funaki and H. Osada. March, 2004

Volume 40 Representation Theory of Algebraic Groups and Quantum Groups.
Edited by T. Shoji, M. Kashiwara, N. Kawanaka, G. Lusztig and K. Shinoda. March, 2004

Volume 41 Stochastic Analysis and Related Topics in Kyoto
In honour of Kiyosi Itô.
Edited by H. Kunita, S. Watanabe and Y. Takahashi.
March, 2004

Volume 42 Complex Analysis in Several Variables
– Memorial Conference of Kiyoshi Oka's Centennial Birthday, Kyoto/Nara 2001.
Edited by K. Miyajima, M. Furushima, H. Kazama, A. Kodama, J. Noguchi, T. Ohsawa, H. Tsuji and T. Ueda. June, 2004

Volume 43 Singularity Theory and Its Applications.
Edited by S. Izumiya, G. Ishikawa, H. Tokunaga, I. Shimada and T. Sano. January, 2007

Volume 44 Potential Theory in Matsue.
Edited by H. Aikawa, T. Kumagai, Y. Mizuta and N. Suzuki. August, 2006

Volume 45 Moduli Spaces and Arithmetic Geometry (Kyoto, 2004).
Edited by S. Mukai, Y. Miyaoka, S. Mori, A. Moriwaki and I. Nakamura. January, 2007

Volume 46 Singularities in Geometry and Topology 2004.
Edited by J.-P. Brassele and T. Suwa. February, 2007

Volume 47 Asymptotic Analysis and Singularities
1—Hyperbolic and dispersive PDEs and fluid mechanics
2—Elliptic and parabolic PDEs and related problems.
Edited by H. Kozono, T. Ogawa, K. Tanaka, Y. Tsutsumi and E. Yanagida. October, 2007

Volume 48 Finsler Geometry, Sapporo 2005
— In Memory of Makoto Matsumoto.
Edited by S. V. Sabau and H. Shimada.
November, 2007

Volume 49 Probability and Number Theory — Kanazawa 2005.
Edited by S. Akiyama, K. Matsumoto, L. Murata and H. Sugita. December, 2007

Volume 50 Algebraic Geometry in East Asia — Hanoi 2005.
Edited by K. Konno and V. Nguyen-Khac.
February, 2008

Volume 51 Surveys on Geometry and Integrable Systems.
Edited by M. Guest, R. Miyaoka and Y. Ohnita.
November, 2008

TO BE CONTINUED